HANDBOOK OF ENVIRONMENTAL MANAGEMENT AND TECHNOLOGY

Second Edition

HANDBOOK OF ENVIRONMENTAL MANAGEMENT AND TECHNOLOGY

SECOND EDITION

Gwendolyn Burke

MTA Bridges & Tunnels
New York, New York

Ben Ramnarine Singh

U.S. Environmental Protection Agency
New York, New York

Louis Theodore

Manhattan College
Riverdale, New York

A JOHN WILEY & SONS, INC. PUBLICATION

New York · Chichester · Weinheim · Brisbane · Singapore · Toronto

This book is printed on acid-free paper. ⊚

Copyright © 2000 by John Wiley & Sons, Inc. All rights reserved.

Published simultaneously in Canada.

For ordering and customer service, call 1(800) CALL-WILEY.

Library of Congress Cataloging-in-Publication Data:

Burke, Gwendolyn, 1963–
    Handbook of environmental management and technology / Gwendolyn
Burke, Ben Ramnarine Singh, Louis Theodore. — 2nd ed.
        p. cm.
    Includes bibliographical references and index.
    ISBN 0-471-34910-0
    1. Pollution. 2. Environmental management. I. Singh, Ben.
II. Theodore, Louis. III. Title.
TD191.5B87   2000
628 — dc21   99-43061
                                                                         CIP

Printed in the United States of America

10 9 8 7 6 5 4 3 2 1 0

To John and Jessica,
  and our new bundle of sunshine—
  you make my life complete.
                    — Gwen

To my wife, Miriam,
  I will always be grateful.
To my sons, Joshua and Joeseph,
  to my daughter Kayla—
  who light up my life.
To my mother, Sonia,
  without whom this would not be possible.
                    — Ben

To Mary O'Dowd—
  A truly wonderful and caring woman
  who has brought much joy to us through her song and spirit;
  Continued success with The Frances Pope Memorial
  Foundation.
                    — Lou

CONTENTS

PREFACE

In the last three decades, people have become aware of a wide range of environmental issues. All sources of air, land, and water pollution are under constant public scrutiny. Increasing numbers of professionals are being confronted with problems related to pollution control. Because many of these issues are of relatively new concern, individuals must develop a proficiency and an improved understanding of technical and scientific, as well as regulatory, issues regarding pollution prevention and remediation in order to cope with these challenges. Although this is not the first professional book to treat this particular subject, it is one of the few books which attempt to highlight all aspects of the spectrum of environmental management and control.

This handbook is intended primarily for engineers, industrial hygienists, health and safety officers, and plant engineers and managers. Lawyers, news media personnel, and regulatory officials can also benefit from this text. The authors' aim is to offer the reader a historical perspective on pollution problems and solutions, and to provide an introduction to the specialized literature in this and related areas. The readers are encouraged, through the reference lists at the end of each chapter, to continue their own development beyond the scope of this book.

As is the case in preparing a book, the problem of what to include and what to omit has been particularly difficult. However, every attempt has been made to offer material to individuals with a limited technical background at a level that should enable them to better cope with some of the complex problems encountered in environmental management today.

The book is divided into ten parts. Part I provides an introduction to background issues such as regulatory approaches, international effects of pollution, and sources of pollutants. Part II covers issues related to air pollution. Dispersion and control of pollutants are discussed, as well as the popular topics of acid rain, the greenhouse effect, indoor air quality, and noise pollution. Part III is devoted to dispersion of pollutants in water systems and control and treatment of wastewaters. Solid, radioactive, and hazardous waste issues are covered in Parts IV and V. The general topics of municipal, medical, radioactive, and hazardous waste control programs are introduced, as well as five individual hazardous pollutants: underground storage tanks, asbestos, household hazardous waste, used oil, and metals. Part VI addresses methods of pollution prevention, including domestic and architectural considerations, waste reduction, and energy conservation. Part VII is primarily devoted to issues faced daily by management. Concerns regarding worker training and safety, emergency management, and monitoring of background levels of contaminants are addressed. Economic considerations are discussed, because in many instances this factor weighs heavily in the ultimate determination of how an environmental management program will proceed. Electromagnetic fields are also introduced in this section. Part VIII introduces three new processes for pollution control and waste treatment: bioremediation, soil vapor extraction, and biofiltration. Part IX discusses risk-related topics. Three types of risk assessment are described, as well as a process for determining an appropriate corrective action plan for a given level of risk and methods for environmental risk communication. Finally, Part X looks into four new environmental topics: The ISO 14000 Quality Standards Certification Plan, how to conduct an environmental audit, suggestions for determining ethical actions, and the regulations behind environmental justice (nondiscrimination policies required for environmental planning).

During the preparation of this book, the authors were ably assisted by a number of individuals. These people contributed much time and energy researching and writing parts of various sections of this book. We gratefully acknowledge their invaluable assistance; the names of these individuals are listed under the titles of the chapters to which they contributed. The authors are particularly indebted to Elizabeth Ruyack and Miriam Ramnarine Singh for their endless support, their impeccable typing skills, and their extra sets of eyes when it came time for proofreading. And, of course, to John Burke, for if he hadn't spent several hours playing with baby Jessica, this second edition would never have been completed.

GWENDOLYN BURKE
BEN RAMNARINE SINGH
LOUIS THEODORE

December 1999

I

INTRODUCTION TO THE ISSUES

1

ENVIRONMENTAL MANAGEMENT OVERVIEW

1.1 THE PAST—TRAGEDIES AND REGULATIONS

The past three decades have been filled with environmental tragedy as well as with a heightened environmental awareness. The oil spills of the *Exxon Valdez* in 1989 and in the Gulf War of 1991 showed how delicate our oceans and their ecosystems truly are. The disclosures of Love Canal in 1978 and Times Beach in 1979 made the entire nation aware of the dangers of hazardous chemical waste. The discovery of acquired immunodeficiency syndrome virus (AIDS) and the beach washups of 1985 brought the issue of medical waste disposal to the forefront of public consciousness. A nuclear accident placed the spotlight on Chernobyl, and to this day we are still seeing the effects of that event.

However, in the face of these disasters, great advances have occurred in the area of environmental regulations. The Clean Water Act of 1977 enforced safe drinking water standards. Discharges to natural water systems were also addressed by this act and recent amendments to the act. The Clean Air Act, passed in 1970, was a first step toward reducing the level of pollutants being released into the air we breathe. This act has been amended several times to address additional sources of air emissions. The Resource Conservation and Recovery Act (RCRA) was developed in 1976 to regulate solid and hazardous waste facilities. As an extension of RCRA, Superfund was enacted in 1988 for cleanup of uncontrolled hazardous waste sites. Thirteen hundred ninety-six hazardous sites requiring remediation were identified. Recent legislation regarding solid waste has emphasized the practices of reuse, reduction, and recycling in order to decrease the amount of waste currently requiring landfill or incineration.

In order to make advances in improving the condition of our environment, we must learn from these past disasters and review the results of these legislative policies to determine how best to proceed. The past must be our guide.

1.2   THE PRESENT—PROBLEMS AND SOLUTIONS

The purpose of this book is to present recent events and problems, in addition to past and present regulatory efforts, in several environmental realms. Attempts will be made to suggest corrections or improvements to our present environmental status. Each section of the book deals with a particular topic.

Introduction to the Issues discusses background topics such as international concerns, environmental regulations, and generators of pollutants. Degradation of the environment is not a problem that is restricted to the United States, or even to developed countries. On the contrary, underdeveloped countries are struggling with several environmental issues that have already been resolved in many developed countries. In the United States, the Environmental Protection Agency (EPA) as well as the individual states are working hard to implement regulations addressing areas of environmental concern. Generators and sources of pollutants are being identified so that solutions may be targeted to specific areas.

Air Pollution Management Issues looks into several different areas related to air pollutants and their control. Atmospheric dispersion of pollutants can be mathematically modeled to predict where pollutants emitted from a particular source, such as a combustion facility stack, will fall to the ground and at what concentration. Pollution control equipment can be added to various sources to reduce the amount of pollutants before they are emitted into the air. Acid rain, the greenhouse effect, and global warming are all indicators of adverse effects to the air, land, and sea which result from the excessive amount of pollutants being released into the air. One topic that few people are aware of is the issue of indoor air quality. Inadequate ventilation systems in homes and businesses directly affect the quality of health of the people within the buildings. For example, the episode of Legionnaires' disease which occurred in Philadelphia in the 1970s was related to microorganisms which grew in the cooling water of the air-conditioning system. Noise pollution, although not traditionally an air pollution topic, is included in this section. The effects of noise pollution are not generally noticed until hearing is impaired. And although impairment of hearing is a commonly known result of noise pollution, few people realize that stress is also a significant result of excessive noise exposure. The human body enacts its innate physiologic defensive mechanisms under conditions of loud noise, and the fight to control these physical instincts causes tremendous stress on the individual.

Pollutant dispersion in water systems and wastewater treatment are discussed in *Water Pollution Management Issues.* Pollutants entering our rivers, lakes, and oceans come from a wide variety of sources, including stormwater runoff, industrial discharges, and accidental spills. It is important to understand how these substances disperse in order to determine how to control them. Municipal and industrial wastewater treatment systems are designed to reduce or eliminate problem substances before they are introduced into natural water systems, industrial use systems, drinking water supply, and other water systems. Often, wastewater from industrial plants must be pretreated before it can be discharged into a municipal treatment system.

Solid and Radioactive Waste Management Issues addresses treatment and disposal methods for municipal, medical, and radioactive wastes. Programs to reduce and dispose of municipal waste include reuse, reduction, recycling, and composting, in addition to incineration and landfilling. Potentially infectious waste generated in medical facilities must be specially packaged, handled, stored, transported, treated, and disposed of to ensure the safety of both the waste handlers and the general public. Radioactive waste may have far more serious impacts on human health and the environment, and treatment and disposal requirements for radioactive substances must be strictly adhered to.

Similar items to which attention must be given are considered under *Hazardous Waste Management Issues.* Incineration has been a typical treatment method for hazardous waste for many years. Superfund was enacted to identify and remedy uncontrolled hazardous waste sites. It also attempts to place the burden of cleanup on the generator rather than on the federal government. Asbestos, household hazardous wastes, used oil, metals, and underground storage tanks either contain, or inherently are, hazardous materials which require special handling and disposal. Further, it is important to realize that small quantity generators of hazardous wastes are regulated as well as large generators.

Pollution Prevention covers domestic and industrial means of reducing pollution. This can be accomplished through (a) proper residential and commercial building design; (b) proper heating, cooling, and ventilation systems; (c) energy conservation; (d) reduction of water consumption; and (e) attempts to reuse or reduce materials before they become wastes. Domestic and industrial solutions to environmental problems result from considering ways to make our homes and workplaces more energy-efficient as well as ways to reduce the amount of wastes generated within them.

The section *Additional Environmental Concerns and Management Considerations* includes a discussion of electromagnetic fields which emanate from power distribution systems. It then moves on to discuss a variety of management considerations that are important concerns in every industry. Items related to both worker and community health and safety and training are brought to the forefront by the increasingly stringent regulations developed by the Occupational Safety and Health Administration (OSHA) and other federal and state regulatory agencies. The best way to prevent a dangerous situation is to be informed of the possible outcomes ahead of time and to be prepared to respond to an emergency situation. Guidelines on how to monitor the results of an environmental action are needed to determine how well an existing cleanup effort is proceeding or how present background levels will affect discharges from new facilities. Economic considerations also play a large role in the implementation of an environmental strategy.

Three new waste remediation technologies are discussed in *New Technologies and Approaches.* Bioremediation is a process that utilizes microorganisms to transform harmful substances into nontoxic compounds. It may be used to treat contaminated soil or groundwater, and it is one of the most promising new technologies for treating chemical spills and hazardous wastes. Soil vapor extraction is used in to remove volatile organic compounds from soil. A vacuum is applied to the soil, causing the movement of vapors toward extraction wells. Volatiles are then readily removed from the subsurface of the soil through the extraction wells. Biofiltration is a process which exploits the ability of microorganisms to remove biodegradable substances in air (gas) streams. In the past, it has been used successfully in Europe to remove odors from wastewater treatment plants and compost factories, and it is now being used to remove volatile organic compounds.

Managers also need to be informed on how to make decisions about associated risks and how to communicate these risks and their effects on the environment to the public. *Risk Related Topics* looks at short-term and long-term threats to human health and the environment. Risk assessment is the most important consideration for remediation of harmful effects stemming from the presence of a hazardous substance, and risk-based decision-making is a tool which is now routinely being used to select a clean-up alternative. The section provides an explanation of both how to estimate and how to avoid environmental, health, and hazard risks.

The section *Recent Developments* discusses four topics that are relatively new in the area of environmental management. ISO 14000 is an international certification standard for an organization's environmental management system. It ensures that the objectives, targets, procedures, and systems of the environmental management system are part of the organization's routine operations. *Environmental audits* provide a means of assessing the environmental condition of the organization, to prevent health risks. *Environmental justice* is a new term for describing the disproportionate distribution of environmental risks in minority and low-income communities. Federal attention is now focused on environmental and human health conditions in these areas, with the goal of achieving equality of environmental protection for all communities. And finally, environmental ethics, which we will discuss in more detail in the next section, relates to rules of proper environmental conduct.

This book is not intended to be all-encompassing. Rather, it is to be used as a starting point. References are provided in each chapter which provide more detailed information on each topic.

1.3 THE FUTURE—A QUESTION OF ETHICS

Now that a regulatory framework and general guidelines are in place, it is up to the individual to follow these procedures and strive to provide a safer, cleaner environment for the future. *Ethics* is defined as the branch of philosophy dealing with the rules of right conduct. When defining environmental ethics, right conduct with respect to the environment must be evaluated. The American Consulting Engineers Council Professional and Ethical Conduct Guidelines state that "consulting engineers shall hold paramount to the safety, health and welfare of the public in the performance of their professional duties. Consulting engineers shall at all times recognize that their primary obligation is to protect the safety, health, property and welfare of the public. If their professional judgment is overruled under circumstances where the safety, health, property or welfare of the public are endangered, they shall notify their client and such other authority as may be appropriate. Consulting engineers shall approve only engineering work which, to the best of their knowledge and belief, is safe for public health, property and welfare and in conformity with accepted standards."

Ethics responsibilities will be discussed further in Chapter 47. For now, it is important to realize that this statement of ethics can be extended to apply to all individuals. Whether your particular area of responsibility lies with the government, industry, or your family, or whether you are an engineer, attorney, politician, or scientist, your actions should aim to promote the general health and well-being of the community in which you work and live and the surrounding environment. If every individual makes even a small contribution to the reduction of pollution and the protection of the environment, the sum of the contributions will be great.

2

INTERNATIONAL ENVIRONMENTAL CONCERNS

2.1 INTRODUCTION

Environmental pollution has transcended national boundaries and is threatening the global ecosystem. International environmental concerns such as stratospheric ozone depletion, the greenhouse effect, global warming, deforestation, acid rain, and mega-disasters such as the devastating nuclear accident at Chernobyl and the toxic methyl isocyanate gas accidentally released by a subsidiary of Union Carbide in Bhopal, India, have set the stage to address global pollution problems. The potential effects of global environmental pollution necessitate global cooperation in order to secure and maintain a livable global environment.

Pollution that crosses political boundaries, such as acid rain, has caused friction between countries for at least a decade. Now, however, people are beginning to recognize a class of pollution problems that can affect not just one region, but the entire planet (1).

Over twenty years have passed since 113 governments agreed in Stockholm, Sweden, to cooperate in attacking a new threat to human welfare: the degradation of the global ecosystem from environmental pollution, over-population, and mismanagement of the natural resource base. Since then our awareness of how human and biological systems interlock has increased significantly, resulting in a far more sophisticated grasp of what must be done.

Is the goal of an environmentally healthy world realistic? To answer that question, one has only to examine the extent and significance of what has been accomplished to date. Governments have responded to the environmental challenge at national, regional, and global levels with a broad spectrum of institutional and programmatic initiatives. Indeed, despite many false starts, setbacks, continuing constraints, and the emergence of new hazards, the spirit of international cooperation has steadily grown stronger (2).

Some pollution is found throughout the world's oceans, which cover about two-thirds of the planet's surface. Marine debris, farm runoff, industrial waste, sewage, dredge material, stormwater runoff, and atmospheric deposition (acid rain) all contribute to ocean pollution. Litter and chemical contamination occur across the globe, including in such remote places as Antarctica and the Bering Sea. But the level of pollution varies a good deal from region to region and from one locality to another.

The open ocean is generally healthy, especially in comparison to the coastal waters and semi-enclosed seas that are most directly affected by human activities. The pressures from those activities are immense; some 50–75% of the world's population probably will live within 50 miles of a coastline within the next 10 years.

The EPA is working through various federal laws and international agreements to reduce marine pollution. For example, the EPA is helping to carry out the London Dumping Convention, MARPOL (the International Convention for the Prevention of Pollution from Ships), the Great Lakes Agreement, the Caribbean Regional Sea Program, and other marine multilateral and bilateral agreements. The EPA is also helping to develop regional and international programs to control discharges to the oceans from the land. In a major domestic initiative, the EPA is working to clean up major estuaries and coastal areas that have suffered the most from pollution (1).

A major global problem, depletion of the ozone layer, is linked to a group of chemicals called *chlorofluorocarbons* (CFCs). These chemicals are used widely by industry as refrigerants and by consumers in polystyrene products. Once released into the air, they rise into the stratosphere and eat away at the earth's protective ozone layer. The ozone layer shields all life on the planet from the sun's hazardous ultraviolet radiation, a leading cause of skin cancer.

The United States, which has been in the forefront of efforts to reduce CFC emissions, outlawed the use of CFCs in aerosol spray products more than a decade ago. We also have joined other nations, in a treaty called the Montreal Protocol, in pledging to eliminate the use of CFCs by the year 2000. Equally important is the commitment by industry to develop products and processes that don't use CFCs, and to share these substitutes with other countries. The EPA evaluates all possible substitutes to make sure they do not present new health or environmental problems.

The Clean Air Act of 1990 contains many measures to protect the ozone layer. Most important, the law requires a gradual end to the production of chemicals that deplete the ozone layer. CFC refrigerants found in car air conditioners, household refrigerators, and dehumidifiers—also known as R-12—were no longer produced after 1995. Hydrochlorofluorocarbon (HCFC) refrigerants for windows and central air-conditioning units also known as R-22—will be produced until 2020. The production of halons ended after 1993, while methyl bromide production will be limited beginning in 1994 and phased out by 2001. The Clean Air Act also bans the release of ozone-depleting refrigerants during the service, maintenance, and disposal of air conditioners and all other equipment that contains these refrigerants (3).

The threat to the ozone layer illustrates an important principle: It is not enough simply to outlaw an environmental problem. One also must work toward a comprehensive and economically acceptable solution.

Except for solar, nuclear, and geothermal power, the production of energy requires that something be burned. That something is usually a fossil fuel such as oil, gasoline, natural gas, or coal. But it also can be other fuels — for example, wood or municipal waste.

Burning any of these substances uses up oxygen and creates carbon dioxide gas. Bodies also create and exhale carbon dioxide. Plants, algae, and plankton, on the other hand, take in carbon dioxide and produce oxygen.

Modern industrial society and its need for power create far more carbon dioxide than the planet's vegetation can consume. As this excess carbon dioxide rises into the atmosphere, it acts as a kind of one-way mirror, trapping the heat reflected from the Earth's surface. Many leading scientists expect that this "greenhouse" effect from increased levels of carbon dioxide and other heat-trapping gases eventually will cause an increase in global temperatures. Some predict that temperatures will rise significantly within the twenty-first century, and that global climate patterns could be dramatically disrupted.

If these experts are correct, areas in the United States that are now crop land could become desert, and ocean levels could rise by three feet or more. The EPA is working with other federal agencies to improve the understanding of the likely amount and possible effects of global climate change. The EPA also is looking at ways to reduce carbon dioxide and other greenhouse gas emissions. This effort, like the agreement to eliminate CFCs, will require a major commitment to international cooperation by all the countries of the world.

Tropical rainforests, by absorbing large quantities of carbon dioxide, help to retard global warming. The rapid depletion of these tropical forests as well as those in the temperate zone has become a pressing global concern in recent years.

New data suggest that tropical forests are being lost twice as fast as previously believed; at present rates of destruction, many forests will disappear within 10 to 15 years. In July 1990, concern for the rapid loss of the great forest systems worldwide led the United States to propose a global forest convention at an economic summit attended by most industrialized nations. The agreement would address all forests — north temperate, temperate, and tropical — as well as mapping and monitoring research, training, and technical assistance (1).

2.2   THE GREENHOUSE EFFECT (4)

The threat of global warming now forces a more detailed evaluation of the environment. It forces consideration of the sacrifices which must be made to ensure an acceptable quality of the environment for the future.

As an environmental problem, global warming must be considered on an entirely different scale from that of most other environmental issues: The effects of climate change are long-term, global in magnitude, and largely irreversible. Because of the enormity of the problem and the uncertainties involved — it may take decades to determine with absolute certainty that global warming is under way — the difficult questions faced today are how and when one should react.

Fossil-fuel burning and forestry and agricultural practices are responsible for most of the man-made contributions to the gases in the atmosphere that act like a greenhouse to raise the earth's temperature; hence the term *greenhouse effect*. Most of the processes that produce greenhouse gases are common everyday activities such as driving cars, generating electricity from fossil fuels, using fertilizers, and using wood-burning stoves. Because so many of these activities are so ingrained in our society, reducing emissions could be a difficult task.

Environmentally, the potential effects of climate change are extensive. The earth's ecosystems, water resources, and air quality could all experience profound impacts; agriculture and forestry could be seriously affected. Politically, global climate change has the potential to become a very sensitive issue among countries if nations cannot agree on a comprehensive solution — and if climate change shifts the relative advantages among them.

Global temperatures in 1998 were the warmest in the past 119 years. Over the past century, the average surface temperature has increased by 1°F. Evidence of global warming is confirmed by melting glaciers, decreased snow cover in the Northern Hemisphere, and even warming below ground. Rising global temperatures are expected to rise sea level and change precipitation and other local climate conditions contributing to numerous natural disasters. A rapid reversal in the sea surface temperature anomaly pattern manifested itself in the east equatorial Pacific as warm anomalies (El Nino) transitioned to cold anomalies (La Nina) during the latter half of 1998. Accompanying these conditions, large portions of central North America experience increased storminess, increased precipitation, and increased significant cold air outbreaks.

And now that the search for solutions has begun, there is a growing concern that the costs of reducing emissions may be too high. But to put cost concerns in proper perspective, one must ask what kind of future one wants on this planet and how much does one value the environment and the cultural heritage that depends on it (4).

A consensus has emerged in the scientific community that a global warming will occur. Scientists are certain that the concentrations of carbon dioxide (CO_2) and other greenhouse gases in the atmosphere are increasing, and they generally agree that these gases will warm the earth. Two questions remain to be answered: how much will the temperature rise, and when.

Recent estimates indicate that if the concentrations of these gases in the atmosphere continue to increase, the earth's average temperature could rise by as much as 1.5–4.5°C in the next century. While this may not sound like a tremendous increase, one must keep in mind that during the last ice age 18,000 years ago, when glaciers covered much of North America, the earth's average temperature was only 5°C cooler than today.

Certainly global cooperation is an important consideration when addressing global warming issues. No single country contributes more than a fraction of greenhouse gases, and only a concerted effort can reduce emissions. In the future, as developing nations grow and consume more energy, their share of greenhouse-gas emissions will steadily increase. It is important for other nations to offer technological assistance so that these developing nations can grow in an energy-efficient manner (4).

The sources of greenhouse gases are so numerous and diverse that no single source contributes more than a tiny fraction of total emissions. Similarly, no single country contributes more than a fraction of emissions.

Unlike other environmental problems that the EPA could address with the stroke of a regulation, potential climate change is a problem that needs innovative global solutions. Future trends of emissions will depend on a wide range of factors, from population and economic growth to technological development and policies to reduce emissions. Past trends show that all countries have been producing greenhouse gases at a growing rate, and many countries will continue to do so for years to come. Based on careful study of the sources and trends of greenhouse emissions around the globe, countries can begin implementing prudent measures for slowing down emissions while increasing economic development.

The United States is responsible for the largest portion of human-made contributions to the greenhouse effect (21%), followed by the countries that comprised the former Soviet Union, (14%), European countries (14%), India (4%) and the rest of the world, 36%. The rate of greenhouse-gas buildup during the next century will depend heavily on future patterns of population and economic growth and technological development; these, in turn, are influenced by the policies of local, state, national, and international private and public institutions.

To assemble a better picture of how emissions may change in the future, the EPA, in conjunction with other countries and the International Panel on Climate Change, is assessing future energy plans of different countries and their implications for emissions of Greenhouse gases. The approach relies on information from individual countries evaluated in comparison to results obtained from large global economic models such as were used in preparing the EPA's recent draft report to Congress titled *Policy Options for Stabilizing Global Climate*.

As with all attempts to forecast into the future, the results become less reliable the farther they extend into the future; however, from the projections summarized in Table 2.1, a certain picture of the future emerges. The analysis suggests that global CO_2 emissions will more than double by the year 2025 (from 5.24 to 12.18

TABLE 2.1 Projected global CO_2 emissions[a]

	Billion Tons Carbon		Percentage Annual Growth	Per-Capita CO_2 Emissions[b] (metric tons/year) in 2025
	1985	2025		
Developed	**3.95**	**6.71**	**1.31**	**4.24**
North America	1.46	2.37	1.23	6.50
Western Europe	0.77	1.11	0.91	2.63
Japan and Australia	0.34	0.63	1.53	2.65
Eastern Europe	1.38	2.60	1.60	5.07
Developing	**1.29**	**5.47**	**3.91**	**0.80**
Centrally Planned Asia	0.55	1.80	3.00	0.98
South and East Asia	0.28	1.55	4.41	0.50
Latin America	0.21	0.65	2.91	1.68
Africa	0.14	0.80	4.53	0.48
Middle East	0.11	0.67	4.72	1.91
Global	**5.24**	**12.18**	**2.61**	**1.41**

[a]Note: These projections assumes no specific international agreements to reduce emissions.
[b]Per-capita CO_2 emissions are calculated for each region based on projected CO_2 emissions and population.

billion tons per year) in the absence of specific government policies to reduce emissions. This estimate is higher than that indicated in EPA's draft report to Congress; most individual countries tend to be optimistic about their future use of energy and do not consider global constraints.

The developed countries, currently the largest CO_2 emitters, will grow in population at approximately 1.0–1.5% per year and are projected to emit 6.7 billion tons of carbon by the year 2025. Developed countries are likely to continue to emit more CO_2 per person than developing countries. For example, the average citizen living in the United States produced six times more CO_2 each year than the average citizen in a developing country. In developing countries, population and economic growth will lead to a substantial increase in CO_2 emissions to over 5 billion tons per year, despite anticipated improvements in efficiency of energy use.

Developing countries now contribute only a small fraction of Greenhouse gases, but their share of emissions is expected to increase significantly in the next 35 years. The table shows the share of CO_2 emissions from Asia (including China), Africa, Latin America, and the Middle East increasing from slightly over one-fourth of the global total in 1985 to nearly one-half the total by 2025. Technologies developed in more industrialized nations to use energy efficiently could help developing nations reduce emissions as they continue to develop, but channels to transfer this technology must be developed.

On a regional basis, energy use in Western European countries is projected to grow at a relatively slow rate because of low population growth and policies that are anticipated to be implemented over the next decade. Several countries, such as Norway, Sweden, and The Netherlands, have already adopted policies specifically designed to slow the growth rate of greenhouse-gas emissions. These measures include special taxes, energy-efficiency programs, and promotion of nuclear energy, natural gas, and renewable energy sources.

The case in Eastern Europe is quite different, largely because many of these countries are among the most energy intensive and most energy inefficient in the world. In Eastern Europe and the countries of the former Soviet Union, energy use and CO_2 emissions are projected to grow considerably over the next 35 years, but policies aimed at restructuring the economy and improving energy efficiency in Russia and the CIS could have a significant impact. If these economies and those of Eastern Europe become more energy-efficient and move from heavy industrial production to production of less energy-intensive consumer goods, they may be able to increase economic growth and enjoy the added benefit of reduced greenhouse-gas emissions.

In the coming years, we must reevaluate how emissions are likely to change. But given this preliminary picture of the future, it is important to take the next step of assessing the specific technologies and policy measures that can reduce emissions now at low costs. Each country will have to examine its unique situation and determine appropriate responses. However, only by acting together will the global community slow the trend toward high emissions in the next century (3). Methods for reducing the greenhouse effect are discussed in more detail in the chapters addressing pollution prevention approaches, waste reduction and energy conservation (Chs. 26, 28, and 29).

2.3 OZONE DEPLETION IN THE STRATOSPHERE (6)

Increasing concentrations of the synthetic chemicals known as CFCs and halons are breaking down the ozone layer, allowing more of the sun's ultraviolet rays to penetrate to the earth's surface. Ultraviolet rays can break apart important biological molecules, including DNA. Increased ultraviolet radiation can lead to greater incidence of skin cancer, cataracts, and immune deficiencies, as well as decreased crop yields and reduced populations of certain fish larvae, phytoplankton, and zooplankton that are vital to the food chain. Increased ultraviolet radiation can also contribute to smog and reduce the useful life of outdoor paints and plastics. Stratospheric ozone protects oxygen at lower altitudes from being broken up by ultraviolet light and keeps most of these harmful rays from penetrating to the earth's surface.

Chlorofluorocarbons (CFCs) are compounds that consist of chlorine, fluorine, and carbon. First introduced in the late 1920s, these gases have been used as coolants for refrigerators and air conditioners, propellants for aerosol sprays, agents for producing plastic foam, and cleaners for electrical parts. CFCs do not degrade easily in the troposphere. As a result, they rise into the stratosphere where they are broken down by ultraviolet light. The chlorine atoms react with ozone to convert it into two molecules of oxygen. In the upper atmosphere ultraviolet light breaks off a chlorine atom from a CFC molecule. The chlorine attacks an ozone molecule, breaking it apart. An ordinary oxygen molecule and a molecule of chlorine monoxide are formed. A free oxygen atom breaks up the chlorine monoxide. The chlorine is free to repeat the process. Chlorine acts as a catalyst and is unchanged in the process. Figure 2.1 illustrates how ozone is destroyed. Consequently, each chlorine atom can destroy as many as 10,000 ozone molecules before it is returned to the troposphere.

Halons are an industrially produced group of chemicals that contain bromine, which acts in a manner similar to chlorine by catalytically destroying ozone. Halons are used primarily in fire extinguishing foam.

Laboratory tests have shown that nitrogen oxides also remove ozone from the stratosphere. Levels of nitrous oxide (N_2O) are rising from increased combustion of fossil fuels and use of nitrogen-rich fertilizers (6).

In the early 1970s, CFCs were primarily used in aerosol propellants. After 1974, U.S. consumption of aerosols had dropped sharply as public concern intensified about stratospheric ozone depletion from CFCs. Moreover, industry anticipated future regulation and shifted to other, lower cost chemicals. In 1978, EPA and other federal agencies banned the nonessential use of CFCs as propellants. However, other uses of CFCs continued to grow, and only Canada and a few European nations followed the United States's lead in banning CFC use in aerosols.

In recognition of the global nature of the problem, 31 nations representing the majority of the CFC-producing countries signed the Montreal Protocol in 1987. The Protocol, which had to be ratified by at least 11 countries before it became official at the start of 1989, requires developed nations to freeze consumption of CFCs at 1986 levels by mid-1990 and to halve usage by 1999. The Protocol came into force, on time, on January 1, 1989, when 29 countries and the EEC representing approximately 82% of world consumption had ratified it. Since then

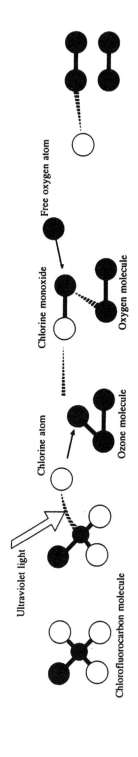

Ultraviolet light

Chlorine atom

Chlorine monoxide

Free oxygen atom

Chlorofluorocarbon molecule

Ozone molecule

Oxygen molecule

In the upper atmosphere ultraviolet light breaks off a chlorine atom from a chlorofluorocarbon molecule.

The chlorine attacks an ozone molecule breaking it apart. An ordinary oxygen molecule and a molecule of chlorine monoxide are formed.

A free oxygen atom breaks up the chlorine monoxide. The chlorine is free to repeat the process.

FIGURE 2.1 How ozone is destroyed.

several other countries have joined. Now 165 countries are parties to the Convention and the Protocol, of which well over 100 are developing countries. The Protocol is constructively flexible and it can be tightened as the scientific evidence strengths without having to be completely renegotiated. Its control provisions were strengthened through 4 adjustments to the Protocol adopted in London (1990), Copenhagen (1992), Vienna (1995), and Montreal (1997). The Protocol aims to reduce and eventually eliminate the emissions of human-made ozone depleting substances.

In addition to implementing the Montreal Protocol, the EPA is working with industry, the military, and other government organizations to reduce unnecessary emissions of CFCs and halons by altering work practices and testing procedures, or by removing institutional obstacles to reductions. The EPA is working with the National Aeronautics and Space Administration, the National Oceanic and Atmospheric Administration, the Department of Energy, the National Science Foundation, and other federal agencies to better understand the effects of global warming and stratospheric ozone depletion.

In 1986 and again in 1987, research teams were sent to investigate the causes and implications of the hole in the ozone over Antarctica. In 1986, the EPA published a multivolume summary with the United Nations on the effects of global atmospheric change. In addition, in 1987 EPA published a major risk assessment of the implications of continued emission of gases that can alter the atmosphere and climate.

In December of 1987, the Agency published proposed regulations for implementing the Montreal Protocol. The provisions of the Protocol would be implemented by limiting the production of regulated chemicals and allowing the marketplace to determine their future price and specific uses.

While the Montreal Protocol represents a major step toward safeguarding the earth's ozone layer, considerable work remains to be done. The major challenge is to develop a better understanding of the effects of stratospheric ozone depletion and global warming on human health, agriculture, and natural ecosystems. Substantial scientific uncertainty still exists. More must be learned about the Antarctic ozone hole and its implications, both for that region and the rest of the earth. More must also be learned about recent evidence of global ozone losses of 2–5% during the past 15 years.

Efforts to develop alternatives to CFCs and halons must be expedited. The Montreal Protocol provides a clear signal for industry to shift away from these chemicals. New technologies and new chemicals that will not deplete the ozone layer and increased conservation and recovery are essential to reducing the economic effects of the Protocol both in the United States and abroad. In the time since the Protocol was signed, major advancements in alternative technologies have been announced for CFC use in food packaging and solvents. Yet these are only a beginning and more must be done.

The EPA plans to continue international efforts to protect the ozone layer and to assess the risks of future climate change. EPA will send advisory teams to several key nations to help them explore options for reducing use of CFCs, such as producing different products, substituting other chemicals, and controlling emissions.

The Clean Air Act of 1990 sets a schedule for ending the production of chemicals that destroy stratospheric ozone. Chemicals that cause the most damage will be phased out first. CFCs, Halons, HCFCs and other ozone-destroying chemicals were listed by Congress in the 1990 Clean Air Act and must be phased out. CFCs from car air conditioners are the biggest single source of ozone-destroying chemicals. By the end of 1993, all car air conditioners systems were required to be serviced using equipment that recycles CFCs and prevents their release into the air. Only specially trained and certified repair persons are allowed to buy the small cans of CFCs used in servicing auto air conditioners. Methly chloroform, also called 1,1,1-trichloroethane, was phased out by 1996. This had been a widely used solvent found in products such as automotive brake cleaners (often sold as aerosol sprays) and spot removers used to take greasy stains off frabics. Replacing methyl chloroform in the workplace and consumer products has led to changes in many products and processes (5).

2.4 OZONE/SMOG (7)

Ozone or "smog" is just one of six major air pollutants that EPA regulates, but it is by far the most complex, intractable, and pervasive. It is also an extremely difficult pollutant to regulate effectively.

Many more Americans live in areas that suffer unhealthy levels of ozone than are affected by any other air pollutant. Adverse health effects have been observed in test animals and in humans even at exposure levels only slightly higher than federal health standards for ozone. In fact, concentrations of the pollutants are often far higher than federal standards in many urban areas of the country. Permanent damage to respiratory systems and other adverse health effects are known to occur from repeated exposure to ozone at such high levels.

Ozone is difficult to control because of the extremely large number of individual sources that can contribute to its formation, and because much of the pollution these sources produce may be transported to areas long distances away. Unlike the other major air pollutants, ozone is not emitted directly by specific sources. Instead, it is formed in the air by chemical reactions from nitrogen oxides and volatile organic compounds (VOCs). In each area, the sources of ozone may consist of literally thousands of large and small stationary sources in addition to motor vehicles — the major mobile source contributor. Sources of VOCs include products of combustion from motor vehicle engines and other machinery; vapors of gasoline emitted by motor vehicles, service station pumps, gasoline refineries, and petroleum storage tanks; and chemical solvent vapors emitted by a host of commercial and industrial sources such as dry-cleaning establishments, solid waste facilities, and metal surface paints.

The reactions that form ozone are stimulated by sunlight, so that ozone reaches peak levels in most of the United States during the summer months — when air particularly is stagnant for extended periods. This type of pollution first gained public attention in the 1940s as Los Angeles "smog." The highest concentrations have long been found in that city, but very high concentrations also began to develop in other areas as motor vehicle travel increased following World War II. Ozone generally affects all areas that have extended periods of abundant sunlight coupled with high emissions from motor vehicles — a major source of both VOCs and nitrogen oxides.

Ozone severely irritates the mucous membranes of the nose and throat, impairs normal functioning of the lungs, and reduces the ability to perform physical exercise. In general, the pollutant's adverse health effects depend on a combination of factors: the amount of ozone in the air, and the frequency and duration of exposure. However, the effects of ozone at any concentration are felt most by people with asthma, chronic obstructive lung disease (such as emphysema), or allergies, and by persons who regularly perform strenuous exercise outdoors. Sensitive individuals may experience adverse health effects from even relatively low concentrations of the pollutant. It also appears that ozone in combination with other pollutants presents greater potential respiratory effects than any single air pollutant alone.

The health effects of ozone have been confirmed in closely controlled and monitored laboratory testing programs and in epidemiological surveys of population groups that are routinely exposed to high concentrations of the pollutant. When ozone levels are up, hospital admissions go up, there is more sickness generally, and physical activity becomes difficult even for healthy individuals. The most vulnerable suffer extreme discomfort and distress.

In addition to a growing body of evidence about the health effects of ozone, there are recent findings about ozone's adverse effects on cash crops, forests, and other forms of vegetation. Since the late 1970s, the EPA has conducted extensive field surveys of ozone's effects on agriculture through the National Crop Loss Assessment Network (NCLAN) study. This study puts the agricultural loss from ozone pollution at between $2 and $3 billion each year. One set of studies showed that even levels of ozone below the health standard can reduce several major cash crops by as much as 10% a year. Additional levels have reduced plant yield in tomatoes by 33%, beans by 26%, soybeans by 20%, and snapbeans by up to 22%.

Conclusive statements about the role of ozone and other air pollutants in damage to forests are not possible at present because data are limited. Many scientists, however, think ozone is a major contributor to the decline in growth of many species of trees. The existing data, though limited, do suggest strongly that ozone pollution has played a role in the loss of at least some forests. One study in the San Bernardino Mountains of California concluded that ozone was the cause of foliar injury, premature leaf drop, decreased radial growth and photosynthetic capacity, and death by bark beetles in ponderosa and Jeffrey pine. Repeated ozone peaks near the standards have been implicated in damage to white pine in the eastern United States and Canada, and reduced growth rates for the red spruce at numerous high elevation sites in the Appalachian Mountains.

It should be noted that the actual chemistry of smog formation is extremely complicated. There are variations according to temperature level, quantity of sunshine, and wind patterns. Also, substances other than VOCs play a role in the photochemical reaction that generates ozone. Prominent among these are nitrogen oxides. The EPA has determined, however, that VOCs should be the principal target of efforts to control ozone-containing smog in most cases.

The fumes from internal combustion engines contain many VOCs that, when released into the atmosphere, interact with other gases in the presence of sunlight to generate the ozone components of urban smog.

Various new technologies, most notably the catalytic converter, have led since the mid-1970s to major reductions in tailpipe emissions of VOCs, as well as the nitrogen oxides also linked to the smog problem. In fact, the use of the catalytic converter on passenger cars and light trucks became virtually universal by the early 1980s as the auto industry scrambled to meet new EPA regulatory deadlines. Unfortunately, its effectiveness has in many cases been undermined by motorists who fouled the devices with leaded gasoline or by mechanics who illegally removed them.

Interest in alternative motor vehicle fuels has come full circle. Alternative fuels can be divided into two distinct groups: those that could completely replace gasoline and those that can be low-level additives to gasoline. The three primary replacement fuels of interest are methanol, ethanol, and compressed natural gas (CNG).

Methanol, ethanol, and CNG all have the potential to significantly reduce the contribution of motor vehicles to ozone formation. This is not so much because these fuels produce fewer hydrocarbon emissions compared to gasoline, but rather because their hydrocarbon emissions have been shown to be far less photochemically reactive than those of gasoline. It has been known for quite some time that methane emissions, the primary hydrocarbon in CNG vehicle exhaust, are very, very slow to react in the atmosphere. More recently, it was learned that methanol emissions also have a low photochemical reactivity, and ethanol has a higher, but still relatively low, reactivity.

Methanol is an excellent engine fuel that can be produced from natural gas, coal, or biomass. It is currently priced at a level fairly close to gasoline on an energy basis. Presently, there are more than 1,000 methanol vehicles operating in California, and those vehicles use engines very similar to those in today's gasoline vehicles. Projections have been made that emissions from current methanol vehicles create 20–50% less ozone than comparable gasoline vehicles. Cold startability and formaldehyde emissions are two areas of concern.

Ethanol is produced in the United States primarily by fermenting grains such as corn. To date, few vehicles here have been designed to operate on pure ethanol, although Brazil's transportation system runs predominantly on ethanol. It is believed that the use of pure ethanol as a motor vehicles fuel would offer emissions benefits somewhat lower but still comparable to methanol. The primary issues associated with ethanol's use are supply and cost.

Most of the vehicles currently operating on CNG use conversion kits to allow the vehicle to operate on either CNG or gasoline. It is estimated that such vehicles, when operated on CNG, would contribute 40–80% less to ozone than gasoline vehicles. If properly performed and maintained, conversion typically provides carbon monoxide emissions reductions as well. Drawbacks associated with CNG conversions include generally higher nitrogen oxide emissions and poorer vehicle performance, due to reduced engineer power and increased weight from the pressurized CNG cylinders. As with methanol, it is believed that CNG is best suited for engines designed specifically for its use. Such vehicles would likely achieve 80–90% reduction in ozone-producing potential and very low carbon monoxide emissions.

The second group of fuels that could reduce motor vehicle emissions includes those composed primarily of gasoline with low levels of additives. Four blends have been approved by the EPA: gasohol, which contains 10% ethanol; DuPont's blend, which includes 5% methanol and 2.5% ethanol; Oxinol, which contains 5% methanol and 5% tertiary butyl alcohol, and methyl *tert*-butyl ether (MTBE), which can be blended up to 15% with gasoline. Currently gasohol accounts for 7% of all gasoline sales and MTBE blends for approximately another 10%.

The primary emission benefit of these additives is lower carbon monoxide emissions from increased air/fuel ratio. Analysis shows that gasohol, DuPont, and Oxinol blends reduce fleet-wide carbon monoxide emissions by around 22%. MTBE, which contains less oxygen, would reduce emissions by 12%. The magnitude of these reductions will decrease somewhat in the future as new cars with computer controls replace older vehicles.

The one emissions concern with respect to oxygenated blends is that the addition of ethanol and/or methanol to gasoline increases the volatility of gasoline. This in turn increases the amount of evaporative hydrocarbon emissions. Using oxygenated blends could thus increase the ozone-producing potential of motor vehicles, unless their use is limited to the winter months when carbon monoxide is typically high and ozone low, or the base gasoline is modified to provide oxygenated blends with the same overall volatility as straight gasoline.

2.5 INTERNATIONAL ACTIVITIES

Several factors are pushing environmental concerns increasingly into the international arena. More and more, pollution is transboundary and even global in scope. Pressures on shared resources, such as river basins and coastal fisheries, are mounting. Resource deterioration in many nations is so extensive that other countries are affected for example, when ecological refugees flee across borders. As international trade increases, commodities and merchandise become the carriers of domestic environmental policies that must be rationalized.

It is not just that there are more environmental problems like ozone depletion that must be dealt with at the international level; it is also that the line between national and international environmental problems is fast disappearing.

Nitrogen oxide emissions, for example, must be regulated locally because of ground-level ozone formation, regionally because of acid rain, and globally because ground-level ozone is an infrared-trapping greenhouse gas. Methane and indirectly, carbon monoxide also contribute to the greenhouse effect.

In these instances, domestic and global environmental concerns push in the same direction. On the other hand, a major move to methanol as a substitute for gasoline could actually increase the global warming risk. A car burning methanol made from coal would result in perhaps twice the carbon dioxide emissions per mile as one burning gasoline.

Environmental diplomacy is the logical outgrowth of the desire to protect one's own national environment, to minimize environment-related conflicts with other countries, and to realize mutual benefits, including economic progress and the protection of the common natural heritage of humankind. As such, it is not entirely new. The register of international conventions and protocols in the field of the environment has grown steadily in this century: the main multilateral treaties today number about 100, many of them having to do with the protection of the marine environment and wildlife.

What is new is the prospect that environmental issues will move from being a secondary to a primary international concern and increasingly crowd the diplomatic agendas of nations. And these diplomatic agendas in turn will increasingly affect domestic environmental policy. U.S. environmental policy will more and more be set in concert with those of other nations. Efforts to give international dimensions a higher priority within the Agency should continue. The EPA has established an Office of International Affairs to address international activities. Even more important is ensuring that domestic and international activities are actually coordinated internally.

The EPA also needs a world-class capacity to follow relevant developments in other countries and in international institutions, to understand and analyze the various approaches to environmental protection being taken abroad, and to anticipate future needs and developments at the international level. Beyond EPA's internal workings, new patterns of relating to other federal agencies seem desirable. Neither global nor local atmospheric issues are likely to be solved unless energy and environmental policy are made together in the future. As environmental diplomacy increases, finding appropriate patterns of interaction with the Department of State will become imperative.

Moreover, the future is likely to bring increasing efforts to link environmental objectives and trade policy. For example, should the United States restrict imports of products that are manufactured by processes that harm the environment, much as one restricts imports of endangered species and harmful products? Should one import copper from countries where smelters operate without serious pollution control?

Much of the EPA's international activity in the past has focused on the Organization of Economic Cooperation and Development and other trans-Atlantic matters. In the future, the North–South and East–West dimensions will rival the North–North ones in importance. It already seems clear that solutions to the most serious global environmental challenges will require a series of vital understandings between the industrial and the developing countries.

For example, the developing countries will expect the industrial countries to take the first and strongest actions on global warming. They will want to see the seriousness of the threat validated, and they will conclude, quite correctly, that the industrial countries are largely responsible for the problem and have the most resources to do something about it.

But a tragic stalemate will occur if this argument is carried too far. Developing countries already account for about a fourth of all greenhouse gas emissions, and their share could double by the middle of the next century. Increasingly, all countries will be pressed to adopt energy and forestry strategies that are consistent with containing the greenhouse effect within tolerable limits.

The United States and the EPA need to build a new set of relationships with developing country officials so that confidence and trust are built for the challenging times ahead. One major step in this direction would be for the United States to initiate a new program of international environmental cooperation with developing countries. Such a program would not be limited to aid-eligible countries but would extend to countries like Brazil and Mexico. It would provide technical assistance, training, access to information and expertise, and planning grants all aimed at increasing the capacity of developing countries to manage their environmental challenges (8).

The EPA's overseas activities includes negotiating international environmental treaties, maintaining liaison with other health and environment organizations, and cooperating with and encouraging the environmental initiatives of other nations, particularly Third World countries. In addition, the EPA works to spare U.S. industry from unfair foreign competitors benefiting from pollution havens.

2.6 INTERNATIONAL PERSPECTIVE

Acid rain is not considered a threat to the global environment. Large parts of the earth are not now, and probably never will be, at risk from the effects of man-made acidity. But concern about acid rain is definitely growing. Although acid rain comes from the burning of fossil fuels in industrial areas, its effects can be felt on rural ecosystems hundreds of miles downwind. And if the affected area is in a different country, the economic interests of different nations can come into conflict.

Such international disputes can be especially difficult to resolve because we do not yet know how to pinpoint the sources in one country that are contributing to environmental damage in another.

Concerns about acid rain tend to be raised whenever large-scale sources of acidic emissions are located unwind of international borders. Japan, for example, has not yet suffered any environmental damage due to acid rain, but the Japanese are worried about the potential downwind effects of China's rapidly increasing industrialization. A similar problem has risen on the U.S.–Mexican border, where some people were worried that Mexico's copper smelter at Nacozari could cause acid rain on the pristine peaks of the Rocky Mountains. Besides scattered instances such as these, acid rain has emerged as a serious international issue only in two places: western Europe and northeastern North America.

Europe

Diplomatic problems related to cross-boundary air pollution first surfaced in Europe in the 1950s, when the Scandinavian countries began to complain about industrial emissions traveling across the North Sea from Great Britain. Since then, acid deposition has been linked to ecological damage in Norway, Sweden, and West Germany, and low-pH rainfall has been measured in a number of other European countries.

The potential and scientific controversies over acid rain are multiplied in Europe because so many countries are involved. Table 2.2 lists the SO_2 emissions of 21 European nations in 1980. Some countries producing very low amounts of SO_2 are nevertheless experiencing low-pH rainfall and high rates of acid deposition. Norway, for example, produced approximately 137,000 metric tons of SO_2 in 1980, yet received depositions of about 300,000 metric tons. Clearly, Norway, like a number of other European nations, is being subjected to acid deposition that originates outside its borders.

Sweden pioneered the development of extensive and consistent monitoring for acid precipitation in the late 1940s. In 1954, the Swedish monitoring program was expanded to include other European countries. The results of this monitoring revealed the high acidity of rainfall over much of western Europe.

Prompted by these findings, the U.N. Conference on the Human Environment recommended a study of the impact of acid rain, and in July 1972, the U.N. Organization for Economic Cooperation and Development (OECD) began an inquiry into "the question of acidity in atmospheric precipitation." In 1979, a U.N. Economic Commission for Europe (ECE) conference in Stockholm approved a multinational convention for addressing the problem of long-range transboundary air pollution. Both the United States and Canada joined the European signatories. Later, a number of European countries, including France, West Germany,

**TABLE 2.2   European SO$_2$ emissions in 1980
(in thousands of metric tons)**

Austria	440
Belgium	809
Bulgaria	1,000
Czechoslovakia	3,100
Denmark	399
Finland	600
France	3,270
Federal Republic of Germany	3,580
Greece	700
Hungary	1,663
Italy	3,800
Netherlands	487
Norway	137
Poland	2,755
Portugal	149
Romania	200
Sweden	450
Switzerland	119
United Kingdom	4,680
USSR	25,500
Yugoslavia	3,000

Source: U.S. Department of State.

Czechoslovakia, and all the Scandinavian countries, agreed to reduce their 1993 SO$_2$ emissions by at least 30% from 1980 levels.

Following the Stockholm conference, ECE members decided in 1985 to broaden their goals to include the control of nitrogen oxides, which have been gaining recognition as important acid rain precursors. Workshops are underway to determine the nature and extent of NO$_x$ pollution in various countries, as well as possible approaches for controlling it.

North America

The United States and Canada share the longest undefended border in the world and billions of dollars in trade every year. They also share a number of environmental problems, foremost among them the problem of acid rain. In both countries, acidic emissions are concentrated relatively close to their mutual border. Canadian emissions originate primarily in southern Ontario and Quebec, while a majority of U.S. emissions originate along the Ohio River Valley. Each country is contributing to acid rain in the other. But because of prevailing wind patterns and the greater quantities of U.S. emissions, the United States sends much more acidity to Canada than Canada sends to it. In 1980, for example, the United States produced over 23 million metric tons of SO$_2$ and over 20 million metric tons of NO$_x$; Canada produced 4.6 million metric tons of SO$_2$ and 1.7 million tons of NO$_x$.

In the early 1970s, Canadian scientists began to report on the adverse environmental effects of acidity in lake water, and to link fish kills in acidic lakes and streams in eastern Canada to U.S. emission. By the late 1970s, acid rain had become a serious diplomatic issue affecting the relationship of the two countries. In 1980, the two countries took their first joint step toward resolving the issue with a Memorandum of Intent that called for shared research and other bilateral efforts to analyze and control acid rain. One of the most spectacular projects was a high-altitude experiment called "CATEX." Trace elements of various chemicals were inserted into SO_2 plumes from coal-fired power plants in the Midwest. Their dispersion was monitored along a path extending across the northeastern United States to Canada. These and other experiments have helped scientists gain new data on the formation and distribution of acid rain.

When Brian Mulroney became prime minister of Canada in 1984, he pressed for more than research; he wanted bilateral action to control acid rain. At the first "Shamrock Summit" in March 1985, Mulroney and President Reagan agreed that Canada and the United States would each appoint a high-level special envoy to study acid rain. The special envoys would be charged with recommending a plan to alleviate both the environmental and the political damage caused by acid rain.

William Davis, former premier of Ontario, and Drew Lewis, former U.S. secretary of transportation, were named special envoys. In January 1986, the two men presented their joint recommendations for U.S.–Canadian action. They proposed a $5 billion U.S. technology demonstration program, ongoing bilateral consultations at the highest diplomatic levels, and cooperative research projects.

Western Europe and North America are highly industrialized, and it is likely that acid rain will continue to be a serious concern in both areas for the foreseeable future. But the nations involved are coming to terms with their common problem. In Europe, several nations have already taken steps to reduce transboundary air pollution. In North America, the president of the United States has endorsed the proposal to invest $5 billion to demonstrate innovative technologies that can be used to reduce transboundary air pollution. And in both Europe and North America, the diplomatic groundwork for long-term cooperative activities has been established (9).

2.7 SUMMARY

1. Environmental pollution has transcended national boundaries and is threatening the global ecosystem. International environmental concerns such as stratospheric ozone depletion, the greenhouse effect, global warming, deforestation, and acid rain have set the stage for addressing global pollution problems. The potential effects of global environmental pollution necessitate global cooperation in order to secure and maintain a livable global environment.

2. As an environmental problem, global warming must be considered on an entirely different scale from that of most other environmental issues; the effects of climate change are long-term, global in magnitude, and largely irreversible.

3. Increasing concentrations of the synthetic chemicals known as CFCs and halons are breaking down the ozone layer, allowing more of the sun's ultraviolet rays to penetrate to the earth's surface. Ultraviolet rays can break apart important biological molecules, including DNA. Increased ultraviolet radiation can lead to greater incidence of skin cancer, cataracts, and immune deficiencies, as well as

decreased crop yields and reduced populations of certain fish larvae, phytoplankton, and zooplankton that are vital to the food chain.

4. Ozone, or "smog," is just one of six major air pollutants that the EPA regulates, but it is by far the most complex, intractable, and pervasive. It is also an extremely difficult pollutant to regulate effectively.

5. Several factors are pushing environmental concerns increasingly into the international arena. More and more, pollution is transboundary and even global in scope. As international trade increases, commodities and merchandise become the vehicles of domestic environmental policies that must be rationalized across borders.

6. Acid rain is not considered a threat to the global environment. Large parts of the earth are not now, and probably never will be, at risk from the effects of man-made acidity. But concern about acid rain is definitely growing.

REFERENCES

1. U.S. EPA, Communications and Public Affairs (A-107), *Preserving Our Future Today*, 2115-1012, October 1991.
2. U.S. EPA, Office of Public Affairs (A-107), *EPA Journal*, 13(7), September 1987.
3. U.S. EPA, Office and Radiation, *Protecting the Ozone Layer*, EPA 430-F-94-007, April 1994.
4. U.S. EPA, "The Greenhouse Effect," *EPA Journal*, 16(2), 20K-9002, March/April 1990.
5. U.S. EPA, *The Plain English Guide To The Clean Air Act, Air, and Radiation*, EPA 400-K-93-001, April 1993.
6. U.S. EPA, *Environmental Progress and Challenges: EPA's Update*, EPA-230-07-88-033, August 1988.
7. U.S. EPA, "The Challenge of Ozone Pollution," *EPA Journal*, 13(8), October 1987.
8. U.S. EPA, "Protecting the Earth—Are Our Institutions Up To It?" *EPA Journal*, 15(4), July/August 1989.
9. U.S. EPA, "Acid Rain: Looking Ahead," *EPA Journal*, 12(6), June/July 1986.

3

MULTIMEDIA ANALYSES

3.1 INTRODUCTION

It is now increasingly clear that some treatment technologies (as briefly introduced in the first two chapters), while solving one pollution problem, have created others. Most contaminants, particularly toxics, present problems in more than one medium. Since nature does not recognize neat jurisdictional compartments, these same contaminants are often transferred across media. Air pollution control devices or industrial wastewater treatment plants prevent wastes from going into the air and water, respectively, but the toxic ash and sludge that these systems produce can themselves become hazardous waste problems. For example, removing trace metals from a flue gas usually transfers the products to a liquid or solid phase. Does this fact in effect exchange what had been an air quality problem for a liquid or solid waste management problem? Waste disposed of on the land or in deep wells may contaminate ground water, and evaporation from ponds and lagoons can convert solid or liquid waste into air pollution problems (1).

The need for a multimedia approach in many pollution prevention programs should be obvious. This chapter introduces the multimedia analysis concept and applies it to three separate situations: (a) a chemical plant, (b) products and services, and (c) a hazardous waste incineration facility.

3.2 HISTORICAL PERSPECTIVE

The EPA's own single-media offices, often created sequentially as individual environmental problems were identified and responded to through legislation, have played a role in impeding development of cost-effective multimedia prevention strategies. In the past, innovative cross-media agreements involving or promoting pollution prevention, as well as involving or promoting voluntary arrangements for overall reductions in the releases of pollutants, have as a result not been encouraged. However, newer initiatives have been characterized by the use of a wider range of tools, including market incentives, public education and information, small business grants, technical assistance, and research and technology applications, as well as the more traditional regulations and enforcement (see Chapter 4 for additional details).

In the past, the responsibility for pollution prevention and/or waste management was delegated to engineers functioning as the equivalent of an environmental control department. The personnel involved were thus skilled in engineering treatment techniques but had almost no responsibility for and thus no control over what went into the plant that generated the waste they were supposed to manage. In addition, most engineers are trained to make a product work, not to minimize or prevent pollution. There is still little emphasis on pollution prevention in the educational curricula of engineers. Business school students, the future business managers, also have not had the pollution prevention ethic instilled in them.

The reader should also note that the federal government, through its military arm, is responsible for some major environmental problems. It has further compounded these problems by failing to apply a multimedia or multiagency approach. The following are excerpts from a front-page article by Keith Schnedier in the August 5, 1991, edition of the *New York Times*:

> A new strategic goal for the military is aimed at restoring the environment and reducing pollution at thousands of military and other government military-industrial installations in the United States and abroad ... the result of environment contamination on a scale almost unimaginable. The environmental projects are spread through four federal agencies and three military services, and are directed primarily by deputy assistant secretaries. Many of the military–industry officials interviewed for this article said the scattered environmental offices were not sharing information well, were suffering at times from duplicated efforts, and might not be supervising research or contractors closely enough. Environmental groups, state agencies, and the Environmental Protection Agency began to raise concerns about the rampant military–industrial contamination in the 1970s, but were largely ignored. The Pentagon, the Energy Department, the National Aeronautics and Space Administration (NASA) and the Coast Guard considered pollution on their property to be a confidential matter. Leaders feared not only the embarrassment from public disclosure, but also that solving the problems would divert money from projects they considered more worthwhile. Spending on military–environmental projects is causing private companies, some of them among the largest contractors for the military industry, to establish new divisions to compete for government contracts, many of them worth $100 million to $1 billion.

This lack of communication and/or unwillingness to cooperate among the various agencies within the federal government has created a multimedia problem that has just begun to surface. The years of indifference and neglect have allowed pollutants/wastes to contaminate the environment significantly beyond what would have occurred had the responsible parties acted sooner.

3.3 THE GREAT LAKES ECOSYSTEM

The Great Lakes offer an excellent example of how society's understanding of environmental problems has expanded, revealing the need for a multimedia and a total systems approach to adequately protect human health and a fragile, inter-dependent ecosystem. Three decades ago, a study by the International Joint Commission (IJC) identified nutrients and toxics problems in the five Great Lakes and found that Lake Ontario and Lake Erie, in particular, were afflicted with eutrophication problems. Since then, the United States and Canada have under-taken cooperative efforts that have successfully reduced nutrient loadings, particu-larly phosphorus, and helped to reverse eutrophication in the most severely affected areas. Since 1972 the U.S. government has spent over $7.6 billion on pollution problems in the Great Lakes, mostly for more than 1000 municipal sewage treatment plants. With point source contributions of phosphorus increasingly under control, the importance of controlling toxic contamination is becoming more evident (see Chapter 12 for a discussion of point sources of pollution). Although some progress has been made, concentrations of persistent toxic substances such as mercury, polychlorinated biphenyls (PCBs), and lead remain unacceptably high in some parts of the Great Lakes, both in water and in sediments. Interestingly, the IJC has found atmospheric deposition to be a major pathway to contamination and has observed airborne sources for 10 or 11 critical toxic pollutants. For example, an estimated 50% of the heavy metals found in the Great Lakes results from atmospheric deposition, not water (or other liquid) discharges. Studies have registered deformities in fish and wildlife exposed to contaminated sediments and other sources of toxic chemicals in the Great Lakes. Although the decline in conventional pollutants has encouraged an increase in fish populations in some areas, all Great Lakes states advise residents to limit or, in some cases, to eliminate their consumption of popular sport fishing species, such as perch, walleye, brown trout, and chinook salmon, because of their contamination by toxics (1).

3.4 MULTIMEDIA APPLICATION: A CHEMICAL PLANT

A multimedia approach can be applied to any type of system, product, service, or process. In the discussion to follow in this section, a relatively simple example involving a chemical plant is discussed. This is followed in the next section by a more detailed analysis of the manufacture, use, and ultimate disposal of a product. The chapter then concludes with a review of potential emissions to the environment from a hazardous waste incineration (HWI) facility.

Regarding application to a chemical plant, the reader is referred to Figure 3.1. This relatively simple method of analysis, as applied to a chemical plant alone, enables one to understand and quantify how various chemical inputs to the plant — chemicals, fuel, additives, and so forth — are chemically transformed within the facility and partitioned through plant components into various chemical outputs-the gaseous, aqueous, and solid discharge streams. The principles of the law of conservation of mass (see Chapter 26) clearly show that if a substance is removed from one discharge stream without chemical change (as opposed to one that undergoes a combustion reaction or is subjected to an incineration process), the source and fate of the remaining chemicals in the process streams of a chemical plant may accordingly be determined and quantified. Chemical changes can be accounted for through chemical kinetic considerations (2).

A multimedia approach to the chemical plant normally starts with a detailed analysis of the discharge and/or emissions from the plant. These potentially

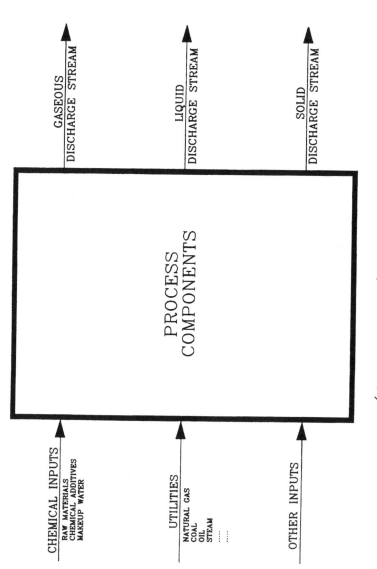

FIGURE 3.1 Multimedia approach.

harmful streams need to be examined not only in terms of the types of control that may be required (and their corresponding discharges) but also in terms of the impact that the changes in feed/inlet streams and design/operating conditions can have. Katin (3) has suggested that the following pollution prevention procedures be employed:

1. Inventory management
2. Raw material substitution
3. Process design and operation
4. Volume reduction
5. Recycling
6. Chemical alteration

A short description of each of these procedures is provided below.

1. Inventory all raw materials and only purchase the minimum amount required. As new plants are built and as processes change, the storeroom is often the last to find out. The original raw materials needed and the quantity required may change over time. There is no need to purchase chemicals that are no longer needed in the process. Likewise, if more is purchased than is needed, the shelf life may expire before the material is completely consumed. By ordering only the amount required, the disposal of unused raw materials as hazardous waste can be minimized. Improve material receiving, storage, and handling procedures. Develop a "first-in, first-out" (FIFO) program, placing newly purchased raw materials at the back of the shelf and bringing the older stock forward. This reduces the probability of a container of raw material needing to be disposed of before it is ever opened because its shelf life has expired.

2. Focus on the one or two streams that have the largest volume or that account for the greatest degree of hazard. Waste streams with the highest costs for treatment and disposal should receive priority. The largest single waste stream is often the most economical to modify. Substituting raw materials can be a very expensive endeavor—a significant amount of labor will be expended, process design changes may be required, and equipment may need to be modified or replaced. Therefore, it is imperative that waste streams be assessed and prioritized. Also, attempt to purchase fewer materials that are hazardous; instead, buy either nonhazardous or less hazardous materials.

3. Install equipment that produces minimal or no waste, and modify equipment to enhance recovery or recycling operations. Redesign equipment or production lines to produce less waste. The generation of paint waste can be minimized by using high-volume, low-pressure paint guns or electrostatic paint equipment. Using proportional paint-mixing equipment to prepare two- or three-component paints can also reduce the generation of hazardous paint waste. Also, maintain a strict preventive maintenance program.

4. Segregate wastes by type for recovery. Users of large quantities of solvents should segregate, not mix, solvents so that they can be recycled.

5. Common industrial solvents can be recycled by distillation. One DuPont plant used 30,000 gal/year of 1.1.1-trichloroethane to degrease parts after forming and plating. A distillation system was installed that saved $148,000 in its first year.

The still bottoms were incinerated, reducing fuel requirements for the process furnace.

6. Hazardous waste can be encapsulated so that it is sealed with a material that makes the exterior of the waste nonhazardous. An example of this is the encapsulation of nuclear waste in borosilicate glass. Adding lime to a spent catalyst that contains clay and phosphoric acid allows the material to be sold as concrete for parking lots and sidewalks. Waste oil can be chemically treated and turned into a saleable aromatic distillate. At a 1-billion-lb/year ethylene plant, 10 million lb/year of waste oil was so treated for a cost savings of $2 million/year.

3.5   MULTIMEDIA APPLICATION: PRODUCTS AND SERVICES

Perhaps a more meaningful understanding of the multimedia approach can be obtained by examining the production and ultimate disposal of a product or service. A flow diagram representing this situation is depicted in Figure 3.2. Note that each of the ten steps in the overall "process" has potential inputs of mass and energy, and may produce an environmental pollutant and/or a substance or form of energy that may be used in a subsequent or later step. Traditional partitioned

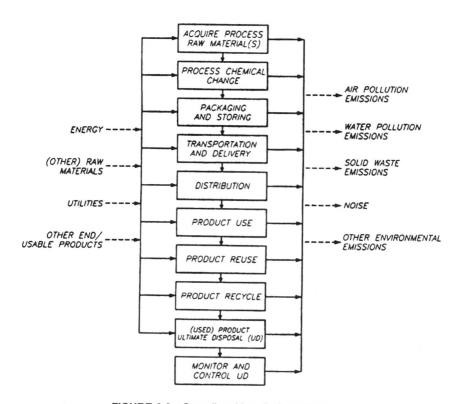

FIGURE 3.2   Overall multimedia flow diagram.

approaches to control can provide some environmental relief, but a total systems approach is required if optimum improvements in terms of pollution/waste reduction are to be achieved.

One should note that a product and/or service is usually conceived to meet a specific market need with little thought given to manufacturing parameters. At this stage of consideration, it may be possible to avoid some significant waste-generation problems in future operations by answering a few simple questions:

1. What are the raw materials used to manufacture the product?
2. Are there any toxic or hazardous chemicals likely to be generated during manufacturing?
3. What performance regulatory specifications must the new product(s) and or service(s) meet? Is extreme purity required?
4. How reliable will the delivery manufacturing/distribution process be? Are all steps commercially proven? Does the company have experience with the operations required?
5. What type of wastes are likely to be generated? What is their physical and chemical form? Are they hazardous? Does the company currently manage these wastes on-site or off-site?

3.6 MULTIMEDIA APPLICATION: A HAZARDOUS WASTE INCINERATION FACILITY

One may also examine a hazardous waste incineration (HWI) facility using a comprehensive, multimedia approach. This system is presented in Figure 3.3. The reader is referred to the literature (4) for more details on incinerators. In addition to the various items of equipment contained in this plant facility, the major environmental emission discharge points are also indicated. The reader is left the exercise of determining other potential emission locations.

Note that emissions can appear from a variety of sources and locations, not only from the incinerator. The major emission points are the fan (noise), stack (gaseous), quench (liquid), water treatment (liquid and solid), and land disposal (solid). A hazard operability study, often referred to as a HAZOP, can be performed in order to locate (potential) emissions and hazardous problems (5).

Looking at a facility such as this from a comprehensive, multimedia perspective allows one to see the forest rather than the trees. It permits the development of cost-effective strategies for managing chemical discharges to minimize risks to public health and the environment while avoiding unnecessarily expensive control requirements.

3.7 SUMMARY

1. It is now increasingly clear that some treatment technologies, while solving one pollution problem, have created others. For this reason, there is a need for a multimedia approach in many pollution prevention programs.

2. The EPA's single-media offices, often created sequentially as individual environmental problems were identified and responded to in legislation, have

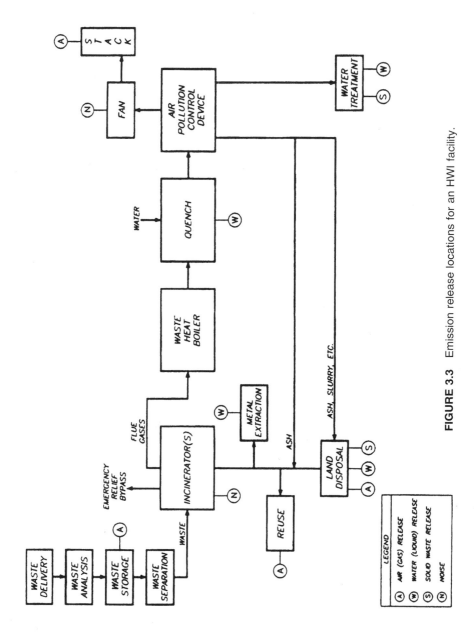

FIGURE 3.3 Emission release locations for an HWI facility.

played a role in impeding the development of cost-effective multimedia strategies. In addition, most engineers are challenged to make a product work, not to prevent pollution.

3. The Great Lakes offer an excellent example of how our understanding of environmental problems has expanded, revealing the need for a multimedia focus and a total systems approach to adequately protect human health and a fragile, interdependent ecosystem.

4. A multimedia approach can be applied to any type of system, product, or process. A multimedia analysis of a chemical plant enables one to understand and quantify how various chemical inputs to the plant are chemically transformed and partitioned through plant components into various chemical outputs.

5. A multimedia approach can also be applied to any product or service. Traditional partitioned approaches to environmental control can provide some environmental relief, but a total systems approach is required if optimum improvements are to be achieved.

6. Emissions can appear from a variety of sources and locations. A hazard operability study (HAZOP) can be performed to locate emissions and hazardous problems.

REFERENCES

1. R. Dupont, K. Ganesan and L. Theodore, *Pollution Prevention: The Waste Management Option for the 21st Century*, CRC Press, Boca Raton, FL., 2000 (in press).
2. L. Theodore, *Chemical Reaction Kinetics*, Theodore Tutorial, AWMA Bookstore, Pittsburgh, PA, 1998.
3. R. A. Katin, *Pollution Prevention at Operating Chemical Plants*, CEP, July 1991.
4. J. Santoleri, J. Reynolds and L. Theodore, *Hazardous Waste Incineration Systems*, Wiley-Interscience, New York, 2000 (in press).
5. L. Theodore, J. Reynolds, and F. Taylor, *Accident and Emergency Management*, Wiley-Interscience, New York, 1989.

4

REGULATORY
FRAMEWORK

4.1 INTRODUCTION

In many ways, the Environmental Protection Agency (EPA) is the most far-reaching regulatory agency in the federal government because its authority is so broad. The EPA is charged by the Congress of the United States of America to protect the nation's land, air, and water systems. Under a mandate of national environmental laws, the EPA strives to formulate and implement actions which lead to a compatible balance between human activities and the ability of natural systems to support and nurture life (1).

The EPA is working with the states and local governments to develop and implement comprehensive environmental programs. Amendments to federal legislation such as the Clean Air Act, the Clean Water Act, the Safe Drinking Water Act, the Resource Conservation and Recovery Act, and the Comprehensive Environmental Response, Compensation, and Liability Act all mandate more involvement by state and local government in the details of implementation.

This chapter presents the regulatory framework governing waste management. It provides an overview of environmental laws and regulations used to protect human health and the environment from the potential hazards of waste disposal. New legislation and reauthorizations of acts have created many changes in the way both government and industry manage their business.

4.2 THE ROLE OF THE STATES

The Resource Conservation and Recovery Act (RCRA) of 1976, like most federal environmental legislation, encourages states to develop and run their own hazardous waste programs as an alternative to EPA management. Thus, in a given state, the hazardous waste regulatory program may be run by the EPA or by a state agency. For a state to have jurisdiction over its hazardous waste program it must receive approval from the EPA by showing that its program is at least as stringent as the EPA program.

States that are authorized to operate RCRA programs oversee the hazardous waste tracking system in their state, operate the permitting system for hazardous waste facilities, and act as the enforcement arm in cases where an individual or a company practices illegal hazardous waste management. If needed, the EPA steps in to assist the states in enforcing the law. The EPA also acts directly to enforce RCRA in states that do not yet have authorized programs. The EPA and the states currently act jointly to implement and enforce the regulations resulting from the 1984 RCRA amendments. Many states assume full responsibility for carrying out these newest sections of the EPA programs (2).

4.3   HAZARDOUS WASTE MANAGEMENT

Defining what constitutes a "hazardous waste" requires consideration of both legal and scientific factors. The basic definitions used in this chapter are derived from: The Resource Conservation and Recovery Act (RCRA) of 1976, as amended in 1978, 1980, and 1986; the Hazardous and Solid Waste Amendments (HSWA) of 1984; and the Comprehensive Environmental Response, Compensation and Liability Act (CERCLA) of 1980, as amended by the Superfund Amendments and Reauthorization Act (SARA) of 1986. Within these statutory authorities a distinction exists between a hazardous waste and a hazardous substance. The former is regulated under RCRA while the latter is regulated under the Superfund program.

Hazardous Waste refers to "... a solid waste, or combination of solid wastes, which because of its quantity, concentration, or physical, chemical or infectious characteristics may [pose a] substantial present or potential hazard to human health or the environment when improperly... managed..." [RCRA, Section 1004(5)]. Under RCRA regulations, a waste is considered hazardous if it is reactive, ignitable, corrosive, or toxic or if the waste is listed as a hazardous waste in Parts 261.31–33 of the *Code of Federal Regulations* (40 CFR) (3).

In addition to hazardous wastes defined under RCRA, there are "hazardous substances" defined by Superfund. Superfund's definition of a hazardous substance is broad and grows out of the lists of hazardous wastes or substances regulated under the Clean Water Act (CWA), the Clean Air Act (CAA), the Toxic Substances Control Act (TSCA), and RCRA. Essentially, Superfund considers a hazardous substance to be any hazardous substance or toxic pollutant identified under the CWA and applicable regulations, any hazardous air pollutant listed under the CAA and applicable regulations, any imminently hazardous chemical for which a civil action has been brought under TSCA, and any hazardous waste identified or listed under RCRA and applicable regulations.

The Resource Conservation and Recovery Act (RCRA) of 1976 completely replaced the previous language of the Solid Waste Disposal Act of 1965 to address the enormous growth in the production of waste. The objectives of this act were to promote the protection of health and the environment and to conserve valuable materials and energy resources by (4,5).

- Providing technical and financial assistance to state and local governments and interstate agencies for the development of solid waste management plans (including resource recovery and resource conservation systems) that promote improved solid waste management techniques (including more effective organizational arrangements), new and improved methods of collection, separation, and recovery of solid waste, and the environmentally safe disposal of nonrecoverable residues.

- Providing training grants in occupations involving the design, operation, and maintenance of solid waste disposal systems.

- Prohibiting future open dumping on the land and requiring the conversion of existing open dumps to facilities that do not pose danger to the environment or to health.

- Regulating the treatment, storage, transportation, and disposal of hazardous wastes that have adverse effects on health and environment.

- Providing for the promulgation of guidelines for solid waste collection, transport, separation, recovery, and disposal practices and systems.

- Promoting a national research and development program for improved solid waste management and resource conservation techniques; more effective organization arrangements; and new and improved methods of collection, separation, recovery, and recycling of solid wastes and environmentally safe disposal of nonrecoverable residues.

- Promoting the demonstration, construction, and application of solid waste management, resource recovery and resource conservation systems that preserve and enhance the quality of air, water, and land resources.

- Establishing a cooperative effort among federal, state, and local governments and private enterprises in order to recover valuable materials and energy from solid waste.

Structurewise, the RCRA is divided into eight subtitles. These subtitles are (A) General Provisions; (B) Office of Solid Waste; Authorities of the Administrator; (C) Hazardous Waste Management; (D) State or Regional Solid Waste Plans; (E) Duties of the Secretary of Commerce in Resource and Recovery; (F) Federal Responsibilities; (G) Miscellaneous Provisions; and (H) Research, Development, Demonstration, and Information.

Subtitles C and D generate the framework for regulatory control programs for the management of hazardous and solid nonhazardous wastes, respectively. The hazardous waste program outlined under Subtitle C is the one most people associate with the RCRA (5).

4.4 MAJOR TOXIC CHEMICAL LAWS ADMINISTERED BY THE EPA

People have long recognized that sulfuric acid, arsenic compounds, and other chemical substances can cause fires, explosions, or poisoning. More recently, researchers have determined that many chemical substances such as benzene and a number of chlorinated hydrocarbons may cause cancer, birth defects, and other long-term health effects. Today, we are evaluating the hazards of new kinds of substances, including genetically engineered microorganisms. The EPA has a

TABLE 4.1   Major toxic chemical laws administered by the EPA

Statute	Provisions
Toxic Substances Control Act	Requires that the EPA be notified of any new Chemical prior to its manufacture and authorizes EPA to regulate production, use, or disposal of a chemical.
Federal Insecticide, Fungicide, and Rodenticide Act	Authorizes the EPA to register all pesticides and specify the terms and conditions of their use, and remove unreasonably hazardous pesticides from the marketplace.
Federal Food, Drug, and Cosmetic Act	Authorizes the EPA in cooperation with FDA to establish tolerance levels for pesticide residues on food and food products.
Resource Conservation and Recovery Act	Authorizes the EPA to identify hazardous wastes and regulate their generation, transportation, treatment, storage, and disposal.
Comprehensive Environmental Response, Compensation, and Liability Act	Requires the EPA to designate hazardous substances that can present substantial danger and authorizes the cleanup of sites contaminated with such substances.
Clean Air Act	Authorizes the EPA to set emission standards to limit the release of hazardous air pollutants.
Clean Water Act	Requires the EPA to establish a list of toxic water pollutants and set standards.
Safe Drinking Water Act	Requires the EPA to set drinking water standards to protect public health from hazardous substances.
Marine Protection, Research, and Sanctuaries Act	Regulates ocean dumping of toxic contaminants.
Asbestos School Hazard Act	Authorizes the EPA to provide loans and grants to schools with financial need for abatement of severe asbestos hazards.
Asbestos Hazard Emergency Response Act	Requires the EPA to establish a comprehensive regulatory framework for controlling asbestos hazards in schools.
Emergency Planning and Community Right-to-Know Act	Requires states to develop programs for responding to hazardous chemical releases and requires industries to report on the presence and release of certain hazardous substances.

number of legislative tools to use in controlling the risks from toxic substances (Table 4.1).

The Federal Insecticide, Fungicide, and Rodenticide Act of 1972 (FIFRA) encompasses all pesticides used in the United States. When enacted in 1947, FIFRA was administered by the U.S. Department of Agriculture and was intended to protect consumers against fraudulent pesticide products. When many pesticides were registered, their potential for causing health and environmental problems was unknown. In 1970, the EPA assumed responsibility for FIFRA, which was amended in 1972 to shift emphasis to health and environmental protection. Allowable levels of pesticides in food are specified under the authority of the

Federal Food, Drug, and Cosmetic Act of 1954. FIFRA contains registration and labeling requirements for pesticide products. The EPA must approve any use of a pesticide, and manufacturers must clearly state the conditions of that use on the pesticide label. Some pesticides are listed hazardous wastes and are subject to RCRA rules when discarded.

The Toxic Substances Control Act (TSCA) of 1976 authorizes EPA to control the risks that may be posed by the thousands of commercial chemical substances and mixtures (chemicals) that are not regulated as either drugs, food additives, cosmetics, or pesticides. Under TSCA, the EPA can, among other things, regulate the manufacture and use of a chemical substance and require testing for cancer and other effects. TSCA regulates the production and distribution of new chemicals and governs the manufacture, processing, distribution, and use of existing chemicals. Among the chemicals controlled by TSCA regulations are PCBs, chloroflurocarbons, and asbestos. In specific cases, there is an interface with RCRA regulations. For example, PCB disposal is generally regulated by TSCA. However, hazardous wastes mixed with PCBs are regulated under RCRA. Under both TSCA and FIFRA, the EPA is responsible for regulating certain biotechnology products, such as genetically engineered microorganisms designed to control pests or assist in industrial processes.

The Clean Air Act (CAA), under Section 112, authorizes the EPA to list various hazardous air pollutants. Currently included are asbestos, beryllium, vinyl chloride, benzene, arsenic, radionuclides, mercury, and coke oven emissions. The CAA also sets certain emission standards for many types of air emission sources, including RCRA-regulated incinerators and industrial boilers or furnaces.

The Clean Water Act (CWA) lists substances to be regulated by effluent limitations in 21 primary industries. The CWA substances are incorporated into both RCRA and CERCLA. In addition, the CWA regulates discharges from publicly owned treatment works (POTWs) to surface waters, and indirect discharges to municipal wastewater treatment systems (through the pretreatment program). Some hazardous wastewaters which would generally be considered RCRA regulated wastes are covered under the CWA because of the use of treatment tanks and a National Pollutant Discharge Elimination System (NPDES) permit to dispose of the wastewaters. Sludges from these tanks, however, are subject to RCRA regulations when they are removed from these tanks.

The Safe Drinking Water Act (SDWA) regulates underground injection systems, including deep-well injection systems. Prior to underground injection, a permit must be obtained which will impose conditions which must be met to prevent the endangerment of underground sources of drinking water.

The Marine Protection, Research, and Sanctuaries Act of 1972 has regulated the transportation of any material for ocean disposal and prevents the disposal of any material in oceans which could affect the marine environment. Amendments enacted in 1988 were designed to end ocean disposal of sewage sludge, industrial waste and medical wastes (3).

4.5 LEGISLATIVE TOOLS FOR CONTROLLING WATER POLLUTION

Congress has provided the EPA and the states with three primary statutes to control and reduce water pollution: the Clean Water Act; the Safe Drinking Water

Act; and the Marine Protection, Research, and Sanctuaries Act. Each statute provides a variety of tools that can be used to meet the challenges and complexities of reducing water pollution in our nation.

Clean Water Act

Under the Clean Water Acts, the states adopt water quality standards for every stream within their respective borders. These standards include a designated use such as fishing or swimming and prescribe criteria to protect that use. The criteria are pollutant-specific and represent the permissible levels of substances in the waters that would enable the use to be achieved. Water quality standards are the basis for nearly all water quality management decisions. Depending upon the standard adopted for a particular stream, controls may be needed to reduce the pollutant levels. Water quality standards are reviewed every three years and revised as needed.

National Pollutant Discharge Elimination System   Under the Clean Water Act, the discharge of pollutants into the waters of the United States is prohibited unless a permit is issued by the EPA or a state under the National Pollutant Discharge Elimination System (NPDES). These permits must be renewed at least once every five years. There are approximately 48,400 industrial and 15,300 municipal facilities that currently have NPDES permits.

An NPDES permit contains effluent limitations and monitoring and reporting requirements. Effluent limitations are restrictions on the amount of specific pollutants that a facility can discharge into a stream, river, or harbor. Monitoring and reporting requirements are specific instructions on how sampling of the effluent should be done to check whether the effluent limitations are being met. Instructions may include required sampling frequency (i.e., daily, weekly, or monthly) and the type of monitoring required. The permittee may be required to monitor the effluent on a daily, weekly, or monthly basis. The monitoring results are then regularly reported to the EPA and state authorities. When a discharger fails to comply with the effluent limitations or monitoring and reporting requirements, the EPA or the state may take enforcement action.

Congress recognized that it would be an overwhelming task for the EPA to establish effluent limitations for each individual industrial and municipal discharger. Therefore, Congress authorized the Agency to develop uniform effluent limitations for each category of point sources such as steel mills, paper mills, and pesticide manufacturers. The EPA develops these effluent limitations on the basis of many factors, most notably efficient treatment technologies. Once the EPA proposes an effluent limit and public comments are received, the EPA or the states issue all point sources within that industry category NPDES permits using the technology-based limits. Sewage treatment plants also are provided with effluent limitations based on technology performance.

Limitations that are more stringent than those based on technology are sometimes necessary to ensure that state-developed water quality standards are met. For example, several different facilities may be discharging into one stream, creating pollutant levels harmful to fish. In this case, the facilities on that stream must meet more stringent treatment requirements, known as *water-quality-based limitations*. These limits are developed by determining the amounts of pollutants that the stream can safely absorb and calculating permit limits such that these amounts are not exceeded (7).

The EPA and the U.S. Army Corps of Engineers implement jointly a permit program regulating the discharge of dredged or fill material into waters of the United States, including wetlands. As part of this program, the EPA's principal responsibility as set forth in the Clean Water Act is to develop the substantive environmental criteria by which permit applications are evaluated. The EPA also reviews the permit applications and, if necessary, can veto permits that would result in significant environmental damage.

The National Estuary Program is also regulated under the Clean Water Act. States nominate and the EPA selects estuaries of national significance that are threatened by pollution, development, or overuse. The EPA and the involved state(s) form a management committee consisting of numerous workgroups to assess the problems, identify management solutions, and develop and oversee implementation of plans for addressing the problems.

Grants   The CWA authorizes the EPA to provide financial assistance to states to support programs such as the construction of municipal sewage treatment plants; water quality monitoring, permitting, and enforcement; and implementation of non-point-source controls. These funds also may support development and implementation of state ground-water protection strategies. In addition, the EPA provides grants to states for the creation of State Water Pollution Control Revolving Funds. States may use these funds for loans and other types of financial assistance to local governments for the construction of municipal wastewater treatment plants, implementation of non-point-source programs, and the development and implementation of estuary protection programs.

Safe Drinking Water Act

The EPA establishes standards for drinking water quality through the Safe Drinking Water Act. These standards represent the Maximum Contaminant Levels (MCL) allowable, and consist of numerical criteria for specified contaminants. Local water supply systems are required to monitor their drinking water periodically for contaminants with MCLs and for a broad range of other contaminants as specified by the EPA. Additionally, to protect underground sources of drinking water EPA requires periodic monitoring of wells used for underground injection of hazardous waste, including monitoring of the ground water above the wells.

States have the primary responsibility for the enforcement of drinking water standards, monitoring, and reporting requirements. States also determine requirements for environmentally sound underground injection of wastes. The Safe Drinking Water Act authorizes EPA to award grants to states for developing and implementing programs to protect drinking water at the tap and ground-water resources. These several grant programs may be for supporting state public water supply, wellhead protection, and underground injection programs, including compliance and enforcement.

The Clean Water Act and the Safe Drinking Water Act place great reliance on state and local initiatives in addressing water problems. With the enactment of the 1986 Safe Drinking Water Act amendments and the 1987 Water Quality Act, significant additional responsibilities were assigned to the EPA and the states.

Faced with many competing programs and limited resources, the public sector will need to set priorities. With this in mind, EPA is encouraging states to address their water quality problems by developing State Clean Water Strategies. These strategies are to set forth state priorities over a multiyear period. They will help target the most valuable and/or most threatened water resources for protection.

Success in the water programs is increasingly tied to state and local leadership and decision-making and to public support. The EPA will work with state and local agencies, industry, environmentalists, and the public as we develop our environmental agenda in the following three areas:

- *Protection of Drinking Water.* Although more Americans are receiving safer drinking water than ever before, there are still serious problems with contamination of drinking water supplies and of ground water that is or could be used for human consumption. Contaminated ground water has caused well closings. The extent and significance of contamination by toxics has not been fully assessed for most of the nation's rivers and lakes, which are often used for drinking water supply. All of these issues are areas for continued work and improvement.

- *Protection of Critical Aquatic Habitats.* Contamination or destruction of previously under protected areas such as oceans, wetlands, and near coastal waters must be addressed.

- *Protection of Surface-Water Resources.* The EPA and the states will need to establish a new phase of the federal–state partnership in ensuring continuing progress in addressing conventional sources of pollution (7).

Marine Protection, Research, and Sanctuaries Act (Title I)

EPA designates recommended sites and times for ocean dumping. Actual dumping at these designated sites requires a permit. The EPA and the Corps of Engineers share this permitting authority, with the Corps responsible for the permitting of dredged material (subject to an EPA review role), and the EPA responsible for permitting all other types of materials. The Coast Guard monitors the activities and the EPA is responsible for assessing penalties for violations. The EPA also is responsible for designating sites and times for the ocean dumping activities.

4.6   THE SUPERFUND AMENDMENTS AND REAUTHORIZATION ACT OF 1986

The 1986 amendments to the Comprehensive Environmental Response, Compensation, and Liability Act (CERCLA), known as the Superfund Amendments and Reauthorization Act (SARA), authorized $8.5 billion for both the emergency response and longer term (or remedial) cleanup programs. The Superfund amendments focused on:

- *Permanent Remedies.* The EPA must implement permanent remedies to the maximum extent practicable. A range of treatment options will be considered whenever practicable.

- *Complying with Other Regulations.* Applicable or relevant and appropriate standards from other federal, state, or tribal environmental laws must be met at Superfund sites where remedial actions are taken. In addition, state standards that are more stringent than federal standards must be met in cleaning up sites.
- *Alternative Treatment Technologies.* Cost-effective treatment and recycling must be considered as an alternative to the land disposal of wastes. Under RCRA, Congress banned land disposal of some wastes. Many Superfund site wastes, therefore, will be banned from disposal on the land; alternative treatments are under development and will be used where possible.
- *Public Involvement.* Citizens living near Superfund sites have been involved in the site decision-making process for over five years. They will continue to be a part of this process. They also will be able to apply for technical assistance grants that may further enhance their understanding of site conditions and activities.
- *State Involvement.* States and tribes are encouraged to participate actively as partners with EPA in addressing Superfund sites. They will assist in making the decisions at sites, can take responsibility in managing cleanups, and can play an important role in oversight of responsible parties.
- *Enforcement Authorities.* Settlement policies already in use were strengthened through Congressional approval and inclusion in SARA. Different settlement tools, such as *de minimis* settlements (settlements with minor contributors), are now part of the Act.
- *Federal Facility Compliance.* Congress emphasized that federal facilities "are subject to, and must comply with, this Act in the same manner and to the same extent . . . as any non-government entity." Mandatory schedules have been established for federal facilities to assess their sites, and if listed on the National Priority List (NPL), to clean up such sites. EPA will be assisting and overseeing federal agencies with these new requirements.

The amendments also expand research and development, especially in the area of alternative technologies. They also provide for more training for state and federal personnel in emergency preparedness, disaster response, and hazard mitigation.

Major Provisions of Title III of SARA

- *Emergency Planning:* Title III establishes a broad-based framework at the state and local levels to receive chemical information and use that information in communities for chemical emergency planning.
- *Emergency Release Notification:* Title III requires facilities to report certain releases of extremely hazardous chemicals and hazardous substances to their state and local emergency planning and response officials.
- *Hazardous Chemical Inventory Reporting:* Title III required facilities that already had prepared Material Safety Data Sheets (MSDS) under Occupational Safety and Health Administration worker Right-to-Know rules to submit those sheets to state and local authorities by October 1987. It also required them to report by March 1988 and annually thereafter information

on chemicals on their premises to local emergency planning and fire protection officials, as well as state officials.

- *Toxic Release Inventory Reporting:* Title III requires facilities to report annually on routine emissions of certain toxic chemicals to the air, land, or water. Facilities must report if they are in Standard Industrial Classification code 20 through 39 (i.e., manufacturing facilities) with 10 or more employees; manufacture or process more than 75,000 pounds of a specified chemical, and use more than 10,000 pounds in one calendar year of specific toxic chemicals or chemical compounds. The reporting thresholds for manufacturing and processing dropped to 50,000 pounds for reports covering 1988, and to 25,000 pounds for 1989 and thereafter. EPA is required to use these data to establish a national chemical release inventory database, making the information available to the public through computers, via telecommunications, and by other means.

4.7   THE CLEAN AIR ACT AMENDMENTS OF 1990

On November 15, 1990, the president signed the Clean Air Act Amendments. Several progressive and creative new themes are embodied in the Amendments, themes necessary for effectively achieving the air quality goals and regulatory reform expected from these far-reaching amendments. Specifically, the new law

- Encourages the use of market-based principles and other innovative approaches, like performance-based standard and emission banking and trading.
- Promotes the use of clean low-sulfur coal and natural gas, as well as innovative technologies to clean high-sulfur coal through the acid rain program.
- Reduces enough energy waste and creates enough of a market for clean fuels derived from grain and natural gas to cut dependency on oil imports by one million barrels/day.
- Promotes energy conservation through an acid rain program that gives utilities flexibility to obtain needed emission reductions through programs that encourage customers to conserve energy.

Title I: Provisions for Attainment and Maintenance of National Ambient Air Quality Standards

Although the Clean Air Act of 1977 brought about significant improvements in our nation's air quality, the urban air pollution problems of ozone (smog), carbon monoxide (CO), and particulate matter (PM-10) persist. In 1995, approximately 70 million U.S. residents were living in counties with ozone levels exceeding the EPA's current ozone standard.

The Clean Air Act Amendments of 1990 created a new, balanced strategy for the nation to address the problem of urban smog. Overall, the new law reveal the Congress's high expectations of the states and the federal government. While it gave states more time to meet the air quality standard (up to 20 years for ozone in Los Angeles), it also required states to make constant formidable progress in reducing emissions. It requires the federal government to reduce emissions from cars, trucks, and buses; from consumer products such as hair spray and window-washing compounds; and from ships and barges during loading and unloading of petroleum products. The federal government must also develop the technical guidance that states need to control stationary sources.

The 1990 law addressed the urban air pollution problems of ozone (smog), carbon monoxide (CO), and particulate matter (PM-10). Specifically, it clarified how areas are designated and redesignated "attainment." It also allows the EPA to define the boundaries of "nonattainment" areas — geographical areas whose air quality does not meet federal air quality standards designed to protect public health. The 1990 law also established provisions defining when and how the federal government can impose sanctions on areas of the country that have not met certain conditions.

For the pollutant ozone, the 1990 law established nonattainment area classifications ranked according to the severity of the area's air pollution problem. These classifications are marginal, moderate, serious, severe, and extreme. The EPA assigns each nonattainment area one of these categories, thus triggering varying requirements the areas must comply with in order to meet the ozone standard.

As mentioned, nonattainment areas have to implement different control measures, depending upon their classification. Marginal areas, for example, are the closest to meeting the standard. They are required to conduct an inventory of their ozone-causing emissions and institute a permit program. Nonattainment areas with more serious air quality problems must implement various control measures. The worse the air quality, the more controls areas will have to implement.

The 1990 law also established similar programs for areas that do not meet the federal health standard for carbon monoxide and particulate matter. Areas exceeding the standards for these pollutants are divided into "moderate" and "serious" classifications. Depending upon the degree to which they exceed the carbon monoxide standard, areas are then required to implement programs introducing oxygenated fuels and/or enhanced emission inspection programs, among other measures. Depending upon their classification, areas exceeding the particulate matter standard have to implement either reasonably available control measures (RACM) or best available control measures (BACM), among other requirements.

Title II: Provisions Relating to Mobile Sources

While motor vehicles built today emit fewer pollutants (60–80% less, depending on the pollutant) than those built in the 1960s, cars and trucks still account for almost half the emissions of the ozone precursors volatile organic carbons (VOC) and nitrogen oxides (NO_x), and up to 90% of the CO emissions in urban areas. The principal reason for this problem is the rapid growth in the number of vehicles on the roadways and the total miles driven. This growth has offset a large portion of the emission reductions gained from motor vehicle controls.

In view of the unforeseen growth in automobile emissions in urban areas combined with the serious air pollution problems in many urban areas, the Congress has made significant changes to the motor vehicle provisions on the 1977 Clean Air Act. The Clean Air Act of 1990 established tighter pollution standards for emissions from automobiles and trucks. These standards were set so as to reduce tailpipe emissions of hydrocarbons, carbon monoxide, and nitrogen oxides on a phased-in basis beginning in model year 1994. Automobile manufacturers also were required to reduce vehicle emissions resulting from the evaporation of gasoline during refueling.

Fuel quality is also controlled. Scheduled reductions in gasoline volatility and sulfur content of diesel fuel, for example, will be required. New programs requiring cleaner (so-called "reformulated") gasoline was initiated in 1995 for the nine cities with the worst ozone problems. Other cities can "opt in" to the reformulated gasoline program. Higher levels (2.7%) of alcohol-based oxygenated fuels were to be produced and sold in 41 areas during the winter months that exceed the federal standard for carbon monoxide.

The 1990 law also established a clean fuel car pilot program in California, requiring the phase-in of tighter emission limits for 150,000 vehicles in model year 1996 and 300,000 by the model year 1999. These standards could be met with any combination of vehicle technology and cleaner fuels. The standards become even stricter in 2001. Other states can "opt in" to this program, though only through incentives, not sales or production mandates.

Title III: Air Toxics

Toxic air pollutants are those pollutants which are hazardous to human health or the environment but are not specifically covered under another portion of the Clean Air Act. These pollutants are typically carcinogens, mutagens, and reproductive toxins. The Clean Air Act Amendments of 1977 failed to result in substantial reductions of the emissions of these very threatening substances. In fact, over the history of the air toxics program only seven pollutants have been regulated.

The toxic air pollution problem is widespread. Information generated from The Superfund "Right to Know" rule (SARA Section 313) indicated that more than 2.7 billion pounds of toxic air pollutants are emitted annually in the United States. The EPA studies indicate that exposure to such quantities of air toxics may result in 1,000–3,000 cancer deaths each year.

The Clean Air Act of 1990 offers a comprehensive plan for achieving significant reductions in emissions of hazardous air pollutants from major sources. Industry reports in 1987 suggested that an estimated 2.7 billion pounds of toxic air pollutants were emitted into the atmosphere, contributing to approximately 300–1,500 cancer fatalities annually. The 1990 law was designed to improve the EPA's ability to address this problem effectively and dramatically accelerate progress in controlling major toxic air pollutants.

The 1990 law includes a list of 189 toxic air pollutants of which emissions were to be reduced. It stated EPA must publish a list of source categories that emit certain levels of these pollutants within one year after enactment of the law. The list of source categories must include (1) major sources emitting 10 tons/yr of any one, or 25 tons/yr of any combination of those pollutants; and (2) area sources (smaller sources, such as dry cleaners).

The EPA then must issue maximum achievable control technology (MACT) standards for each listed source category according to a prescribed schedule. These standards are based on the best are demonstrated control technology or practices within the regulated industry, and EPA must issue the standards for 40 source categories within 2 years of enactment. The remaining source categories would be controlled according to a schedule that ensures all controls will be achieved within 10 years of enactment. Companies that voluntarily reduce emissions according to certain conditions can get a six-year extension from the original date for meeting

MACT requirements. Eight years after MACT is installed on a source, EPA must examine the risk levels remaining at the regulated facilities and determine whether additional controls are necessary to reduce unacceptable residual risk.

The 1990 law also established a Chemical Safety Board to investigate accidental releases of extremely hazardous chemicals. Further, the 1990 law required the EPA to issue regulations controlling air emissions from municipal, hospital, and other commercial and industrial incinerators.

Title IV: Acid Deposition Control

Acid rain occurs when sulfur dioxide and nitrogen oxide emissions are transformed in the atmosphere and return to the earth in rain, fog, or snow. Approximately 20 million tons of sulfur dioxide are emitted annually in the United States, mostly from the burning of fossil fuels by electric utilities. Acid rain damages lakes, harms forests and buildings, contributes to reduced visibility, and is suspected of damaging health.

The Clean Air Act of 1990 will result in a permanent 10 million-ton reduction in sulfur dioxide (SO_2) emissions from 1980 levels. To achieve this, the EPA allocated allowances in two phases, permitting utilities to emit one ton of sulfur dioxide. The first phase, which became effective January 1, 1995, required 110 power plants to reduce their emissions to a level equivalent to the product of an emissions rate of 2.5 lbs of SO_2/mm Btu × an average of their 1985–1987 fuel use. Emissions data indicate that 1995 SO_2 emissions at these units nationwide were reduced by almost 40% below the required level. Plants that use certain control technologies to meet their Phase I reduction requirements may receive a 2-year extension of compliance until 1997. The new law also allowed for a special allocation of 200,000 annual allowances per year each of the 5 years of phase I to powerplants in Illinois, Indiana and Ohio.

The second phase, becoming effective January 1, 2000, will require approximately 2,000 utilities to reduce their emissions to a level equivalent to the product of an emissions rate of 1.2 lbs of SO_2/mm Btu × the average of their 1985–1987 fuel use. In both phases, affected sources are required to install systems that continuously monitor emissions in order to track progress and assure compliance.

The 1990 law allowed utilities to trade allowance within their systems and/or buy or sell allowance to and from other affected sources. Each source must have sufficient allowances to cover its annual emissions. If not, the source is subject to a $2,000/ton excess emissions fee and a requirement to offset the excess emissions in the following year.

Nationwide, plants that emit SO_2 at a rate below 1.2 lbs/mm Btu were able to increase emissions by 20% between a baseline year and 2000. Bonus allowances are distributed to accommodate growth by units in states with a statewide average below 0.8 lbs/mm Btu. Plants experiencing increases in their utilization in the second five years of phase I also receive bonus allowances. A total of 50,000 bonus allowances per year is allocated to plants in 10 midwestern states that make reductions in Phase I. Plants that repower with a qualifying clean coal technology may receive a four-year extension of the compliance date for Phase II emission limitations.

The 1990 law also included specific requirements for reducing emissions of nitrogen oxides, based on the EPA regulations promulgated April 13, 1995, and December 19, 1996.

Title V: Permits

The 1990 law introduced an operating permits program modeled after a similar program under the federal National Pollution Elimination Discharge System (NPDES) law. The purpose of the operating permits program is to ensure compliance with all applicable requirements of the Clean Air Act and to enhance the EPA's ability to enforce the Act. Air pollution sources subject to the program must obtain an operating permit, states must develop and implement the program, and the EPA must issue permit program regulations, review each state's proposed program, and oversee the state's effort to implement any approved program. The EPA must also develop and implement a federal permit program when a state fails to adopt and implement its own program.

This program — in many ways the most important procedural reform contained in the new law — will greatly strengthen enforcement of the Clean Air Act. It will enhance air quality control in a variety of ways. First, adding such a program updates the Clean Air Act, making it more consistent with other environmental statutes. The Clean Water Act, the Resource Conservation and Recovery Act, and the Federal Insecticide, Fungicide, and Rodenticide Act all require permits. The 1977 Clean Air laws also requires a construction permit for certain pollution sources, and about 35 states have their own laws requiring operating permits.

Title VI: Stratospheric Ozone and Global Climate Protection

The 1990 law builds on the market-based structure and requirements currently contained in the EPA's regulations to phase out the production of substances that deplete the ozone layer. The law requires a complete phase-out of CFCs and halons, with interim reductions. Under these provisions,the EPA must list all regulated substances along with their ozone-depletion potential, atmospheric lifetime and global warming potentials within 60 days of enactment.

In addition, the EPA must ensure that Class I chemicals be phased out on a schedule similar to that specified in the Montreal Protocol — CFCs, halons, and carbon tetrachloride by 2000; methyl chloroform by 2002 — but with more stringent interim reductions. Class II chemicals (HCFCs) will be phased out by 2030. Regulations for Class I chemicals will be required within 10 months, and Class II chemical regulations will be required by December 31, 1999. The law also requires EPA to publish a list of safe and unsafe substitutes for Class I and II chemicals and to ban the use of unsafe substitutes.

The law requires nonessential products releasing Class I chemicals to be banned within 2 years of enactment. In 1994 a ban went into effect for aerosols and noninsulating foams using Class II chemicals, with exemptions for flammability and safety. Regulations for this purpose were required within 1 year of enactment, and effective 2 years afterwards.

Title VII: Provisions Relating to Enforcement

The Clean Air Act of 1990 contains provisions for a broad array of authority to make the law more readily enforceable, thus bringing it up to date with the other major environmental statutes. EPA has new authorities to issue administrative penalty orders up to $200,000, and field citations up to $5,000 for lesser infractions.

Civil judicial penalties are enhanced. Criminal penalties for knowing violations are upgraded from misdemeanors to felonies, and new criminal authorities for knowing and negligent endangerment will be established. In addition, sources must certify their compliance, and EPA has authority to issue administrative subpoenas for compliance data. The EPA will also be authorized to issue compliance orders with compliance schedules of up to one year.

The citizen suit provisions have also been revised to allow citizens to seek penalties against violators, with the penalties going to a U.S. Treasury fund for use by the EPA for compliance and enforcement activities. The government's right to intervene is clarified and citizen plaintiffs will be required to provide the United States with copies of pleadings and draft settlements.

Other Titles

The Clean Air Act Amendments of 1990 continue the federal acid rain research program and contain several new provisions relating to research, development, and air monitoring. They also contain provisions to provide additional unemployment benefits through the Job Training Partnership Act to workers laid off as a consequence of compliance with the Clean Air Act. The Act also contains provisions to improve visibility near National Parks and other parts of the country (8).

4.8 THE POLLUTION PREVENTION ACT OF 1990

The Pollution Prevention Act, along with the Clean Air Act Amendments passed by Congress on the same day in November 1990, represents a clear breakthrough in this nation's understanding of environmental problems. The Pollution Prevention Act calls pollution prevention a "national objective" and establishes a hierarchy of environmental protection priorities as national policy.

Under the Pollution Prevention Act, it is the national policy of the United States that pollution should be prevented or reduced at the source whenever feasible; where pollution cannot be prevented, it should be recycled in an environmentally safe manner; in the absence of feasible prevention and recycling opportunities, pollution should be treated; disposal should be used only as a last resort.

Among other provisions, the Act directs the EPA to facilitate the adoption of source reduction techniques by businesses and federal agencies, to establish standard methods of measurement for source reduction, to review regulations to determine their effect on source reduction, and to investigate opportunities to use federal procurement to encourage source reduction. The Act also authorizes an $8 million state grant program to promote source reduction, with a 50% state match requirement (9).

The EPA's pollution prevention initiatives are characterized by their use of a wide range of tools, including market incentives, public education and information, small business grants, technical assistance, research and technology applications, as well as the more traditional regulation and enforcement. In addition, there are other significant behind-the-scenes achievements: identifying and dismantling barriers to pollution prevention; laying the groundwork for a systematic prevention focus; and creating advocates for pollution prevention that serve as catalysts in a wide variety of institutions.

The EPA's Pollution Prevention Strategy establishes EPA's future direction in pollution prevention. The strategy indicates how pollution prevention concepts will be incorporated into the EPA's ongoing environmental protection efforts and it set up the "33/50 Program" (Industrial Toxics Project), under which the EPA is seeking substantial voluntary reductions of 17 targeted high-risk industrial chemicals that offer significant opportunities for prevention. The goal was to reduce environmental releases of these chemicals by at least 50% by the end of 1995. The ambitious reduction goals raise several issues which are addressed below.

Although the goals of a one-third reduction by 1992, and a 50% reduction by 1995 were ambitious in 1990, there may well be certain cases — individual chemicals, sources, or types of releases — where even greater reduction targets would be appropriate. As new programs come into being such as those envisioned in the Clean Air Act amendments, the EPA will reevaluate the magnitude and timing of its reduction targets.

Voluntary reduction efforts can be a cost-effective and environmentally effective means of achieving these national goals. Many companies have already made significant progress in reducing their toxic emissions, and have found that their pollution prevention measures often save, rather than cost, money. Establishing national reduction goals will spur additional activity. Where appropriate, the EPA will use its enforcement and regulatory authorities to promote pollution prevention of these chemicals. However, achieving these goals through voluntary programs will be an effective demonstration of environmental progress through nonregulatory means.

Progress will initially be measured by reliance on the Toxics Release Inventory (TRI), with 1988 as a baseline year. Achievement of the goals will be documented by downward trends in the TRI data. The goals are independent of any increasing levels of production; toxic releases can be reduced even as economic activity increases. In effect, industries will have had four years to achieve the initial target of a one-third reduction, and seven years to reach the 50% mark. Those that have been actively pursuing pollution prevention should have little difficulty in achieving these goals; others may have to work more aggressively.

The 17 chemicals are by no means an exhaustive list of the EPA's concerns. These are our nation's principal starting points for achieving major reductions, and the authors believe these targets to be achievable and beneficial. Other targets may be set in the future as information on other chemicals raises concerns. Reductions in these 17 chemicals are anticipated to have a spillover effect in fostering across-the-board reductions in toxics. In all cases, the EPA's existing toxic chemical control programs, aimed at thousands of substances, will be continued and strengthened.

The EPA intends these goals to apply to all sources of releases of these chemicals. Ultimately, this will entail reducing releases of toxic pollutants in the home, office, in farming, in motor vehicles, and elsewhere throughout society. However, in order to document progress in the near term, the EPA will rely chiefly on the TRI to track reductions from manufacturing sources. For some chemicals and sources, it may be necessary to develop separate means of documenting reductions. As substantial progress is made in this sector, the EPA will expand its targeting effort to include other sources as well.

Pollution prevention is the primary means of achieving these national goals. The thrust of this initiative is not only to reduce releases, but to do so by minimizing the quantities of wastes generated in the first place, either by replacing toxic materials with nontoxic substitutes or by running processes more efficiently so as to produce less wastes. Processes that rely on destruction of wastes after they are generated are not as effective in achieving either the environmental or economic benefits of pollution prevention.

Environmental releases as well as off-site transfers of waste are targeted for reductions. It is not the EPA's intent to shift toxic chemical wastes from one disposal route to another. The best reduction option by far is to avoid generating wastes in the first place by eliminating the use of toxic chemicals wherever possible, minimizing the quantities needed, and making operations more efficient so that less toxics end up in waste streams. This goal is best realized by documenting reductions in all forms of waste generation.

Not all facilities will be able to achieve the same level of reductions. The EPA recognizes that facilities will differ in their potential for reducing their waste generation for these particular toxic chemicals. The goals set are national goals, and will not automatically be applied to specific chemicals or facilities. Doubtless, some facilities will be able to exceed them, while others may find it takes a longer time to implement pollution prevention measures in order to achieve the goals. Although the reductions are intended to apply across-the-board to the TRI data, the EPA will focus particular attention on the largest sources of releases of each of the 17 chemicals; the facilities involved can effectively contribute to national reductions by setting reduction goals that exceed those established by the EPA (10).

4.9 SUMMARY

1. Regulatory framework provides an overview of environmental laws and regulations used to protect human health and the environment from potential hazards of waste disposal.

2. The Resource Conservation and Recovery Act (RCRA) of 1976, like most federal environmental legislation, encourages states to develop and run their own hazardous waste programs as an alternative to EPA management.

3. Defining what constitutes a "hazardous waste" requires consideration of both legal and scientific factors.

4. The EPA is evaluating the hazards of new kinds of substances, including genetically engineered microorganisms. The EPA has a number of legislative tools to use in controlling the risks from toxic substances.

5. Congress has provided the EPA and the states with three primary statutes to control and reduce water pollution: the Clean Water Act, the Safe Drinking Water Act, and the Marine Protection, Research, and Sanctuaries Act.

6. The 1986 amendments of CERCLA, known as the Superfund Amendments and Reauthorization Act (SARA), authorized $8.5 billion for both the emergency response and longer term (or remedial) cleanup programs.

7. On November 15, 1990, President Bush signed the Clean Air Act Amendments. Several progressive and creative new themes are embodied in the Amendments, these themes are necessary for effectively achieving the air quality goals and regulatory reform expected from these far-reaching amendments.

8. The Pollution Prevention Act calls pollution prevention a "national objective" and establishes a hierarchy of environmental protection priorities as national policy.

REFERENCES

1. U.S. EPA, *EPA J.* 14(2), March 1988.
2. Office of Solid Waste, *Solving the Hazardous Waste Problem*, EPA/530–SW–86–037.
3. U.S. EPA, Solid Waste and Emergency Response, *The Waste System*, November 1988.
4. P. N. Cheremisinoff, and F. Ellerbusch, Solid Waste Legislation, Resource Conservation & Recovery Act, A Special Report, Washington, D.C., 1979.
5. Bureau of National Affairs, Washington, D.C., Resource Conservation and Recovery Act of 1976, *International Environmental Reporter*, October 21, 1976.
6. Hazardous Waste Management System; Final Codification Rule, *Federal Register*, 40 CFR, Washington, D.C., Monday, July 15, 1985.
7. U.S. EPA, *Environmental Progress and Challenges: EPA's Update*, EPA–230–07–88–033, August 1988.
8. The Clean Air Act Amendments of 1990 Summary Materials, November 15, 1990.
9. Office of Pollution Prevention, *Pollution Prevention News*, October 1991.
10. Office of Pesticides and Toxic Substances, *Overview of the Industrial Toxics Project*, December 5, 1990.

5

MOBILE AND STATIONARY SOURCES OF POLLUTION

Contributing Author: Abdool Jabar

5.1 INTRODUCTION

Pollution of air, water, and land comes from either mobile or stationary sources. Mobile sources such as automobiles, buses, trucks, locomotives, and airplanes contribute to our nation's air and land pollution problems, whereas ships, oil tankers, and barges contribute to our water pollution problems. Stationary or fixed sources of pollution are frequently encountered in everyday life. Stationary sources generate pollutants mainly by burning fuel for energy and as by-products of industrial and chemical processes. Pollutants from stationary sources contribute to the pollution of all three media. Electric utilities, factories, and residential and commercial buildings that burn coal, oil, natural gas, wood, and other fuels are principal sources of such pollutants such as sulfur dioxide, nitrogen oxides, carbon monoxide, particulates, and volatile organic compounds (VOCs). Hazardous waste generated from chemical and industrial processes is the most well-known source of pollution. The problem of past hazardous waste mismanagement caught the nation's attention in a series of incidents. One of these incidents involved the Love Canal in Niagara Falls, New York, where people were evacuated from their homes after hazardous waste buried over 25 years seeped to the surface and into the basements. Times Beach, Missouri, represents another prominent story of hazardous waste mismanagement.

5.2 SOURCES OF AIR POLLUTION

Air pollution presents one of the greatest risks to human health and the environment in the United States. The long list of health problems caused or aggravated by air pollution includes lung diseases such as chronic bronchitis and pulmonary emphysema; cancer, particularly lung cancer; neural disorders, including brain damage; bronchial asthma and the common cold, which are most persistent with highly polluted air; and eye irritations. Environmental problems range from damage to crops and vegetation to increased acidity of lakes that makes then uninhabitable for fish and other aquatic life.

More than half of the nation's air pollution comes from mobile sources. Mobile sources can be classified as any nonstationary source of air pollution, such as passenger cars, trucks, trains, motorcycles, boats, and aircrafts. Exhaust from such sources contains carbon monoxide, VOCs, nitrogen oxide, particulates, and lead. VOCs, along with nitrogen oxides, are the principal contributors to the formation of ground-level ozone and an aerosol which inhibits visibility. The ozone and aerosol are collectively referred to as "smog."

Stationary sources of air pollution are commonly thought of as those emanating from any point that is fixed or stationary. These sources range from large processing plants to neighborhood dry cleaners and gas stations. Some of the common sources are steel plants, coke plants, power plants, gas stations, dry cleaners, and printers.

Stationary sources generate air pollutants mainly by combusting fuel for energy and as by-products of industrial processes. Electrical utilities, factories, and residential and commercial buildings that burn coal, oil, natural gas, wood, and other fuels are principal sources of pollutants such as sulfur dioxide, nitrogen oxides, carbon monoxide, particulates, VOCs, and lead.

Hazardous air pollutants also come from a variety of industrial and manufacturing processes. Fuel oils contaminated with toxic chemicals, hazardous waste disposal facilities, municipal incinerators, landfills, and electrical utilities are potential sources of toxic air pollutants.

This section deals with pollutants from mobile and stationary sources and how the EPA has been dealing with the problem, and the progress made in lowering the emissions of these pollutants. The major pollutants discussed are carbon dioxide, ozone, sulfur dioxide, lead, airborne particulates, air toxics, and acid deposition.

Carbon Monoxide

Carbon monoxide is an invisible, odorless product of incomplete fuel combustion. When inhaled, it replaces the oxygen in the bloodstream and can impair vision, alertness, and other mental and physical capacities. It has particularly severe health effects for people with heart and lung problems.

The main source of carbon monoxide is motor vehicles, especially when their engines are burning fuel inefficiently as they do when starting up in the morning, idling, or moving slowly in congested traffic. Other sources are wood stoves, incinerators, and industrial processes.

Carbon monoxide levels have declined in most parts of the country since 1970 and long-term improvements continued between 1988 and 1997. Ambient carbon monoxide concentrations decreased 38%, and the estimated number of exceedences of the national standards decreased 95%. Emissions from motor vehicle exhaust contributes about 60% of all carbon monoxide emissions nationwide. High concentrations of carbon monoxide generally occur in areas with heavy traffic congestion. In cities, as much as 95% of all carbon monoxide emissions may come from automobile exhaust. While carbon monoxide emissions from highway vehicles alone decreased 29%, total carbon monoxide emissions decreased only 25% overall. Long-term air quality improvement in carbon monoxide occurred despite a 25% increase in vehicle miles traveled in the United States during this 10-year period. Between 1996 and 1997, ambient carbon monoxide concentrations decreased 7%, and carbon monoxide emissions decreased 3%. Transportation sources including highway and off-highway vehicles now account for 77% of the national total carbon monoxide emissions.

Volatile Organic Compounds

Volatile organic compound emissions from large stationary sources such as chemical plants, oil refineries, and industrial sources have been substantially controlled through the efforts of EPA, state, and local regulatory and enforcement efforts. However, many smaller sources such as paint manufacturers, dry cleaners, and gas stations have not been widely controlled. The EPA is assessing methods to control a number of these stationary sources. New emission standards for industries producing synthetic organic chemicals, paints and other surface coatings, and pesticides are part of this effort (1).

Ozone

Ozone is one of the most intractable and widespread environmental problems. Although significant efforts were made, including controls on oil refineries and motor vehicles, many urban areas in the country are still not in attainment of health-based standards. Ozone, which is a major component of smog, can cause serious respiratory problems such as breathing difficulty, asthma, and reduced resistance to infection.

Chemically, ozone is formed with three atoms of oxygen instead of the usual two found in a regular oxygen molecule. This makes it very reactive, so that it combines with practically every material it comes into contact with. This reactivity causes health problems because it tends to break down biological tissues and cells. In the upper atmosphere, where ozone is needed to protect the skin from ultraviolet radiation, it is being destroyed by man-made chemicals such as chlorofluorocarbons (CFCs). At ground level, ozone is a harmful pollutant.

Ozone is produced in the atmosphere when sunlight triggers chemical reactions between naturally occurring atmospheric gases and pollutants such as VOCs and nitrogen oxides. VOCs and nitrogen oxides are produced by combustion sources such as automobiles. Therefore, ozone levels are highest during the day, usually after the heavy morning traffic has released large amounts of VOCs and nitrogen oxides. Motor vehicle traffic is growing so fast and is such an essential aspect of everyday life in many places that even strenuous efforts may not sufficiently reduce emissions. In just four years, between 1980 and 1984, Americans increased their driving by almost two billion vehicle-miles. Individuals, as well as state and local government, must face tough choices if we are to make an adequate reduction in ozone (1).

Sulfur Dioxide

The ambient standards for sulfur dioxide are exceeded in several areas of the United States. The resulting conditions pose serious health and environmental problems. Excessive levels of sulfur dioxide in the ambient air are associated with significant increases in acute and respiratory diseases. Sulfur dioxide can be transported long distances in the atmosphere because it bonds to particles of dust or aerosols. When sulfur dioxide combines with the water vapor in the atmosphere, it forms sulfuric acid. Because of this, sulfur dioxide emissions constitute one of the major contributors to acid rain.

Sulfur dioxide is emitted to the atmosphere primarily by the combustion of coal and fuel oils. Up until the 1950s, the burning of coal by railroad locomotives was a major source of sulfur dioxide pollution. Emissions from industrial sources grew rapidly between 1940 and 1970, as a result of increased production. Since 1970, there has been a decrease in industrial emissions because of controls on nonferrous smelters and sulfuric acid plants. Today two-thirds of all sulfur dioxide emissions come from electric power plants, with those that are coal-fired responsible for 95% of all power plant emissions.

Other sources of sulfur dioxide include refineries, pulp and paper mills, smelters, steel and chemical plants, and energy facilities related to oil shale synfuels and oil and gas production. Home furnaces and wood-burning stoves are sources that more directly affect residential neighborhoods.

Prior to the EPA's establishment in 1970, some states recognized the problems posed by excessive emission of sulfur dioxide and limited the emissions from power plants and factories. One of the first acts of the EPA was to set National Ambient Air Quality Standards for sulfur dioxide.

To meet the EPA's standards, state environmental authorities developed control plans for the various facilities emitting sulfur dioxide. Many utilities installed equipment to wash excessive sulfur from their emissions, while some of these facilities converted sulfur emissions into commercial products such as sulfuric acid.

One technique used to attain ambient standards has proven to be shortsighted. Power plants and factories were allowed to use tall stacks as an alternative to further reduce emissions. This action reduced the local impact of the sulfur dioxide emissions, but the stacks dispersed the pollutant hundreds of miles away. As a result, sulfur dioxide emission in the midwest are contributing to New England's acid rain problem.

Efforts to control sulfur dioxide on a national level have been reasonably successful. Ambient levels decreased by 37% between 1977 and 1986, with a 2% reduction between 1986 and 1987. An even greater improvement was observed in the number of violations of the ambient standard which decreased by 98% during the period. However, emissions decreased by only 21%. Controls on existing plants in urban areas and construction of new power plants in rural areas account for the difference between reductions in ambient levels and emission levels.

The decrease in sulfur dioxide emissions in industrial areas is due to the use of fuels with lower average sulfur content, the introduction of scrubbers to remove sulfur oxides from flue gases, and control on industrial process. The decrease in sulfur dioxide levels in residential and commercial areas is due to a combination of energy conservation measures and the use of cleaner fuels (2).

Lead

The amount of lead emitted in the atmosphere has decreased dramatically in past decades and is one of the EPA's most important success stories. Lead is a heavy metal which can cause serious physical and mental impairment. Children are particularly vulnerable to the effects of high lead levels. Lead has been used in gasoline to increase octane levels to avoid engine-knocking. Two efforts begun over 20 years ago are responsible for a 95% decrease of lead in gasoline.

The EPA, recognizing the health risks posed by lead, required in the early 1970s, that the lead content of all gasoline must be reduced over time. The lead content of leaded gasoline was reduced in 1985 to an average of 0.5 g/gal from 1.0 g/gal and was reduced further in 1986 to 0.1 g/gal.

In addition to the phasing down of lead in the gasoline, the EPA's overall automotive emission control program required the use of unleaded gas in any car beginning in 1975. Currently about 70% of the gas sold is unleaded.

These efforts, together with reductions in lead emissions from stationary sources (such as battery plants and nonferrous smelters), have substantially reduced ambient lead levels. This achievement is one of the greatest contributions the EPA has made to the nation's health (1).

Airborne Particulates

Particulates in air such as dust, smoke, and aerosols may have both short-term and long-term health and environmental effects. These effects range from irritating the eyes and throat and reducing resistance to infection, to being the cause of chronic respiratory diseases. When fine particulates about the size of cigarette smoke particulates are inhaled deeply and lodged in the lungs, they can cause temporary or permanent damage.

Some particulates emitted from diesel engines are suspected of causing cancer, whereas others such as wind-blown dust can carry toxic substances such as polychlorinated biphenyls (PCBs) and pesticides. Particulates are also responsible for corrosion of building materials, damage to vegetation, and severe reduction in visibility.

Major sources of particulates include steel mills, power plants, cotton gins, cement plants, smelters, and diesel engines. Other sources are grain storage elevators, industrial haul roads, construction work, and demolition. Wood-burning stoves and fireplaces also can be significant emitters of particulates. Urban areas are likely to have wind-blown dust from roads, parking lots, and construction activity.

In 1991, the EPA issued National Ambient Air Quality standards for total suspended particulates covering all kinds and sizes. In 1987, the EPA published new standards based on particulate matter smaller than 10 microns. The smaller inhalable particulates present a more serious health threat because they tend to lodge in the lungs and remain in the body for a long time.

Some particulates can be controlled by conventional means, but others need more creative approaches. The EPA and states have sought to meet particulates standards by limiting emissions from industrial facilities and other sources. In order to meet emission standards, industry installed pollution control equipment such as electrically charged plates and huge filters. In 1999, the EPA signed legislation requiring maximum achievable control technology, or MACT, which affects incinerators. The emissions limits for particulates are twice as stringent as those previously allowed. The EPA has also set emission standards for diesel automobiles. These standards took effect in the 1988 model year and became progressively more stringent in the 1991 and 1994 model years. Improved paving, better street cleaning, limits on agricultural and forest burning practices, and bans on backyard burning in urban areas are also responsible for reductions in ambient particulate concentration. There has been a 23% decrease in ambient particulate level from 1977 to 1986 (1).

Air Toxics

One of today's most serious emerging problems is that caused by the toxic pollutants found in all environmental media. Regardless of their low concentrations, emissions of toxic chemicals into the air by human activities may have long- and short-term effects on human health and the environment. The sources that emit toxic chemicals into the atmosphere are numerous. These include industrial and manufacturing processes, solvent uses, sewage treatment plants, hazardous waste handling and disposal sites, municipal waste sites, incinerators, and motor vehicles. Smelters, metal refiners, manufacturing processes, and stationary fuel combustions emit toxic metals such as cadmium, lead, arsenic, chromium, mercury, and beryllium. Toxic organics such as vinyl chloride and benzene are emitted by a variety of sources such as chemical and plastic manufacturing plants. Chlorinated dioxins are emitted by some chemical processes and high-temperature burning of plastics in incinerators.

The most common exposure of these contaminants is by inhalation after they are emitted from smokestacks or tailpipes. After the toxics are airborne and fall to the earth, they are taken up by crops, animals, or fish that are consumed by humans. Airborne toxics may also contaminate drinking water. Toxics enter the body through these routes and are accumulated over time. They may become highly concentrated in human fatty tissue and breast milk. The data on direct human effects of airborne toxics comes from studies of industrial workers. Exposure to airborne toxics is much higher in the workplace than in ambient areas (2).

5.3 SOURCES OF WATER POLLUTION

Protecting the nation's drinking water, coastal zone waters, and surface waters is made complex by the variety of sources that affect them. Groundwater is being contaminated by leaking underground storage tanks, fertilizers and pesticides, uncontrolled hazardous wastes sites, septic tanks, drainage wells, and other sources, threatening 50% of the nation's drinking water supplies for half of this nation's population. Many coastal towns along the Atlantic Coast and Gulf of Mexico have to close beaches one or more times during summer months because of shoreline pollution. In Puget Sound, fecal coliform bacteria contaminate oysters, and the harbor seals have higher concentrations of PCBs than do almost any other seal population in the world.

In March 1989, the *Exxon Valdez* ran aground on Bligh Reef in Prince William Sound, flooding one of the nation's most pristine and sensitive environments with approximately 11 million gallons of crude oil. The spill, the largest in U.S. history, spread over 700–800 miles of shoreline, damaging the area's diverse wildlife and directly affecting the lives of many Alaskans.

Pollution of groundwater by pesticides and nitrates due to the application of agricultural chemicals is a major environmental concern in many parts of the country, particularly in the Midwest. In Iowa, where agricultural chemicals are used in 60% of the state, some private and public drinking water wells have exceeded public health standards for nitrates. Pesticides also have been found in groundwater. About 30 towns in Nebraska have excessive amounts of nitrate in their drinking water. Bottled water must be provided to infants, and monthly well testing is required.

The United States is losing one of its most valuable, and perhaps irreplaceable, resources — the nation's wetlands. Once regarded as wastelands, wetlands are now recognized as an important resource to people and the environment. Wetlands are among the most productive of all ecosystems. Wetland plants convert sunlight into plant material or biomass, which in turn serve as food for many types of aquatic and terrestrial animals. The major food value of wetland plants occurs as they break down into small particles to form the base of an aquatic food chain.

Wetlands are habitats for many forms of fish and wildlife. Approximately two-thirds of this nation's major commercial fisheries use estuaries and coastal marshes as nurseries or spawning grounds. Migratory waterfowl and other birds also depend on wetlands, some spending their entire lives in wetlands and others using them primarily as nesting, feeding, or resting grounds.

The roles of wetlands in improving and maintaining water quality in adjacent water bodies is increasingly being recognized in the scientific literature. Wetlands remove nutrients such as nitrogen and phosphorus, and thus help prevent over-enrichment of water (eutrophication). Also they filter harmful chemicals, such as pesticides and heavy metals, and trap suspended sediments, which otherwise would produce turbidity (cloudiness) in water. This function is particularly important as a natural buffer for nonpoint pollution sources (see Chapter 2 for a discussion of point versus nonpoint pollution sources).

Wetlands also have socioeconomic values. They play an important role in flood control by absorbing peak flows and releasing water slowly. Along the coast, they buffer land against storm surges resulting from hurricanes and tropical storms. Wetland vegetation can reduce shoreline erosion by absorbing and dissipating wave energy and encouraging the deposition of suspended sediments. Also, wetlands contribute $20–40 billion annually to the nation's economy — for example, through recreational and commercial fishing, hunting of waterfowl, and the production of cash crops such as wild rice and cranberries. Unfortunately, our natural heritage of swamps, marshes, bogs, and other types of wetlands is rapidly disappearing. Once there were over 200 million acres of wetlands in the lower 48 states; by the mid-1970s, only 99 million acres remained. Between 1955 and 1975, more than 11 million acres of wetlands — an area three times the size of the state of New Jersey — were lost entirely. The average rate of wetland loss during this period was 458,000 acres per year: 440,000 acres of inland wetlands and 18,000 acres of coastal wetlands. Agricultural development involving drainage of wetlands was responsible for 87% of the losses during those two decades. Urban and other development caused 8% and 5% of losses, respectively. In addition to the physical destruction of habitat, wetlands are also threatened by chemical contamination and other types of pollution.

Ocean dumping of dredged material, sewerage sludge, and industrial wastes are major sources of ocean pollution. Sediments dredged from industrialized urban harbors are often highly contaminated with heavy metals and toxic synthetic organic chemicals such as PCBs and petroleum hydrocarbons. When these sediments are dumped in the ocean, the contaminants can be taken up by marine organisms.

However, persistent disposal of plastics from land and ships at sea has become a serious problem. The most severe impact of this biodegradable debris floating in the ocean is in causing the injury and death of fish, marine mammals, and birds. Debris on beaches from sewer and storm drain overflows or from mismanagement of trash poses public safety and aesthetic concerns and can result in major economic losses for coastal communities during the tourist season (3).

5.4 SOURCES OF LAND POLLUTION

Historically, land has been used as the dumping ground for wastes, including those removed from air and water. Improper handling, storage, and disposal of chemicals can cause serious problems. Most of us are familiar with the examples that follow.

In 1984, gasoline that leaked out of an underground storage tank at a service station seeped down through the soil to the water table and spread out across the surface of the groundwater. When heavy spring rains cause the water table to rise, the moisture that seeped into the gas station's basement carried gasoline with it. The fumes eventually reached explosive levels. A spark from an air compressor that controlled the lift ignited the vapors, and the gas station building was destroyed in an explosion.

In another case, a leak from an underground storage tank at a local service station in 1984 contaminated groundwater, affecting 14 wells. Property values dropped dramatically. Twelve families were forced to rely on bottled water for months. Some residents were affected by headaches, nausea, and eye and skin irritations.

When a homeowner in a residential neighborhood noticed a pool of gasoline near the foot of his driveway, an investigation uncovered a pinhole leak in the petroleum pipeline. Fortunately, most of the neighborhood was connected to the municipal water system and did not rely on wells for drinking water. But the private wells that did exist in the neighborhood had to be sealed as a precautionary measure. Three recovery systems, involving about 60 monitoring wells and more than 100 recovery wells, had to be established in the neighborhood as part of the cleanup operation.

In December 1984, an accidental release of methyl isocyanate from a pesticide facility killed 2800 people in Bhopal, India. This incident focused international attention on the seriousness of chemical accidents. It also created an awareness of the possibility of a major chemical accident occurring in the United States. Six months later, a chemical accident in the United States brought even greater attention to the problem. A release of methylene chloride and aldicarb oxime occurred in Institute, West Virginia, causing concern among many of the town's residents and indicating that better chemical emergency response procedures were needed.

Chemical accidents can be caused by human error, equipment malfunction, explosion, highway accidents, and other factors. The extent to which each community is vulnerable to a serious chemical emergency depends on these factors and on its particular meteorological and topographical conditions. These conditions will determine how chemicals might disperse after an explosion or an accidental release.

Communities at greater risk are (a) those near facilities that produce or use toxic chemicals, (b) a transportation corridor where large-volume chemicals are moved from one facility to another, or (c) a waterway where ships carry or dock and unload chemical cargoes.

In the Love Canal (Niagara Falls, New York) and in Times Beach (Missouri), improper disposal of hazardous waste resulted in contaminated land and water in surrounding communities. In addition, hundreds of drinking water wells have been contaminated by improper waste disposal throughout the United States.

A barge carrying 90 tons of garbage from the northeastern United States traveled for several months in the Caribbean and Gulf of Mexico as its operators searched for a place to dispose the cargo in an appropriate manner.

Municipal wastes include household and commercial wastes, demolition materials, and sewer sludge. Solvents and other harmful household and commercial wastes are generally so intermingled with other materials that specific control of each is virtually impossible. Leachate resulting from rainwater seeping through municipal landfills may contaminate underlying groundwater. While the degree of hazard presented by this leachate is often relatively low, the volume produced is so great that it may contaminate groundwater.

Uncontrolled disposal sites containing hazardous wastes and other contaminants present some of the most serious environmental problems our nation has ever faced. These sites can contaminate groundwater, lead to explosions, and present other dangers to people and the nearby environment. In many cases, the people who disposed the waste were unaware of the problems that the sites eventually would create for public health and the environment. Most of the abandoned or inactive sites and many of the active hazardous waste facilities where hazardous wastes have escaped are linked in some fashion to the chemical and petroleum industries (1).

5.5 SUMMARY

1. Pollution of our air, water, and land comes from mobile or stationary sources.

2. Air pollution presents one of the greatest risks to human health and the environment in the United States. The long list of health problems caused or aggravated by air pollution includes respiratory problems, cancer, and eye irritations. Environmental problems range from damage to crops and vegetation to increased acidity of lakes, which makes them uninhabitable for fish and other aquatic life.

This section of the chapter dealt with the air pollutants that are associated with mobile and stationary sources. The pollutants discussed are carbon monoxide, volatile organic compounds, ozone, sulfur dioxide, lead, airborne particulates, air toxics, and acid deposition.

3. Protecting the nation's drinking water, coastal zone waters, and surface waters is made complex by the variety of sources of pollution that affects them. Groundwater is being contaminated by leaking underground storage tanks, fertilizers and pesticides, uncontrolled hazardous waste sites, septic tanks, drainage wells, and other sources, threatening 50% of the U.S. drinking water supplies.

4. Historically, land has been used as the dumping ground for wastes, including those removed from air and water. Improper handling, storage, and disposal of chemicals can cause serious problems.

REFERENCES

1. U.S. EPA, *Environmental Progress and Challenges, EPA's Update,* August 1988.
2. Clean Air Amendments—Summary, 1990.
3. U.S. EPA, *EPA Journal,* 17(1), January/February 1991.

II

AIR POLLUTION MANAGEMENT ISSUES

6

ATMOSPHERIC DISPERSION

6.1 INTRODUCTION

This chapter focuses on some of the practical considerations of dispersion in the atmosphere. Both continuous and instantaneous discharges are of concern to individuals involved with environmental management. However, the bulk of the material presented here is for continuous emissions from point sources — for example, a stack. This has traditionally been an area of much concern in the air pollution field because stacks have long been one of the more common industrial methods of "disposing" waste gases. The concentrations that humans, plants, animals, and structures are exposed to at ground level can be reduced significantly by emitting the waste gases from a process at great heights. This permits the pollutants to be dispersed over a much larger area and will be referred to as *control by dilution*. Although tall stacks may be effective in lowering the ground-level concentration of pollutants, they still do not in themselves reduce the amount of pollutants released into the atmosphere. However, in certain situations, it can be the most practical and economical way of dealing with an air pollution problem.

Air quality models describe the fate of airborne gases and particles. As these pollutants travel over their pathways, physical and chemical reactions may occur. The categories of mechanisms are nonreactive, reactive (photochemical and non-photochemical), gas-to-particle conversions, gas/particle processes, and particle/particle processes. In addition, the gases and particles may be radioactive, in which case the models must contain some provisions for accounting for radioactive decay and the production of subsequent radioactive elements. In addition, to adequately assess the significance of the air quality impact of a source, background concentrations must be considered. Background air quality relevant to a given source includes those pollutant concentrations due to natural sources and also distant, unidentified man-made sources. For example, it is commonly assumed that the annual mean background concentration of particulate matter is $30-40\ \mu g/m^3$ over much of the eastern United States. Typically, air quality data are used to establish background concentrations in the vicinity of the source under consideration.

A four-step procedure is recommended in performing dispersion calculations, particularly for health effect studies:

1. Estimate the rate, duration, and location of the release into the environment.
2. Select the best available model to perform the calculations.
3. Perform the calculations and generate "downstream" concentrations, including isopleths (i.e., lines of constant concentration) resulting from the source emission(s).
4. Determine what effect, if any, the resulting discharge has on the environment, including humans (particularly and often the only concern), animals, vegetation, and materials of construction. These calculations often include estimates of the so-called venerability zones—that is, regions that may be adversely affected because of the emissions. This is treated in more detail in the chapter on Health and Safety (Chapter 33).

The U.S. EPA's *Guideline on Air Quality Models* (1, 2) specifically addresses atmospheric dispersion calculations. This guideline is used by the EPA, by the states, and by private industry in the review and preparation of Prevention of Significant Deterioration (PSD) permits and State Implementation Plans (SIP) revisions. The guideline serves as a means of maintaining consistency in air quality analyses. On September 9, 1986 (51 FR 32180), the EPA proposed to include four changes to this guideline: (i) addition of a specific version of the Rough Terrain Diffusion Model (RTDM) as a screening model, (ii) modification of the downwash algorithm in the Industrial Source Complex (ISC) model, (iii) addition of the Offshore and Coastal Dispersion (OCD) model to the EPA's list of preferred models, and (iv) addition of the AVACTA 11 model as an alternative model in the guideline. Other minor modifications have been introduced since then.

6.2 DISPERSION CONSIDERATIONS

Perhaps the most important consideration in dispersion applications is first to determine the acceptable ground-level concentration of the waste pollutant(s). The topography of the area must also be considered. Awareness of the meteorological conditions prevalent in the area, such as the prevailing winds, humidity, and rainfall, is also essential. Finally, an accurate knowledge of the constituents of the waste gas and its physical and chemical properties is paramount (3).

Elementary Principles of Dispersion

The ground-level concentration (GLC) is defined as the amount of solid, liquid, or gaseous material per unit volume of air from 0 to 2 m above the ground. Determination of the acceptable GLC of a particular pollutant will vary from state to state and community to community, depending on the local (air pollution) regulations. Current standards may be found by consulting the local environmental control agencies. Sulfur dioxide, for example, may be limited by pollution regulations to concentrations below 0.5 ppm, whereas hydrogen chloride might be permitted in a particular area up to a concentration of 5 ppm (4).

When a small, concentrated puff of gaseous pollutant is released into the atmosphere, it tends to expand in size due to the dynamic action of the atmosphere. In doing so, the concentration of the gaseous pollutant is decreased because the same amount of pollutant is now contained within a larger volume. This natural process of high concentration spreading out to lower concentrations is the process of *diffusion*. Atmospheric diffusion is primarily accomplished by the wind movement of pollutants, but the character of the source of pollution requires that this action of the wind be taken into account in different ways.

These sources can be conveniently grouped into three classes: point sources, line sources, and area sources. In practice, the first two classes must be further divided into instantaneous and continuous sources. Continuous point sources (the smoke plume from a factory, the pall from a burning dump) are the most familiar, the most conspicuous, and the most studied of pollution sources.

The dilution of air contaminants is also a direct result of atmospheric turbulence and molecular diffusion. The rate of turbulent diffusion is so many thousand times greater than the rate of molecular diffusion that the latter effect can be neglected in atmospheric diffusion problems. Atmospheric turbulence and hence atmospheric diffusion vary widely with weather conditions and topography. Atmospheric turbulence is discussed below in the next two subsections.

Vertical Turbulent Diffusion

For all intents and purposes, rapid atmospheric diffusion in the vertical is always bounded: on the bottom by the surface of the earth and at the top by the tropopause. The tropopause — the demarcation between the troposphere, where temperature decreases with altitude, and the stratosphere, where the temperature is relatively constant or increases with altitude — is lowest over the poles, at about 8 km. The detection of radon products throughout the troposphere is conclusive evidence of the eventual availability of the full depth of the troposphere for vertical dispersion, since the radon source is exclusively at the earth's surface. Utilization of this total vertical dimension can take place at very different rates, depending on the thermally driven vertical wind. These rates are intimately related to the vertical temperature profile. On the average (and if one neglects the effects of the phase change of water in the air), enhanced turbulence is associated with a drop in temperature with height of 10°C per kilometer or greater. (This is the dry adiabatic rate.) If the temperature change with height is at a lesser rate, turbulence tends to be decreased, and if the temperature increases with height (an "inversion"), turbulence is very much reduced. The temperature profiles, particularly over land, show a large diurnal variation. Shortly after sunrise, the heating of the land surface by the sun results in rapid warming of the air near the surface; the reduced density of this air causes it to rise rapidly. Cooler air from aloft replaces the rising air "bubble," to be warmed and to rise in turn. This vigorous vertical interchange creates a "superadiabatic" lapse rate (a temperature decrease of more than 10°C per vertical kilometer), and vertical displacements are accelerated. The depth of this well-mixed layer depends on the intensity of solar radiation and the radiation characteristics of the underlying surface. This vigorous mixing may extend well above 3 km over the deserts, whereas the layer may be only 100 or 200 meters thick over forested lake country. Obviously, this effect is highly dependent on the season;

in winter, the lesser insulations and unfavorable radiation characteristics of snow cover greatly inhibit vertical turbulence.

In contrast, with clear or partly cloudy skies the temperature profile at night is drastically changed by the rapid radiational cooling of the ground and the subsequent cooling of the layers of air near the surface. This creates an inversion of the daytime temperature profile, since there is now an increase in temperature with height. In such a situation the density differences rapidly damp out vertical motions, tend to reduce vertical turbulence, and stabilize the atmosphere.

Two other temperature configurations, on very different scales, have important effects on vertical turbulence and the dilution of air pollution. At the smaller end of the scale, the heat capacity of urban areas and, to a lesser extent, the heat generated by fuel consumption act to modify the temperature profile. The effect is most marked at night, when the heat stored by day in the buildings and streets warms the air and prevents the formation of the surface-based temperature inversion typical of rural areas. Over cities it is rare to find an inversion in the lowest 100 m, and the city influence is still evident 200–300 m above the surface. The effect is a function of city size and building density, but not enough observations are yet available to provide any precise quantitative relations. Although the effect — even for the largest cities — is probably insignificant above 1 km, this locally produced vertical mixing is quite important. Pollution, instead of being confined to a narrow layer near the height of emission, perhaps only 100 m in thickness, can be freely diluted in more than double the volume of air, with the concentrations being reduced by a similar factor.

On a much larger scale the temperature profile can be changed over thousands of square kilometers by the action of large-scale weather systems. In traveling storm systems (cyclones) the increased pressure gradients and resulting high winds, together with the inflow of air into the storm, create relatively good vertical mixing conditions. On the other hand, the flat pressure patterns, slower movement, and slow outflow of surface air in high-pressure cells (anticyclones) result in much less favorable vertical mixing. This is primarily due to the gradual subsidence of the air aloft as it descends to replace (mass-continuity requirement) the outflow at the surface. During this descent the air warms adiabatically, and eventually a temperature inversion aloft, inhibiting the upward mixing of pollution above the inversion level, is created. As the anticyclone matures and persists, this subsidence inversion may lower to very near the ground and persist for the duration of the particular weather pattern.

Horizontal Turbulent Diffusion

The most important difference between the vertical and horizontal dimensions of diffusion is that of scale. In the vertical, rapid diffusion is limited to about 10 km. But in the horizontal, the entire surface of the globe is eventually available. Even when the total depth of the troposphere is considered, the horizontal scale is larger by at least three orders of magnitude; and the difference, say during a nocturnal inversion which might restrict the vertical diffusion to a few tens of meters, is even greater since the lateral turbulence is reduced less than the vertical component. Mechanically produced horizontal turbulence is, on a percentage basis, much less important than the thermal effects; its effects are of about the same order of magnitude as the vertical mechanical effects.

The thermally produced horizontal turbulence is not so neatly related to horizontal temperature gradients as vertical turbulence is to the vertical temperature profile. The horizontal temperature differences create horizontal pressure fields, which in turn drive the horizontal winds. These are acted upon by the earth's rotation (the Coriolis effect) and by surface friction, so that there is no such thing as a truly steady-state wind near the surface of the earth. Wind speeds may vary from nearly zero near the surface at night in an anticyclone to 100 m/s under the driving force of the intense pressure gradient of a hurricane. The importance of this variation, even though in air pollution one is concerned with much more modest ranges, is that for continuous sources the concentration is inversely proportional to the wind speed.

The variation of turbulence in the lateral direction is perhaps the most important factor of all, and certainly one of the most interesting. Within a few minutes, the wind may fluctuate rapidly through 908 or more. Over a few hours it may shift, still with much short-period variability, through 1808, and in the course of a month it will have changed through 3608 numerous times. Over the seasons, preferred directional patterns will be established depending upon latitude and large-scale pressure patterns. These patterns may be very stable over many years, and thus establish the wind climatology of a particular location.

The emitted pollution travels with this ever-varying wind. The high-frequency fluctuations spread out the pollutant, and the relatively steady average direction carries it off—for example, toward a suburb or a business district. A gradual turning of direction transports material toward new targets and gives a respite to the previous ones. Every few days the cycle is repeated, and over the years the prevailing winds can create semipermanent patterns of pollution downwind from factories or cities.

6.3 EFFECTIVE HEIGHT OF EMISSION

For an emission height, the calculational sequence begins by first estimating the "effective" height of the emission, employing an applicable plume rise equation. The maximum GLC may then be determined using an appropriate atmospheric diffusion equation (to be considered in the next section).

The effective height of an emission rarely corresponds to the physical height of the source or the stack. If the plume is caught in the turbulent wake of the stack or of buildings in the vicinity of the source or stack, the effluent will be mixed rapidly downward toward the ground. If the plume is emitted free of these turbulent zones, a number of emission factors and meteorological factors will influence the rise of the plume.

The remainder of this section will focus on the effective height of stack emissions that develops because of plume rise considerations. This effective height depends on a number of factors. The emission factors include (a) the gas flow rate and temperature of the effluent at the top of the stack and (b) the diameter of the stack opening. The meteorological factors influencing plume rise are wind speed, air temperature, shear of the wind speed with height, and atmospheric stability. No theory on plume rise presently takes into account all of these variables, and indeed it almost appears as if the number of formulas known to a given researcher for calculating plume rise varies inversely with the relative degress of understanding of the overall process possessed by that individual. Most of the equations that have been formulated for computing the effective height of an emission are, however, semiempirical in nature. Hence when considering any of these plume rise equations, it is important to evaluate each in terms of the assumptions made and the circumstances existing at the time the particular correlation was formulated. Depending on the circumstances, some equations may definitely be more applicable than others.

The effective stack height (equivalent to the effective height of the emission) is usually considered as the sum of the actual stack height, the plume rise due to the velocity (momentum) of the issuing gases, and the buoyancy rise, which latter factor is a function of both the temperature of the gases being emitted and the surrounding atmospheric conditions.

Several plume rise equations have been described in the literature. The David-son–Bryant method (5) is empirical because it is based on Bryant's wind tunnel experiments. It is restricted to gases below 125°F and to stacks of moderate height and larger. It does not give maximum plume rise because it is only a function of momentum and not buoyancy. Values for plume rise are often low. The method applies when the atmosphere neither resists nor assists the vertical motion of the plume and when wind velocities are 20 mph or higher (because above this velocity the vertical motion is insignificant compared with the horizontal). The Holland equation (6) is valid for effluent gases hotter than 125°F and for neutral conditions. Briggs (7) developed a series of equations that are currently used by the U.S. EPA. The Tennessee Valley Authority (TVA) plume rise equations (8) resulted from an extensive investigation of stack emissions from large coal-fired power plants indicating that for particular ranges of atmospheric stability expressions may be determined for estimating the effective plume rise at various distances from the stack. The stack heights ranged from 152 to 244 m, with inside stack diameters ranging from 8 to 9 m. The temperature of flue gas leaving the stacks was approximately 140°C.

Perhaps the simplest of these equations is that of Holland (6), which is shown below.

$$\Delta h = d_s(v_s/u)[1.5 + 2.68 \times 10^{-3}P(\Delta T/T_s)d_s] \tag{6.3.1}$$

where d_s = inside stack diameter, m

                y_s = stack exit velocity, m/s

                u = wind speed, m/s

                P = atmospheric pressure, mbar

           T_s, T = stack gas and ambient temperature, respectively, K

                $T = T_s - T$

                h = plume rise, m

6.4 THE PASQUILL–GIFFORD MODEL (6,9,10)

Having estimated the effective height of an emission, the next step is to study its path downwind using the appropriate atmospheric dispersion formula. There are many dispersion equations presently available, most of them semiempirical in nature. It is not the intent of this chapter to develop each in detail, but rather to look at the one which has found the greatest applicability today. [In the authors' opinion, the best atmospheric dispersion workbook published to date is that by Turner (6).]

The coordinate system (see Figure 6.1) used in making atmospheric dispersion estimates, as suggested by Pasquill and modified by Gifford, is employed and developed in the following discussion. The origin is at ground level at or beneath the point of emission, with the x axis extending horizontally in the direction of the mean wind. The y axis is in the horizontal plane perpendicular to the x axis, and the z axis extends vertically. The plume travels along or parallel to the x axis (in the mean wind direction). The concentration, c, of gas or aerosols at (x, y, z) from a continuous source with an effective emission height, H_e, is given by

$$c(x, y, z; H_e) = \frac{Q'}{2\pi\sigma_y\sigma_z u} \exp\left[-\frac{1}{2}\left(\frac{y}{\sigma_y}\right)^2\right]\left\{\exp\left[-\frac{1}{2}\left(\frac{z - H_e}{\sigma_z}\right)^2\right]\right.$$

$$\left. + \exp\left[-\frac{1}{2}\left(\frac{z + H_e}{\sigma_z}\right)^2\right]\right\} \tag{6.4.1}$$

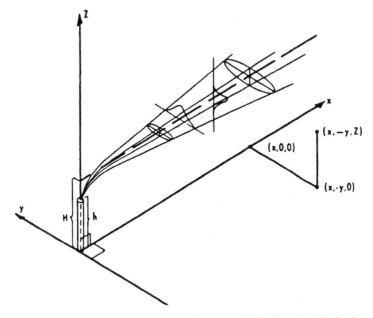

FIGURE 6.1 Coordinate system showing Gaussian distributions in the horizontal and vertical.

where H_e = effective height of emission [sum of physical stack height, H_s, and the plume rise, Δh(m)]

$\quad u$ = mean wind speed affecting the plume (m/s)

$\quad Q'$ = emission rate of pollutants (g/s)

$\quad \sigma_y, \sigma_z$ = stability parameters (m)

$\quad c$ = concentration of gas (g/m³)

$\quad x, y, z$ = coordinates (m)

The assumptions made in the development of Equation 6.4.1 are as follows: (a) The plume spread has a Gaussian distribution in both the horizontal and vertical planes, with standard deviations of plume concentration distribution in the horizontal and vertical of σ_y and σ_z, respectively; (b) there is a uniform emission rate of pollutants, Q'; (c) total reflection of the plume takes place at the earths surface; and (d) the plume moves downwind with mean wind speed, u. Although any consistent set of units may be used, the cgs system is preferred.

Equation 6.4.2 is valid where diffusion in the direction of the plume travel can be neglected; that is, where there is no diffusion in the x direction. This is a valid assumption if the release is continuous or if the duration of release is equal to or greater than the travel time (x/u) from the source to the location of interest.

For concentrations calculated at ground level ($z = 0$), Equation 6.4.1 simplifies to

$$c(x, y, o; H_e) = \frac{Q'}{\pi \sigma_y \sigma_z u} \exp\left[-\frac{1}{2}\left(\frac{y}{\sigma_y}\right)^2 \right] \exp\left[-\frac{1}{2}\left(\frac{H_e}{\sigma_z}\right)^2 \right] \qquad (6.4.2)$$

Where the concentration is to be calculated along the centerline of the plume ($Y = 0$), further simplification gives

$$c(x, 0, 0; H_e) = \frac{Q'}{\pi \sigma_y \sigma_z u} \exp\left[-\frac{1}{2}\left(\frac{H_e}{\sigma_z}\right)^2 \right] \qquad (6.4.3)$$

The values of σ_y and σ_z, which are provided in Figures 6.2 and 6.3—vary with the turbulent structure of the atmosphere, height above the surface, surface roughness, sampling time over which the concentration is to be estimated, wind speed, and distance from the source. Stability categories for σ_y and σ_z are given in Table 6.1.

Cota (11) has produced a program that has computerized many of the above calculations for a point source. Baasel (12) has developed a simple technique for determining the maximum GLC resulting from discharges from elevated point sources. Other approaches and equations (13,14) are available for performing dispersion calculations.

6.5   LINE AND AREA SOURCES

The entire development of the topic of atmospheric dispersion that is possible has been limited here to emissions from a "point" (e.g., stack) source. Although most dispersion applications involve point sources, there are instances where the location

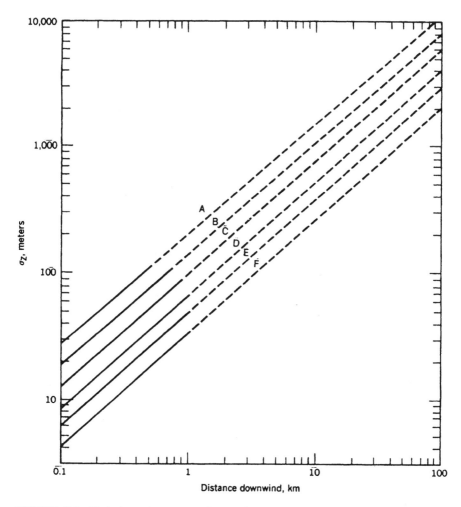

FIGURE 6.2 Variation of concentration profile downwind from points source; A–F designate stability categories listed in Table 6.1.

of the emission can be more accurately described physically and mathematically by either a line source or an area source.

Line sources are generally confined to roadways and streets along which there are well-defined movements of motor vehicles. For these types of line sources, data are required on the width of the roadway and its center strip, the types and amounts (grams per second per meter) of pollutant emissions, the number of lanes, the emissions from each lane, and the height of emissions. In some situations (e.g., a traffic jam at a tollbooth or a series of industries located along a river, or heavy traffic along a straight stretch of highway), the pollution problem may be modeled as a continuously emitting infinite line source (6). Concentrations downwind of a continuously emitting infinite line source, when the wind direction is normal to the

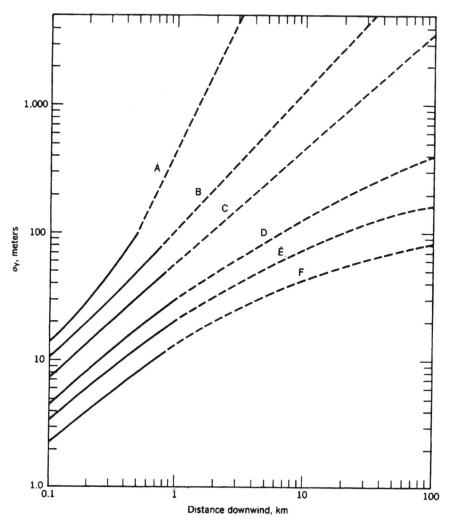

FIGURE 6.3 Vertical dispersion coefficient as a function of downwind distance from the source; A–F designate stability categories listed in Table 6.1.

line, can be calculated from

$$c(x, y, 0; H_e) = \frac{2q}{\sqrt{2\pi}\,\sigma_z u} \exp\left[-\frac{1}{2}\left(\frac{H_e}{\sigma_z}\right)^2\right] \qquad (6.5.1)$$

Here q is the source strength per unit distance, for example, g/sec-m. Note that the horizontal dispersion parameter, σ_y, does not appear in this equation, since it is assumed that lateral dispersion from one segment of the line is compensated by dispersion in the opposite direction from adjacent segments. Also, y does not appear because the concentration at a given x is the same for any value of y.

TABLE 6.1 Stability categories[a]

Surface Wind Speed at 10 m (m/sec)	Day			Night	
	Incoming Solar Radiation			Thinly Overcast or $\geqslant 4/8$ Low Cloud	$\leqslant 3/8$ Cloud
	Strong	Moderate	Slight		
2	A	A–B	B		
2–3	A–B	B	C	E	F
3–5	B	B–C	C	D	E
5–6	C	C–D	D	D	D
6	C	D	D	D	D

[a]Class A is the most unstable, and Class F is the most stable. The neutral class, D, should be assumed for overcast conditions during the day or the night, regardless of wind speed. Night refers to the period from 1 hr before sunset to 1 hr after sunrise.

Concentrations from infinite line sources, when the wind is not perpendicular to the line, can also be approximated. If the angle between the wind direction and line source is ϕ, one obtains

$$c(x, y, 0; H_e) = \frac{2q}{\sin \phi \sqrt{2\pi} \, \sigma_z u} \exp\left[-\frac{1}{2}\left(\frac{H_e}{\sigma_z}\right)^2 \right] \qquad (6.5.2)$$

This equation should not be used where ϕ is less than 45°. When the continuously emitting line source is reasonably short in length (i.e., "finite"), one can account for the edge effects caused by the two ends of the source. If the line source is perpendicular to the wind direction, then it is convenient to define the x axis in the direction of the wind and also passing through the sampling point downwind. The ends of the line source, then, are at two positions in the crosswind direction, y_1 and y_2, where y_1 is less than y_2. The concentration along the x axis at ground level is then given by the expression

$$c(x, 0, 0; H_e) = \frac{2q}{\sqrt{2\pi} \, \sigma_z u} \left\{ \exp\left[-\frac{1}{2}\left(\frac{H_e}{\sigma_z}\right)^2 \right] \right\} \times \int_{p_1}^{p_2} \frac{1}{\sqrt{2\pi}} \exp\left(-\frac{1}{2}p^2 \right) dp \quad (6.5.3)$$

where $p_1 = y_1/\sigma_y$ and $p_2 = y_2/\sigma_y$. Once the limits of integration are established, the value of the integral may be determined from standard tables of integrals (3,4).

Area sources (3,4) include the multitude of minor sources with individually small emissions that are impractical to consider as separate point or line sources. Area sources are typically treated as a grid network of square areas, with pollutant emissions distributed uniformly within each grid square. Area source information required includes types and amounts of pollutant emissions, the physical size of the area over which emissions are prorated, and representative stack height for the area. In dealing with dispersion of pollutants in areas having large numbers of sources — for example, as in fugitive dust from (coal) piles, a large number of automobiles in a parking lot, or a multistack situation — there may be too many sources to consider each source individually. Often an approximation can be made by combining all of the emissions in a given area and treating this area as a single source.

6.6 ATMOSPHERIC DISPERSION EQUATION FOR INSTANTANEOUS SOURCES (6)

Unfortunately, little information is available on instantaneous or "puff" sources. Only Turner's workbook (6) provides a simplified equation that may be used for estimation purposes. This approach is presented below.

Only sources emitting continuously or for time periods equal to or greater than the travel times from the source to the point of interest were treated earlier. Cases of instantaneous releases (as from an explosion) or of short-term releases (on the order of seconds) are often of practical concern. To determine concentrations at any position downwind, one must consider the time interval after the time of release and diffusion in the downwind direction as well as lateral and vertical diffusion. The determination of the path or trajectory of the "puff" is of considerable importance, but very difficult to accomplish in practice. It is most important if concentrations are to be determined at specific points. Determining the trajectory is less important if knowledge of the magnitude of the concentrations for particular downwind distances or travel times is required, but without a need to know exactly at what points these concentrations occur. An equation that may be used for estimates of ground-level concentrations downwind from a release from height, H, is

$$c(x, y, 0; H) = \frac{2Q_T}{(2\pi)^{1.5}\sigma_x\sigma_y\sigma_z} \exp\left[-\frac{1}{2}\left(\frac{x - ut}{\sigma_x}\right)^2\right]$$
$$\times \exp\left[-\frac{1}{2}\left(\frac{H}{\sigma_z}\right)^2\right]\exp\left[-\frac{1}{2}\left(\frac{y}{\sigma_y}\right)^2\right] \qquad (6.6.1)$$

The numerical value of $(2\pi)^{3/2}$ is 15.75. The notations have the usual meaning, with the important exceptions that Q_T represents the total mass of the release and the σ's are not necessarily those evaluated with respect to the dispersion of a continuous source at a fixed point in space. This equation can be simplified for centerline concentrations and ground-level emissions by setting $y = 0$ and $H = 0$, respectively.

The σ's in Equation 6.6.1 refer to dispersion statistics following the motion of the expanding puff. The σ_x is the standard deviation of the concentration distribution in the puff in the downwind direction, and t is the time after release. Note that there is no dilution in the downwind direction by wind speed. The speed of the wind mainly serves to give the downwind position of the center of the puff, as shown by examination of the exponential involving σ_x. Wind speed may influence the dispersion indirectly because the dispersion parameters σ_x, σ_y, and σ_z may be functions of wind speed. The σ's and σ_z's for an instantaneous source may be assumed (as a first approximation) to be equal to those given earlier. The problem that remains is to make best estimates of σ_x. Much less is known of diffusion in the downwind direction than is known of lateral and vertical dispersion. In general, one should expect the σ_x value to be about the same as σ_y.

6.7 DESIGN SUGGESTIONS (3,4)

As experience in designing stacks has accumulated over the years, several rules of thumb have evolved:

1. Stack heights should be at least 2.5 times the height of any surrounding buildings or obstacles so that significant turbulence is not introduced by these factors.

2. The stack gas exit velocity should be greater than 60 ft/sec so that stack gases will escape the turbulent wake of the stack. In many cases, it is good practice to have the gas exit velocity on the order of 90 or 100 ft/sec.

3. A stack located on a building should be located in a position which will assure that the exhaust escapes the wakes of nearby structures.

4. Gases from stacks with diameters less than 5 ft and heights less than 200 ft will hit the ground part of the time, and ground concentrations may be excessive. In this case, the plume becomes unpredictable.

5. The maximum ground concentration of stack gases subjected to atmospheric diffusion occurs about five to ten effective stack heights downwind from the point of emission.

6. When stack gases are subjected to atmospheric diffusion and building turbulence is not a factor, GLC on the order of 0.001–1% of the stack concentration are possible for a properly designed stack.

7. Ground concentrations can be reduced by the use of higher stacks. The ground concentration varies inversely as the square of the effective stack height.

8. Average concentrations of a contaminant downwind from a stack are directly proportional to the discharge rate. An increase in discharge rate by a given factor increases GLC at all points by the same factor.

9. In general, increasing the dilution of stack gases by the addition of excess air in the stack does not affect GLC appreciably. Addition of diluting air will increase the effective stack height, however, by increasing the stack exit velocity. This effect may be important at low wind speeds. On the other hand, if the stack temperature is decreased appreciably by the dilution, the effective stack height may be reduced. Stack dilution will have an appreciable effect on the concentration in the plume close to the stack.

The nine rules of thumb listed above represent the basic design elements of a stack system. The engineering approach suggests that each element be evaluated independently and as part of the whole control system. However, the engineering design and evaluation must be an integrated part of a complete air pollution control program.

6.8 SUMMARY

1. Both continuous and instantaneous discharges are of concern to individuals involved with environmental management. This has traditionally been an area of much concern in the air pollution field because stacks have long been one of the more common industrial methods of "disposing" waste gases. The concentrations that humans, plants, animals, and structures are exposed to at ground level can be reduced significantly by emitting the waste gases from a process at great heights.

2. Perhaps the most important consideration in dispersion applications is first to determine the acceptable GLC of the waste pollutant(s). The topography of the

area must also be considered. Awareness of the meteorological conditions prevalent in the area, such as the prevailing winds, humidity, and rainfall, is also essential. Finally, an accurate knowledge of the constituents of the waste gas and its physical and chemical properties is paramount.

3. The effective stack height (equivalent to the effective height of the emission) is usually considered as the sum of the actual stack height, the plume rise due to the velocity (momentum) of the issuing gases, and the buoyancy rise, which is a function of the temperature of the gases being emitted and the atmospheric conditions.

4. Having determined the effective height of an emission, the next step is to study its path downwind using the appropriate atmospheric dispersion formula. There are many dispersion equations presently available, most of them semi-empirical in nature. The Pasquill–Gifford model or some modification of it is usually employed in performing downwind concentration calculations.

5. Although most dispersion applications involve point sources, there are instances where the location of the emission can be more accurately described physically and mathematically by either a line source or an area source.

6. Cases of instantaneous releases (as from an explosion) or of short-term releases (on the order of seconds) are often of practical concern. To determine concentrations at any position downwind, one must consider the time interval after the time of release and diffusion in the downwind direction, as well as lateral and vertical diffusion.

7. There are numerous design suggestions for stacks. One of the key recommendations is that stack heights should be at least 2.5 times the height of any surrounding buildings or obstacles so that significant turbulence is not introduced by these factors.

REFERENCES

1. U.S. EPA, *Guideline on Air Quality Models*, Publication No. EPA–450/2–78–027, Research Triangle Park, NC, 1978 (OAQPS No. 1.2–08).

2. U.S. EPA, *Industrial Source Complex (ISC) Dispersion Model Users Guide*, second edition, Vols. 1 and 2, Publication Nos. EPA–450/4–86–005a, and EPA–450/4–86–005b, Research Triangle Park, NC, 1986 (NRTIS PB86 234259 and 23467).

3. L. Theodore et al., *Accident and Emergency Management Student Manual*, U.S. EPA APTI Course 503, 1989.

4. L. Theodore, J. Reynolds, and F. Taylor, *Accident and Emergency Management*, John Wiley & Sons, New York, 1989.

5. W. F. Davidson, *Transactions of the Conference on Industrial Wastes, 14th Annual Meeting of the Industrial Hygienics Foundation of America*, 1949, p. 38.

6. D. B. Turner, *Workbook of Atmospheric Dispersion Estimates*, Environmental Protection Agency, Publication No. AP-26, Research Triangle Park, NC, 1970 (revised).

7. G. A. Briggs, *Plume Rise*, AEC Critical Review Series, USAEC, Division of Technical Information.

8. T. L. Montgomery et al., *Journal of the Air Pollution Control Association*, 22(10), 779, 1972.

9. F. Pasquill, *Meteorological Magazine*, 90(33), 1063, 1961.

10. F. A. Gifford, Nuclear Safety 2(4), 47, 1961.

11. H. Cota, *Journal of the Air Pollution Control Association*, 31(8), 253, 1984.

12. W. Baasel, *Journal of the Air Pollution Control Association*, 38(8), 866, 1981.

13. O. G. Sutton, *Quarterly Journal of the Royal Meteorological Society*, 73, 257, 1947.

14. C. H. Bosanquet, and J. L. Pearson, *Transactions of the Faraday Society*, 32, 1249, 1936.

7

AIR POLLUTION CONTROL EQUIPMENT

7.1 INTRODUCTION

Controlling the emission of pollutants from industrial and domestic sources is important in protecting the quality of air. Air pollutants can exist in the form of particulate matter or as gases. Air cleaning devices have been reducing pollutant emissions from various sources for many years. Originally, air cleaning equipment was used only if the contaminant was highly toxic or had some recovery value. Now, with recent legislation, control technologies have been upgraded and more sources are regulated in order to meet the National Ambient Air Quality Standards (NAAQS). In addition, state and local air pollution agencies have adopted regulations that are in some cases more stringent than federal emission standards.

Equipment used to control particulate emissions are gravity settlers (often referred to as *settling chambers*), mechanical collectors (cyclones), electrostatic precipitators (ESPs), scrubbers (venturi scrubbers), and fabric filters (baghouses). Techniques used to control gaseous emissions are absorption, adsorption, combustion, and condensation. The applicability of a given technique depends on the physical and chemical properties of the pollutant and the exhaust stream. More than one technique may be capable of controlling emissions from a given source. For example, vapors generated from loading gasoline into tank trucks at large bulk terminals are controlled by using any of the above four gaseous control techniques. Most often, however, one control technique is used more frequently than others for a given source–pollutant combination. For example, absorption is commonly used to remove sulfur dioxide (SO_2) from boiler flue gas.

The material presented in this chapter regarding air pollution control equipment contains, at best, an overview of each control device. Equipment diagrams and figures, operation and maintenance procedures, etc., have not been included in this discussion. More details, including calculational predictive and design procedures, are available in the literature (1, 2). Applications of control equipment to specific processes can be found in other chapters in this book.

7.2   AIR POLLUTION CONTROL EQUIPMENT FOR PARTICULATES

Gravity Settlers

Gravity settlers, or gravity settling chambers, have long been utilized industrially for the removal of solid and liquid waste materials from gaseous streams. Advantages accounting for their use are simple construction, low initial cost and maintenance, low pressure losses, and simple disposal of waste materials. Gravity settlers are usually constructed in the form of a long, horizontal parallelepiped with suitable inlet and outlet ports. In its simplest form the settler is an enlargement (large box) in the duct carrying the particle-laden gases; the contaminated gas stream enters at one end, while the cleaned gas exits from the other end. The particles settle toward the collection surface at the bottom of the unit with a velocity at or near their terminal settling velocity. One advantage of this device is that the external force leading to separation is provided freely by nature. Its use in industry is generally limited to the removal of the larger particles (i.e., those greater than 40 μm microns (or micrometers).

Cyclones

Centrifugal separators, commonly referred to as *cyclones*, are widely used in industry for the removal of solid and liquid particles (or particulates) from gas streams. Typical applications are found in mining and metallurgical operations, the cement and plastics industries, pulp and paper mill operations, chemical and pharmaceutical processes, petroleum production (cat-cracking cyclones), and combustion operations (fly ash collection).

Particles suspended in a moving gas stream possess inertia and momentum and are acted upon by gravity. Should the gas stream be forced to change direction, these "properties" can be utilized to promote centrifugal forces to act on the particles. In a conventional unit the entire mass of the gas stream with the entrained particles enters the unit tangentially and is forced into a constrained vortex in the cylindrical portion of the cyclone. Upon entering the unit, a particle develops an angular velocity. Because of its greater inertia, it tends to move outward across the gas streamlines in a tangential rather than a rotary direction; thus, it attains a net outward radial velocity. By virtue of its rotation with the carrier gas around the axis of the tube (main vortex) and its higher density with respect to the gas, the entrained particle is forced toward the wall of the unit. Eventually the particle may reach this outer wall, where it is carried by gravity and assisted by the downward movement of the outer vortex and/or secondary eddies, toward the dust outlet at the bottom of the unit. The flow vortex is reversed in the lower (conical) portion of the unit, leaving most of the entrained particles behind. The cleaned gas then passes up through the center of the unit (inner vortex) and out of the collector.

Multiple-cyclone collectors (multicones) are high-efficiency devices that consist of a number of small-diameter cyclones operating in parallel with a common gas inlet and outlet. The flow pattern differs from a conventional cyclone in that instead of bringing the gas in at the side to initiate the swirling action, the gas is brought in at the top of the collecting tube and swirling action is then imparted by a stationary vane positioned in the path of the incoming gas. The diameters of the collecting tubes usually range from 6 to 24 in. Properly designed units can be constructed and operated with a collection efficiency as high as 90% for particulates in the 5- to 10-μm range. The most serious problems encountered with these systems involve plugging and flow equalization.

Electrostatic Precipitator

Electrostatic precipitators (ESPs) are satisfactory devices for removing small particles from moving gas streams at high collection efficiencies. They have been used almost universally in power plants for removing fly ash from the gases prior to discharge.

The two major types of high-voltage ESP configurations currently used are tubular and plate. Tubular precipitators consist of cylindrical collection tubes with discharge electrodes located on the axis of the cylinder. However, the vast majority of ESPs installed are of the plate type. Particles are collected on flat, parallel collecting surfaces spaced 8–12 in. apart, with a series of discharge electrodes located along the center line of adjacent plates. The gas to be cleaned passes horizontally between the plates (horizontal-flow type) or vertically up through the plates (vertical-flow type). Collected particles are usually removed by rapping.

Depending on operating conditions and the required collection efficiency, the gas velocities in an industrial ESP are usually between 2.5 and 8.0 ft/sec. A uniform gas distribution is of prime importance for precipitator operation, and it should be achieved with a minimum expenditure of pressure drop. This is not always easy, since gas velocities in the duct ahead of the precipitator may be 30–100 ft/sec in order to prevent dust buildup. It should be clear that the best operating condition for a precipitator will occur when the velocity distribution is uniform. When significant maldistribution occurs, the higher velocity in one collecting plate area will decrease efficiency more than a lower velocity at another equal area will increase the efficiency of that area. Small particles tend to follow flow streamlines better than do large particles, so a maldistribution in particle size distribution in areas of the precipitator will also occur. Gross flow maldistribution also contributes to reentrainment losses and sneakage of raw gases around collecting plates. Uniformity of the bulk flow entering the precipitator is therefore important to maximize precipitator performance, and a customary practice in precipitator design is lto conduct a complete model study prior to installation, so as to define the flow control devices necessary to ensure compliance with stipulated uniformity criteria. The economic and technical advantages of conducting three-dimensional flow model studies for this purpose are readily apparent because providing good gas distribution is an art; model studies are the best means available for obtaining the necessary confidence in the design for these high-cost installations. Suitable distribution models must be tested; the results must be related to the full-size unit considering the laws of dynamic similarities.

The maximum voltage at which a given field can be maintained depends on the properties of the gas and the dust being collected. These parameters may vary from one point to another within the precipitator, as well as with time. In order to keep each section of the precipitator working at high efficiency, a high degree of sectionalization is recommended. This means that many separate power supplies and controls will produce better performance on a precipitator of a given size than if there are only one or two independently controlled sections. This is particularly true if high efficiencies are required.

Modern precipitators have voltage control devices which automatically limit precipitator power input. A well-designed automatic control system tends to keep the voltage level at approximately the value needed for optimum particle charging by the corona current. The controls operate according to the following principles. The higher the voltage, the greater the spark rate between the discharge electrodes and the collecting plate. As the spark rate increases, however, a greater percentage of input power is wasted in the spark current; consequently, less useful power is applied to dust collection. There is, however, an optimal sparking rate where the gains in particle charging from increased voltage are just offset by corona current losses from sparkover. Measurements on commercial precipitators have determined that this optimal sparking rate is between 50 and 150 sparks per minute per electrical section. The objective in corona power control is to maintain corona power input at this optimal sparking rate. This is usually accomplished by momentarily reducing precipitator power input whenever excessive sparking occurs.

Venturi Scrubbers

Wet scrubbers have found widespread use in cleaning contaminated gas streams because of their ability to effectively remove particulate and gaseous pollutants. Specifically, wet scrubbing involves the technique of bringing a contaminated gas stream into intimate contact with a liquid. Wet scrubbers include all the various types of gas absorption equipment. The term "scrubber" will be restricted to those systems which utilize a liquid, usually water, to achieve or assist in the removal of particulate matter from a carrier gas stream. The use of wet scrubbers to remove gaseous pollutants from contaminated streams is considered in the next section.

To achieve high collection efficiency of particulates by impaction, a small droplet diameter and high relative velocity between the particle and droplet are required. In a venturi scrubber this is often accomplished by introducing the scrubbing liquid at right angles to a high-velocity gas flow in the venturi throat (vena contracta). Very small water droplets are formed, and high relative velocities are maintained until the droplets are accelerated to their terminal velocity. Gas velocities through the venturi throat typically range from 12,000 to 24,000 ft/min. The velocity of the gases alone causes the atomization of the liquid.

The size of the droplets generated in scrubber units affects both the collection efficiency and pressure drop; that is, small droplet sizes requiring high-pressure atomization give greater collection efficiencies. Various correlations are available in the literature to estimate the mean liquid drop diameter from different types of atomizers under different operating conditions. These correlations are applicable to fluids within a certain range of operating conditions and properties such as the volume ratio of liquid to gas, the relative velocity of gas to liquid, the type of nozzle, the surface tension of the liquid, and so on. In using one of these correlations to estimate droplet diameter, it is important to select a correlation which takes these factors into consideration.

Another important design consideration for venturi scrubbers, (as well as absorbers) is concerned with suppressing the steam plume. Water scrubber systems removing pollutants from high-temperature processes (i.e., combustion) can generate a supersaturated water vapor which becomes a visible white plume as it leaves the stack. Although not strictly an air pollution problem, such a plume may be objectionable for aesthetic reasons. Regardless, there are several ways to avoid or eliminate the steam plume. The most obvious way is to specify control equipment which does not use water in contact with the high-temperature gas stream (i.e., ESP, cyclones, or fabric filters). Should this not be possible or practical, a number of suppression methods are available:

1. Mixing with heated and relatively dry air.
2. Condensation of moisture by direct contact with water, then mixing with heated ambient air.
3. Condensation of moisture by direct contact with water, then reheating scrubber exhaust gases.

Baghouses

The basic filtration process may be conducted in many different types of fabric filters in which the physical arrangement of hardware and the method of removing collected material from the filter media will vary. The essential differences may be related, in general, to:

1. Type of fabric
2. Cleaning mechanism
3. Equipment
4. Mode of operation

Gases to be cleaned can be either "pushed" or "pulled" through the baghouse. In the pressure system (push through) the gases may enter through the cleanout hopper in the bottom or through the top of the bags. In the suction type (pull through) the dirty gases are usually forced through the inside of the bag and exit through the outside.

Baghouse collectors are available for either intermittent or continuous operation. Intermittent operation is employed where the operational schedule of the dust-generating source permits halting the gas cleaning function at periodic intervals (regularly defined by time or by pressure differential) for removal of collected material from the filter media (cleaning). Collectors of this type are primarily utilized for the control of small-volume operations such as grinding, polishing, and so on, and for aerosols of a very coarse nature. For most air pollution control installations and major dust control problems, however, it is desirable to use collectors which allow for continuous operation. This is accomplished by arranging several filter areas in a parallel flow system and cleaning one area at a time according to some preset mode of operation.

Baghouses may also be characterized and identified according to the method used to remove collected material from the bags. Particle removal can be accomplished in a variety of ways, including shaking the bags, blowing a jet of air on the bags from a reciprocating manifold, or rapidly expanding the bags by a pulse of compressed air. In general, the various types of bag cleaning methods can be divided into those involving fabric flexing and those involving a reverse flow of clean air. In pressure-jet or pulse-jet cleaning, a momentary burst of compressed air is introduced through a tube or nozzle attached to the top of the bag. A bubble of air flows down the bag, causing bag walls to collapse behind it. In the pulse-jet bag cleaning method a momentary burst of compressed air is introduced in the discharge nozzle of a filter bag, which inflates the bag in the opposite direction.

A wide variety of woven and felted fabrics are used in fabric filters. Clean felted fabrics are more efficient dust collectors than are woven fabrics, but woven materials are capable of giving equal filtration efficiency after a dust layer accumulates on the surface. When a new woven fabric is placed in service, visible penetration of dust within the fabric may occur. This normally takes from a few hours to several days for industrial applications, depending on dust loadings and the nature of the particles.

Baghouses are constructed as single units or compartmental units. The single unit is generally used on small processes that are not in continuous operation such as grinding and paint-spraying processes. Compartmental units consist of more than one baghouse compartment and are used in continuous operating processes with large exhaust volumes such as electric melt steel furnaces and industrial boilers. In both cases, the bags are housed in a shell made of rigid metal material.

7.3   AIR POLLUTION CONTROL EQUIPMENT FOR GASEOUS POLLUTANTS

Absorption

Absorption is a mass transfer operation in which a gas is dissolved in a liquid. A contaminant (pollutant exhaust stream) contacts a liquid, and the contaminant diffuses from the gas phase into the liquid phase. The absorption rate is enhanced by (a) high diffusion rates of the contaminant in both the liquid and gas phase, (b) high solubility of the contaminant, (c) large liquid–gas contact area, and (d) good mixing between liquid and gas phases (turbulence).

The liquid most often used for absorption is water because it is inexpensive, is readily available, and can dissolve a number of contaminants. Reagents can be added to the absorbing water to increase the removal efficiency of the system. Certain reagents merely increase the solubility of the contaminant in the water. Other reagents chemically react with the contaminant after it has been absorbed (such as a lime slurry used to scrub SO_2 flue gases). In reactive scrubbing the absorption rate is much higher, so in some cases a smaller, more economical system can be used. However, the reactions can form a precipitate which could cause plugging problems in the absorber or in associated equipment.

Gas absorbers or wet scrubbers are designed to provide good mixing of the gas and liquid phases. The devices used for gas absorption are often the same as those used in particulate emission scrubbing. These include packed towers, plate towers, spray columns, and venturi scrubbers. However, a wet particulate emission scrubber is not operated the same way as a gas absorber. In the case of a wet scrubber, particulate matter removal efficiency is a function of the pressure drop: the higher the pressure drop, the greater the removal efficiency. For gas absorption, a higher pressure drop can result in shorter contact time between the phases, thereby limiting the amount of absorption that can occur. Therefore, optimizing both gas and particulate pollutant removal in one device is difficult.

If a gaseous contaminant is very soluble, almost any of the wet scrubbers will adequately remove this contaminant. However, if the contaminant is of low solubility, the packed tower or the plate tower is more effective. Both of these devices provide long contact time between phases and have relatively low pressure drops. The packed tower, the most common gas absorption device, consists of an empty shell filled with packing. Liquid flows down over the packing, exposing a large film area to the gas flowing up countercurrently through the packing. Plate towers consist of horizontal plates placed inside the tower. Gas passes up through orifices in these plates, while liquid flows down across the plate, thereby providing the desired contact. The major disadvantage in using these devices is that they both are susceptible to plugging when the exhaust stream contains high concentrations of particulate matter. When an exhaust stream contains a high level of particulate matter, venturis, spray devices, or combinations of systems can be used.

Adsorption

Adsorption is a mass transfer process that involves removing a gaseous contaminant by adhering it to the surface of a solid. Adsorption can be classified as physical or chemical. In physical adsorption, a gas molecule adheres to the surface of the solid due to an imbalance of natural forces (electron distribution). In chemisorption, once the gas molecule adheres to the surface, it reacts chemically with it. The major distinction is that physical adsorption is readily reversible whereas chemisorption is not.

All solids physically adsorb gases to some extent. Certain solids, called *adsorbents*, have a high attraction for specific gases; they also have a large surface area, which gives them a large capacity. By far the most important adsorbent for air pollution control is activated carbon. Because of its surface, activated carbon will preferentially adsorb hydrocarbon vapors and odorous organic compounds from an airstream. Most other adsorbents (molecular sieves, silica gel, and activated aluminas) will preferentially adsorb water vapor, which may render them useless to remove other contaminants.

For activated carbon, the amount of hydrocarbon vapors that can be adsorbed depends on the physical and chemical characteristics of the vapors, their concentration in the gas stream, system temperature, system pressure, humidity of the gas stream, and molecular weight of the vapor. Physical adsorption is a reversible process; the adsorbed vapors can be released by increasing the temperature, decreasing the pressure, or using a combination of both. Vapors are normally desorbed by heating the adsorber with steam.

Adsorption can be a very useful removal technique, since it is capable of removing very small quantities (a few parts per million) of vapor from an airstream. The vapors are not destroyed; instead, they are stored on the adsorbent surface until they can be removed by desorption. The desorbed vapor stream is highly concentrated. Depending on the nature of this material, it can be condensed and recycled or burned.

The most common adsorption system is the fixed bed adsorber. These systems consist of two or more adsorber beds operating on a timed adsorbing and desorbing cycle. One or more beds are adsorbing vapors, while the other bed is being regenerated. If particulate matter or liquid droplets are present in the vapor-laden airstream, this stream is sent to pretreatment to remove them. If the temperature of the inlet vapor stream is high (much above 120°F), cooling may also be provided. Since all adsorption processes are exothermic, cooling coils in the carbon bed itself may be needed to prevent excessive heat buildup. Carbon bed depth is usually limited to a maximum of 4 ft, and vapor velocity through the adsorber is held below 100 ft/min to prevent an excessive pressure drop.

Combustion

Combustion is defined as rapid, high-temperature gas-phase oxidation. Simply stated, the contaminant (a carbon–hydrogen substance) is burned with air and converted to carbon dioxide and water vapor. The operation of any combustion source is governed by the three T's of combustion; temperature, turbulence, and time. For complete combustion to occur, each contaminant molecule must come in contact with oxygen (turbulence) at a sufficient temperature and be held at this temperature for an adequate time. These three variables are totally dependent on each other. For example, if a higher temperature were to be used, less mixing of the contaminant and combustion air or shorter residence time may then be used. If adequate turbulence cannot be provided, a higher temperature or longer residence time may be required for complete combustion.

Combustion devices can be categorized as flares, thermal incinerators, or catalytic incinerators. Flares are direct combustion devices used to dispose of large quantities or emergency releases of combustible gases. Flares are normally elevated (from 100 to 400 ft) to protect the surroundings from heat and flames. Flares are often designed for steam injection at the flare tip. The steam provides sufficient turbulence to ensure complete combustion; this prevents smoking. Flares are very noisy, which can cause problems for adjacent neighborhoods.

Thermal incinerators are also called *afterburners*, *direct flame incinerators*, or *thermal oxidizers*. These are devices in which the contaminant airstream passes around or through a burner and into a refractory-line residence chamber where oxidation occurs. To ensure complete combustion of the contaminant, thermal incinerators are designed to operate at a temperature of 700–800°C (1300–1500°F) and a residence time of 0.3–0.5 sec. Ideally, as much of the fuel value as possible is supplied by the waste contaminant stream. This reduces the amount of auxiliary fuel needed to maintain the proper temperature.

In catalytic incineration the contaminant-laden stream is heated and passed through a catalyst bed, which promotes the oxidation reaction at a lower temperature. Catalytic incinerators normally operate at 370–480°C (700–900°F). This reduced temperature results in a continuous fuel savings. However, it may be offset by the cost of the catalyst. The catalyst, which is usually platinum, is coated on a cheaper metal or ceramic support base. The support can then be arranged to expose a high surface area, which provides a sufficient number of active sites, on which the reactions occur. Catalysts are subject to both physical and chemical deterioration. Halogens and sulfur-containing compounds act as catalyst suppressants and decrease the catalysts' usefulness. Certain heavy metals such as mercury, arsenic, phosphorus, lead, and zinc are particularly poisonous.

Condensation

Condensation is a process in which the volatile gases are removed from the contaminant stream and changed into a liquid. Condensation is usually achieved by reducing the temperature of a vapor mixture until the partial pressure of the condensable component equals its vapor pressure. Condensation requires low temperatures to liquefy most pure contaminant vapors. Condensation is affected by the composition of the contaminant gas stream. The presence of additional gases which do not condense at the same conditions, such as air, hinders condensation.

Condensers are normally used in combination with primary control devices. Condensers can be located upstream of (before) an incinerator, adsorber, or absorber. These condensers reduce the volume of vapors that the more expensive equipment must handle. Therefore, the size and the cost of the primary control device can be reduced. Similarly, condensers can be used to remove water vapors from a process stream with a high moisture content upstream of a control system. A prime example is the use of condensers in rendering plants to remove moisture from the cooker exhaust gas. When the condensers are used alone, refrigeration is required to achieve the low temperatures required for condensation. Refrigeration units are used to successfully control gasoline vapors at large bulk gasoline dispensing terminals.

Condensers are classified as either contact condensers or surface condensers. Contact condensers cool the vapor by spraying liquid directly on the vapor stream. These devices resemble a simple spray scrubber. Surface condensers are normally shell-and-tube heat exchangers. Coolant flows through the tubes, while the vapor is passed over and condenses on the outside of the tubes. In general, contact condensers are more flexible, simpler, and less expensive than surface condensers. However, surface condensers require much less water and produce nearly 20 times less wastewater that must be treated than do contact condensers. Surface condensers also have an advantage in that they can directly recover valuable contaminant vapors.

7.4 HYBRID SYSTEMS

Hybrid systems are defined as those types of control devices that involve combinations of control mechanisms—for example, fabric filtration combined with electrostatic precipitation. Four of the major hybrid systems found in practice today are briefly reviewed below. These include wet electrostatic precipitators, ionizing wet scrubbers, dry scrubbers, and electrostatically augmented fabric filtration.

Wet Electrostatic Precipitators

The wet electrostatic precipitator (WEP) is a variation of the dry ESP design. The two major added features in a WEP system are

1. A preconditioning step, where inlet sprays in the entry section are provided for cooling, gas absorption, and removal of coarse particles.
2. A wetted collection surface, where liquid is used to continuously flush away collected materials.

Particle collection is achieved by (a) the introduction of evenly distributed liquid droplets to the gas stream through sprays located above the electrostatic field sections and (b) migration of the charged particles and liquid droplets to the collection plates. The collected liquid droplets form a continuous downward-flowing film over the collection plates, and keep them clean by removing the collected particles.

The WEP overcomes some of the limitations of the dry ESP. Its operation is not influenced by the resistivity of the particles. Furthermore, since the internal components are continuously being washed with liquid, buildup of tacky particles is controlled and there is some capacity for removal of gaseous pollutants. In general, applications of the WEP fall into two areas: removal of fine particles and removal of condensed organic fumes. Outlet particulate concentrations are typically in the 10^{-3} to 10^{-2} gr/ft^3 range (3). Data on the capability of the WEP to remove acid gases are very limited. This device has been installed to control HF emissions. Using a liquid-to-gas (L/G) ratio of 5 gal/1000 acf and a liquid pH between 8 and 9, fluoride removal efficiencies of >98% have been measured; outlet concentrations of HF were found to be <1 ppm (4).

Some of the advantages of a WEP include:

- Simultaneous gas absorption and dust removal.
- Low energy consumption.
- No dust resistivity problems.
- Efficient removal of fine particles.

Disadvantages of the WEP are the following:

- Low gas absorption efficiency.
- Sensitive to changes in flow rate.
- Dust collection is wet.

Ionizing Wet Scrubbers

The ionizing wet scrubber (IWS) is a relatively new development in the technology of removal of particulate matter from a gas stream. These devices have been incorporated in commercial incineration facilities (5,6). In the IWS, high-voltage ionization in a charge section places a static electrical charge on the particles in the gas stream, which then passes through a crossflow packed-bed scrubber. The packing is normally polypropylene in the form of circular-wound spirals and gearlike wheel configurations, providing a large surface area. Particles with sizes of 3 "m or larger are trapped by inertial impaction within the packed bed. Smaller charged particles pass close to the surface of either the packing material or a scrubbing water droplet. An opposite charge in that surface is induced by the charged particle, which is then attracted to an ion attached to the surface. All collected particles are eventually washed out of the scrubber. The scrubbing water also functions to absorb gaseous pollutants. According to Ceilcote (the IWS vendor), collection efficiency of a two-stage IWS is greater than that of a baghouse or a conventional ESP for particles in the 0.2–0.6-μm range. For 0.8 μm and above, the ESP is as effective as the IWS (3). Scrubbing water can include caustic soda or soda ash when needed for efficient absorption of acid gases. Corrosion resistance of the IWS is achieved by fabricating its shell and most internal parts from fiberglass-reinforced plastic (FRP) and thermoplastic materials. Pressure drop through a single-step IWS is around 5 in. H_2O (primarily through the wet scrubber section). All internal areas of the ionizer section are periodically deluge-flushed with recycled liquid from the scrubber recycle system. The advantages of an IWS are those of the combination of a WEP and a packed-bed absorber. The disadvantages include the need for separation of particulates from the scrubbing medium, potential scaling and fouling problems (aggravated by recycle neutralizing solution), possible damage to the scrubber if scrubber solution pumps fail, and the need for a downstream mist eliminator. Despite some of these limitations, the IWS is a particulate control device that has worked efficiently.

Dry Scrubbers

The success of fabric filters in removing fine particles from flue gas streams has encouraged the use of combined dry-scrubbing/fabric-filter systems for the dual purpose of removing both particulates and acid gases simultaneously. Dry scrubbers offer potential advantages over their wet counterparts, especially in the areas of energy savings and capital costs. Furthermore, the dry-scrubbing process design is relatively simple, and the product is a dry waste rather than a wet sludge.

There are two major types of so-called dry scrubber systems: spray drying and dry injection. The first process is often referred to as a *wet–dry* system. When compared to a conventional wet scrubber, it uses significantly less scrubbing liquid. The second process may be referred to as a *dry–dry* system because no liquid scrubbing is involved. The spray-drying system is predominantly used in utility and industrial applications.

The method of operation of the spray dryer is relatively simple, requiring only two major equipment items: (a) a spray dryer similar to those used in the chemical food-processing and mineral-preparation industries and (b) a baghouse (or ESP) to collect the fly ash and entrained solids. In the spray dryer, the sorbent solution, or slurry, is atomized into the incoming flue gas stream to increase the liquid–gas interface and to promote the mass transfer of the SO_2 from the gas to the slurry droplets, where it is absorbed. Simultaneously, the thermal energy of the gas evaporates the water in the droplets to produce a dry powdered mixture of sulfite-sulfate and some unreacted alkali. Because the flue gas is not saturated and contains no liquid carryover, potentially troublesome mist eliminators are not required. After leaving the spray dryer, the solids-bearing gas passes through the fabric filter (or ESP), where the dry product is collected and where a percentage of the unreacted alkali reacts with SO_2 for further removal. The cleaned gas is then discharged through the fabric-filter plenum to an induced draft (ID) fan and to the stack.

Among the inherent advantages that the spray dryer enjoys over wet scrubbers are

1. Lower capital costs.
2. Lower draft losses.
3. Reduced auxiliary power.
4. Reduced water consumption, with liquid-to-gas (L/G) ratios significantly lower than those of wet scrubbers.
5. Continuous, two-stage operation, from liquid feed to dry product.

The sorbent of choice for most spray-dryer systems is a lime slurry. One system under development uses a sodium carbonate solution. Although the latter will generally achieve a higher level Of SO_2 removal than will a lime slurry at similar operating conditions, the significant cost advantage that lime has over sodium carbonate makes it the overwhelming favorite. Also, when sodium alkalis are used, the products are highly water-soluble and may create disposal problems.

Dry-injection processes generally involve pneumatic introduction of a dry, powdery alkaline material, usually a sodium-based sorbent, into the flue gas stream with subsequent fabric-filter collection. The injection point in such processes can vary from the boiler–furnace area all the way to the flue gas entrance to the baghouse, depending on operating conditions and design criteria.

Electrostatically Augmented Fabric Filtration

Advanced electrostatic simulation of fabric filtration is another example of a hybrid system. It combines the efficiency of fabric filters with the efficiency and low pressure drop of an ESP. This combination is accomplished by placing a high-voltage electrode coaxially inside a filter bag to establish an electric field between the electrode and the bag surface. The electric field alters the dust deposition pattern within the bag, yielding a much lower pressure drop than is found in conventional bags. The bags can also operate at much higher air-to-cloth ratios.

Another system combines fabric filtration with a surface electric field. This version is similar to the above system but uses an array of wires separated by insulating spacers mounted on the clean side of the fabric. These electrostatic forces can be utilized by allowing the natural particle charges to accumulate on the fabric and then collecting the particles, and by applying an electric field at the fabric surface.

Numerous attempts have been made to use electrostatic effects to improve upon the performance of fabric filters. These efforts have met with varying degrees of success. Presently, these techniques are only available in the pilot plant stage. The implementation of some of these new techniques is being generally resisted by most industries. As with any new technology, there is room for improvement, but the results may be promising enough to continue research and development (7).

7.5 FACTORS IN CONTROL EQUIPMENT SELECTION

There are a number of factors to be considered prior to selecting a particular piece of air pollution control hardware (8, 9). In general, they can be grouped into three categories: environmental, engineering, and economic.

Environmental

1. Equipment location
2. Available space
3. Ambient conditions
4. Availability of adequate utilities (i.e., power, water, etc.) and ancillary system facilities (i.e., waste treatment and disposal, etc.)

5. Maximum allowable emissions (air pollution regulations)

6. Aesthetic considerations (i.e., visible steam or water vapor plume, impact on scenic vistas, etc.)

7. Contribution of air pollution control system to wastewater and solid waste

8. Contribution of air pollution control system to plant noise levels

Engineering

1. Contaminant characteristics (i.e., physical and chemical properties, concentration, particulate shape and size distribution, etc.; in the case of particulates, chemical reactivity, corrosivity, abrasiveness, toxicity, etc.)

2. Gas stream characteristics (i.e., volume flow rate, temperature, pressure, humidity, composition, viscosity, density, reactivity, combustibility, corrosivity, toxicity, etc.)

3. Design and performance characteristics of the particular control system (i.e., size and weight, fractional efficiency curves, etc.; in the case of particulates, mass transfer and/or contaminant destruction capability, etc.; in the case of gases or vapors, pressure drop, reliability and dependability, turndown capability, power requirements, utility requirements, temperature limitations, maintenance requirements, flexibility toward complying with more stringent air pollution regulations, etc.)

Economic

1. Capital cost (equipment, installation, engineering, etc.)

2. Operating cost (utilities, maintenance, etc.)

3. Expected equipment lifetime and salvage value

Prior to the purchase of control equipment, experience has shown that the following points should be emphasized (8,9):

1. Control equipment should not be purchased without reviewing certified independent test data on its performance under a similar application. The manufacturer should be asked to provide performance information and design specifications.

2. In the event that sufficient performance data are unavailable, the equipment supplier can often provide a small pilot model for evaluation under existing conditions.

3. Participation of the local control authorities in the decision-making process is strongly recommended.

4. A good set of specifications is essential. A strong performance guarantee from the manufacturer should be obtained to ensure that the control equipment will meet all applicable local, state, and federal codes at specific process conditions.

5. Process and economic fundamentals should be closely reviewed. Assess the possibility for emission trade-offs (offsets) and/or applying the "bubble concept." The bubble concept permits a plant to find the most efficient way to control its emissions as a whole, rather than having the EPA regulate the emissions from

individual sources. Reductions at a source where emissions can be reduced for the least cost can offset emissions of the same pollutant from another source in the plant.

6. A careful material balance study should be made before authorizing an emission test or purchasing control equipment.

7. Equipment should not be purchased until firm installation cost estimates have been added to the equipment cost. Escalating installation costs are the rule rather than the exception.

8. Operation and maintenance costs should be given high priority on the list of equipment selection factors.

9. Equipment should not be purchased until a solid commitment from the supplier(s) is obtained. Make every effort to ensure that the new system will utilize fuel, controllers, filters, motors, and so on, that are compatible with those already available at the plant.

10. The specification should include written assurance of prompt technical assistance from the equipment supplier. This, together with a complete operating manual (with parts list and full schematics), is essential and is too often forgotten in the rush to get the equipment operating.

11. Schedules, particularly on projects being completed under a court order or consent judgment, can be critical. In such cases, delivery guarantees should be obtained from the manufacturer(s) and penalties should be identified.

12. The air pollution equipment should be of fail-safe design with built-in indicators to show when performance is deteriorating.

13. Withhold 10–15% of the purchase price until compliance is clearly demonstrated.

The usual design/procurement/construction/startup problems can be further compounded by any one or combination of the following (8):

1. Unfamiliarity of process engineers with air pollution engineering
2. New and changing air pollution regulations
3. New suppliers with unproven equipment
4. Lack of industry standards in some key areas
5. Interpretations of control agency field personnel
6. Compliance schedules that are too tight
7. Vague specifications
8. Weak guarantees for the new control equipment
9. Unreliable delivery schedules
10. Variability, unreliable process operation

Proper selection of a particular system for a specific application can be extremely difficult and complicated. In view of the multitude of complex and often ambiguous pollution control regulations, it is in the best interest of the prospective user to work closely with regulatory officials as early in the process as possible. Finally, previous experience on a similar application cannot be overemphasized.

7.6   COMPARING CONTROL EQUIPMENT ALTERNATIVES (9)

The final choice in equipment selection is usually dictated by the need for equipment that is capable of achieving compliance with regulatory codes at the lowest uniform annual cost (amortized capital investment plus operation and maintenance costs). In order to compare specific control equipment alternatives, knowledge of the particular application and site is essential. A preliminary screening, however, may be performed by reviewing the advantages and disadvantages of each type of air pollution control equipment. For example, if water or a waste treatment system is not available at the site, this may preclude use of a wet scrubber system and instead focus particulate removal on dry systems, such as cyclones, baghouses, and/or ESP. If auxiliary fuel is unavailable on a continuous basis, it may not be possible to combust organic pollutant vapors in an incineration system. If the particulate-size distribution in the gas stream is relatively fine, cyclone collectors probably would not be considered. If the pollutant vapors can be reused in the process, control efforts may be directed to adsorption systems. There are many more situations where knowledge of the capabilities of the various control options, combined with common sense, will simplify the selection procedure. General advantages and disadvantages of the most popular types of air pollution control equipment for gases and particulates are presented in Tables 7.1 to 7.9.

7.7   FUTURE TRENDS (10)

The basic design of air pollution control equipment has remained relatively unchanged since it was first used in the early part of the twentieth century. Some modest equipment changes and new types of devices have appeared in the last twenty years, but all have employed essentially the same capture mechanisms as those used in the past. One area that has recently received some attention is hybrid systems (see earlier section)—equipment that can in some cases operate both at

TABLE 7.1   Advantages and disadvantages of cyclone collectors

Advantages

1. Low cost of construction.
2. Relatively simple equipment with few maintenance problems.
3. Relatively low operating pressure drops (for degree of particulate removal obtained) in the range of an approximately 2–6 in. water column.
4. Temperature and pressure limitations imposed only by the materials of construction used.
5. Dry collection and disposal.
6. Relatively small space requirements.

Disadvantages

1. Relatively low overall particulate collection efficiencies, especially on particulates below 10 μm in size.
2. Inability to handle tacky materials.

Table 7.2   Advantages and disadvantages of wet scrubbers

Advantages

1. No secondary dust sources.
2. Relatively small space requirements.
3. Ability to collect gases as well as particulates (especially "sticky" ones).
4. Ability to handle high-temperature, high-humidity gas streams.
5. Capital cost is low (if wastewater treatment system is not required).
6. For some processes, the gas stream is already at high pressures (so pressure drop considerations may not be significant).
7. Ability to achieve high collection efficiencies on fine particulates (at the expense, however, of pressure drop).

Disadvantages

1. May create water disposal problem.
2. Product is collected wet.
3. Corrosion problems are more severe than with dry systems.
4. Steam plume opacity and/or droplet entrainment may be objectionable.
5. Pressure drop and horsepower requirements may be high.
6. Solids buildup at the wet–dry interface may be a problem.
7. Relatively high maintenance costs.

Table 7.3   Advantages and disadvantages of electrostatic precipitators

Advantages

1. Extremely high particulate (coarse and fine) collection efficiencies can be attained (at a relatively low expenditure of energy).
2. Dry collection and disposal.
3. Low pressure drop (typically less than 0.5-in. water column).
4. Designed for continuous operation with minimum maintenance requirements.
5. Relatively low operating costs.
6. Capable of operation under high pressure (to 150 psi) or under vacuum conditions.
7. Capable of operation at high temperatures (to 1300°F).
8. Relatively large gas flow rates can be effectively handled.

Disdvantages

1. High capital cost.
2. Very sensitive to fluctuations in gas stream conditions (in particular, flows, temperatures, particulate and gas composition, and particulate loadings).
3. Certain particulates are difficult to collect because of extremely high or low resistivity characteristics.
4. Relatively large space requirements required for installation.
5. Explosion hazard when treating combustible gases and/or collecting combustible particulates.
6. Special precautions are required to safeguard personnel from the high voltages.
7. Ozone is produced by the negatively charged discharge electrode during gas ionization.
8. Relatively sophisticated maintenance personnel required.

Table 7.4   Advantages and disadvantages of fabric-filter systems

Advantages

1. Extremely high collection efficiency on both coarse and fine (submicron) particulates.

2. Relatively insensitive to gas stream fluctuation. Efficiency and pressure drop are relatively unaffected by large changes in inlet dust loadings for continuously cleaned filters.

3. Filter outlet air may be recirculated within the plant in many cases (for energy conservation).

4. Collected material is recovered dry for subsequent processing or disposal.

5. No problem with liquid waste disposal, water pollution, or liquid freezing.

6. Corrosion and rusting of components are usually not problems.

7. There is no hazard of high voltage, simplifying maintenance and repair and permitting collection of flammable dusts.

8. Use of selected fibrous or granular filter aids (precoating) permits the high-efficiency collection of submicron smokes and gaseous contaminants.

9. Filter collectors are available in a large number of configurations, resulting in a range of dimensions and inlet and outlet flange locations to suit installation requirements.

10. Relatively simple operation.

Disadvantages

1. Temperatures much in excess of 550°F require special mineral or metallic fabrics that are still in the developmental stage and can be very expensive.

2. Certain dusts may require fabric treatments to reduce dust seeping or, in other cases, assist in the removal of the collected dust.

3. Concentrations of some dusts in the collector (~ 50 g/m^3) may represent a fire or explosion hazard if a spark or flame is admitted by accident. Fabrics can burn if readily oxidizable dust is being collected.

4. Relatively high maintenance requirements (bag replacement, etc.).

5. Fabric life may be shortened at elevated temperatures and in the presence of acid or alkaline particulate or gas constituents.

6. Hygroscopic materials, condensation of moisture, or tarry adhesive components may cause crusty caking or plugging of the fabric or may require special additives.

7. Replacement of fabric may require respiratory protection for maintenance personnel.

8. Medium pressure-drop requirements, typically in the range 4–10-in. water column.

higher efficiency and more economically than conventional devices. Tighter regulations and a greater concern for environmental control by society has placed increased emphasis on the development and application of these systems. The future will unquestionably see more activity in this area.

Several specialty air pollution control devices, particularly in the scrubber area, are receiving more attention. Although they have been on the market for many years, their unique collection mechanism has recently attracted increased interest. Four such units include the Dynawave Unit, the Multi-Micro-Venturi (MMV), the condensing scrubber, and the collision scrubber. These are briefly described in the next paragraph.

Table 7.5 Advantages and disadvantages of absorption systems (packed and plate columns)

Advantages

1. Relatively low pressure drop.
2. Standardization in fiberglass-reinforced plastic (FRP) construction permits operation in highly corrosive atmospheres.
3. Capable of achieving relatively high mass-transfer efficiencies.
4. Increasing the height and/or type of packing or number of plates can improve mass transfer without the need to purchase a new piece of equipment.
5. Relatively low capital cost.
6. Relatively small space requirements.
7. Ability to collect particulates as well as gases.

Disadvantages

1. May create water (or liquid) disposal problem.
2. Product is collected wet.
3. Particulates deposition may cause plugging of the bed or plates.
4. When FRP construction is used, it is sensitive to temperature.
5. Relatively high maintenance costs.

The DynaWare™ unit operates in a manner that creates a "froth" zone through which contaminated gas must pass. There is extreme turbulence in the zone and efficient collection of small particles occurs. The MMV Scrubber consists of a "stacked" bank of staggered tubular elements arranged in a pattern similar to a bank of heat exchanger tubes. The thin spacing causes successive micro-venturi flows at the "pinch" sections, leading to high efficiencies. In condensing scrubbers, water vapor in the flue gas stream is condensed onto particulates, causing the particle to grow to a size which enables easier collection by conventional means. Finally, in the collision scrubber two gas streams are made to collide head-on. The collision action shreds the water drops into finer ones which can more effectively collect submicron particles. The process also produces a larger liquid surface area for gas absorption.

Table 7.6 Comparison of plate and packed columns

Packed Column

1. Lower pressure drop.
2. Simpler and cheaper to construct.
3. Preferable for liquids with high foaming tendencies.

Plate Column

1. Less susceptible to plugging.
2. Less weight.
3. Less of a problem with channeling.
4. Temperature surge will result in less damage.

Table 7.7 Advantages and disadvantages of adsorption systems

Advantages

1. Product recovery may be possible.
2. Excellent control and response to process changes.
3. No chemical disposal problem when pollutant (product) is recovered and returned to process.
4. Capability of systems for fully automatic, unattended operation.
5. Capability to remove gaseous or vapor contaminants from process streams to extremely low levels.

Disadvantages

1. Product recovery may require an exotic, expensive distillation (or extraction) scheme.
2. Adsorbent progressively deteriorates in capacity as the number of cycles increases.
3. Adsorbent regeneration requires a steam or vacuum source.
4. Relatively high capital cost.
5. Prefiltering of gas stream may be required to remove any particulate capable of plugging the adsorbent bed.
6. Cooling of the gas stream may be required to get to the usual range of operation (less than 120°F).
7. Relatively high steam requirements to desorb high-molecular-weight hydrocarbons.

Other recent advances in this field have been primarily involved in the treatment of metals. A dry scrubber followed by a wet scrubber has been employed in the United States to improve the collection of fine particulate metals in hazardous waste incinerators; the dry scrubber captures metals that condense at the operating temperature of the unit, and the wet scrubber captures residue metals (particularly mercury) and dioxin/furan compounds. Another recent application in Europe involves the injection of powdered activated carbon into a flue gas stream from a hazardous waste incinerator at a location between the spray dryer (the dry scrubber) and the baghouse (or electrostatic precipitator). The carbon mixes with the lime particulates in the gas stream from the dry scrubbing system and adsorbs

Table 7.8 Advantages and disadvantages of combustion systems

Advantages

1. Simplicity of operation.
2. Capability of steam generation or heat recovery in other forms.
3. Capability for virtually complete destruction of organic contaminants.

Disdvantages

1. Relatively high operating costs (particularly associated with fuel requirements).
2. Potential for flashback and subsequent explosion hazard.
3. Catalyst poisoning (in the case of catalytic incineration).
4. Incomplete combustion can create potentially worse pollution problems.

Table 7.9 Advantages and disadvantages of condensers

Advantages

1. Pure product recovery (in the case of indirect-contact condensers).
2. Water used as the coolant in an indirect contact condenser (i.e., shell-and-tube heat exchanger) does not contact the contaminated gas stream and can be reused after cooling.

Disadvantages

1. Relatively low removal efficiency for gaseous contaminants (at concentrations typical of pollution control applications).
2. Coolant requirements may be extremely expensive.

the mercury vapors and residual dioxin/furan compounds; they are then separated from the gas stream by a particulate control device. More widespread use of these types of systems is anticipated in the future.

7.8 SUMMARY

1. Controlling the emission of pollutants from industrial and domestic sources is important in protecting the quality of air. Air pollutants can exist in the form of particulate matter or as gases. Air cleaning devices have been reducing pollutant emissions from various sources for many years. Originally, air cleaning equipment was used only if the contaminant was highly toxic or had some recovery value.

2. Typs of equipment used to control particulate emissions are gravity settlers (often referred to as *settling chambers*), mechanical collectors (cyclones), electrostatic precipitators (ESPs), scrubbers (venturi scrubbers), and fabric filters (baghouses).

3. Techniques used to control gaseous emissions are absorption, adsorption, combustion, and condensation.

4. Hybrid systems are defined as those types of control devices that involve combinations of control mechanisms — for example, fabric filtration combined with electrostatic precipitation. Four of the major hybrid systems found in practice today include wet electrostatic precipitators, ionizing wet scrubbers, dry scrubbers, and electrostatically augmented fabric filtration.

5. There are a number of factors to be considered prior to selecting a particular piece of air pollution control hardware. In general, they can be grouped into three categories: environmental, engineering, and economic.

6. The final choice in equipment selection is usually dictated by that equipment capable of achieving compliance with regulatory codes at the lowest uniform annual cost (amortized capital investment plus operation and maintenance costs). In order to compare specific control equipment alternatives, knowledge of the particular application and site is essential. A preliminary screening, however, may be performed by reviewing the advantages and disadvantages of each type of air pollution control equipment.

7. The basic design of air pollution control equipment has remained relatively unchanged since first used in the early part of the twentieth century. Some modest

equipment changes and new types of devices have appeared in the last twenty years, but all have essentially employed the same capture mechanisms used in the past.

REFERENCES

1. L. Theodore and P. Feldman, *Air Pollution Control Equipment for Particulates*, Theodore Tutorials, East Williston, NY, 1992.

2. L. Theodore, J. P. Reynolds, and R. Richman, *Air Pollution Control Equipment for Gaseous Pollutants*, Theodore Tutorials, East Williston, NY, 1992.

3. The Ceilcote Corporation, *IWS Bulletin*, Berea, OH, 1983.

4. J. W. MacDonald, private communication, 1996.

5. U.S. EPA, *Engineering Handbook for Hazardous Waste Incineration*, prepared by Monsanto Research Corporation, Dayton, OH, EPA Contract No. 68–03–3025, September 1982.

6. U.S. EPA, *Revised Engineering Handbook for Hazardous Waste Incineration*, unpublished.

7. Assorted technical literature. Environmental Testing Services (ETS), Roanoke, VA, 1998.

8. L. Theodore and A. J. Buonicore, *Industrial Air Pollution Control Equipment for Particulates*, CRC Press, West Palm Beach, FL, 1988.

9. L. Theodore and A.J. Buonicore, *Selection, Design, Operation and Maintenance: Air Pollution Control Equipment*, ETS, Roanoke, VA, 1982.

10. Section 7.7 is adapted with permission from M. K. Theodore and L. Theodore, *Major Environmental Issues Facing the 21st Century*, Prentice-Hall, Upper Saddle River, NJ, 1998.

8

GREENHOUSE EFFECT AND GLOBAL WARMING

Contributing Author: Indira H. Sweeny

8.1 INTRODUCTION

The greenhouse effect appears to be a completely man-made phenomenon in the world today—one that some scientists feel is leading the planet to the brink of disaster (1). The term "greenhouse effect" describes two separate but dependent occurrences: (a) the increase of trace "greenhouse gases" (carbon dioxide, nitrous oxides, methane, tropospheric ozone, and chlorofluorocarbons) in the earths atmosphere and (b) the absorption and re-emission of long-wave radiation by these gases. The increased concentration of greenhouse gases (most especially carbon dioxide) in the atmosphere since the Industrial Revolution is a well-documented fact; however, the predicted effects of this increase are still in debate among scientists in the environmental field. In theory, the greenhouse gases act like the glass in a botanical greenhouse, trapping heat and warming the planet. Current debate centers around questions such as (2): (a) Have greenhouse gases affected global weather as yet? (b) How much will temperature rise once the greenhouse gases in the atmosphere reach concentrations double their current levels? (c) How long does it take for changes in greenhouse gases to affect global climate?

Climatic Changes

Global mean surface temperatures have increased 0.6–1.2°F since the late nineteenth century. The 10 warmest years in the twentieth century have all occurred in the last 15 years. Of these, 1998 was the warmest year on record. The snow cover in the Northern Hemisphere and floating ice in the Arctic Ocean have decreased. Globally, sea level has risen 4–10 in. over the past century. Worldwide precipitation over land has increased by about 1%. The frequency of extreme rainfall events has increased throughout much of the United States.

Increasing concentrations of greenhouse gases are likely to accelerate the rate of climate change. Scientists expect that the average global surface temperature could

103

rise 1.6–6.3°F with significant regional variation by the year 2100. Evaporation will increase as the climate warms, which in turn will increase average global precipitation. Soil moisture is likely to decline in many regions, and intense rainstorms are likely to become more frequent. Sea level is likely to rise 2 ft along most of the U.S. coast (3).

The Carbon Cycle in Nature

Carbon dioxide has been the earth's natural temperature regulator since the time of the formation of the planet, even though there is a comparatively small amount present in the atmosphere (0.03% by weight or by volume). Energy radiated from the sun to the earth is absorbed by the atmosphere, and is balanced by a comparable amount of long-wave energy emitted back to space from the earth's surface. Carbon dioxide molecules absorb some of the long-wave energy radiation from the planet, and at the same time allow solar energy radiation to the planet to penetrate to the surface of the earth. This exchange causes the surface of the earth to increase in temperature. In this way, the thermal balance of the planet, and thus global climate, is dependent on the amount of carbon dioxide in the atmosphere (4). As described above, the effect of carbon dioxide (and other greenhouse gases) in the atmosphere has been compared to that of the glass panes of a greenhouse: The panes let sunlight pass into the greenhouse during the day and the interior of the structure is warmed by the solar radiation. At night some heat radiates out of the structure, but the "glass" decreases the heat loss and keeps the greenhouse warmer (5).

There are various natural cycles in the world that play a major part in determining the concentration of carbon dioxide in the atmosphere. As winds blow over the large surface area of the ocean, carbon dioxide is exchanged across the water–air interface until an equilibrium between the partial pressures of the carbon dioxide in the atmosphere and the carbon dioxide in the surface ocean water is reached. The carbon dioxide is then transferred to deeper water by convection cycles which lower the concentration of carbon dioxide at the surface, thereby allowing more of the gas from the atmosphere to diffuse into solution. The most widely accepted models of oceanic mixing estimate that the oceans have taken up 23–30 gigatons of carbon dioxide between 1958 and 1980, which is equal to 26–34% of the fossil fuel carbon emitted to the atmosphere during that time span (6).

Another important natural influence on the amount of atmospheric carbon dioxide is the terrestrial carbon cycle. Photosynthetic plants remove inorganic carbon (such as carbon dioxide) from the air and convert it into organic compounds. The cycle is completed when carbon is returned to the atmosphere by respiration, decay, and the burning of hydrocarbons. Scientists have estimated that the net uptake of carbon dioxide by vegetation is 62 gigatons per year, which is assumed to be balanced, over several years, by an equal amount of carbon returned to the atmosphere by the decomposition of trash and organic matter (6). Of course, there are wide variations in both the oceanic and terrestrial cycles, and thus it is difficult to measure exactly how effective nature is at storing the increased amount of carbon dioxide being produced in the modern age. All ecological processes (respiration, photosynthesis, etc.) are sensitive to environmental conditions, especially temperature, moisture, and even fluctuations in carbon dioxide concentration.

For the most part, greenhouse gases trap heat because of their chemical makeup and, in particular, their triatomic nature (7). They are transparent to the visible and near-infrared wavelengths in the sunlight, but they absorb and re-emit back down a large percentage of the longer infrared wavelengths previously emitted outward by the earth. Incoming solar radiation is essentially shortwave; 25% of it is reflected back to space, and another 25% is absorbed by atmospheric gases, whose temperatures are then raised. The remaining 50% reaches the surface of the earth, and 5% of this is reflected back to space. An additional 55% is transferred to the atmosphere by thermals, and 24% is converted to atmospheric warming by evaporation of water and condensation within clouds (8). Because of the greenhouse heat trapping effect, the atmosphere itself radiates a large amount of long-wavelength energy downward to the surface of the earth and makes the earth twice as warm as it would have been if warmed by solar radiation alone.

Evidence of Increasing Trace Gases

Having discussed the vital role of carbon dioxide in planetary climate control, it is necessary to examine whether or not industry is affecting the natural climate cycles of the earth, and if it is, what the future consequences of these trends will be. One result of current greenhouse effect research that appears irrefutable is the observation that the levels of naturally occurring atmospheric trace gases (such as carbon dioxide, methane, and nitrous oxides) have risen over preindustrial levels, and also that chlorofluorocarbons and other halocarbons released by industrial societies have become a new class of significant trace gases (8).

The primary man-made contributor to the greenhouse effect is carbon dioxide, taking approximately 50% of the blame for global warming. The U.S. Environmental Protection Agency (EPA) concluded that energy consumption in general contributed almost 60% to the greenhouse effect between 1980 and 1990 (9). Accurate measurement of carbon dioxide started in 1957 with a study by C. D. Feeling of the Scientific Committees on the Problem of the Environment at Mauna Loa, Hawaii. In 1957, monthly averages of continuous carbon dioxide samples totaled 315 parts per million (ppm), which had risen to 343 ppm by 1984 (1). These data are supported in similar findings from the South Pole, Alaska, and Samoa, and illustrate that in rural areas, where the local atmospheric conditions reflect a global air circulation, carbon dioxide concentrations have risen at a measurable rate (10). From analysis of air trapped in arctic ice, it is known that preindustrial levels of carbon dioxide were at about 280 ppm; thus, carbon dioxide levels have risen at least 25% in the last 150 years (8). This rise of carbon dioxide in the atmosphere is largely due to the burning of fossil fuels in addition to worldwide deforestation. Deforestation adds carbon dioxide to the atmosphere in two ways. Firstly, deforestation produces land that ends up as barren soil or grassland, which absorbs much less carbon dioxide than would a growing rainforest. A rainforest can hold 1–2 kg of carbon per square meter per year, as compared to a field of crops, which can fix less than 0.5 kg of carbon per square meter, most of which is recycled every year. Secondly, most of the trees that are destroyed by deforestation are either burnt or decomposed by bacteria, and thus give up carbon dioxide directly to the atmosphere as they are destroyed. The current rate of atmospheric increase in carbon dioxide is 0.4%, which is equivalent to 10^{10} metric tons per year (11).

Although methane and nitrous oxides are present in the atmosphere in much lower quantities than carbon dioxide, they trap 60% as much total low-wavelength radiation as does carbon dioxide (8). This is the case because of their wider range of electromagnetic absorption ($7-13\,\mu$m). Their atmospheric levels are rising rapidly (methane by 2% per year and nitrous oxide by 0.2% per year) and their capacities to trap heat are also much higher than that of carbon dioxide (25 times higher for methane and 259 times higher for nitrous oxide), and thus it is expected that the effects of these trace gases will hasten the greenhouse effect, leading to an "effective doubling" of carbon dioxide levels by 2030 (11).

Over 50% of methane released to the atmosphere comes from the action of anaerobic bacteria on vegetation, such as in the stomachs of ruminants, in the guts of termites, and in the rice paddies and wetlands across the globe. It also has sources in natural gas pipeline leaks, the decomposition of organic matter, and the incomplete combustion of vegetation. Nitrous oxide is emitted to the atmosphere from biomass burning and artificial fertilizers at a rate of 5 million tons per year. This gas is a danger for two reasons: (a) It contributes to global warming by increasing the greenhouse effect, and (b) it destroys tropospheric ozone, which helps to shield the earth from the full effects of solar radiation (8).

The concentration of chlorofluorocarbons (CFCs), which was 0% 150 years ago, is rising at a faster rate than that of any other greenhouse gas–nearly 5% per year (11). This is due to the large-scale industrial uses of CFCs, which include aerosol propellants (CFC-11), refrigerants (CFC-12), and a wide range of solvents. Like nitrous oxides, CFCs trap heat 20,000 times as efficiently as does carbon dioxide and also destroy tropospheric ozone and thus pose a serious twofold environmental problem. A worldwide phaseout of CFCs was undertaken under the Montreal Protocol (see Section 8.6), and as of the end of the twentieth century, production of these harmful chemicals is minimal.

It is clear that because of industrial activity, the atmospheric concentrations of greenhouse gases are increasing at a steady rate. What is in dispute is what effect these trends are having on the climate. Scientists are looking to computer modeling to simulate the climate of the earth and are trying to predict the future concentrations of greenhouse gases and their probable effect on weather patterns. Climate data from the past is also being analyzed to evaluate present-day weather in a valid setting.

8.2 THE GREENHOUSE DEBATE

Although scientists are in agreement that concentrations of trace greenhouse gases in the atmosphere are increasing, the effects of this increase are undergoing wide debate. Some scientists believe there is strong enough evidence to show that global warming has already begun, whereas others feel that the planet is actually entering another ice age. The current trends in weather could be natural fluctuations or could be the results of global climate patterns that run on cycles of thousands of years. Or, on the other hand, they could be the result of industrial activity on an unprecedented scale—activity that could lead to climatic chaos. For scientists, understanding and predicting how increasing carbon dioxide concentrations might alter the earth's climate requires in-depth analysis of observational data, which includes past climatic change and also the production of theoretical "warm-world" scenarios (1).

Analysis of Natural Weather Cycles

The analysis of past climate helps in the study of the greenhouse effect by supplying data on the natural variations in weather, which must be eliminated from present data in order to determine the pure impact of increases in carbon dioxide levels. The most significant climatic cycle the earth experiences is the regular ice age, which occurs about every 100,000 years. The main factor that influences the onset of an ice age appears to be minute changes in global climate, but its effects are too long-term to be counted in the measurement of the greenhouse effect over a relatively shorter period of time. Short-term weather cycles are much more important in evaluating the greenhouse effect. Historical weather observations have shown that the Northern Hemisphere experiences a cooling cycle every 180 years. Because the first cooling period occurred in the early 1800s, it was expected that the 1980s would bring in record cold spells. Although the winters in the 1970s were extremely severe, those in the 1980s were characteristically warm, suggesting that something broke the cycle. Since this change in weather cycles was not brought on by natural means, some scientists have pinpointed the greenhouse effect as the culprit. By analyzing data on changes in upper atmospheric temperature, surface temperature, and precipitation over dozens of years, the general conclusion is that the temperature of the earth has risen anywhere from 0.5°C to 0.7°C over the last century because of increased carbon dioxide levels from industrial activity (12).

The ocean also provides evidence of interruptions in natural cycles, apparently due to the greenhouse effect. In the El Niño–La Niña wind cycle, the El Niño warms tropical waters, resulting in warmer weather around the globe about 6 months later. The La Niña winds perform a completely opposite task, lowering the temperature of the tropical seas and, by extension, the air temperature. A La Niña occurred in the last half of 1988, and therefore should have cooled the air temperature during 1989 (13). This cooling was not experienced, and in fact the temperatures during 1989 increased from those in 1988. Scientists from the British Meteorology Office and the University of East Anglia's Climate Research unit reported in 1989 that the six warmest years on record were 1988, 1987, 1983, 1981, 1989, and 1980 respectively (14). This fact points toward a warming trend of some magnitude occurring in the decade of the 1980s that is out of sync with the natural climate trends witnessed in the past.

In fact, today, scientists believe that it is possible that global warming has had an effect on the El Niño–La Niña cycle. Since the end of the nineteenth century, El Niños are occurring more often and are warmer each time. Four of the strongest El Niños of the 1900s occurred since 1980, with the last occurring in 1997–1998. However, it is not clear if pollutants of human origin are actually the cause of either global warming or the changes seen in the El Niño–La Niña cycle. (15)

Opposition to the Greenhouse Effect Theory

There is much evidence that the whole concept of greenhouse effect could just be an illusion. Firstly, many researchers remain unconvinced that there is concrete evidence that combustion-produced gases accumulating in the atmosphere have produced any rise in the average earth temperatures during the past century. Three reputable scientists from the George C. Marshall Institute in Washington, D.C., wrote a 35-page report on this lack of evidence, which was submitted to Congress in 1989 for analysis (16). In their report, *Scientific Perspectives on the Greenhouse Problem*, they claimed that any warming of the earth over the last 100 years is better explained by variation in natural climate and solar energy output than by increased concentrations of carbon dioxide in the atmosphere. According to this theory, the most probable source of global warming appears to be variations in solar activity.

The amount of solar radiation reaching the earth, coupled with the planet's changing orientations to the sun, has been the major cause of climatic change throughout the earth's history. If the electromagnetic radiation from the sun were to decline even 5%, the globe would be covered by ice in less than 100 years (8). Unfortunately, the variations in solar radiation are not included in current climate models. In the past one million years, the earth has undergone 10 major and 40 minor ice ages which have been controlled by three orbital elements varying cyclically over time. The first orbital element is the tilt of the earth's axis, which varies from 22° to 24.5° and back again every 41,000 years. The second element is the month of the year in which the earth is closest to the sun, which also varies over cycles of 19,000 and 24,000 years. This "month-of-closest approach factor" can make a difference of 20% in the amount of solar radiation reaching a certain point at any given time. Finally, the third element is the shape of the orbit of the earth, which, over a period of 100,000 years, changes from being more elliptical to being almost fully circular. Studies of ocean cores have revealed that these orbital changes are the primary causes of ice ages, although the precise way in which the earth responds to these changes are still uncertain. Nor do these orbital changes act alone to cause vast climatic shifts: Dust in the atmosphere, carbon dioxide and methane concentrations, and changes in the earth's reflectivity all act together with orbital changes to contribute to global warming and cooling. There are also changes in the sun that occur over days, months, or even years. Scientists both in Germany and in the United States have linked an 11-yr sunspot cycle to weather patterns on a global scale (8).

One of the scientists from the George C. Marshall Institute, Dr. Jastrow, has stated (16), the following:

> Changes in the earth's temperature have followed changes in solar activity over the last 100 years. When solar activity increased from the 1880s to the 1940s, global temperatures increased. When it declined from the 1940s to the 1960s, temperatures also declined. When solar activity and sunspot numbers started to move up again in the 1970s and 1980s, temperatures did the same.

Jastrow also pointed out that the flaws in the greenhouse effect models make the conclusions that result from them very uncertain, and certainly not something around which to center legislation. He states that the observed global temperature rise of 1°F over the last 100 years actually took place from 1880 to 1930, before 67% of global carbon dioxide emissions had even occurred. Also, from 1940 to 1970, world temperatures decreased significantly and even affected farming areas in northern Europe. This is particularly hard to explain as a greenhouse effect phenomenon because the increased emissions of greenhouse gases should have created a period of accelerated temperature rise. According to the report, an ice age is to be expected in the twenty first century, with cooling of as much as $2°F$, which is expected to offset any greenhouse effect.

In Defense of the Greenhouse Effect Theory

The leading herald of the greenhouse effect and the dire consequences of global warming is Dr. James Hansen, a climate modeler and director of NASA's Goddard Institute of Space Studies in New York City. He claims that climate models have been reliable enough to conclude that the greenhouse effect is causing global warming. Hansen gave three points as the crux of what he believes is happening (17):

> Firstly, I believe the earth is getting warmer, and I can say that with 99% confidence.... Secondly, with a high degree of confidence, we can associate the warming and the greenhouse effect. Thirdly, in our climate model, by the 1980s and early 1990s, there is a noticeable increase in the frequency of drought.

Hansen has many critics who take special exception to the second point — that the warming of $1°F$ of the planet is definitely due directly to an increase in atmospheric carbon dioxide. "The one thing that has the greatest impact on my thinking," says Hansen, "is the increase in atmospheric carbon dioxide from 280 ppm in the 19th century to its present 350 ppm. It is just inconceivable that that is not affecting our climate. There is no model that would not say that it is affecting it right now." Hansen believes that most scientists in the field agree with him but are not willing to voice their opinions until there is more conclusive evidence. However, for Hansen, time is of the essence in trying to reverse the damaging industrial trends being followed today.

If Hansen's models are correct, there will have been an increased frequency of drought in the 1990s provided that there were no large volcanic eruptions, which serve to cool the atmosphere. Without such data to review, scientists linked the 1988 drought in the United States to naturally occurring climatic forces in the Pacific Ocean that had nothing to do with the greenhouse effect. It was considered to be the result of the El Niño-La Niña cycle mentioned previously. The drought originated in unusually cold water along the equator pushing the tropical convergence zone (the zone where tradewinds collide) farther north than usual. When the cooler-than-usual water met the warmer-than-usual water, there was thunderstorm activity that disturbed the atmosphere. These newly created high and low pressures pushed the jet stream far to the north and created a large, dry high-pressure system over the United States which brought weeks of drought (18).

8.3 EFFECTS OF GLOBAL WARMING

Although the scientific world is not unanimous with regard to the causes of global warming, the need to evaluate and prepare for future situations has led the Environmental Protection Agency (EPA) and other environmental groups to create theoretical scenarios that try to give a better picture of what would happen to the planet after a 2–5°C warming. Through the use of sophisticated computer systems, these groups have shown that a "reasonable chance" of climatic change exists, and have already begun to define the potential environmental implications of such changes. Hansen predicts that several North American cities can expect a dramatic rise in summer temperatures; for example, New York City would experience 48 days over 90°F in 2030 as compared to 15 days in 1987 (19). However, the effects will not be restricted to populated regions, or solely to rising temperatures. The EPA, operating on a mandate from Congress, has used climate change scenarios to study future consequences of rising temperatures on water sources, forests, farms, and shorelines. The premise for this study is a predicted 1.5–4.5°C rise in temperature over the next 100 years (12). According to this report, a warmer climate will harm several North American ecological systems which will not adapt quickly to changes in rainfall and temperature. The last 5°C warming to occur globally was during the last ice age, and progressed gradually over 18,000 years. The evolutions and migrations that took place over that wide period of time cannot be expected to occur over 100 years.

Environmental Effects

The range over which forests may survive will be reduced. The southern forest boundary will move north approximately 400 miles, while the northern boundary will move south 60 miles. The composition of forests will also change as soaring temperatures reduce soil moisture levels. Certain species, such as the Michigan sugar maple and oak, will be replaced by grassland, while Minnesota's mixed boreal will give way to northern hardwood. These changes will begin to occur after a 1°C rise in temperature (30–60 years from now) and will continue for centuries. The greenhouse effect is also expected to threaten 400 species of mammals, 460 species of reptiles, 650 species of birds, 660 species of freshwater fish, and the tens of thousands of species of invertebrates and plants in the United States. The EPA predicts an increase in extinction rate for some of these species as well as changes in their migration patterns. Many fish and shellfish will migrate northward along coastlines to cooler waters, although much of this population will be eliminated due to the loss of coastal wetlands from rising sea levels. The increase in salinity in coastal estuaries may reduce the abundance of freshwater fish species and increase populations of marine species. Higher temperatures may also lead to more algae and stratification in lakes, making them less hospitable for freshwater fish. In short, many aquatic habitats will shift north or disappear altogether (12).

Effects on Population

What is more important to determine is the direct effect global warming will have on humanity and society. A 1-m rise in sea level could destroy 25–80% of the coastal wetlands of the United States if left unhindered. This, coupled with the fact that the quality of water in water reserves will deteriorate due to increased salinity, spells big problems for agricultural and urban water needs in the future. An EPA report done in 1986 identified four major policy issues related to the 0.5–2-m rise in the sea level projected for the twenty-first century. The report states that responses to the expected rise in sea level (construction of sea walls, sand pumping, etc.) can be postponed until the last minute, because they can be effected rapidly. However, current engineering and land-use decisions should incorporate the expected sea level rise, because this foresight will cost little in comparison to the expense of future protective measures. The report also predicted that wetland ecosystems will only survive sea level rises if coastal states begin to adopt land-use measures. One example of this is that the government of Maine has issued laws allowing wetlands to shift inland by ordering the removal of any man-made structures as the sea level rises. In its report, the EPA also pointed out the implications of these coastal occurrences on future insurance programs. This observation is especially pertinent in light of a Department of Housing and Urban Development Authorization Act, which has required that the U.S. government buy ocean-front property that is susceptible to erosion (12).

Seasonal Effects

The length of agricultural growing seasons will change, as will the frequency of heat waves. However, high carbon dioxide concentrations atmospherically could have a beneficial effect on crops, which may reduce the otherwise negative effects of the greenhouse on agriculture and produce a steadier supply of agricultural commodities. The demand for electricity will increase in the summer and decrease in the winter. The total annual electrical generation in 2055 will jump 4–6% from present levels due to climatic change, an increase which will require an estimated $33–73 billion to meet the excess demand. Changes in weather patterns and atmospheric water vapor levels will reduce air quality, and increased temperatures will speed up the reaction rates among chemicals in the air, causing increased air pollution. Summertime will be extended, and because summer usually is a time of high air pollution activity, ozone levels are expected to increase by up to 20% as well. Contagious diseases, such as influenza and pneumonia, and allergenic diseases, such as asthma, will be affected by the weather and become more prolific. The life cycles of mosquitos and other disease-carrying insects are also extended in warmer weather (12).

The study showed exceptionally dire predictions for California, the Great Lakes Region, and the southeastern United States. California will experience severe water problems. The present-day Central Valley reservoir system does not have the capacity to store more winter runoff and still provide adequate flood protection. Thus, when the greenhouse effect turns winter snow into winter rain, there will be less water in the system for late spring and summer deliveries, when the runoff will be lower. The reduction in net water deliveries from the Central Valley may be 7–16%. The inland salinity levels are expected to increase during this time, and the salt front will move upstream 2.5–6 mi in the Sacramento-San Joaquin River Δ. The Great Lakes Region contains 95% of the nation's surface freshwater supply. The EPA report projects a decrease in water levels for Lake Superior and Lake Michigan of 0.5 m and 1.25 m respectively, by the middle of the twenty-first century. The Southeast is noted for its warm temperatures, abundant rainfall, large coastal plains, and marine fisheries. The EPA predicts that up to 50% of the total agricultural acreage in this area will become sterile and that the coastal fisheries will be severely threatened (12).

8.4 COMBATTING THE GREENHOUSE EFFECT

In light of the uncertainties in the greenhouse effect models and lack of unanimity among scientists with regard to the expected consequences of increased carbon dioxide levels, the chances for effective legislation and solutions to the problems are seriously reduced.

Approaches to Political Response

There are three available avenues of response on the political level, ranging from efforts to drastically reduce carbon dioxide output to suggestions that nothing be done at all. The first approach is the "wait-and-see" approach (20). "Wait-and-see" theorists contend that a global warming would not necessarily be harmful and that countermeasures are inappropriate unless a huge adverse climatic change is established beyond a doubt. However, even "wait-and-see" theorists generally suggest that extensive research ought to be done in the meantime in order to help detect the first climatic changes associated with increased levels of carbon dioxide. Only when the existence of the latter is actually established should counter-measures be pursued.

The second course of action is the "adaptation-to-incurable-changes" approach (20). In this view, attempts at prevention of global warming are seen as useless, and instead humanity is urged to rely on its innate ability to evolve in response to changes in the environment. This theory is based on the assumption that there will be plenty of time in which to decide and act on climatic change. There will be a time lag before science is certain as to the effects of the greenhouse phenomenon; there will also be a time lag before any international agreement is made on a course of action; and, finally, implementing the course of action will involve decades of work. Thus, no serious prevention of the greenhouse process could be activated before the end of the twenty first century, and by that time, changes in climate may already have occurred. Thus, it is up to humanity to adapt to and make fullest use of any benefits of a changing environment.

The third approach to the global warming question is the "act-now" approach which is the only one which demands an immediate legislative (and industrial) response in the formation of national policy and the implementation of federal control and regulation of emissions (20). This approach is based on the idea that even in the absence of conclusive scientific evidence, the threat of the greenhouse is enough to warrant the initiation of preventive measures. Preventive action is less expensive than any form of adaptation in the future, and the "act-now" approach sees wisdom in putting limits on carbon dioxide emissions, stopping deforestation, and generally increasing fuel efficiency, even if the greenhouse effect is proven to be a false alarm. The problem now lies in the legal and economic systems, which normally respond only to immediate and certain threats. The "act-now" policy includes the installation of air pollution control devices to remove carbon dioxide and nitrous oxides from industrial streams before they reach the atmosphere.

Air pollution control equipment such as absorbers and adsorbers can be used to control greenhouse gas emissions. However, the author of this chapter, as well as the other authors of this volume, do not view the control equipment approach—with associated disposal problems—as a viable option for managing greenhouse gas emissions to the atmosphere.

8.5 RESPONSE TO THE GREENHOUSE EFFECT ON THE NATIONAL LEVEL

As indicated in Section 8.1, the primary contributor of human origin to the greenhouse effect is carbon dioxide, and EPA data have shown that energy consumption in general has contributed almost 60% to the global warming effect in the 1980s (9). In 1985, the United States was the highest worldwide emitter of carbon dioxide (26%), followed by the countries of the then-Soviet Union (21%) and Western Europe (17%) (21). Thus, while it is impossible to point fingers in the greenhouse effect debate to any great effect, it is obvious that the United States contributes at least its fair share of the problem.

America's policy toward the greenhouse effect was best illustrated by the Clean Air Act Amendments of 1990, which represented the first revisions of the original Clean Air Act since 1977. The bill set forth technical standards for 191 toxic air pollutants and called for significant reductions in tailpipe emissions. It mandated a 10 million-ton, two-phase reduction in sulfur dioxide and a two million-ton reduction in oxides of nitrogen (NO_x) emissions, providing a flexible plan for the 111 affected coal-fired utility plants (22). The cost for all of this emission control is hefty for industry ($25–35 billion per year), but the passage of the Clean Air Act was a triumph for environmentalists (23). Auto manufacturers now have to build cars that run on alternate fuels, such as electricity, natural gas, or methanol, and utilities have to install coal scrubbers or switch to low-sulfur coal burning. For the first time, small businesses were also required to invest in pollution control equipment. CFCs will be phased out by 2000, and research into the manufacture of substitutes for the product was almost completely accelerated (24). These amendments were to be implemented in a 15-yr plan in order to give industry a chance to catch up with the phenomenal costs associated with the Act. While none of the new amendments directly affect carbon dioxide emissions, the next generation of environmental restraints will accelerate the search for other forms of energy, including nuclear.

During the State of the Union Address on January 19, 1999, President Clinton endorsed the development of an early action credit program in the United States to encourage the private sector to make voluntary reductions in greenhouse gases. On March 4, 1999, Senator Chafee introduced the Credit for Voluntary Reductions Act (S.547) that would authorize the president to enter agreements with industry and businesses to award tradable credits for voluntarily reducing greenhouse gas emissions. Under such agreements, credits earned or purchased from another industry or business could be used to comply with any regulatory limits that may be placed on greenhouse gases in the future. (25)

Having determined that the United States is swiftly becoming more open to taking an approach of active prevention against the greenhouse effect, the field of policy and legislation to combat global warming can be categorized in three main subdivisions: (a) the reduction of the amount of greenhouse gases emitted to the atmosphere by the burning of fossil fuels, (b) the total elimination of fossil fuels as energy sources, and (c) the reduction of the effects of greenhouse gas emissions by reforestation or alternate methods of carbon dioxide disposal. Governmental policy calls for shifts that could immediately affect the cost and availability of all types of energy in most aspects of society. Modern life is very industrial, and the magnitude of the changes could easily dwarf the conservation efforts of the early 1970s which were sparked by the oil embargo. The industry that will be most affected is the energy industry, especially coal producers and users. Combustion provides almost 90% of the United State's energy needs, and there are more than 300 years of coal reserves in the nation (21). United States electrical utilities alone contribute 7.5% of the worldwide carbon dioxide output. Assuming that power from combustion cannot be done away with immediately, there are two avenues for the United States to follow to reduce emissions of carbon dioxide and NO from energy utilities. The first is to do nothing about the actual emissions themselves and merely increase energy efficiency, and the second is to reduce the amount of emissions of greenhouse gases to the atmosphere.

8.6   THE INTERNATIONAL ASPECTS OF THE GREENHOUSE EFFECT

The United States can spend trillions of dollars on reducing greenhouse gas emissions within its own borders, but it is wasted money unless the rest of the world follows suit. The greenhouse effect is a global problem, and the whole world must consent to an international agreement to constrain the emission of the greenhouse gases and combat the effects of a warming. The problems associated with any international agreement on a given issue are both political and social. At present, both national and international politicians are beginning to notice the important implications of the greenhouse effect, and plans are being made to deal with the problem. However, the drive to solve the greenhouse effect may well falter as a consequence of the time lag between emission and effect. Ten years of cool weather could also test the strength political motivation that will be required in order to deal with the problem. The most serious obstacle in realizing a worldwide agreement on greenhouse gas emissions is the conflict between developing and developed nations. Developing nations see the problem as a result of industrialized nation's activities and do not want their fledgling economies restrained (26). As a result of this, they are resistant to international legal control on industry in which developed nations are anxious to involve them. The developed countries should keep this in mind when meeting with the underdeveloped nations, and they should also be open to forgiving Third World debts, sharing technology, and financing energy projects in foreign countries for the greater good of preventing world catastrophe.

The solutions on an international level are basically the same as those proposed on the national level. Carbon dioxide emissions should be reduced, either by increasing global efficiency, scrubbing the flue gases from energy plants, or using new energy sources. Reforestation is more vital on a global scale, because the slash-and-burn agriculture common in the tropics is devastating for the environment. The tropical rainforests are far more important in the control of carbon dioxide in the atmosphere than are the forests of the Northern Hemisphere. The EPA recommends substituting "agro-foresting" as a solution to this problem: Trees and crops are planted together, and branches and leaves from the trees are cut for fertilizer. By contrast, slash-and-burn agriculture uses the ashes from burnt forests to fertilize crops. The EPA plans to undertake large-scale tree replanting programs on denuded, abandoned lands; it also plans to establish plantations of fast growing trees in the Third World villages that could be harvested on a rotating basis, instead of villagers cutting down wild trees.

International Committees Working on Global Warming Solutions

The initial undertaking of solutions to the greenhouse effect is clear on the international scale: It is their implementation and protection that are the real problems. The Intergovernmental Panel on Climatic Change (IPCC), which was formed in 1988 under the auspices of the United Nations Environmental Program and the World Meteorological Organization (9), is one of the many committees that have agreed to work on the problem of the greenhouse effect on a global scale. In 1990, there were three Working Groups (WG) within the IPCC: WG1, known as the science group, was chaired by the United Kingdom and was responsible for assessing climatic change-causing elements; WG2 was chaired by the then-Soviet Union and was responsible for issuing the environmental and socioeconomic aspects of climatic change; WG3 was chaired by the United States Assistant Secretary of State for Oceans and International Environmental Issues, and its task was the understanding and evaluation of society for response to global climate change. The IPCC still meets today, and is essentially an informal gathering of committees, and it is up to the United Nations and the United Nations Environmental Program to determine how regulations for a treaty should proceed. In an International Conference on the Changing Environment in Toronto in June 1988, 300 government representatives, scientists, and environmentalists from 46 countries made plans to reduce carbon dioxide emissions by 20% by 2005 (27). There was a great deal of talk about switching from coal and gasoline plants to natural gas plants, and there was also a large discussion regarding funding for solar, wind, geothermal, and nuclear energy efforts. The conference also discussed reforestation of the Third World countries, the halting of deforestation, and the need to eliminate CFCs by 2000.

Such committees, while making progress in general, are not without their setbacks. Great Britain, Greece, Spain, and Portugal blocked moves in June 1990 to set the year 2000 as a deadline for stabilizing carbon dioxide emissions and wanted to postpone such a stabilization to the year 2005 (28). The current international trend is to decide the limit of carbon dioxide emissions for a country on a per capita basis, which would require the more industrialized nations to adopt stiffer controls. In November 1989, a similar event took place in an international conference in Norway, when the United States and the then-Soviet Union refused to join 39 other nations in stabilizing greenhouse gases by 2000 at the latest. It is believed that the former Soviet Union and Japan would have joined the other nations if the United States had taken the initiative at this important first ministerial level meeting on climatic change. President Bush defended his decision by claiming that the United States was not yet able to double its efficiency standards, increase reforestation, and improve home insulation in so short a period of time.

The Montreal Protocol — A Legislative Precedent

Despite the problems inherent in coming to any global agreement on pollution, it has been done in the past. In 1987, the Montreal Protocol was signed which called for a freeze on the production and use of CFCs at 1986 levels, with a subsequent reduction of 20% and 50% in the early 1990s (26). Eleven countries ratified the Protocol by January 1989, when it went into force by the signatories. Part of the Protocol involves the gradual cessation of all imports of substances containing

CFCs from nonparticipating nations. For example, a country not a party to the Protocol could not export computer chips or foam products made with CFCs to the United States. Signatories agreed to ban exports of controlled substances to nonparticipating nations entirely after January 1993. The Montreal Protocol is considered to be the state-of-the-art agreement among many nations. The question is whether an agreement of this kind can be implemented for carbon dioxide control. One difference between the two situations is that CFCs are only made by 37 countries worldwide and that controlling their production is relatively simple compared to trying to control the infinite number of sources of carbon dioxide. But the existence of international cooperation on the scale illustrated by the Montreal Protocol is indicative of what can be done between nations to solve long-term environmental problems. It allowed consideration for both developing and developed nation's needs: Developed countries were to freeze consumption levels of CFCs immediately, whereas developing countries may delay compliance for 10 years. Also, developed countries were to stabilize production levels as soon as possible, whereas developing nations could increase them by 10% in the next 5 years. However, in the case of carbon dioxide, there cannot be double standards for different nations. An international treaty to solve this problem could include different efforts for different countries: The goal of one nation could be to decrease emissions, whereas that of another could be an intense reforestation plan. In the case of CFCs, a substitute can be found; but the worldwide control of carbon dioxide is going to require major changes and innovative approaches.

The Kyoto Protocol

On December 11, 1997, 160 parties, including the United States, completed negotiations on the Kyoto Protocol to the United Nations Framework Convention on Climate Change. The Unites States signed the Protocol on November 12, 1998. If ratified by the Senate, it would commit the United States to reducing greenhouse gases by 7% below 1990 levels. While nations would be allowed to trade emission reductions to assist them in meeting their commitments, numerous issues must be resolved before trading could be implemented. First, uniform standards would be necessary to ensure accurate monitoring and effective enforcement in each nation. Second, trading would need to extend beyond large stationary sources and include smaller sources that are more difficult to monitor because sources of greenhouse gases are more diverse than other pollutants. Third, some participating nations might lack the resources necessary to implement trading programs effectively. During a conference held in Buenos Aires, Argentina in November 1998, the parties to the Kyoto Protocol established a deadline for resolving such issues by late 2000. [For more on trading programs, see Chapter 9 (25)].

Problems with International Environmental Law

Under the customary international law, the orthodox approach to international environmental pollution is to apply the rules of state responsibility (26). This approach is common in the solving of international disputes, but some uncertainties in it raise particular problems in the context of the greenhouse effect. The general rules of state responsibility are as follows: State responsibility arises when (a) conduct consisting of an action or omission is attributable to a state under international law and (b) that conduct constitutes a breach of an international obligation of that state. The generally accepted way of stating obligation toward pollution on the global level is as follows: "....the international obligation is to prevent environmental harm within the territory of another state," or, alternatively, "...states are under an obligation not to allow pollution that might reasonably be prevented to damage foreign nations" (26). The legal and practical problems arising in the situation of so far-reaching a problem as the greenhouse effect are extremely complicated. Hypothetically, the case would be State A bringing action against State B for damage to State A's land due to the greenhouse effect. State A would have to successfully establish that the damage inflicted by the greenhouse effect is the fault of State B. Would State A make a claim only against State B? And since the commercial industries of a nation are in the hands of private owners, can State B as a nation take the blame for all of the activity of industry that go on within its borders? Can State A prove that State B's carbon dioxide emissions were a cause of the greenhouse effect itself? At the present level of scientific knowledge, these questions would be impossible to answer. There is also the issue of reparation and compensation for damage and/or economic loss.

Finally, the greatest difficulty in formulating an international agreement on the greenhouse effect is how to enforce it. No mechanism for this exists presently, and compliance is dependent only on the will of the state itself. From the current trends in carbon dioxide emissions, it appears that developing nations may well become the primary contributors of carbon dioxide in the future and it is unlikely that they will be able to pay for damage caused by the greenhouse effect.

8.7 FUTURE TRENDS

The theory behind the greenhouse effect, combined with the trends of modern society, certainly points toward a climatic phenomenon that would, if allowed to proceed unchecked, create a global warming; however, when the physical evidence for climatic change is closely investigated, the situation does not appear to be changing as quickly or catastrophically as is predicted. This does not rule out the possibility that certain gas emissions in society are causing severe atmospheric problems, but it does mean that there is time to evaluate the world situation without creating an environmental panic. The steps that have to be taken in response to the problems can be more fully investigated before laws are made and money is spent. Today's environmental problems are closely interlinked, planetary in scale, and deadly serious. They cannot be addressed issue by issue, or by one nation, or even a by a small group of nations acting alone. Obviously, problems such as the greenhouse effect and global warming must be faced with a unified effort by the world as a whole, because the atmosphere is not separated by borders. Each nation must do its part with internal legislation to control the emissions of greenhouse gases and investigations into other forms of energy. On the international level, nations must be willing to cooperate, overlook debts, and share technology with other countries. It is important to remember that individual nations have successfully dealt with such climate problems on a small scale, and this indicates that the planet as a whole can combat the harsh potential consequences of the greenhouse effect.

8.8 SUMMARY

1. The greenhouse effect appears to be a phenomenon of completely human origin in the world today — one that some scientists feel is leading the planet to climatic chaos. The term "greenhouse effect" describes two separate but dependent occurrences: (a) the increase of trace "greenhouse gases" (carbon dioxide, nitrous oxides, methane, tropospheric ozone, and chlorofluorocarbons) in the earth's atmosphere and (b) the absorption and re-emission of long-wave radiation by these gases.

2. Although scientists are in agreement that concentrations of trace greenhouse gases in the atmosphere are increasing, the effects of this increase are in wide debate. Some scientists feel that there is strong evidence to show that global warming has already begun, and others feel that the planet is actually entering another ice age.

3. Although the scientific world is not unanimous with regard to the causes of global warming, the need to evaluate and prepare for future situations has led the Environmental Protection Agency (EPA) and other environmental groups to create theoretical scenarios that try to give a better picture of what would happen to the planet after a 2–5°C warming.

4. In light of the uncertainties in the greenhouse effect models, and lack of unanimity among scientists with regard to the expected consequences of increased carbon dioxide levels, the chances for effective legislation and solutions to the problems are seriously reduced.

5. America's policy toward the greenhouse effect was best illustrated by the Clean Air Act Amendments of 1990, which represented the first revisions of the original Clean Air Act since 1977. The bill set forth technical standards for 191 toxic air pollutants and called for significant reductions in tailpipe emissions. It mandated a 10 million-ton, two-phase reduction in sulfur dioxide and a two million-ton reduction in oxides of nitrogen (NO_x) emissions, providing a flexible plan for the ill affected coal-fired utilities plants.

6. At present, both national and international politicians are beginning to notice the important implications of the greenhouse effect, and plans are being made to deal with the problem. However, the drive to solve the greenhouse effect may well falter as a consequence of the time lag between emission and effect.

7. The theory behind the greenhouse effect, combined with the trends of modern society, certainly points toward a climatic phenomenon that would, if allowed to proceed unchecked, create a global warming; however, when the physical evidence for climatic change is closely investigated, the situation does not appear to be changing as quickly or catastrophically as is predicted.

REFERENCES

1. A. Nadakavukaren, *Man & Environment*, 3rd ed., Waveland Press, Prospect Heights, IL, 1990.
2. G. Cowley, "Is It All Hot Air?" *Newsweek*, November 20, 1989.
3. Energy Information Administration, U.S. Department of Energy, *Greenhouse Gases, Global Climate Change, and Energy*.

4. R. Rotty, *Engineering Progress*, 3(4), 253, November 1984.

5. R. Bailey, *Chemistry of the Environment*, Academic Press, New York, 1978, p. 385.

6. W. Post, T. Peng, and E. Emanuel, *American Scientist*, 310–326, July–August 1990.

7. A. Rind, *A Character Sketch of Greenhouse*, 15 EPA 5, 1989.

8. B. Hileman, *Chemical and Engineering News*, 26, March 13, 1989.

9. L. Rodgers, *Public Utilities Fortnightly*, 29, March 15, 1990.

10. L. Soloman and B. Friedburg, The Greenhouse Effect: A Legal and Policy Analysis, 20 Environmental Law 83.

11. W. Stevens, "With Cloudy Crystal Balls, Scientist Race to Assess Global Warming," *New York Times*, February 7, 1989.

12. Office of Policy Planning and Evaluation, Office of Research and Development, United States Environmental Protection Agency (Draft Report), *The Potential Effects of Global Climate Change in the United States*, 1988, pp. 2–20.

13. R. Kerr, *Science*, 521, February 21, 1990.

14. J. Shabecoff, "Global Warming in '88 is Found to Set a Record," *New York Times*, Al, January 26, 1988.

15. C. Suplee, "El Niño–La Niña: Nature's Vicicus Cycle," *National Geographic*, March 1999.

16. "Scientists Bare Cold Facts about Greenhouse Effect," *Electrical World*, 9, August 1989.

17. J. Hansen, *Science*, 1041, June 2, 1989.

18. W. Stevens, "Scientists Link '88 Drought to Natural Cycle in Tropical Pacific," *New York Times*, C1, January 3, 1989.

19. S. Appel, "Groups Use Computer Models to Forecast Global Climate," *Christian Science Monitor*, A5, April 13, 1987.

20. D. Einhorn, "Carbon Dioxide and the Greenhouse Effect: Possibilities for Legislative Action," *Columbia Journal of Environmental Law*, 1986.

21. "Fighting the Greenhouse Effect," *New York Times,* C1, August 28, 1988.

22. "Senate Approves Clean Air Act," *Public Utilities Fortnightly*, 8, April 26, 1990.

23. J. Main, "Here Comes the Big Clean-Up," *Time*, 31, November 5, 1990.

24. "Clean Air Act Conference Held," *Public Utilities Fortnightly*, 8, April 26, 1990.

25. D. Bearden, *Air Quality and Emissions Trading: An Overview of Current Issues*, Congressional Research Service, Report for Congress, updated March 9, 1999.

26. F. Taylor, "International Legal Control of the Greenhouse Effect," *Victoria University of Wellington Law Review*, Wellington, New Zealand, 45, January 1990.

27. International Conference on the Changing Atmosphere, *Current*, 4, February 1989.

28. "Britain Blocks Move on Global Warming," *New York Times*, June 8, 1990

9

ACID RAIN*

9.1 INTRODUCTION

Few environmental problems have caused so much controversy, and so much confusion, as acid rain. People were worrying about pollution problems related to acid rain hundreds of years ago; now countries in several parts of the world are working together to control it. Research into many different aspects of acid rain is advancing and so is the technology to reduce it.

This chapter will look at the precursors to acid rain and explain how this phenomenon occurs. It will also examine how acid rain affects the environment and what can be done to control it.

9.2 AN ACID RAIN CHRONOLOGY

The following synopsis provides a historical perspective of how acid rain was first identified and highlights what steps have been taken to monitor and reduce acid rain depositions (1):

1661-1662

English investigators John Evelyn and John Graunt publish separate studies speculating on the adverse influence of industrial emissions on the health of plants and people. They mention the problem of transboundary exchange of pollutants between England and France. They also recommend remedial measures such as locating industry outside of towns and using taller chimneys to spread "smoke" into "distant parts."

1734

Swedish scientist C. V. Linne describes a 500-year old smelter at Falun, Sweden: "...we felt a strong smell of sulfur...rising to the west of the city...a poisonous, pungent sulfur smoke, poisoning the air wide around...corroding the earth so that no herbs can grow around it."

*This chapter was adapted from U.S. EPA, "Acid Rain: Looking Ahead," *EPA Journal*, 12(6), June/July 1986.

1872

English scientist Robert Angus Smith coins the term "acid rain" in a book called *Air and Rain: The Beginnings of a Chemical Climatology.* Smith is the first to note acid rain damage to plants and materials. He proposes detailed procedures for the collection and chemical analysis of precipitation.

1911

English scientists C. Crowther and H. G. Ruston demonstrate that acidity of precipitation decreases the further one moves from the center of Leeds, England. They associate these levels of acidity with coal combustion at Leeds factories.

1923

American scientists W. H. MacIntyre and I. B. Young conduct the first detailed study of precipitation chemistry in the United States. The focus of their work is the importance of airborne nutrients to crop growth.

1948

Swedish scientist Hans Egner, working in the same vein of agricultural science as MacIntyre and Young, set up the first large-scale precipitation chemistry network in Europe. Acidity of precipitation is one of the parameters tested.

1954

Swedish scientists Carl Gustav Rossby and Erik Eriksson help to expand Egner's regional network into the continent-wide European Air Chemistry Network. Their pioneering work in atmospheric chemistry generates new insights into the long-distance dispersal of air pollutants.

1972

Two Canadian scientists, R. J. Beamish and H. H. Harvey, report declines in fish populations due to acidification of Canadian lake waters.

1975

Scientists gather at Ohio State University for the First International Symposium on Acid Precipitation and the Forest Ecosystem.

1977

The U.N. Economic Commission for Europe (ECE) sets up a Cooperative Program for Monitoring and Evaluating the Long-Range Transmission of Air Pollutants in Europe.

1979

The U.N.'s World Health Organization establishes acceptable ambient levels for SO_2 and NO_x. Thirty-one industrialized nations sign the Convention on Long-Range Transboundary Air Pollution under the aegis of the ECE.

1980

The U.S. Congress passes an Acid Deposition Act providing for a ten-year acid rain research program under the direction of the National Acid Precipitation Assessment Program.

1980

The United States and Canada sign a Memorandum of Intent to develop a bilateral agreement on transboundary air pollution, including "the already serious problem of acid rain."

1985

The ECE sets 1993 as the target date to reduce SO, emissions, or their transboundary fluxes by at least 30% from 1980 levels.

1986

On January 8, the Canadian and U.S. Special Envoys on Acid Rain present a joint report to their respective governments calling for a $5 billion control technology demonstration program.

1986

In March, President Ronald Reagan and Prime Minister Brian Mulroney of Canada endorse the Report of the Special Envoys and agree to continue to work together to solve the acid rain problem.

1990

The Clean Air Act Amendments, which were signed on November 15, 1990, established federal standards for the emissions that cause acid rain. The amendments established specific time deadlines for compliance with pollutant reduction criteria by the various states, and more significantly, shifted the burden and responsibility for proof of compliance onto the owner/ operator of the facility (2).

9.3   THE ACID RAIN PHENOMENON

All rainfall is by nature somewhat acidic. Decomposing organic matter, the movement of the sea, and volcanic eruptions all contribute to the accumulation of acidic chemicals in the atmosphere, but the principal factor is atmospheric carbon dioxide, which causes a slightly acidic rainfall (pH of 5.6) even in the most pristine of environments. (See Figure 9.1 for an explanation of pH.)

In some parts of the world, the acidity of rainfall has fallen well below 5.6. In the northeastern United States, for example, the average pH of rainfall is 4.6, and it is not unusual to have rainfall with a pH of 4.0, which is 1000 times more acidic than distilled water. Although precipitation in the western United States tends to be less acidic than in the East, incidents of fog with a pH of less than 3.0 have been documented in southern California.

There is no doubt that man-made pollutants accelerate the acidification of rainfall. Man-made emissions of sulfur dioxide (SO_2) and nitrogen oxides (NO_x) are transformed into acids in the atmosphere, where they often travel hundreds of miles before falling as acidic rain, snow, dust, or gas. Table 9.1 lists the precursors to acid rain. All these wet and dry forms of acid deposition are known loosely as "acid rain," which is now recognized as a potentially serious long-term air pollution problem for many industrialized nations.

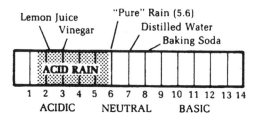

FIGURE 9.1 How "acid" is acid rain? The pH scale ranges from 0 to 14. A value of 7.0 is neutral. Readings below 7.0 are acidic; readings above 7.0 are alkaline. The more pH decreases below 7.0, the more acidity increases. Because the pH scale is logarithmic, there is a tenfold difference between one number and the one next to it. Therefore, a drop in pH from 6.0 to 5.0 represents a tenfold increase in acidity, whereas a drop from 6.0 to 4.0 represents a hundredfold increase. All rain is slightly acidic. Only rain with a pH below 5.6 is considered "acid rain."

Emissions and Deposition

Before the Clean Air Act was passed in 1970, U.S. SO_2 and NO_x emissions were increasing dramatically (see Table 9.2). Between 1940 and 1970, annual SO_2 emissions had increased by more than 55%. Over the same period, NO_x emissions had almost tripled.

The Clean Air Act helped to curb the growth of these emissions. By 1984, annual SO_2 emissions had declined by 24%, and NO_x emissions had increased by only 9%. These reductions in historical growth rates took place despite the fact that the U.S. economy and the combustion of fossil fuels grew substantially over the same period. In Title IV of the 1990 Clean Air Act Amendments, (Title IV) Congress set out to decrease the adverse effects of acid rain through reductions in annual emissions of SO_2 and NO_x from utilities burning fossil fuels. The SO_2 reductions were to be implemented in two phases: Phase I began on January 1, 1995; and Phase II on January 1, 2000. Phase I affects the highest emitting electric generating units and those units choosing to comply early, and Phase II includes the remaining electric generating units (3). Title IV set a cap on national emissions of SO_2 at 10 million metric tons (roughly 50%) below levels generated in 1980 by

Table 9.1 Acid rain precursors (1986)

Nitrogen Oxides (NO_x) 19.7 Million Metric Tons NO_x				
Transportation (44%)	Electrical utilities (34%)	Industrial processes and fuel combustion (18%)	Commercial, industrial, residential (3%)	Other (1%)
Sulfur Dioxide (SO_2) 21.4 Million Metric Tons SO_2				
Transportation (4%)	Electrical utilities (68%)	Industrial processes and fuel combustion (25%)	Commercial, industrial, residential (3%)	Other

Table 9.2   Historic U.S. SO_2 and NO_x emissions (in millions of tons)

	1940	1950	1960	1970	1980	1984	1995
SO_2	19.8	22.4	22.0	31.1	25.6	23.6	18.3
NO_x	7.5	10.3	14.1	20.0	23.3	21.7	21.8

2010. NO_x emissions will be reduced by approximately 2 million tons from 1980 levels through Titles I, II, and IV of the 1990 Amendment. By the end of 1995, after the implementation of the first phase of Title IV, SO_2 emissions were reduced dramatically (3). In 1997, monitoring data indicated that total emissions were roughly 23% below the allowable level of 7.15 million tons of SO_2.

Acid-forming emissions are not spread evenly over the United States. In the mid-1980s, ten states — Missouri, Illinois, Indiana, Tennessee, Kentucky, Michigan, Ohio, Pennsylvania, New York, and West Virginia — produced 53% of total U.S. SO_2 emissions and 30% of total U.S. NO_x emissions.

Table 9.3 lists the top ten SO_2- and NO_x-emitting states in 1984. SO_2 emissions were concentrated along the Ohio River Valley in Ohio, Indiana, Pennsylvania, Illinois, and West Virginia. These five states, along with Missouri and Tennessee, produced 44% of all SO_2 in the United States. The order of ranking changed somewhat as Title IV took effect. For instance, in 1995, as a result of Phase I of Title IV, Ohio, Missouri, Indiana, Illinois, and Tennessee decreased SO_2 emissions by 3.4 million tons from 1980 levels. However, Texas emissions increased in 1995, as it has newer coal-fired power plants that will not be regulated until Phase II of Title IV is implemented in January 2000.

U.S. NO_x emissions tend to be more evenly distributed, but again, states along the Ohio River are especially high producers. Four of the five highest SO_2-producing states — Ohio, Indiana, Pennsylvania, and Illinois — are also among the top ten NO_x-producing states. Thus, the Ohio River Valley and the states immediately adjacent to it lead the United States in emissions of both major components of acid rain.

Table 9.3   Top ten SO_2- and NO_x-producing states in 1984 (in millions of tons)

Rank	State	SO_2	Rank	State	NO_x
1	Ohio	2.58	1	Texas	3.25
2	Indiana	1.67	2	California	1.17
3	Pennsylvania	1.60	3	Ohio	1.14
4	Illinois	1.38	4	Illinois	0.99
5	Texas	1.24	5	Pennsylvania	0.92
6	Missouri	1.18	6	Indiana	0.83
7	West Virginia	1.02	7	Florida	0.70
8	Florida	0.99	8	Michigan	0.69
9	Georgia	0.93	9	Louisiana	0.68
10	Tennessee	0.92	10	New York	0.62

Although long-term trends in acid deposition can't be exactly determined, it is possible to draw conclusions about current patterns. A comparison of the pH of U.S. rainfall with the states producing the greatest SO_2 and NO_x emissions clearly shows the solid link between acidic emissions and acidic deposition. Data collected by several different monitoring networks show that the areas of the United States receiving the most acid rainfall are downwind and northeast of those states with the highest SO_2 and NO_x emissions.

Effects of Acid Rain

The environmental effects of acid rain are usually classified into four general categories: aquatic, terrestrial, materials, and human health. Although there is evidence that acid rain can cause certain effects in each category, the extent of those effects is very uncertain. The risks these effects may pose to public health and welfare are also unclear and very difficult to quantify.

The extent of damage caused by acid rain depends on the total acidity deposited in a particular area and the sensitivity of the area receiving it. Areas with acid-neutralizing compounds in the soil, for example, can experience years of acid deposition without problems. Soils like this are common throughout the midwestern United States. On the other hand, the thin soils of the mountainous Northeast have very little acid-buffering capacity, making them vulnerable to damage from acid rain.

The adverse effects of acid rain are seen most clearly in aquatic ecosystems. The most common impact appears to be on reproductive cycles. When exposed to acidic water, female fish, frogs, salamanders, and so on may fail to produce eggs or may produce eggs that fail to develop normally.

Low pH levels also impair the health of fully developed organisms. Some scientists believe that acidic water can kill fish and amphibian reptiles by altering their metabolism, but we have little evidence that this is happening now.

It is known, however, that acid rain plays a role in what scientists call the "mobilization" of toxic metals. These metals remain inert in the soil until acid rain moves through the ground. The acidity of this precipitation is capable of dissolving and "mobilizing" metals such as aluminum, manganese, and mercury. Transported by acid rain, these toxic metals can then accumulate in lakes and streams, where they may threaten aquatic organisms.

Some lakes in areas of high acid deposition and low buffering capacity have been found to be both highly acidic and lifeless. Yet other lakes in similarly sensitive areas have not. Different lakes vary in the time it takes to reach an acidic condition, and rates of recovery from acidification also seem to vary.

Scientists are using field studies, long-term water quality data, studies of fish population declines, and lake sediment studies to analyze the acidification of various lakes. However, both the data and the theoretical models currently available are unproven in their ability to make an accurate prediction of the effects of continued acidic emissions.

Less is known about acid rain's effects on forests and crops than about effects on aquatic systems. The most extreme form of damage some have attributed to acid rain is the phenomenon known as "dieback." Dieback is a term applied to the unexplained death of whole sections of a once-thriving forest. At this time, however, little direct evidence links acid rain to forest dieback.

Scientists do agree that acid rain can lead to other, less extreme effects on soil and forest systems. It can leach nutrients from soil and foliage while inhibiting photosynthesis. Acid rain can also kill certain essential microorganisms. The toxic metals it mobilizes when passing through soil can be harmful not just to aquatic life but to trees and crops as well. But, again, little evidence is available which actually shows that such damage is occurring now because of acid rain.

Some experts even point to data indicating that acid deposition may actually benefit certain trees and crops. For example, some pitch pine seedlings have grown better when treated with increasingly acidic water, and exposure to combinations of acid rain and mist has stimulated red spruce growth. It is possible that nitrates derived from the nitrogen oxides in acid rain confer some nutritional benefits on trees and plants.

Acid rain can also damage man-made materials, such as those used in construction and sculpture. Photographs of statues that are losing their features and shape are familiar, with acid rain often cited as the culprit.

The problem is far more than aesthetic. Building materials, too, can be degraded by acidity. For example, limestone, marble, carbonate-based paints, and galvanized steel all can be eroded and weakened by the kind of dilute acids found in acid deposition.

Because materials naturally deteriorate with time, it is difficult to differentiate the effects of acid rain from damage caused by normal weathering. It is also hard to identify the specific damage caused by given pollutants or combinations of pollutants. As a result, the particular role played by acid rain in the deterioration of materials is still unknown.

So far, no known human health problems result from direct contact with acid rain. Inhaling acidic particles in acid fog may possibly carry some health risk, but more research is needed to confirm whether this constitutes a real risk.

Acid rain may also indirectly affect human health when it mobilizes toxic trace metals such as aluminum and mercury. When dissolved in acidic water, these materials can be ingested by fish and animals, thereby building up within the human food chain. Acidic water could also leach lead out of pipe solder and into drinking water supplies.

But these are only possibilities. No one has established that current emissions of SO_2 and NO_x are actually causing such damage, or that such damage will continue or increase in the future if SO_2 and NO_x emissions are not reduced.

9.4 ACID RAIN RESEARCH

Despite intensive research into most aspects of acid rain, scientists still have many areas of uncertainty and disagreement. That is why the United States emphasizes the importance of further research into acid rain.

Scientific research into acid rain accelerated significantly in the 1980s. With a dozen federal agencies involved, acid rain research can be complicated organizationally as well as scientifically. To prevent duplication of effort and foster creative cooperation among the agencies, the National Acid Precipitation Assessment Program (NAPAP) was set up in 1980.

The NAPAP was initially chaired jointly by the EPA, the President's Council on Environmental Quality, the National Oceanic and Atmospheric Administration, and the Departments of Agriculture, Energy, and the Interior. The National Aeronautics and Space Administration is now also a member of the program.

The EPA played a major role in several of the early NAPAP key research initiatives:

- Expansion of the National Trends Network, which gathered definitive acid rain data at monitoring stations throughout the nation. This network monitored wet deposition at 150 locations around the country, and was extended to include 100 dry deposition monitoring stations.

- Investigations into "source–receptor relationships," i.e., the relation between changes in emissions and changes in deposition levels at distant locations. The EPA's Atmospheric Processes program developed an ambitious Regional Acid Deposition Model to enable scientists to predict the amounts of acid rain resulting from given levels of emissions. With the model's predictive powers, policy-makers were to be able to weigh the benefits and drawbacks of different regulatory scenarios.

- The Delayed/Direct Response Project, which was working to determine the rate at which lakes acidify and to identify factors that hasten or retard that process, such as the acid-neutralizing capacity of surrounding soil. A "delayed" response is one that takes 10 years or longer. A "direct" response is acidification occurring in fewer than 10 years. Under this program, the EPA sampled 145 watersheds in New England with the help of the Soil Conservation Service.

In 1982, the federal agencies involved in the NAPAP budgeted $14.4 million for acid rain research. For 1987, the President requested $85 million for acid rain research, a more than fivefold increase in as many years. The increased funding has shown results. Scientists today have a much greater understanding of the chemistry of acid rain than they did in 1980. But they are still seeking a better grasp of the effects of acid rain on lakes, streams, forests, and construction materials.

Title IV mandated NAPAP to continue the coordination of federal acid rain research and monitoring. In addition, Congress asked NAPAP to evaluate the costs, benefits, and effectiveness of Title IV and to assess what further reductions in deposition rates are needed to prevent adverse ecological effects. The first report to Congress was prepared in May 1998.

National Surface Water Survey

The National Surface Water Survey is the EPA's primary source of data on the impact of acid rain on America's lakes and streams. Plans for the project began in 1983, with the first of three planned phases to be completed by the fall of 1984. The goal of Phase I was to measure the acidity of U.S. lakes and streams. It was not feasible to sample all the lakes and streams in potentially susceptible areas, so methods of statistical sampling were used to make the final selection. Phase I data collection was divided into three components: eastern lakes, western lakes, and eastern streams. Preliminary findings from the Eastern Lakes Survey were made public in August 1985.

Many people expected that more acidic lakes would be found in the Northeast than in other parts of the United States. They based this expectation on the fact that northeastern states are downwind of the major generators of acid rain precursors in the Ohio River Valley. Eastern Lake Survey teams took samples at 763 northeast lakes. On the basis of those samples, EPA scientists estimated that only 3.4% of the lakes sampled in the Northeast had pH values of 5.0 or less. The comparable figure for the Upper Midwest was also low: 1.5%. Surprisingly, Florida — far to the south of industrial sources of acid rain — had a much higher percentage of acidic lakes than did the Northeast and the Upper Midwest. Over 12% of lakes sampled in Florida had pH levels of 5.0 or less. The EPA believes that it is too early to attribute this high Florida figure to the impact of acid rain. Natural processes or land-use practices may also contribute substantially to the acidity of many Florida lakes.

Expertise gained during Phase I of the National Surface Water Survey proved useful in Phase II, which was initiated in the Northeast at the end of 1985. Phase II researchers were looking for variations in surface water chemistry from region to region and from season to season. They also calculated the fish population at selected lakes and streams surveyed in Phase 1. This data will be valuable as scientists try to evaluate the impact of acid rain on aquatic life. For Phase III, the EPA modified a long-term monitoring project already in progress. The goal of Phase III was to identify trends in surface water chemistry using long-term monitoring data. The work, which is planned to continue indefinitely, is being designed to be adaptable to other surface water pollution problems as well as acid rain. A series of databases was compiled between 1986 and 1990, based on the result of these surveys.

Title IV required that a National Acid Lakes Registry be created. The registry lists all lakes known to be acidified due to acid depository. Title IV also authorized appropriations to the U.S. Fish and Wildlife Service for research related to acid deposition and the monitoring of high altitude mountain lakes in the Wind River Reservation in Wyoming, and also to do a study to be conducted in conjunction with the University of Wyoming on various buffering and neutralizing agents used to restore lakes and streams damaged by acid deposition.

Materials Effects Research

Scientists who specialize in the materials effects of acid rain still don't know how wet and dry acid deposition affects the natural process of decay. One way to answer this question is to measure tombstones. In the 1980s, the EPA sponsored research into the rates of deterioration of headstones at 18 Veterans Administration cemeteries.

Two of the cemeteries provided particularly valuable data. One was located in an industrial area close to New York City, while the other was in a semirural area of Long Island. New York University had previously traced changes in the thickness of tombstones at both cemeteries, as well as the depth of their emblem inscriptions. Using these data to calculate weathering rates at the two cemeteries, scientists compared them with estimates of rates of increase in SO_2 in New York City from 1880 to 1980. They found what is known as a "linear" relation between the two rates. In other words, increased SO_2 concentrations were directly proportional to increased weathering rates.

This correlation enabled scientists to develop a formula for calculating the damage caused to materials in the New York area by SO_2: 10 mm of fine grain marble will be worn away every century for every part per million of SO_2 in the air.

This study was the first statistically significant proof of damage to stone from an acid rain precursor. It would be difficult to carry out other experiments of this kind, because historical data on air pollution levels are extremely rare. But it is clear that decay accelerated by acid deposition has ramifications far beyond the graveyard.

Some acid rain concerns are primarily cultural. For example, the rapid deterioration of the Acropolis in modern times prompted the EPA in the mid-1980s to join a North Atlantic Treaty Organization (NATO) pilot study on the conservation and restoration of monuments. Scientists from 10 countries monitored acid rain damage to monuments, developed formats and procedures for documenting acid rain damage, and evaluated various means of conserving and restoring damaged monuments.

But acid rain threatens more than cultural artifacts. Though experts cannot yet fix an exact dollar value to the materials damage caused by acid rain, they agree that it damages homes, commercial buildings, highways, bridges, and other structures vital to our everyday lives. The EPA is working with the U.S. Army Corps of Engineers to develop a list of materials subject to acid rain damage. This inventory will draw together the data needed to assess the magnitude of acid rain-induced materials damage.

Forest Response Program

In the early 1980s, experts began to see unexplained growth reductions and foliage damage in U.S. forests. The evidence was first spotted in New York and New England, but similar problems have now been detected in the Appalachians and the Carolinas. Even worse forest deterioration has occurred in Europe, where whole stands of European trees, especially on mountain peaks, have gone into an unprecedented decline.

Scientists are still uncertain of acid rain's role in such instances. Many factors other than acid rain could be responsible for forest damage. Changes in soil or climate could play a role, as could changes in insect or pathogen activity. For these reasons among others, the evidence for acid rain damage to forests is thought to be weaker than corresponding evidence of damage to aquatic systems.

To clarify the effects of acid rain on trees and other vegetation, the EPA began the Forest Response Project (FRP) in 1985. The FRP scientists were studying the role of acid rain and other pollutants in causing or contributing to forest damage in the United States. They were also trying to determine (a) the mechanisms causing the damage and (b) the relationship between various "doses" of acid deposition and the "responses" they are suspected of causing.

Initial research involved two types of U.S. forest that experienced damage or decline. The first type of forest, common to New England and New York, contained spruce and fir. The second, known as "Southern commercial," included several species of pines valuable to the economy of the southeastern United States. At two sites in New England and three sites in the Southeast, trees were classified and checked for height and radial growth. Scientists also conducted field experiments to compare the growth of trees in open-top chambers with that of trees in rain-exclusion chambers. Control chambers in laboratories permitted comparable experiments with seedlings, although it is still difficult to extrapolate from seedlings to mature trees.

The EPA also set up a "Mountain Cloud" data-gathering network to study the effects of various acid rain patterns on forests at differing elevations. Mountain Cloud sites were co-located with biological stations that measured (a) plant growth and productivity and (b) soil chemistry. This work and other studies that were planned for eastern hardwood forests and western conifers began to provide a clearer idea of the kind of threat that acid rain poses to the $38.5 billion forest products industry.

The Future

Many challenges confront acid rain scientists. There is still a need to increase scientific understanding of the effects of acid rain, as well as the rate at which those effects occur. As yet, scientists lack reliable methods of extrapolating on a regional level what is known about the effects of acid rain in small-scale environments. They also need to determine the level of acid deposition that is realistically compatible with protecting our valuable resources. As these and other questions are answered, a much clearer understanding of the type of control program needed to protect all the resources at risk from acid rain will result.

9.5 CONTROL TECHNOLOGIES

As mentioned above, Title IV calls for the reduction of SO_2 by 10 million tons and NO_x by 2 million tons from 1980 levels. To achieve that level of control, many existing sources of SO_2 and NO_x—especially utility and industrial coal-fired boilers—would have to be retrofitted with control equipment. But the availability, cost, and technical complexity of existing retrofit controls leave much to be desired. Therefore, utilities have pursued a combination of compliance options, including scrubber installation, fuel switching, energy efficiency, and emissions trading (emissions trading will be discussed in detail in Section 9.6).

Existing Control Options

A number of different methods of equipping new boilers with NO_x controls have been developed and tested. But, overall, NO_x control technologies have not been commercially retrofitted on existing boilers as extensively as SO_2 controls.

At present, there are three techniques available for reducing the amount of SO_2 emitted from existing coal-fired boilers: coal-switching, coal-cleaning, and flue gas desulfurization. Unfortunately, each of these techniques has drawbacks that limit their ability to reduce SO_2 emissions by 10 million tons per year.

A coal-burning facility could cut down on SO_2 emissions by switching from a high-sulfur to a low-sulfur coal. Deregulation of the railroads in the 1980s and changes in the low-sulphur coal markets have made this possible. However, this fuel shift could damage some kinds of boiler equipment.

A second option is for sulfur to be cleaned from coal before it is burned. Physical coal-cleaning technologies are available commercially today. A substantial amount of coal already is being cleaned because of the savings that result from lower shipping costs, lower boiler-maintenance costs, and the higher energy content of the cleaned coal. However, coal is cleaned primarily to rid it of ash and other noncombustibles. Not enough SO_2 could be cleaned from coal to hit the emissions reduction target of a large-scale acid rain control program.

Currently, there is only one technology available that could reduce SO_2 emissions to the extent required by an ambitious acid rain control program: flue gas desulfurization (FGD), a process better known as "scrubbing." FGD uses sorbents such as limestone to soak up (or scrub) SO_2 from exhaust gases. This technology, which is capable of reducing SO_2 emissions by up to 95%, can be added to existing coal-fired boilers.

FGD does have several drawbacks. The control equipment is very expensive and unreliable. Smaller facilities do not always have the capital or the space needed for FGD equipment. Even some larger power plants would find it technically very difficult to retrofit FGD systems on older cramped facilities.

Expanding Control Options

The *Report of the Special Envoys on Acid Rain*, presented to President Reagan on January 8, 1986, recognized the political and economic problems that stem from having only a limited menu of pollution control options. The report stated: "The availability of cheaper, more efficient control technologies would improve our ability to formulate a national response that is politically and economically acceptable." The Special Envoys went on to recommend a $5 billion U.S. program to fund the commercial demonstration of control technologies that promise greater emissions reductions, lower costs, or applicability to a wider range of existing sources. They also recommended that special consideration be given to projects that have the potential to reduce SO_2 emissions from existing facilities that burn high-sulfur coal.

Over the past several years, millions of dollars have been spent researching a variety of innovative approaches to the control Of SO_2 and NO_x emissions from existing coal-fired utility and industrial boilers. Major federal research programs are being funded by the EPA, the Department of Energy (DOE), the national laboratories (Argonne, Brookhaven, Lawrence Berkeley, and Oak Ridge), and the Tennessee Valley Authority. In addition, the Electric Power Research Institute is cooperating with different electric utilities to improve the control of utility boilers. This research and testing have already generated a number of attractive candidates for the kind of commercial demonstrations recommended in the *Report of the Special Envoys on Acid Rain*.

The four technologies described below—Limestone Injection Multistage Burning, In-Duct Spraying, Reburning, and Fluidized Bed Combustion—represent just a few of the wide range of potential candidates for funding as commercial demonstration projects. The purpose of these projects will be to determine whether technologies such as these can be proven to work in existing commercial facilities.

The limestone injection multistage burner (LIMB) is an emerging control apparatus that can be retrofitted on a large portion of existing coal-fired boilers, both utility and industrial. Its broad applicability makes it an attractive candidate for funding under the proposed commercial demonstration program. In a LIMB system, an SO_2 sorbent (e.g., limestone) is injected into a boiler equipped with low-NO_x burners. The sorbent absorbs the SO_2 and the low-NO_x burners limit the amount of NO_x formed. Thus, a LIMB is capable of reducing both SO_2 and NO_x by about 50–60%.

LIMB technology will not be applied widely until a number of technical problems are solved. The sorbent injected into the boiler tends to increase slagging and fouling, which in turn increase operation and maintenance costs. Because boilers retrofitted with LIMB tend to produce more particulates of smaller size, particulate control becomes more difficult. Furthermore, technical questions remain as to what sorbents are most effective in a LIMB system, and how and where to inject the sorbents.

LIMB controls SO_2 and NO_x emissions during the combustion process itself. It is also possible to control SO_2 after combustion by cleaning it out of the exhaust gases. The scrubbers now in use apply this kind of post-combustion technology. If ways could be found to reduce the technical complexity and economic costs of scrubbing, post-combustion controls would become a more attractive method of reducing SO_2 emissions. The EPA, the DOE, and private industry are involved in efforts to improve FGD technology. Much of the research focuses on the development of more effective sorbent materials. In addition, the possibility of injecting a sorbent directly into existing exhaust ductwork is being investigated.

An in-duct spray-drying FGD system would improve on traditional scrubbers in several ways. Current scrubbers require the construction of very large reaction vessels where the exhaust gases and sorbent can mix to extract the SO_2. These vessels are very expensive, and sometimes the space they demand simply isn't available at existing facilities. If, however, the sorbent could be injected into existing ductwork, the cost of the reaction vessel could be eliminated, and it would be much easier to retrofit controls on a wider range of sources. Space constraints would no longer be a limiting factor.

Another relatively new technology, known as reburning or fuel staging, is capable of reducing NO_x emissions in existing boilers. In a coal-fired boiler, reburning is accomplished by substituting 15–20% of the coal with natural gas or low-sulfur oil and burning it at a location downstream of the primary combustion zone of the boiler. Oxides of nitrogen formed in the primary zone are reduced to nitrogen and water vapor as they pass through the reburn zone. Additional air is injected downstream of the reburn zone to complete the combustion process at a lower temperature. In general, NO_x reductions of 50% or more are achievable by reburning. When combined with other low-NO_x technologies (such as low NO_x burners), NO_x reductions of up to 90% may be achievable.

Fluidized bed combustion (FBC) is an innovative approach to SO_2 and NO_x control in both utility and industrial boilers. In a FBC boiler, pulverized coal is burned while suspended over a turbulent cushion of injected air. This technique is promising from an economic perspective, because FBC boilers allow improved combustion efficiencies and reduced boiler fouling and corrosion. Such boilers also are capable of burning different kinds of low-grade fuels like refuse, wood bark, and sewage sludge. In addition, FBC offers a number of environmental advantages. If the coal is mixed with limestone or some other sorbent material during combustion, the SO_2 is captured and retained in the ash.

FBC boilers have another environmental advantage over typical coal-fired boilers: They have the potential to control NO_x as well as SO_2. FBC boilers must operate within a narrow temperature range (1500–1600°F) that is substantially lower than typical boiler temperatures. Lower combustion temperatures inherently limit the formation of NO_x. Thus, FBC boilers may be able to control NO_x by 50–75% at the same time as they control SO_2 by up to 90%. An FBC system does have one major drawback: It requires the construction of a new boiler. Thus, it is more of a replacement technology than a retrofit. The number of existing boilers that could be replaced with FBC boilers at reasonable cost is limited, and its promise is more likely to be realized on new sources.

A Less Limited Future

Limestone injection multistage burning, in-duct spraying, reburning, and fluidized bed combustion, as well as other technologies, are capable of expanding the current, rather limited set of acid rain control options. If they can be proven to work on existing commercial facilities, state and federal lawmakers will have much more latitude as they frame legislation for controlling acid rain. Clearly, it would be inefficient and ineffective to try to implement a major acid rain control program before technically viable and economically affordable technologies are available.

9.6 EMISSIONS TRADING

Emissions trading is a market-based alternative to conventional regulation in which sources facing high costs to control pollution have the flexibility to meet their emission limits by purchasing excess reductions from other sources that can afford to lower their emissions further than federal or state regulations require. Currently, a trading program is being implemented at the federal level to reduce acid rain. At the state level, California is operating a trading program to assist the Los Angles area in complying with the federal air quality standard for ozone. Although specific savings are difficult to estimate, initial monitoring data indicate that these programs likely have reduced overall emissions at lower costs than conventional regulations have done. Their early success has encouraged the consideration of trading to address similar air quality issues.

Conventional air quality regulations place fixed limits on emissions from individual sources and require them to install specific technologies to control pollution. However, compliance may be more economically feasible for some sources than it is for others. Trading has the potential to improve air quality in cases where a pollutant disperses over a broad geographic area and the environmental objective is to control total emissions rather than limit local emissions from individual sources.

Trading programs use either credits or allowances. Credits are emission reductions that a pollution source has achieved in excess of required amounts. Sources that have earned credits can sell them to others that need additional reductions. Allowances differ from credits in that they represent the amount of a pollutant that a source is permitted to emit during a specified time in the future. If a source estimates that its emissions will be less than its allowances, it can sell its excess allowances to other sources that need them. Some trading programs also allow sources to save credits or allowances for meeting limits on emissions in future years. This practice is commonly referred to as banking.

For diffuse pollutants, decreasing total emissions over large areas through trading has the potential to improve overall air quality. For example, trading is well suited for SO_2 and NO_x because they can drift far from their points of release. However, trading is less suited for pollutants that remain locally concentrated because excess reductions in one area would not improve air quality elsewhere, and localized problems with air quality could arise from trading if sources clustered in one area are permitted to offset substantial increases in pollution with credits or allowances.

To decrease levels of acid rain, Title IV established an allowance program to reduce emissions of SO_2. Phase I of the program began in 1995 and currently involves over 400 coal-fired electric utility generating units. Phase II begins in 2000, and the number of participating units will increase to over 2,000. The EPA reports that the program has achieved full compliance with each unit holding enough allowances to cover its emissions. In 1997, monitoring data indicated that total emissions were roughly 23% below the allowable level of 7.15 million tons of SO_2. However, actual emissions were below the established cap primarily because many utilities had banked a significant portion of their allowances. Consequently, emissions could rise in coming years if utilities decide to use their banked allowances to achieve compliance. As the program enters its second phase, the substantial increase in the number of participating sources also could make the program more difficult to administer and perhaps reduce its effectiveness.

A second example of a trading program is the Regional Clean Air Incentives Market (RECLAIM), which was implemented by California's South Coast Air Quality Management District in the Los Angeles area in January 1994. RECLAIM is an allowance program aimed at reducing emissions of SO_2 and NO_x. The program's goal is to achieve an 80% reduction in emissions by 2003. Stationary sources that annually emit at least four tons of SO_2 or NO_x receive allowances, which are based on past peak production and existing emission requirements. The total number of allowances equals a cap on emissions, and the state reports that actual emissions have been below the cap from 1994 to 1996. However, some environmentalists argue that the program has been less successful in improving air quality than monitoring data indicate because emission allowances were initially high.

In the future, while traditional regulation remains the most common measure to control air pollution, the early success of the federal acid rain program and RECLAIM in California has encouraged the consideration of trading to address similar air quality issues. Domestically, trading may be used to reduce NO_x emissions in the eastern U.S. and to control possible increases in pollution if the electric utility industry is restructured. Internationally, trading may also be used to assist nations in reducing greenhouse gases thought to contribute to global warming (see Chapter 8 for more on global warming) (4).

9.7 SUMMARY

1. Problems associated with acid rain were noted as early as 1861. The phrase "acid rain" was first used in 1872.

2. All rainfall is slightly acidic, with a pH of approximately 5.6. Only rain with a pH below 5.6 is considered acid rain.

3. Man-made emissions of SO_2 and NO_x are transformed into acids in the

atmosphere, where they often travel hundreds of miles before falling as acid rain, snow, dust, or gas.

4. The environmental effects of acid rain are usually classified into four general categories: aquatic, terrestrial, materials, and human health. The extent of the effects of acid rain is very difficult to prove and measure. Ongoing research attempts to verify and quantify these effects.

5. The United States is attempting to substantially reduce SO_2 and NO_x emissions. Existing control options include coal-switching, coal-cleaning, and flue gas desulfurization. Potential control options include limestone injection, multi-stage burning, in-duct spraying, reburning, and fluidized bed combustion.

6. Emission trading is a market-based alternative to conventional regulation in which sources facing high costs to control pollution have the flexibility to meet their emission limits by purchasing excess reductions from other sources that can afford to lower their emissions further than federal or state regulations require.

REFERENCES

1. E. B. Cowling, "Acid Precipitation in Historical Perspective," *Environmental Science and Technology* 16(2), 1982.
2. G. Urbanowicz, "The Impact of Title V Regulations on Hospitals and Healthcare Facilities," paper presented at the 1995 Annual Conference and Technical Exhibition of the American Society for Hospital Engineering, Las Vegas, Nevada, June 20, 1995.
3. National Acid Precipitation Assessment Program Biennial Report to Congress: An Integrated Assessment, National Science and Technology Council Committee on Environment and Natural Resources, May 1998.
4. D. Bearden, *Air Quality and Emissions Trading: An Overview of Current Issues*, Congressional Research Service, Report for Congress, updated March 9, 1999.

10

INDOOR AIR QUALITY

10.1 INTRODUCTION

Indoor Air Quality is rapidly becoming a major environmental concern due to the fact that a significant number of people spend a substantial amount of time in a variety of different indoor environments. These indoor environments include, but are not limited to, homes, offices, hotels, restaurants, government buildings, factories, warehouses, and vehicles, including cars, planes, buses, and trains. In these "environments," people are exposed to pollutants emanating from a wide array of sources, including smoking, building materials and furnishings, combustion appliances, pesticides, cleaning and deodorizing agents, and other commercial and consumer goods. Pollutants which individuals may be exposed to indoors include gases like carbon monoxide, carbon dioxide, and nitrogen dioxide; metals and other inorganic compounds like mercury, chlorine, and sulfur; particulates and fibers like asbestos, tobacco smoke, and various microbes; radioactive pollutants like radon; and a wide array of compounds known as volatile organic compounds (VOCs). Among the more well-known VOCs are formaldehyde, benzene, and carbon tetrachloride. Adverse health effects caused by these pollutants range from eye and respiratory irritations to more severe toxic effects and cancer.

The serious concern over pollutants in indoor air is due largely to the fact that indoor pollutants are not as easily dispersed or diluted as pollutants outdoors. Thus, the concentrations indoors are often many times higher than those outdoors. Research by EPA, called the Total Exposure Assessment Methodology (TEAM) studies, has documented the fact that levels for some pollutants indoors may exceed outdoor levels by 200–500%. In addition, because people spend as much as 90% or more of their time indoors, these high levels of indoor air pollutants can present a serious health concern (1).

The degree of risk associated with exposure to indoor pollutants depends on how well buildings are ventilated and the type, mixture, and amount of pollutants present in the building. Improperly designed and operated ventilation systems can cause "sick building syndrome," with complaints of eye, nose, and throat irritations, fatigue, lethargy, headaches, nausea, irritability, or forgetfulness. Long-term health effects range from impairment of the nervous system to cancer (2).

This chapter discusses the most common specific sources of pollutants, their potential health effects, and ways to reduce the levels of pollutants in the home.

10.2 HEALTH EFFECTS OF INDOOR AIR POLLUTION

Health effects stemming from indoor pollutants fall into two categories: those that are experienced immediately after exposure and those that do not show up until years later. Immediate effects, which may occur after either a single exposure or repeated exposures, include irritation of the eyes, nose, and throat, headaches, dizziness, and fatigue. These immediate effects are usually of short-term duration and are treatable by some means. Sometimes the treatment consists simply of eliminating the person's exposure to the source of pollution, if it can be identified. Symptoms of some diseases, including asthma, hypersensitivity, pneumonitis, and humidifier fever, can also show up soon after exposure to some indoor air pollutants.

The likelihood of an individual developing immediate reactions to indoor air pollutants depends on several factors. Age and preexisting medical conditions are two important influences. In other cases, whether a person reacts to a pollutant can be determined by individual sensitivity, which varies tremendously from person to person. Some people can become sensitized to biological pollutants after repeated exposures, and it appears that some people can become sensitized to chemical pollutants as well.

Certain immediate effects are similar to those from colds or other viral diseases, so it is often difficult to determine if the symptoms are a result of exposure to indoor air pollution. For this reason, it is important to pay attention to the time and place the symptoms occur. If the symptoms fade or go away when a person is away from the home and return when the person returns home, an effort should be made to identify indoor air sources that may be possible causes. Some effects may be made worse by an inadequate supply of outside air or from the heating, cooling, or humidity conditions prevalent in the home.

Other health effects may show up either years after exposure has occurred or only after long or repeated periods of exposure. These effects, which include emphysema and other respiratory diseases, heart disease, and cancer, can be severely debilitating or fatal. More information on potential health effects from particular indoor air pollutants is provided in the Section 10.3, "Common Indoor Air Pollutants."

While pollutants commonly found in indoor air can be responsible for many harmful effects, there is considerable uncertainty about what concentrations or periods of exposure are necessary to product specific health effects. People also react very differently to exposure to indoor air polllutants. Further research is needed to better understand which health effects can occur after exposure to the average pollutant concentrations found in homes and which can occur from the higher concentrations that occur for short periods of time (3).

10.3 COMMON INDOOR AIR POLLUTANTS

This section takes a look, source by source, at the most common indoor air pollutants, their potential health effects, and ways to reduce their levels in the house.

Radon

Radon is a colorless, odorless gas that occurs naturally and is found everywhere at very low levels. It is when radon becomes trapped in buildings and concentrations build up in indoor air that exposure to radon becomes of concern.

The most common source of indoor radon is uranium in the soil or rock on which homes are built. As uranium naturally breaks down, it releases radon gas, and radon gas breaks down into radon decay products (also called *radon daughters* or *radon progeny*). Radon gas enters homes through dirt floors, cracks in concrete walls and floors, floor drains, and sumps. A second entry route for radon in some areas of the country is through well water. In some unusual situations, houses are made of radon-containing construction materials; in such cases, those materials can release radon into the indoor air.

Studies by the EPA indicate that as many as 10% of all American homes, or about 8 million homes, may have elevated levels of radon, and the percentage may be higher in geographic areas with certain soils and bedrock formations. Radon can be detected only by the use of measurement instruments call radon detectors.

The only known health effect associated with exposure to elevated levels of radon is lung cancer. The EPA estimates that about 5,000 to 20,000 lung cancer deaths a year in the United States may be attributed to radon. (The American Cancer Society estimated there would be a total of about 139,000 lung cancer deaths in 1988 from all causes.)

Exposure to Radon in the home can be reduced by the following steps:

1. *Measure levels of radon in the home.* Two types of radon detectors are most commonly used in homes: charcoal canisters that are exposed for 2 to 7 days; and alpha track detectors that are exposed for one month or longer. (Some states recommend that residents use only the alpha track monitors.)

2. *The state radiation protection office can provide you with information on the availability of detection devices or services.* Ask for materials specifically developed for your state and for the EPA's *A Citizen's Guide to Radon.* States may also provide you with EPA's *Radon Measurement Proficiency Report* for your state. This publication lists firms and laboratories that have demonstrated their ability to accurately measure radon in homes.

3. *Refer to the EPA guidelines in deciding whether and how quickly to take action based on test results.* The guidelines are given in the booklet *A Citizen's Guide to Radon.* The higher the radon level in the home, the faster action should be taken to reduce exposure. The EPA believes that radon levels in homes can be reduced to about 4 picocuries per liter of air and sometimes less.

4. *Learn about control measures.* An effective radon mitigation plan may include one or more of the following actions: sealing cracks and other openings in basement floors, ventilating crawl spaces, installing sub-slab or basement ventilation, or

installing air-to-air heat exchangers. The EPA booklet *Radon Reduction Methods: A Homeowner's Guide* describes some possible reduction measures.

5. *Take precautions not to draw larger amounts of radon into the house.* Increasing ventilation can be an effective means of reducing exposure to many indoor air pollutants; in homes with elevated concentrations of radon, however, increasing ventilation may increase infiltration through the foundation and result in drawing even larger amounts of radon into the home. The benefits of increased ventilation can be achieved without increasing radon exposure by opening windows evenly on all sides of the home. Opening windows is particularly important when you are using outdoor-vented exhaust fans.

6. *Select a qualified contractor to draw up and implement a radon mitigation plan.* The EPA suggests that all but the most experienced "do-it-yourselfer" get professional help in selecting and installing radon reduction measures. The EPA booklet *Radon Reduction Methods: A Homeowner's Guide* offers advice about how to select a contractor and how to evaluate proposals for radon mitigation. The EPA does not certify contractor competency for planning or executing radon mitigation measures.

7. *Stop smoking, and discourage smoking in your home.* Scientific evidence indicates that smoking may increase the risk of cancer associated with exposure to radon.

8. *Treat radon-contaminated well water by aerating or filtering through granulated-activated charcoal.* Contact the state's radiation protection office or drinking water office for more information concerning radon in drinking water in a particular community.

Environmental Tobacco Smoke

Environmental tobacco smoke is composed of sidestream smoke (the smoke that comes from the burning end of a cigarette) and smoke that is exhaled by the smoker. It is a complex mixture of over 4,700 compounds, including both gases and particles. Nonsmokers' exposure to environmental tobacco smoke is often called "passive smoking," "second-hand smoking," or "involuntary smoking."

According to reports issued in 1986 by the Surgeon General and the National Academy of Sciences, environmental tobacco smoke is a cause of disease, including lung cancer, in both smokers and healthy nonsmokers. Studies indicate that exposure to tobacco smoke may increase the risk of lung cancer by an average of 30% in the nonsmoking spouses of smokers. Very young children exposed to smoking at home are more likely to be hospitalized for bronchitis and pneumonia. Recent studies suggest that environmental tobacco smoke may also contribute to heart disease.

The following steps can reduce exposure to environmental tobacco smoke in the home:

1. *Give up smoking and discourage smoking in your home or ask smokers to smoke outdoors.* The 1986 Surgeon General's report concluded that physical separation of smokers and nonsmokers in a common air space, such as different rooms, within the same house, may reduce — but will not eliminate — nonsmokers' exposure to environmental tobacco smoke.

2. *Ventilation, a common method of reducing exposure to indoor air pollutants, also will reduce but not eliminate exposure to environmental tobacco smoke.* Because smoking produces such large amounts of pollutants, natural or mechanical ventilation techniques do not remove them from the air in your home as quickly as they build up. In addition, the large increases in ventilation it takes to significantly reduce exposure to environmental tobacco smoke can also increase energy costs substantially. Consequently, the most effective way to reduce exposure to environmental tobacco smoke in the home is to eliminate smoking there.

Biological Contaminants

Biological contaminants ("biologicals") include bacteria, mold and mildew, viruses, animal dander and cat saliva, mites, cockroaches, and pollen. There are many sources for these pollutants. For example, pollens originate from plants; viruses are transmitted by people and animals; bacteria are carried by people, animals, and soil and plant debris; and household pets are sources of saliva and animal dander. The protein in urine from rats and mice is a potent allergen; when it dried, it can become airborne. Contaminated central air handling systems can become breeding grounds for mold, mildew, and other biological contaminants and can then distribute these contaminants through the home.

By controlling the relative humidity level in a home, the growth of biologicals can be minimized. A relative humidity of 30–50% is recommended for homes. Standing water, water-damaged materials, or wet surfaces can serve as a breeding ground for molds, mildews, bacteria, and insects. House dust mites, one of the most powerful biologicals in triggering allergic reactions, can grow in any damp, warm environment.

Some biological contaminants trigger allergic reactions, including hypersensitivity pneumonitis, allergic rhinitis, and some types of asthma. Some transmit infectious illnesses, such as influenza, measles, and chicken pox. And some biologicals, such as certain molds and mildews, release disease-causing toxins. Symptoms of health problems caused by biological pollutants include sneezing, watery eyes, coughing, shortness of breath, dizziness, lethargy, fever, and digestive problems.

Many allergic reactions caused by biological allergens occur immediately after exposure; other allergic reactions are the result of previous exposures that a person may not have been aware of. As a result, people who have noticed only mild allergic reactions, or no reactions at all, may suddenly find themselves very sensitive to particular allergens. Some diseases, like humidifier fever, have generally been associated with exposure to toxins from microorganisms that can grow in large building ventilation systems. However, these diseases can also be traced to microorganisms that grow in home heating and cooling systems and humidification devices. Children, elderly people, and people with breathing problems, allergies, and lung diseases are particularly susceptible to disease-causing biological agents in the indoor air.

The following steps can reduce exposure to biological contaminants in the home:

1. *Install and use exhaust fans that are vented to the outdoors in kitchens and bathrooms and vent clothes dryers outdoors.* These actions can eliminate much of the moisture that builds up from everyday activities. There are exhaust fans on the

market that produce little noise, an important consideration for some people. Another benefit to using kitchen and bathroom exhaust fans is that they can reduce levels of organic pollutants that vaporize from hot water used in showers and dishwashers.

2. *Ventilate the attic and crawl spaces to prevent moisture buildup.* Keeping humidity levels in these areas between 30–50% can prevent water condensation on building materials.

3. *If using cool mist or ultrasonic humidifiers, clean water trays and fill with fresh, distilled water daily.* Because these humidifiers can become breeding grounds for biological contaminants, they have the potential for causing diseases such as hypersensitivity pneumonitis. Evaporation trays in air conditioners, dehumidifiers, and refrigerators should also be cleaned frequently.

4. *Thoroughly dry and clean water-damaged carpets and building materials (within 24 hours if possible).* Water-damaged carpets and building materials can harbor mold and bacteria. If health problems persist after you have tried to dry these materials, consider replacing them. It can be very difficult to completely rid such materials of biological contaminants.

5. *Keep the house clean.* House dust mites, pollens, animal dander, and other allergy-causing agents can be reduced, although not eliminated, through regular cleaning. People who are allergic to these pollutants should not vacuum (and may even need to leave the house while vacuuming is occurring), because vacuuming can actually increase levels of airborne mite allergens and other biological contaminants. Using central vacuum systems that are vented to the outdoor may reduce allergic reactions to biologicals.

6. *Take steps to minimize biological pollutants in basements.* Clean and disinfect the basement floor drain regularly. Do not finish a subsurface basement unless all water leaks are patched and outdoor ventilation and adequate heat to prevent condensation are provided. Operate a dehumidifier in the basement if needed to keep relative humidity levels between 30–50%.

Stoves, Heaters, Fireplaces, and Chimneys

In addition to environmental tobacco smoke, other sources of combustion products are unvented kerosene and gas space heaters, woodstoves, fireplaces, and gas stoves. The major pollutants released from these sources are carbon monoxide, nitrogen dioxide, and particles. In addition, woodstoves, fireplaces, and unvented kerosene space heaters emit polycyclic aromatic hydrocarbons. Unvented kerosene heaters may also generate acid aerosols.

Other sources of combustion gases and particulates are chimneys and flues that are improperly installed or maintained and cracked furnace heat exchangers. Pollutants from fireplaces and woodstoves with no outside air supply vent can be "down-drafted" from the chimney back into the living space, particularly in "weather-tight" homes.

Carbon monoxide is a colorless, odorless gas that interferes with the delivery of oxygen throughout the body. At low concentrations, it can cause fatigue in healthy people and episodes of increased chest pain in people with chronic heart disease. At higher concentrations, carbon monoxide can cause headaches, dizziness, weakness, nausea, confusion, and disorientation. The symptoms of carbon monoxide poisoning are sometimes confused with the flu or food poisoning. At very high concentrations carbon monoxide can cause unconsciousness and death. Fetuses, infants, pregnant women, elderly people, and people with anemia or with a history of heart or respiratory disease can be especially sensitive to carbon monoxide exposures.

Nitrogen dioxide can irritate mucous membranes in the eye, nose, and throat and cause shortness of breath after exposure to high concentrations. There is evidence that high concentrations or continued exposure to low levels of nitrogen dioxide can increase the risk of respiratory infection; there is also evidence from animal studies that repeated exposures to elevated nitrogen dioxide levels may lead, or contribute, to the development of lung disease such as emphysema. People at particular risk from exposure to nitrogen dioxide include children and individuals with asthma and other respiratory diseases.

Respirable particles, released when fuels are incompletely burned, can lodge in the lungs and irritate or damage lung tissue. A number of pollutants, including radon and benzo(*a*)pyrene, both of which can cause cancer, attach to small particles that are inhaled and then carried deep into the lung.

The following steps can reduce exposure to combustion products in the home:

1. *Take special precautions when operating fuel-burning unvented space heaters.* Consider potential effects of indoor air pollution when you decide to use an unvented kerosene or gas space heater. Follow the manufacturer's directions, especially instructions on the proper fuel and keeping the heater properly adjusted. A persistent yellow-tipped flame is generally an indicator of maladjustment and increased pollutant emissions. While a space heater is in use, open a door from the room where the heater is located to the rest of the house and open a window slightly.

2. *Install and use exhaust fans over gas cooking stoves and ranges and keep the burners properly adjusted.* Using a stove hood with a fan vented to the outdoors can greatly reduce exposure to pollutants during cooking. Improper adjustment, often indicated by a persistent yellow-tipped flame, can result in increased pollutant emissions. Ask the gas company to adjust the burner so that the flame tip is blue. If you purchase a new gas stove or range, consider buying one with pilotless ignition because they do not have a pilot light that burns continuously. Never use a gas stove to heat the home and always make certain the flue in the gas fireplace is open when the fireplace is in use.

3. *Keep woodstove emissions to a minimum.* Choose properly sized new stoves that are certified as meeting EPA emission standards. Make certain that doors in old woodstoves are tight-fitting. Use aged or cured (dried) wood only, and follow the manufacturer's directions for starting, stoking, and putting out the fire in woodstoves. Chemicals are used to pressure-treat wood; such wood should never be burned indoors. (Because some old gaskets in woodstove doors contain asbestos, when replacing gaskets refer to the instructions in the EPA and Consumer Product Safety Commission (CPSC) booklet, *Asbestos in Homes*, to avoid creating an asbestos problem. New gaskets are made of fiberglass.)

4. *Have central air handling systems — including furnaces, flues, and chimneys — inspected annually and promptly repair cracks or damaged parts.* Blocked, leaking, or damaged chimneys or flues can release harmful combustion gases and particles

and even fatal concentrations of carbon monoxide. Strictly follow all service and maintenance procedures recommended by the manufacturer, including those that tell how frequently to change the filters. If the manufacturer's instructions are not readily available, change filters once every month or two during periods of use. Proper maintenance is important even for new furnaces, because they can also corrode and leak combustion gases, including carbon monoxide.

Household Products

Organic chemicals are widely used as ingredients in household products because of their many useful characteristics, such as the ability to dissolve substances and evaporate quickly. Paints, varnishes, and wax all contain organic solvents, as do cleaning, disinfecting, cosmetic, degreasing, and hobby products. Fuels are made up of organic chemicals. All of these products can release organic compounds while they are being used and, to some degree, when they are stored.

In research conducted by the EPA, called the Total Exposure Assessment Methodology (TEAM) studies, levels of about a dozen common organic pollutants were found to be 2 to 5 times higher inside homes than outside, regardless of whether the homes were located in rural or highly industrial areas. Additional TEAM studies indicate that while people are using products containing organic chemicals, they can expose themselves and others to very high pollutant levels, and also that elevated concentrations can persist in the air long after the activity is completed. Three out of four specific organic compounds mentioned later in this chapter — benzene, perchloroethylene, and paradichlorobenzene — are among the most prevalent organic compounds identified by the TEAM studies. The fourth organic compound, methylene chloride, is used widely in consumer goods.

The ability of organic chemicals to cause health effects varies greatly, ranging from those that are highly toxic to those with no known health effect. Eye and respiratory tract irritation, headaches, dizziness, visual disorders, and memory impairment are among the immediate symptoms that some people have experienced soon after exposure to some organics. At present not much is known about what health effects occur as a result of the levels of organics found in homes. Many organic compounds are known to cause cancer in animals; some are suspected of causing, or are known to cause, cancer in humans.

The following steps can reduce exposure to household chemicals:

1. *Follow label instructions carefully.* Products often have warnings aimed at reducing exposure to the user. For example, if a label says to use the product in a "well-ventilated" area, go outdoors or in areas equipped with an exhaust fan to use the product, if possible. Otherwise, open windows to provide the maximum amount of outdoor air possible.

2. *Throw away partially full containers of old or unneeded chemicals safely.* Because gases can leak even from closed containers, this single step could do much to lower concentrations of organic chemicals in your home. Be sure that materials you decide to keep are stored not only in a well-ventilated area but are also safely out of reach of children. Do not simply toss these unwanted products in the garbage can. Find out if the local government or any organization in the community sponsors special days for the collection of toxic household wastes. If such days are available, use them to dispose of the unwanted containers safely. If no such collection days are available, think about organizing one.

3. *Buy limited quantities.* In the future, if products such as paints, paint strippers, and kerosene for space heaters or gasoline for lawn mowers are used only occasionally or seasonally, buy only as much as will be used right away.

4. *Keep exposure to emissions from products containing methylene chloride to a minimum.* Consumer products that contain methylene chloride include paint strippers, adhesive removers, aerosol spray paints, and pesticide "bombs." Methylene chloride is known to cause cancer in animals. Also, methylene chloride is converted to carbon monoxide in the body and can cause symptoms associated with exposure to carbon monoxide. Carefully read the labels containing health hazard information and cautions on the proper use of these products. Use methylene chloride-containing products outdoors when possible; use them indoors with as much ventilation as possible.

5. *Keep exposure to benzene to a minimum.* Benzene is a known human carcinogen. The main indoor sources of this chemical are environmental tobacco smoke, stored fuels and paint supplies, and automobile emissions in attached garages. Actions which will reduce benzene exposure include eliminating smoking within the home, providing for maximum ventilation during painting, and discarding paint supplies and special fuels that will not be used immediately.

6. *Keep exposure to perchlorethylene emissions from newly dry-cleaned materials to a minimum.* Perchloroethylene is the chemical most widely used in dry cleaning. In laboratory studies, it has been shown to cause cancer in animals. Recent studies indicate that people breathe low levels of this chemical both in homes where dry-cleaned goods are stored and as they wear dry-cleaned clothing. Dry cleaners, recapture the perchlorethylene during the dry cleaning process so that they can save money by reusing it, and they remove more of the chemical during the pressing and finishing processes. Some dry cleaners, however, do not remove as much perchloroethylene as possible all the time.

Taking steps to minimize your exposure to this chemical is prudent. If dry-cleaned goods have a strong chemical odor when you pick them up, do not accept them until they have been properly dried. If goods with a chemical odor are returned to you on subsequent visits, try a different dry cleaner.

Formaldehyde

Formaldehyde is an important chemical used widely by industry to manufacture building materials and numerous household products. It is also a by-product of combustion and certain other natural processes. Thus is may be present in substantial concentrations both indoors and outdoors.

Sources of formaldehyde in the home include smoking, household products, and the use of unvented, fuel-burning appliances, like gas stoves or kerosene space heaters. Formaldehyde, by itself or in combination with other chemicals, serves a number of purposes in manufactured products. For example, it is used to add permanent-press qualities to clothing and draperies, as a component of glues and adhesives, and as a preservative in some paints and coating products.

In homes, the most significant sources of formaldehyde are likely to be pressed wood products made using adhesives that contain urea–formaldehyde (UF) resins. Pressed wood products made for indoor use include particleboard (used as subflooring and shelving and in cabinetry and furniture), hardwood plywood paneling (used for decorative wall covering and in cabinets and furniture), and medium-density fiberboard (used for drawer fronts, cabinet doors, and furniture tops). Medium-density fiberboard contains a higher resin-to-wood ratio than any other UF pressed wood product, and is generally recognized as being the highest formaldehyde-emitting pressed wood product.

Other pressed wood products, like softwood plywood and flake or oriented strand-board, are produced for exterior construction use and contain the dark, or reddish-black colored phenol–formaldehyde (PF) resin. Although formaldehyde is present in both types of resins, pressed woods that contain PF resin generally emit formaldehyde at considerably lower rates than those containing UF resin.

Since 1985, the Department of Housing and Urban Development (HUD) has only permitted the use of plywood and particleboard that conform to specified formaldehyde emission limits in the construction of prefabricated and mobile homes. In the past, some of these homes had elevated levels of formaldehyde because of the large amount of high-emitting pressed wood products used in their construction and because of their relatively small interior space.

The rate at which products like pressed wood or textiles release formaldehyde emissions can change. Formaldehyde emissions will generally decrease as products age. When the products are new, high indoor temperatures or humidity can cause increased releases of formaldehyde from these products.

Formaldehyde, a colorless, pungent-smelling gas, can cause watery eyes, burning sensations in the eyes and throat, nausea, and difficulty in breathing in some humans exposed at elevated (>0.1 ppm) levels. High concentrations may trigger asthma attacks in people with asthma. There is some evidence that some people can develop chemical sensitivity after exposure to formaldehyde. Formaldehyde has also been shown to cause cancer in animals and may cause cancer in humans.

The following steps can reduce exposure to formaldehyde in the home:

1. *Ask about the formaldehyde content of pressed wood products, including building materials, cabinetry, and furniture before purchasing them.* If you experience adverse reactions to formaldehyde, you may want to avoid the use of pressed wood products and other formaldehyde-emitting goods. Even if you do not experience such reactions, you may wish to reduce your exposure as much as possible by purchasing exterior-grade products, which emit less formaldehyde. For further information and consumer products, call the EPA Toxic Substance Control Act (TSCA) assistance line (202/554-1404).

Some studies suggest that coating pressed wood products with polyurethane may reduce formaldehyde emissions for some period of time. To be effective, any such coating must cover all surfaces and edges and remain intact. Increase the ventilation and carefully follow the manufacturer's instructions while applying these coatings. If you are sensitive to formaldehyde, check the label contents before purchasing coating products to avoid buying formaldehyde-containing products, as they will emit the chemical for a short time after application.

2. *Maintain moderate temperature and humidity levels and provide adequate ventilation.* The rate at which formaldehyde is released is accelerated by heat and may also depend somewhat on the humidity level. Therefore, the use of dehumid-

ifiers and air conditioners to control humidity and to maintain a moderate temperature can help reduce formaldehyde emissions. Drain and clean dehumidifier collection trays frequently so that they do not become a breeding ground for microorganisms. Increasing the rate of ventilation in your home will help in reducing formaldehyde levels.

Pesticides

According to an EPA survey, nearly 9 out of 10 U.S. households use pesticides. One study by EPA suggests that 80–90% of most people's exposure to pesticides in the air occurs indoors and that measurable levels of up to a dozen pesticides have been found in the air inside homes. The amount of pesticides found in homes appears to be greater than can be explained by recent pesticide use in those households; other possible sources include contaminated soil or dust that floats or is tracked in from outside, stored pesticide containers, and household surfaces that collect and then release the pesticides.

The EPA registers pesticides for use and requires manufacturers to put information on the label about when and how to use the pesticide. It is important to remember the "-cide" in pesticides means "to kill." These products are dangerous if not used properly.

In addition to the active ingredient, pesticides are also made up of ingredients which are used to carry the active agent. These carrier agents are called "inerts" in pesticides because they are not toxic to the targeted pest; nevertheless, some inerts are capable of causing health problems. For example, methylene chloride, discussed under "Household Products," is used as an inert.

Pesticides used in and around the home include products to control insects (insecticides), termites (termiticides), rodents (rodenticides), and fungi (fungicides). They are sold as sprays, liquids, sticks, powders, crystals, balls, and foggers or "bombs."

Chlordane, and three other related termiticides (the "cyclodienes")—heptachlor, aldrin, and dieldrin—deserve special attention because of their ability to remain active for long periods of time. In recent studies, air samples taken in homes soon after well-applied termiticide treatments contained residues of these chemicals. As a result of these studies, the EPA has taken a series of actions that led to the removal of these chemicals from the marketplace. All use of aldrin and dieldrin has been banned, while chlordane and heptachlor cannot be used until an application method that will not result in any measurable exposure to household occupants is successfully developed. Alternative termiticides are currently available, and the EPA anticipates that manufacturers will soon apply to register others.

Both the active and inert ingredients in pesticides can be organic compounds; therefore, both can add to the levels of airborne organics inside homes. However, there is little understanding at present about what concentrations are necessary to produce these effects.

Exposure to high levels of cyclodienes, commonly associated with misapplication, has produced various symptoms, including headaches, dizziness, muscle twitching, weakness, tingling sensations, and nausea. In addition, the EPA is concerned that cyclodienes might cause long-term damage to the liver and the central nervous system, as well as increased risk of cancer.

The following steps can reduce exposure to pesticides in the home:

1. *Read the label and follow the directions.* It is illegal to use any pesticides in any manner inconsistent with the directions on its label. Unless you have had special training and are certified, never use a pesticide that is restricted to use by state-certified pest control operators. Such pesticides are simply too dangerous for application by a noncertified person. Use only the pesticides approved for use by the general public and then only in recommended amounts; increasing the amount does not offer more protection against pests and can be harmful to you and your plants or pets.

2. *Use in well-ventilated areas.* Open windows when applying pesticides. Mix or dilute pesticides outdoors or in a well-ventilated area and only in the amounts that will be immediately needed. If possible, take plants or pets outside when applying pesticides to them.

3. *Use alternative nonchemical methods for pest control.* Since pesticides can be found far from the site of their original application, it is prudent to reduce the use of chemical pesticides outdoors as well as indoors. Depending on the site and pest to be controlled, one or more of the following steps can be effective: use of biological pesticides, such as *Bacillus thuringiensis* for the control of gypsy moths; selection of disease-resistant plants; and frequent washing of indoor plants or pets. Termite damage can be reduced or prevented by making certain that wooden building materials do not come into direct contact with the soil and by storing firewood away from the home. By appropriately fertilizing, watering, and aerating lawns, the need for chemical pesticide treatments of lawns can be dramatically reduced.

4. *If you decide to use a pest control company, chose one carefully.* Ask for an inspection of your home and get a written control program for evaluation before you sign a contract. The control program should list specific names of pests to be controlled and chemicals to be used; it should also reflect any of your safety concerns. Insist on a proven record of competence and customer satisfaction.

5. *Dispose of unwanted pesticides safely.* If you have unused or partially used pesticide containers you want to get rid of, dispose of them according to the directions on the label or on special hazardous waste collection days. If there are no such collection days in your community, work with others to organize them.

6. *Keep exposure to moth repellents to a minimum.* One pesticide often found in the home is paradichlorobenzene, a commonly used active ingredient in moth repellents. This chemical is known to cause cancer in animals, but substantial scientific uncertainty exists over what may be the effects, if any, of long-term human exposure to paradichlorobenzene. The EPA requires that products containing paradichlorobenzene bear warnings such as "Avoid breathing vapors," to warn users of potential short-term toxic effects. Where possible, paradichlorobenzene, and items to be protected against moths, should be placed in trunks or other containers that can be stored in areas that are separately ventilated from the home, such as attics and detached garages. Paradichlorobenzene is also the key active ingredient in many air fresheners (in fact, some labels for moth repellents recommend that these same products be used as air fresheners or deodorants). Proper

ventilation and basic household cleanliness will go a long way toward preventing unpleasant odors, thereby reducing or eliminating the need for air refresheners.

7. *Call the National Pesticide Telecommunications Network (NPTN).* The EPA sponsors the NPTN (800/858-PEST) to answer questions about pesticides and to provide EPA publications on pesticides. Two such EPA publications are *A Citizen's Guide to Pesticides* and *Termiticides Consumer Information.*

Asbestos

Asbestos is a mineral fiber that has been used commonly in a variety of building construction materials for insulation as a fire-retardant. The EPA and CPSC have banned several asbestos products. Manufacturers have also voluntarily limited use of asbestos. Today asbestos is most commonly found in older homes in pipe and furnace insulation materials, asbestos shingles, millboard, textured paints and other coating materials, and floor tiles.

Elevated concentrations of airborne asbestos can occur after asbestos-containing materials are disturbed by cutting, sanding, or other remodeling activities. Improper attempts to remove these materials can release asbestos fibers into the air in homes, increasing asbestos levels and endangering people living in those homes.

The most dangerous asbestos fibers are too small to be visible. After they are inhaled, they can remain and accumulate in the lungs. Asbestos can cause lung cancer, mesothelioma (a cancer of the chest and abdominal linings) and asbestosis (irreversible lung scarring that can be fatal). Symptoms of these diseases do not show up until many years after exposure began. Most people with asbestos-related diseases were exposed to elevated concentrations on the job; some developed disease from exposure to clothing and equipment brought home from job sites.

The following steps can reduce exposure to asbestos in the home:

1. *Learn how asbestos problems are created in homes.* Read the booklet, Asbestos in the Home, issued by the CPSC and the EPA. To order a copy of this publication send a postcard to : Asbestos in the Home, Washington, DC 20207.

2. *If you think your home may have an asbestos problem, ask a trained professional to help determine whether your home has asbestos-containing materials and whether those materials are damaged or deteriorating.* Look to trained contractors, the manufacturer or installer of particular products or materials in your home, your state asbestos program (if there is one), and your state health department for information on how to identify and remedy asbestos problems in your home.

3. *Do not cut, rip, or sand asbestos-containing materials.* Leave undamaged materials alone, and to the extent possible, prevent them from being disturbed, damaged, or touched. Inspect periodically for damage or deterioration.

4. *When you need to remove or clean up asbestos, use a professional, trained contractor.* Select a contractor only after careful discussion of the problems in your home and the steps the contractor will take to clean up or remove them. Consider the option of sealing off the materials instead of removing them. Call the EPA TSCA assistance line (202/554-1404) to find out whether your state has a training and certification program for asbestos removal contractors and for information on the EPA's asbestos programs.

Lead

Lead has long been recognized as a harmful environmental pollutant. There are many ways in which humans are exposed to lead, including air, drinking water, food, and contaminated soil and dust. Airborne lead enters the body when an individual breathes lead particles or swallows lead dust once it has settled. Until recently, the most important airborne source of lead was automobile exhaust.

Lead-based paint has long been recognized as a hazard to children who eat lead-containing paint chips. A 1988 National Institute of Building Sciences (NIBS) Task Force report found that harmful exposures to lead can be created when lead-based paint is removed from surfaces by sanding or open-flame burning. The NIBS Task Force called for development of better and safer techniques to remove lead-based paints and effective clean-up methods. High concentrations of airborne lead particles in homes can also result from the lead dust from outdoor sources, contaminated soil tracked inside, use of lead in activities such as soldering, electronics repair, and stained glass art work.

Lead is toxic to many organs within the body at both low and high concentrations. Lead is capable of causing serious damage to the brain, kidneys, peripheral nervous system (the sense organs and nerves controlling the body), and red blood cells. Even low levels of lead may increase high blood pressure in adults.

Fetuses, infants, and children are more vulnerable to lead exposure than adults since lead is more easily absorbed into growing bodies, and the tissues of small children are more sensitive to the damaging effects of lead. In addition, an equal concentration of lead is more damaging because of a child's smaller body weight. Children may also have higher exposures since they are more likely to get lead dust on their hands and then put their fingers or other lead-contaminated objects in their mouths. The effects of lead exposure on fetuses and young children include delays in physical and mental development, lower IQ levels, shortened attention spans, and increased behavioral problems.

The following steps can reduce exposure to lead:

1. *If you suspect that paint in your home contains lead, have it tested.* It has been estimated that lead paint was used in about two-thirds of the houses built before 1940; one-third of the houses built from 1940 to 1960; and some housing built since 1960. Consult your state health or housing department for suggestions on which private laboratories or public agencies may be able to help test your home for lead in paint.

2. *Leave lead-based paint undisturbed if it is in good condition and there is little likelihood that it will be eaten by children — do not sand or burn off paint that may contain lead.* Ordinary household cleaning methods are ineffective at removing lead dust produced by sanding or burning; vacuuming does not sufficiently reduce lead dust levels because the particles pass through the filtering system in ordinary vacuums. Repainting areas covered with lead-based paint is not recommended because steps to prepare the surface area, such as sanding or removing cracked paint, produce lead dust.

If paint is cracked or peeling, cover with wallpaper or some other building material or replace the painted surface. Also, consider having painted woodwork such as doors and molding taken out of the house and sent off-site for chemical removal.

If on-site removal of lead-based paint cannot be avoided, then everyone not involved in doing the removal should leave the building during the period that removal takes place — no matter whether it involves sanding, burning, or chemical stripping. Workers should be protected and thorough cleanup should follow removal.

3. *People who may have been exposed to lead dust recently should have the lead levels in their blood tested by their doctor or local health department.* If exposure occurred some time previously, a blood test may not be a reliable indicator of exposure and it may be advisable for exposed children or adults to have neurological tests done. If either test shows lead exposure has occurred, follow the advice of your doctor or health department.

4. *Keep surface areas clean.* Frequent cleaning of smooth surfaces, especially food preparation areas, with a wet cloth or mop can reduce the amount of lead dust that drifts or is tracked in from outdoors. However, lead dust will remain in carpeting and on furnishings.

5. *Choose well-ventilated areas to engage in activities that involve the use of lead.* As with other activities, increasing ventilation can reduce potential health effects by reducing the concentrations of indoor air pollutants. Consider using "no-lead" solder.

6. *Have the drinking water in your home tested for lead.* Homes most likely to have high lead levels in their water are those with lead-soldered plumbing that is less than 5 years old or those that have water service connections or interior plumbing made of lead. Send for the EPA pamphlet, Lead and Your Drinking Water, for more information about what you can do if you have lead in your drinking water (3).

10.4   IMPROVING THE AIR QUALITY IN THE HOME

Three basic strategies for improving the air quality within the home are source control, ventilation improvements, and air cleaners. Usually the most effective way to improve indoor air quality is to eliminate individual sources of pollution or to reduce their emissions. Some sources, like those that contain asbestos, can be sealed or enclosed; others, like gas stoves, can be adjusted to decrease the amount of emissions. In some cases, source control is also a more cost-efficient approach to protecting indoor air quality than increasing ventilation because increasing ventilation can increase energy costs.

Another approach to lowering the concentrations of indoor air pollutants in your home is to increase the amount of outside air coming indoors. Opening windows and doors, when the weather permits, increases the natural ventilation rate. Turning on local bathroom or kitchen exhaust fans, if they are vented to the outdoors, can lower pollution levels by removing contaminants from the room where the fan is located. Where radon may be a problem, a window should be opened while bathroom or kitchen exhaust fans are in use. This keeps the amount of radon entering the house from increasing.

It is particularly important to take as many of these steps as possible while you are involved in short-term activities that can generate high levels of pollutants — for example, painting, paint stripping, heating with kerosene heaters, cooking with gas stoves that are not vented to the outdoors, or engaging in maintenance and hobby activities such as welding, soldering, or sanding. You might also choose to do some of these activities outside, if you can and if weather permits.

Another way to increase the mechanical ventilation rate is to install heat recovery ventilators (also known as air-to-air heat exchangers) in homes. These devices, which can be installed in windows or as part of a central air system, increase ventilation by drawing outside air into the home and conserve energy by recovering the heat from air that is exhausted to the outdoors. Heat recovery ventilators are most easily installed in central air systems in new homes or during extensive remodeling; window units can be installed in existing homes.

Before you buy a mechanical ventilation device for your home, you should read books and articles on these devices and consult a mechanical engineer. Write to Renewable Energy Information, P.O. Box 8900, Silver Spring, MD 20907 for the U.S. Department of Energy fact sheet. For additional information look in the yellow pages of a telephone directory under "Engineers" or write the American Society of Heating, Refrigerating, and Air Conditioning Engineers (ASHRAE) for the name of the president of the local ASHRAE organization in your community.

The third basic strategy for improving air quality is by utilizing an air cleaner. There are many types and sizes of air cleaners on the market, ranging from relatively inexpensive tabletop models to sophisticated and expensive whole-house systems. Some air cleaners are highly effective at particle removal, while others, including most tabletop models, are much less efficient. Air cleaners are generally not designed to remove gaseous pollutants.

How well an air cleaner works depends on how well it collects pollutants from indoor air (expressed as a percentage efficiency rate) and how much air it draws through the cleaning or filtering element (expressed in cubic feet per minute). A very efficient collector with a low air-circulation rate will not be effective, nor will a cleaner with a high air-circulation rate but a less efficient collector. The long-term performance of any air cleaner depends on maintaining it in accordance with the manufacturer's directions.

Another important factor in determining the effectiveness of an air cleaner is the strength of the pollutant source. Tabletop air cleaners, in particular, may not remove satisfactory amounts of pollutants from strong nearby sources. People with a sensitivity to particular sources may find that air cleaners are helpful only in conjunction with concerted efforts to remove the source.

At present, the EPA does not recommend using air cleaners to reduce levels of radon and its decay products. The effectiveness of these devices is uncertain because they only partially remove the radon decay products and do not diminish the amount of radon entering the home. EPA plans to do additional research on whether air cleaners are, or could become, a reliable means of reducing the health risk from radon (3).

Some air cleaners may be installed in the ducts which are part of central heating or air-conditioning systems in homes. Portable air cleaners stand alone in a room. Types of air cleaners include

- *Mechanical filters* similar to, and including, the typical furnace filter.
- *Electronic air cleaners* (for example, electrostatic precipitators) which trap charged particles using an electrical field.
- *Ion generators* which act by charging the particles in a room. The charged

particles are then attracted to walls, floors, draperies, etc., or to a charged collector.

- *"Hybrid" devices,* which contain two or more of the particle removal devices discussed above (4).

At a minimum, you should consider the following major factors affecting the performance of the air cleaner:

- The percentage of the particles removed as they go through the device (that is, the efficiency).
- The amount of air handled by the device. For example, an air cleaner may have a high efficiency filter, but it may process only 10 ft^3 of air each minute. Suppose that the air cleaner is put in a room of typical size, containing 1,000 ft^3 of air. In this room, it will take a long time for all the air to be processed. In some cases, pollutants may be generated more quickly than they are removed.
- The effective volume of the air to be cleaned. A single portable unit used in a room within a large building in which the air flows between several apartments or offices would be of little or no value.
- The decrease in performance which may occur between maintenance periods and if periodic maintenance is not performed on schedule.

Additional factors to consider include the following:

- Ion generators and electronic air cleaners may produce ozone, particularly if they are not properly installed and maintained. Ozone can be a lung irritant.
- Gases and odors from particles collected by the devices may be redispersed into the air.
- The odor of tobacco smoke is largely due to gases in the smoke, rather than particles. Thus, you may smell a tobacco odor even when the smoke particles have been removed.
- Some devices scent the air to mask odors, which may lead you to believe that the odor-causing pollutants have been removed.
- Ion generators, especially those that do not contain a collector, may cause soiling of walls and other surfaces.
- You may be bothered by noise from portable air cleaners, even at low speeds.
- Maintenance costs, such as costs for the replacement of filters, may be significant. You should consider these costs in addition to the initial cost of purchase. In general, the most effective units are also the most costly.

Several key elements are required to obtain adequate performance fron an air cleaner. First, follow the manufacturer's directions to assure that the air cleaner works properly. To avoid any electrical or mechanical hazards, be sure the unit is listed with Underwriters Laboratories (UL) or another recognized independent safety testing laboratory. Second, perform routine maintenance, as required. Generally speaking, air cleaners require frequent cleaning and filter replacement to function properly. Finally, be sure the element is properly located: Place portable air cleaners so they are near a specific pollutant source, if one exists that they force the cleaned air into occupied areas, and the inlet and outlet are not blocked by walls, furniture, or other obstructions. For in-duct devices, assure that the inlets and outlets of the heating or cooling system are not blocked by furniture and other obstructions.

Several standards are used for comparing types of air cleaners. One common method of rating high efficiency filters uses a procedure in Military Standard 282. This procedure measures how well small particles of a specific chemical are removed by the filter. The Federal government has not published guidelines or standards that can be used to determine how well low to medium efficiency air cleaners work. However, standards that have been developed by private standard-setting trade associations may be useful.

For further information on standards for **in-duct air cleaners**, contact your local heating or air-conditioning contractor or write to

> Air-Conditioning and Refrigeration Institute (ARI)
> 1501 Wilson Blvd., 6th Floor
> Arlington, VA 22209

For further information on standards for **portable air cleaner**, send a stamped, self-addressed envelope to

> Association of Home Appliance Manufacturers (AHAM)
> Air Cleaner Certification Program
> 20 North Wacker Drive
> Chicago, IL 60606

An in-depth analysis of air cleaners can be found in the EPA document *Residential Air-Cleaning Devices: A Summary of Available Information.* For this document and other EPA indoor air pollutants, contact

> Public Information Center
> U.S. Environmental Protection Agency
> Mail Code PM-211B
> 401 M St., SW
> Washington, DC 20460

10.5 SUMMARY

1. Indoor Air Quality is rapidly becoming a major environmental concern due to the fact that a significant number of people spend a substantial amount of time in a variety of different indoor environments.

2. Health effects from indoor pollutants fall into two categories: those that are experienced immediately after exposure and those that do not show up until years later.

3. A source-by-source look at the most common indoor air pollutants focuses on their potential health effects, and ways to reduce levels in the house.

4. Three basic strategies improve indoor air quality; One method is source control, another is through ventilation improvements, and the third is the utiliz-ation of some sort of mechanical device such as air cleaners.

REFERENCES

1. U.S. EPA, Indoor Air Facts No. 1, *EPA and Indoor Air Quality*, June 1987.
2. U.S. EPA Environmental Progress and Challenges, *EPA's Update*, August 1988, EPA-230-07-88-033.
3. U.S. EPA, *The Inside Story — A Guide to Indoor Air Quality*, September 1988, EPA/400/1-188/004.
4. U.S. EPA, Residential Air Cleaners Indoor Air Facts No. 7, *Air and Radiation*, 20A-4001, February 1990.

11

NOISE POLLUTION

11.1 INTRODUCTION

Noise pollution is traditionally not placed among the top environmental problems facing the nation; however, it is one of the more frequently encountered sources of pollution in everyday life. Noise pollution can be defined simply by combining the meaning of both environmental terms. Noise is typically defined as unwanted sound and pollution is generally defined as the presence of matter or energy whose nature, location, or quantity produces undesired environmental effects. Noise is typically thought of as a nuisance rather than a source of pollution. This is true in part because noise does not leave a visible impact on the environment like other sources of pollution.

The damage done by the pollution of our air and water is widely recognized. The evidence is right before our eyes, in contaminated water, oil spills and dying fish, and in smog that burns the eyes and sears the lungs. Noise is a more subtle pollutant. Aside from sonic booms that can break windows, noise usually leaves no visible evidence, although it also can pose a hazard to our health and well-being. An estimated 14.7 million Americans are exposed to noise that poses a threat to their hearing on the job. Another 13.5 million of us are exposed to dangerous noise levels without knowing it from trucks, airplanes, motorcycles, hi-fi's, lawnmowers, and kitchen appliances.

Recent scientific evidence shows that relatively continuous exposures to sound exceeding 70 decibels — expressway traffic, for instance — can be harmful to hearing. More than that, noise can cause a temporary stress reaction which includes increases in heart rate, blood pressure, blood cholesterol levels, and effects in the digestive and respiratory systems. With persistent, unrelenting noise exposure, it is possible that these reactions become chronic stress diseases such as high blood pressure or ulcers (1).

11.2 WHAT IS SOUND?

Sound travels in waves through the air like waves through water. The higher the wave, the greater its power. The greater the number of waves a sound has, the greater is its frequency or pitch.

The strength of sound or sound level is measured in decibels (dB). The decibel scale ranges from 0, which is regarded as the threshold of hearing for normal, healthy ears, to 194, regarded as the theoretic maximum for pure tones. Because the decibel scale, like the pH scale, is logarithmic, 20 dB is 100 times louder than 0, 30 dB is 1,000 times louder, 40 dB 10,000 times louder, and so forth. Thus at high levels, even a small reduction in level values can make a significant different in noise intensity.

The frequency is measured in Hertz (Hz) (cycles per second). It can be described as the rate of vibration. The faster the movement, the higher the frequency of the sound pressure waves created. The human ear does not hear all frequencies. Our normal hearing ranges from 20 Hz to 20,000 Hz or, roughly, from the lowest note on a great pipe organ to the highest note on a violin.

The human ear also does not hear all sounds equally. Very low and very high notes sound more faint to our ear than 1,000 Hz sounds of equal strength. This is the way our ears function. The human voice in conversation covers a median range of 300 to 4,000 Hz. The musical scale ranges from 30 to 4,000 Hz. Noise in these ranges sound much louder to us than very low- or very high-pitched noises of equal strength.

Because hearing also varies widely between individuals, what may seem loud to one person may not to another. Although loudness is a personal judgement, precise measurement of sound is made possible by use of the decibel scale. This scale, shown below, measures sound pressure or energy according to international standards. Table 11.1 compares some common sounds and shows how they rank in potential harm. Note that 70 dB is the point at which noise begins to harm hearing. To the ear, each 10-dB increase seems twice as loud.

11.3 HEARING PRINCIPLE

Sound waves enter the auditory canal, causing the eardrum to vibrate. The three small bones of the middle ear transmit these vibrations to the inner ear, through which they move as fluid-pressure waves. The Organ of Corti, running the length of the cochlea, converts these vibrations to nerve impulses, which are then carried to the brain by the auditory nerve.

Noise results in hearing loss through its destructive effect on the delicate hair cells in the Organ of Ciorti within the cochlea of the inner ear. These hair cells convert fluid vibrations in the inner ear into nerve impulses, which are carried by the auditory nerve to the brain, resulting in the sensation of sound.

Progressive destruction of the hair cells and a correlated reduction in the number of associated nerve fibers can be caused either by aging (deterioration of hearing due to old age is referred to as *presbycusis*) or by exposure to noise (noise-induced hearing loss is termed *sociocusis*). Prolonged exposure of excessive noise levels results in damage to, or the complete collapse of, individual hair cells, thus affecting the transmission of nerve impulses. Although there are many

Table 11.1 Sound levels and human response

Common Sounds	Noise Level (dB)	Effect
Carrier deck jet operation Air raid siren	140	Painfully loud
Jet takeoff (200 ft) Thunderclap	130	
Discothèque Auto horn (3 ft)	120	Maximum vocal effort
Pile drivers	110	
Garbage truck	100	
Heavy truck (50 ft) City traffic	90	Very annoying Hearing damage (8 hr)
Alarm clock (2 ft) Hair dryer	80	Annoying
Noisy restaurant Freeway traffic Man's voice (3 ft)	70	Telephone use difficult
Air conditioning unit (20 ft)	60	Intrusive
Light auto traffic (100 ft)	50	Quiet
Living room Bedroom Quiet office	40	
Library Soft whisper (15 ft)	30	Very quiet
Broadcasting studio	20	
	10	Just audible
	0	Hearing begins

thousands of hair cells in the Organ of Corti, when a large enough number are damaged, hearing ability is inevitably affected. This type of hearing loss is irreversible and cannot be restored by the use of hearing aids. Since the ear, unlike the eye, has no automatic defense mechanism—no "earlid" to prevent unwanted sounds from penetrating—the only way to protect oneself against noise-induced hearing loss it to limit one's exposure to a noisy environment as much as possible.

Americans are daily being exposed to levels of noise which are permanently damaging to their ability to hear. Most people are familiar with the temporary deafness and ringing in the ears which occurs after sudden exposure to a very loud noise such as a cap gun or firecracker exploding close to one's head. This type of partial hearing loss generally lasts a few hours at the most and is referred to as *temporary threshold shift* (TTS). However, it is not widely recognized that regular exposure to levels of noise commonly encountered in everyday life can, over a period of time, result in permanent hearing loss. Because damage to the ear is usually painless and seldom visible, few people recognize the injury they are incurring until it is too late. Research has shown that when daily noise levels average about 85 dB or more, permanent hearing loss can occur; for particularly sensitive individuals, average sound levels as low as 70 dB may be dangerous (2). The noise of city traffic, jet planes, chainsaws, power lawnmowers, some household appliances, babies screaming, and even people shouting all exceed the decibel level considered as "safe" (3).

11.4 SOURCES OF NOISE

The sources of noise pollution are extremely diverse and are constantly increasing as more and more noise-generating products become available to consumers. The greater variety of cars, home appliances, motorized tools and equipments both for home and work related use, and pleasure and hobby accessories are increasingly contributing to the wider variety of sources of noise pollution. Noise from commonly encountered motor vehicles such as cars, trucks, buses, motorcycles, emergency vehicles with sirens such as police cars, ambulances, and fire trucks represent some of the most intrusive sounds. Noise levels near major airports have become so intolerable that residents sometimes are forced to relocate and property value sometimes depreciates because of noise pollution. It is the most common source of noise pollution, producing an immediate effect ranging from temporary deafness to a prolonged irritation. Home appliances and home shop tools can be grouped into four categories based on the noise levels they produce. Machines in the first group, which include quieter major appliances such as refrigerators and clothes dryers, usually produce sound levels lower than 60 dB. Although the level is relatively low, such noise may be objectionable to a few people.

The second group includes clothes washers, food mixers, many dishwashers, and sewing machines that produce noise from 65–75 dB. Exposure time tends to be brief and infrequent, but the resulting noise may disrupt the understanding of speech and may be disturbing to neighbors in multifamily dwellings. The third group includes vacuum cleaners, noisy dishwashers, food blenders, electric shavers and food grinders. They usually produce 75 to 85 dB. The risk of hearing damage from them is small since use is not continuous or cumulative. Generally the noise from such appliances is annoying. Appliances in the fourth group produce the highest noise levels in the home environment — above 85 dB. They include millions of yard-care and shop tools. Any amount of exposure to such equipment will probably interfere with activities, disrupt your neighbor's sleep, cause annoyance and stress, and may contribute to hearing loss. Both gasoline and electric walk-behind lawn mowers range from about 87 to 92 dB at the operator's ear, and even 50 feet away ranges up to 72 dB; some riding mowers reach 83 dB at 50 feet (2). Table 11.2 illustrates sound levels produced by typical sources around the home.

11.5 HEALTH EFFECTS OF NOISE POLLUTION

An estimated 20 million Americans are exposed to noise that poses threat to their hearing. Everyone at sometime or another has experienced the effects of noise pollution. Many people are unaware that the sounds which cause them so much annoyance may also be affecting their hearing. Hearing loss as a result of noise pollution is one of the most serious of health threats.

Table 11.2   Noise levels at home

Noise Source	Sound Level for Operator (in dB)
Refrigerator	40
Floor fan	38–70
Clothes dryer	55
Washing machine	47–78
Dishwasher	54–85
Hair dryer	59–80
Vacuum cleaner	62–85
Sewing machine	64–74
Electric shaver	75
Food disposal (grinder)	67–93
Electric lawn edger	81
Home shop tools	85
Gasoline power mower	87–92
Gasoline riding mower	90–95
Chain saw	100
Stereo	Up to 120

Hearing disability caused by noise can range in severity from difficulty in comprehending normal conversation to total deafness. In general, the ability to hear high-frequency sounds is the first thing to be affected by noise exposure; for this reason, tests for early detection of hearing loss should pay special attention to hearing ability in the 4000-Hz range. People affected often have difficulty hearing such sounds as a clock ticking or telephone ringing and cannot distinguish certain consonants, particularly s, sh, ch, p, m, t, f, and th. Individuals suffering hearing loss not only have trouble with the volume of speech, but also with its clearness. They frequently accuse people with whom they are speaking of mumbling, particularly when talking on the telephone or when background noises interfere with conversation; listening to the radio or television may become impossible. The psychological impact of such difficulties — the fear of being laughed at for misunderstanding questions or comments, the frustration of not being able to follow a conversation, the feeling of isolation or alienation experienced as friends unconsciously avoid trying to converse — are frequently as severe as the physical disability. Individuals experiencing hearing loss tend to become suspicious, irritable, and depressed; their careers suffer and social life become severely restricted, sometimes to the point of causing complete withdrawal (4). In addition to these problems, a person with partial hearing loss may suffer sharp pain in the ears when exposed to very loud noise and may experience repeated bouts of *Tinnitus* — a ringing or buzzing sound in the head which can drive the victim to distraction, interfering with sleep, conversation, and normal daily activities (3).

Stressful reaction to loud or sudden noise undoubtedly represents an evolutionary adaption to warnings signaling approaching danger. Bodily responses start to the snarl of a predatory beast or the rumble of a boulder crashing down a mountainside and cause a surge of adrenalin, an increase in heart and breathing rates, and the tensing of muscles — all physiological preparations for fight or flight which have important survival value. However, these same metabolic responses are today being triggered repeatedly by the innumerable artificial noises of modern society. As a result, our bodies are subjected to a constant state of stress which, far from being advantageous, is literally making millions of people sick — even though they may not attribute their problems to noise. While many individuals insist that they have "gotten used to noise" and are no longer bothered by it, the truth is that no one can prevent the automatic biological changes which noise provokes. Research into the association between noise exposure and stress-related disease has produced findings such as the following:

- Workers exposed to high levels of occupational noise were found to exhibit up to five times as many cases of ulcers as would normally be expected among people in quieter surroundings.

- A 5-yr study of factory workers revealed that employees assigned to noisy areas of the plant had a higher frequency of diagnosed medical problems, including respiratory ailments, than did workers in quieter sections of the same plant.

- Exposure of a laboratory population of rhesus monkeys to noise levels typically experienced on a daily basis by factory workers resulted in a 30% elevation of the monkeys' blood pressure, a level which persisted for a long period after the experiment ended. Such findings indicate that adverse noise-induced health effects cannot be reversed quickly simply by removing the noise source (5).

Contrary to earlier beliefs, it is now known that outside noise can penetrate the womb and provoke responses in the developing unborn child. A fetus responds to loud noise with an increase in heart rate and kicking. Pregnant women have reported that they felt considerably more fetal movement while listening to music in a concert hall, the kicking reaching a peak when the audience began to applaud! Although this type of direct fetal response to sound may not present any serious concern, indirect responses caused by noise-induced maternal stress are more worrisome.

A study conducted in Japan suggests that expectant mothers living in noisy environments are more likely to give birth to underweight babies than are women from quieter areas. It is assumed that stress caused by high noise levels affected the production of certain maternal hormones responsible for fetal growth. Other studies have demonstrated that stress experienced by an expectant mother can cause blood vessels in the uterus to constrict, thereby reducing the supply of oxygen and nutrients in the developing child. A preliminary investigation among people living near a major airport indicate a higher-than-expected incidence of harelip, cleft palate, and spina bifida among children born in this area. Much more research is needed to define the extent of the relationship between noise and birth defects, as well as to establish how high noise levels must be to produce teratogenic effects. Lacking definitive information, some doctors recommend that pregnant women try to avoid noise exposure to the greatest extent possible, one such expert offering the tongue-in-cheek advice that "any expectant mother should get out of New York."

Noisy surroundings at home and school can adversely affect children's language development and their ability to read. High noise levels interfere with a youngster's capacity to distinguish certain sounds, such as "b" and "v," for example, and can foster a tendency to drop the endings of words, thereby distorting speech. Research has shown that reading skills are seriously impaired when the student's surroundings are noisy. One study focused on children living in a noisy apartment complex found that the longer they had resided in that environment, the poorer was their reading development. The investigators conducting the study concluded that a noisy home environment has more of an impact on reading skills than do such factors as parents' educational level, number of children in the family, or the child's grade level. Another study which examined the effect of classroom exposure to noise revealed that in one school located adjacent to an elevated railway, students whose classrooms faced the track scored significantly lower on reading tests than did those whose rooms were on the opposite side of the school.

A comparable situation relating to decreased efficiency on the job faces millions of American workers. Several years ago the National Institute for Occupational Safety and Health conservatively estimated that over 2.5 million U.S. industrial workers were exposed to harmful levels of noise. Aside from the health aspects of this exposure, noise hinders the performance of tasks requiring high levels of accuracy (total quantity of work, as opposed to quality, does not appear to suffer appreciably). Very loud or sporadic noises seem to be the most disruptive, distorting time perception, increasing the variability in work performance, disturbing concentration, and making it more difficult to remain alert. The effects of working all day in a noisy environment frequently carry over into domestic life, making the worker more prone to aggravation and frustration when he or she comes home. Pent-up stress from daytime noise exposure may prevent relaxation in the evening, and if the home environment is noisy also, the worker may remain in a constant state of tension and irritability (3).

11.6 OCCUPATIONAL NOISE EXPOSURE

The Occupational Safety and Health Administration (OSHA) has established standards for occupational noise exposure. Determining occupational noise exposure has been defined by OSHA under 29 CFR Part 1926.52 as codified in the *Code of Federal Regulations* under the U.S. Department of Labor. The established procedure to determine occupational noise exposure are detailed as follows.

Protection against the effects of noise exposure are to be provided when the sound levels exceed those shown in Table 11.3 when measured on the A-scale of a standard sound level meter at slow response. When employees are subjected to sound levels exceeding those listed in Table 11.3, feasible administrative or engineering controls are to be utilized. If such controls fail to reduce sound levels within the levels of the table, personal protective equipment is to be provided and used to reduce sound levels within the levels of the table. In all cases where the sound levels exceed the values shown in Table 11.3, a continuing, effective hearing conservation program will be administered. Exposure to impulsive or impact noise should not exceed 140 dB peak sound pressure level (6).

If the variations in noise level involve maxima at intervals of 1 second or less, the noise is considered to be continuous. When the daily noise exposure is composed of two or more periods of noise exposure of different levels, their combined effect should be considered, rather than the individual effect of each. Exposure to different levels for various periods of time are computed according to

Table 11.3   Permissible noise exposure

Duration per Day (hr)	Sound Level (dB), Slow Response
8	90
6	92
4	95
3	97
2	100
1.5	102
1	105
0.5	110
0.25 or less	115

the following formula.

$$F_e = (T_1/L_1) + (T_2/L_2) + \cdots + (T_n/L_n)$$

where F_e is the equivalent noise exposure factor. T is the period of noise exposure at any essentially constant level, and L is the duration of the permissible noise exposure at the constant level (from Table 11.3). If the value of F_e exceeds unity (1), the exposure exceeds permissible levels.

A sample computation showing an application of this formula is as follows. An employee is exposed at these levels for these periods:

110 db 1/4 hr
100 db 1/2 hr
 90 db 1 1/2 hs

$$F_e = \frac{0.25}{0.5} + \frac{0.5}{2} + \frac{1.5}{8}$$

$$= 0.500 + 0.25 + 0.188$$

$$= 0.938$$

Since the value of F_e does not exceed unity, the exposure is within permissible limits.

11.7   NOISE ABATEMENT

Noise abatement measures are under the jurisdiction of local government except for occupational noise abatement efforts. In today's mechanized world it is virtually impossible for an active person to avoid exposure to potentially harmful sound levels. For this reason, hearing specialists now recommend that individuals get into the habit of wearing protectors, not only to guard against hearing loss but to reduce the annoying effects of noise.

There are two basic types of hearing protectors: muffs worn over the ears and inserts worn in the ears. Well-fitting protective muffs are more effective, but inserts also do a good job if properly fitted. Since ear canals are rarely the same size, inserts should be separately fitted for each ear. Cotton plugs are virtually useless. Protective muffs should be adjustable to provide a good seal around the ear, proper tension of the cups against the head, and comfort. Both types of protectors are available at many sports stores and drugstores. They are well worth the small inconvenience they cause for the wearer. Hearing protectors are recommended at work, especially if employed in the construction, lumber, mining, steel, or textiles industries. They are also recommended during recreational and home activities such as target shooting and hunting, power-tool use, lawn mowing and sowmobile riding.

There are several factors relating to noise to consider when choosing a new house or apartment. Be aware of major noise sources near any residence you are considering, including airport flight paths, heavy truck routes, and high-speed freeways. Ask prospective neighbors if there is a local noise problem. When buying a home, check the area zoning master plan for projected changes. In some places, you can't get FHA loans for housing in noisy locations. Use the HUD "walk-away test." By means of this method, a couple can assess background noise around a house. Simply have one person stand with some reading material at chest level and begin reading in a normal voice while the other person slowly backs away. If the listener cannot understand the words closer than 7 ft, the noise is clearly unacceptable. At 7–25 ft, it is normally unacceptable; at 26–70 ft, normally acceptable; and over 70, clearly acceptable.

Look for wall-to-wall carpeting, especially in the apartment above and in the corridors. Find out about the wall construction. Staggered-stud interior walls provide better noise control. Studs are vertical wooden supports located behind walls. Staggering them breaks up the pattern of sound transmission. Check the electrical outlet boxes. If they are back-to-back, noise will pass through the walls. Check the door construction. Solid or core-filled doors with gaskets or weather stripping provide better noise control. Make sure sleeping areas are well away from rooms with noise-producing equipment. Check the heating and air-conditioning ducts. Insulation makes them quieter.

Several helpful hints to make a home quieter include the use of carpeting to absorb noise, especially in areas where there is a lot of foot traffic. Hang heavy drapes over windows closest to outside noise sources. Put rubber or plastic treads on uncarpeted stairs. Also, covered stairs are safer than uncovered stairs, as the covering provides resistance to slipping. Use upholstered rather than hard-surfaced furniture to deaden noise. Install sound-absorbing ceiling tile in the kitchen. Wooden cabinets will vibrate less than metal ones. Use a foam pad under blenders and mixers. Use insulation and vibration mounts when installing dishwashers. Install washing machines in the same room with heating and cooling equipment, preferably in an enclosed space away from bedrooms. If you use a power mower, operate it at reasonable hours. The slower the engine setting, the quieter it will operate. When listening to a stereo, keep the volume down. Place window air conditioners where their hum can help mask objectionable noises. However, try to avoid locating them facing your neighbor's bedrooms. Use caution in buying children's toys that can make intensive or explosive sounds. Some can cause permanent ear injury. Compare, if possible, the noise outputs of different makes of an appliance before making your selection.

Noise problems are worse in dwellings where the construction is of a type that relies on thinner and lighter materials. These materials do not effectively block noise and vibration from outside or between rooms, and in some cases actually can amplify sound. Poor siting also may add to the noise problem. Housing developments often are built near the landing pattern of major airports, and apartment houses located near high-speed highways. Poor housing placement is on the increase in many communities across the country. To cope with the problem of lightweight construction and poor planning, the U.S. Department of Housing and Urban Development (HUD) has developed "Noise Assessment Guidelines" to aid in community planning, construction, modernization and rehabilitation of existing buildings. In addition, the Veterans Administration requires disclosure of information to prospective buyers about the exposure of existing VA-financed houses to noise from nearby airports.

For the community, the control of noise around the home involves proper land use, zoning, and building regulations. For the construction industry it means better engineering. For the homeowner, it means insistence on quieter appliances and equipment, and the initiative to create less noisy dwellings. One of the most effective actions residents can take regarding noise in the home is to make appliance dealers and manufacturers aware of their desire for quieter products and to influence their local governments to enact and enforce the necessary building codes. Beyond that, persons with noisy appliances and equipment should try to schedule use of these items when the least amount of disturbance is created. Discretion should be used in controlling the volume of TVs and stereos. Hearing protectors should be worn when operating very noisy equipment such as chain saws and power lawnmowers.

The EPA has under preparation a model building code for various building types. The code will spell out extensive acoustical requirements and will make it possible for cities and towns to regulate construction in a comprehensive manner to produce a quieter local environment.

The Noise Control Act of 1972 provides the EPA with authority to require labels on all products, both domestic and imported, that generate noise capable of adversely affecting public health or welfare and on those products sold wholly or in part for their effectiveness in reducing noise (such as acoustic tile, some types of carpeting, certain building materials, etc.) The EPA is initiating a study to rate home appliances and other consumer products by the noise generated and the impact of the noise on users and other persons normally exposed to it. Results of the study will be used to determine whether noise labeling or noise emission standards are necessary (6).

11.8 SUMMARY

1. Noise pollution is traditionally not placed among the top environmental problems facing the nation; however, it is one of the more frequently encountered sources of pollution in everyday life.

2. Sound travels in waves through the air like waves through water.

3. Sound waves enter the auditory canal, causing the eardrum to vibrate.

4. The sources of noise pollution are extremely diverse and are constantly

increasing as more and more noise-generating products become available to consumers.

5. An estimated 20 million Americans are exposed to noise that poses a threat to their hearing. Everyone at some time or another has experienced the effects of noise pollution.

6. The Occupational Safety and Health Administration (OSHA) has established OSHA standards for occupational noise exposure.

7. In today's mechanized world it is virtually impossible for an active person to avoid exposure to potentially harmful sound levels.

REFERENCES

1. U.S. EPA, *Noise and its Measurements*, OPA 22/1, January 1981.
2. U.S. EPA, *Noise: A Health Problem*, Office of Noise Abatement and Control, August 1978.
3. A. Nadakavukaren, *Man & Environment, A Health Perspective*, 3rd ed.
4. C. Perham, "The Sound of Silence" (A-107), *EPA Journal* 5(9), October 1979.
5. J. Stansbury, "Noise in the Workplace" (A-107), *EPA Journal* 5(9), 1979.
6. U.S. EPA, *Occupational Noise Exposure*, 29 CFR Part 1926.52, revised as of July 1, 1991.

WATER POLLUTION
MANAGEMENT

12

WATER POLLUTION MODELING AND CONTROL

12.1 INTRODUCTION

The basic objective of the field of water quality engineering is the determination of the environmental controls that must be instituted to achieve a specific environmental quality objective. The problem arises principally from the discharge of the residues of human and natural activities that result, in some way, in an interference of a desirable use of water. What constitutes a desirable use is, of course, a matter of considerable discussion and interaction between the social–political environment and the economic ability of a given region to live with or otherwise improve its water quality.

The principal desirable uses of water are:

1. Water supply — municipal and industrial.
2. Recreational — swimming, boating, and aesthetics.
3. Fisheries — commercial and sport.
4. Ecological balance.

Table 12.1 shows the major water quality problems that have been perceived through interferences with various uses of the water and subsequently confirmed through water quality sampling and analysis. For example, it has long been noted that the problem of low dissolved oxygen in a stream interferes with the fish life of that stream, and the manifestation of that interference is indicated by fish kills in the stream and associated aesthetic nuisances. The basic purpose of water quality engineering is to diagnose the type of problem shown in Table 12.1, relate that problem to the water use interference and the manifestation of that interference, and then make a judgment on which water quality variables need to be controlled and the means available for control.

Table 12.1   Principal pollution problems, affected uses, and associated water quality variables

Manifestation of Problem	Water-Use Interference	Water Quality Problem	Water Quality Variables

1. Fish kills
    Nuisance odors,
    H_2S "Nuisance"
    organisms
    Radical change in
    ecosystem
	Fishery
	Recreation
	Ecological health
		Low Do (dissolved
		    oxygen)
			BOD
			NH_3, organic N
			Organic solids
			Phytoplankton
			DO

2. Disease transmission
    Gastrointestinal
    disturbance, eye
    irritation
	Water supply
	Recreation
		High bacterial levels	Total coliform bacteria
			Fecal coliform bacteria
			
			Fecal streptococci
			Viruses

3. Tastes and odors—
    blue green algae
    Aesthetic beach
    nuisances—algal
    mats "Pea soup"
    Unbalanced ecosystem
	Water supply
	Recreation
	Ecological health
		Excessive plant
		    growth
		    (Eutrophication)
			Nitrogen
			Phosphorus
			Phytoplankton

4. Carcinogens in water
    supply
    Fishery closed—
    unsafe toxic levels
    Ecosystem upset:
    mortality,
    reproductive
    impairment
	Water supply
	Fishery
	Ecological health
		High toxic chemical
		    levels
			Metals
			Radioactive substances
			Pesticides
			Herbicides
			Toxic product
			    chemicals

In general then, the role of the water quality engineer and scientist is to analyze water quality problems by dividing the problem into its principal components:

1. Inputs—that is, the discharge of residue into the environment from man's and nature's activities.
2. The reactions and physical transport (that is, the chemical and biological transformations) and water movement that result in different levels of water quality at different locations in time in the aquatic ecosystem.
3. The output (that is, the resulting concentration of a substance, such as dissolved oxygen or nutrients, at a particular location in the water body) during a particular time of the year or day.

The inputs are discharged into an ecological system such as a river, lake, estuary, or oceanic region. As a result of chemical, biological, and physical phenomena (such as bacterial biodegradation, chemical hydrolysis, and physical sedimentation), these inputs result in a specific concentration of the substance in the given water body. Concurrently, through various mechanisms of public hearings, legislation, and evaluation, a desirable water use is being considered or has been established for the particular region of the water body under study. Such a desirable water use is translated into public health and/or ecological standards, and such standards are then compared to the concentration of the substance resulting from the discharge of the residue. This desired versus actual comparison may result in the need for an environmental engineering control, if the actual or forecasted concentration is not equal to that desired. Environmental engineering controls are then instituted on the inputs to provide the necessary reduction to reach the desired concentration. The presentation of various environmental engineering control alternatives to reach the same objective has a central role in the decision-making process of water quality management.

The intent of this chapter is not to make the reader an expert in the field of water quality mathematical modeling. It is intended as an introduction to the array of variables present in a model of a specific body of water. For further information, the reader is referred to *Principles of Surface Water Quality Modeling and Control* by Thomann and Mueller (1).

12.2 WASTE LOAD ALLOCATION PRINCIPLES

The central problem of water quality management is the assignment of allowable discharges to a water body so that a designated water-use and quality standard is met using basic principles of cost-benefit analysis. There are several components to the overall waste load allocation (WLA) problem for dissolved oxygen, including the determination of desirable water use standards, the relationship between load and water quality, and selection of projected conditions. It is generally not sufficient to simply make a scientific engineering analysis of the effect of waste load inputs on water quality. The analysis framework must also include economic impacts which, in turn, must also recognize the sociopolitical constraints that are operative in the overall problem context.

The principal steps in the WLA process are:

1. A designation of a desirable water use or uses—for example, recreation, water supply, and agriculture.

2. An evaluation of water quality criteria that will permit such uses.

3. The synthesis of the desirable water use and water quality criteria to a water qualitystandard promulgated by a local, state, interstate, or federal agency.

4. An analysis of the cause–effect relationship between present and projected waste load inputs and water quality response through use of: a. Site-specific field data or data from related areas and a calibrated and verified mathematical model. b. A simplified modeling analysis based on the literature, other studies, and engineering judgment.

5. A sensitivity analysis and a projection analysis for achieving water quality standards under various levels of waste load input

6. Determination of the "factor of safety" to be employed through, for example, a set-aside of reserve waste load capacity.

7. For the residual load, an evaluation of: a. The individual costs to the

dischargers. b. The regional cost to achieve the load and the concomitant benefits of the improved water quality.

8. Given all of the above, a complete review of the feasibility of the designated water use and water quality standard.

9. If both are satisfactory, a promulgation of the waste load allocated to each discharger.

Within the above framework, it is assumed that a calibrated and verified water quality model is available. There are several points at which careful judgments are required to provide a defensible WLA. For example, the determination of design conditions including flow and parameters must be evaluated for a WLA. The specification or projection of flow and parameter conditions under a given design event is a most critical step and is a blend of engineering judgment and sensitivity analysis.

There are several other issues that must be addressed to answer the basic WLA question, which is: "What is the permissible equitable discharge of residuals that will not exceed a water quality standard?" These questions include:

1. What does "permissible" mean? Is "permissible" meant in terms of maximum daily load, 7-day average load, or 30-day average load?

2. What does "exceed" mean? Does it mean "never" or 95% of the time and for all locations?

3. What are the design conditions to be used for the analysis?

4. How credible is the water quality model projection of expected responses due to the WLA — that is, what is the "accuracy" of the model calculations and how should the level of the analysis be reflected, if at all, in the WLA?

From a water quality point of view, the basic relationship between waste load input and the resulting response is given by a mathematical model of the water system. The development and applications of such a water quality model in the specific context of a WLA involve a variety of considerations, including the specifications of parameters and model conditions.

The principal inputs can be divided into two broad categories: point sources and non-point sources. The point sources are those inputs that are considered to have a well-defined point of discharge which, under most circumstances, is usually continuous. A discharge pipe or group of pipes can be located and identified with a particular discharger. The two principal point source groupings are: (a) municipal point sources that result in discharges of treated and partially treated sewage [with associated bacteria and organic matter, biochemical oxygen demand (BOD), nutrients, and toxic substances] and (b) industrial discharges which also result in the discharge of nutrients, BOD, and hazardous substances.

The principal non-point sources are agricultural, silviculture, atmospheric, urban and suburban runoff, and groundwater. In each case, the distinguishing feature of the non-point source is that the origin of the discharge is diffuse. That is, it is not possible to relate the discharge to a specific well-defined location. Furthermore, the source may enter the given river or lake via overland runoff as in the case of agriculture or through the surface of the land and water as an atmospheric input. The urban and suburban runoff may enter the water body through a large number of smaller drainage pipes not specifically designed for the carriage of wastes but for the carriage of storm runoff. In some instances in urban runoff, the discharge may be a large pipe draining a similarly large area. Other non-point sources include pollution due to groundwater infiltration, drainage from abandoned mines and construction activities, and leaching from land disposal of solid wastes.

In addition to the fact that the non-point sources result from diffuse locations, non-point sources also tend to be transient in time, although not always. For example, agriculture, silviculture, and urban and suburban runoff tend to be transient, resulting from flows due to precipitation at various times of the year. Other inputs such as the atmospheric input and leaching of substances out of solid waste disposal sites are more or less continuous.

One of the more important aspects of water quality engineering is the determination of the input mass loading — that is, the total mass of a material discharged per unit time into a specific body of water. The mass input depends on both the input flow and the input concentration and, for atmospheric inputs, includes the airborne deposition on an area-wide basis.

In any given problem context, reductions in waste inputs may be required to mitigate existing water quality violations. The choice of reducing point or non-point sources will depend not only on the feasibility and economics of existing engineering controls but also on the relative magnitudes of the sources. Early identification of the major waste sources will allow concentration of available resources on the most significant inputs. To this end, published values of point and non-point sources may be used in a preliminary screening effort, although caution should be exercised because of the wide variability in such data. Data from areas similar to the study areas should be used as much as possible.

Point Source Mass Loading Rates

For those defined sources with continuous flow, the input load is determined by the concentration of the input and the input flow. It should be recognized that effluents from sources such as waste treatment plants can vary significantly over time. Although average values of various constituents can be specified, the temporal variability of the load often needs to be incorporated either directly or indirectly into the analysis. There may be significant diurnal variation in both flow and quality constituent. In streams that are dominated by the effluent, these variations may be reflected in downstream water quality.

Tributary Mass Loading Rates

For some sources, the flow is continuously measured (as for example in the tributary to a larger river or lake), but estimates of the concentration of the water quality parameter are only available at certain intervals. If the waste load is determined only at those times when both flow and concentration are available, the actual loading from the source may be significantly underestimated. If, for example, the concentration is not measured during times of peak runoff, a major component of the load may be missed. Therefore, some other approach is required to estimate the average load.

One approach is to plot the available concentration data as a function of the river flow, usually both as logarithms, and estimate the relationship between concentration and flow. For each flow where the concentration was not measured, this relationship is used to estimate the concentration and, hence, the load. The difficulty with this approach is that the relationship may depend on whether the concentration was measured when the flow was rising (when bottom material is suspended and tributaries and land runoff are discharging into the river) or when the flow was declining.

Intermittent Mass Loading Rates

Loading for intermittent sources depends on a number of factors that may influence both the flow and the concentration. The flow from urban runoff is usually highly transient, resulting from variable precipitation, so at times there is no load being discharged. For discharges from combined urban sewers, therefore, several input loads can be estimated:

1. Equivalent annual (or other interval) loading rate (comparable to continuous load).
2. Average load discharged per event of overflow.
3. Distribution of load within an event of overflow.

A statistical procedure has been developed which allows one to make an estimate of the mean load and the expected high load events with a given probability (2).

12.3 MODELING OF RIVERS AND STREAMS

For anyone who has been thrilled by the excitement of canoeing on a river or quietly paddling on a stream, it is quite clear that the distinguishing feature of describing water quality in rivers and streams is the movement of the water, more or less rapidly, in a downstream direction. From a water quality engineering point of view, rivers have been studied more extensively and longer than other bodies of water, probably reflecting the fact that many people live close to or interact with rivers and streams. Hydrologically then, our interest in rivers begins with the analysis of river flows. The magnitude and duration of flows, coupled with the chemical quality of the waters, determine (to a considerable degree) the biological characteristics of the stream. The river is an extremely rich and diverse ecosystem, and any water quality analysis must recognize this diversity. The river system may therefore be considered from the physical, chemical, and biological perspective. The principal physical characteristics of rivers that are of interest include:

1. Geometry: width, depth.
2. River slope, bed roughness, "tortuosity".
3. Velocity.
4. Flow.
5. Mixing characteristics (dispersion in the river).
6. River water temperature.
7. Suspended solids and sediment transport.

For river water quality management, the important chemical characteristics are:

1. Dissolved oxygen (DO) variations, including associated effects of oxidizable nitrogen on the DO regime.
2. pH, acidity, alkalinity relationships in areas subjected to such discharges—for example, drainage from abandoned mines.
3. Total dissolved solids and chlorides in certain river systems—for example, natural salt springs in the Arkansas–White–Red River basins.
4. Chemicals that are potentially toxic.

Biological characteristics of river systems that are of special significance in water quality studies are:

1. Bacteria and viruses.
2. Fish populations.
3. Rooted aquatic plants.
4. Biological slimes; *Sphaerotilus.*

As with all water quality analyses, the objective in river water quality engineering is to recognize and quantify, as much as possible, the various interactions between river hydrology, chemistry, and biology.

The study of river hydrology includes many factors of water movement in river systems, including precipitation, stream flow, droughts and floods, groundwater, and sediment transport (3). The most important aspects of river hydrology are the river flow, velocity, and geometry. Each of the characteristics are used in various ways in the water quality modeling of rivers. Measurements of river flows focus on those times when the flow is "low" due to the factor of dilution. If a discharge is running into a stream, then conditions will probably be most critical during the times when there is less water in the channel. The flow at a given point in a river will depend on:

1. Watershed characteristics such as the drainage area of the river or stream basin up to the given location.
2. Geographical location of the basin.
3. Slope of the river.
4. Dams, reservoirs, or locks which may regulate flow.
5. Flow diversions into or out of the river basin.

The flow in the river can be obtained by several methods. A direct measurement of river velocity and cross-sectional area at a specific location can give an estimate of the flow at that location and time. River velocities are measured either directly by current meters or indirectly by tracking the time for objects in the water to travel a given distance. Since the velocity of a river varies with width and depth due to frictional effects, the mean vertical velocity must be estimated.

With an estimate of the velocity at hand, a first approximation can be made to the time of travel between various points on the river. For example, the travel time, in days, to cover a given distance, in miles, can be estimated by knowing the velocity, in miles per day. This relationship ignores dispersion or mixing in the river and any effects of "dead" zones such as deep holes or side channel coves.

With the flow and hydraulic properties of the river system defined and the estimates of these properties at hand, some first approaches to describing the discharge of residual substances into rivers and streams can be examined. Such residuals may include discharges from waste treatment plants, from combined sewer overflows, or from agricultural and urban runoff. First, consider point sources — that is, those sources that enter the river from a fixed discharge point such as an effluent pipe or tributary stream.

The basic idea in describing the discharge of material into a river is to write a mass balance equation for various reaches of the river. Begin by examining the mass balance right at the point of discharge. The first key assumption is that the river is homogeneous with respect to water quality variables across the river (laterally) and with respect to depth (vertically).

Consideration of how to compute the distance from an outfall to complete mixing is a separate, more complicated topic. However, the order of magnitude of the distance from a single point source to the zone of complete mixing is obtained by knowing the average stream velocity, width, and depth (4).

The second key assumption to be made in this analysis of water quality in streams is that there is no mixing of water in the longitudinal downstream direction. Each element of water and its associated quality flows downstream in a unique and discrete fashion. There is no mixing of one parcel with another due to dispersion or velocity gradients. This condition is referred to as an *advective system*, a *plug flow system*, or *maximum gradient system*. A pulse discharge retains its identity, and any spreading of the pulse is assumed to be negligible. In actual streams and rivers, however, true plug flow is never really reached. Lateral and vertical velocity gradients, "dead" zones in the river (coves, deep holes, backwater regions), produce some mixing and retardation of the material in a stream. For many purposes, however, the nondispersive assumption is a good one.

Mass Balance at Discharge Point

For the mass balance at the outfall, the principal equation is:

Mass rate of substance upstream + mass rate added by outfall = mass rate of substance immediately downstream from outfall assuming complete mixing

A similar equation can also be written for the balance of the flows, that is, flow continuity:

Flow rate upstream + flow rate added by outfall = flow rate immediately downstream from the outfall

The upstream conditions of flow and concentration are often known or can be measured, and typically some information is available on the effluent conditions. For example, the discharge may be a proposed industrial treatment plant, and an estimate is available of the flow and concentration of the waste substance to be expected upon completion of the plant. Interest then centers on estimating the concentration of the substance in the river at the outfall after mixing of the effluent with the upstream concentration. The downstream concentration is thus dependent on the upstream and downstream flows and the concentrations of the upstream and effluent inputs. If the upstream concentration of the substance is zero, then the downstream concentration is equivalent to the effluent concentration reduced by the ratio of influent flow to total river flow. This is a dilution effect. If the river flow is increased for a given mass discharge, this will result in a decreased concentration in the river due to the diluting effect of the increased river flow. These two relationships contain a considerable amount of information regarding water quality engineering controls, including reduction of the mass loading and/or upstream concentration of the substance (e.g., a toxic substance, bacteria, or BOD) at the outfall. Although the concentration at the outfall may be reduced due to increased river flow, the situation becomes more complex as one proceeds downstream from the point of discharge.

Water Quality Downstream of Point Source

Attention will now be given to the downstream behavior of water quality. Consider first a substance that is conserved — that is, there are no losses due to chemical reactions or biochemical degradations. Such substances may include, for example, total dissolved solids, chlorides, and certain metals during times of the year when transport is in the dissolved form. The assumption of complete mixing will be retained and applied to any new entering point sources and tributaries. In order to simplify the analysis, it is also assumed that there is no change of flow between any point waste inputs or tributaries; flow into or out of the river due to groundwater effects is therefore excluded. Finally, it is assumed that the magnitudes of the waste inputs and flows are temporarily invariant — the so-called steady-state condition.

Because the substance is conservative, there is no change in concentration between tributaries or waste inputs. The concentration changes only at the entrances of new sources of the substance, with the associated changes in flow. Therefore, the concentration at each tributary or waste input is computed, and the concentration remains at that level until the next tributary or waste source. Imagine a portion of a river upstream with two tributaries and a waste input. The river is divided into three reaches, and the upstream reach above the first tributary is considered as the upstream boundary reach. At the first tributary, the concentration is reduced due to the dilution effect of the tributary, where the tributary is assumed to have a negligible concentration of the substance in question. The concentration remains at this level until the next tributary. The concentration at the entrance of the secondary tributary is calculated and is greater than the upstream concentration. The new level is held constant (no change is possible because of the conservation assumption) until the waste input, where once again the outfall mass balance equation is applied. In this way, the complete downstream profile can easily be generated. If, however, it is known or suspected that the substance is not conservative, then an additional consideration must be included.

For nonconservative substances, the substance decays with time due to chemical reactions, bacterial degradation, radioactive decay, or perhaps settling of particulates out of the water column. Many substances exhibit decay or nonconservative behavior, including oxidizable organic matter, nutrients, volatile chemicals, and bacteria. A very useful assumption is that the substance decays according to a first-order reaction — that is, the rate of loss of the substance is proportional to the concentration at any time. The mass balance equation, at steady state, is a first-order linear differential equation, which is beyond the scope of this book (5).

Streams are often subjected to sources or sinks of a substance which are distributed along the length of the stream. An example of an external source is runoff from agricultural areas, where oxygen-demanding material distributed over the bottom of the stream exemplifies an instream, or internal, source. An estimate of these sources would also be incorporated in the water quality model.

Effect of Spatial Flow Variations on Water Quality

Distributed flow changes in a stream can alter the spatial distributions of conservative and nonconservative parameters. If flow increases in a stream as a result of uncontaminated groundwater infiltration, instream concentrations will decrease because of the additional dilution. For contaminated groundwater, the instream concentration will increase—if the groundwater concentration is higher than the instream value—or decrease—if the groundwater is of lesser concentration than that instream. Decreases in inflow without removal of the contaminant, such as those caused by evaporation, will result in increased instream concentrations.

For the conservative substance, neglect of the inflow might lead an analyst to mistakenly conclude that some type of removal mechanism was operating in the stream when in fact dilution was the cause of decreasing concentrations. Similarly, for nonconservative substances, appropriate care in estimating the flow increase is required so as not to overestimate the reaction rate.

Time-Variable Analysis

For some problem contexts, it is important to be able to describe the time-variable behavior of water quality in a river downstream of an outfall or tributary input. Such problems include describing the downstream transport of a peak in a wastewater discharge load, an accidental spill of a chemical, or the day-to-day variation in water quality due to day-to-day changes in waste load inputs.

The basic principle of the time-variable response in a river or stream can be quickly seen by making the initial assumption that there is no mixing in the longitudinal direction. If there is no mixing of the water parcels that are moving downstream, then each parcel does not interact with the parcel in front of it or behind it. Each parcel of water retains its identity, and hence as noted earlier this type of condition is called plug flow. For a discharge pulse at time $t = 0$, the "slice" of water passing by the discharge pipe (assumed to mix completely in the cross section, as before) receives the pulse discharge, and the resulting concentration in the slice is simply the mass of the discharge (i.e., the pulse discharge rate times the time interval of the pulse) divided by the volume of water in the slice over the interval.

At some time interval later, say t_1, the slice of the river that contains the discharged pulse is now a distance x_1 downstream. If a person were therefore stationed at location x_1 the pulse would be seen for the same time interval only. There would be an equal concentration in the slice of water as it passed by, and there would be sharp edges to the pulse in the river. If decay were to occur in the slice of river water, then the concentration in the slice would be reduced to a degree proportional to the travel time.

Engineering Controls

There are several points at which the water quality in a system can be controlled. The initial concentration at the outfall can be controlled by:

1. Reducing the effluent concentration of the waste input by:
   a. Wastewater treatment.
   b. Industrial in-plant process control and/or "housekeeping."
   c. Eliminating effluent constituents by pretreatment prior to discharge to municipal sewer systems or by different product manufacturing for an industry.

2. Reducing the upstream concentration by upstream point and non-point source controls.

Reduction of effluent flow and/or augmentation of stream river flow reduces the initial concentration and hence may achieve water quality standards. Thus the initial concentration can be controlled by the following:

3. Reducing the effluent volume by:
   a. Reduction in infiltration into municipal sewer systems.
   b. Reduction of direct industrial discharge volumes into the sewer system.
   c. Reduction, for industry, of waste volumes through process modifications.

4. Increasing the upstream flow by low flow augmentation—that is, releases from upstream reservoir storage or from diversions from nearby bodies of water.

However, these latter two controls on effluent flow and upstream river flow also may affect the downstream transport through the velocity. A reduction in concentration at the outfall due to increased dilution from low flow augmentation may result in an increase in the concentration downstream because of an increased velocity. A reverse situation may occur if the flow is reduced. The concentration profile also depends on the decay rate. Thus, a final general control point is to:

5. Increase the environmental, instream degradation rate of the substance.

The latter control can be accomplished by a redesign of the chemical to result in a more rapid breakdown of the chemical by the natural heterotrophic bacteria in the stream. Examples include the redesign of synthetic detergents to reduce foaming and downstream transport through increased biodegradation rate. Also, the thrust in contemporary manufacture of potentially toxic chemicals is to attempt as much as possible to increase biodegradation rates so that a chemical buildup does not occur.

The choice of the mix of the above controls involves issues of

1. The costs of the controls—locally, regionally, and nationally.

2. The expected benefits of the resulting water quality in terms of water use.

3. The technological bounds (e.g., available storage for low flow augmentation) on the controls.

12.4 MODELING OF ESTUARIES, BAYS, AND HARBORS

The region between the free flowing river and the ocean is a fascinating, diverse, and complex water system: the coastal regime of estuaries, bays, and harbors. The ebb and flow of the tides, the incursion of salinity from the ocean, and the influx of nutrients from the upstream drainage all contribute to the generation of a unique aquatic ecosystem. The estuarine and wetland regions are considered to be crucial to the maintenance of major fish stocks such as the striped bass and blue fish, which to varying degrees utilize the estuarine areas as spawning and nursery grounds.

The movement of the tides into and out of estuaries, the associated density effects created by the incursion of salinity, are of particular importance in describing the water quality of such bodies of water. Many major cities are located along estuaries primarily as a result of the historical need for ready access to national and international commerce routes. For many years, such cities discharged large quantities of waste, but, because of the large volumes of the estuaries, effects were not immediately felt. Later, however, especially in the 1950s, the load on estuaries became very great, quality deteriorated rapidly, and great interest centered on the analysis of water quality in estuaries.

Several distinct zones of a river can be defined. The *tidal river* is that region of a river where there is some current reversal but sea salts have not penetrated to this region so that the tidal river is still "fresh." The *estuary* is the "drowned" part of a river system due to incursion of the ocean landward with marked current reversal and brackishness due to the saline water. (Note that if a river discharges to a large lake such as one of the Great Lakes, a condition similar to that in an estuary can be created through the incursion of lake water up into the mouth of the river.)

Dwellers by their shores have always been fascinated by the movement of water into and out of estuaries and bays along coastal regions. No coastline is without tides, and over the many centuries of observation, a great degree of regularity in the vertical and horizontal motion of water along the coast has been noticed. Tides are the movement of water above and below a datum plane, usually mean sea level. Tidal currents are the associated horizontal movement of the water into and out of an estuary. Tides and tidal currents are due to the attractive forces of the moon and sun on the waters of the earth. There is a "pulling and tugging" which raises the water at certain locations and lowers it at other locations. These motions occur on a more or less regular cyclical basis reflecting the regularity of the lunar and solar cycles (6–8). Tides are also generated in lakes and seas, produced principally by winds blowing across the lake surface and "piling up" the water, which, in turn, sets the lake into an oscillatory motion or seiche. The approximately regular motion of the lake results in a motion in lake tributaries similar to estuarine tides. The National Oceanic and Atmospheric Administration (NOAA) publishes annually tidal height and current predictions.

Tidal excursion is the approximate distance a particle will travel along the main axis of an estuary in going from low to high water, or vice versa. The *tidal flow* is the total volume of water passing a given point in the estuary over time.

The tidal currents in open offshore waters behave in an interesting fashion due to the lack of physical boundaries. The tidal current tends to move about a point in a rotary-type current. This type of current therefore will tend to move any wastes discharged offshore in an elliptical motion on which may be superimposed a net current drift. The current structure in offshore waters is therefore quite complex and is of particular significance in the transport of wastes discharged at sea.

An important characteristic of estuarine hydrology is the net flow through the estuary over a tidal cycle or a given number of cycles. This is the flow that, over a period of several days or weeks, flushes material out of the estuary and is a significant parameter in the estimation of the distribution of estuarine water quality. If the estuary is well mixed from top to bottom and from side to side (i.e., no significant gradients in velocity), then the net flow at any location in the estuary is approximately equal to the sum of the upstream external flow inputs to the estuary, assuming no other significant net hydrologic inputs or losses. This is so since it is known that the estuary is not overflowing due to the flow inputs. Therefore, this flow must, on balance, be leaving the estuary at any cross section.

Estimating the time and spatial behavior of water quality in estuaries is complicated by the effects of tidal motion. The upstream and downstream currents produce substantial variations of water quality at certain points in the estuary, and the calculation of such variation is indeed a complicated problem. Some simplifications can, however, be made which provide some remarkably useful results in estimating the distribution of estuarine water quality. The simplifications can be summarized through the following assumptions:

1. Estuary is one-dimensional (i.e., it is subject to reversals in direction of the water velocity, and only the longitudinal gradient of a particular water quality parameter is dominant).
2. Water quality is described as a type of average condition over a number of tidal cycles.
3. Area, flow, and reaction rates are constant with distance.
4. Estuary is in a steady-state condition.

12.5 MODELING OF LAKES

A major portion of our water-based recreational activities centers about the thousands of lakes, reservoirs, and other small, relatively quiescent bodies of water. In addition, these waters serve as a source of water for municipal and industrial use, including water released from reservoirs for agricultural purposes, water quality control, and fisheries management. The ecosystems and quality of lakes throughout the world are therefore of primary concern in water quality management. Indeed, for those who have drifted across a quiet lake at sundown, half attempting to fish, but marveling at the beauty and complexity of the scene, the study of the water quality of lakes is of particular interest.

Lakes and reservoirs vary from small ponds and dams to the magnificent and monumental large takes of the world such as Lake Superior, one of the Great Lakes, and Lake Baikal in Russia, the deepest lake in the world (5310 ft). The ecosystems supported by this broad range of water bodies vary from the very attractive local sport fishes such as bass and perch to the large top predators of both sport and commercial value such as lake trout and landlocked and migratory salmon. Lake Baikal, for example, is the habitat for a fresh water seal, an aquatic animal of significant ecological symbolism (9).

Limnology is the study of the physical, chemical, and biological behavior of lakes. Recreation, sport fishing (and for the larger lakes, commercial fishing), and water supply for municipal and industrial uses are all intimately related to the quality of these water bodies. The distinguishing physical features of lakes include relatively low flow-through velocities and development of significant vertical gradients in temperature and other water quality variables. Lakes therefore often become sinks for nutrients, toxicants, and other substances in incoming rivers. As a result, eutrophication is one of the more significant water quality problems of lakes.

The principal physical features of a lake are length, depth, area (both of the water surface and of the drainage area), and volume. The overall physical relationships for a lake can be summarized in area—depth and volume—depth curves. The relationship between the flow out of a lake or reservoir and the volume is also an important characteristic. The ratio of the volume to the flow represents the hydraulic detention time, that is, the time it would take to empty out the lake or reservoir if all inputs of water to the lake ceased. The hydraulic detention times, as a function of the ratio of lake drainage area to surface area for northern U.S. lakes and reservoirs range from 1 day to about 6000 days (16 years) (10). A long detention time does not necessarily indicate a large lake. A small lake with a small flow may still have a long detention time.

As with rivers and estuaries, an understanding of the water balance and circulation of lakes is of considerable importance in water quality analysis and engineering (11). A general and simple hydrologic balance equation for a given body of water is:

The net flows into and out of the lake due to river and/or groundwater flow + precipitation directly on the lake + lake evaporation = the change in the lake volume over a period of time

Inflows may include surface inflow, subsurface inflow, and water imported into the lake. Outflows may include surface and subsurface outflow from the reservoir and exported water. The change in storage in the lake or reservoir may also include subsurface storage or "bank" storage of water. In determining the hydrologic balance of a lake, the change in volume and surface inflow and outflow can usually be measured easily. Precipitation can also be measured without difficulty except for large lakes, where it must be estimated for the open water. The remaining unknowns include subsurface water movements and evaporation.

12.6 INDICATOR BACTERIA, PATHOGENS, AND VIRUSES

The exploration of an individual water quality problem begins with an investigation of the impact of bacteria and the organisms which may cause communicable diseases. This is appropriate because:

1. The kinetics of bacterial die-away are usually considered to be first-order.
2. It is the oldest of the water pollution problems in the discovery of the links between contaminated water and communicable disease.
3. The problem is still quite relevant today and is manifested in continuing high

levels of waterborne diseases in some countries, closing of bathing beaches, and restrictions on water consumption.

The transmission of waterborne diseases (e.g., gastroenteritis, amoebic dysentery, cholera, and typhoid fever) has been a matter of concern for many years. The impact of high concentrations of disease-producing organisms on water uses can be significant. Bathing beaches may be closed permanently or intermittently during rainfall conditions when high concentrations of pathogenic bacteria are discharged from urban runoff and combined sewer overflows. Diseases associated with water used for drinking purposes continue to occur. Data were collected regarding the average annual number of waterborne disease outbreaks in the United States for the period 1920–1975. The outbreaks for 1971–1975 involved a total of 27,829 people and included acute gastrointestinal illness, hepatitis A, shigellosis, giardiasis (caused by the parasite *Giardia*), and four cases of typhoid fever (12). Virtually all of the population affected used water from municipal or semipublic systems.

The modes of transmission of pathogens are (a) ingestion of contaminated water and food and (b) exposure to infected persons or animals. Infections of the skin, eyes, ears, nose, and throat may result from immersion in the water while bathing. The water uses impacted by pathogenic bacteria, viruses, and parasites are

1. Drinking water: municipal, domestic, industrial, and individual supplies.
2. Primary contact recreation such as bathing and water skiing.
3. Secondary contact recreation such as boating and fishing.
4. Shellfishing such as harvesting for clams and oysters.

Measurement approaches include analysis for indicator bacterial groups that reflect the potential presence of pathogens, the pathogenic bacteria directly, viruses, and intestinal parasites. The various measures can be summarized as in Table 12.2 (13). Although there are a variety of indicators and direct enumeration

Table 12.2   Examples of communicable disease indicators and organisms

Indicators	Viruses
Bacteria	Hepatitis A
Total coliform	Enteroviruses
Fecal coliform	Polioviruses
Fecal streptococci	Echoviruses
Obligate anaerobes (*Clostridium perfringins*)	Coxsackieviruses
Bacteriophages (bacterial viruses)	
	Pathogenic protozoa and helminths
Pathogenic bacteria	**(intestinal worms)**
Vibrio cholerae	*Giardia lambia*
Salmonella species	*Entamoeba histolytica*
Shigella species	Facultatively parasitic amoebae
	(*Naegloria* and *Hartmanella*)
	Nematodes

of pathogens is possible, primary emphasis in water quality has been placed in the past on the coliform group of bacteria. This is due principally to the fact that the coliform groups meet many of the criteria for a suitable indicator organism; for example, they are easily detected by simple laboratory tests, they are generally not present in unpolluted waters, and the number of indicator bacteria tends to be correlated with extent of contamination. The National Academy of Sciences indicates that although other indicators are under study (such as the bacteriophases) (14), they should not yet be substituted for the coliform, group. Also, the World Health Organization recommends guidelines for drinking water based on the coliform group and for virological quality, which is expressed in terms of water turbidity and disinfection (15,16).

Indication Bacteria

The total coliform (TC) bacteria group is a large group of bacteria that has been isolated from both polluted and nonpolluted soil samples as well as from the feces of humans and other warm-blooded animals. This group was widely used in the past as a measure of health hazard and continues to be used in some areas. The fecal coliform (FC) bacteria group are indicative of organisms from the intestinal tract of humans and other animals. The fecal streptococci (FS) bacteria group includes several species or varieties of streptococci, and the normal habitat of these bacteria is the intestines of humans and animals.

For primary water contact recreation, increased emphasis has been placed on the FC group as the bacterial indicator group to be used for regulatory purposes. It should be noted, however, that the issue of bacteriological standards for bathing waters has been the subject of considerable controversy. The origins of such standards was sometimes on an aesthetic basis rather than on the basis of impact on public health (17). The wide range of standards and conditions for primary water contact recreation reflects a range of interpretation of the relative health risk.

The stringent condition in the allowable concentration in shellfish (TC < 70/ 100 mL and FC < 14/100 mL) is aimed at protecting the consumer of clams and oysters from communicable diseases such as hepatitis and gastrointestinal disorders. Since shellfish filter the overlying water and concentrate bacteria as part of the feeding process, the low concentration of bacteria in the water column is intended to result in an acceptable level in the organism itself.

Pathogenic Bacteria

It is clear that in spite of the contemporary concern with chemicals and other substances in water, the occurrence of diseases associated with pathogenic bacteria continues to be a problem especially on the worldwide scene. A summary of the number of notified disease cases for industrialized countries is shown in Table 12.3 (18). The principal waterborne pathogens of concern include: *Vibrio cholerae*, resulting in the disease of cholera; *Salmonella* species, causing typhoid and paratyphoid; and *Shigella* species, causing dysentery.

Determination of specific pathogenic bacteria is severely limited in practice especially for routine water quality surveys. The limitations are due principally to the high degree of expertise needed to perform the specific isolations especially in small numbers. The procedures are accurate only for *Salmonella* species (13). Thus,

Table 12.3 Notified infectious disease cases per 100,000 population

Cholera	0.0–0.03
Typhoid	0.0–9.1
Paratyphoid	0.0–2.5
Other *Salmonella*	0.3–54.0
Shigella	0.02–22.0

Source: From Bonde (18). Reprinted by permission of Elsevier Science Publishers.

the emphasis continues to be on the coliform group, the indicator organisms of most utility in water quality.

Viruses

As summarized by the National Academy of Sciences, viruses are "submicroscopic, inert particles that are unable to replicate or adapt to environmental conditions outside a living host" (13). The viruses are thus parasites, and those that are of most significance in water are the enteric viruses, that is, those that inhabit the intestinal tract. The enteric virus particles are relatively persistent in the environment. Viruses ingested from water can result in a variety of diseases, including hepatitis (a disease of continuing importance in the United States), and diseases of the central nervous system caused by the polioviruses, coxsackieviruses, and echoviruses.

Pathogenic Protozoan—*Giardia lambia*

As noted in Table 12.2, communicable diseases can also be transmitted in parasitic protozoa and helminths (intestinal worms). The eggs and cysts of the organisms may be ingested, protozoa may reproduce, and helminths mature in the human host. Amoebic dysentery (from *Entamoeba histolytica*) is a particularly severe disease in the world. Helminths include the whipworm, hookworm, and dwarf tapeworm. Of particular concern is the protozoa *Entamoeba histolytica* and *Giardia lambia*, which may survive public water treatment. In the United States in 1974, the largest number of cases in a waterborne disease outbreak was due to *Giardia lambia* (13).

The existence of the intestinal parasite and protozoan *Giardia lambia* has been known for several hundred years (19,20). The disease caused by this organism, called *giardiasis*, is characterized by (a) diarrhea, nausea, and intestinal cramps in the chronic or low-level stage and (b) vomiting and anorexia at the acute stage. While normally associated with travelers from abroad or with hikers and campers drinking from streams, there is an apparent increase in the incidence of the disease even in protected supplies and systems. From 1965 to 1977, a total of 7009 cases of giardiasis were reported in 23 waterborne outbreaks. Although there is uneasiness over the increased incidence of this disease, most of the outbreaks appear to be associated with individual breakdowns of water supply systems.

Inputs of Organisms

The principal sources of organisms associated with waterborne communicable disease are point sources from domestic, municipal, and some industrial sources; combined sewer overflows; runoff from urban and suburban land through separate sewers; and municipal waste sludges disposed of on land or in water bodies. Variations in point source inputs can be expected on a diurnal, weekly, and seasonal basis depending on the water-use patterns of the community (21). The variation of an order of magnitude can be noted within a day.

Extensive bacteriological data on discharges from both combined and separate sewers have been collected as a result of increased concern about the relative impact of these sources on bacteriological quality. For example, studies were conducted on the effects of bypasses during 1964 in the Detroit and Ann Arbor, Michigan areas and in Syracuse, New York (22,23). The order of magnitude of the results bears testimony to the bacteriological problem created by combined sewer overflows. The values of coliforms in some instances equaled the order expected in raw sewage. The maximum observed density of TC in Detroit was 160×10^6 organisms/10 mL. Across 25 cities, a geometric mean of 6×10^6/100 mL of TC bacteria has been reported (24).

Combined sewer overflows therefore can contribute high waste loads (relative to dry weather loads), but for short periods of time and at randomly distributed intervals. This "impulse"-type load can be a point source, or the load can be almost uniformly distributed along the length of a stream or estuary where in some large cities bypass locations exist at every block. The short-term nature of this type of load requires particular attention in the mathematical modeling of the effects of combined sewer discharges on water quality.

The increasing urbanization of the environment has also affected the quality of runoff that is physically separated from the sanitary sewerage system. Field sampling and analysis of the water quality characteristics of separate storm water runoff have been conducted in several cities (22). A geometric mean for 20 cities of 3×10^5/100 mL of TC has been reported by the U.S. EPA representing the discharge from storm sewers and unsewered areas (24).

Organism Decay Rate

The survival, fate, and distribution of bacteria and other organisms in natural waters depend on the particular type of water body (i.e., stream, estuary, lake) and associated phenomena that influence the growth, death, and other losses of the organisms. The factors that influence the kinetic behavior of the communicable disease organisms after discharge to a water body are sunlight, temperature, salinity, predation, nutrient deficiencies, toxic substances, settling of the organism population after discharge, resuspension of particulates with associated sorbed organisms, and aftergrowth—that is, the growth of organisms in the body of water. These factors may be present in varying degrees depending on the specific situation. The resultant distribution of the organism concentration will then reflect the net decay (or increase) of the organism as a function of location in the body of water.

Sediment may be a significant source of organisms. It has been known for quite some time that organisms in the microscopic range can adhere to particles dispersed in water and wastewater. Thus, the discharge of bacteria and viruses to natural waters may result in the sorption of such organisms to particles, settling into bottom sediments along with subsequent resuspension of contaminated particles. Microorganisms can apparently survive in the sediments for longer periods of time than in the overlying water column. Since the sediment may include large concentrations of microorganisms, the resuspension of such sediment and subsequent desorption may be an important source of contamination in the overlying waters. In the context of modeling, the bacteria in the sediment can be considered as a distributed source of organisms.

Environmental Controls

There are three broad categories for control of pathogenic bacteria, viruses, and parasites: control at the input source of the microorganism, control at the area of water use, and control of the product that is affected by contamination. Point sources of municipal wastes are the usual principal inputs of communicable disease organisms, and such inputs can be reduced by treatment of wastes without disinfection, and disinfection by chlorination, ultraviolet radiation, ozonation, and chlorine dioxide. Controls of the area of water use would include bathing restrictions on a transient basis (that is, during and after storms, resulting in combined sewer overflows), construction of dikes, and diversion structures to protect a given area. Controls of the product would include chlorination or other forms of disinfection of water used for municipal water supply, treatment plants for contaminated shellfish to allow depuration of bacteria prior to marketing, and distribution of high-quality bottled water during emergencies.

12.7   DISSOLVED OXYGEN

The problem of dissolved oxygen (DO) in surface waters has been recognized for over a century. The impact of low DO concentrations or of anaerobic conditions was reflected in an unbalanced ecosystem, fish mortality, odors, and other aesthetic nuisances. While coliform bacteria was the surrogate variable for communicable disease and public health, DO is a surrogate variable for the general health of the aquatic ecosystem.

The discharge of municipal and industrial waste of urban and other non-point-source runoff will necessitate a continuing effort in understanding the DO resources of surface waters. The DO problem can thus be summarized as the discharge of organic and inorganic oxidizable residues into a body of water, which, during the processes of ultimate stabilization of the oxidizable material (in the water or sediments) and through interaction of aquatic plant life, results in the decrease of DO to concentrations that interfere with desirable water uses.

The major components of the DO problem can be summarized as follows. The principal inputs include the biochemical oxygen demand (BOD) of municipal and industrial discharges, the oxidizable nitrogen forms, and nutrients which may stimulate phytoplankton or rooted aquatic plant growth. The nature of the aquatic ecosystem then determines the DO level through such processes as reaeration, photosynthesis, or sediment oxygen demand. For a given desirable water use, or set of uses, a determination of the DO level consistent with that use must be made and compared to the observed or predicted DO level.

The relationships of the level of DO to specific uses has been a continued subject of debate. The principal use affected is fish preservation, including survival and reproduction. This also affects recreational and commercial fishing. Much research has been conducted relating specific levels of DO to fish behavior (25–28). The U.S. EPA, based on a review of laboratory and field data on the impact of varying levels of DO in freshwater ecosystems, found the assignment of risk to the fishery to be valid and useful to decision-making. The four levels of protection or risk are:

1. *No production impairment.* Representing nearly maximal protection of fishery resources.

2. *Slight production impairment.* Representing a high level of protection of important fishery resources, risking only slight impairment of production in most cases.

3. *Moderate production impairment.* Protecting the persistence of existing fish populations but causing considerable loss of production.

4. *Severe production impairment.* For low-level protection of fisheries of some value but whose protection in comparison with other water uses cannot be a major objective of pollution control.

The U.S. EPA suggested national DO criteria at 0.5 mg/L above the "slight production impairment" values, or approximately 5.5 mg/L.

The DO problem begins, of course, with the input of oxygen-demanding wastes into a water body. In the water body itself, the sources of DO are reaeration from the atmosphere, photosynthetic oxygen production, and DO in incoming tributaries or effluents. Internal sinks of DO are oxidation of carbonaceous waste material (CBOD), oxidation of nitrogenous waste material (NBOD), oxygen demand of sediments of water body (SBOD), and use of oxygen for respiration by aquatic plants. With these inputs and sources and sinks, the following general mass balance equation for DO in a segment volume can be written as:

reaeration + (photosynthesis − respiration)

> − oxidation of CBOD, NBOD (from inputs)

> − sediment oxygen demand + oxygen inputs

> ± oxygen transport (into and out of segment)

> = the change with time of DO in a specific volume of water

This equation is applied to a specific water body where the transport and sources and sinks are unique to that aquatic system. These source–sink components can then be examined in order to develop further the mass balance equation.

There are several points at which engineering control can be utilized to improve the DO in streams, estuaries, or lakes. These points can be grouped as follows:

1. Point and non-point reduction of sources of CBOD and NBOD through reduction of effluent concentration and/or effluent flow.

2. Aeration of the effluent of a point source to improve initial value of DO.

3. Increase in river flow through low-flow augmentation to increase dilution.

4. Instream reaeration by turbines and aerators.

5. Control of SOD through dredging or other means of inactivation.

6. Control of nutrients to reduce aquatic plants and resulting DO variations.

12.8   EUTROPHICATION

Even the most casual observer of water quality has probably had the dubious experience of walking along the shores of a lake that has turned into a sickly green "pea soup." Or perhaps one has walked the shores of a slow-moving estuary or bay and had to step gingerly to avoid rows of rotting, matted, stringy aquatic plants. These problems have been grouped under a general term called *eutrophication*. The unraveling of the causes of eutrophication, the analysis of the impact of human activities on the problem, and the potential engineering controls that can be exercised to alleviate the condition have been a matter of special interest for the past several decades.

Aquatic plants can be thought of in two very broad categories: those that move freely with the water (planktonic aquatic plants) and those that remain fixed (i.e., attached or rooted in place). The first category includes the microscopic phytoplankton as well as free-floating water weeds or certain types of plants such as the blue-green algae which may float to the surface and move with the surface current. The second category includes rooted aquatic plants of various sizes and attached microscopic plants (the benthic algae). Algae therefore is an all-inclusive designation of simple plants, mostly microscopic, which includes both the free-moving plants, the phytoplankton, and the attached benthic algae. In all cases, the plants obtain their primary energy source from sunlight through the photosynthesis process.

Eutrophication is the excessive growth of aquatic plants, both attached and planktonic, to levels that are considered to be an interference with desirable water uses. The growth of a aquatic plants results from many causes. One of the principal stimulants, however, is an excess level of nutrients such as nitrogen and phosphorus. In recent years, this problem has been increasingly acute due to the discharge of such nutrients by municipal and industrial sources, as well as agricultural and urban runoff. It has often been observed that there is an increasing tendency for some water bodies to exhibit increases in the production of aquatic plants, apparently as a result of elevated levels of nutrients.

This increased production of aquatic plants has several consequences regarding water uses:

1. Aesthetic and recreational interferences — algal mats, decaying algal clumps, odors, and discoloration — may occur.

2. Large diurnal variations in dissolved oxygen (DO) can result in low levels of DO at night, which, in turn, can result in the death of desirable fish species.

3. Phytoplankton and weeds settle to the bottom of the water system and create a sediment oxygen demand (SOD), which, in turn, results in low values of DO in the hypolimnion of lakes and reservoirs as well as in the bottom waters of deeper estuaries.

4. Large diatoms (phytoplankton that require silica) and filamentous algae can

clog water treatment plant filters and result in reduced time between backwashing.

5. Extensive growth of rooted aquatic macrophytes (larger plant forms) interfere with navigation, aeration, and channel-carrying capacity.

6. Toxic algae have sometimes been associated with eutrophication in coastal regions and have been implicated in the occurrence of "red tide," which may result in paralytic shellfish poisoning.

The eutrophication problem can then be further defined as the input of organic and inorganic nutrients into a body of water which stimulates the growth of algae or rooted aquatic plants, resulting in the interference with desirable water uses of aesthetics, recreation, fish maintenance, and water supply. The growth of a proliferation of aquatic plants is a result of the utilization and conversion of inorganic nutrients into organic plant material through the photosynthesis mechanism. The fundamental driving force for the process is the incoming solar radiation. Therefore, the eutrophication of a given water body may vary depending on the geographical location of the surface water, the degree of penetration of the solar radiation to different depths, the magnitude and type of nutrient inputs, and the particulars of the water movement through flow transport and dispersion. Since the aquatic plants may vary widely in species composition, the impact of each of the aforementioned factors may differ widely. For example, some species may require significantly less light and nutrients for growth than others. Some forms may remain rooted (as noted earlier), others may be buoyant, while still other forms may sink under different physiological conditions.

In spite of the complexities, an engineering–scientific predictive modeling approach for eutrophication control can be constructed. The principles of the approach are discussed here with the emphasis initially on the phytoplankton. The basic phenomena underlying the process of phytoplankton growth in the North Temperate regions are as follows. Increasing solar radiation provides the energy source for the photosynthesis reaction. Phytoplankton biomass then begins to increase as water temperature increases; as a result, nutrients in dissolved form are utilized by the plankton. This mechanism continues until nutrients reach levels that will no longer support growth and at which the increase in phytoplankton biomass ceases. A decline is then observed, often due to zooplankton predation and often a late summer-to-early fall bloom may be observed again due to nutrient recycling. Biomass then declines as solar radiation and temperature decrease to the lower levels of late fall and early winter.

The principal variables of importance then in the analysis of eutrophication are:

1. Solar radiation at the surface and with depth.
2. Geometry of water body; surface area, bottom area, depth, volume.
3. Flow, velocity, dispersion.
4. Water temperature.
5. Nutrients.
   a. Phosphorus.
   b. Nitrogen.
   c. Silica.
6. Phytoplankton — chl *a*.

It should be recognized that the nutrients above are present in several forms in a body of water, and not all forms are readily available for uptake of the phytoplankton. Total phosphorus is composed of two principal components: the total dissolved form and the total particulate form. The dissolved form, in turn, is composed of several forms, one of which is the dissolved reactive phosphorus. This form of phosphorus is available for phytoplankton growth. The particulate phosphorus forms include inorganic soil runoff phosphorus particulates and the organic particulate phosphorus, which includes detritus and the phytoplankton phosphorus.

Total nitrogen is composed of four major components, namely, the organic, ammonia, nitrite, and nitrate forms. The latter three forms make up the total inorganic nitrogen, which is the form utilized by the phytoplankton for growth. The organic form of nitrogen represents both a dissolved and a particulate component. The particulate form, in turn, is composed of organic detritus particles and the phytoplankton.

The principal external sources of nutrient inputs are municipal wastes, industrial wastes, agricultural runoff, forest runoff, urban and suburban runoff, and atmospheric fallout. The control of the eutrophication of various water bodies can take a variety of forms. The engineering control techniques essentially aim at changing the components of the governing equations in the direction of decreasing plant biomass. As with engineering controls for other water quality problems, principal interest first centers on control of the inputs — in this case, the nutrient inputs. There are really three principal control areas: control of inputs, alteration of system kinetics, and instream treatment and flow control. For the eutrophication problem, then, the engineering techniques include reduction or elimination of external nutrient inputs; reduction of internal sources and/or concentrations and/or cycling of nutrients; acceleration of nutrient transport through the system and selective withdrawal; and direct biological and chemical control of aquatic plants. Each of these control techniques has relative advantages and disadvantages when applied to a specific problem context.

There are five means of reducing the direct input of nutrients:

1. Wastewater treatment of municipal and industrial point sources and combined sewer overflows, including phosphorus and nitrogen removal by physical, chemical, and biological treatment processes.

2. Alteration of non-point nutrient inputs through land conservation practices and changes in agricultural use of nutrients. These practices would include reduction of erosion, testing of soil before fertilizer application, reduction in winter fertilizing, and use of buffer strips.

3. Diversion of a nutrient input to a different water body where the impact is less.

4. Treatment and control of a nutrient at the inflow point, including treatment of the entire river inflow and building of upriver detention basins or storage reservoirs.

5. Modification of product to reduce generation of nutrients at the source (e.g., nonphosphate detergents).

12.9 TOXIC SUBSTANCES

The issue of the release of chemicals into the environment in concentrations that are toxic is an area of intense concern in water quality and ecosystem analyses. Passage of the Toxic Substances Control Act of 1976 in the United States, unprecedented monetary fines, and continual development of data on lethal and sublethal effects attest to the expansion of control on the production and discharge of such substances. However, as illustrated by pesticides, the ever-present potential for insect and pest infestations with attendant effects on humans and livestock results in a continuing demand for product development. As a result of these competing goals, considerable effort has been devoted in recent years to the development of predictive schemes that would permit an a priori judgment of the fate and effects of a chemical in the environment.

The uniqueness of the toxic substances problem lies in the potential transfer of a chemical to humans with possible attendant public health impacts. This transfer can occur through two principal routes: ingestion of the chemical from the drinking water supply and ingestion of the chemical from contaminated aquatic foodstuffs (e.g., fish and shellfish) or from food sources that utilize aquatic foodstuffs as a feed. The concern of the general public lies principally in the apparent unknown long-term effects of such chemical ingestion on the whole person, both physical and psychological. Concern over potential cancers, tumors, and birth defects is real and most profound and reflects a certain sense of "mystery" associated with toxic substances. Such substances are therefore unique in water quality in the sense that the potential impact on the public health is perceived to be direct and individual-ized. This is in contrast to such water quality issues as dissolved oxygen or eutrophication where the impact is primarily on the aquatic ecosystem and the effect on the public health is minimal or nonexistent. The discharge of bacteria and parasites is the closest "conventional" issue within the toxic substances concern; however, more is known relatively about waterborne diseases than about the public health implication of toxic substances.

Numerous examples of the interferences with desirable water uses by toxic substances have been documented, such as:

1. Metal and organic chemical accumulation in fish, resulting in the banning of the fishery for human consumption.

2. Trihalomethane formation in water treatment and in finished water, resulting in a potential interference with the use of water for drinking.

3. Leaching and runoff from pesticides in agriculture and subsequent accumu-lation in the food chain or in water supplies.

4. Leaching and runoff from chemical waste disposal sites, resulting in inter-ferences with water supplies and ecosystem accumulation.

In all cases, the interest in the problem is particularly heightened by the potential transfer of chemicals to humans, with subsequent short-term or long-term impact on the public health. Particularly extreme cases have occurred such as greatly elevated cadmium and polychlorinated biphenyl (PCB) concentrations leading to severe toxic effects in humans.

The toxic substances water quality problem can therefore be summarized as the discharge of chemicals into the aquatic environment which results in concentrations in the water or aquatic food chain at levels that are determined to be toxic, in a public health sense or to the aquatic ecosystem itself, and thus may interfere with the use of the water body for water supply or fishing or contribute to ecosystem instability.

The potentially toxic substance can be grouped into several broad categories:

1. Metals — for example, mercury, cadmium, lead, or selenium — resulting from industrial activities such as electroplating, battery manufacturing, mining, smelting, and refining.
2. Industrial chemicals — thousands of chemical compounds for thousands of uses from which several broad subcategories can be identified as:
   a. Plasticizers — for example, alkyl phthalates; used in plastic products.
   b. Solvents — for example, chlorinated benzenes; used in cleaning operations.
   c. Waxes — for example, chlorinated paraffins; used in commercial and residential waxes and cleaners.
   d. Miscellaneous — for example, polychlorinated biphenyls (PCB); originally used in transformers.
3. Hydrocarbons resulting from oil refining and combustion of fuels; examples include the polycyclic aromatic hydrocarbons (PAHs) from coal conversion processes.
4. Agricultural chemicals — a large class of chemicals designed for specific environmental controls such as pesticides, insecticides, herbicides, and weedicides; examples include DDT, malathion, chlordane, atrazine, and heptachlor.
5. Radioactive substances resulting from production of nuclear energy, nuclear weapons, and production of radioactive materials for industrial use; examples include strontium-90, plutonium-239/240, and cesium-137.
6. Miscellaneous — examples include ammonia and chlorine from municipal and industrial discharges.

It should, of course, be recognized that a chemical becomes a "toxic substance" when, at some concentration, the chemical has some deleterious impact. The judgement of what constitutes a deleterious impact is usually subject to considerable debate. Nevertheless, the mere presence of a chemical does not constitute the presence of a toxic substance. Therefore, chemicals should really be viewed as "potentially toxic" at some concentration in various sectors of the water system, especially with regard to the effect that the concentration has on the ecosystem or on man. The analysis of the fate and transport is focused on calculating the expected concentration in the water body and its ecosystem and does not make any statements about the significance of the resulting concentration. The analysis of the effects of a chemical usually begins with a given range of concentrations, followed by determination of the impact of the concentration range on the ecosystem or on a specified organism of the ecosystem. Such impact may be registered in terms of mortality (acute effects) or the reproductive behavior, growth, or other physiologic characteristic of the organism (chronic effects). It is the effects of the chemical that generally form the basis for the specification of water quality criteria.

Chemical criteria and guidelines for drinking water quality have been given by the U.S. EPA (*U.S. Federal Register*, 1980 and 1984) and the World Health Organization (15,16). In the former case, risk levels are associated with the chemical where risk is measured in terms of the number of additional cases of cancer in a population of one million. Thus for carbon tetrachloride, 0.4 μg/L in treated drinking water is proposed for a 10^{-6} risk level (i.e., one additional case per 10^6 people), 0.04 μg/L for a 10^{-7} risk, and 4 μg/L for a 10^{-5} risk. Drinking water criteria and guidelines may therefore vary by orders of magnitude depending on the risk assigned to the public health impact. For some chemicals, the drinking water criteria may be more stringent than for protection of the aquatic ecosystem.

Considering next the focus of the specification of a chemical water quality criterion and standard to be the effect on the aquatic ecosystem, then several factors must be considered:

1. The concentration level that results in acute effects of mortality.

2. The concentration level that shows some specified longer-term chronic effect.

3. The tendency for the chemical to bioaccumulate in the aquatic food chain.

4. The time of exposure of the organism or ecosystem to the chemical at varying concentrations.

5. The return period of the design conditions (e.g., river flow) for use in a chemical waste load allocation.

The questions then are:

1. What are the water quality criteria for a given chemical under both acute and chronic conditions?

2. What should be the target concentration in the water body — that is, the water quality standard or the target concentration of the chemical in organisms of the ecosystem?

3. What river flow, temperature, or other environmental condition (e.g., pH) should be used in arriving at an allowable effluent concentration and mass discharge level?

There are several points at which control can be exercised to achieve a desirable water use. These control points include:

1. Control at the source, including:
   a. Waste treatment to reduce chemical input using processes such as (29):
      i. Adsorption to activated carbon.
      ii. Chemical oxidation.
      iii. Ultraviolet photolysis and catalyzation.
      iv. Air stripping.
      v. Biological treatment.
   b. Control of agricultural runoff containing pesticides and urban stormwater runoff containing a variety of chemicals.
   c. Modification of the product to:
      i. Increase biodegradation.
      ii. Decrease solids partitioning.
      iii. Decrease bioconcentration and food chain magnification.
2. Control within the water body, including:

a. Dredging of contaminated sediment with subsequent treatment and disposal of dredge spoils.

b. Covering or in-place fixation of contaminated sediments.

3. Control at point of use, including:

a. Treatment of municipal water supply before distribution.

b. Consumption of only a certain size or weight class of fish where chemical concentrations are acceptable.

12.10 SUMMARY

1. The basic objective of the field of water quality engineering is the determination of the environmental controls that must be instituted to achieve a specific environmental quality objective. The problem arises principally from the discharge of the residues of human and natural activities that result, in some way, in an interference of a desirable use of water.

2. The central problem of water quality management is the assignment of allowable discharges to a water body so that a designated water-use and quality standard is met using basic principles of cost–benefit analysis. It is generally not sufficient to simply make a scientific engineering analysis of the effect of waste load inputs on water quality. The analysis framework must also include economic impacts which, in turn, must also recognize the sociopolitical constraints that are operative in the overall problem context.

3. Water quality in rivers and streams can be controlled by reducing the effluent concentration of the waste input, reducing the upstream concentration, reducing the effluent volume, and increasing the upstream flow.

4. There are three broad categories for control of pathogenic bacteria, viruses, and parasites: control at the input source of the microorganism, control of the area of water use, and control of the product that is affected by the contamination.

5. There are several points at which engineering control can be utilized to improve the DO in streams, estuaries, or lakes. These points can be grouped as follows: point and non-point reduction of sources of CBOD and NBOD through reduction of effluent concentration and/or effluent flow; aeration of the effluent of a point source to improve initial value of DO; increase in river flow through low-flow augmentation to increase dilution; instream reaeration by turbines and aerators; control of SOD through dredging or other means of inactivation; and control of nutrients to reduce aquatic plants and resulting DO variations.

6. For the eutrophication problem, the engineering techniques include reduction or elimination of external nutrient inputs; reduction of internal sources and/or concentrations and/or cycling of nutrients; acceleration of nutrient transport through the system and selective withdrawal; and direct biological and chemical control of aquatic plants.

7. When dealing with toxic inputs, several points can be controlled to achieve a desirable water use. These include: reducing the chemical input; control of pesticides in runoff; control of urban stormwater runoff; dredging or fixation of contaminated sediment; and treatment of municipal water supply before distribution.

REFERENCES

1. R. V. Thomann and J. A. Mueller, *Principles of Surface Water Quality Modeling and Control*, Harper & Row, New York, 1987.

2. D. M. Di Toro, *Statistics of Receiving Water Response to Runoff.* Paper presented at the Urban Stormwater and Combined Sewer Overflow Impact on Receiving Water Bodies, National Conference, University of Central Florida, Orlando, FL, November 1979, 35 pp.

3. R. K. Linsley, Jr., M. A. Kohler, and J. L. H. Paulhus, *Hydrology for Engineers*, 3rd ed., McGraw-Hill, New York, 1982, 508 pp.

4. Yotsukura, 1968. As referenced in preliminary report by F. A. Kilpatrick, L. A. Martens, and J. F. Wilson, *Techniques of Water Resources Investigations of the U.S. Geological Survey, Measurement of Time of Travel and Dispersion by Dye Tracing*, Book 3, Chapter A9, 1970.

5. D. J. O'Connor, "The Temporal and Spatial Variation of Dissolved Oxygen in Streams," *Water Resources Research*, 3(l), 65–79, 1967.

6. A. Defant, *Ebb and Flow, the Tides of Earth, Air and Water,* University of Michigan Press, Ann Arbor, MI, 1958, 121 pp.

7. G. Neumann and W. J. Pierson, *Principles of Physical Oceanography,* Prentice-Hall, Englewood Cliffs, NJ, 1966, 545 pp. + xii.

8. A. Ippen, *Estuary and Coastline Hydrodynamics*, McGraw-Hill, New York, 1966, 744 pp. + xvii.

9. M. M. Kochov, *The Biology of Lake Baikal*, Academy of Sciences of the USSR, Moscow, 1962, 315 pp.

10. A. F. Bartsch and J. H. Gakstatter, "Management Decisions for Lake Systems on a Survey of Trophic Status, Limiting Nutrients, and Nutrient Loadings," in *American-Soviet Symposium on Use of Mathematical Models to Optimize Water Quality Management, 1975*, EPA-600/9-78-024, U.S. Environmental Protection Agency, Office of Research and Development, Environmental Research Laboratory, Gulf Breeze, FL, 1978, pp. 372–394.

11. R. K. Linsley, M. A. Kohler, and J. L. H. Paulhus, *Hydrology for Engineers*, McGraw-Hill, New York, 1958, 340 pp.

12. G. F. Craun, Impact of the Coliform Standard on the Transmission of Disease, in *Evaluation of the Microbiology Standards for Drinking Water,* EPA-570/9-78-OOC, C. W. Hendricks, ed., U.S. EPA, Washington, D.C., Office of Drinking Water, 1978, pp. 21–35.

13. National Academy of Sciences, *Drinking Water and Health*, Safe Drinking Water Committee, National Research Council, Washington, D.C., 1977, 939 pp.

14. P. V. Scarpino, Bacteriophage Indicators, in *Indicators of Viruses in Water and Food*, G. Berg, Ed., Ann Arbor Science Publishers, Ann Arbor, MI, 1978, pp. 201–227.

15. World Health Organization, *Guidelines for Drinking Water Quality, Vol. 1: Recommendations*, Geneva, Switzerland, 1984, 130 pp.

16. World Health Organization, *Guidelines for Drinking Water Quality, Vol. 2: Health Criteria and Other Supporting Information,* Geneva, Switzerland, 1984, 335 pp.

17. H. Salas, Historia y Applicacion de Normas Microbiologicas de Calidad de Agua en el Medio Marino, en: *Hojas de Divulgation Tecnica*, CEPIS, OPS/OMS, No. 29, Lima, Peru, 1985, 15 pp.

18. G. J. Bonde, Salmonella and Other Pathogenic Bacteria, *The Science of the Total Environment,* Vol. 18, Elsevier, New York, 1981, pp. 1–11.

19. N. D. Levine, *Giardia lambia:* Classification, Structure, Identification, in *Waterborne Transmission of Giardiasis*, Proceedings of a Symposium, EPA-600/9-79-00 1, W. Sakubowski and J. C. Hoff, Eds., U.S. EPA, ORD, ERC, Cincinnati, OH, pp. 2–8.

20. G. F. Craun, Waterborne Outbreaks of Giardiasis, *Waterborne Transmission of Giardiasis,* Proceedings of a Symposium, EPA-600/9-79-001, W. Jakubowski and J. C. Hoff, Eds., U.S. EPA, ORD, ERC, Cincinnati, OH, 1979, pp. 127–149.

21. Hazen and Sawyer, Engineers, Storm/CSO Laboratory Analysis, NYC 208 Task Report 223, Vol. 11, prepared for NYC Department of Environmental Protection, March 10, 1978.

22. W. J. Benzie and R. J. Courchaine, Discharges from Separate Storm Sewers and Combined Sewers, *J. Water Pollut. Control Fed.*, 38(3), 410–421, 1966.

23. F. J. Drehwing, A. J. Oliver, D. A. MacArthur, and P. E. Moffa, *Disinfection/Treatment of Combined Sewer Overflows, Syracuse, New York,* EPA-600/2-79-134, U.S. EPA, MERL, Cincinnati, OH, 1979, 244 pp.

24. U.S. EPA, *Areawide Assessment Procedures Manual*, Vol. 1, EPA-600/9-76/014, MERL, ORD, 1976, 6 Chapters.

25. P. Doudoroff and D. L. Shumway, *Dissolved Oxygen Requirements of freshwater Fishes,* Food and Agricultural Organization of the United Nations, FAO Fisheries Technical Paper No. 86, 291 pp.; pp. 225–275 reprinted in Warren et al. (26).

26. C. E. Warren, P. Doudoroff, and D. L. Shumway, *Development of Dissolved Oxygen Criteria for Freshwater Fish,* EPA-R3-73-019, U.S. EPA, ORD, Washington, D.C., 1973, 121 pp.

27. J. S. Alabaster and R. Lloyd, *Water Quality Criteria for Freshwater Fish*, Butterworths, London, 1980, 297 pp.

28. U.S. EPA, "Ambient Water Quality Criterion for Dissolved Oxygen," *Federal Register*, 50(76), 1985.

29. T. J. Mulligan, J. A. Mueller, and O. K. Shabbily, *Treatment of Toxic Waste waters,* presentation at May 1981 American Society of Civil Engineers' International Convention and Exposition, New York, 12 pp.

13

MUNICIPAL WASTEWATER TREATMENT

Contributing Author: Michael H. Reid

13.1 INTRODUCTION

In the United States, the treatment and disposal of wastewater did not receive much attention in the late 1800s because the extent of the nuisance caused by the discharge of untreated wastewater into the relatively large bodies of water was not severe, and large areas of land were available for disposal purposes. During the early 1900s, nuisance and declining health conditions created the need for more effective means of wastewater management. Land space was no longer readily available, especially in larger cities. This led to the planning, design, construction, and operation of a higher level of pollution technology in wastewater treatment facilities.

If untreated wastewater is allowed to accumulate, the decomposition of the organic materials it contains can lead to the production of offensive odors and gases. In addition, untreated wastewater contains numerous pathogenic (i.e., disease-causing) microorganisms, released from the human intestinal system. It contains nutrients which can stimulate the growth of aquatic life, and it may also contain toxic compounds. For these reasons, the immediate removal of wastewater from its sources of generation, followed by treatment and disposal, is imperative.

Today, not only must a wastewater treatment plant satisfy effluent quality requirements, it must also satisfy many other environmental conditions. Some of these conditions are designed to meet requirements for aesthetics and the minimization of obnoxious odors at treatment and disposal sites; to prevent contamination of water supplies from physical, chemical, and biological agents; to prevent destruction of fish, shellfish, and other aquatic life; to prevent degradation of water quality of receiving waters due to overfertilization; to prevent impairment of beneficial uses of natural water (recreation, agriculture, commerce, or industry); to protect against the spread of disease from crops grown on sewage irrigation or sludge disposal; to prevent decline in land values and, therefore, to not restrict the level of community growth and development; and to encourage other beneficial uses of effluent.

The purpose of any wastewater treatment plant is to convert the components in raw wastewater, with its inherently harmful characteristics, into a relatively harmless final effluent for discharge to a receiving body of water and to safely dispose of the solids (sludge) produced in the process. The planning, design, construction, and operation of wastewater treatment facilities is a complex problem. This chapter presents a brief review of the various factors that lead to the development and operation of a wastewater treatment plant.

13.2 WASTEWATER REGULATIONS

The Federal Water Pollution Control Act of 1972 established levels of treatment, deadlines for meeting these levels, and penalties for violators. The Act mandated that requirements for treatment performance based on percent removal of solids may no longer be used as the primary design standards. Receiving water quality criteria must also be used to define limitations of material concentrations resulting from the commingled effluent and receiving water.

Through this Act the federal government established certain minimum effluent criteria, as listed in Table 13.1, as a first step in upgrading water quality. If these minimum effluent requirements are not sufficient to result in attainment of acceptable water quality, the water quality standards of the receiving waters will dictate the level of treatment. The individual states reserve the right to impose more stringent effluent requirements than the federal requirements outlined in Table 13.1.

For wastewater treatment plant discharge to surface waters, effluent requirements are expressed in permits issued by the National Pollutant Discharge Elimination System (NPDES). The NPDES permits have generally mandated the upgrading of existing treatment plants or the construction of new plants to provide higher levels of treatment and reliability.

The Clean Water Act of 1977 contains two major provisions for wastewater solids (those solids removed during treatment): those affecting utilization and disposal, respectively. Future guidelines and regulations intended to limit the quantity and kind of toxic materials reaching the general public may set limits on the quantity and quality of sludge from wastewater treatment distributed for public use or applied to lands where crops are grown for human consumption. The

TABLE 13.1   Secondary effluent criteria for publicly owned treatment facilities

Parameter	Monthly Average	Weekly Average
BOD,[a] mg/L	30	45
Suspended solids, mg/L	30	45
Fecal coliform bacteria, number/100 mL	200	400
pH	Within range of 6.0 to 9.0	

[a]BOD = biochemical oxygen demand.

methods by which sludge is applied to land are expected to be controlled to meet aesthetic requirements. Groundwater protection will probably be required at wastewater solids disposal sites. The Resource Conservation and Recovery Act (RCRA) of 1976 requires that solid wastes, including sludge from wastewater treatment plants, be utilized or disposed of in a safe and environmentally acceptable manner. The Marine Protection, Research, and Sanctuaries Act 1977 amendments prohibited disposal of "sewage sludge" by barging to the ocean after December 31, 1981.

State and municipal requirements, which are more stringent than federal regulations, govern in many areas. This is commonly noted in regulations regarding sludge disposal. Many municipalities that apply sludge to land on which food crops are grown analyze their sludges more frequently than is required by federal guidelines, or limit sludge application rates more stringently.

13.3   WASTEWATER CHARACTERISTICS AND SOLIDS PRODUCTION

Municipal wastewater is composed of a mixture of dissolved, colloidal, and particulate organic and inorganic materials. The total amount of the substances accumulated in a body of waste water is referred to as the mass loading. The concentration of any individual component is constantly changing as a result of sedimentation, hydrolysis, and microbial transformation and degradation of organic compounds.

Wastewater characteristics are described in terms of water flow conditions and chemical quality. The characteristics depend largely on the types of water usage in the community and industrial and commercial contributions. During wet weather, a significant quantity of infiltration or inflow may also enter the municipal collection system. This will significantly change the characteristics of wastewater.

The characteristics of a wastewater may be obtained from the plant flow records and laboratory data at the municipal wastewater treatment plant. The data describing the wastewater characteristics should include minimum, average, and maximum dry weather flows, peak wet weather flows, sustained maximum flows, and chemical parameters such as biological oxygen demand (BOD), total suspended solids, total dissolved solids, pH, total nitrogen, phosphorus, and toxic chemicals. It is important that reliable estimates of the wastewater characteristics be made, since this is what the municipal wastewater treatment plant will be treating.

Quality of Wastewater

Municipal wastewater contains 99.9% water. The remaining materials include suspended and dissolved organic and inorganic matter as well as microorganisms. These materials make up the physical, chemical, and biological qualities that are characteristic of residential and industrial waters.

The physical quality of municipal wastewater is generally reported in terms of its temperature, color, odor, and turbidity. The temperature of wastewater is slightly higher than that of the water supply. The latter is an important parameter because of its effect upon aquatic life and the solubility of gases. The temperature varies slightly with the seasons, normally remaining higher than air temperature during most of the year and falling lower only during the hot summer months. The color of a wastewater is usually indicative of age. Fresh water is usually gray; septic wastes impart a black appearance. Odors in wastewater are caused by the decomposition of organic matter that produces offensive-smelling gases such as hydrogen sulfide. Wastewater odor generally can provide a relative indication of its condition.

Turbidity in wastewater is caused by a wide variety of suspended solids. Suspended solids are defined as the matter which can be removed from water by filtration through prepared membranes. Volatile suspended solids for the most part represent organics and may affect oxygen resources of the stream; however, they are not a direct measure of total organics. Suspended solids may cause the undesirable conditions of increased turbidity and silt load in the receiving water.

Chemical characteristics of wastewater are expressed in terms of organic and inorganic constituents. Different chemical analyses furnish useful and specific information with respect to the quality and strength of wastewater.

Organic components in wastewater are the most significant factor in the pollution of many natural waters. The principal groups of organic substances found in municipal wastewater are proteins (40–60%), carbohydrates (25–50%), and fats and oils (10%). Carbohydrates and proteins are easily biodegradable. Fats and oils are more stable and are decomposed by microorganisms. In addition, wastewater may also contain small fractions of synthetic detergents, phenolic compounds, and pesticides and herbicides. These compounds, depending on their concentration, may create problems such as nonbiodegradability, foaming, or carcinogenicity. The concentrations of these toxic organic compounds in wastewaters are very small. Their sources are usually industrial wastes and surface runoff.

The inorganic compounds most often found in wastewater are chloride, hydrogen ions, alkalinity-causing compounds, nitrogen, phosphorus, sulfur, and heavy metals. Trace concentrations of these compounds can significantly affect organisms in the receiving water through their growth-limiting characteristics. Algae and macroscopic plant forms are capable of using inorganics as substrate in their metabolism. The major elements which serve as inorganic metabolites are carbon, ammonia–nitrogen, and phosphorus.

Gases commonly found in raw wastewater include nitrogen, oxygen, carbon dioxide, hydrogen sulfide, ammonia, and methane. Of all these gases mentioned, the ones that are considered most in the design of a treatment facility are oxygen, hydrogen sulfide, and methane concentrations. Oxygen is required for all aerobic life forms either within the treatment facility or in the receiving water. During the absence of aerobic conditions, oxidation is brought about by the reduction of inorganic salts such as sulfates or through the action of methane-forming bacteria. The end products are often very obnoxious. To avoid such conditions it is important that an aerobic state be maintained.

The quality and species of micro- and macroscopic plants and animals which make up the biological characteristics in a receiving body of water may be considered as the final test of wastewater treatment effectiveness. Within the treatment facility, the wastewater provides the perfect medium for good microbial growth, whether it be aerobic or anaerobic. Bacteria and protozoa are the keys to the biological treatment process used at most treatment facilities, and to the natural biological cycle in receiving waters. In the presence of sufficient dissolved oxygen, bacteria convert the soluble organic matter into new cells and inorganic elements.

This causes a reduction of organics loading through the buildup of more complex organisms and/or removal.

Within a wastewater treatment plant handling domestic wastes, bacteria (with concentrations ranging from 10^5 to 10^8/mL) will be the dominant plant/animal species, with other organisms achieving varying degrees of importance depending on the individual, or unit, processes used for treatment and the plant design. Consequently, wastewater treatment is directed towards using and removing the common bacteria along with organic and inorganic components.

Water quality in a receiving body of water is strongly influenced by the biological interactions that take place. The discharge (effluent) to the receiving waters becomes a normal part of the biological cycle, and its effect on aquatic organisms is the ultimate consideration of treatment plant operation. Typically, the species and organisms found in biological examination of the receiving waters include zooplankton and phytoplankton, peryphyton, macroinvertebrates, macrophytes, and fish.

Because of the increasing awareness that enteric viruses can be waterborne, attempts have been made to identify and quantify virus contributions to receiving waters via wastewater treatment plants. Different physical and chemical analytical techniques have been used in an effort to identify such viruses. Virus removal is mandated in connection with reclaimed wastewater discharged for indirect reuse in groundwater, and into lakes and rivers used for body-contact purposes.

Sludge Characteristics

Most wastewater treatment plants use primary sedimentation to remove readily settleable solids from raw wastewater. In a typical plant, the dry weight of primary sludge solids (those removed by filtration, settling or other physical means) is roughly 50% of that for the total sludge solids. Primary sludge is usually easier to manage than biological and chemical sludges — which are produced in the advanced or secondary stages of treatment — for several reasons. First, primary sludge is readily thickened by gravity, either within a primary sedimentation tank or within a separate gravity thickener. In comparison with biological and many chemical sludges, primary sludge with low conditioning requirements can rapidly be mechanically dewatered. Furthermore, the dewatering device will produce a drier cake and give better solids capture than it would for most biological and chemical sludges.

Primary sludge always contains some grit, even when the wastewater has been processed through degritting. Typically, it also contains different anaerobic and facultative species of bacteria, such as sulfate-reducing and oxidizing bacteria. Primary sludge production is typically within the range of 800–2500 lbs per million gallons (100–300 mg/L) of wastewater. A basic approach to estimating primary sludge production for a particular plant is to compute the quantity of total suspended solids (TSS) entering the primary sedimentation tanks, assuming efficiency of removal.

Biological sludges are produced by secondary treatment processes such as activated sludge, trickling filters, and rotating biological contactors. Quantities and characteristics of biological sludges vary with the metabolic and growth rates of the various microorganisms present in the sludge. Biological sludge that contains debris such as grit, plastics, paper, and fibers is produced at plants lacking primary treatment. Plants with primary sedimentation normally produce a fairly pure biological sludge. Biological sludges are generally more difficult to thicken and dewater than are primary sludge and most chemical sludges.

Biochemical Oxygen Demand Test and Chemical Oxygen Demand Test

The biochemical oxygen demand (BOD) test was developed in an attempt to reflect the depletion of oxygen that would occur in a stream due to utilization by living organisms as they metabolize organic matter. Often the BOD is used as the sole basis for determining the efficiency of the treatment plant in stabilizing organic matter. Effluent ammonia-nitrogen poses an analytical problem in measuring BOD. At 20°C (68°F), the nitrifying bacteria in raw domestic wastewater usually are significant in number and normally will not grow sufficiently during the 5-day BOD test to exert a measurable oxygen demand. To obtain a true measure of the treatment plant performance in removing organic matter, the BOD test may require correction for nitrification.

The chemical oxygen demand (COD) analysis is more reproducible and less time-consuming. The COD test and BOD test can be correlated, but the correlation ultimately gives a qualitative value. The COD test measures the nonbiodegradable as well as the ultimate biodegradable organics. A change in the ratio of biodegradable to nonbiodegradable organics affects the correlation between COD and BOD. Such correlation is specific for a particular waste but may vary considerably between treatment plant influent and effluent. For additional information on the BOD and COD tests, see Chapter 14.

13.4 WASTEWATER PLANT DESIGN

Facility Plan and Studies

A significant amount of time is involved in the planning and design of a wastewater treatment facility. The initial phase consists of a facility plan, which must be prepared before grants from the federal and state governments can be obtained.

A facility plan is prepared to identify the water pollution problems in a specific area, develop design data, evaluate alternatives, and recommend a solution. Most of the data developed in a facility plan are used in the preparation of design plans, specifications, and cost estimates of the wastewater treatment facility.

Steps in the facility plan include:

1. *List effluent limitations:* The plan should list the effluent limitations applicable to the facility being planned. Publicly owned wastewater treatment plants built after June 30, 1974 must achieve "best practicable waste treatment technology" (equivalent of secondary treatment).
2. *Assess current situation:* Describe briefly the existing conditions to be considered when examining alternatives during planning. The following conditions should be described:
   a. Planning area
   b. Organizational context
   c. Demographic data
   d. Water quality

    e. Other existing environmental conditions

    f. Existing wastewater flows and treatment systems, including system performance

    g. Infiltration and inflow

3. *Assess future conditions:*

    a. Twenty-year planning period beyond start-up of the facility. Phased construction should be considered.

    b. Land use; must be carefully coordinated with the state, municipal, and regional regulations, policies, and plans.

    c. Demographic and economic projections.

    d. Forecasts of flow and wasteloads; includes projections of economic and population growth, infiltration/inflow estimates, analysis of pollutant content and flows in the existing system, sewer overflow data, industrial wasteload data projections, and pollution-reducing possibilities.

    e. Future environment of the planning area without the project.

4. *Develop and evaluate alternatives:* Alternative treatment systems and their impact on the environment, long-range sewer plans for the planning area, sludge utilization and/or disposal, and facility location.

5. *Select plan:* The public is provided with alternative proposals and hearings are held to explain each proposal.

6. *Preliminary design of treatment works:* The following items are included: a schematic flow diagram; unit processes; plant site plans; sewer pipe plans and profiles; design data regarding detention items, flow rates, and sizing of units; operation and maintenance summary; cost estimates; and a completion schedule.

7. *Arrangements for implementation:* Following selection of plan and design, existing institutional arrangements should be reviewed and a financial program developed, including preliminary allocation of costs among various classes of users of the system.

Wastewater treatment plants utilize a number of unit operations and processes to achieve the desired degree of treatment. The collective treatment schematic is called a *flowsheet*. Many different flow schemes can be developed from various unit operations and processes for the desired degree of treatment. However, the most desirable flow scheme is the one that is the most cost-effective.

Wastewater treatment plants are designed to process the liquid and solid portions of the wastewater. Treatment systems and solids disposal systems must be put together so as to assure the most efficient utilization of resources such as money, materials, energy, and work force in meeting treatment requirements. Logic dictates what the process elements must be and the order in which they go together.

A methodical process of selection must be followed in choosing a resource-efficient and environmentally sound system from the numerous treatment and disposal options available. The basic selection mechanism used is the "principle of successive elimination," an iterative procedure in which less effective options are progressively eliminated until only the most suitable system or systems for the particular site remain.

The concept of a "treatment train" is a result of a systems approach to problem solving. However, this concept is useful only if all components of the train are considered. This includes not only sludge treatment and disposal components, but also wastewater treatment options and other critical linkages such as sludge transportation, storage, and side stream treatment.

The general sequence of events in system selection is:

1. Selecting relevant criteria.
2. Identifying options.
3. Narrowing the list of candidate systems.
4. Selecting a system.

The basis of unit process design is the initial and future volume and characteristics of the wastewater, anticipated variations, and statutory requirements of regulatory agencies. Data acquisition required will be, in part, determined by the treatment processes considered and largely determined by the treatment requirements. The plant process design is usually a function of peak and minimum loading conditions and not a function of average or median conditions. Wastewater flows for design purposes are estimated by:

1. Gaging flow in existing systems and making corrections appropriate to increased future requirements.
2. Estimating and totaling the various components of the flow.

For updating an existing wastewater facility, the first method of estimating flow is more reliable. In the case of a new facility, the second method is preferred. A commonly used basis for wastewater treatment plant design is the adoption of a maximum flow of two to four times the average dry-weather flow. The value depends on local factors, including population.

Several flow rates are used for design of various elements in a wastewater treatment plant. The average daily flow rate for the period of design is determined by totaling the 24-hr average of all components. This rate is used to determine such items as pumping and chemical costs, sludge solids, and organic loading. The design average flow rate is generally used for mass loading of treatment units. Peak design rate, usually 2 to 2.25 times the design average flow rate, is used for hydraulic sizing.

The method of solids disposal usually controls the selection of solids treatment systems and not vice versa. Thus, the system selection procedure normally begins when the solids disposal option is specified. The process selection procedure consists of developing treatment/disposal systems which are compatible with one another and appear to satisfy local relevant criteria, and choosing the best system or systems by progressive elimination. Proper selection of the sludge processing equipment is important for trouble-free operation of a wastewater treatment plant. Sludge is quite odorous and may cause environmental concerns. Therefore such factors as solids captured, chemical quality of return flows, ability to handle variable quality of sludge, ease of operation, and odors are often given serious considerations.

Treatment

Wastewater treatment plants utilize a number of individual or unit operations and processes to achieve the desired degree of treatment. The collective treatment schematic is called a *flow scheme*, a *flow diagram*, a *flow sheet*, a *process train*, or a *flow schematic*. Many different flow schemes can be developed from various unit operations and processes for the desired level of treatment. Unit operations and processes are grouped together to provide what is known as primary, secondary, and tertiary (or advanced) treatment. The term *primary* refers to physical unit operations, *secondary* refers to chemical and biological unit processes, and *tertiary* refers to combinations of all three.

Treatment methods in which the application of physical forces predominate are known as *physical unit operations*. These were the first methods to be used for wastewater treatment. Screening, mixing, flocculation, sedimentation, flotation, and filtration are typical unit operations.

Treatment methods in which the removal or conversion of contaminants is brought about by the addition of chemicals or by other chemical reactions are known as *chemical unit processes*. Precipitation, gas transfer, adsorption, and disinfection are the most common examples used in wastewater treatment.

Treatment methods in which the removal of contaminants is brought about by biological activity are known as *biological unit processes*. Biological treatment is used primarily to remove the biodegradable organic substances (colloidal or dissolved) in wastewater. Basically these substances are converted into gases that can escape to the atmosphere or into biological cell tissue that can be removed by settling. Biological treatment is also used to remove the nitrogen in wastewater. With proper environmental control, wastewater can be treated biologically in most cases.

Municipal wastewater treatment plants are mandated by the federal government to provide secondary level treatment in order to comply with the final effluent discharge to the receiving waters. As a result of this mandate, secondary level treatment is considered the minimum acceptable design criterion for a wastewater treatment plant. Secondary treatment implies that chemical and biological processes are utilized in the overall treatment process in addition to physical treatment. An overview of a typical secondary wastewater treatment plant may be described as being grouped in the following sections:

1. Primary treatment and handling facilities.

2. Activated sludge treatment/secondary facilities.

3. Sludge treatment, storage, and disposal facilities.

As influent plant flow, primary raw sewage enters the plant by means of interceptors; then the flow enters the wet wells which are located at the lowest elevation in the plant. The purpose of the low elevation is to provide gravity flow and to create sufficient pressure differential for the use of sewage pumps leading into the collection systems. Prior to the pumps, the raw sewage first enters screens such as a bar screen rack mechanism. This device physically separates large objects such as paper cups, rags, and sticks from the raw sewage flow. Grinders and shredders are also frequently incorporated into the screening facilities. The next phase of processing involves grit removal; here materials such as sand, coffee grounds, and cigarette filter tips are separated in grit tanks or chambers. Typically a grit tank is rectangular in shape, equipped with velocity control devices, raking devices, and aeration piping. Grit tanks are designed to do the following:

1. Protect moving mechanical equipment from abrasion and abnormal wear.

2. Reduce clogging in pipes and sludge hoppers.
3. Prevent accumulations in aeration tanks and sludge digesters and the consequent loss of usable volume.

From the grit tanks, the raw sewage enters the primary settling tanks. The purpose of the primary settling tanks is to provide the detention time needed to separate the settleable and floatable solids from the wastewater for appropriate handling. The objectives of primary settling are

1. Removal of finely dispersed solids by floc formation with larger particles.
2. Removal of colloidal material via adsorption to larger particles.

At this stage of the processing and handling of the primary effluent, further treatment will now employ biological and chemical processes (secondary treatment). The most common secondary treatment process is the activated sludge process. The aeration tank is the heart of the activated sludge process. Here, oxygen is introduced into the system, along with the effluent from the primary treatment process, to satisfy the requirements of the organisms in the sludge and to keep the sludge dispersed in the aeration liquor. These organisms break down the waste material remaining in the primary effluent.

The activated sludge process is used to convert nonsettleable substances from finely divided, colloidal, and dissolved form to biological floc. This newly formed biological floc, known as *sludge*, is removed from the system through sedimentation, thereby providing a high degree of secondary treatment. The biological floc is developed in aeration tanks and settled out in final settling tanks.

There are different types of aeration systems; the two most common types are subsurface diffusion and mechanical aeration. In the diffused air system, compressed air is introduced at the bottom of the tank near one side. This causes the tank's contents to be circulated by the air-lift effect. The floating or fixed bridge aerators are common mechanical aeration devices. Some mechanical aerators use a blade to agitate the surface of the tank and disperse air bubbles into the aeration liquor. Others circulate the aeration liquor by an updraft or downdraft pump or turbine. This action produces surface and subsurface turbulence, while at the same time diffusing air through the liquid mass.

The basic function of separating the newly formed activated sludge from the treated wastewater cannot be accomplished without proper operation of the final settling tanks. The final settling tanks' function is liquid–solid separation, solids concentration, and solids return.

Secondary treatment as described above can remove up to 99% of the bacteria from raw wastewater. Tertiary treatment is capable of an even greater removal. In some instances, however, regulatory agencies may require removal or inactivation of pathogenic organisms in excess of that achieved by secondary or tertiary treatment. In those cases, disinfection is required. Chlorine is the disinfectant used at the vast majority of wastewater treatment plants. It is effective and reliable and may be the least costly disinfection alternative. Disinfection usually must be performed year round if the receiving stream is used as a source of drinking water by downstream communities or if it flows into shellfish areas. Seasonal disinfection may be required if the main usage is bathing or other primary contact recreation. After disinfection, the treated water is now ready for use by the receiving community.

The problem of sludge treatment still needs to be addressed. The principal sources of sludge at a municipal wastewater treatment plant are the primary sedimentation tanks and the final settling tanks. Additional sludge may come from chemical precipitation, nitrification facilities, screening, and grinder devices. Sludge contains large volumes of water. The small fraction of solids in the sludge is highly offensive, thus requiring treatment such as conditioning. Common sludge management processes include thickening, stabilization, dewatering, and disposal.

Sludge thickening is used to concentrate solids and reduce volume. Thickened sludge requires less tank capacity and chemical dosage for stabilization, and smaller piping and pumping equipment for transport. Common methods of sludge thickening used at medium-to-large plants are gravity thickening, dissolved air flotation, and centrifugation.

The main purpose of sludge stabilization is to reduce pathogens, eliminate offensive odors, and control the potential for putrification of organic matter. Sludge stabilization can be accomplished by biological, chemical, or physical means. Selection of any method depends largely on the ultimate sludge disposal method. Various methods of sludge stabilization are: anaerobic or aerobic digestion (biological); chemical oxidation or lime stabilization (chemical); and thermal conditioning (physical). The anaerobic digestion process is the most widely selected stabilization process used in most plants.

The principal purpose of sludge digestion is twofold:

1. To reduce in volume the solids content from the treatment process.
2. To decompose highly putrescible organic matter to relatively stable or inert organic and inorganic compounds.

The digestion process is carried out in the digester or digestion tanks. In an anaerobic digestion process, the tanks are covered to exclude air and oxygen, to prevent release of offensive odors, and to collect digester gases. In the process of decomposition, gases are formed by bacterial action. These gases are comprised mostly of carbon dioxide and methane. Anaerobic digesters are most commonly operated at 35–42°C.

Anaerobic sludge digesters are of two types, namely, standard-rate and high-rate. In the standard-rate digestion process the digester contents are usually unheated and unmixed. The digester period may vary from 30 to 60 days. In a high-rate digestion process, the digester contents are heated and completely mixed. The required detention period is 10–20 days. Often a combination of standard- and high-rate digestion is achieved in two-stage digestion. The second-stage digester mainly separates the digested solids from the supernatant liquor, although additional digestion and gas recovery may also be achieved.

13.5 OVERVIEW OF ADVANCED WASTEWATER TREATMENT TECHNOLOGY (TERTIARY TREATMENT)

Advanced wastewater treatment technology is designed to remove those constituents that are not adequately removed in the secondary treatment plants. These include nitrogen, phosphorus, and other soluble organic and inorganic compounds. Nitrogen and phosphorus are nutrients that accelerate the growth of plants in the receiving waters. Ammonia is toxic to fish, exerts nitrogenous oxygen demand, and increases chlorine demand. Heavy metals, hydrogenated hydrocarbons, and phenolic compounds are toxic to fish and other aquatic life, concentrate in the food chain, and may create taste and odor problems in water supplies. Many of these constituents must be removed to meet stringent water quality standards and to allow reuse of the effluent for municipal, industrial, irrigation, recreation, and other water needs.

The most commonly used advanced wastewater treatment processes are chemical precipitation of phosphorus, nitrification, denitrification, ammonia stripping, breakpoint chlorination, filtration, carbon adsorption, ion exchange, reverse osmosis, and electrodialysis. Some of these unit operations and processes are discussed below.

Coagulation and Flocculation

Coagulation involves the reduction of surface charges and the formation of complex hydrous oxides. Flocculation involves combining the coagulated particles to form settleable floc. The coagulant (alum, ferric chloride, ferrous sulfate, ferric sulfate, etc.) is mixed rapidly and then stirred in order to encourage formation of floc prior to settling. The objective is to improve the removal of BOD, total suspended solids (TSS), and phosphorus. To accomplish this, chemicals are added prior to primary treatment, in biological treatment processes, or in separate facilities following the biological treatment processes. Polymers are also used in conjunction with these chemicals. The chemical dosage is adjusted to give the desired amount of floc formation and BOD, TSS, or phosphorus removal. The approximate average alum and ferric chloride dosages in municipal wastewater are 170 and 80 mg/L, respectively. Lime may be needed in conjunction with iron salts.

Major equipment for a coagulation–flocculation system includes: chemical storage; chemical feeders, pipings, and control systems; flash mixer; flocculator; and sedimentation basin. Proper mixing of chemicals at the point of addition, along with proper flocculation prior to clarification, is essential for maximum effectiveness. Flocculation may be accomplished in a few minutes to half an hour in basins equipped with mixers, paddles, or baffles. The coagulants react with alkaline to produce insoluble metal hydroxide for floc formation. If sufficient alkalinity is not present, lime or soda ash (sodium carbonate) is added in the desired dosages.

In biological treatment units the addition of coagulants has a marked influence on biota. Population levels of protozoans and higher animals are adversely affected. However, the BOD, TSS, and phosphorus removal is significantly improved. Overdosing of chemicals may cause toxicity to microorganisms necessary to the treatment process. Large quantities of sludge are produced from chemical precipitation. The chemical sludges may cause serious handling and disposal problems.

Lime Precipitation

Lime reacts with bicarbonate alkalinity and orthophosphate, causing flocculation. The objectives of lime addition are to increase removal of BOD, TSS, and phosphorus. Lime is added prior to primary sedimentation, in biological treatment, or in a separate facility after secondary treatment. Lime addition may be in a single stage or in two stages. Major equipment for a lime precipitation process includes: lime storage; lime feeders, pipings, and control systems; flash mixer; flocculator; and sedimentation basin. For a two-stage system, a CO_2 source such as an incinerator or an internal combustion engine is required.

Lime in a single stage is used in primary biological treatment or after secondary treatment. The procedure is the same as in coagulation. The pH of the wastewater is raised to about 10, and the wastewater is flocculated and then settled. Normal lime dosage is about 180–250 mg/L as CaO. The actual dosage depends primarily on phosphorus concentration, hardness, and alkalinity. The biological system is not adversely affected by lime addition in moderate amounts (80–120 mg/L as CaO). The microbial production of carbon dioxide is sufficient to maintain a pH near neutral. High dosage may upset the biological process.

Lime addition prior to primary and after secondary treatment may be achieved in two stages. First, the pH of wastewater is raised to greater than 11, flocculated, and settled. Lime dosage up to 450 mg/L as CaO may be necessary. Next, the effluent is carbonated by adding CO_2 to lower the pH, and then it is flocculated and settled. Higher BOD, TSS, and phosphorus removal is achieved in a two-stage lime process. However, use of excess lime causes scale formation in tanks, pipes, and other equipment. Also, handling and disposal of large quantities of lime sludge is a problem.

Nitrification

Nitrification converts ammonia to nitrate form, thus eliminating toxicity to fish and other aquatic life and reducing the nitrogenous oxygen demand. Ammonia oxidation to nitrite and then to nitrate is performed by autotrophic bacteria. The reactions are shown by Equations 13.5.1 and 13.5.2:

$$NH_3 + \tfrac{2}{3}O_2 \rightarrow NO_2 + H^+ + H_2O + \text{biomass} \qquad (13.5.1)$$

$$NO_2 + \tfrac{1}{2}O_2 \rightarrow NO_3 + \text{biomass} \qquad (13.5.2)$$

Temperature, pH, dissolved oxygen, and the ratio of BOD to total Kjeldahl nitrogen (TKN) are important factors in nitrification.

Denitrification

Nitrite and nitrate are reduced to gaseous nitrogen by a variety of facultative heterotrophs in an anaerobic environment. An organic source, such as acetic acid, acetone, ethanol, methanol, or sugar, is needed to act as hydrogen donor (oxygen acceptor) and to supply carbon for synthesis. Methanol is preferred because it is least expensive. Equations (13.5.3) through (13.5.5) express the basic reactions:

$$3O_2 + 2CH_3OH \rightarrow 2CO_2 + 4H_2O \qquad (13.5.3)$$

$$6NO_3 + 5CH_3OH \rightarrow 3N_2 + 5CO_2 + 7H_2O + 6OH \qquad (13.5.4)$$

$$2NO_2 + CH_3OH \rightarrow N_2 + CO_2 + H_2O + 2OH \qquad (13.5.5)$$

Biological Nutrient Removal

Biological phosphorus and nitrogen removal has received considerable attention in recent years. Basic benefits reported for biological nutrient removal include monetary saving through reduced aeration capacity and the obviated expense for chemical treatment. Biological nutrient removal involves anaerobic and anoxic treatment of return sludge prior to discharge into the aeration basin. Based on the anaerobic, anoxic, and aerobic treatment sequence and internal recycling, several processes have been developed. Over 90% phosphorus and high nitrogen removal (by nitrification and denitrification) has been reported by biological means.

Ammonia Stripping

Ammonia gas can be removed from an alkaline solution by air stripping as expressed by Equation (13.5.6):

$$NH_4 + OH \rightarrow NH_3\uparrow + H_2O \qquad (13.5.6)$$

The basic equipment for an ammonia-stripping system includes chemical feed, a stripping tower, a pump and liquid spray system, a forced air draft, and a recarbonation system. This process requires raising the pH of the wastewater to about 11, formation of droplets in the stripping tower, and providing air–water contact and droplet agitation by countercurrent circulation of large quantities of air through the tower. Ammonia-stripping towers are simple to operate and can be very effective in ammonia removal, but the extent of their efficiency is highly dependent on air temperature. As the air temperature decreases, the ammonia removal efficiency drops significantly. This process, therefore, is not recommended for use in a cold climate. A major operational disadvantage of stripping is the need for neutralization and prevention of calcium carbonate scaling on the tower. Also, there is some concern over discharge of ammonia into the atmosphere.

13.6   SLUDGE DISPOSAL

Ensuring the safe disposal of municipal sludge and other residues, such as screenings, grits, and skimmings, is considered an integral part of good planning, design, and management of municipal wastewater treatment facilities. Acceptable sludge disposal practices include conversion processes such as: incineration; wet oxidation; pyrolysis and composting; and land disposal by land application and landfilling. Table 13.2 summarizes the various processes available for sludge disposal.

Incineration

Incineration involves drying the sludge cake followed by complete combustion of the organic matter. A minimum temperature of 700°F is needed to deodorize the stack emissions. Excess air of 50–100% over the stoichiometric air requirements is necessary. Natural gas or fuel oil is provided as auxiliary fuel for ignition and to maintain the proper temperature. Often sludge is incinerated with municipal solid waste and other residues. Two major incineration systems are the multiple-hearth furnace and the fluidized bed reactor. The multiple-hearth furnace contains several hearths arranged in a vertical stack. Sludge cake enters the top and proceeds downward through the furnace from hearth to hearth. The fluidized bed incinerator utilizes a hot sand reservoir in which hot air is blown from below to expand and fluidize the bed. Air pollution control equipment is needed in both cases to clean the flue gases before they are emitted.

TABLE 13.2 Summary of conversion processes for disposal of municipal sludges

Conversion Process	Recommended Pretreatment	Codisposal with Other Residue	Additional Processing Requirements
Incineration	Thickening and dewatering	Yes	Ash landfilling
Wet oxidation	Thickening	No	Separation of ash, treatment of returned liquid, landfilling of ash
Pyrolysis	Thickening	No	Utilization of by-products (gas, liquid, carbon, etc.), disposal of residues
Recalcining	Thickening and dewatering	No	Recovery of lime, landfilling of ash
Composting by heating and drying	Thickening and dewatering	No	Utilization or sale of compost
Composting by microbial action	Thickening, digestion, and dewatering	No	Utilization or sale of compost

Composting by Heat Drying

The purpose of heat drying is to remove the moisture from the wet sludge and partially combust the organics. Sludge drying occurs at temperatures of approximately 370°C. At this temperature, part of the volatile matter is removed and the sludge is stable enough for use as compost. The gases evolved in the drying process are reheated to approximately 750–800°C to eliminate odors. The most common types of heat-drying systems for sludge are flash dryers, spray dryers, rotary dryers, multiple-hearth dryers, and the Carver–Greenfield process. After it is dried, the sludge is processed into soil conditioner. Drying permits grinding, weight reduction, and prevention of continued microbial decomposition.

Composting by Microbial Action

Microbial composting is a process in which organic matters undergo biological decomposition to produce a stable end product that is acceptable as a soil conditioner. There are two methods of composting, namely, open windrow and mechanical processing. Windrow composting uses long, narrow piles approximately 1.2 m high and 2.5 m wide. These piles are turned every few days. Moisture is maintained at 55–70%. It takes 4–6 weeks to produce stable compost. Mechanical systems of composting produce stable compost in 5–10 days. There has been a limited market for compost in this country.

13.7 SUMMARY

1. The ultimate goal of wastewater treatment is the protection of the environment in a manner commensurate with economic, social, and political concerns.

2. The Federal Water Pollution Control Act Amendments of 1972 require municipalities to prevent, reduce, or eliminate pollution of surface waters and groundwaters. The planning and design of a wastewater treatment plant must achieve these criteria. Presently, municipal wastewater plants are being upgraded and expanded to meet all government regulations.

3. Primary treatment, which includes filtration and settling tanks, removes paper, rags, sand, coffee grounds, and other solid waste materials from the wastewater.

4. Secondary treatment typically utilizes a biological process to further remove substances from the water.

5. Sludge is produced as a waste product during the wastewater treatment process. Sludge management and disposal issues are currently being tackled to provide a safer and cleaner environment.

6. Advanced wastewater treatment, known as *tertiary treatment*, is designed to remove those constituents that may not be adequately removed by secondary treatment. This includes removal of nitrogen, phosphorus, and heavy metals.

REFERENCES

R. S. Burd, *A Study of Sludge Handling and Disposal*, National Technical Information Service, Springfield, VA, 1968.

Malcolm Pirnie, Inc., and Michael Baker, Jr., of New York, Inc., J.V., 201 Facilities Plan for WP-287, the Coney Island WPCP, December 1980.

New England Interstate Water Pollution Control Commission, *Guides for the Design of Wastewater Treatment Works*, 1980.

Perry, Chemical Engineering Handbook, 7th edition, 1997, "Industrial Wastewater Management", pg 25-58 to 25-80.

S. R. Qasim, *Wastewater Treatment Plants*, CBS College Publishing, New York, 1985.

U.S. Environmental Protection Agency, *Process Design Manual for Sludge Treatment and Disposal*, Technology Transfer, EPA 625/1-79-011, Washington, D.C., September 1979.

U.S. Environmental Protection Agency, *Design Guides for Biological Wastewater Treatment Process*, Project No. 11010 ESQ, Austin, TX, August 1971.

U.S. Environmental Protection Agency, *Cost to the Consumer for Collection and Treatment of Wastewater*, Water Pollution Control Research Series 17090-07/70, Washington, D.C., July 1970.

Water Pollution Control Federation, *Wastewater Treatment Plant Design*, Lancaster Press, Lancaster, PA, 1977.

14

INDUSTRIAL WASTEWATER TREATMENT

Contributing Authors: Robert Politzer and Thomas Ryan

14.1 INTRODUCTION

In a very real sense, many of the environmental problems faced by today's world are of relatively recent occurrence. For eons, effective nutrient and end-product cycling in natural systems eliminated any excess, useless residues. Industrial production, however, has evolved in a more linear mode, with the input of raw materials and output of "useful" and seemingly "useless" (waste) products.

For much of human history, "the solution to pollution is dilution" was the general approach in dealing with wastes. In the case of wastewater, when the sewage of previous years was dumped into waterways, the natural processes of purification began. First, the sheer volume of clean water in the stream diluted the small amounts of wastes. Bacteria and other small organisms in the water consumed the sewage or other organic matter, turning it into new bacterial cells, carbon dioxide, and other products. But the bacteria normally present in water must have oxygen in order to do their part in breaking down the sewage. Water acquires this all-important oxygen by absorbing it from the air and from plants that grow in the water itself. These plants use sunlight to turn the carbon dioxide present in water into oxygen.

The life and death of any body of water depend mainly upon its ability to maintain a certain amount of dissolved oxygen (DO). This DO is what fish breathe. If only a small amount of sewage is dumped into a stream, fish are not affected and the bacteria can do their work; the stream can quickly restore its oxygen loss from the atmosphere and from plants. Trouble begins when the sewage load is excessive. If carried to the extreme, the water could lose all of its oxygen, resulting in the death of fish and beneficial plant life.

It was not until the early 1800s that any sort of organized system for collection of wastewater was conceived. And it took nearly a century thereafter for systematic treatment to commence. Development of the germ theory by Koch and Pasteur in the latter half of the nineteenth century marked the beginning of a new era in sanitation. Before that time, the relationship of pollution to disease had been only faintly understood, and the science of bacteriology, then in its infancy, had not been applied to the subject of wastewater treatment (1).

Both municipal and industrial wastewater treatment plants were originally designed in part to speed up the natural processes by which water purifies itself. However, these natural processes, even though they were accelerated in a waste treatment plant, were not sufficient to remove other contaminants such as disease-causing germs, excessive nutrients such as phosphates and nitrates, and chemicals and trace elements. New pollution problems of modern industry placed additional burdens on wastewater treatment systems, with the result that these pollutants have become ever more difficult to remove from the wastewater.

Beginning in the 1970s, several pieces of legislation were designed to address the increasing wastewater crisis. The EPA was given the authority to promulgate and enforce the various regulations that derived from the Federal Water Pollution Control Act, as amended (1972); the Clean Water Act Amendments (1977); the Resource Conservation and Recovery Act (1976); and the Toxic Substances Control Act (1977). Specific industry pretreatment regulations were also promulgated by the EPA in the early 1980s which limited the discharge concentrations of targeted pollutant parameters.

In response to these dictates, scientists and engineers have developed more advanced technologies for the treatment of both municipal and industrial wastewaters. Such relatively new techniques range from extensions of biological treatment capable of removing nitrogen and phosphorus nutrients to physical–chemical separation, utilizing adsorption and ion-exchange. For smaller industries, wastewaters usually undergo in-plant pretreatment and reuse and are then discharged into the municipal system. Larger industries often employ various combinations of strategies for complete on-site treatment. These processes in various combinations can achieve virtually any degree of pollution control desired, and, as waste effluents are purified to higher and higher degrees by such treatment, the point is reached where effluents become "too good to throw away."

Unfortunately, some of the same advanced techniques that have helped to further purify wastewater effluent have led to an increase in the production of solid waste. Sludge is the combination of precipitated chemicals, original turbidity, and suspended material. With the outlawing of ocean dumping and with the emerging landfill crisis, alternatives to disposal such as composting, land application, and other recycling techniques are presently being researched and applied on a limited basis.

To meet wastewater management objectives, the following four major areas of concern must be addressed:

1. Sources and characteristics of wastes and wastewaters must be ascertained.

2. The particular type of wastewater treatment process, or combination of processes, must be determined.

3. Wastewater effluent must be effectively controlled.

4. Sludge and other solid wastes must be properly managed.

These four areas of concern will be the major focus of this chapter.

14.2 SOURCES AND CHARACTERIZATION OF INDUSTRIAL WASTEWATERS

Types of Pollutants

The various types of pollutants found in industrial wastewaters can be classified into eight categories: common sewage and other oxygen-demanding wastes; disease-causing agents; plant nutrients; synthetic organic chemicals; inorganic chemicals and other mineral substances; sediments; radioactive substances; and heat (2).

Oxygen-demanding wastes are the traditional organic wastes, ammonia, iron, or any other oxidizable compound contributed by industrial wastes. Such wastes result from food processing, paper mill production, tanning, and other manufacturing processes. These wastes are usually destroyed by bacteria if there is sufficient oxygen present in the water. Because fish and other aquatic life depend on oxygen for life, the oxygen-demanding wastes must be controlled — otherwise the fish will die.

The disease-causing agents include infectious organisms which are carried into surface water and groundwater by certain kinds of industrial wastes, such as tanning and meat-packing plants. Humans or animals may come in contact with these microbes either by drinking the water or through swimming, fishing, or other activities. Although modern disinfection techniques have greatly reduced the danger of this type of pollutant, the problem must be carefully monitored.

Plant nutrients are the substances in the food chain of aquatic life (such as algae and water weeds) which support and stimulate plant growth. Carbon, nitrogen, and phosphorus are the three chief nutrients present in natural waters. Large amounts of these nutrients are present in sewage, certain industrial wastes, and drainage from fertilized land. Biological waste treatment processes do not remove the phosphorus and nitrogen to any substantial extent — in fact, they convert the organic forms of these substances into mineral form, making them more usable by plant life. The problem starts when an excess of these nutrients overstimulates the growth of water plants (particularly algae), a condition called *eutrophication*. When the plants and algae eventually die, the decomposition of their organic matter can lead to a severe depletion of dissolved oxygen in the body of water.

The synthetic organic chemicals include pesticides, synthetic industrial chemicals, and wastes from their manufacture. Many of these substances are toxic to aquatic life and are possibly harmful to humans. They cause taste and odor problems in drinking water, and they resist waste treatment. Some are known to be poisonous at very low concentrations.

A vast array of metal salts, acids, solid matter, and many other chemical compounds are included in the category of inorganic pollutants. Their sources include mining and manufacturing processes, oil field operations, and agricultural practices. While a wide variety of acids are discharged as waste by industry, the largest single source comes from mining operations and mines that have been abandoned.

Many of these types of chemicals are being created each year. If untreated, they can interfere with natural stream purification, destroy fish and other aquatic life, and cause excessive hardness of water supplies. Moreover, such chemicals can corrode expensive water treatment equipment and increase the cost of waste control. The EPA has developed a list of 129 priority pollutants that include both toxic organic and inorganic chemicals. The heightened danger of these chemicals has necessitated the drafting of specific limitations in most treatment plant permits.

While they are not as dangerous as some other types of pollutants, sediments are a major problem because of the sheer magnitude of the amount reaching our waterways. Sediments fill stream channels and harbors, requiring expensive dredging; they also fill reservoirs, reducing their capacity and useful life. They erode power turbines and pumping equipment, and they reduce fish and shellfish populations by blanketing fish nests and food supplies.

Radioactive pollution results from the mining and processing of radioactive ores, from the use of refined radioactive materials in power reactors, and from their use in industrial, medical, and research purposes. Because radiation bioaccumulates, control of this type of pollution must take into consideration total exposure in the environment.

While not directly toxic, heat reduces the capacity of water to absorb oxygen. Tremendous volumes of water are used by power plants and industry for cooling. Most of the water, with the added heat, is returned to natural bodies of water, raising their temperatures. Also having less oxygen, the water is not as efficient in assimilating oxygen-consuming wastes and in supporting fish and aquatic life. Unchecked discharges of waste heat can seriously alter the ecology of a lake, a stream, or even part of the sea.

To complicate matters, most industrial wastes are a mixture of the eight types of pollution. Such mixing makes the problems of treatment and control that much more difficult and costly.

Characterization of Wastewater

The volume and strength of industrial wastewaters are usually defined in terms of units of production. For example, for a pulp-and-paper mill, waste is often measured in gallons per ton of pulp. However, because of variations in the production process as well as differences in housekeeping and water reuse, considerable variation may occur in the flow and characteristics of wastewaters of similar industries. In fact, very few industries utilize identical process operations. Thus, in the development of an effective strategy for wastewater control, an industrial waste survey is usually required to determine the character of a particular industry's waste load (3).

The industrial waste survey involves a definite procedure designed to develop a flow and material balance of all processes using water and producing wastes and to establish the variation in waste characteristics from specific process operations as well as from the plant as a whole. The results of the survey should establish possibilities for water conservation and reuse, and the variation in flow and strength of the effluents that need to undergo wastewater treatment.

The general procedure to be followed in developing the necessary information with a minimum of effort can be summarized in four steps:

1. Develop a sewer map from consultation with the plant engineer and an inspection of the various process operations. This map should indicate possible sampling stations and a rough order of magnitude of the anticipated flow.

2. Establish sampling and analysis schedules. Continuous samples with composites weighted according to flow are the most desirable, but these either are not always possible or do not lend themselves to the physical sampling location. The

period of sample composite and the frequency of sampling must be established according to the nature of the process being investigated.

3. Develop a flow-and-material-balance diagram. After the survey data are collected and the samples analyzed, a flow-and-material-balance diagram should be developed that considers all significant sources of waste discharge. How closely the summation of the individual sources checks the measured total effluent provides a check on the accuracy of the survey.

4. Establish statistical variation in significant waste characteristics. The variability of certain waste characteristics is significant for waste-treatment plant design. These data should be prepared as a probability plot showing frequency of occurrence.

Data from industrial waste surveys are highly variable and are usually susceptible to statistical analysis. Statistical analysis of variable data provides the basis for treatment plant process design.

Determination of Wastewater Composition

As implied in the industrial waste survey, the objective of water quality management is to control the discharge of pollutants so that water quality is not degraded to an unacceptable extent below the natural background level. Direct measurement of the pollutants must be made. The impact of the pollutant on water quality must be predicted and compared to the background water quality which would be present without human intervention. Based on this, a decision can be made on the levels acceptable for the intended uses of the water (4).

The *biochemical oxygen demand (BOD) test* is used to determine the amount of biodegradable contents in a sample of wastewater. As discussed in Chapter 13, the BOD test measures the amount of oxygen consumed by living organisms (mainly bacteria) as they metabolize the organic matter present in a waste. The test simulates conditions as close as possible to those that occur naturally (5). As is true for any bioassay, the success of a BOD test depends on the control of such environmental and nutritional factors as pH and osmotic conditions, essential nutrients, constant temperature, and population of organisms representative of natural conditions.

Although it theoretically takes an infinite amount of time for all the oxidizable material in a sample of water to be consumed, it has been empirically determined that a period of 20 days is required for near-completion. Because the 20-day waiting period is too long to wait in most cases, a 5-day incubation period (the time in which 70–80% of available material is usually oxidized) has been adopted as standard procedure. This shorter test is referred to as BOD5. The BOD5 test also serves the purpose of avoiding the contribution of nitrifying bacteria to the overall DO measurement, since these bacteria populations do not become large enough to make a significant oxygen demand until about 8–10 days from the start of the BOD test (6).

While the BOD test is the only test presently available that gives a measure of the amount of biologically oxidizable organic material present in a body of water, it does not provide an accurate picture of the total amount of overall oxidizable material and thus the total oxygen demand. For such measurements, the *chemical oxygen demand (COD) test* is used. In this test, a very strong oxidizing chemical, usually potassium dichromate, is added to samples of different dilution. To ensure full oxidation of the various compounds found in the samples, a strong acid and a chemical catalyst are added. One of the products of this oxidation–reduction reaction is the chromate ion which gives a very sharp color change that can be easily detected in a spectrophotometer. Thus, the greater the absorbance measured (inverse of transmittance), the greater the amount of oxidation that has taken place; also, the more the oxidizable material originally present, and the greater the oxygen demand of that material. To find the corresponding concentration of DO in the sample, the absorbance measured is related to an absorbance-concentration graph of a standard whose oxygen concentration is known.

A third method for measuring the organic matter present in a wastewater is the *total organic carbon (TOC) test*. The test is performed by placing a sample into a high-temperature furnace or chemically oxidizing environment. The organic carbon is oxidized to carbon dioxide. The carbon dioxide that is produced can then be measured. While the TOC test does directly measure the concentration of organic compounds, it does not provide a direct measurement of the rate of reaction or the degree of biodegradability. For this reason, the TOC test has been accepted as a monitoring technique but has not been utilized in the establishment of treatment regulations.

The BOD, COD, and TOC tests provide estimates of the general organic content of a wastewater. However, because the particular composition of the organics remains unknown, these tests do not reflect the response of the wastewater to various types of biological treatment technologies. It is therefore necessary to separate the wastewater into its specific components.

The *solids analysis* focuses on the quantitative investigation and measurement of the specific content of solid materials present in a wastewater. The *total solids* are the materials that remain after the water from the solution has evaporated. For any solution, the total solids consist of *suspended* and *dissolved* particles. Dissolved particles are classified according to size as either *soluble* or colloidal.

When a solution is poured through a filter (usually 0.45 μm in size), those solids that are too large to pass through the filter openings will be retained. These solids are called *suspended solids*. The filter containing the suspended solids is heated to remove the water and to produce the *total suspended solids (TSS)*. The TSS are further heated in a muffle furnace at 550°C. At such high temperatures, the *volatile suspended solids (VSS)*, mostly organics, will be vaporized, leaving behind the remaining *fixed suspended solids*. Those solids that are small enough to pass through the filter are called *dissolved (soluble) solids*. By heating the solution to remove the water, the *total dissolved solids* are obtained. As with the suspended solids, these solids are further heated at extreme temperatures to remove the *volatile dissolved solids* (VDS), leaving the *fixed dissolved solids*.

The assumption is made that all volatile solids (VSS and VDS) are organic compounds. A distinction is made as to whether such solids may or may not be adsorbed by a specific treatment medium. The soluble organics that are nonsorbable are further separated into degradable and nondegradable constituents.

Additional measurements of pH, nitrogen, and phosphorus levels in wastewaters are often required when complex pollution conditions exist. Various measuring devices are available for direct measurement of these inorganic components.

14.3 WASTEWATER TREATMENT PROCESSES

Numerous technologies exist for treating industrial wastewater. These technologies range from simple clarification in a settling pond to a complex system of advanced technologies requiring sophisticated equipment and skilled operators. Finding the proper technology or combination of technologies adequate to treat a particular wastewater to meet federal and local requirements and yet still be cost effective can be a challenging task.

Treatment technologies can be divided into three broad categories: physical, chemical, and biological. Many treatment processes combine two or all three categories to provide the most economical treatment. There are a multitude of treatment technologies for each of these categories. Therefore, although the technologies selected for discussion in this chapter are among the most widespread in the industrial field, they represent only a fraction of the available technologies. Figure 14.1 matches various types of industry with candidate treatment technologies (7,8,9).

Physical Treatment Processes

Clarification (Sedimentation). When an industrial wastewater containing a suspension of solid particles that have a higher specific gravity than the transporting liquid is in a relatively calm state, the particles will settle out because of the effects of gravity. This process of separating the settleable solids from the liquid is called *clarification* or *sedimentation*. In some treatment systems employing two or more stages of treatment and clarification, the terms *primary*, *secondary*, and *final* clarification are used. *Primary clarification* is the term normally used for the first clarification process in the system. This process is used to remove the readily settleable solids prior to subsequent treatment processes, particularly biological treatment. This treatment step results in significantly lower pollutant loadings to down stream processes and is appropriate for industrial wastewaters containing a high suspended solids content.

The actual physical sizing of the clarifier (depth, surface area, inlet structure, etc.) is highly dependent upon the quantity and composition of the wastewater flow to be treated. Because these criteria will vary substantially among industries, sizing must be done following a detailed study conducted on a site-by-site basis.

Clarification units can be either circular or rectangular and are normally designed to operate on a continuous, flow-through basis. Circular units are generally called *clarifiers*, whereas rectangular units are commonly referred to as *sedimentation tanks*.

Generally, clarifiers are designed on the basis of the type of suspended solids to be removed from the waste stream. There are three types of suspended solids. Class I solids are discrete particles that will not readily flocculate and are typical of raw influents. Class II particles are characterized by a relatively low solids concentration of flocculated material. These types of particles are usually found in wastewaters that have been subjected to chemical addition. Either of the above two particles may be found in industrial wastewaters. Class I particles are typical of physically manufactured operations, whereas Class II particles are more common to chemically manufactured operations. The last type are Class III particles. These

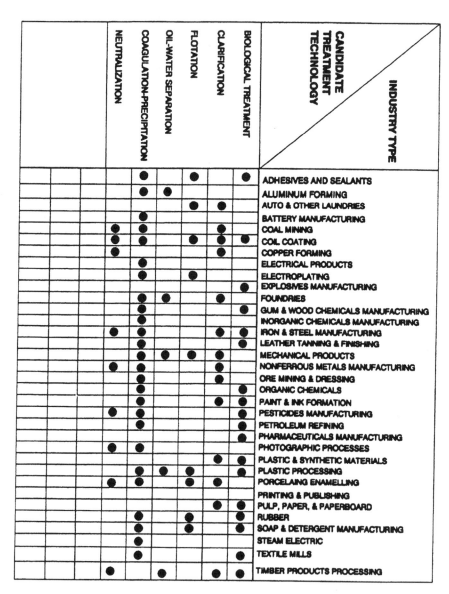

FIGURE 14.1 Candidate treatment technologies for various types of industries.

are normally the solids generated from a biological treatment process. Because of their poor settling properties, special care must be exercised in the design of their removal. Clarification is effective in removing a substantial portion of the suspended solids and BOD in an industrial wastewater. Not only does this aid in the performance of other downstream processes, but it also protects them from negative effects, such as clogging. This technology is relatively receptive to modifications, such as chemical addition, which allows for alterations to compensate for changes in wastewater characteristics. Finally, clarification is a relatively simple operation, which requires neither sophisticated operation nor highly skilled operators. The disadvantages of clarification as a component of an industrial wastewater treatment system include high capital cost and limitations on specific pollutant loadings and flows.

Flotation. In contrast to clarification, which separates suspended particles from liquids by gravitational forces, flotation separates these particles by their density, through the introduction of air into the system. Fine bubbles adhere to, or are absorbed by, the solids, which are then lifted to the surface. There are several methods of achieving flotation. In one flotation method, dissolved air flotation, small gas bubbles are generated as a result of the precipitation of a gas from a solution supersaturated with that gas. Supersaturation occurs when air is dispersed through the sludge in a closed pressure tank. When the sludge is removed from the tank and exposed to atmospheric pressure, the previously dissolved air leaves solution in the form of fine bubbles.

Flotation separator tanks can be either rectangular or circular in shape and constructed of either concrete or steel. Tanks are constructed with equipment to provide uniform flow distribution at the inlet, to provide pressurized gas, and to skim off float material. In designing a dissolved air flotation separator system, the following variables are typically considered: full, partial, or recycled pressurization; feed characteristics; surface area; float characteristics; hydraulic loading; chemical usage; type of pressurization equipment; and operating pressure.

Flotation separation is an appropriate technology for treating suspended solids and oil and grease in industrial waters. Generally, the process will achieve 40–65% suspended solids removal and 60–80% of oil and grease removal. The reduction of BOD levels will also be realized through the removal of suspended solids. The addition of chemicals prior to actual flotation can significantly improve the performance of this technology.

In addition to the pollutant removals mentioned, flotation has other advantages. The resulting float material may be used as an auxiliary fuel source. Minimized construction costs may result because the higher overflow rates and shorter detention times require smaller tanks. Also, because of the short detention periods and the presence of DO in the wastewater, odor problems are kept at a minimum. Some disadvantages associated with this type of treatment process are high operations and maintenance costs, inability to remove "heavy" suspended solids, and the requirement of more skilled personnel than for conventional settling processes.

Oil–Water Separation. In practically all manufacturing industries, oil and grease can be found in a plant's wastewater. They generally result from equipment lubrication, accidental spills, and similar incidence. However, for some industries, such as petroleum and edible oil refining, oil compounds can represent a significant constituent of their flows. This is due to the use of these oils as the raw materials in production. Not only can the oil have a detrimental effect on the performance of a wastewater treatment system, but it also represents a valuable raw material being wasted. Thus, specific treatment processes have been developed to separate the major portion of the oil from the waste stream. These are called oil–water separators.

The configuration of the separator is that of a flow-through tank. The basic principle by which oil–water separators work is the differential between the specific gravities of water and the oils to be removed. Since the oils have lower specific gravities, they will rise to the top of the unit while the heavier water sinks to the bottom. An important consideration in the separator design is the oil globule size, since Stokes' law for terminal velocity of spheres in a liquid medium will determine the rate at which the oil rises. Retaining baffles and skimmers capture the oil compounds as the separated water leaves the tank. Sludge collectors can be used to scrape the bottom of the tank to remove any deposited solids.

The major advantage of oil–water separators is their ability to treat wastewater that is heavily laden with oil compounds. Because of the design of these separators, they represent a very simple treatment operation that minimizes personnel requirements. Since this technology relies on natural forces rather than on chemicals or aeration, its operating costs are minimized. Also, it results in a more "pure" oil, which can make recycling much easier.

The disadvantages of oil–water separators are as follows. The ability to remove only specific size oil globules can result in very small particles passing through the system. Also, if the wastewater has a high solids content and the oil and solids become mixed, the resultant globules may be too heavy to float to the surface and thus remain suspended. Finally, improperly maintained units are subject to odor problems.

Chemical Treatment Processes

Coagulation–Precipitation. The nature of an industrial wastewater is often such that conventional physical treatment methods will not provide an adequate level of treatment. Particularly, ordinary settling or flotation processes will not remove ultrafine colloidal particles and metal ions. In these instances, natural stabilizing forces (such as electrostatic repulsion and physical separation) predominate over the natural aggregating forces and mechanisms, namely, van der Waals forces and Brownian motion, which tend to cause particle contact. Therefore, to adequately treat these particles in industrial wastewaters, coagulation–precipitation may be warranted.

The first and most important part of this technology is coagulation, which involves two discrete steps. Rapid mixing is employed to ensure that the chemicals are thoroughly dispersed throughout the wastewater flow for uniform reaction. Next, the wastewater undergoes flocculation, which provides for particle contact so that the particles can agglomerate to a size large enough for removal. The final part of this technology involves precipitation. This is really the same as settling and thus can be performed in a unit similar to a clarifier.

Coagulation–precipitation is capable of removing from industrial wastewater pollutants such as BOD, COD, and TSS. In addition, depending upon the specifics of the wastewater being treated, coagulation–precipitation can remove additional pollutants such as phosphorus, nitrogen compounds, and metals. This technology is attractive to industry because a high degree of clarifiable and toxic pollutants removal can be combined in one treatment process. A disadvantage of this process is the substantial quantity of sludge generated, which presents a sludge disposal problem.

Neutralization. In virtually every type of manufacturing industry, chemicals play a major role. Whether they result from the raw materials or from the various processing agents used in the production operation, some residual compounds will ultimately end up in a process wastewater. Thus, it can generally be expected that most industrial waste streams will deviate from the neutral state (i.e., will be acidic or basic in nature).

Highly acidic or basic wastewaters are undesirable for two reasons. First, they can adversely impact the aquatic life in receiving waters; second, they might significantly affect the performance of downstream treatment processes at the plant site or at a publicly owned treatment works. Therefore, in order to rectify these potential problems, one of the most fundamental treatment technologies, neutralization, is employed at industrial facilities. Neutralization involves adding an acid or a base to a wastewater to offset or neutralize the effects of its counterpart in the wastewater flow, namely, adding acids to alkaline wastewaters and bases to acidic wastewaters.

The most important considerations in neutralization treatment are a thorough understanding of the wastewater constituents so that the proper neutralizing chemicals are used, and proper monitoring to ensure that the required quantities of these chemicals are used and that the effluent is in fact neutralized. For acid waste streams, lime, soda ash, and caustic soda are the most common base chemicals used in neutralization. In the case of alkaline waste streams, sulfuric, hydrochloric, and nitric acid are generally used for neutralization. Some industries have operations that separate acid and alkaline waste streams. If properly controlled, these waste streams can be mixed to produce a neutralized effluent without the need for adding neutralizing chemicals.

Neutralizing treats the pH level of a wastewater flow. Although most people do not think of pH as a pollutant, it is in fact designated by the EPA as such. Since many subsequent treatment processes are pH-dependent, neutralization can be considered as a preparatory step in the treatment of all pollutants.

Eliminating the adverse impacts on water quality and wastewater treatment system performance is not the only benefit of neutralization. Acidic or alkaline wastewaters can be very corrosive. Thus, by neutralizing its wastewaters, a plant can protect its treatment units and associated piping. The major disadvantage of neutralization is that the chemicals used in the treatment process are themselves corrosive and can be dangerous.

Biological Treatment Processes

For many industrial wastewaters, one of the most important concerns has to do with those constituents which can exert an oxygen demand and have an impact on receiving waters. Although most industries are discussed in terms of the toxics they discharge, they are also significant sources of biochemical oxygen demand (BOD) and chemical oxygen demand (COD). In many instances, the most appropriate industrial treatment technology for removing this oxygen-demanding pollutant is biological treatment. Discussion in this section will be limited to those processes frequently used in the industrial field: aerobic suspended growth processes (activated sludge), aerobic contact processes, aerated lagoons (stabilization ponds), and anaerobic lagoons.

Aerobic Suspended Growth Processes (Activated Sludge). An aerobic suspended growth process is one in which the biological growth products (microorganisms) are kept in suspension in a liquid medium consisting of entrapped and suspended colloidal and dissolved organic and inorganic materials. This biological process uses the metabolic reactions of the microorganisms to attain an acceptable effluent quality by removing those substances exerting an oxygen demand. Depending on the type of material in the raw wastewater stream, this process may be preceded by one or more other treatment technologies (i.e., clarification, oil and grease removal, etc.) to improve removal efficiencies.

In the suspended growth processes, wastewater enters a reactor basin, concrete–steel–earthen tank(s) where microorganisms are brought into contact with the organic components of the wastewater by some type of mixing device. This mixing device not only maintains all material in suspension but also promotes transfer of oxygen to the wastewater, thus providing oxygen necessary for sustaining the biological activities in the reactor basin. The organic matter in the wastewater serves as a carbon and energy source for microbial growth and is converted into microbial cell tissue and oxidized end products, mainly carbon dioxide. Contents of the reactor basin are referred to as *mixed liquor suspended solids* (MLSS) and consist mainly of microorganisms and inert and nonbiodegradable matter.

When the MLSS are discharged from the reactor basin, a means of separating them is normally provided. Concentrated microbial solids are recycled back to the reactor basin to maintain a concentrated microbial population for degradation of the wastewater. Because microorganisms are usually synthesized in the process, a means must be provided for wasting some of the microbial solids. Wasting of the solids is usually done from the settling basin, although wasting from the reactor basin is an alternative.

The first suspended growth process, now called the *conventional activated sludge process*, was developed to achieve carbonaceous BOD removal. However, since its inception, many modifications to the basic process have taken place. The variations in the activated sludge process are too numerous to be discussed in detail here. However, some of these are step aeration, contact stabilization, aerated lagoon, and deep tank aeration.

Aerobic suspended growth systems can be adapted to treat a wide range of industrial wastewaters. The process can be easily expanded to accommodate increased flows. However, these systems do have some drawbacks. Suspended growth systems generally perform best under uniform hydraulic and pollutant loadings. For some industries it is extremely difficult to maintain these conditions because of their manufacturing operations. A common event is a "shock" loading of high strength entering the treatment process, with the result being poor pollutant treatment. Also, certain toxic pollutants can kill microorganisms in the reactor basin, causing a loss of treatment from this part of the overall system. Operational skills and controls required to effectively operate an aerobic suspended growth system are higher than for most other biological processes. Finally, high energy costs can be expended in providing mixing and oxygen in the process to sustain microbial growth.

Excluding those waste streams with very high concentrations of toxic pollutants, aerobic suspended growth systems can be used to treat any wastewater containing biodegradable matter. Removal efficiencies in excess of 90% for carbonaceous BOD have been achieved through this process.

Aerobic Attached Growth Processes. An aerobic attached growth process is one in which the biological growth products (microorganisms) are attached to some type of medium (i.e., rock, plastic sheets, plastic rings, etc.) and where either the wastewater trickles over the surface or the medium is rotated through the wastewater. The process is related to the aerobic suspended growth process in that both depend upon biochemical oxidation of organic matter in the wastewater to carbon dioxide, with a portion oxidized for energy to sustain and promote the growth of microorganisms. There are three general types of aerobic attached growth systems: conventional trickling filters, roughing filters, and rotating biological contactors (RBCs).

There are several advantages for attached growth processes over other biological processes. First, microorganisms growth can be easily reinstituted in the case of an accidental kill. Second, since oxygen is supplied naturally the need for air- or oxygen-generating equipment is eliminated. This, along with a much more simple operation, lowers the requirements for highly skilled operational personnel. Both can result in substantial cost savings. The disadvantages are that attached growth treatment processes experience operating difficulty in cold climates. Enclosing the units for temperature protection can lead to other problems such as condensation. These units are also susceptible to clogging if dense media are used and/or high solid loadings are applied.

Excluding those waste streams with very high concentrations of toxic pollutants, aerobic attached growth processes can be used to treat any wastewater containing biodegradable matter. In general, the aerobic attached growth processes are not quite as efficient as the aerobic suspended growth processes in removing BOD suspended solids, and toxic pollutants.

Aerobic Lagoons (Stabilization Ponds). Aerobic lagoons are large, shallow earthen basins that are used for wastewater treatment by utilizing natural processes involving both algae and bacteria. The objective is microbial conversion of organic wastes into algae. Aerobic conditions prevail throughout the process.

In aerobic photosynthesis, the oxygen produced by the algae through the process of photosynthesis is used by the bacteria in the biochemical oxidation and degradation of organic waste. Carbon dioxide, ammonia, phosphate, and other nutrients released in the biochemical oxidation reactions are, in turn, used by the algae, forming a cyclic–symbiotic relationship. Aerobic lagoons are used for treatment of weak industrial wastewater containing negligible amounts of toxic and/or nonbiodegradable substances.

Anaerobic Lagoons. Anaerobic lagoons are earthen ponds built with a small surface area and a deep liquid depth of 8–20 ft. Usually these lagoons are anaerobic throughout their depth, except for an extremely shallow surface zone. Once greases form an impervious layer, completely anaerobic conditions develop. In a typical anaerobic lagoon, raw wastewater enters near the bottom of the lagoon (often at the center) and mixes with the active microbial mass in the sludge blanket, which is usually about 6 ft deep. The discharge is located near one of the sides of the lagoon, submerged below the liquid surface. Excess undigested grease floats to the top, forming a heat-retaining and airtight cover. Excess sludge is washed out with the effluent. Anaerobic lagoons are effective prior to aerobic treatment of high-strength organic wastewater that also contains a high concentration of solids. Under optimal operating conditions, BOD removal efficiencies of up to 85% are possible.

The advantage in using either aerobic or anaerobic lagoons are low cost (excluding land if not readily available); simplicity of operation; low operation and maintenance cost; and when designed properly, high reliability. The disadvantages in using any lagoon process are high land requirements; possible odor emissions; and the potential for seepage of wastewater into groundwater unless the lagoon is adequately lined. In addition, in most locales of the United States there are seasonal changes in both available light and temperature. Typically in the winter, biological activity decreases because of a reduction in temperature.

14.4   CONTROL OF TREATED EFFLUENT

The development of advanced treatment technologies, along with an increasing scarcity of fresh water, has led to marked changes in effluent management. Numerous strategies for purified wastewater reuse are presently being employed in ways appropriate to the particular industrial operation. The combination of scarcity of water with increasingly stringent regulations has made effluent disposal into natural receiving bodies the option of last resort.

Water Reuse and In-Plant Wastewater Control

While it is theoretically possible to reuse ("close up") all the water of many industrial processes, inevitable limits will be encountered as a result of quality control. In a paper mill, for example, a closed system will result in increasing concentrations of dissolved organic solids. Such a buildup will lead to an increase in the costs of slime control, greater downtime on the paper machines, and possible discoloration of the paper stock. The goal is to maximize reuse while maintaining the integrity of the production process.

In the design of water reuse strategies, the particular water requirements of the process must be addressed. For example, water use by hydropulpers in a paper mill does not have to be treated for suspended solid removal. However, if these solids are not removed from shower water used on the paper machines, clogging of the shower nozzles will result. In the production of produce, such as tomatoes, protection against microbial contamination through chlorination is of greater importance than overall purification.

In the case where two or more waste streams are generated, it is often necessary that they be separated. This is all the more important when mixing can result in health hazards. In a plating plant where both acidic metal rinses and cyanide streams are generated, the lack of separation can lead to the production of toxic hydrogen cyanide.

It is often the case that only a portion of the waste flow contains the majority of the suspended solids loading. Treatment for solids removal would then only be necessary on this part of the waste stream. Such is the case in a tannery where nearly 60% of the flow from the beam house yields 90% of the suspended solids.

Where little, if any, contamination of waste streams has occurred, such as in cooling water, the segregation of these effluents can allow for direct reuse or discharge into a receiving water.

Stormwater Control

In most industrial wastewater treatment plants, containment and control of pollutional discharges from stormwater is deemed essential. Present strategies include adequate diking around process sewers. The design of holding basins for temporary storage of contaminated stormwater is based on storm frequency for the particular region. Once collected, the water is passed through the treatment plant at a controlled rate.

Effluent Disposal

That portion of wastewater that has not been reused must be disposed of in the environment, where it reenters the hydrologic cycle. Disposal thus may be considered a first step in the long-term, global process of reuse.

The most common strategy for disposal of treated wastewater effluent is release into ambient waters. Where streams or rivers of sufficient volume or flow are not available, it may be necessary to directly discharge treated wastewater into lakes or reservoirs. Because of their tremendous assimilative capacity, oceans and large lakes (such as the Great Lakes) are used by many communities for wastewater disposal. Discharge in these conditions is accomplished with the use of pipes or tunnels laid on, or buried in, the floor of the water body.

In very dry regions, land application is often utilized. Where successful, effluent will seep into the ground and replenish the underground aquifers. At the same time, a portion of the wastewater will return to the hydrologic cycle through evaporation.

Of principal importance in wastewater disposal is the consideration of environmental impact. Environmental regulations, criteria, policies, and reviews protect ambient waters from negative impacts of treated wastewater discharges. When selecting discharge locations and outfall structures, as well as the level of treatment required, one must bear this in mind. Both treatment and disposal of an industrial effluent must be considered when designing a wastewater system for the facility (10).

14.5 SLUDGE MANAGEMENT

The utilization and disposal of sludge is jointly addressed under the Clean Water Act and the Resource Conservation and Recovery Act. Both of these federal laws emphasize the need to employ environmentally sound sludge management techniques and to use sludge beneficially whenever possible. At the same time, the national requirements for improved wastewater treatment will result in the production of a greater quantity of residuals, and possibly more concentrated forms of contaminants will be present in these residuals than ever before. The permits required for effluent discharge from sewage treatment plants will be affected by, and will contain, provisions related to the sludge management techniques employed by the facilities.

Prior to utilization or disposal, sludge may be stabilized to control odors and reduce the number of disease-causing organisms. The sludge may also be dewatered to reduce the volume to be transported or to prepare it for final processing. The liquid sludge, which contains 90–98% water, can be partially dewatered by a number of processes. Vacuum filtration, pressure filtration, and centrifugation are three of the more common dewatering technologies presently being employed.

Digestion of sludge is accomplished in heated tanks where the material can decompose aerobically and the odors can be controlled. Anaerobic sludge digestion has the added benefit of producing methane gas, which may be used by the same treatment plant as an energy source.

Until recently, all but 20% of generated sludge had been incinerated, landfilled, or dumped into the ocean. Currently, much more attention is being given to sludge utilization by application to land as a soil conditioner or fertilizer, and by combustion in facilities to recover energy.

Liquid digested sludge has been used successfully as a fertilizer and for restoring areas disrupted by strip mining. Under this sludge management approach, digested sludge in semiliquid form is transported to the spoiled areas. The slurry, containing nutrients from the wastes, is spread over the land to give nature a hand in returning grass, trees, and flowers to barren land. Restoration of the countryside will also help control the flow of acids that drain from mines into streams and rivers, endangering fish and other aquatic life and adding to the difficulty of reusing the water.

14.6 RECENT DEVELOPMENTS IN INDUSTRIAL WASTEWATER TREATMENT

With the tremendous increase in recent years in the overall generation of industrial pollutants, along with the escalating costs of pollution control, increasing efforts are being made with regard to pollution prevention. The policies of the U.S. EPA now give the highest priority to source reduction, followed by recycling, with pollution control as the final alternative.

One example of this is the National Pretreatment Program, which is intended to eliminate the discharge of pollutants into the nation's waters. The major objective of the program is to protect publicly owned treatment works (POTWs) and the environment from the hazards of various toxic and other dangerous pollutants. Water quality-based toxics limits and monitoring requirements are becoming a common provision in the National Pollution Discharge Elimination System (NPDES) permits of POTWs. Pretreatment of toxic pollutants is necessary in order to comply with these water quality-based toxics requirements (11).

Industry is responding in new, innovative ways to these mandates. The Union Carbide plant at South Charleston, West Virginia, is an example of what is happening in the organic chemicals industry. Through various in-plant changes, the average waste flow of 11.1 million gallons per day (MGD) and waste load of 55,700 pounds of BOD per day was reduced to a flow of 8.3 MGD and load of 37,000 pounds of BOD per day. These changes were accomplished principally by replacement of scrubbers, segregation, collection and incineration of wastewater streams, and additional processing of collected tail streams.

Responding to the difficulty of many dairy industries to meet present federal wastewater standards, Search, Inc. of Norman, Oklahoma, has developed the vertically integrated environmental waste (VIEW) treatment system. The VIEW system utilizes nothing other than basic wastewater treatment technologies. In this system, a vertical unit has been constructed through the combination of an aeration basin, a trickling filter, and a clarifier. There, BOD levels are reduced to EPA standards through the continuous recirculation of plant effluent.

The largest water reclamation effort in California was initiated in the early 1990s by the Metropolitan Water District of Southern California and the West Basin Municipal Water District. By the end of the decade, the project is expected to be reclaiming nearly 70,000 acres of wastewater. District officials are planning to use the water for irrigation of schools and golf courses. In addition, the local Chevron and Mobil Refineries are to receive water for use in cooling towers (12).

14.7 SUMMARY

1. To meet wastewater management objectives, four major areas of concern must be addressed: (a) Sources and characteristics of wastes and wastewaters must be ascertained; (b) the particular type of wastewater treatment process, or combination of processes, must be determined; (c) wastewater effluent must be effectively controlled; and (d) sludge and other solid wastes must be properly managed.

2. All the various types of pollutants found in industrial wastewaters can be classified into eight categories: common sewage and other oxygen-demanding wastes; disease-causing agents; plant nutrients; synthetic organic chemicals; inorganic chemicals and other mineral substances; sediments; radioactive substances; and heat.

3. In the development of an effective strategy for wastewater control, an industrial waste survey is usually required to determine the character of a particular industry's waste load. The general procedure can be summarized in four steps: (a) Develop a sewer map from consultation with the plant engineer and an inspection of the various process operations; (b) establish sampling and analysis schedules; (c) develop a flow-and-material-balance diagram; and (d) establish statistical variation in significant waste characterization.

4. Controlling waste discharges must be a quantitative endeavor. Direct measurement must be made of the pollutants, along with (a) a prediction of the impact of the pollutant on water quality and (b) a determination of acceptable levels for the intended uses of the water.

5. The biochemical oxygen demand (BOD) test is used to determine the amount of biodegradable contents in a sample of wastewater. The chemical oxygen demand (COD) test measures the oxygen demand of the overall oxidizable material. A third method for measuring the organic matter in a wastewater is the total organic carbon (TOC) test. The BOD, COD, and TOC tests provide estimates of the general organic content of a wastewater. The solids analysis focuses on the

quantitative investigation and measurement of the specific content of solid materials present in a wastewater. Additional measurements of pH, nitrogen, and phosphorus levels of wastewaters are often required when complex pollution conditions exist.

6. Industrial users discharge tremendous volumes of wastewater, and hence considerable pollutant loadings, to municipal treatment plants and receiving waters. Therefore, effective industrial wastewater treatment is required to attain acceptable discharge concentrations and protect the water quality of receiving waters and of the surrounding ecosystems.

7. There are a variety of industrial technologies available. This chapter has addressed the technologies which are currently among the most frequently used. Selection of the appropriate technology can be difficult and must be made on a site-by-site basis.

8. The constituent composition of an industry's wastewater is likely to narrow down the selection to technologies which remove pollutants present in that wastewater. Necessary pollutant removals which are required to meet discharge standards will further reduce the list of applicable technologies. It is often necessary to combine two or more of the technologies to achieve the desired pollutant removal.

9. The development of advanced treatment technologies, along with an increasing scarcity of fresh water, has led to marked changes in effluent management. Numerous strategies for purified wastewater reuse are presently being employed in ways appropriate to the particular industrial operation.

10. In most industrial wastewater treatment plants, containment and control of pollutional discharges from stormwater is deemed essential. Present strategies include adequate diking around process areas, storage tanks, and liquid transfer points, with drainage into a process sewer.

11. That portion of wastewater that has not been reused must be disposed of in the environment, where it reenters the hydrologic cycle. Disposal thus may be considered a first step in the long-run, global process of reuse.

12. The utilization and disposal of sludge is jointly addressed under the Clean Water Act and the Resource Conservation and Recovery Act. Both of these federal laws emphasize the need to employ environmentally sound sludge management techniques and to beneficially use sludge whenever possible.

13. With the tremendous increase in recent years in the overall generation of industrial pollutants, along with the escalating costs of pollution control, increasing efforts are being made at attaining pollution prevention. One example of this is the National Pretreatment Program which is intended to eliminate the discharge of pollutants into the nation's waters.

REFERENCES

1. Metcalf & Eddy, Inc., *Wastewater Engineering: Treatment, Disposal, and Reuse*, 3rd edition, McGraw-Hill, New York, 1991, p. 2.

2. U.S. EPA, *Primer for Wastewater Treatment*, EPA Publication, Washington, D.C., 1980.

3. W. W. Eckenfelder, Jr. *Industrial Water Pollution Control*, 2nd edition, McGraw-Hill, New York, 1989.

4. D. A. Cornwell and D. L. Mackenzie, *Introduction to Environmental Engineering*, 2nd ed., McGraw-Hill, New York, 1991, pp. 265–266.

5. P. L. McCarty and C. N. Sawyer, *Chemistry for Environmental Engineering*, 3rd ed., McGraw-Hill, New York, 1978, p. 416.

6. Metcalf & Eddy, Inc., *Wastewater Engineering: Treatment, Disposal, and Reuse*, 3rd ed., McGraw-Hill, New York, p. 82.

7. U.S. EPA, *Pretreatment of Industrial Wastes*, EPA Publication, Washington, D.C., 1978.

8. U.S. EPA, *Innovative and Alternative Technology Assessment Manual*, EPA Publication, Washington, D.C., 1980.

9. Perry, Chemical Engineering Handbook, 7th edition, 1997. "Industrial Wastewater Treatment", pg 25-58 to 25-80.

10. Metcalf & Eddy, Inc., *Wastewater Engineering: Treatment, Disposal, and Reuse*, 3rd ed., McGraw-Hill, New York, p. 1195.

11. J. S. Kishwara, The National Pretreatment Program, Clearwaters, *Journal of the New York Water Pollution Control Association*, 8, Spring 1990.

12. *U.S. Water News*, 1 and 13, September 1991.

IV

SOLID AND RADIOACTIVE WASTE MANAGEMENT ISSUES

15

MUNICIPAL WASTE MANAGEMENT

15.1 INTRODUCTION

The United States has a population of almost 250 million people. On average, each American generates approximately 3–5 lbs of solid waste every day. That is equal to 1 billion lbs of waste requiring disposal every day. In addition to the waste generated by the residents of this country, waste is trucked in daily across the Canadian border. The municipal waste management problem is reaching crisis levels in several areas of this country, especially in highly urbanized areas where available open land for landfilling is limited.

Problems associated with waste disposal include siting a disposal facility, transporting the waste, and costs associated with implementation and operations. The major factors related to siting include lack of space and the "Not in My Back Yard" (NIMBY) syndrome. Most people realize we are facing a waste disposal crisis, but they do not feel that the solution should be located within their community.

This chapter looks at the problems associated with solid waste management and available disposal alternatives. For the purposes of this chapter, solid waste excludes sludges, industrial wastes, hazardous wastes, and radioactive wastes. In most instances, separate regulations and disposal requirements govern these materials. Instead, the focus here will be on wastes commonly generated by residential homes and commercial establishments, such as restaurants and retail stores.

15.2 REGULATORY STATUS

In the 1980s, the Environmental Protection Agency (EPA) reviewed federal and state solid waste regulatory programs to identify any areas of inadequacy. In

October 1988, the EPA submitted the results of the study to Congress in a report entitled *Report to Congress: Solid Waste Disposal in the United States* (1). The results of this study confirmed that the United States was in the midst of a municipal solid waste disposal crisis. EPA data showed that in 1988 the nation generated nearly 180 million tons of municipal solid waste and that this quantity would likely grow to 216 million tons by the year 2000. This growing volume of waste was coupled with a steadily decreasing availability of disposal capacity. Since the mid-1980s, almost three-quarters of the nation's municipal landfills have closed as regulations governing land disposal have tightened (2). Today's disposal capacity crisis is further compounded by the difficulty in siting new solid waste management facilities.

In response to the growing national concern about the solid waste disposal crisis, the EPA developed a national strategy for addressing municipal solid waste management problems. This strategy was set out in a document entitled *The Solid Waste Dilemma: An Agenda for Action*, which the EPA issued in final form in February 1989 (3). The strategy described a wide range of activities that were to be undertaken by various parties, including government, industry, and the general public, to bring municipal solid waste management problems under control.

The cornerstone of the strategy was "integrated waste management," where the following solid waste reduction and management options worked together to form an effective system: source reduction, recycling, combustion, and landfilling. In keeping with the EPA's policy of pollution prevention, which is discussed below, the strategy strongly encouraged the use of source reduction (i.e., reduction of the quantity and toxicity of materials and products entering the solid waste stream) followed by recycling as first steps in a solid waste management system. These techniques could then be complemented by environmentally sound combustion and landfilling.

The strategy set out three national goals for municipal solid waste management: Increase source reduction and recycling, increase disposal capacity and improve secondary material markets, and improve the safety of solid waste management facilities. To promote the attainment of the first goal, the EPA established a national goal of 25% source reduction and recycling of municipal solid waste by 1992.

The EPA identified a series of actions or activities that were to be carried out to achieve the above national goals. These activities sought to increase use of source reduction, increase recycling, and improve the design and management of municipal waste combustors and landfills. The following describes some of the most significant actions completed in implementing these strategies.

The highest priority in the EPA's strategy for addressing the nation's solid waste problems is increasing source reduction. The EPA has taken several steps to promote the reduction of the quantity and toxicity of materials entering the municipal solid waste stream. First, the EPA convened a steering committee of national source reduction experts to evaluate and develop recommendations on specific opportunities for source reduction, along with incentives for promoting source reduction. The results of this project were published in a report entitled *Getting at the Source: Strategies for Reducing Municipal Solid Waste* (4). The EPA also completed a review and analysis of economic incentives, including volume-based pricing schemes, to promote increased source reduction.

With regard to toxicity reduction, the EPA completed a report identifying the sources of lead and cadmium in the waste stream and will issue a report identifying potential substitutes for these constituents in products (5). The EPA is also examining mercury in the municipal waste stream. In March 1990, the EPA completed a comprehensive report to Congress on methods for managing plastic waste (6). This report examined the full range of options for addressing plastic wastes, including source reduction.

To increase recycling nationwide, the EPA undertook a number of efforts to stimulate markets for secondary materials and to promote increased separation, collection, processing, and recycling of waste. In the area of markets for secondary materials, the EPA produced a report examining disincentives to recycling and has conducted a series of market studies on various components of municipal solid waste (paper, glass, aluminum, tires, and compost). To improve federal procurement of recycled materials, the EPA finalized four procurement guidelines (retread tires, building insulation products, paper and paper products containing recovered materials, and lubricating oils containing re-refined oil) in 1988 and 1989 and began examining future candidate materials (other building and construction materials) for additional procurement guidelines. The EPA also funded the establishment of a National Recycling Institute, composed of high-level representatives from business and industry, to identify and resolve issues in recycling. Recycling and composting have been the fastest growing methods of waste management, accounting for 27% of waste management in 1996, up from 10% in 1986 (2).

In 1990, the EPA took a major step forward in improving the design and management of municipal waste combustion facilities. The EPA published a final municipal waste combustion rule for air emission standards for new and existing municipal waste combustors on February 11, 1991. The rule included requirements for good combustion practices and air emission control of particulates, organics, NO_x, and acid gases (see 56 FR 5488). Municiple combustion facilities which generate steam or electricity, also known as waste-to-energy facilities, have increased their capacity to manage waste tenfold during the 1980s and early 1990s. They now manage 17% of the nation's municiple solid waste (MSW) (2).

On October 9, 1991, the EPA published a final rule which set forth the requirements for owners and operators of municipal solid waste landfills. The ruling included requirements for facility design, operating criteria, groundwater monitoring, and facility closures. It also restricted the siting of new landfills or landfill expansions in areas that are especially vulnerable to contamination (7).

In most areas of the country, the lead role in transforming solid waste management has been played by state and local governments. These agencies generally decide how waste will be managed — whether by landfill, incineration, recycling, composting, waste reduction, or a combination thereof. States set standards for the resulting facilities, and funding for MSW programs comes overwhelmingly from state and local sources. As discussed above, the federal government has also played an important role in MSW management in recent years, setting minimum national landfill standards under the Resource Conservation and Recovery Act (RCRA), setting incinerator and landfill emission standards under the Clean Air Act, and promoting recycling through the use of federal procurement policy (2).

15.3 DETERMINING WASTE GENERATION

Before a determination can be made of the type of waste management plan which would be most effective, the quantity and type of waste generated within the area to be managed must be estimated to the highest degree of accuracy possible. This is necessary to properly size and design a waste management facility, whether it be a composting, recycling, resource recovery, or landfill facility. It is also necessary in order to estimate the costs of implementing and operating the program, as well as the amount of labor which will be required. The facility should be able to handle both current waste generation loads and any anticipated increased in generation.

Three methods are commonly used to estimate the quantity of waste which is generated. The most accurate means of estimating the waste quantity is to weigh the waste requiring disposal. This is most easily done at the transfer or disposal facility. All the waste entering the facility should be weighed for a defined period of time, for example, one year. This total annual quantity can then be multiplied by a ratio representing any projected increases in population to determine the anticipated generation rate for future years.

The second method is to determine the volume of waste which is being generated and use known density factors to convert this into the associated weight. A record of the volume of waste generated is usually kept by the waste hauler and/or the disposal facility. As a rule of thumb, it is generally assumed that a mixture of uncompacted solid waste has a density of approximately $100 \, \text{lbs/yd}^3$, whereas compacted waste has a density of approximately $300-500 \, \text{lbs/yd}^3$. Table 15.1 lists density data for various waste streams and individual waste items. A growth factor should be applied to determine future waste quantities. Waste volume is also important in determining the useful life of a landfill.

The third, and least accurate, method for determining waste generation rates in a region is to determine the population of the area and then multiply this by typical waste generation factors. As stated in the introduction to this chapter, it is generally assumed that each person generates $3-5 \, \text{lbs}$ of waste per day. This factor can be used to determine a very rough estimate of waste generation.

Facility sizing will also have to take into account seasonal variations. For example, winter recreational areas, such as ski resorts, will tend to generate peak quantities of waste during the winter time, whereas beach-front communities will generate peak quantities during the summer. Agricultural areas will generate peak waste quantities during harvest season. Minimum, maximum, and average waste generation should all be determined.

In addition to the quantity of waste generated, it is also necessary to determine the type of waste generated. When implementing a recycling program, the types of materials generated must be known to select processing equipment and to determine markets for the recovered materials. In sizing a resource recovery facility, knowing the type of waste disposed of is critical to determining the size of the furnace and the steam and electrical generation equipment.

The most accurate method of determining the type of waste generated is to conduct an audit of the waste stream. This is done by taking a representative sample of the waste stream and sorting it into its individual components. The typical components include batteries, cardboard, construction debris, food waste, glass, metals, paper, plastic, rubber, textiles, tires, wood, yard waste, and miscellan-

TABLE 15.1   Industrial solid waste density data[a]

Waste	Density (lb/yd^3) As Discarded
Department store waste	80
Hospital waste (not research)	100
School waste including lunch program	110
Supermarket waste	100
Bakelite	600
Bitumen waste	1500
Brown paper	135
Cardboard	180
Cork	320
Corn cobs	300
Corrugated paper (loose)	100
Disposable hospital plastics	120
Grass, green	75
Hardboard	900
Latex	1200
Magazines	945
Meat scraps	400
Milk cartons, coated	80
Nylon	200
Paraffin — wax	1400
Plastic-coated paper	135
Polyethylene film	20
Polypropylene	100
Polystyrene	175
Polyurethane (foamed)	55
Resin bonded fiberglass	990
Rubber — synthetics	1200
Shoe leather	540
Tar paper	450
Textile waste (nonsynthetic)	280
Textile waste (synthetic)	240
Vegetable food waste	375
Wax paper	150
Wood	300

[a]This table shows the various weights of materials commonly encountered in incinerator applications. The values given are approximate and may vary based on their exact characteristics or moisture content.

eous fines. For the purposes of developing a recycling program, certain categories may be broken down further. For example, glass is typically sorted by color: clear, green, brown, and miscellaneous.

Another method of determining waste type is to multiply the quantity of waste by known ratios for common waste type distributions. Data, such as provided in Table 15.2, are available in reference books for various waste streams. The accuracy of this method will be affected by factors such as the age and affluence of the community.

TABLE 15.2 Municipal solid waste composition

Component	Percent by Weight
Food wastes	15
Paper	38
Cardboard	5
Plastics	5
Leather and textiles	2
Yard wastes	10
Wood	3
Miscellaneous metals and glass	18
Dirt, ashes, etc.	4

15.4 REUSE, REDUCTION, AND RECYCLING

As mentioned above, the EPA instituted a national waste reduction and recycling goal of 25% by the year 1992. Many states implemented programs of their own which were even more ambitious. For example, the State of New York intended to reduce their waste stream by 40% by the year 1997. The three strategies involved in waste minimization are reuse, reduction, and recycling.

The first strategy, which is most effective on an individual basis, is reuse. People must make a concentrated effort to reuse existing materials, rather than to purchase new goods. For example, items such as aluminum or plastic food containers can be reused several times. Plastic and paper bags can be returned to the store and reused when grocery shopping, or cloth sacks can be used in place of these items. Electronics and major appliances should be repaired if possible, rather than discarded. Clothing and furniture can be donated to charities such as Goodwill, the Salvation Army, local churches, and shelters for the abused or the homeless. Books can be donated to the local library. Many construction materials can often be reused.

The second strategy is waste reduction. Again, the local grocer is a good starting point. When purchasing only one or two items, carry the items out of the store without bagging them. Purchase items in bulk quantities, because this cuts down on the amount of packaging requiring disposal. At the manufacturing level, vendors can limit the amount of packaging used on their products. In some instances, the one-time use of disposable items can be replaced with reusable items. However, this often has other associated adverse environmental effects. The controversy over disposable versus reusable diapers is a good example of this.

The third strategy for waste minimization is recycling. Many materials, including aluminum, batteries, cardboard, glass, metals, newspaper, paper, tin cans, and tires, have the potential to be recycled. Many states have instituted a Bottle Bill to encourage recycling of glass and aluminum cans. Separation of the recyclables from the waste stream can be accomplished on an individual basis at the point of generation or can be done at a municipal facility.

A number of mechanical devices are employed at municipal separation facilities. Conveyors are used to move waste past workers who pick out specified recyclables by hand. This is a very unsophisticated method for recovering materials; however,

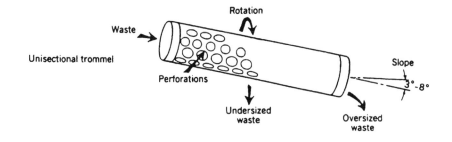

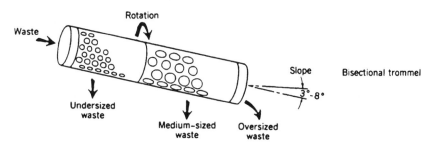

FIGURE 15.1 Trommel.

its main disadvantage is worker safety. The use of magnets is fairly successful in recovering ferrous materials from the waste stream. Magnetic drums or conveyors are often used for this purpose. Air classifiers are used to separate heavy and light fractions of a waste stream. Air is blown through the waste stream. The lighter fraction, which may be comprised of such items as paper and plastics, would be blown upwards, whereas the heavier fraction, which may be comprised of such items as glass and textiles, would fall to the bottom. Air classifiers are not very reliable. Trommels, such as the one depicted in Figure 15.1, are used to sort materials by size. The trommel rotates and tumbles the waste. As the material tumbles through the unit, holes of increasing size allow materials to pass through. The most commonly encountered problem associated with the use of trommels is blockage of the holes. Various technologies are also employed for separating glass, paper, or plastic. The systems are usually very complex and expensive, with limited reliability.

Composting is another process commonly associated with recycling. Composting is the microbiological decay of organic materials in an aerobic environment. Materials which potentially could be composted include agricultural waste, grass clippings, leaves and other yard waste, food waste, and paper products. Many municipalities have implemented leaf composting programs. Three methods for leaf composting are the leaf pile, the windrow method, and forced aeration (8).

The pile method typically requires an acre of land for every 8000–12,000 yd³ of leaves to be composted. Buffer zones are also required. The site surface must be earthen. The leaf pile should be maintained at a height of 6 ft. Piles may be combined after initial shrinkage. The compost process time is generally 2–3 yr. Generation of odors may be high upon initial disturbance of the piles. The end product of this process may have limited marketability because of its appearance; however, shredding of the material may improve its resale.

The windrow method typically requires an acre of land for every 3500–8000 yd^3 of leaves to be composted. Buffer zones are also required. The site surface can be earthen or paved. The leaf pile, or windrow, should be maintained at a height of 6 ft. Piles may be combined after initial shrinkage. Air is pulled through the piles, and the entire pile is turned in accordance with moisture and temperature data. Specialized mechanical equipment can be used for turning the windrows. The compost process time varies with the frequency of turning of the windrows, but typically is 6–12 mo. There may be some generation of odors upon initial disturbance of the piles; however, this can usually be reduced through proper management. The end product of this process is generally of acceptable quality; however, screening of the compost will give a more uniform product.

The forced aeration method typically requires an acre of land for every 5000–10,000 yd^3 of leaves to be composted. Buffer zones are also required. The site surface can be earthen or paved. Air is blown through the pile of leaves. Plastic piping is used to direct the air through the piles. An organic material, such as wood chips or compost, is used as a pile cover for insulation. The frequency and time of aeration is controlled by a timer switch or temperature controller. The compost process time typically takes 4–6 mo. Odor generation is minimal if the system is properly designed, installed, and operated. Field experimental data available on the quality of the end product of this process are limited.

These methods can be adapted to compost other organic materials as well. In addition, in-vessel composting applications have been successfully attempted for composting mixed organic wastes. The finished compost product can be used as a soil additive (topsoil), mulch, or soil conditioner, and it can also be used for landscaping. Compost is typically not used as a fertilizer because of its low nitrogen content.

A major problem with the implementation of any recycling program is the public perception of associated costs. Many people believe that recycling is free or, at the very least, inexpensive. However, in most instances, that is not the case. Costs are associated with every aspect of the program, including collection of the materials, processing of the materials and disposing of any residues. Purchase of new equipment or the retrofitting of existing equipment which is used to separate or recycle materials, or to incorporate recycled materials into a process, is often very expensive. Direct operational costs include labor and utilities. With the exceptions of glass and aluminum, it is usually more cost effective to use virgin materials rather than recycled materials in manufacturing processes. Many times markets are not available for sorted materials. The plastics industry is representative of this problem. Although much research has gone into plastics recycling in the past few years, markets for both the segregated material and the end products are very limited. The public needs to realize that recycling is not cheap; and many times the cost of recycling is only offset by the avoided cost of disposal, rather than by any profits generated.

15.5 COMBUSTION

As space for new landfills and the capacity of existing systems becomes increasingly diminished, waste disposal problems continue to grow. Although incineration is often viewed unfavorably by the general public, it has several advantages. It provides for a significant reduction of both the volume and weight of solid wastes, which in turn extends the available life of existing landfills. Municipal incineration systems, or resource recovery facilities, also can provide steam and electricity generation for the surrounding community. However, as many people are aware, the disadvantages of incineration include high capital and operational expenditures, and requirements for skilled operators. Improper equipment or operations can lead to problems associated with air pollution and emissions deposition.

Municipal waste can be combusted in bulk form or in reduced form. Shredding, pulverizing, or any other size reduction method which can be used before incineration decreases the amount of residual ash, due to better contact of the waste material with oxygen during the combustion process (9). Shredded waste used as fuel is generally referred to as *refuse-derived fuel* (RDF) and is sometimes combined with other fuel types. Table 15.3 lists the American Society for Testing and Materials (ASTM) classification for RDF (10).

Good combustion depends on three principles, known as the three Ts: time, temperature, and turbulence. *Time* refers to providing adequate residence time of the combustible matter within the system. *Temperature* refers to the optimum temperature for complete combustion. *Turbulence* refers to the proper mixing of the flowing gases through the system. Every incinerator must be designed to optimize

TABLE 15.3   ASTM classification for refuse-derived fuel (RDF)

ASTM Classification	RDF Nomenclature	RDF Description
RDF-1	Raw	Solid waste used as a fuel in as-discarded form, without oversize bulky waste.
RDF-2	Coarse	Solid waste processed to a coarse particle size, with or without ferrous metal extraction, such that 95% by weight passes through a 6-in.^{2} mesh screen.
RDF-3	Fine or fluff	Solid waste processed to a particle size such that 95% by weight passes through a 2-in.^{2} mesh screen and from which the majority of metals, glass, and other inorganics have been extracted.
RDF-4	Powder	Solid waste processed into a powdered form such that 95% by weight passes through a 10-mesh (0.035-in.^{2}) screen and from which most metals, glass, and other inorganics have been extracted.
RDF-5	Densified	Solid waste that has been processed and densified into the form of pellets, slugs, cubettes, or briquettes.
RDF-6	Liquefied	Solid waste that has been processed into a liquid fuel.
RDF-7	Gaseous	Solid waste that has been processed into a gaseous fuel.

Source: American Society for Testing and Materials, Philadelphia, PA.

these three variables according to the waste type in order to provide complete combustion.

An incineration system is comprised of several components. It must have a waste feeding system, also referred to as a *loading* or *charging* system, to ensure uniform loading of the incinerator. The incinerator itself generally consists of a primary chamber, a secondary chamber, an auxiliary fuel system, air supply systems, a hearth or a grate area, and either moving grates or rams to move the waste and the ash through the unit. The incineration system must also have an ash removal system; both wet and dry ash removal systems are available. Air pollution control equipment will most likely be required on all new incineration systems. Many municipal incinerators also are equipped with steam and/or electricity generators.

The types of incinerators used in municipal waste combustion include fluidized bed incinerators, rotary waterwall combustors, reciprocating grate systems, and modular incinerators. The basic variations in the design of these systems are related to the waste feed system, the air delivery system, and the movement of the material through the system. As an illustration of typical system configurations, Figure 15.2 depicts a modular incinerator, and Figure 15.3 depicts a rotary combustor. Both are equipped with a heat recovery boiler. Several reference texts are available which provide further details on the various system designs (11,12).

Gross electric power output from a resource recovery system ranges from 340–600 kW · h per ton of raw solid waste incinerated. Output is dependent on the type of incineration technology utilized and the type of waste fed. Electricity generated

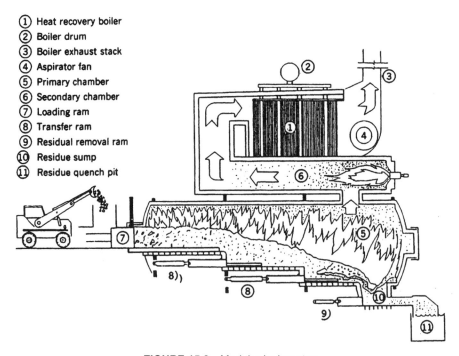

① Heat recovery boiler
② Boiler drum
③ Boiler exhaust stack
④ Aspirator fan
⑤ Primary chamber
⑥ Secondary chamber
⑦ Loading ram
⑧ Transfer ram
⑨ Residual removal ram
⑩ Residue sump
⑪ Residue quench pit

FIGURE 15.2 Modular incinerator.

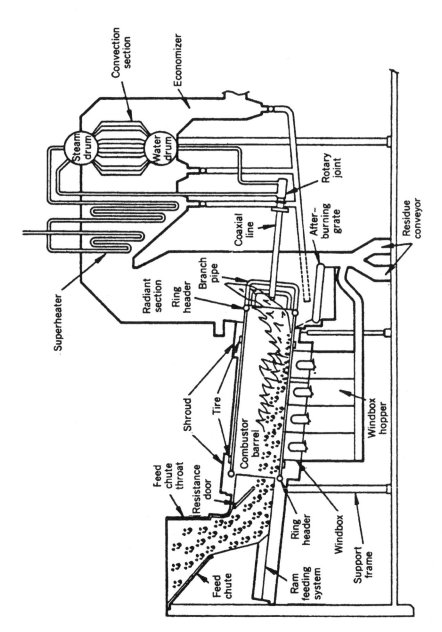

FIGURE 15.3 Westinghouse—O'Connor water-cooled rotary combustor and boiler.

by a resource recovery facility will usually be used to supply the total electrical need for in-house power consumption, which ranges from 10% to 15% of the gross amount generated. The remaining 85–90% can be sold to the local utility.

In general, 99% of incineration effluent gases consist of carbon dioxide, oxygen, nitrogen, and water. Their exact amounts depend on the proportion of excess air employed in combustion. Compared to other combustion processes which contribute gaseous pollutants to the atmosphere, the contribution of pollutants due to incineration of typical solid wastes is low. A prime example would be the total nitrogen and sulfur oxides produced during incineration of municipal wastes compared to the amounts generated from fossil fuel combustion in power plants. Municipal wastes generally make up only 1–10% (on a per-ton basis) of the amounts generated by power plants. The amount of sulfur in fossil fuel is thus 5–30 times the amount in solid wastes. The emissions per ton of nitrogen oxides from fossil fuels are 10 times greater than nitrogen oxide emissions from incineration. The emissions from incinerators which are of concern include acid gases (such as nitrogen oxides and sulfur oxides), heavy metals, dioxins, and furans. All of these emissions can be safely reduced through the use of air pollution control equipment, such as wet or dry scrubbers. The undesirable aspect of vapor plumes may accompany incineration processes when water is used in the air pollution control equipment process. These plumes generally condense and disappear within a few feet of the stack and are not usually harmful in of themselves. Further details on air pollution control equipment are provided in Chapter 7.

Because of the high degree of public opposition to these facilities, siting of the facility is a critical stage of implementation. It is generally recommended that these facilities be located in a commercial or industrial zone, away from residential areas. The facility should be designed to blend in with the surrounding facilities.

15.6 LANDFILLING

Because it is not possible to reuse, recycle, or incinerate the entire solid waste stream, a portion of the material must be landfilled. Landfills have been a common means of waste disposal for centuries. A process that originally was nothing more than open piles of waste has now evolved into a sophisticated facility. The criteria provided herein for municipal waste landfills are as set forth by the EPA (7). State-specific criteria may be even more stringent.

A municipal solid waste landfill (MSWLF) is defined as a discrete area of land or an excavation that receives household waste. It may also receive other types of Resource Conservation and Recovery Act (RCRA) subtitle D wastes, such as commercial solid waste, nonhazardous sludge, small-quantity generator waste, and industrial solid waste. It may not accept hazardous waste regulated under subtitle C of RCRA.

When siting an MSWLF, several restrictions must be considered. Any MSWLF located within 10,000 ft of any airport runway end used by turbojet aircraft or within 5,000 ft of any airport runway end used by only piston-type aircraft must demonstrate that the unit is designed and operated so that the facility does not pose a bird hazard to aircraft. Any MSWLF located in 100-yr floodplains must demonstrate that the facility will not restrict the flow of the 100-yr flood, reduce the temporary water storage capacity of the floodplain, or result in washout of solid waste so as to pose a hazard to human health and the environment. With a few exceptions, a new MSWLF cannot be located in wetlands, within 200 ft of a fault that has had displacement, or in seismic impact zones.

An MSWLF must be designed with a composite liner and a leachate collection system that is designed and constructed to maintain less than a 30-cm depth of leachate over the liner. A composite liner is a system consisting of two components; the upper component consists of a minimum 30-mm flexible membrane liner (FML), and the lower component consists of at least a 2-ft layer of compacted soil with a hydraulic conductivity of no more than 1×10^{-7} cm/sec. The FML component is installed in direct and uniform contact with the compacted soil component.

Unless the owner or operator of a landfill can adequately demonstrate that there is no potential for migration of hazardous constituents from the MSWLF unit to the uppermost aquifer during the active life of the unit and the postclosure period, the facility is required to have a groundwater monitoring system. The groundwater monitoring system consists of a sufficient number of wells, installed at appropriate locations and depths, to yield groundwater samples from the uppermost aquifer that represent that the quality of background groundwater has not been affected by leakage from the landfill. The groundwater monitoring program must include consistent sampling and analysis procedures that are designed to ensure monitoring results that provide an accurate representation of groundwater quality at the background and down-gradient wells.

Disposed solid waste must be covered with 6 in. of earthen material at the end of each operating day, or more frequently if necessary to control disease vectors, fires, odors, blowing litter, and scavenging. The concentration for methane gas generated by the facility cannot exceed 25% of the lower explosive limit for methane in the facility structure or exceed the lower explosive limit for methane at the facility boundary. Open burning of solid waste is prohibited at the facility.

Upon facility closure, a final cover system must be installed that is designed to minimize infiltration and erosion. Postclosure care must be conducted for 30 yr. This includes maintaining the integrity and effectiveness of the final cover, maintaining and operating the leachate collection system, groundwater monitoring, and maintaining and operating the gas monitoring system.

15.7 SUMMARY

1. The U.S. generates approximately 1 billion lbs of municipal waste every day.

2. In 1989, the EPA proposed an integrated waste management system of source reduction, recycling, combustion, and landfilling, and established a national goal of 25% source reduction and recycling of municipal waste.

3. The quantity and type of wastes generated within a community must be estimated before an appropriate waste management plan can be developed.

4. Three strategies employed in waste minimization are reuse, reduction, and recycling. Composting is generally considered to be a form of recycling.

5. Resource recovery facilities are advantageous because they provide for significant reduction of both the volume and weight of solid wastes, which in turn

extends the available life of existing landfills. These facilities can also provide steam and/or electricity for the surrounding community.

6. Landfills are not open dumps. Criteria for liners, groundwater monitoring, and soil cover are defined by the EPA.

REFERENCES

1. U.S. EPA, OSWER, *Report to Congress: Solid Waste Disposal in the United States*, EPA/530-SW-88-011B, October 1988.
2. E. McCarthy, IB10002: Solid Waste Issues in the 106th Congress, CRS Issue Brief for Congress, April 9, 1999.
3. U.S. EPA, OSWER, *The Solid Waste Dilemma: An Agenda for Action*, EPA/530-SW89-019, February 1989.
4. World Wildlife Fund and The Conservation Foundation, *Getting at the Source: Strategies for Reducing Municipal Solid Waste*, 1991.
5. U.S. EPA, OSW, *Characterization of Products Containing Lead and Cadmium in Municipal Solid Waste in the United States, 1970 to 2000*, EPA/503-SW-89-015, January 1989.
6. U.S. EPA, OSWER, *Report to Congress: Methods to Manage and Control Plastic Wastes*, EPA/530-SW-89-051, February 1990.
7. *U.S. Federal Register*, 40 CFR Parts 257 and 258, Solid Waste Disposal Facility Criteria; Final Rule. October 9, 1991.
8. CDEP, *Guidance on Leaf Composting*.
9. G. Geiger, "Incineration of Municipal and Hazardous Waste," *National Environmental Journal*, 1(2), Nov./Dec. 1991.
10. ASTM, Philadelphia, PA.
11. H. L. Hickman, et al., *Thermal Conversion Systems for Municipal Solid Waste*, Noyes Publications, Park Ridge, NJ, 1984.
12. APCA, Thermal Treatment of Municipal, Industrial, and Hospital Wastes, in *Proceedings of an APCA International Specialty Conference*, Pittsburgh, PA, November 1987.

16

MEDICAL WASTE MANAGEMENT

16.1 INTRODUCTION

Phenomenal progress has been made in methods and equipment for the care of hospital patients. Hundreds of single-service items have been marketed to reduce the possibility of hospital-acquired infections. Yet hospitals generally have been slow to improve their techniques for the handling and disposing of the waste materials, which are increasing in quantity as a result of more patients and higher per-patient waste loads.

What the effect has been on health and safety has not been measured, but without proper management, wastes containing contaminated materials, dangerous chemicals, or discarded needles are a potential hazard to millions of patients, health care workers, and visitors. Furthermore, the health of the entire community can be jeopardized if these wastes are temporarily but inadequately stored outside the hospital or thrown onto open dumps.

Medical waste comes in a wide variety of forms. These forms include packaging, such as wrappers from bandages and catheters; disposable items, such as tongue depressors and thermometer covers; and infectious wastes, such as blood, tissue, sharps, cultures, and stocks of infectious agents.

There is an equally wide variety of sources. While hospitals, clinics, and health care facilities may generate the vast majority of medical waste, both infectious and noninfectious waste is also generated by private practices, home health care, veterinary clinics, and blood banks. In New York and New Jersey alone, there are some 150,000 sources producing over 240 million lbs/yr.

The beach closures along coastal New Jersey in 1987 and along the south shore of Long Island in 1988 focused attention on medical wastes. Their volume is relatively small (probably less than 1% of the total), but as with sewage wastes, concern centers around the issue of public health. Why these wastes are appearing more frequently is not certain. However, there are several possible contributing factors. Among these are the following:

1. Marked increase in disposable medical care materials.

2. Increase in the use of medically associated equipment on the streets as drug paraphernalia.
3. Increase in illegal disposal of medical wastes as a consequence of the increased costs of disposal.

16.2   MEDICAL WASTE REGULATIONS AND STANDARDS

The "regulatory environment" relating to the management of medical waste was in a constant state of change in the late 1980s through the late 1990s. The dynamics of this process and the frequent regulatory changes resulted in significant confusion and misunderstanding on the part of the health care industry. Uncertainty regarding the future regulatory environment made it difficult to develop long-range plans for the management of medical waste. However, despite these uncertainties, the deadlines imposed by existing laws and regulations required that health care professionals make significant decisions regarding the future of medical waste management at their facility. In view of the above, health care professionals had to keep a watchful eye on the impact of new or revised regulations.

On December 18, 1978, the Environmental Protection Agency proposed comprehensive regulations under the Solid Waste Disposal Act, as amended by the Resource Conservation and Recovery Act (RCRA) of 1976, for hazardous waste management. Wastes listed or designated by the Environmental Protection Agency (EPA) as "hazardous" under Section 3001 of RCRA are subjected to stringent "cradle-to-grave" regulations under Subtitle C of RCRA, which include, among other provisions, tracking of such wastes via a national uniform manifest, restrictions on land disposal, and permits for all treatment, storage, and disposal facilities.

Congress defined hazardous waste generally to mean "a solid waste, or combination of solid wastes, which, because of its quantity, concentration, or physical, chemical, or infectious characteristics, may either (a) cause or significantly contribute to an increase in mortality or an increase in serious irreversible, or incapacitating reversible, illness or (b) pose substantial present or potential hazard to human health or the environment when improperly treated, stored, transported, disposed of, or otherwise managed" (1).

During the public comment period for this rule-making, the EPA received approximately 60 comments which specifically addressed the infectious waste provisions of the proposed regulations. On May 19, 1980, the EPA published the first phase of the hazardous waste regulations. The EPA stated in the preamble to the regulations that the sections on infectious waste would be published when work on treatment, storage, and disposal standards was completed. While the EPA had evaluated management techniques for infectious waste, considerable evidence that these wastes cause harm to human health and the environment was needed to support federal rule-making (2).

In response to numerous requests for technical information and guidance on infectious waste management, the EPA published its findings in September 1982 as a guidance manual, the *Draft Manual for Infectious Waste Management* (WS-957). After consideration of comments on the Draft Manual, the EPA decided to revise and finalize the manual. An ad hoc working group consisting of a cross section of health care professionals from government, academia, private industry, and trade associations was convened to provide expertise and assistance. The EPA believed that this guidance document represented environmentally sound practices for the handling, treatment, and disposal of infectious waste and also reflected current thinking and practices in infectious waste management.

On November 1, 1988, the Medical Waste Tracking Act of 1988 (the Act) was signed into law. The Act amended the Solid Waste Disposal Act by adding "Subtitle J-Demonstration Medical Waste Tracking Program." Among other things, this Act required the EPA to establish a 2-yr demonstration program for tracking medical waste generated in New York, New Jersey, Connecticut, Rhode Island, and the Commonwealth of Puerto Rico, even if the wastes are taken out of state for treatment or disposal. The regulations establishing this program were to include a list of medical wastes to be tracked and minimum standards for segregation from other wastes, packaging, and labeling before transport to treatment and/or disposal facilities (3). The only effective dates for the demonstration program were June 22, 1989 through June 22, 1991. Three reports were prepared as a result of this program. The reports addressed a wide variety of topics, such as the quantities of medical wastes generated, types and number of generators, cost of mismanaged medical waste, and the costs of complying with the regulations. In addition, Congress asked for information on current handling methods, available and potential threats to human health, and existing state requirements. Finally, Congress asked the EPA to provide an assessment of the demonstration program's success and to evaluate the possibility of using existing state regulations or the hazardous waste regulations to control medical waste. A final report was never submitted to Congress, as newly developed OSHA standards on Occupational Exposure to Bloodborne Pathogens and Federal Hospital/Medical/Infectious Waste Incinerator Regulations resulting from the 1990 Clean Air Act Amendments addressed problems related to medical waste handling and disposal.

16.3 DEFINING REGULATED MEDICAL WASTE

On March 24, 1989, the EPA published regulations in the Federal Register as required under the Medical Waste Tracking Act. The term "medical waste" was defined as any solid waste which is generated in the diagnosis, treatment, or immunization of human beings or animals, in research pertaining thereto, or in the production or testing of biologicals. Medical waste can be either infectious or noninfectious. The term "medical waste" does not include any hazardous or household waste, defined in regulations under Subtitle C.

Infectious waste is waste which contains pathogenic microorganisms. In order for a disease to be transmitted, the waste must contain a sufficient quantity of the pathogen which causes the disease, there must be a method of transmitting the disease from the waste material to the recipient, and there must be a point of entry for the pathogens to afflict the recipient. Many pathogenic organisms cannot survive long periods outside of the human body, and therefore they are rendered innocuous over time.

Medical waste that has not been specifically excluded in the EPA provisions (for example, household waste), and is either a listed medical waste or a mixture of a listed medical waste and a solid, is known under the demonstration program as "regulated medical waste." Seven classes of listed wastes are defined by the EPA as regulated medical waste. Details on these seven classes are provided below. The

Centers for Disease Control, as well as each individual state, has its own classifications of medical waste. For the most part, these other classifications are similar to the EPA definitions. Regulated medical wastes require special handling and treatment in most states.

1. *Cultures and Stocks.* Cultures and stocks of infectious agents and associated biologicals, including: cultures and stocks of infectious agents from research and industrial laboratories; wastes from the production of biologicals; discarded live and attenuated vaccines; and culture dishes and devices used to transfer, inoculate, and mix cultures.

2. *Pathological Waste.* Human pathological wastes, including (a) tissues, organs, body parts, and body fluids that are removed during surgery, autopsy, or other medical procedures and (b) specimens of body fluids and their containers.

3. *Human Blood and Blood Products.* (a) Liquid waste human blood; (b) products of blood; (c) items saturated and/or dripping with human blood that are now caked with dried human blood, including serum, plasma, and other components; (d) containers which were used or intended for use in either patient care, testing, laboratory analysis, or the development of pharmaceuticals (intravenous bags are also included in this category).

4. *Sharps.* Sharps that have been used in animal or human patient care or treatment, in medical research, or in industrial laboratories, including hypodermic needles, syringes (with or without the attached needles), pasteur pipettes, scalpel blades, blood vials, needles with attached tubing, and culture dishes (regardless of presence of infectious agents). Also included are other types of broken or unbroken glassware that were in contact with infectious agents, such as used slides and coverslips.

5. *Animal Waste.* Contaminated animal carcasses, body parts, and bedding of animals that were known to have been exposed to infectious agents during research (including research in veterinary hospitals), production of biologicals, or testing of pharmaceuticals.

6. *Isolation Wastes.* Biological waste and discarded materials contaminated with blood, excretion, exudates, or secretions from either (a) humans who are isolated to protect others from certain highly communicable diseases or (b) isolated animals known to be infected with highly communicable diseases.

7. *Unused Sharps.* The following unused, discarded sharps: hypodermic needles, suture needles, syringes, and scalpel blades.

In addition to the seven classes of waste listed above, waste materials which are trace contaminated with antineoplastic agents require special handling and treatment as well. A wide variety or antineoplastic, or chemotherapeutic, drugs are used to treat cancer patients by destroying the diseased cell tissues. Due to their inherent nature, these chemicals are extremely toxic. In fact, seven antineoplastic agents are defined under RCRA as hazardous chemicals. Items which may be trace contaminated with these agents include empty vials, syringes, gloves, gauze, and other materials contaminated during the preparation and administration of these drugs.

A final category of waste generated within a hospital which requires special handling is "low-level radioactive wastes." Most hospitals generate a very small quantity of this type of waste. The materials are generally allowed to decay on-site in a protected "hot lab" until the materials have reached acceptable ambient levels.

The decayed material can then be disposed as regulated or nonregulated medical waste, depending on the contact of the material with infectious agents. Contaminated waste materials of this nature typically include gloves, gauze, syringes, and other patient contact materials. For more on radioactive waste, see Chapter 19.

16.4 PACKAGING AND STORAGE

Infectious waste should be separated from the general waste stream to ensure that these wastes will receive appropriate handling and treatment. Also, segregation of the waste ensures that the added costs of special handling will not be applied to noninfectious waste (4). Infectious waste should be segregated from the general waste stream at the point of generation (i.e., the point at which the material becomes a waste). It is best accomplished by those who generate the waste and therefore are best qualified to assess the hazards associated with the waste.

The EPA believes that it is generally necessary to segregate sharps (including sharps containing residual fluids) and fluids in quantities greater than 20 mL from all other medical waste, as well as to segregate these wastes from each other. The EPA believes that sharps and fluids pose special problems in waste handling and are best managed if placed in special containers. If not segregated, these items can contaminate other medical waste. Segregating sharps and fluids should increase the integrity and safety of all medical waste packaging and protect those persons handling the waste. Used needles, syringes, and other sharps, including the residual fluids contained therein, must meet more stringent packaging requirements than all other waste. In addition, fluids in quantities greater than 20 mL must meet certain packaging requirements. Other regulated medical waste must be segregated, to the extent practical, from other waste (e.g., hazardous, radioactive, or general refuse), as well as from sharps and fluids.

When regulated medical waste cannot be segregated from other waste (i.e., it is not practical), the generator must ensure that the waste is packaged and marked according to the applicable packaging requirements. For example, if general refuse is placed in the same container as fluids, then the packaging must meet the requirements for fluids. Furthermore, if "untreated" regulated medical waste is mixed and co-packaged with "treated" regulated medical waste, the package must be labeled and identified as "untreated medical waste." The scheme outlined above provides incentives for segregation when it is advantageous for the generator to do so, but still does not preclude co-packaging when the generator determines it to be necessary or appropriate (5).

Infectious waste should be discarded directly into containers or plastic bags which are clearly identifiable and distinguishable from the general waste stream. Containers should be marked with the universal biological hazard symbol (Figure 16.1) (6–10). Plastic bags also should be distinctively colored or marked with the universal biological hazard symbol. Red or orange plastic bags generally are used to identify infectious waste.

Infectious waste should be packaged in order to protect waste handlers and the public from possible injury and disease that may result from exposure to the waste. Infectious waste should be contained from the point of origin up to the point of final disposal. Therefore, the integrity of the packaging must be preserved through handling, storage, transportation, and treatment.

FIGURE 16.1 The biological hazard symbol. The symbol is fluorescent orange or orange-red. The background may be any color that provides sufficient contrast for the symbol to be clearly defined.

The following factors should be considered in selecting appropriate packaging:

* Waste type.
* Handling and transport of the packaged waste (before treatment).
* Treatment technique.
* Special considerations for plastic bags.
* Package identification.

All regulated medical waste must be placed in a container or combination of containers that are rigid and leak-resistant, impervious to moisture, strong enough to prevent tearing or bursting under normal handling, and sealed securely. To provide adequate waste containment, packaging should be appropriate for the type of waste. For example, liquid infectious waste should be placed in capped or tightly stoppered bottles or flasks resistant to breaking. Containment tanks may be used for large quantities of liquid waste. Solid or semisolid wastes such as pathological wastes, animal carcasses, and laboratory wastes may be placed in plastic bags.

Sharps should be placed directly into impervious, rigid, and puncture-resistant containers to eliminate the hazard of physical injury. Clipping of needles is not recommended, unless the clipping device effectively contains needle parts and aerosols which might otherwise become airborne and pose a hazard.

If the waste is to be moved within the facility for treatment or storage, single plastic bags may not effectively contain the waste. If necessary, additional packaging should be used to preserve the integrity of the bags and to ensure containment of the waste. One option is to place single-bagged waste within a rigid or semirigid container such as a bucket, box, or carton. Plastic bags may be used as liners for such containers. Another suitable practice is double-bagging, that is, placing the sealed plastic bag within another bag which is subsequently sealed. Containers of sharps and liquids may be placed within other containers (e.g., boxes) for ease of handling. The containers should be covered with secure lids during transport and storage.

Whenever plastic bags of infectious waste are handled and especially when they are transferred, care must be exercised to prevent tearing of the bags. For example, plastic bags containing infectious waste should not be transported through a chute or dumbwaiter. In general, mechanical devices should not be used for transport or loading of plastic bags containing this type of waste. Carts are suitable for moving packaged infectious waste within the facility. These carts should be disinfected frequently. Carts used to transport infectious waste should not be used to transport other materials prior to decontamination.

When infectious waste is transported off-site for treatment, plastic bags—single or double—should be placed within rigid or semirigid containers to maintain the integrity of the packaging and prevent spillage. Depending on the management system, these containers can be recycled or disposed of. Recycled containers (such as heavy plastic barrels) which are used repeatedly for transport and treatment of bagged waste should be disinfected after each use. Single-use containers (such as strong cartons) are usually destroyed as part of the treatment process (e.g., incineration).

Infectious waste should not be compacted prior to treatment because compacting may actually disperse the infectious waste by destroying the integrity of the packaging. In addition, compaction may interfere with the effectiveness of the treatment process.

While it is preferable to treat infectious waste as soon as possible after generation, same-day treatment is not always possible or practical. For example, treatment equipment may not be available because of insufficient capacity, malfunction, or unavailability of personnel, or providers of off-site treatment may not pick up the waste on a daily basis. In such cases it may be necessary to store the waste. Stored medical waste must be:

- Protected from water, rain, and wind.
- Kept nonputrescent.
- Kept in locked dumpsters, sheds, tractor trailers, or other secure containers, if outdoors.
- Protected from unauthorized employees, animals, and pests.

There are four important factors to be considered when storing infectious waste: the integrity of the packaging; storage temperature; the duration of storage; and the location and design of the storage area.

The packaging should provide containment of the waste throughout the waste management process. In addition, the packaging should deter rodents and vermin, which can be vectors in disease transmission.

Storage temperature and duration are important considerations. As temperature increases, the rates of microbial growth and putrefaction also increase. This results in the unpleasant odors typically associated with wastes containing decaying organic matter.

There is no unanimity of opinion on optimum storage time and temperature. Some states, however, establish storage requirements as a function of time and temperature. For example, regulations in California permit storage for a maximum of 4 days at temperatures above 32°F. Massachusetts allows infectious waste to be stored for 1 day at room temperature (64–77°F) or for 3 days in a refrigerator (34–45°F). These requirements are for total storage time prior to treatment, regardless of whether the waste is stored at the generating facility or at a separate treatment facility. The EPA recommends that storage times be kept as short as possible.

The storage area should be a specially designated area located at or near the treatment or pick-up site. For security reasons, the storage area should have a limited access that restricts the entry of unauthorized personnel. It should be kept free of rodents and vermin. In addition, the universal biohazard label should be posted where appropriate (e.g., doors, waste containers, refrigerators, and freezers).

16.5 TREATMENT AND DISPOSAL OF INFECTIOUS WASTE

Since packaging integrity cannot be ensured during landfilling, loss of containment may result in dispersal of potentially infectious waste into the environment. Therefore, to ensure protection from the potential hazards posed by these wastes, the EPA recommends that all infectious waste be treated prior to disposal. "Treatment," when used to refer to waste management, means any method, technique, or process designed to change the biological character or composition of any regulated medical waste so as to substantially reduce or eliminate its potential for causing disease.

The purpose of treating infectious waste is to reduce the hazard associated with the presence of infectious agents. To be effective, treatment must reduce or eliminate pathogens present in the waste so that they no longer pose a hazard to persons who may be exposed to them.

Infectious waste which has been effectively treated is no longer biologically hazardous. However, it is still subjected to tracking requirements under many state requirements. For a regulated medical waste to be removed from the tracking system, it must be both treated and destroyed. Destroyed regulated medical waste is defined as regulated medical waste that has been ruined, torn apart, or mutilated through processes such as thermal treatment, melting, shredding, grinding, tearing, or breaking, so that it is no longer generally recognized as medical waste but has not necessarily been treated. It does not mean compacted regulated medical waste.

Treated liquid wastes may be poured down the drain to the sewer system. Sewer disposal is also an option for certain treated solid wastes (e.g., pathological waste) that are amenable to grinding and flushing to the sewer system. Sewer disposal, however, is subject to the approval of the local sewer authority. Treated solid wastes and incinerator ash can be disposed of in a sanitary landfill.

Additional processing of some wastes may be necessary or advisable before disposal. For example, some states require that needles and syringes be rendered nonusable before disposal. Treated sharps can be ground up, incinerated, or compacted. For aesthetic reasons, body parts should not be recognizable when disposed of.

Hospital wastes are disposed of in a number of ways, usually by the hospital's maintenance or engineering department. Eventually, almost two-thirds of the wastes leave the hospitals and go out into the community for disposal. About 35% by weight, principally combustible rubbish and biological materials, are disposed of in hospital incinerators. Noncombustibles are usually separated and, along with the incinerator residue, leave the hospital to be disposed of on land.

Infectious waste management methods include incineration, autoclaving, sanitary landfilling, sewer systems, chemical disinfection, thermal inactivation, ionizing radiation, and gas vapor sterilization. Table 16.1 lists typical treatment/disposal methods for each waste type as determined by a survey conducted by Odette in

TABLE 16.1 Percent of hospitals using treatment/disposal methods for each waste type

Typical Treatment/Disposal Methods[a]	Blood and Blood Products	Body Fluids	Lab Waste[a]	Pathological Wastes	Sharps	Animal Wastes	Disposal Materials[b]	Cytotoxic/ Antineoplastic Wastes[c]
Incineration	58%	58%	61%	92%	79%	81%	29%	Unknown
Sanitary landfill	12%	32%	16%	4%	16%	2%	54%	Unknown
Steam sterilization	25%	11%	33%	4%	14%	2%	4%	Unknown
Sewer	23%	6%	3%	2%	0%	0%	6%	Unknown

Source: Odette (11).

[a]Treatment/disposal methods are not equal to 100% because hospitals use more than one method or insufficient data for some responses.
[b]The waste-type categories used in this table do not directly correspond to Odette's categories. Therefore, some percentages are averages of those given in Odette's paper.
[c]Data not specified.

1988 (11). Odette's survey of hospitals revealed that 55% of hospitals that segregate infectious from noninfectious waste only incinerate their infectious waste. Eighteen percent of the hospitals which segregate treat infectious waste by steam sterilization and then incinerate or landfill the waste. Three percent of the hospitals surveyed dispose of infectious waste in sanitary landfills without prior treatment (12). Because of the special treatment requirements for medical waste, disposal costs are 10–20 times greater than those for nonregulated wastes.

Incineration

Incineration is not a new technology; it has been routinely used for treating waste for many years in both the United States and Europe. It is currently the preferred treatment method for hospital wastes. The major benefit of incineration is that the process actually destroys most of the waste rather than disposing of or storing it. It can be used for a variety of specific wastes and is reasonably competitive in cost compared to other disposal methods (13).

Hospital waste incineration involves the application of combustion processes under controlled conditions to convert wastes containing infectious and pathological material to inert mineral residues and gases. The three types of incinerators used most frequently for hospital waste treatment in the United States are multiple-hearth, rotary-kiln, and controlled-air (14). All three types can use primary and secondary combustion chambers to ensure maximum combustion of the waste. Many hospitals also may have small (usually older) incinerators used only for pathological wastes. Most, probably over 90%, of the hospital incinerators installed during the last three decades have been controlled-air units, which tend to be modular (15).

Multiple-hearth (multiple-chamber) incinerators consist of two or more combustion chambers. The primary chamber is for solid-phase combustion, whereas the secondary chamber is for gas-phase combustion (see Figure 16.2). These incinerators are often referred to as *excess-air incinerators* because they operate with excess air levels well above stoichiometric in both the primary and secondary combustion chambers.

Almost all of the existing hospital incinerators that were installed over 35 years ago are of this type; however, very few are being built for this purpose today. These systems are not preferable because the chambers are designed for very high excess air levels. Because of the high air flow, these systems emit high levels of particulates and are unable to comply with recent regulations regarding hospital waste incinerator emissions. Air pollution control equipment would have to be added to these systems, which can be very difficult and costly.

Rotary-kiln process incinerators were originally designed for lime processing. The rotary kiln is a cylindrical refractory-lined shell that is mounted at a slight incline from the horizontal plane to facilitate mixing the waste materials with circulating air (see Figure 16.3). The kiln accepts most types of solid and liquid waste materials. Residence time of the wastes is controlled by the rotational speed of the kiln and the angle at which it is positioned. The residence times of liquids and volatilized combustibles are controlled by the gas velocity in the incineration system. Thus, the residence time of the waste material can be controlled to provide complete burning of the combustibles. This is a critical factor in limiting release of some air pollutants from incineration devices.

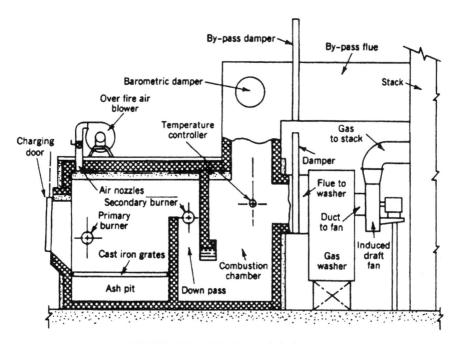

FIGURE 16.2 Multiple-hearth incinerator.

Rotary-kiln systems usually have a secondary combustion chamber after the kiln to ensure complete combustion of the wastes. Airtight seals close off the high end of the kiln; the lower end is connected to the secondary combustion chamber or mixing chamber. The kiln acts as the primary chamber to volatilize and oxidize combustibles in the wastes. Inert ash is then removed from the lower end of the kiln. The volatilized combustibles exit the kiln and enter the secondary chamber where additional oxygen is available and ignitable liquid wastes or fuel can be introduced. Complete combustion of the waste and fuel occurs in the secondary chamber. Both the secondary combustion chamber and kiln are usually equipped with an auxiliary fuel-firing system to bring the units up to the desired operating temperatures.

Most of the incinerators built for medical waste treatment in the last 15–20 years have been controlled-air incinerators (sometimes referred to as starved-air incinerators). These incinerators burn waste into two or more chambers under conditions of both low and excess stoichiometric oxygen requirements. In the primary chamber, waste is dried, heated, and burned at 40–80% of the stoichiometric oxygen requirement. Combustible gas produced by this process is mixed with excess air and burned in the secondary chamber. Excess air is introduced into the secondary chamber at usually between 100–150% of the stoichiometric requirement. A supplementary fuel burner is used to maintain elevated gas temperatures and provide for complete combustion (see Figure 16.4).

One advantage of using low levels of air in the primary chamber is that there is very little entrainment of particulate matter in the flue gas. For example, multiple-

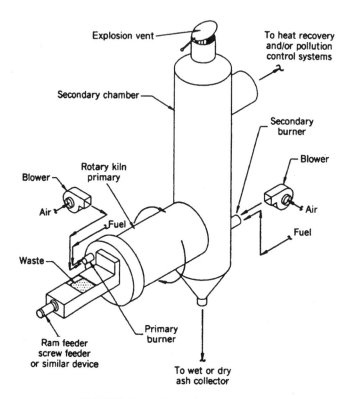

FIGURE 16.3 Rotary-kiln incinerator.

chamber air incinerators have average particulate emission factors of 7 lbs per ton, compared with 1.4 lbs per ton for controlled air units. Available data indicate that many controlled-air incinerators can be operated to meet existing particulate standards that are at or below 0.08 grains per dry standard cubic foot (gr/dscf), corrected to 12% carbon dioxide (16). Many states, however, are adopting lower standards (e.g., 0.015 gr/dscf) for medical waste incinerators, which probably would require additional air pollution control technologies. Additional controls raise capital costs and require expansion space (which may or may not be available). Additional controls, however, would capture finer particulates and some other pollutants.

Advantages of the controlled-air system include high thermal efficiency as a result of lower stoichiometric air use, higher combustion efficiencies, and low capital costs (which may increase as more controls are required). As with all types of incinerators, disadvantages include potential incomplete combustion under poor operating conditions, high capital, and objectionable public perception.

Autoclaving

Steam sterilization (autoclave) systems are designed to bring steam into direct contact with waste in a controlled manner and for sufficient duration to kill

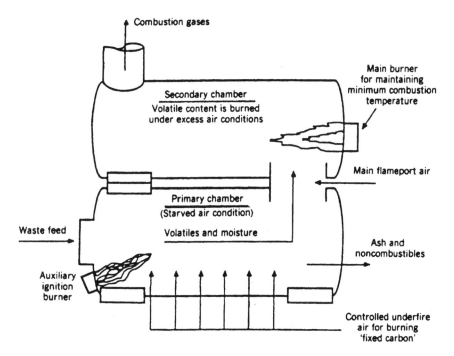

FIGURE 16.4 The principle of the controlled-air incinerator

pathogenic microorganisms that may be contaminating the waste. There are several different types and designs of autoclave systems. These design differences basically achieve improved steam contact efficiencies such that increasing waste volumes can be treated within the shortest possible time periods. Without full steam penetration and contact into the most densely packed wastes, decontamination will not be complete and sterilization cannot be achieved.

The three basic types of steam autoclave systems are gravity, pre-vacuum, and retort systems. Gravity-type autoclaves, in which pressure alone is used to evacuate air from the treatment chamber, typically operate with steam temperatures of about 250°F. These systems require a typical cycle time of approximately 60–90 min in order to achieve full steam penetration into the most densely packed waste loads. Pre-vacuum-type autoclave systems evacuate air from the treatment chamber using vacuum pumps. This enables them to reduce cycle times to about 30–60 min, as the time to heat the air within the chamber is eliminated. Pre-vacuum systems typically operate at about 270°F. Retort-type autoclaves basically comprise large-volume treatment chambers designed for much higher steam temperatures and pressures, and therefore their cycle times can be substantially less than those of the other systems. The treatment capacities of gravity- and pre-vacuum-type systems typically range between 0.25 and 2 yd^3 per cycle (50–200 lb per cycle). Retort-type autoclave systems have been used on large-scale, commercial medical waste treatment applications in capacities of up to 20 tons per day per unit.

Microwaving

One medical waste treatment process uses a combination of shredding and microwaving. Waste is loaded into the system and shredded while being injected with steaming. Waste is loaded into the system and shredded while being injected with steam. The shredded and steam-moistened waste is transported by an auger conveyor past a series of magnetrons, or microwave generators. In the process, shredded medical waste is reportedly held within the system for approximately 30 min at a temperature of approximately 205°F. The system is manufactured in nominal capacities of 220 and 550 lb/hr.

Decontamination is accomplished by the heat generated in the waste by microwave radiation. Wetting the waste by steam injection facilitates the waste heat-up process because the water molecules absorb microwave energy much more efficiently. The shredding process not only serves to disfigure the waste but also serves to prepare it for distribution through the auger conveyor, as well as break up sealed items, such as closed containers, so that they will not rupture or explode when heated by the microwaves. Shredding also reportedly provides up to an 80% volume reduction.

Chemical Disinfection Processes

Chemical treatment is synonymous with chemical disinfection. There are reportedly in excess of 8000 registered disinfectants. Currently, most medical waste treatment systems use chlorine compounds, but other disinfectants, such as mercurial compounds, phenolic compounds, iodine, alcohols, hexachlorophene, formaldehydes, iodine–alcohol combinations, and formaldehyde–alcohol combinations, could also be used. Most chemical disinfectants are used as aqueous solutions. Water is needed to bring the chemicals and microorganisms together as necessary to achieve inactivation.

Several manufacturers have developed, or are planning to develop, chemical disinfection systems for treating medical waste. Most of these systems start with a shredding step in order to provide sufficient contact between the waste and disinfectants. The shredding step, in turn, disfigures the waste such that recognizability is not a problem. Disinfectant is sprayed onto the waste while it is being shredded.

The shredded waste is usually transported from below the shredder via a dewatering system that allows liquids to drain away from the solid residues. The solid residues, in turn, are deposited into collection containers, and the free liquids are typically recycled back into the process or discharged to a sanitary sewer. The solids, which are wet from the disinfectant solution, are held for sufficient time to provide an acceptable degree of disinfection. Afterwards, they are typically discharged into trash containers for disposal.

Irradiation Processes

Irradiation is synonymous with electromagnetic or ionizing radiation. Two irradiation processes, gamma radiation and electron beam radiation, have been developed or considered for medical waste treatment.

Gamma radiation primarily involves the use of radioisotope cobalt-60. The radiation from this is powerful and penetrating and therefore appears to offer substantial advantages in the treatment of medical waste. Its application would require lead and thick concrete shielding. Cobalt-60 is readily available but very costly. Experience has shown that cobalt-60 replacement is required approximately every 4 years.

Electron beam radiation is used to sterilize food and medical supplies, but it is still under development as a medical waste treatment process. Electron beam radiation involves the use of a linear accelerator, or electron beam gun. The advantages of this are that a direct beam from the gun can be accurately focused and swept across the waste. Medical waste is passed through the electron beam by a conveyor system and remains recognizable after treatment. A secondary process, such as a shredder or grinder, can be added to the system to render the waste nonrecognizable.

Plasma Systems

A plasma is basically a material in which the temperatures are so high that some of its electrons are separated from its atoms. Natural plasmas include lightning and auroras. Direct and alternating current electrical arcs are used to create plasmas. Plasma temperatures reaching in excess of 20,000°C have been produced in laboratories. In industrial applications, plasmas are typically in the range of 2500°C to 10,000°C. An electric arc welder operates in the range of 15,000°C to 20,000°C. There are several firms and entrepreneurs attempting to develop medical waste treatment systems using plasma burners or torches. These systems are, in many respects, very similar to incineration systems except that conventional combustion does not take place. They are also very similar to pyrolysis systems because plasma-fired chambers operate in an oxygen-deficient mode, and off-gases need to be combusted separately. The residue produced is a glass-like substance, rather than a particulate ash typical of incinerators.

16.6 SELECTION OF WASTE MANAGEMENT OPTIONS

In an infectious waste management system, several options are available within each area of management. Options for treatment techniques for the various types of infectious waste, types of treatment equipment, treatment sites, and various waste handling practices all need to be carefully evaluated. The selection of available options at a facility depends upon a number of factors such as the nature of the infectious waste, the quantity of infectious waste generated, the availability of equipment for treatment on-site and off-site, regulatory constraints, and cost considerations. These factors are presented here in order to provide assistance in the development of an infectious waste management program.

Since treatment methods vary with waste type, the waste must be evaluated and categorized with regard to its potential to cause disease. Such characteristics as chemical context, density, water content, bulk and so on, are known to influence waste treatment decisions. For example, many facilities use a combination of techniques for the different components of the infectious waste stream, such as stream sterilization for laboratory cultures and incineration for pathological waste.

The quantity of each category of infectious waste generated at the facility may also influence the method of treatment. Decisions should be made on the basis of the major components of the infectious waste stream. Generally, it would be desirable and efficient to handle all infectious waste in the same manner. However, if a selected option is not suitable for treatment of all wastes, then other options must be included in the waste management plan.

Regulation at the federal, state, and local level may have impact on the treatment of infectious waste. For example, federal air pollution regulations were published in 1997 which apply to infectious waste incineration. These regulations are presently being implemented on the state level (17). A hospital may also be required, under Title V of the Clean Air Act Amendments of 1990, to obtain an operating permit for all combined emissions sources at the facility. Emissions from incinerators and other waste treatment technologies would have to be considered in the Title V analysis (18). Furthermore, many localities apply particulate standards to all incinerators regardless of type. Water quality regulations may also be applicable. For example, regulations and standards pertaining to chemical pollutants, thermal discharges, organic loading (biological oxygen demand), and particulates (total suspended solids) are relevant to the medical waste treatment systems that utilize chemicals, thermal inactivation, and grinding, respectively, as part of the treatment process. In addition, several states have regulations that specify requirements for the disposal of pathological waste and sharps. Therefore, it is imperative that state and local regulations be carefully considered when developing a hospital waste treatment plan.

Another important factor in the selection of options for infectious waste management is the availability of on-site and off-site treatment. On-site treatment of infectious waste provides the advantage of a single facility or generator maintaining control of the waste. When selecting an on-site treatment system, both the treatment technology and the vendor providing the system should be carefully evaluated. For some facilities, however, off-site treatment may offer the most cost-effective option. Off-site treatment alternatives include such options as crematories operated by morticians (for pathological wastes), a shared treatment unit at another institution, and commercial or community treatment facilities. With off-site treatment, precautions should be taken in packaging and transporting to ensure containment of the infectious waste. In addition, generators should comply with all state and local regulations pertaining to the transport of regulated medical waste and should ensure that the waste is being handled and treated properly at the off-site treatment facility.

It is also important to consider prevailing community attitudes in such matters as site selection for off-site treatment facilities. These include local laws, ordinances, and zoning restrictions as well as unofficial public attitudes which may result in changes in local laws.

Cost considerations are the determining factor in the selection of infectious waste management options. Cost factors include personnel, equipment cost (capital expense, annual operating, and maintenance), hauling costs (for infectious waste and the residue from treatment), and, if applicable, service fees for the off-site treatment option.

16.7 SUMMARY

1. Sources of medical waste include dentists, morticians, veterinary clinics, home health care, blood banks, and private practices, as well as hospitals and clinics.

2. Regulated medical waste includes the following categories: cultures and stocks; pathological waste; human blood and blood products; sharps; animal waste; isolation waste; and unused sharps. Chemotherapy waste should also be segregated from noninfectious waste.

3. Infectious waste should be packaged and stored in a safe manner which will protect waste handlers and the general public from possible injury and/or disease that may result from exposure to the waste.

4. Treatment methods substantially reduce or eliminate the potential for a waste to spread disease. Incineration, autoclaving, microwave processing, chemical disinfection, plasma systems, and irradiation systems are currently used to treat infectious waste.

5. Factors to consider when selecting a waste management system include: evaluation and categorization of the types of waste to be treated; selection of an on-site or off-site system or a combination of both; selection of a type of equipment and a vendor; current and proposed federal, state, and local regulations; community relations; and cost.

REFERENCES

1. U.S. EPA, EPA *Guide for Infectious Waste Management*, EPA/530-SW-86-014, May 1986, p. v.

2. U.S. EPA, *Hospital Waste Incinerator Field Inspection and Source Evaluation Manual*, EPA-340/1-89-001, 1989, pp. 7–30.

3. Rules and Regulations, *Federal Register* 54(56), 12326, 1989.

4. R. Silvagni et al., *A Waste Reduction Project Within the University of Minnesota Hospitals*, Physical Plant Department, University of Minnesota Hospitals, Minneapolis, MN, March 15, 1980.

5. Standards for the Tracking and Management of Medical Waste; Interim Final Rule and Request for Comments, 40 CFR Parts 22 and 259, March 24, 1989.

6. U.S. Department of Health and Human Services, National Institutes of Health (NIH), *A Supplement to the NIH Guidelines for Recombinant DNA Research*, Laboratory Safety Monograph, NIH, Office of Research Safety, National Cancer Institute, and the Special Committee of Safety and Health Experts, Bethesda, MD, July 1978.

7. U.S. Department of Health and Human Services, Interstate Shipment of Etiologic Agents: Transportation of Materials Containing Certain Etiologic Agents; Minimum Packaging Requirements, in *Code of Federal Regulations*, 42 CFR 72.3, U.S. Government Printing Office, Washington, D.C., 1984.

8. U.S. Department of Labor, Occupational Safety and Health Administration, Specifications for Accident Prevention Signs and Tags. Biological Hazard Signs, in *Code of Federal Regulations*, 29 CFR 1910.145(e)(4), U.S. Government Printing Office, Washington, D.C., 1979.

9. U.S. Department of Transportation, Etiologic Agent Label, in *Code of Federal Regulations*, 49 CFR 172.444, U.S. Government Printing Office, Washington, D.C., 1984.

10. U.S. Department of Transportation, Labeling of Packages Containing Etiologic Agents, in *Code of Federal Regulations*, 49 CFR 173.388, U.S. Government Printing Office, Washington, D.C., 1984.

11. Odette, 1988.

12. U.S. Environmental Protection Medical Waste Meeting, November 14–16, 1988, Annapolis, MD, 1988, pp. 13–14.

13. L. Theodore, *Air Pollution Control and Waste Incineration for Hospitals and Other Medical Waste Facilities*, Van Nostrand Reinhold, New York, 1990.

14. U.S. Environmental Protection Agency, Hospital Waste Combustion Study, data gathering phase, final draft (prepared by the Radian Corp.), October 1987.

15. L. Doucet, "State-of-the-Art Hospital and Institutional Waste Incineration: Selection, Procurement and Operations," paper presented at the 75th Annual Meeting of the Association of Physical Plant Administrators of Universities and Colleges, Washington, D.C., July 24, 1988.

16. L. Doucet, *Update of Alternative and Emerging Medical Waste Treatment Technologies*, AHA Technical Document Series, 1991.

17. L. Doucet, "EPA's Final Hospital/Medical/Infectious Waste Indinerator Regultions: Requirements, Compliance Decisions, and Related Environmental Issues," paper presented at the 35th Annual Conference And Technical Exhibition of the American Society for Healthcare Engineering, Denver Colorado, July 15, 1998.

18. G. Urbanowicz, "The Impact of Title V Regulations on Hospitals and Healthcare Facilities," paper presented at the 1995 Annual Conference & Technical Exhibition of the American Society for Hospital Engineering, Las Vegas, Nevada, June 20, 1995.

17

ASBESTOS

17.1 INTRODUCTION

Asbestos became a popular commercial product because it is strong, won't burn, resists corrosion, and insulates well. In the United States, its commercial use began in the early years of the twentieth century and peaked in the period from World War II into the 1970s. Under the Clean Air Act of 1970 the EPA has been regulating many asbestos-containing materials which, by EPA definition, are materials with more than 1% asbestos. The Occupational Safety and Health Administration's (OSHA) asbestos construction standard in Section K, "Communication of Hazards to Employees," specifies labeling many materials containing 0.1% or more asbestos. In the mid-1970s several major kinds of asbestos materials, such as spray-applied insulation, fireproofing, and acoustical surfacing material, were banned by the EPA because of growing concern about health effects, particularly cancer, associated with exposures to such materials.

The term *asbestos* describes six naturally occurring fibrous minerals found in certain types of rock formations. Of that general group, the minerals chrysotile, amosite, and crocidolite have been most commonly used in building products. When mined and processed, asbestos is typically separated into very thin fibers. When these fibers are present in the air, they are normally invisible to the naked eye. Asbestos fibers are commonly mixed during processing with a material which binds them together so that they can be used in many different products. Because these fibers are so small and light, they may remain in the air for many hours if they are released from asbestos-containing material (ACM) in a building. When fibers are released into the air they may be inhaled by people in the building.

Asbestos fibers can cause serious health problems. If inhaled, they can cause diseases which disrupt the normal functioning of the lungs. Three specific diseases — asbestosis (a fibrous scarring of the lungs), lung cancer, and mesothelioma (a cancer of the lining of the chest or abdominal cavity) — have been linked to asbestos exposure. These diseases do not develop immediately after inhalation of asbestos fibers; it may be 20 years or more before symptoms appear.

In general, as with cigarette smoking and the inhalation of tobacco smoke, the more asbestos fibers a person inhales, the greater the risk of developing an asbestos-related disease. Most of the cases of severe health problems resulting from asbestos exposure have been experienced by workers who held jobs in industries such as shipbuilding, mining, milling, and fabricating, where they were exposed to very high levels of asbestos in the air, without benefit of the worker protections now afforded by law. Many of these same workers were also smokers. These employees worked directly with asbestos materials on a regular basis and, generally, for long periods of time as part of their jobs. Additionally, there is an increasing concern for the health and safety of construction, renovation, and building maintenance personnel, because of possible periodic exposure to elevated levels of asbestos fibers while performing their jobs (1).

Intact and undisturbed asbestos materials do not pose a health risk. The mere presence of asbestos in a building does not mean that the health of building occupants is endangered. ACM which is in good condition, and is not somehow damaged or disturbed, is not likely to release asbestos fibers into the air. When ACM is properly managed, release of asbestos fibers into the air is prevented or minimized, and the risk of asbestos-related disease can be reduced to a negligible level.

However, asbestos materials can become hazardous when, due to damage, disturbance, or deterioration over time, they release fibers into building air. Under these conditions, when ACM is damaged or disturbed—for example, by maintenance repairs conducted without proper controls—elevated airborne asbestos concentrations can create a potential hazard for workers and other building occupants (1).

17.2   REGULATIONS ADDRESSING ASBESTOS

The Asbestos National Emission Standards for Hazardous Air Pollutant (NESHAP) regulation has been in existence since 1973 and has been amended several times. The following is a summary of the main provisions contained in the original stipulations and subsequent revisions:

- March 31, 1971—Asbestos listed as a hazardous air pollutant under Section 112.
- April 6, 1973—Original promulgation developed regulations for:
  - Asbestos mills and manufacturing sources;
  - Asbestos-containing spray-on materials;
  - Use of tailings in roadways;
  - Demolition of buildings containing friable asbestos-containing fireproofing and insulating material;
  - The spraying of ACM on buildings and structures for fireproofing and insulating purposes.
- May 3, 1974—Regulations were revised to include:
  - Addition of clarifying definitions.
  - Clarification of demolition provisions.

- Clarification of the no visible emission standard to exclude condensed uncombined water vapor.
- October 14, 1975 — Substantial changes were made including:
  - Addition of fabricators.
  - Inclusion of renovation projects as regulated activities.
  - Prohibition of the use of wet applied and molded insulation (i.e., pipe lagging).
  - Expansion of the scope of the regulation to cover asbestos-containing waste handling and disposal.
  - Inclusion of inactive waste disposal sites that were operated by milling, manufacturing, and fabricating sources.
- March 2, 1977 — Minor changes, mostly addressing definitions.
- June 19, 1978 — Important changes made include:
  - Expansion of the coverage of spraying restriction to prohibit application of asbestos-containing materials for decorative purposes;
  - Adoption of provisions to exempt bitumen- or resin-based (i.e., contains material such as tar or asphalt) materials from the spraying restrictions;
  - Repromulgation of certain work practice provisions.
- April 5, 1984 — Repromulgation to make sure that work practice standards could be enforced. In 1978, the United States Supreme Court ruled that EPA's authority to enforce emission standards under the Clean Air Act (CAA) of 1970 did not extend to work practice standards. The CAA Amendments of 1977 gave EPA clear authority to enforce work practice standards. By repromulgating the standard, EPA removed any doubt that the work practice standards could be enforced. The standard was also reorganized and clarified, and placed in Subpart M.
  - October 1990 — Promulgated revisions to clarify standard, promote compliance, and aid enforcement, including:
  - Requirements for milling, manufacturing, and fabricating to monitor and inspect control devices and keep records of monitoring activities;
  - Renotification requirements and other clarifying revisions for demolition and renovation; and
  - Requirements to keep records of waste shipments and waste disposal (2).

The EPA and OSHA have major responsibility for regulatory control over exposure to asbestos. Emissions of asbestos to the ambient air are controlled under Section 112 of the Clean Air Act, which establishes the National Emission Standards for Hazardous Air Pollutants (NESHAP). The regulations specify control requirements for most asbestos emissions, including work practices to be followed to minimize the release of asbestos fibers during handling of asbestos waste materials. These regulations do not identify a safe threshold level for airborne asbestos fibers. For additional information about the NESHAPs regulations for asbestos, refer to the *Code of Federal Regulations* (40 CFR Part 61, Subpart M).

The OSHA regulations are established to protect workers handling asbestos or asbestos-containing products. The current OSHA regulations include (a) a maximum workplace airborne asbestos concentration limit of 2 fibers/cc on an 8-hr time weighted average basis, and (b) a ceiling limit of 10 fibers/cc in any 15-min period. The standard includes (a) requirements for respiratory protection and other safety equipment, and (b) work practices to reduce indoor dust levels. For details regarding the OSHA regulations, refer to the *Code of Federal Regulations* (29 CFR Part 1910).

The EPA has implemented a separate regulation under the Toxic Substances Control Act (TSCA) to handle the problem of asbestos construction materials used in schools. This regulation requires that all schools be inspected to determine the presence and quantity of asbestos and that the local community be notified as well as the building posted. Corrective actions, such as asbestos removal or encapsulation, are currently left to the discretion of the school administrators. The EPA provides technical assistance under this program through Regional Asbestos Coordinators or the toll-free TSCA hotline: (800) 424-9065 (554-1404 in Washington, DC). The specific details of the TSCA program are contained in the *Code of Federal Regulations* (40 CFR Part 763, Subpart F).

The Asbestos School Hazard Abatement Act of 1984 (ASHAA) establishes a $600 million grant and loan program to assist financially needy schools with asbestos abatement projects. The program also includes (a) the compilation and distribution of information concerning asbestos, and (b) the establishment of standard for abatement projects and abatement contractors. Under this program, centers to train contractors on asbestos handling and abatement have been established at the Georgia Institute of Technology (Atlanta, GA), Tufts University (Medford, MA), and the University of Kansas (Lawrence, KN). Additional information can be obtained through the toll-free ASHAA hotline: (800) 835-6700 (554-1404 in Washington, D.C.).

Wastes containing asbestos are not hazardous wastes under the Resource Conservation and Recovery Act (RCRA). However, because state regulations can be more restrictive than the federal regulations under RCRA, some states may have listed asbestos-containing wastes as hazardous wastes. Since this will greatly impact on transportation and disposal of the waste, the state hazardous waste agency should be contacted. A list of state hazardous waste agencies may be obtained by calling the RCRA hotline: (800) 424-9346 (382-3000 in Washington, DC). Current nonhazardous waste regulations under RCRA pertain to facility siting and general operation of disposal sites (including those that handle asbestos). Details concerning these RCRA requirements are contained in the *Code of Federal Regulations* (40 CFR Part 257).

Other federal authorities and Agencies controlling asbestos include the Clean Water Act, under which EPA has set standards for asbestos levels in effluents to navigable waters; the Mine Safety and Health Administration, which oversees the safety of workers involved in the mining of asbestos; the Consumer Product Safety Commission; the Food and Drug Administration; and the Department of Transportation. State and local agencies may have more stringent standards than the federal requirements; these agencies should be contacted prior to any asbestos removal or disposal operations (3).

17.3   DEFINITIONS

In the initial Asbestos NESHAP rule promulgated in 1973, a distinction was made between building materials that would readily release asbestos fibers when damaged or disturbed and those materials that were unlikely to result in significant fiber release. The terms "friable" and "nonfriable" were used to make this distinction. EPA has since determined that, if severely damaged, otherwise nonfriable materials can release significant amounts of asbestos fibers.

Friable asbestos-containing material (ACM), is defined by the Asbestos NESHAP, as any material containing more than 1% asbestos as determined using the method specified in Appendix A, Subpart F, 40 CFR Part 763, Section 1, *Polarized Light Microscopy* (PLM), that, when dry, can be crumbled, pulverized or reduced to powder by hand pressure (Sec. 61.141).

Nonfriable ACM is any material containing more than 1% asbestos as determined using the method specified in Appendix A, Subpart F, 40 CFR Part 763, Section 1, Polarized Light Microscopy (PLM),that when dry, cannot be crumbled, pulverized, or reduced to powder by hand pressure. The EPA also defines two categories of nonfriable ACM, Category I and Category II nonfriable ACM, which discussed are Section 17.5.

Regulated Asbestos-Containing Material (RACM) is (a) friable asbestos material, (b) Category I nonfriable ACM that has become friable, (c) Category I nonfriable ACM that will be or has been subjected to sanding, grinding, cutting or abrading, or (d) Category II nonfriable ACM that has a high probability of becoming or has become crumbled, pulverized, or reduced to powder by the forces expected to act on the material in the course of demolition or renovation operations (4).

17.4 FRIABLE ASBESTOS CONTAINING-MATERIALS

Because of their high tensile strength, incombustibility, corrosion and friction resistance, and other properties such as acoustical and thermal insulation abilities, asbestos fibers have been incorporated into over 3,600 commercial products. Thermal system, fireproofing, and acoustical insulation materials have been used extensively in the construction industry.

Thermal system applications include steam or hot water pipe coverings and thermal block insulation found on boilers and hot water tanks. Fireproofing insulation may be found on building structural beams and decking. Acoustical insulation (soundproofing) commonly has been applied as a troweled-on plaster in school and office building stairwells and hallways. Unfortunately, with time and exposure to damaging forces (e.g., severe weather, chemicals, mechanical forces, etc.), many ACMs may become crumbled, pulverized or reduced to powder, thereby releasing asbestos fibers, or may deteriorate to the extent that they may release fibers if disturbed. Since inhalation of asbestos fibers has been linked to the development of respiratory and other diseases, any material which is friable, or has a high probability of releasing fibers, must be handled in accordance with the Asbestos NESHAP.

The following work practices should be followed whenever demolition/renovation activities involving RACM occur:

- Notify the EPA of intention to demolish/renovate.
- Remove all RACMs from a facility being demolished or renovated before any disruptive activity begins or before access to the material is precluded.

- Keep RACMs adequately wet before, during, and after removal operation.
- Conduct demolition/renovation activities in a manner which produces no visible emissions to the outside air.
- Handle and dispose of all RACM in an approved manner (4).

17.5   NONFRIABLE ASBESTOS-CONTAINING MATERIALS

Because of the resilient nature of asbestos, it is used in materials exposed to a wide variety of stressful environments. These environments can cause the deterioration of binding materials and cause nonfriable materials to become friable. For example, nonfriable asbestos-containing packings and gaskets used in thermal systems may be found to be in poor condition as a result of the heat they have encountered. In petrochemical handling facilities, which may have miles of transfer pipes and fittings which contain asbestos gaskets and/or packings, profound degradation of the ACM may occur due to exposure to organic-based liquids and gases or to corrosive agents used to chemically clean these lines.

When nonfriable ACM is subjected to intense mechanical forces, such as those encountered during demolition or renovation, it can be crumbled, pulverized, or reduced to powder, and thereby release asbestos fibers. When nonfriable materials are damaged or are likely to become damaged during such activities, they must be handled in accordance with the Asbestos NESHAP.

As previously mentioned, the two categories of nonfriable materials are Category I Nonfriable ACM and Category II Nonfriable ACM. Category I nonfriable ACM is any asbestos-containing packing, gasket, resilient floor covering, or asphalt roofing product which contains more than 1% asbestos as determined using polarized light microscopy (PLM) according to the method specified in Appendix A, Subpart F, 40 CFR Part 763 (Sec. 61.141). Category I nonfriable ACM must be inspected and tested for friability if it is in poor condition before demolition to determine whether or not it is subject to the Asbestos NESHAP. If the ACM is friable, it must be handled in accordance with the NESHAP. Asbestos-containing packing, gaskets, resilient floor coverings, and asphalt roofing materials must be removed before demolition only if they are in poor condition and are friable.

Category II nonfriable ACM is any material, excluding Category I nonfriable ACM, containing more than 1% asbestos as determined using polarized light microscopy according to the methods specified in Appendix A, Subpart F, 40 CFR Part 763 that, when dry, *cannot* be crumbled, pulverized, or reduced to powder by hand pressure (Sec. 61.141). Category II nonfriable ACMs (cement siding, transite board shingles, etc.) subjected to intense weather conditions such as thunderstorms, high winds, or prolonged exposure to high heat and humidity may become "weathered" to a point where they become friable.

The Asbestos NESHAP further requires that if a facility is demolished by intentional burning, all of the facility's ACM, including Category I and II nonfriable ACM, be considered RACM and be removed prior to burning (Sec. 61.145(c)(10)). Additionally, if Category I or Category II nonfriable ACM is to be sanded, ground, cut, or abraded, the material is considered RACM and the owner or operator must abide by the following (Sec. 61.145(c)(1)):

1. Adequately wet the material during the sanding, grinding, cutting, or abrading operations.

2. Comply with the requirements of 61.145(c)(3)(i) if wetting would unavoidably damage equipment or present a safety hazard.

3. Asbestos material produced by the sanding, grinding, cutting, or abrading should be handled as asbestos-containing waste material subject to the waste handling and collection provisions of Section 61.150 (4).

17.6 PRODUCTS CONTAINING ASBESTOS (3)

There are several products in use containing asbestos. It is used in friction products such as brake linings for automobiles, buses, trucks, railcars, and industrial machinery, and in vehicle or industrial clutch linings, drum brake linings, disc brake pads, and brake blocks. In the past, asbestos linings have accounted for up to 99% of this market. Friction materials are generally tough and nonfriable, but they release asbestos dust during fabrication. In addition, accumulated dust in a brake drum from lining wear contains high levels of asbestos. Brake installation facilities (e.g., city bus service centers, tire and brake shops) may generate significant quantities of asbestos waste. Substitute nonasbestos brake linings have been developed and are beginning to replace asbestos lining in some applications.

Plastic products containing asbestos include resilient vinyl and asphalt floor coverings, asphalt roof coatings, and traditional molded plastic products such as a cooking pot handle or plastic laboratory sink. The products in this category are usually tough and inflexible. The asbestos in these products is tightly bound and is not released under typical conditions of use. However, any sawing, drilling, or sanding of these products during installation or removal would result in the release of asbestos dust.

Vinyl (linoleum) and asphalt flooring are used in many types of construction. Vinyl–asbestos flooring has about a 90% share of the resilient floor covering market. These materials are not friable, and asbestos is released primarily through sawing or sanding operations during installation, remodeling, and removal. Asphalt–asbestos coatings, used primarily as roof sealants, generally remain flexible and nonfriable, but can become friable or brittle as they age.

Asbestos–cement (A–C) pipe has been widely used for water and sewer mains, and is occasionally used for electrical conduits, drainage pipe, and vent pipes. Asbestos–cement sheet, manufactured in flat or corrugated panels and shingles, has been used primarily for roofing and siding, but also for cooling tower fill sheets, canal bulkheads, laboratory tables, and electrical switching gear panels. Asbestos–cement products are dense and rigid with gray coloration, unless the material is lined or coated. The asbestos in these products is tightly bound, and would not be released to the air under typical conditions of use. However, any sawing, drilling, or sanding of these products during installation or renovation would result in release of asbestos dust. In addition, the normal breakage and crushing involved in the demolition of structures can release asbestos fibers from these materials. For this reason they are subject to the NESHAPs regulation during demolition operations. Also, normal use of A–C pipe for water or sewer mains has been shown to release asbestos fibers to the fluid being carried.

By the late 1970s, A–C pipe had a 40% share of the water main market and a 10% share of the sewer main market. However, since A–C pipe has only been in existence for 50 years, it only accounts for a small fraction of the total pipe in place in the United States.

Roofing felts, gaskets, and other paper products are manufactured on conventional papermaking equipment using asbestos fibers instead of cellulose. The raw asbestos paper produced in this process has a high asbestos content ($\sim 85\%$), but is typically coated or laminated with other materials in the final product. The asbestos fibers in most paper products are sufficiently bound to prevent their release during normal product use. Cutting or tearing the material during installation, use, or removal would result in the release of asbestos dust.

Asbestos-containing roofing felt has been widely used for application of "built-up" roofs. Built-up roofing is used on a flat surface, and consists of alternating layers of roofing felt and asphalt. The roofing felt consists of asbestos paper, saturated and coated with asphalt. Asphalt–asbestos roofing shingles for residential structures, made from roofing felt coated with asphalt, were reportedly used for only a short time between 1971 and 1974.

Other asbestos-containing paper products include pipeline wrap, millboard, rollboard, commercial insulating papers, and a variety of specialty papers. Pipeline wrap is used to protect underground pipes from corrosion, particularly in the oil and gas industry. Millboard and rollboard are laminated paper products used in commercial construction such as walls and ceilings. Commercial insulating papers are used for high temperature applications in the metals and ceramics industries, for low-grade electrical insulation, and for fireproofing steel decks in building construction. Corrugated asbestos paper was used for pipe coverings, block insulation, and specialty panel insulation. Although these uses have generally been discontinued, significant amounts are typically found in older structures. These products are generally considered friable.

Asbestos yarn, cloth, and other textiles are made using conventional textile manufacturing equipment. These materials are used to manufacture fire-resistant curtains or blankets, protective clothing, electrical insulation, thermal insulation, and packing seals. The raw textile products have a high asbestos content ($\sim 85\%$). However, they are typically coated or impregnated with polymers before assembly into a final product, which is not required to be labeled as containing asbestos and typically is not labeled. These products may release asbestos dust if cut or torn, or for some products, during normal use. There still remains a significant quantity of noncoated fabrics in use, especially in schools and fire departments.

Asbestos-containing thermal insulation generally refers to sprayed and troweled asbestos coatings, and molded or wet-applied pipe coverings. These materials generally have an asbestos content of 50–80%. The coatings were commonly applied to steel I-beams and decks, concrete ceilings and walls, and hot water tanks and boilers. The coatings were applied primarily for thermal insulation, although in many cases the coating also provided acoustical insulation and a decorative finish. Sprayed coatings typically have a rough, fluffy appearance, while troweled coatings have a smooth finish and may be covered with a layer of plastic or other nonasbestos material. Both sprayed and troweled asbestos coatings are considered friable in most applications. Most spray-applied asbestos coatings were banned for fireproofing/insulating in 1973, and for decorative purposes in 1978.

Asbestos insulation board was used as a thermal/fireproofing barrier in many types of walls, ceilings, and ducts or pipe enclosures. This material looks like A–C sheet, but is less dense and much more friable. High asbestos dust levels have been measured for many board handling operations, including simple unloading of uncut sheets.

Preformed pipe coverings having an asbestos content of about 50% were used for thermal insulation on steam pipes in industrial, commercial, institutional, and residential applications. This product is usually white and chalky in appearance and was typically manufactured in 3-ft long, half-round sections, joined around the pipe using plaster-saturated canvas or metal bands.

This covering was applied on straight pipe sections, while wet-applied coatings were used on elbows, flanges, and other irregular surfaces. The preformed pipe coverings may be slightly more dense than the insulating coatings, but are still very friable. The installation of wet-applied and preformed asbestos insulations was banned in 1975; however, significant amounts are typically found in older structures.

Preformed block insulation was used as thermal insulation on boilers, hot water tanks, and heat exchangers in industrial, commercial, institutional, and residential applications. The blocks are commonly chalky white, 2 in. thick, from 1–3 ft in length and held in place around the boiler by metal wires and/or expanded metal lath. A plaster-saturated canvas was often utilized as a final covering or wrap. Asbestos block insulation is friable and rapidly deteriorates in a high humidity environment or when exposed to water. The installation of this type of asbestos insulation was banned by the EPA in 1975.

Other uses of asbestos have included exterior siding shingles, shotgun shell base wads, asphalt paving mix, spackle and joint patching compounds, artificial fireplace logs for gas-burning fireplaces, and artificial snow. The use of asbestos as artificial logs in gas-burning fireplace systems was banned in 1977, while the use of asbestos as an ingredient in spackle and joint compounds was banned in 1978.

17.7 TRAINING FOR OPERATIONS AND MAINTENANCE

Training of custodial and maintenance workers is one of the keys to a successful operations and maintenance (O&M) program. If building owners do not emphasize the importance of well-trained custodial and maintenance personnel, asbestos O&M tasks may not be performed properly. This could result in higher levels of asbestos fibers in the building air and an increased risk faced by both building workers and occupants.

The OSHA and the EPA require a worker training program for all employees exposed to fiber levels (either measured or anticipated) at or above the action level [0.1 f/cc, 8-hr *time-weighted average* — (TWA)] and/or the excursion limit (1.0 f/cc, 30-min TWA). According to the EPA regulations governing schools, all school staff custodial and maintenance workers who conduct any activities that will result in the disturbance of ACM must receive 16 hr of O&M training. Some states and municipalities may also have specific training requirements for workers who may be exposed to asbestos, or who work in a building with ACM present.

With proper training, custodial and maintenance staff can successfully deal with ACM in place, and greatly reduce the release of asbestos fibers. Training sessions should provide basic information on how to deal with all types of maintenance activities involving ACM. However, building owners should also recognize that O&M workers in the field often encounter unusual, "nontextbook" situations. As a result, training should provide key concepts of asbestos hazard control. If these concepts are clearly understood by workers and their supervisors, workers can develop techniques to address a specific problem in the field. Building owners who need to provide O&M training to their custodial and maintenance staff should contact an EPA environmental assistance center or equally qualified training organization for more information (1).

At least three levels of maintenance worker training can be identified:

1. Awareness training for custodians involved in cleaning and simple maintenance tasks where ACM may be accidentally disturbed—for example, fixing a light fixture in a ceiling covered with surfacing ACM. Such training may range from two to eight hours, and may include such topics as:

   • Background information on asbestos.
   • Health effects of asbestos.
   • Worker protection programs.
   • Locations of ACM in the building.
   • Recognition of ACM damage and deterioration.
   • The O&M program for that building.
   • Proper response to fiber release episodes.

2. Special O&M training for maintenance workers involved in general maintenance and asbestos material repair tasks—for example, a repair or removal of a small section of damaged Thermal System Insulation (TSI), or the installation of electrical conduit in an air plenum containing ACM or ACM debris. Such training generally involves at least 16 hours. This level of training usually involves more detailed discussions of the topics included in Level 1 training as well as:

   • Federal, state, and local asbestos regulations.
   • Proper asbestos-related work practices.
   • Descriptions of the proper methods of handling ACM, including waste handling and disposal.
   • Respirator use, care, and fit-testing.
   • Protective clothing donning, use, and handling.
   • Hands-on exercises for techniques such as glovebag work and EPA vacuum use and maintenance.
   • Appropriate and proper worker decontamination procedures.

3. Abatement worker training for workers who may conduct asbestos abatement—for example, conducting a removal job, constructing an enclosure, or encapsulating a surface containing ACM. This work involves direct, intentional contact with ACM. The recognized "abatement worker" training courses approved by EPA or states, under the EPA AHERA model accreditation plan for schools, which involves 24 to 32 hours of training, would fulfill this level of training. If this level of training is provided to in-house staff, it may save time and money in the long run to use these individuals to perform such activities. This level of training is much more involved than Levels 1 and 2, although it should include some of the same elements (e.g., health effects of asbestos). It will typically include a variety of specialized topics, such as:

- Pre-asbestos abatement work activities.
- Work area preparation.
- Establishing decontamination units.
- Personal protection, including respirator selection, use, fit-testing, and protective clothing.
- Worker decontamination procedures.
- Safety considerations in abatement work area
- A series of practical hands-on exercises.
- Proper handling and disposal of ACM wastes.

The Asbestos Program Manager should consider conducting the training program for Levels 1 and 2 if he or she has sufficient specific asbestos knowledge and training. If the Asbestos Program Manager does not conduct the training, the building owner should hire an outside consultant or send workers to an appropriate O&M training course. A trained (preferably certified) industrial hygienist or equally qualified safety and health professional should conduct the training on respirator use and fit-testing. A health professional should conduct the training on health effects. OSHA or EPA Regional Offices, as well as state and local agencies and professional associations, may be able to suggest courses or direct you to listings of training providers for each of the three levels.

Where custodial and maintenance services are performed by a service company under contract, or where some installation or repairs are performed by employees of trade or craft contractors and subcontractors, those workers may need to have training at the different levels described above as appropriate for their work. The Asbestos Program Manager or building owner should verify that these employees receive appropriate training before they begin any work.

17.8 FEDERAL, STATE, AND LOCAL REGULATIONS AFFECTING O&M PROGRAMS

A variety of federal, state, and local regulations govern the way building owners must deal with ACM in their facilities. State and local regulations may be more stringent than federal standards and often change rapidly. Building owners should periodically check with the appropriate Federal, State, and local authorities to determine whether any new asbestos regulations have been developed or whether current regulations have been amended. Specific federal regulations that may affect asbestos-related tasks and/or workers are highlighted here:

- OSHA Construction Industry Standard for Asbestos (29 CFR 1926.58).
- OSHA General Industry Standard for Asbestos (29 CFR 1910.1001).
- OSHA Respiratory Protection Standard (29 CFR 1910.134).
- EPA Worker Protection Rule (40 CFR 763 Subpart G).
- EPA National Emission Standards for Hazardous Air Pollutants (NESHAP) (40 CFR 61 Subpart M).
- EPA Asbestos Hazard Emergency Response Act (AHERA) Regulations (40 CFR 763 Subpart E).
- EPA Asbestos Ban and Phaseout Rule (40 CFR 763 Subpart I).

Appendix G, which is specified as a nonmandatory section to the OSHA regulation 29 CFR 1926.58, may become mandatory under certain circumstances where "small-scale, short-duration" asbestos projects are conducted. These projects are not precisely defined in terms of either size or duration, although their nature and scope are illustrated by examples presented in the text of the regulation. Properly trained maintenance workers may conduct these projects. Examples may involve removing small sections of pipe insulation or covering for pipe repair, replacing valves, installing electrical conduits, or patching or removing small sections of drywall. OSHA issued a clarification of the definition of a "small-scale, short-duration" (SS/SD) project in a September 1987 asbestos directive. The directive focuses on intent, stating that in SS/SD projects, the removal of ACM is not the primary goal of the job. If the purpose of a small-scale, short-duration project is maintenance, repair, or renovation of the equipment or surface behind the ACM — not abatement of ACM — then the appendix provisions may apply. If the intent of the work is abatement of the ACM, then the full-scale abatement control requirements apply.

In any event, this appendix section of the OSHA construction standard outlines requirements for the use of certain engineering and work practice controls such as glovebags, mini-enclosures, and special vacuuming techniques. Similar information on these procedures may be found in the EPA's AHERA regulations for schools. (See final AHERA rule, Appendix B, for SS/SD projects.)

Notification

The EPA or the state (if the state has been delegated authority under NESHAP) must be notified before a building is demolished or renovated. The following information is required on the NESHAP notice:

1. Name and address of the building owner or manager.
2. Description and location of the building.
3. Estimate of the approximate amount of friable ACM present in the facility.
4. Scheduled starting and completion dates of ACM removal.
5. Nature of planned demolition or renovation and method(s) to be used.
6. Procedures to be used to comply with the requirements of the regulation.
7. Name, address, and location of the disposal site where the friable asbestos waste material will be deposited.

The notification requirements do not apply if a building owner plans renovation projects which will disturb less than the NESHAP limits of 160 square feet of friable ACM on facility components or 260 linear feet of friable ACM on pipes (quantities involved over a one-year period). For renovation operations in which the amount of ACM equals or exceeds the NESHAP limits, notification is required as soon as possible.

Emissions Control and Waste Disposal

The NESHAP asbestos rule prohibits visible emissions to the outside air by requiring emission control procedures and appropriate work practices during collection, packaging, transportation or disposal of friable ACM waste. All ACM must be kept wet until it is sealed in a leak-tight container that includes the appropriate label.

Under expanded authority of RCRA, a few states have classified asbestos-containing waste as a hazardous waste, and require stringent handling, manifesting, and disposal procedures. In those cases, the state hazardous waste agency should be contacted before disposing of asbestos for approved disposal methods and recordkeeping requirements, and for a list of approved disposal sites. Friable asbestos is also included as a hazardous substance under EPA's CERCLA regulations. The owner or manager of a facility (e.g., building, installation, vessel, landfill) may have some reporting requirements. Check with your EPA Regional Office for further information.

The Asbestos Hazard Emergency Response Act Regulations (AHERA)

In October 1987, EPA issued final regulations to carry out the Asbestos Hazard Emergency Response Act of 1986 (AHERA). The AHERA regulatory requirements deal only with public and private elementary and secondary school buildings. The regulations require schools to conduct inspections, develop comprehensive asbestos management plans, and select asbestos response actions to deal with asbestos hazards. The AHERA rule does not require schools to remove ACM.

A key element of the AHERA regulations requires school to develop an O&M program if friable ACM is present. The AHERA O&M requirements also cover nonfriable ACM which is about to become friable. For example, drilling through an ACM wall will likely result in friable ACM. Under the AHERA O&M provisions, schools must carry out specific O&M procedures which provide for the clean-up of any ACM releases and help ensure the general safety of school maintenance and custodial workers, as well as other school building occupants. The AHERA regulation's O&M requirements mandate that schools employ specific work practices including wet wiping, HEPA vacuuming, proper waste disposal procedures, and specific training for custodial and maintenance employees who work in buildings with ACM.

U.S. EPA Asbestos Ban and Phaseout Rule

Bans on some uses and applications of asbestos are included under the Clean Air Act. In July 1989, under the Toxic Substances Control Act (TSCA), EPA promulgated an Asbestos Ban and Phaseout Rule. The complete rule was published in the *Federal Register* on July 12, 1989. Beginning in 1990 and taking effect in three stages, the rule prohibits the importation, manufacture, and processing of 94% of all remaining asbestos products in the United States over a period of seven years (1).

17.9 SUMMARY

1. Asbestos became a popular commercial product because it is strong, does not burn, resists corrosion, and insulates well.

2. The Asbestos NESHAP regulation has been in existence since 1973 and has been amended several times.

3. In the initial Asbestos NESHAP rule promulgated in 1973, a distinction was made between building materials that would readily release asbestos fibers when damaged or disturbed and those materials that were unlikely to result in significant fiber release.

4. Friable asbestos-containing materials. Since inhalation of asbestos fibers has been linked to the development of respiratory and other diseases, any material which is friable, or has a high probability of releasing fibers, must be handled in accordance with the Asbestos NESHAP.

5. When nonfriable ACM is subjected to intense mechanical forces, such as those encountered during demolition or renovation, it can be crumbled, pulverized, or reduced to powder, and thereby release asbestos fibers. When nonfriable materials are damaged or are likely to become damaged during such activities, they must be handled in accordance with the Asbestos NESHAP.

6. There are several products in use today containing asbestos. It is used in friction products such as brake linings for automobiles, buses, trucks, railcars, and industrial machinery, and in vehicle or industrial clutch linings.

7. Training of custodial and maintenance workers is one of the keys to a successful O&M program.

8. A variety of federal, state, and local regulations govern the way building owners must deal with ACM in their facilities.

REFERENCES

1. U.S. EPA, *Managing Asbestos in Place: A Building Owners Guide to Operations and Maintenance Programs for Asbestos-Containing Materials*, 20T-2003, July 1990.
2. U.S. EPA, *A Guide to the Asbestos NESHAP*, As revised November 1990, EPA 340/1-90-015, November 1990, Air and Radiation (EN-341).
3. U.S. EPA, *Asbestos–Waste Management Guidance*, Office of Solid Waste, EPA/530-SW-85-007, May 1985.
4. U.S. EPA, *Asbestos/NESHAP Regulated Asbestos-Containing Materials Guidance*, EPA 340/1-90-018, December 1990.

18

USED OIL

18.1 INTRODUCTION

Used oil is a valuable resource, but it can be an environmental problem and financial liability if improperly disposed of. Used oils pose hazards to human health and the environment, and therefore need to be managed safely. The mismanagement of used oil can contaminate air, water, and soil. Contamination primarily occurs from improper storage in containers and tanks, disposal in unlined impoundments or landfills, burning of used oil mixed with hazardous waste, improper storage practices of used oil handling sites and associated facilities, and road oiling for dust suppression.

In 1988, 1.3 million gallons of used oil were generated. Fifty-seven percent of the 1.3 million gallons generated entered the used oil management system and was recycled. Of the remaining used oil, the do-it-yourselfer (DIY) generator population (i.e., generated by homeowners) disposed of approximately 183 million gallons of mostly automotive crankcase oil, while nonindustrial and industrial generators dumped/disposed of 219 million gallons. The Environmental Protection Agency (EPA) believes that the majority of the remaining 43% of used oil that was generated could and should be recycled in an effort to protect the nation's groundwater, to meet the nation's petroleum needs and to converse natural resources (1). In addition to preserving a natural resource, it would lessen our dependence on foreign oil.

18.2 USED OIL INDUSTRY

Much of the used oil that is generated nationally is as a result of routine replacement of deteriorated lubricating oils or what can be classified as *viscosity breakdown*. Frequently, lubricants are replaced because they no longer meet their performance standards. Lubricants and oil are made of two basic components: (a) a base stock or material and (b) additives comprising up to 20% of the volume.

These additives greatly influence the specific performance of the finished product which eventually becomes a trademark of various lubricants. Typical additives include color stabilizer, viscosity improvers, corrosion inhibitators, rust inhibitators, and detergents. These additives contain specific metal and chemical compounds which results in specific performance standards. Typical lubricating oil additives and their functions are represented in Table 18.1.

Lubricants and industrial oils are typically manufactured from chemical feedstocks, mainly petroleum or petroleum products which are derived from crude oil. They contain a wide assortment of hydrocarbons in addition to chemical compounds such as sulfonates, sulphur, chlorine, and nitrogen compounds; they also contain metals such as barium, zinc, and chromium as a result of additives.

Used oils are generated from literally thousands of different resources. These can be broken down into two main categories: (a) automotive oils, which include engine crankcase oils, transmission fluids, diesel engine oils, and automotive hydraulic fluids, and (b) industrial oils, which covers oils and lubricants generated from industrial sources such as metal working processes, hydraulic equipment and machinery, refrigeration equipment, quenching oils, and turbine lubricating oils. In all automotive and industrial applications, the performance of the oil deteriorates over a period of time as additives break down and as contaminants build up in the oil. The oil must be replaced with new oil which results in the steady generation of what is commonly referred to as "used oil." Used oil can therefore be defined as lubricating oil which, through use, has been contaminated by physical or chemical impurities. Used oil is considered to be a waste product because it has served its original, intended purpose and must be discarded. However, it is a unique type of waste in that it can be recycled or reused as another product instead of merely being discarded or destroyed (2).

Much of the used oil that is generated nationally as a result of routine replacement of deteriorated lubricating oil enters the "used oil management

TABLE 18.1 Typical lubricating oils additives and their function

Name of Additive	Chemical Composition	Function
Corrossion inhibitor	Metal sulfonates and sulfurized terpenes, and barium dithiophosphates	To react with metal surfaces to form a zinc corrosion-resistant film
Rust inhibitor	Sulfonates, alkylamines, or amine phosphates	To react chemically with steel surfaces to form an impervious film
Antiseptic	Alcohols, phenols, and chlorine compounds	To inhibit microorganisms
Antioxidant	Sulfides, phosphates, phenols	To inhibit oxidation of oil
Detergent	Sulfonates, phosphites, alkyl substituted salicyates combined with barium, magnesium, zinc, and calcium	To neutralize acids in crankcase oils to form compounds suspended in oil
Color stabilizer	Amine compounds	To stabilize oil color

system." The used oil management system consists of companies that are involved in the generation, collection, transport, processing and reuse of used oils. These companies interact to provide a mechanism for used oil to flow from its point of generation to its reuse or disposal. Figure 18.1 illustrates that fate of used oil. Many companies are involved in just one used oil function while others participate in more than one activity. An example of this is a reprocessor or rerefiner which collects, transports, and recycles used oil.

Independent collectors in the United States collect about 25% of the oil which passes from generators into the used oil management system. There are several hundred independent companies in the United States which collect used oil and sell it to end-use markets and/or reprocessors and rerefiners. Such collectors may store oil, but they do not process the oil to improve quality. A typical company operates approximately two collection vehicles and uses two to five storage tanks. Collectors prefer to sell their used oil to users for fuel or as a dust suppressant because they can command a higher price for their oil. Prices average 30–80% higher than those received from a reprocessor or rerefiner. Thirty-five percent of the used oil collected nationally by independent companies is sold directly to end-users — 15% for road oil and 20% for fuel. The remaining 65% is sold to reprocessors or rerefiners. The vast majority of the used oil (75%) flowing into the used oil management system is collected directly by companies involved in reprocessing or rerefining used oil. They maintain collection services in order to ensure a steady supply.

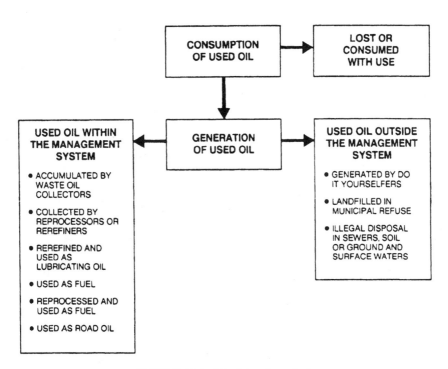

FIGURE 18.1 The fate of used oil.

The National Association of Oil Recyclers (NORA) states that the prices paid for or charged for used oil can vary greatly by region, depending on its urban, suburban, and rural character. In rural areas where used oil is a listed hazardous waste, such as Northern California and upper New York State, prices charged for collection have ranged from $0.00 to $0.25 per gallon. However, in urban areas such as New York City or Chicago, processors and collection firms may pay for the used oil in the range of $0.05 to $0.15 per gallon. The single most important factor affecting the profitability of the industry is the price of virgin crude oil, although other factors such as transportation and delivery cost and processing costs do contribute to a lesser degree.

From 1983 until August 1990 the price of virgin fuel oil prices underwent a general decline. Although the winters of 1983 and 1989 produced short-lived price spikes in the Northeastern United States, the overall impact has been to cause a dramatic drop in the salvage value of used oil.

This price structure resulted in a generator paid system of collection in most parts of the United States. The effects on the collection and reprocessing segments of the industry has resulted in a severe contraction. Collectors declined from 700 to 376 and the number of processors declined by 20%.

Transportation cost, which consists of collection and delivery, can vary between 7 and 14 cents per gallon depending on the degree of enforcement in the operation area and the size and geographical dispersion of the collection points. The most costly factor of the transportation component is insurance for the large tank trucks. These costs range from $100,000 in less regulated areas to over a $1.2 million in annual costs or highly regulated states such as New Jersey and California. Premiums may run between $2,000 to $4,000 per vehicle and ultimately claim ten percent of the recylcers' revenue. The larger used oil transporters try to improve utilization rates by spreading the fixed vehicle costs and operating fuel costs across a larger volume basis.

A system which is currently being used by a major player in the used oil industry is composed of two tier system: a 3,500 gallon collector known as a "street truck" and a 6,000–8,000-gallon long haul tanker. These trucks are the foot soldiers in the collection system. The street trucks will service any small quantity generator such as gasoline retail service stations where the average pickup is 250 gallons, and they will also service small industrial establishments. These street trucks will transport their load 100 miles to a remote bulk storage facility which contains 15,000–20,000 gallons.

The long haul tanker of 6,000–8,000 gallons will then transport used oils from the remote terminals to a rerefinery or reprocessing facility, depending on the quality of the used oil. The maximum distance a large tanker will travel from dispersed large industrial customers is 400 miles from the rerefinery. Sending the long haul tank trucks direct from the customers to the rerefinery is based on the truck being able to backhaul rerefined lubrication stock, without washing the tank of contaminants. Additional modes of transportation are dedicated rail tank cars and barges. The cost of washing these tank vessels is prohibitive and the waste stream must be suitable for the production of rerefined lube stock. The greatest distance that the rerefinery can accommodate used oil for refining by barge or rail is 600 miles from the facility.

The evolution of the collection network is analogous to casting a net from the refinery. As the market net develops, the mesh to the net becomes finer to ensnare smaller and smaller customers. Currently the economics of the collection system dictate that the generator of the oil pay for the collection of the oil for recycling. This is based on the price of virgin crude but also on the costs associated with collecting small amounts from numerous sources.

18.3 USED OIL RECYCLING AND REUSE

Refining

Used oil can be rerefined into a base lubricating oil by employing a variety of techniques and processes. With the additives restored, the rerefined oil can be marketed as lubricating oil for industrial and commercial application. Rerefining used oil results in well-defined, marketable products regardless of the type of technology employed. It can produce a refined lubricating base, a distilled light fuel fraction and distillation bottoms for use as asphalt extenders. The methods of rerefining differ as to the waste material generated, the percent recovered, and by-product marketability.

Common Rerefining Technologies

1. **Acid/Clay Method.** This process involves three steps.

Step 1. Filtering the used oil to remove water and other solids.

Step 2. Acid Treatment — using sulphuric acid treatment to remove toxic impurities.

Step 3. Clarification — the material is clarified to remove odor and color impurities by filtering through clay.

Product: Acceptable base with which additives and virgin oil can be blended. Efficiency 50% recovery.

Disadvantages

1. Process is costly.
2. Batch Operation which results in high operating costs.
3. Environmental Problems. Technique yields a considerable amount of acid sludge and clay-like residue from sulfuric acid treatment and clay refilteration. Associated Waste disposal costs.

2. **Distillation — Clay Treatment Process.** The distillation treatment process involves a five-step process:

Step 1. Screening to remove solids.

Step 2. Evaporation to remove water.

Step 3. Flash vacuum distillation to recover lowboiling components as a distillate fuel.

Step 4. High-temperature, high-vacuum distillation in a thin film evaporator to separate lubricating oil fraction from the residue and depleted additive, with controlled partial condensation of the distillate to separate lubricating oil into light and heavy fractions.

Step 5. Final purification of lubricating oil using clay treatment and filtration.

Product:

1. Lubricating oil base.
2. High ash content fuel by-product which can be burned.

Advantages:

1. Higher recovery 60% to 75%.
2. Continuous operation.
3. Manageable residue by-product.

Distillation–hydrotreating is a type of rerefining that is quite similar to the distillation–clay process. However, it substitutes hydrotreating for clay treating as a final step. Hydrotreating involves mixing heating oil with recirculating hydrogen to remove impurities. The end product is a lubricating oil base and several by-products which have secondary market value. Recovery can be expected to approach 99%.

Phillips Petroleum developed Phillips Rerefining Oil Process (PROP), a re-refined oil process which combines hydrotreating with chemical demetallization to produce 90% yields of based oil. PROP is a two-stage process. In the first stage the metal is recovered from the used oil through the use of chemicals. The oil is then filtered to remove the remaining metals. The demetallized and dehydrated oil is hydrotreated in the second major processing stage. Hydrotreating removes unwanted sulfur, nitrogen, oxygen, and chlorine compounds and improves color.

Reprocessing

Used oil can also be reprocessed to yield fuel oil that can be burned in industrial incinerators or boilers. Reprocessing involves filtering sediment and water from used oil. The technology varies according to the degree of sophistication employed and from facility to facility. Most processing firms produce only fuel.

Reprocessing is a less costly recycling alternative. It does not involve as substantial an investment in its operation as rerefining does. However, the product is of a lesser quality and contains toxic metals and other chemicals that refining techniques can remove.

Minor reprocessors employ fairly simple technology to recycle used oil as a fuel. It includes in-line filtering and gravity settling to remove solids and water. In addition, it may include the addition of a heat source to decrease viscosity and improve gravity settling.

Minor reprocessors market their product by making it available directly to fuel users, fuel oil dealers, road oilers (where allowed) or for purchase by major waste oil reprocessors. In addition, some choose to burn a portion of the oil produced on site to generate heat to induce gravity settling, for space heating, or for some other fuel-consuming process operated at the site.

Major reprocessors utilize comparatively sophisticated processing technology. They go beyond merely filtering and settling used oil and employ treatment devices to further increase oil quality. Some of the devices to improve oil quality include (a) distillation towers to separate and capture light fuel fractions as well as remove water, (b) centrifuges to separate fine solids, and (c) agitators to mix emulsion-breaking chemicals into the oil.

A major portion of used oil which is reprocessed for fuel is blended with virgin fuel oil. Most of this blending is done by virgin fuel oil dealers. However, a small fraction of the blending is done by major reprocessors. The criteria used for blending vary greatly. Some blenders mix used and virgin oil to one part used oil. Others blend to a desired viscosity, moisture content, or any number of other factors, including heat content and percent solids. The criteria are thus a function of the product specifications or characteristics which the blender or his customer have established.

18.4 REGULATIONS GOVERNING USED OIL

On December 18, 1978, the EPA initially proposed guidelines and regulations for the management of hazardous waste as well as specific rules for the identification and listing of hazardous waste under Section 3001 of the Resource Conservation and Recovery Act (RCRA) (43FR 58946). At that time, the EPA proposed to list waste lubricating oil and waste hydraulic and cutting oil as hazardous wastes on the basis of their toxicity. In addition, the EPA proposed recycling regulations to (a) regulate the incineration or burning of used lubricating, hydraulic, transformer, transmission, or cutting oil that was hazardous and (b) the use of waste oils in a manner that constituted disposal.

In May 19, 1980 regulations (45 FR33084), the EPA decided to defer promulgation of the recycling regulations for waste oils in order to consider fully whether waste- and use-specific standards may be implemented in lieu of imposing the full set of Subtitle C regulations on potentially recoverable and valuable materials. At the same time, the EPA deferred the listing of waste oil for disposal so that the entire waste oil issue could be addressed at one time. Under the same regulation, however, any waste oil exhibiting one of the characteristics of hazardous waste (ignitability, corrossivity, reactivity, and toxicity) that was disposed of or accumulated, stored, or treated prior to disposal, became regulated as a hazardous waste subject to all applicable subtitle C regulations.

In an effort to encourage the recycling of used oil and, in recognition of the potential hazards posed by its mismanagement, Congress passed the Used Oil Recycling Act (UORA) on October 15, 1980 (P.L. 96-463). UORA defined used oil as "any oil which has been refined from crude oil, used, and as result of such use, contaminated by physical or chemical impurities." Among other provisions, UORA required the EPA to make a determination as to the hazardousness of used oil and report the findings to Congress with a detailed statement of the data and other information upon which the determination was based. In addition, the EPA was to establish performance standards and other requirements under Section 7 of UORA as "may be necessary to protect public health and the environment from hazards associated with recycled oil" as long as such regulations "do not discourage the recovery or recycling of used oil." These provisions are now included in Section 3014 of RCRA.

In January 1981, the EPA submitted to Congress the used oil report mandated by Section 8 of the UORA. In the report, the EPA indicated its intention to list both used and unused waste oils as hazardous under Section 301 of RCRA based on the presence of a number of toxicants in crude or refined oil (e.g., benzene, naphthalene, and phenols), as well as the presence of contaminants in used oils as a result of use (e.g., lead, chromium, and cadmium). In addition, the report cited the environmental and human health threats posed by these waste oils, including the potential threat of rendering ground water unpotable through contamination.

On November 8, 1984, the Hazardous and Solid Waste Amendments (HSWA) were signed into law. In addition to many other requirements, HSWA mandated that the protection of human health and the environment was to be of primary concern in the regulation of hazardous waste. Specific to used oil, the Administrator was required to "promulgate regulations ... as may be necessary to protect human health and the environment from hazards associated with recycled oil. In developing such regulations, the Administrator shall conduct an analysis of the economic impact of the regulations on the oil recycling industry. The Administrator shall ensure that such regulations do not discourage the recovery or recycling of used oil consistent with the protection of human health and the environment." This altered EPA's mandate with respect to the regulation of used oil by requiring that protection of human health and environment be a prime consideration, even if such regulation may tend to discourage the recovery or recycling of used oil.

The HSWA required the EPA to propose whether to identify or list used automobile and truck crankcase oil by November 8, 1985, and to make a final determination as to whether to identify or list any or all used oils by November 8, 1986. On November 29, 1985 (50 FR 49258), the EPA proposed to list all used oils as hazardous waste, including petroleum-derived and synthetic oils, based on the presence of toxic constituents at levels of concern from adulteration during and after use. Also on November 29, 1985, the EPA proposed management standards for recycled used oil (50 RF 49212) and issued final regulations, incorporated at 40 CFR Part 266, Subpart E, prohibiting the burning of off-specification used oil in nonindustrial boilers and furnaces (50 FR 49164). Marketers of used oil fuel and industrial burners of off-specification fuel are required to notify EPA of their activities and to comply with certain notice and recordkeeping requirements. Used oils that meet the fuel oil specification are exempt from most of the 40 CFR Part 266, Subpart E regulations.

On March 10, 1986 (51 FR 8206), the EPA published a supplemental notice requesting comments on additional aspects of the proposed listing of used oil as hazardous. In particular, commenters to the November 29, 1985 proposal suggested that the EPA consider a regulatory option of only listing used oil as a hazardous waste when disposed of while promulgating special management standards for used oil that is recycled. The supplemental notice also contained a request for comments on additional issues to the "mixture rule" (40 CFR Section 261.3(a)(2)(iii)), on test methods for determining halogen levels in used oils, and on new data on the composition of used oil and used oil processing residuals.

On November 19, 1986, the EPA issued a decision not to list as a hazardous waste used oil that is being recycled (51 FR 41900). At that time, it was the Agency's belief that the stigma associated with a hazardous waste listing might discourage recycling of used oil, thereby resulting in increased disposal of used oil in uncontrolled manners. The EPA stated that several residues, waste waters, and sludges associated with the recycling of used oil may be evaluated to determine if a hazardous waste listing was necessary, even if used oil was not listed. The EPA also outlined a plan that included (a) making the determination whether to list used oil being disposed of as hazardous waste and (b) promulgation of special management standards for recycled oil.

The EPA's decision not to list used oil as a hazardous waste based on the potential stigmatic effects was challenged by the Hazardous Waste Treatment Council, the Association of Petroleum Rerefiners, and the Natural Resources Defense Council. The petitioners claimed that the language of RCRA indicated that in determining whether to list used oil as a hazardous waste, EPA may consider technical characteristics of hazardous waste, but not the "stigma" that listing might involve, and that Congress intended EPA to consider the effects of listing on the recycled oil industry only after the initial listing decision.

On October 7, 1988, the Court of Appeals for the District of Columbia found that EPA acted contrary to law in its determination not to list used oil under RCRA §3001 based on the stigmatic effects. The court ruled that the EPA must determine whether to list any used oils based on the technical criteria for waste listings specified in the statute.

After the 1988 court decision, the EPA began to reevaluate its basis for making a listing determination for used oil. The EPA reviewed the statute, the proposed rule, and the many comments received on the proposed rule. Those comments indicated numerous concerns with the proposed listing approach. One of the most frequent concerns voiced by commenters related to the quality and "representativeness" of the data used by the EPA to characterize used oils in 1985. Numerous commenters indicated that "their oils" were not represented by the data and, if they were represented, those oils were characterized when mixed with other more contaminated oils or other hazardous wastes. Many commenters submitted data demonstrating that their oils, particularly industrial used oils, did not contain high levels of toxicants of concern.

In addition, the EPA recognized that much of the information in the 1985 used oil composition data was more than five years old, as most of the information was collected prior to 1985. Since the time of that data-gathering effort, used automotive oil composition may have been affected by the phase-down of lead in gasoline. The Agency also recognized the need to collect analytical data addressing specific classes of used oils as collected and stored at the point of generation (i.e., at the generator's facility).

Finally, the promulgation of the toxicity characteristic (TC) (55 FR 11798, March 29, 1990) is known to identify certain used oils as hazardous. Due to the possibility of changes in used oil composition described above and the new TC, the EPA recognized that additional data on used oil characterization may be needed prior to making a listing determination. The EPA believed it is important to consider the effects of the TC before taking final action on the listing determination and used oil standards in accordance with its mandate Section 3014(b) of RCRA to "list or identify" used oil as a hazardous waste.

On September 23, 1991, the EPA published a notice in the Federal Register providing information on proposed used oil management standards for recycled oil under section 3014 of RCRA. In addition, the EPA specifically requested public comments on proposed used oils and residuals to be listed as hazardous, on a number of specific aspects of the newly available data, on specific aspects of the Agency's approach for used oil management standards, and on several aspects of the hazardous waste identification program as related to used oil. The EPA's overall approach to used oil consists of three major components. First, the EPA identifies approaches for making a determination whether to list or identify used crankcase oil and other used oil as hazardous wastes, as required by Section

3014(b). Second, the EPA proposes a number of alternatives relating to management standards to ensure proper management of used oils that are recycled. The EPA management standards will be issued in two phases. Phase I will consist of basic requirements for used oil generators, transporters, road oilers, and recyclers, including burners and disposal facilities to protect human health and the environment from potential hazards caused by mismanagement of used oil. Once Phase I standards are in place, the EPA may decide to evaluate the effectiveness of these standards in reducing the impact on human health and the environment. Upon such evaluation, the EPA will consider whether or not more stringent regulations are necessary to protect human health and the environment, and propose these regulations as Phase II standards. The third part of the EPA's general approach to used oil is the consideration of nonregulatory incentives and other nontraditional approaches to encourage recycling and mitigate any negative impacts the management standards may have on the recycling of used oil, as provided by Section 3014(a).

The notice presents supplemented information gathered by the EPA and provided to the EPA by individuals commenting on previous notices on the listing of used oil and used oil management standards. Numerous commenters on the 1985 listing of all used oils unfairly subjects them to stringent Subtitle C regulations because their oils are not hazardous. Based on those comments, the EPA has collected a variety of additional information regarding various types of used oil, their management, and their potential health and environmental effects when mismanaged. This notice presents that new information to the public and requests comment on that information, particularly if and how this information suggests new concerns that the EPA may consider in deciding whether to finalize all or part of its 1985 proposal to list used oil as a hazardous waste.

The EPA intends to amend 40 CFR Section 261.32 by adding four waste streams from the reprocessing and rerefining of used oil to the list of hazardous wastes from specific sources. The EPA noted its intention to include these residuals in the definition of used oil in its November 29, 1985 proposal to list used oil as hazardous. The wastes from the reprocessing and rerefining of used oil include process residuals from the gravitational or mechanical separation of solids, water, and oil; spent polishing media used to finish used oil; distillation bottoms; and treatment residues from primary wastewater treatment.

The notice also includes a description of some of the management standards (in addition to or in place of those proposed in 1985) that the EPA is considering promulgating with the final used oil listing determination. The EPA, under various RCRA authorities, is considering management standards for used oils, whether or not the oil is classified as hazardous waste. When promulgated, the standards may: prohibit road oiling; restrict used oil storage in surface impoundments; limit disposal of nonhazardous used oil; require inspection, reporting, and cleanup of visible released of used oil around used oil storage containers and aboveground tanks and during used oil pickup, delivery, and transfer; impose spill cleanup requirements and allow for limited CERCLA liability exemptions; institute a tracking mechanism to ensure that all used oils reach legitimate recyclers; and require reporting of used oil recycling activities. The used oil burner standards included in 40 CFR Part 266 Subpart E will continue to regulate the burning of used oil for energy recovery. All of the requirements (including those in Part 266, Subpart E) are placed in a new Part (e.g., 40 CFR Part 279). Used oils that are hazardous (either listed or characteristic) that cannot be recycled are not included in these provisions, but are instead subject to 40 CFR Section 261-270.

On September 10, 1992, the EPA promulgated both a final listing decision for recycled used oil and management standards for used oil pursuant to RCRA Section 3014 (57 FR 42566). Part 279, *Standards for the Management of Used Oil*, was added to codify the management standards. In this rule, the EPA stated that it assumes all used oil will be recycled until the used oil is disposed of or sent for disposal (57 FR 41578). Used oil that is disposed of will need to be characterized like any other solid waste and will need to be managed as hazardous if it exhibits a characteristic of hazardous waste or if it is mixed with a listed hazardous waste.

Standards for the management of used oil (40 CFR Part 279) are a comprehensive set of requirements centered around the various entities involved in the management of used oil. The different subparts incorporates specific requirements for those entities. Of particular importance is the requirements under the following subparts:

- Subpart C. Standards for used oil Generators
- Subpart D. Standards for Used Oil Collection Centers and Aggregration Points.
- Subpart E. Standards for Used Oil Transporter and Transfer Facilities.
- Subpart F. Standards for Used Oil Processors and Rerefiners.
- Subpart G. Standards for Used Oil Burners Who Burn-Off Specification Used Oil for Energy Recovery.
- Subpart H. Standards for Used oil Fuel Marketers.
- Subpart I. Standards for Use as a Dust Suppressant and Disposal of Used Oil.

18.5 FACTS ABOUT USED OIL

The following are some well known facts about used oil:

- A gallon of used oil from a single oil change can ruin a million gallons of fresh water — a year's supply for 50 people (4).
- It takes only 1 gallon of used oil to yield the same 2.5 quarts of lubricating oil provided by 42 gallons of crude oil (4).
- Americans who change their own oil throw away 120 million gallons of recoverable motor oil every year (4).
- If this oil were recycled, it would save the United States 1.3 million barrels of crude oil per day. This will reduce our dependence on foreign oil (4).
- The damage used oil causes comes from mismanagement.
- Rerefining used oil takes only about one-third the energy of refining crude oil to lubricant quality.
- If all used oil improperly disposed of by do-it-yourselfers were recycled, it could produce enough energy to power 360,000 homes each year or could provide 96 million quarts of high-quality motor oil.
- One gallon of used oil used as fuel contains about 140,000 Btu of energy.

- Concentrations of 50 to 100 parts per million (ppm) of used oil can foul sewage treatment processes.
- Films of oil on the surface of water prevent the replenishment of dissolved oxygen, impair photosynthetic processes, and block sunlight.
- Oil dumped onto land reduces soil productivity.
- Toxic effects of used oil on freshwater and marine organisms vary, but significant long-term effects have been found at concentrations on 310 ppm in several freshwater fish species and as low as 1 ppm in marine life forms.
- Publicity about used oil recycling can triple do-it-yourselfer participation (3).

18.6   SETTING UP A RECYCLING PROGRAM

In many cases, local recycling programs are cooperative efforts between local governments (towns, cities, and counties) and one or more private or semi-private sponsors, such as (a) environmental or civic groups, or (b) service organizations. Local governments often assist in collecting used oil through collection centers or curbside pickup. Sponsors often help governments design and organize their programs, run the publicity campaigns and outreach, and enlist the help of resourceful and committed volunteers.

Other arrangements can be equally successful such as those run entirely by local governments or by private sponsors. Private companies can also help — used oil haulers and recyclers may act as business sponsors; Car dealerships or local oil retailers also reap benefits from the publicity and customer goodwill these programs generate.

Consider the following basic pointers when setting up a program:

1. Learn the facts about used oil in the state: Call the state DIY used oil recycling coordinator for information on the status of DIY used oil recycling in the state.

2. Bring the most effective participants together: If a local government is thinking of sponsoring a program, seek out community sponsorship. If a community group is willing to sponsor a program, identify the most appropriate local government agency with which one can work and secure the maximum support from local business.

3. Design and implement the program as a group: Work together with the other participants to decide how the program will run — the type of pickup it will use, who will collect and recycle the used oil, how the program may link with other local recycling efforts, how it will be publicized, and so on. General issues may include enlisting additional volunteers, soliciting funds, finding haulers and recyclers and assessing their performance, running collection operations, and tracking progress and accomplishments.

Recycling used oil is an ideal way for interested groups to get constructively involved in environmental action because it deals with an important environmental problem that is best addressed at the local level. A successful program demands commitment, energy, and sustained involvement. Before beginning, make sure that with the following fundamental needs are considered:

1. Ensuring Adequate Resources: Used oil recycling programs are not expensive to run and can rely heavily on volunteer labor and in-kind contributions. They do need money, however, for purposes like equipping pickup or collection operations and designing, printing, and mailing publicity materials. States may offer financial assistance, but each local program will probably have to raise money on its own as well.

2. Properly Managing Used Oil Risks: Programs must prevent other materials from being mixed with used oil. Mixing can be environmentally damaging and also may prevent haulers or recyclers from accepting your used oil. The key point do-it-yourselfers must understand is never to mix used oil with gasoline, solvents, pesticides, or other household chemicals before recycling. Small businesses and consumers also must never use collection centers as dump sites for solvents or other hazardous materials.

3. Paying Adequate Attention to Haulers' and Recyclers' Performance: The most obvious and dramatic environmental damage caused by used oil in recent years has been traced to unsafe hauling and recycling operations. One of the most important contributions to environmental quality local programs can make is to conduct a "safety assessment" of the performance of current and prospective haulers and recyclers in their areas.

Once the basic framework of the program has been set up, the most important next step it to make the public aware of the program. The typical do-it-yourselfer is usually a male between 16 and 45 years old (people older than 45 usually have their oil changed for them). Many of those younger than 16 will be driving someday and may become do-it-yourselfers. A campaign should have three targets — current do-it-yourselfers, young people in school, and the general public.

Promotional activities for a used oil recycling program should have two goals — first, to educate the public about the used oil problem and to encourage more responsible oil management and, second, to tell do-it-yourselfers exactly how to use the program to recycle oil. Educational efforts should raise awareness of the damage used oil can do, its value as a resource, and how to change auto oil in an environmentally sound manner. Emphasis that used oil that is re-refined or made back into a motor oil is as good as regular oil and that purchasing recycled oil helps support the used oil re-refining industry. Encourage the purchase of re-refined oil where it is available. The publicity portion should alert do-it-yourselfers about the location of collection points, the availability of curbside collection (if any), how to obtain appropriate containers, and any other elements of your program aimed directly at the do-it-yourselfer.

Promoting a used oil program involves taking advantage of all possible opportunities to bring the message to the public, educating them about the importance of the used oil issue and how to manage their oil properly, and telling them how to take advantage of the program's services. Since do-it-yourselfer activity is seasonal, the promotions may not have to run the full year, but education of the general public and young people can be a year-round activity.

The program should be in full operation during the time when do-it-yourselfers are most likely to change their oil — the spring through summer months. Have all collection sites in operation by the time warm weather arrives. Promotion should be in high gear one to two months beforehand to give do-it-yourselfers plenty of time to take advantage of new services. For instance, in the Northeast, a program might begin its publicity in March, when winter weather is over. Publicity would peak in May and June, the spring months when most DIYs would be changing their oil, and again in September, the beginning of cooler weather. In the warmest U.S. climates, seasonal variations may be minor and one will want consistent, year-round publicity.

Below are some suggestions of ways to promote a program. Although they introduce proven approaches, one should be creative and invent more ways.

An open meeting is one way to kick-off a program by combining public education and publicity to recruit more volunteers and increase participation among DIYs, potential collection centers, and local area leaders.

Time:	Pick two hours on a weekday evening or a weekend day.
Invitation:	Invite any community service organizations already interested, as well as representatives of business and government.
Press Coverage:	Meet with a reporter from a local newspaper two to three weeks in advance. Provide the reporter with background information about the problem, the program, and the groups involved.
Announcements:	Send public service announcements to local radio and TV stations stating the purpose of the meeting and its date, time, and location.
Press Release:	One week before the meeting, send out a press release to local newspapers.

This first meeting will serve to get people involved. Stress the basics about the nature of the used oil problem and its solution. By the time the meeting is over, one should have a list of the names and phone numbers of additional volunteers.

If a state has a used oil recycling coordinator, he or she would be an excellent speaker at the kickoff meeting. This is also a time to call on local celebrities or community leaders to ask them to lend their influence to the program.

The used oil program should, if possible, have a publicly advertised, local telephone "hotline" that people can dial during normal business hours (and if possible on weekends) to get information regarding collection center locations, how to obtain suitable used oil containers, and how to participate in the program as a volunteer. This might be provided by the civic group sponsor, but could also be run by the local government. In addition, if a state has its own used oil hotline, that fact should be advertised locally as part of the program.

Public service announcements are a good way to get the message out through newspapers, television, magazines, and radio. There is usually no charge. One can use them as regular reminders to do-it-yourselfers to change their oil properly and take advantage of collection centers. They are also invaluable for publicizing special events. Use public service announcements as a vehicle for outside endorsements from business and community leaders.

Full-length articles and editorials are another way to promote a program through newspapers, community newsletters, and local consumer publications. These may include feature articles by environmental editors or correspondents, editorials supporting the program, letters to the editor from prominent people in the community, and so forth. Solicit this type of coverage and be prepared to supply background material as necessary. Keep a list of press and media contacts for the area so that one can reach them quickly.

Where possible, generate news coverage of the program through announcements of special events, progress made, major contributions, new endorsements or testimonials, newly established collection sites, or tie-ins with other environmental and energy groups, businesses, or local government. Send out press releases and call reporters with developments as they occur. Radio and television offer special opportunities for publicity and education through participation of program members or supporters in public affairs shows.

A press release should answer the basic reporting questions of "who, what, when, where, and why." This information should be found in the first sentence or two of the release so the reporter or the news department can quickly learn what the press release is about and decide whether it deserves coverage. Learn local press schedules and send releases so they reach reporters three or four days before the events are covered.

Printed materials of all kinds can be distributed through many outlets. Posters with the program logo should be prominently displayed at all collection centers and, where possible, at points of purchase. Brochures and leaflets can be distributed wherever motor oil is sold—especially at discount stores, supermarkets, and department stores catering to do-it-yourselfers. Handouts can be both educational and promotional, warning against pollution, teaching proper management techniques, and publicizing local collection programs. Try to distribute these materials to everyone who may be a do-it-yourselfer by persuading stores selling lubricating oil to place them where the oil is displayed or near the cash register, or to insert them into each bag carried away. The local office of a state motor vehicle department may be willing to distribute them with licenses or registrations.

Bumper stickers are also effective, with very high visibility to exactly the right audience. They can be distributed (perhaps at the collection centers) to everyone who actively participates in or supports the program. Local motor vehicle fleets can be asked to put a bumper sticker on each of their vehicles to help promote the program.

Regular or special-purpose mailings are another powerful technique for education and publicity. Often local businesses, such as banks, department stores, insurance companies, or utility companies, can be convinced to include inserts or brochures from a program in their mailings as a public service. These can be used to remind people of collection center locations, as educational tools to instruct do-it-yourselfers on proper oil changing and oil management techniques, and so forth.

High schools are natural places to present short programs on the benefits of used oil recycling. Future do-it-yourselfers can be reached with information on how the damages caused by used oil, how to change automobile oil properly, and how to participate in a local collection program—either as a recycler or as a volunteer helping run the program. Drivers' education classes are a perfect place to include this information. One may be able to persuade a state to include used oil recycling in motor vehicle examinations or study guides.

Beyond education and an appeal to public concern for the environment, incentive programs offering money and other prizes can be very useful for increasing participation. Such incentives include:

- Merchandise discount coupons given with the original purchase of motor oil, redeemable on return of used oil.

- Instant prizes issued on the return of used oil, redeemable for merchandise.

- Large-prize contest coupons, issued at the point of purchase and entered into a drawing when oil is returned to a participating collection center. Prize drawings could be held at regular intervals, such as quarterly, with winning numbers posted at participating collection points.

- Inexpensive kickoff prizes, such as funnels or used oil containers, offered at collection sites to all participants during the first days or weeks of the program.

Be aware of any legal issues relating to health, safety, and environmental performance that could affect the activities. There are a few federal requirements affecting used oil management; state requirements vary. EPA Regional Offices can provide information on current federal regulations. States may have their own laws and regulations governing used oil recycling; the state used oil program would be the authority on these and any other requirements.

Generally, the most significant legal issue is to keep used oil from being mixed with any hazardous waste. The easiest way is to prevent mixing used oil with any other substances. Since preventing mixing will be as important to a reputable hauler as it would be to a program, all participants should be willing to cooperate on this issue.

Other important legal requirements include making sure of compliance with local zoning, health, safety, environment, and fire laws. Contact the pertinent local agencies for advice (5).

18.7 SUMMARY

1. Used oil is a valuable resource, but it can be an environmental problem and financial liability if improperly disposed. Used oils pose hazards to human health and the environment, and therefore need to be managed safely. The mismanagement of used oil can contaminate air, water, and soil.

2. Much of the used oil that is generated nationally as a result of routine replacement of deteriorated lubricating oil enters the "used oil management system." The used oil management system consists of companies that are involved in the generation, collection, transport, processing, and reuse of used oils.

3. Used oil can be rerefined into a base lubricating oil by employing a variety of techniques and processes. With the additives restored, the re-refined oil can be marketed as lubricating oil for industrial and commercial application.

4. The regulations governing waste oils were first introduced under the Resource Conservation and Recovery Act. At that time, EPA proposed to list waste lubricating oil and waste hydraulic and cutting oil as hazardous wastes on the basis of their toxicity. In addition, the Agency proposed recycling regulations to regulate (1) the incineration or burning of used lubricating, hydraulic, transformer, transmission, or cutting oil that was hazardous and (2) the use of waste oils in a manner that constituted disposal.

5. Some interesting facts about used oil for example one gallon of used oil can ruin a million gallons of fresh water.

6. Recycling used oil can be a rewarding experience. It is an ideal way for interested groups to get constructively involved in environmental action because it deals with an important environmental problem that is best addressed at the local level.

REFERENCES

1. Hazardous Waste Management System; Identification and Listing of Hazardous Waste; Used Oil; Supplemental Notice of Proposed Rulemaking, 40 CFR Parts 261 and 266, Volume 56, No. 184, September 23, 1991.

2. *Used — But Useful: A Review of the Used Oil Management Program in New York State,* Legislative Commission on Toxic Substances and Hazardous Wastes, October 1986.

3. U.S. EPA, *How to Set up a Local Program to Recycle Used Oil,* EPA/530-SW-89-039A.

4. U.S. EPA, *Recycling Used Oil, 10 Steps to Change Your Oil,* EPA/530-SW-89-039C.

5. U.S. EPA, *How to Set up a Local Program to Recycle Used Oil,* EPA/530-SW-89-039A, May 1989.

19

RADIOACTIVE WASTE MANAGEMENT

Contributing Author: James B. Mernin

19.1 INTRODUCTION

As with many other types of waste disposal, radioactive waste disposal is no longer a function of technical feasibility but rather a question of social or political acceptability. The placement of facilities for the permanent disposal of municipal solid waste, hazardous chemical waste, and nuclear wastes alike has become an increasingly large part of waste management. Today a large percentage of the money required to build a radioactive waste facility will be spent on the siting and licensing of the facility.

Nuclear or radioactive waste can be loosely defined as something that is no longer useful and that contains radioactive isotopes in varying concentrations and forms. Radioactive waste is then further broken down into categories that classify the waste by activity, by generation process, by molecular weight, and by volume.

Radioactive isotopes emit energy as they decay to more stable elements. The energy is emitted in the form of alpha particles, beta particles, neutrons, and gamma rays. The amount of energy that a particular radioactive isotope emits, the timeframe over which it emits that energy, and the type of contact with humans all help determine the hazard it poses to the environment. The major categories of radioactive waste that exist are high-level waste (HLW), low-level waste (LLW), transuranic waste (TRU), uranium mine and mill tailings, mixed wastes, and naturally occurring radioactive materials.

19.2 CURRENT STATUS OF RADIOACTIVE WASTE MANAGEMENT

Radioactive materials are used in many applications throughout today's society. Radioactive materials are used to generate power in nuclear power stations and are used to treat patients in hospitals (see Chapter 16). The generators of radioactive waste in today's society are primarily the federal government, electrical utilities,

TABLE 19.1   Chronology of major events affecting radioactive waste management

Year	Event
1954	The Atomic Energy Act is passed
1963	First commercial disposal of LLW
1967	DOE facilities begin to store TRU wastes retrievably
1970	National Environmental Policy Act becomes effective; Environmental Protection Agency is formed
1974	Atomic Energy Commission divides into the Nuclear Regulatory Commission (NRC) and the Energy Research and Development Administration (ERDA)
1975	WIPP proposed as unlicensed defense TRU disposal facility; West Valley, New York low-level disposal facility closed
1977	President Carter defers reprocessing, pending the review of the proliferation implications of alternative fuel cycles
1979	Three Mile Island, Unit #2 accident; report to the president of the Interagency Review Group of Radioactive Waste Management
1980	Low Level Waste Policy Act is passed; all commercial disposal of TRU wastes ends
1982	Nuclear Waste Policy Act is passed; 10 CFR Part 61 issued as final regulation for LLW
1985	Low-Level Radioactive Waste Policy Act Amendments
1986	The reactor explosion at Chernobyl
1987	Nuclear Waste Policy Act Amendments provide for the characterization of the proposed HLW repository at Yucca Mountain, Nevada
1999	WIPP facility in Carlsbad, NM receives its first shipment

Source: Ref. 1.

private industry, hospitals, and universities. Although, each of these generators uses radioactive materials, the waste that is generated by each of them may be very different and must be handled accordingly. Any material that contains radioactive isotopes in measurable quantities is considered nuclear or radioactive waste. For the purposes of this chapter, the terms *nuclear waste* and *radioactive waste* will be considered synonymous.

Waste management is a field that involves the reduction, stabilization, and ultimate disposal of waste. Waste reduction is the practice of minimizing the amount of material that requires disposal. The object of waste disposal is to isolate the material from the biosphere, and in the case of radioactive waste, allow it time to decay to sufficiently safe levels. Table 19.1 is a chronology of the laws that have affected radioactive waste management practices over the last 50 years.

The federal government has mandated that individual states, or interstate compacts, which are formed and dissolved by Congress, be responsible for the disposal of the LLW generated within the respective states' boundaries. Originally, these states were to bring the disposal capacity on line by 1993. Although access to the few remaining facilities is drawing to an end, none of the states or compacts has a facility available to accept waste. Some states are making progress, but none of the proposed facilities is currently in the construction phase (2).

Waste from the transuranic (TRU) waste program is shipped to the Waste Isolation Pilot Plant (WIPP) in Carlsbad, New Mexico. The WIPP received its first shipment of waste in March 1999. The WIPP facility is a United States

Department of Energy (DOE) research and development facility that is designed to accept 6 million ft^3 of contact-handled TRU waste, as well as 25,000 ft^3 of remote-handled TRU waste. The facility accepts defense generated waste and places it into a retrievable geologic repository. The geologic repository is in the salt formations located near Carlsbad. The facility has a design-based lifetime of twenty-five years. The high-level waste (HLW) program had a site at Yucca Mountain, Nevada. However, it now appears unlikely that the facility will be opened.

19.3 RAMIFICATIONS OF NUCLEAR ACCIDENTS (3)

Nuclear accidents, while being very frightening , have not occurred often. In fact, there have only been a handful of fatal accidents since an understanding of nuclear energy and radiation has been developed. Pioneers of radiation research, including Marie Curie, are known to have died from radiation poisoning because they neglected to effectively control their exposure to this powerful energy source. Today, a better understanding of the risks associated with radioactive materials has led to fewer careless deaths.

The best known U.S. nuclear accident occurred in 1979 on Three Mile Island in Pennsylvania. A series of breakdowns in the cooling system of the plant's number 2 reactor led to a major accident in the early morning hours on March 28. Two days later the Nuclear Regulatory Commission (NRC) warned of possible core meltdown, a catastrophic event that could involve major loss of life. The possible explosion of a hydrogen gas bubble that had formed in the overheated reactor vessel of the crippled plant was also a major threat. Because of concern over the continued emission of radioactive gases, pregnant women and preschool children within a 5-mile radius of the plant were advised to leave the area. On April 2 there was a dramatic reduction in the size of the dangerous gas bubble, as well as further cooling of the reactor core. A week later the bubble had been eliminated.

The accident at Three Mile Island threatened the future of nuclear power in the United States and called into question the safety systems regulated by the NRC and used by the nuclear power industry. At the time of the accident, 72 nuclear reactors provided 13% of the nation's electrical power.

The cause of the incident has been hotly contested by the plant's owners, Metropolitan Edison Company, the NRC, the state of Pennsylvania, and companies that had constructed elements of the reactor system. What apparently happened was failure of a valve in a pump in the primary core cooling system. This interrupted the flow of water used to take heat away from the reactor, which led to the stopping of the steam turbine and a consequent shutdown of the reactor. The reactor continued to generate heat. As a result, the emergency cooling system began operating automatically. However, at some point during the switchover from the primary cooling system to the emergency core cooling system, a plant operator turned off the emergency system and, after a period of time, turned it back on. During that time the core was damaged, since some of the pellets of enriched uranium fuel became so heated that they either melted through or ruptured the zirconium-clad tubes that held them. Some of the water used to cool the core spilled onto the floor of the reactor building. When some of that radioactive water became steam, it was vented into the atmosphere to relieve pressure.

Vented steam was not the only source of radiation leakage, however. Radiation that had escaped by penetrating the 4-ft-thick walls had been traced directly to nuclear materials within the plant. The uranium fuel in the core remained so hot that the plant's managers had to vent more steam into the atmosphere to prevent an explosion in the containment building. The direct result of the venting was the release of small amounts of radioactive iodine, krypton, and xenon. The levels were described as "quite low" and not dangerous to humans. There was no apparent serious exposure for the plant's workers. The long-range effects, however, are not yet known. According to Ernest Sternglass, profession of radiology at the University of Pittsburgh, "It's not a disaster where people are going to fall down like flies. It's a creeping thing.

Federal safety investigators reported that a series of human, mechanical, and design errors had contributed to the Three Mile Island Accident. Metropolitan Edison had closed three auxiliary cooling pumps for maintenance two weeks before the accident and had kept them closed — a major violation of federal regulations. Several other errors contributed significantly to the incident: electrical magnetic relief valves that had opened to release a buildup of water pressure in the reactor had failed to close as planned; plant operators received incorrect readings from the pressure level indicator about the amount of water in the reactor; and, on two occasions after the accident, operators prematurely shut off the emergency core cooling system. Also, the release of slightly radioactive water into the Susquehanna River and the venting of steam into the air had been done without NRC approval (4).

Everyone's worst fears about nuclear power became a reality in the latter part of April 1986. A large Soviet reactor — unit number 4 at Chernobyl, 80 miles north of Kiev and only 3 years old — blew out and burned, spewing radioactive debris over much of Europe. Radiation levels increased from Sweden to Britain, through Poland, and as far south as Italy. The damage caused to the environment far surpassed that due to the accident at Three Mile Island.

The sequence of events at the Soviet reactor seems to have been as follows. First, the reactor suffered a loss of cooling water, which caused the uranium fuel elements to become overheated. The reactor had no containment building to keep in radioactive releases. Therefore, all the radioactivity generated in this part of the accident entered the atmosphere. Eventually, the temperature of the fuel rose to a point at which the graphite casing holding the uranium caught fire. Water could not be used to put the fire out because it would have evaporated, causing plumes of radioactivity to escape.

The explosion was the result of a series of errors by plant operators who were conducting an unauthorized experiment after having shut down the emergency cooling system. Operators were attempting to prove that if a turbine tripped, in the event of a power outage, and was disconnected from the steam supply, they would be able to draw kinetic energy from the still spinning rotor blades to operate emergency coolant pumps until the backup diesel generators began operating. Operators began to reduce the power output on April 25. Tests were conducted at 7% power, a level at which the plant is subject to automatic shutdown. At such low power levels, xenon gas builds up to absorb neutrons and slow the fission process. When this occurred at Chernobyl, it caused loss of control of the reactor. Power dipped to as low as 1% before it was finally stabilized back to 7%. To increase the flow of water to the reactor, two cooling pumps, in addition to the six normally used, were engaged. The cooling water inside the reactor's pressure tubes was already close to boiling or had reached that point because the drop in pressure from the low power output had heated the coolant. When the turbine was tripped, the coolant turned to steam. Heat could not escape because emergency systems had been shut off. Power began to surge as the water dissipated in the reactor. A heat buildup caused the zirconium casing to react with water, releasing hydrogen. Two explosions occurred, blowing the roof off the reactor building, destroying the cooling system, and severely damaging the core.

The accident reportedly killed 31 people, injured 299, and caused the evacuation of 135,000 from the site. The full extent of the damage from this incident probably will not be known for years. It is the long-term effects from exposure to radiation that frighten most people, and these fears may still become a reality at Chernobyl (5–8).

The third accident in recent years occured in September 1999. Two workers at a uranium processing plant in Tokaimura, Japan — located 90 miles northeast of Tokyo — were critically injured after a release due to improper handling techniques. The uncontrolled reaction occured when three workers at the plant mixed 35 pounds of uranium into a tank of nitric acid and water. The mixture was supposed to contain no more than five pounds of uranium. The reaction was stopped by dumping boric acid on the uranium. As a processing plant typically does not set off an atomic reaction, the plant was not shielded to prevent radiation from escaping. Radiation levels were more than 10,000 times higher than normal levels near the site of the accident when it occured, but had dropped off significantly by the next day. Over 50 people were directly exposed to the radiation and people living within 350 yards of the plant were evacuated, while those living within a six-mile radius of the plant were told to remain indoors. It took 20 hours to bring the fission reaction under control. Japan relies heavily on nuclear power, with one-third of the country's power requirements supplied by its 51 nuclear reactors. The incident was a major cause of concern for Japanese officials.

Three of the four largest radiological accidents of the last twenty years were the explosion at Chernobyl, the release at Tokaimura, and the partial core meltdown at Three Mile Island Unit No. 2, discussed above. The fourth largest accident involved the mishandling of a radioactive source in Brazil. The least publicized, but perhaps the most appropriate of these accidents with respect to its implications for waste management, was the incident in Brazil.

An uncontrolled radiotherapy source was overlooked in an abandoned medical clinic in Brazil, and was eventually discarded as scrap. The stainless steel jacket and the platinum capsule surrounding the radioactive cesium were compromised by scavengers in the junkyard. The cesium was distributed among the people for use as "carnival glitter," because of its luminescent properties. The material was spread directly onto individuals' skin and face, as well as their clothing. Severe illness was immediately evident to most of the exposed victims. Four people died from exposure by the spring of 1988, and it was estimated that an additional five persons would die over the next five years. Over 40 tons of material, including clothing, shoes, and housing materials, were contaminated from the release of less than 1 g of radioactive cesium. Although these accidents are very frightening, future accidents of a similar nature can be prevented through proper engineering and waste management controls.

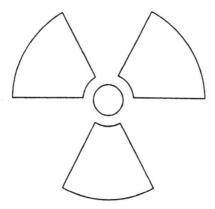

FIGURE 19.1 Radioactive materials warning sign.

19.4 BIOLOGICAL EFFECTS OF RADIATION

Although much still remains to be learned about the interaction between ionizing radiation and living matter, more is known about the mechanism of radiation damage on the molecular, cellular, and organ system level than most other environmental hazards. The radioactive materials warning sign is shown in Figure 19.1. A vast amount of quantitative dose-response data has been accumulated throughout years of studying the different applications of radionuclides. This information has allowed the nuclear technology industry to continue to operate at risks no greater than those faring any other technology. The following subsections provide a brief description of the different types of ionizing radiation and describe the effects that may occur upon overexposure to radioactive materials.

Radioactive Transformations (9)

Radioactive transformations are accomplished by several different mechanisms, most importantly alpha particle, beta particle, and gamma ray emissions. Each of these mechanisms is a spontaneous nuclear transformation. The result of these transformations is the formation of different more stable elements. The kind of transformation that will take place for any given radioactive element is a function of the type of nuclear instability as well as the mass/energy relationship. The nuclear instability is dependent on the ratio of neutrons to protons; a different type of decay will occur to allow for a more stable daughter product. The mass/energy relationship states that for any radioactive transformations the laws of conservation of mass and the conservation of energy must be followed.

An alpha particle is an energetic helium nucleus. The alpha particle is released from a radioactive element with a neutron-to-proton ratio that is too low. The helium nucleus consists of two protons and two neutrons. The alpha particle differs from a helium atom in that it is emitted without any electrons. The resulting daughter product from this type of transformation has an atomic number that is two less than its parent and an atomic mass number that is four less. Equation

19.4.1 shows an example of alpha decay using polonium (Po); polonium has an atomic mass number (protons and neutrons) and atomic number of 210 and 84, respectively.

$$^{210}84 \, Po \rightarrow \, ^{4}2 \, He + \, ^{206}82 \, Pb \qquad (19.4.1)$$

The terms He and Pb represent helium and lead, respectively.

This is a useful example because the lead daughter product is stable and will not decay further. The neutron-to-proton ratio changed from 1.5 to 1.51, just enough to result in a stable element. Alpha particles are known to have a high linear energy transfer (LET). The alphas will only travel a short distance while releasing energy. A piece of paper or the top layer of skin will stop an alpha particle. So, alpha particles are not external hazards, but can be extremely hazardous if inhaled or ingested.

Beta particle emission occurs when an ordinary electron is ejected from the nucleus of an atom. The electron (e), appears when a neutron (n) is transformed into a proton within the nucleus.

$$^{1}0 \, n \rightarrow \, ^{1}1 \, H + \, ^{0}(-1)e \qquad (19.4.2)$$

Note that the proton is shown as a hydrogen (H) nucleus. This transformation must conserve the overall charge of each of the resulting particles. In contrast to alpha emission, beta emission occurs in elements that contain a surplus of neutrons. The daughter product of a beta emitter remains at the same atomic mass number, but is one atomic number higher than its parent. Many elements that decay by beta emission also release a gamma ray at the same instant. These elements are known as betagamma emitters. Strong beta radiation is an external hazard, because of its ability to penetrate body tissue.

Similar to beta decay is positron emission, where the parent emits a positively charged electron. Positron emission is commonly called betapositive decay. This decay scheme occurs when the neutron-to-proton ratio is too low and alpha emission is not energetically possible. The positively charged electron, or positron, will travel at high speeds until it interacts with an electron. Upon contact, each of the particles will disappear and two gamma rays will result. When two gamma rays are formed in this manner the result is called annihilation radiation.

Unlike alpha and beta radiation, gamma radiation is an electromagnetic wave with a specified range of wavelengths. Gamma rays cannot be completely shielded against, but can only be reduced in intensity with increased shielding. Gamma rays typically interact with matter through the photoelectric effect, Compton scattering, pair production, or direct interactions with the nucleus.

Dose-Response (9)

The response of humans to varying doses of radiation is a field that has been widely studied. The observed radiation effects can be categorized as stochastic or non-stochastic effects, depending upon the dose received and the time period over which that dose was received. Contrary to most biological effects, effects from radiation usually fall under the category of stochastic effects. The nonstochastic effects are noted to exhibit three qualities: A minimum dose or threshold dose must be received before the particular effect is observed; the magnitude of the effect increases as the size of the dose increases; and a clearly causal relationship can be determined between the dose and the subsequent effects. Cember uses the analogy between drinking an alcoholic beverage and exposure to a noxious agent (9). For example, a person's intake must exceed a certain amount of alcohol before he or she shows signs of drinking. After that, the effect of the alcohol will increase as the person continues to drink. Finally, if he or she exhibits drunken behavior, there is no doubt that this is a result of his or her drinking.

Stochastic effects, on the other hand, occur by chance. Stochastic effects will be present in a fraction of the exposed population as well as in a fraction of the unexposed population. Therefore, stochastic effects are not unequivocally related to a noxious agent, as the above example implies. Stochastic effects have no threshold; any exposure will increase the risk of an effect, but will not wholly determine if any effect will arise. Cancer and genetic effects are the two most common effects linked with exposure to radiation. Cancer can be caused by the damaging of a somatic cell, while genetic effects are caused when damage occurs to a germ cell that results in a pregnancy.

19.5   SOURCES OF NUCLEAR WASTE

Naturally Occurring Radioactive Materials (10)

Naturally occurring radioactive materials (NORM), are present in the earth's crust in varying concentrations. The major naturally occurring radionuclides of concern are radon, radium, and uranium. These radionuclides have been found to concentrate in water treatment plant sludges, petroleum scale, and phosphate fertilizers.

In United States an estimated 40 billion gallons of water are distributed, through public water supplies, daily. Since water comes from different sources — streams, lakes, reservoirs, and aquifers — it contains varying levels of naturally occurring radioactivity. Radioactivity is leached into ground or surface water while in contact with uranium- and thorium-bearing geologic materials. The predominant radionuclides found in water are radium, uranium, and radon, as well as their decay products.

For reasons of public health, water is generally treated to ensure its quality before consumption by the public. Water treatment includes passing the water through various filters and devices that rely on chemicals to remove any impurities and organisms. If water with elevated radioactivity is treated by one or more of these systems, there exists the possibility of generating waste sludges or brines with elevated levels of radioactive materials. These wastes may be generated even if the original intention of the treatment process was not to remove radionuclides.

Mining of phosphate rock (phosphorite) is the fifth largest mining industry in the United States in terms of quantity of material mined. The southeastern United States is the center of the domestic phosphate rock industry, with Florida, North Carolina, and Tennessee having over 90% of the domestic rock production capacity. Phosphate rock is processed to produce phosphoric acid and elemental phosphorus. These two products are then combined with other materials to produce phosphate fertilizers, detergents, animal feeds, other food products, and phosphorus-containing materials. The most important use of phosphate rock is the production of fertilizer, which accounts for 80% of the phosorite in the United States.

Uranium in phosphate ores found in the United States ranges from 20 to 300 parts per million (ppm), or about 7–100 pCi/g. Thorium occurs at a lower concentration between 1 and 5 ppm, or about 0.1–0.6 pCi/g. The unit picocuries per gram (pCi/g) represents a concentration of each radionuclide based on the activity of that radionuclide. The units of curies represent a fixed number of radioactive transformations in a second. Phosphogypsum is the principal waste by-product generated during the phosphoric acid production process. Phosphate slag is the principle waste by-product generated from the production of elemental phosphorus. Elevated levels of bothnium and thorium as well as their decay products are known to exist at elevated levels in these wastes. Since large quantities of phosphate industry wastes are produced, there is a concern these materials may present a potential radiological risk to individuals who are exposed to these materials if they are distributed in the environment.

Fertilizers are spread over large areas of agricultural land. The major crops that are routinely treated with phosphate-based fertilizer include coarse grains, wheat, corn, soybeans, and cotton. Since large quantities of fertilizer are used in agricultural applications, phosphate fertilizers are included as a NORM material. The continued use of phosphate fertilizers could eventually lead to an increase in radioactivity in the environment and in the food chain.

Currently, there are no federal regulations pertaining directly to NORM containing wastes. The volume of wastes produced is sufficiently large that disposal in a low-level waste facility is generally not feasible. A cost-effective solution must be implemented both to guard industry against large disposal costs and to ensure the safety and health of the public.

Low-Level Radioactive Waste (11)

Low-level radioactive waste (LLRW) is produced by a number of processes and is the broadest category of radioactive waste. Low-level waste is frequently defined for what it is not rather than for what it is. According to the Low-Level Waste Policy Act of 1980, LLRW is defined as: "radioactive waste not classified as high-level radioactive waste, transuranic waste, spent nuclear fuel, or by-product material as defined in Section 1 l(e)(2) of the Atomic Energy Act of 1954."

This definition excludes high-level waste and spent nuclear fuel because of its extremely high activity. Transuranic wastes (those containing elements heavier than uranium) are excluded because of the amount of time needed for them to decay to acceptable levels. Finally, by-product material or mill tailings are excluded because of the very low concentrations of radioactivity in comparison to the extreme volume of waste that is present.

The generators of low-level waste include nuclear power plants, medical and academic institutions, industry, and the government. Low-level waste can be generated from any process in which radionuclides are used. A list of the different waste streams and the possible generators of each is presented in Table 19.2.

Each of the aforementioned generators produces wastes that fall into the category of low-level waste. The waste streams identified in Table 19.2 are categorized by generation process, but may also, in some instances, be identified by the type of generating facility.

TABLE 19.2 Typical waste streams by generator category

Waste Stream	Power Reactors	Medical and Academic	Industrial	Government
Compacted trash or solids	x	x	x	x
Dry active waste	x			
Dewatered ion exchange resins	x			
Contaminated bulk	x		x	x
Contaminated plant hardware	x		x	x
Liquid scintillation fluids		x	x	x
Biological wastes		x		
Absorbed liquids		x	x	x
Animal carcasses		x		
Depleted uranium MgF_2			x	

Source: EG & G Idaho, Inc., 1985.

The disposal of low-level waste is accomplished through shallow land burial. This process usually involves the packaging of individual waste containers in large concrete overpacks. The overpack is designed to reduce the amount of water that may come into contact with the waste. Another function of the overpack is to guard against intruders coming into contact with the waste once institutional control of the facility is lost. When waste is delivered to the facility in drums, boxes, or in high-density polyethylene (HDPE) liners, they are placed in an overpack and sealed with cement before being buried in the landfill.

High-Level Radioactive Waste (11)

High-level waste (HLW) consists of spent nuclear fuel, liquid wastes resulting from the reprocessing of irradiated reactor fuel, and solid waste that results from the solidification of liquid high-level waste. Spent reactor fuel is the fuel that has been used to generate power in a reactor. The spent fuel may be owned by a government reactor, a public utility reactor, or a commercial reactor. The wastes resulting from fuel reprocessing are either governmentally or commercially generated. Only a small fraction of the liquid HLW has been generated commercially. Approximately 600,000 gallons of waste were produced in the nation's only commercial fuel reprocessing facility in West Valley, New York. The remainder of the HLW present in the United States today has been generated by the government in weapon facilities.

Spent nuclear fuel is removed from a reactor and stored in a pool of water on the site. The water in the spent fuel storage pools shields the workers and the environment from the fission products, and also provides cooling to the fuel. The residual heat from a fuel assembly is quantified as approximately 6% of the operating power level of the reactor. Failure to provide additional cooling after the fission reaction has stopped was the reason for the fuel damage at Three Mile Island. Once a geologic repository is constructed, the spent fuel assemblies will be placed in a sealed canister and disposed of.

Most of the liquid high-level waste is stored in underground storage tanks. Many of these tanks are getting old and the availability of a geologic repository in the near future is doubtful. Many methods of solidifying the wastes for transport and ultimate disposal have been investigated. Plans are under way to store HLW in one central location in the United States. The chosen location is Yucca Mountain, Nevada. However, it appears that the Yucca Mountain facility will not be able to meet NRC licensing criteria.

Transuranic Waste

Transuranic wastes are those wastes containing isotopes that are heavier than uranium, U. Generally, transuranic isotopes are not found in nature. These isotopes are manmade, produced by the radiation of heavy elements, such as uranium and thorium. Transuranic wastes are

$$^{238}_{92}\text{U} + {}^{1}_{0}\text{n} \rightarrow {}^{239}_{93}\text{Np} + {}^{0}(-1)\text{e} \leftarrow {}^{239}_{93}\text{Np} \rightarrow {}^{239}_{94}\text{Pu} + {}^{0}(-1)\text{e} \quad (19.5.1)$$

where Np and Pu represent neptunium and plutonium, respectively. They are normally generated by the government, particularly from weapons testing. The transuranic waste is now being shipped from storage at a number of DOE facilities across the country to WIPP in Carlsbad, New Mexico, the permanent disposal facility.

19.6 RADIOACTIVE WASTE TREATMENT AND DISPOSAL (1)

Many treatment processes can be employed to reduce the volume, or increase the stability, of waste that must ultimately be permanently disposed of. Landfill fees for radioactive waste is assessed largely on the volume of the waste to be disposed of. Current trends in the rising cost of waste disposal have led to the generators' implementing one or a number of waste minimization techniques. The physical form of the waste is a critical factor in determining the probability that the waste will remain isolated from the biosphere.

Compacting is a method of directly reducing the volume and increasing the specific weight of the resulting waste. Contaminated materials such as glass vials, protective clothing, and filter media can be compacted to reduce the volume. Compacting does not reduce the environmental hazard of the waste stream — its purpose is purely waste minimization.

Incineration of waste both reduces the volume and provides a more stable waste stream. Many biological wastes, including animal radioactive carcasses, are incinerated. The storage of animal carcasses in drums is generally not cost-effective because of the gas generation of the materials as they decay biologically. A drum packed with animal carcasses must be filled with absorbent material so that the pressure inside the drum does not rise to unsafe levels. Incineration is a very cost-effective waste reduction technique for large generators of combustible materials.

Dewatering or evaporation is another waste minimization and stabilization technique that is practiced by waste generators. Evaporating sludges or slurries can greatly reduce the volume of the radioactive waste stream and stabilize the waste prior to disposal.

19.7 FUTURE TRENDS

Current regulations call for each individual state or interstate compact to store and dispose of all of the LLW generated within its boundaries. An interstate compact consists of a group of states that have joined together to dispose of LLW. Interstate compacts can only be formed and dissolved by Congress. Many regulatory milestones have passed, leaving most states with access restricted to Barnwell, South Carolina. Barnwell is the only remaining LLW disposal facility for such wastes. It is almost certain that the Barnwell facility will close before most states have centralized storage capacity on line. Some states, like New York, have unsuccessfully attempted to sue the federal government, arguing that it is unconstitutional to mandate that states dispose of radioactive waste within their boundaries. Without the individual states or compacts taking immediate action to site and construct a permanent disposal facility or a temporary centralized storage facility, generators of waste will be forced either to store radioactive materials on-site or to stop generating radioactive wastes by ceasing all operations that utilize radioactive materials. While these two options may seem appropriate, neither of them will solve the problem of waste disposal for any extended period of time. Many radioactive waste generators, such as hospitals, do not have the storage space allocated to handle on-site storage for periods exceeding one or two years. Much of the waste generated at hospitals is directly related to patient care, and it is unacceptable to assume that all the processes like chemotherapy that produce radioactive waste will be stopped.

The high-level waste program is limping along because of public concern for the areas surrounding the proposed facilities. The HLW program has met drastic public opposition because of the amount of time that the waste will remain extremely hazardous, a period in the order of thousands of years. Opponents of this facility are arguing that the ability to properly label the disposal facility and guard against future intruders is lacking. Many symbols that would serve as a deterrent to would be intruders, such as thorns or unhappy faces, have been proposed.

The public at large will continue to oppose most activities involving nuclear waste until they are made aware of the unwanted characteristics of current, more acceptable technologies, as well as the positive benefits that radioactive materials have made to society.

19.8 SUMMARY

1. After an individual state or an interstate compact is denied access to the current disposal facilities nationwide, the generators of the state will be forced to store LLW on-site. The only other alternative is to stop generating waste until such time as the state or compact develops and constructs an appropriate disposal or storage facility. Neither of these options constitutes an appropriate choice for generators such as hospitals that offer nuclear medical services.

2. The interaction between ionizing radiation and living matter is one of the most understood environmental hazards. Radioactive isotopes are transformed into more stable elements through the mechanisms of alpha, beta, gamma, and neutron emission.

3. Naturally occurring radioactive materials may be concentrated by many industrial and municipal processes. The individual states and interstate compacts now have the responsibility to site and construct facilities to dispose of low-level waste. Both high-level waste and transuranic waste is being stored on the site of generation until the respective geologic repositories begin to accept waste for disposal.

4. Waste disposal fees are assessed primarily on the basis of the volume of waste. Generators have invested in treatment technologies because of the rising cost of disposal. Many of the treatment technologies also improve the stability of the waste.

5. Generators of radioactive wastes will be forced to store all generated materials on-site until the next generation of disposal facilities comes on line.

REFERENCES

1. R. Berlin and C. Stanton, *Radioactive Waste Management*, John Wiley & Sons, New York, 1989.

2. EG&G Idaho, Inc., *The State by State Assessment of Low Level Radioactive Wastes Shipped to Commercial Disposal Sites*, DOE/LLW50T, December 1985.

3. L. Theodore, J. Reynolds, and F. Taylor, *Accident and Emergency Management*, John Wiley & Sons, New York, 1989.

4. G. M. Ferrara, ed. *The Disaster File: The 1970s*, Facts on File, New York, 1979.

5. C. Norman, "Chernobyl: Errors and Design Flaws," *Science*, 233, 1029–1031 (1986).

6. T. Wilke and R. Milne, "The World's Worst Nuclear Accident," *New Science*, 17–19, May 1, 1986.

7. E. Marshall, "Reactor Explodes Amid Soviet Science," *Science*, 232, 814–815 (1986).

8. S. Cooke, P. Galuszka, and J. J. Kosowatz, "Human Failures Led to Chernobyl," *Engineering News Record*, 10, 11, August 28, 1986.

9. H. Cember, *Introduction to Health Physics*, McGraw-Hill, 1992.

10. U.S. EPA, *Draft: Diffuse NORM Wastes Waste Characterization and Risk Assessment*, May 1991.

11. R. Faw and J. Shultis, *Radiological Assessment Sources and Exposures*, Prentice Hall, Englewood Cliffs, NJ, 1993.

V

HAZARDOUS WASTE MANAGEMENT ISSUES

20

HAZARDOUS WASTE INCINERATION

20.1 INTRODUCTION

Incineration is not a new technology. It has been commonly used for treating organic hazardous waste for many years in Europe and the United States. The major benefits of incineration are that the process actually destroys most of the waste rather than just disposing of or storing it. Incineration can be used for a variety of specific wastes and is reasonably competitive in cost compared to other disposal methods.

Hazardous waste incineration involves the application of combustion processes under controlled conditions to convert wastes containing hazardous materials to inert mineral residues and gases. Four conditions must be present:

1. Adequate free oxygen must always be available in the combustion zone.

2. Turbulence, the constant mixing of waste and oxygen, must exist.

3. Combustion temperatures must be maintained; reactions of combustion must provide enough heat to raise the burning mixture to a sufficient temperature to destroy all organic components.

4. Elapsed time of exposure to combustion temperatures must be adequately long in duration to ensure that even the slowest combustion reaction has gone to completion. In other words, transport of the burning mixture through the high-temperature region must occur over a sufficient period of time.

Thus, four parameters influence the mechanisms of incineration: oxygen, temperature, turbulence, and residence time.

Typically, incinerators operate between 900°C and 1200°C and are fueled by the hazardous wastes, with auxiliary heating supplied by a fuel oil or natural gas when burning wastes with low heating values. Design gas-phase residence times in the high-temperature zones range from 0.2 seconds to several seconds.

Wastes vary significantly in organic content, ash content, contamination, and burning characteristics. Hazardous wastes typically require more oxygen, much higher temperature, vigorous mixing, longer residence time, and more heat input to break the bonds of heavy organic molecules such as halogenated compounds. When hazardous wastes are burned with oxygen, they react to form combustion products such as carbon dioxide (CO_2), carbon monoxide (CO), water (H_2O), and particulate matter. Significant amounts of other species are also normally released during incineration, including oxides of sulfur (SO_x), oxides of nitrogen (NO_x), acids, salts, halogenated organics, free halogen gases, and amines. Products of incomplete combustion (PICs) may contain polycyclic organic compounds, such as benzo(*a*)pyrene (BaP) and 2,3,7,8-tetrachlorodibenzo-*p*-dioxin (TCDD). These two compounds are highly toxic and are considered carcinogenic to humans in significant doses. At the present time, only limited data are available to assess conditions that lead to the generation of these toxic compounds.

Incinerators have the capability to burn a wide range of liquid hazardous wastes. Major attention has been focused on the oxidation of halogenated hydrocarbons because these substances are relatively difficult to completely incinerate. The destruction of complex halocarbons occurs in a sequence of reaction steps. These consist of bond dissociation, creating chemical fragments which undergo further fragmentation and oxidation. The reaction times for the individual steps last on the order of milliseconds at typical incineration temperatures. The availability of hydroxyl (OH) radicals appears to promote the rate of decomposition of many hazardous compounds. Because chlorine serves as an OH radical scavenger, chlorocarbons have been used as fire retardants, thus slowing down oxidation rates. Chemicals with high chlorine-to-hydrogen ratios tend to soot readily. In general, the chlorocarbons must be burned with large quantities of excess air relative to hydrocarbons to prevent soot formation during combustion.

Particulate matter may originate from inorganic or organometallic substances introduced with the waste, auxiliary fuel, combustion air, or some combination of these materials. Inorganic matter, such as salts and trace metals present in the waste and fuel, is called ash and cannot be destroyed by incineration. The ash content is generally low in liquid wastes but can be high in sludge and solid wastes.

A number of wastes present special problems when they are incinerated. For example, when burned, metals become metal oxides which are mostly in the submicron-to-a-few-micron size range that may not be collected by conventional air pollution control equipment. Resins may polymerize, coating surfaces and plugging nozzles in incinerators designed to atomize liquid wastes. Polyolefins and nitrocellulose may detonate rather than burn. Another problem arises when the wastes being incinerated contain significant concentrations of halogenated compounds (say 0.5% by weight) that result in the formation of undesirable combustion products such as hydrogen chloride (HCl), hydrogen fluoride (HF), and hydrogen bromide (HBr). In order to minimize incineration problems, it is also necessary to know the following: (a) whether the waste is gaseous, liquid, or solid, since organics in all three phases can be thermally processed; (b) the fraction of the waste that is organic and whether the inorganic is water, inorganic compounds, or a combination; and (c) why the waste is hazardous and to what degree.

Waste characterization is a major factor in assessing the feasibility of converting a hazardous waste material by process incineration. It affects the design of the process incinerator and its emissions control system, and helps determine the compatibility of a waste with a proposed or available facility. It also plays a part in determining process incineration operating conditions for complete conversion of a specific waste. The reader is referred to the literature (1) for more details on this subject.

20.2 REGULATIONS AND STANDARDS

The current regulatory framework for incineration involves cradle-to-grave regulations of hazardous wastes to be incinerated, provided jointly by the Environmental Protection Agency (EPA) and the Department of Transportation (DOT). The DOT regulates transportation, transfer, and handling of hazardous waste. The EPA regulates all phases of hazardous waste management, but focuses primarily on storage and disposal or treatment of hazardous waste. The EPA also requires companies to provide for emergency response in the event of unexpected releases.

While the control of hazardous waste incineration is based primarily on federal laws and regulations, the responsibility for implementing regulatory programs affecting incineration involves all three levels of government. Complementing the broad federal responsibilities, state and local governments are involved in the areas of enforcement, facility siting, and emergency response to accidents.

Incineration of hazardous wastes under the U.S. Resource Conservation and Recovery Act of 1976 (RCRA) is regulated under the *Code of Federal Regulations*, Title 40, Parts 122, 264, and 265. These regulations establish performance standards for hazardous waste incineration which relate to the following factors:

1. The destruction and removal efficiency (DRE) for each principal organic hazardous constituent (POHC) designated in the permit. The DRE, defined in terms of the mass emission of a POHC in the stack exhaust gas versus the mass feed rate of the same POHC in the waste, must be >99.99%. The DRE performance standard implicitly requires sampling and analysis to quantify the designated POHC(s) in the waste stream and in the stack gas during a trial burn (to be discussed later).

2. A limitation on hydrochloric acid emissions from the stack of an incinerator. This standard implicitly requires, in some cases, sampling and analysis to quantify HCl acid in the stack gas and/or to determine the efficiency of air pollution control devices. A control efficiency of 99% is required for emissions greater than 4 lb/hr.

3. A limitation of stack emissions of particulate material to <180 mg/dscm (<0.08 gr/dscf, or grains per dry standard cubic feet of air), corrected to 7% oxygen (or an approximate excess air level of 50%).This performance standard requires measurement of the particulate emission rate. However, state regulations for particulates are at or below 0.03 gr/dscf.

The above standards were promulgated as a result of the RCRA regulations of 1976 in 40 CFR Parts 264 and 265. The regulations under Parts 264 and 265, Subpart 0, applied to owners and operators of facilities that incinerated hazardous waste. Several revisions were made during the 1980s incorporating continuous emission monitors (CEMs) for carbon monoxide (CO), metals emissions, products of incomplete combustion (PICs), chlorine emissions, and particulate matter (PM). Emission levels of particular interest were metallic compounds such as lead (Pb), cadmium (Cd), arsenic (As), nickel (Ni), copper (Cu), and mercury (Hg). These are present in the flue gas primarily as oxides and chlorides. Most of the metallic compounds are in the vapor phase within the incineration system since these compounds boil or sublime at temperatures around 1800°F. These metallic compounds tend to condense as the flue gas is cooled and adsorb onto fine particulate matter that is usually submicron in size. A portion of the more volatile metals such as mercury and lead may remain in the vapor phase, depending upon temperature conditions. On April 19, 1996, the EPA published a proposed rule, called the "MACT" rule (for maximum achievable control technology), under the joint authority of RCRA and the Clean Air Act to upgrade the emission standards for hazardous waste combustors (61 FR 17358). Specifically, this rule will affect incinerators, cement kilns, and lightweight aggregate kilns. The existing combustion regulations will change significantly because of the proposed new emission standards for dioxins/furans, PM, HCl, CO, HCs (hydrocarbons), Hg, semi-volatile metals including Pb and Cd, and low-volatility metals including berylium, arsenic, and chromium. The standards were signed on July 30, 1999, but have yet to appear in the Federal Register (as of September 1, 1999).

The proposed new standards, which will affect both the existing and new hazardous waste incinerators, may be summarized as follows:

	NEW	EXISTING
Dioxins/Furans	0.02 ngTEQ/dscm	0.02 ngTEQ/dscm
PM	0.015 gr/dscm @ 7% O_2	0.030 gr/dscm @ 7% O_2
HCl/Cl_2	21 ppmv (dry)	77 ppmv (dry)
CO	100 ppmv (dry)	100 ppmv (dry)
HCs	10 ppmv (dry)	10 ppmv (dry)
Mercury	45 μg/dscm	130 μg/dscm
Semi-volatile metals	24 μg/dscm	240 μg/dscm
Low-volatile metals	90 μg/dscm	97 μg/dscm

Note: TEQ = Toxic Equivalent Quotient. All emission rates corrected to seven percent oxygen (7% O_2). In addition, emission standards for cement kilns and lightweight aggregate kilns are similar to those for incinerators, with some exceptions due to differences in equipment and operators.

Normally, a trial burn (1) is required. The information to be reported to regulatory agencies after the burn includes the following [Section 122.2(b)]:

1. Quantitative analysis of POHCs in the waste feed.

2. Determination of the concentration of particulates, POHCs, oxygen, and HCl in the exhaust gas.

3. Quantitative analysis of any scrubber water, ash residues, and other residues to determine the fate of POHCs.

4. A computation of the destruction and removal efficiency (DRE) of the POHCs.

5. Computation of HCl removal efficiency if the HCl emission rate exceeds 1.8 kg of HCl per hour (4 lb/hr).

6. Computation of particulate emissions.

7. Identification of sources of fugitive emissions and their means of control.

8. Measurement of average, maximum, and minimum temperatures and of combustion gas velocities (gas flows).

9. Continuous measurement of CO in the exhaust gas.

10. Any other information the EPA may require to determine compliance.

The Hazardous and Solid Waste Amendments of 1984, or "the New RCRA," has had a significant impact on the management or "treatment" of hazardous wastes. Congressional findings [1002(a)(7)] indicated "...certain classes of land disposal facilities are not capable of assuring long-term containment of certain hazardous wastes, and to avoid substantial risk to human health and the environment (HH&E), *reliance on land disposal should be minimized or eliminated, and land disposal, particularly landfill and surface impoundment, should be the least favored method for managing hazardous wastes.*" The objective and national policy position [1003(a)(6)] is "...to promote the protection of health and the environment ... by ... minimizing the generation of hazardous waste and the land disposal of hazardous waste by encouraging process substitution, materials recovery, properly conducted recycling and reuse, and treatment." This congressional position may have led to the passage of the Pollution Prevention Act of 1990 (2).

The trial burn referred to above is defined as any attempt to incinerate the waste in question for a limited period, and it is designed to establish the conditions at which incineration of waste in a given facility must be carried out to assure protection to public health and the environment (1, 3). A trial burn may be requested when the EPA believes that the information is insufficient to assure protection to public health and the environment, and that a trial burn can provide information necessary to assure such protection (i.e., to verify 99.99% destruction efficiency). A trial burn may comprise either a single burn or a sequence of burns conducted at constant or varied incineration parameters. A trial burn will be typically conducted at a given facility which is applying for a permit to incinerate the waste in question on a more permanent schedule. The facility could be a commercial (full-sized) incinerator, a pilot-scale unit operated by an incinerator vendor, or a pilot-scale unit operated by a vendor specializing in trial burns. A trial burn requires a temporary permit from the EPA and may be conducted in the presence of an EPA official. A trial burn itself should not present any serious threat to the public's or the operators' health and the environment. To prevent any serious hazard to the public or the operators' health and the environment, a trial burn should provide for (a) rapid detection in the incineration effluents of hazardous materials in quantities potentially threatening health and safety of the public and/or the operators and (b) rapid incinerator shutdown upon detection of such quantities.

Some general information on the activities and costs associated with a trial or test burn at an incinerator facility is given below.

1. *Site Survey.* Costs include professional services and travel to a local site for inspection of the facility to be tested and discussion of plans for trial burns. Specific characteristics of wastes to be incinerated are assumed to be already

known or are provided by waste generators and listed on manifest records, or have been analyzed previously. Information on the compatibility of these wastes with the specific facility characteristics is also assumed to be available.

2. *Equipment Preparation.* Sampling equipment may be leased by the facility or provided by a consulting firm. Certain costs are incurred in calibrating and loading the instruments for transport to the facility site, as well in the actual transport itself.

3. *Equipment Setup and Take down.* Installation of equipment includes any scaffolding and securing of ports and proper sampling instruments at facility stack(s) to ensure the necessary monitoring and procurement of trial burn samples.

4. *Stack Sampling.* A minimum of three tests involving 3–6 h per test is assumed. Costs for testing, instrumentation, and adjustments associated with a permit application are considered to be separate costs not attributable to the trial burn activity. Development of sampling procedures and verification of test methods are presumed available and accomplished prior to the trial burns.

5. *Sample Analysis.* Laboratory analysis costs include preservation and transporting of samples to an off-site laboratory. Compounds to be analyzed generally include potential air pollutants. Gas chromatography and mass spectrometry (GC/MS) testing is conducted for the chlorinated hydrocarbons.

6. *Equipment Cleanup.* These costs are the routine costs incurred for cleaning and storing various pieces of sampling and analysis equipment.

7. *Report Preparation.* The written report displays and interprets the trial burn results. Preparation of this report and the information contained therein is considered independent from any information produced for permit negotiations.

It is especially important that appropriate quality assurance (QA) procedures be employed throughout both the sampling and analytical phases of any test, because every test represents a unique situation. All solvents and resins should be checked for purity before the sampling program is begun; filter paper and glass wool should be solvent-cleaned. Appropriate blanks should be included to provide background information. The sampling trains should be all glass and should be thoroughly cleaned to remove all sealant greases and any other residues. Plastics or silicone grease should be used on the organic sample portions of the sampling trains. All sample containers should be glass with Teflon-lined screw caps and should be hexane-rinsed prior to deployment in the field. Ground glass plugs and aluminum foil should be used to cap off the adsorbent trap and other sample-exposed portions of the trains to minimize the potential for contamination. Sample packing lists should be included for each test burn with information regarding source lot numbers and preservation/extraction techniques.

20.3 PROCESS INCINERATOR SYSTEMS

Process incineration has been used over a period spanning many years as a means of disposing of various industrial wastes containing hazardous materials. A complete incineration system may be illustrated by dividing the process into six major components: waste feed, auxiliary fuel feed, combustion, emissions control, monitoring systems, and supporting facilities.

Table 20.1 Incineration processes and their typical ranges

Process	Temperature Range (°C)	Residence Time
Liquid injection	650–1600	0.1–2 sec
Rotary kiln	820–1600	Liquid and gases, seconds; solids, hours
Fluidized bed	450–980	Not applicable
Multiple hearth	320–540 (dry zone)	0.25–1.5 hr
	760–980 (incineration)	0.25–1.5 hr
Co-incineration	150–1600	Seconds to hours

The two most important operating conditions for incineration are temperature and residence time. These two conditions vary with a waste's chemical structure, the physical form of the waste, and the type of process incinerator. Table 20.1 summarizes the typical ranges for the two operating conditions of the five types of process incinerators. (These units are briefly described below.)

Most land-based process incinerators are installed either at industrial plant sites where wastes are generated or at privately owned central disposal facilities. On-site incinerators are usually installed only by large companies having quantities of incinerable materials large enough to justify the expense. On the other hand, for small volumes of hazardous wastes, it is frequently cheaper and certainly easier to pay the shipping and disposal charges to a private central facility than to build and operate an on-site facility.

There are many factors which must be considered in making the decision to site an incinerator, whether on-site or off-site. These factors include expected waste volumes, liabilities associated with transportation and disposal, available technical expertise in operating incinerators, economic feasibility, permit requirements, available space, and potential local opposition. A central off-site plant offers a number of important advantages:

1. It services a large area, which permits greater utilization of equipment.

2. All new techniques and R&D (research and development) activities can be concentrated, with specialized personnel predominant.

3. Responsibility for plant operation is in the hands of specialists.

4. Small industries can direct their attention solely to their normal operations without having to worry about their waste disposal problems.

5. The government rules and regulations can be effectively and economically implemented.

As described in Table 20.1, there are five major types of process incinerators commercially available for burning combustible hazardous wastes: liquid-injection, rotary-kiln, fluidized-bed, multiple-hearth, and co-incineration. Liquid injection and rotary kiln are at present the most highly developed and most commonly used incinerators for hazardous waste incineration. Both exist throughout the United States in full-scale operation. Each type of the process incinerators is briefly described below.

Liquid-Injection Incinerators

Liquid-injection process incinerators (see Figure 20.1) can be used to dispose of almost all combustible liquid wastes with low viscosities. Viscosity can usually be controlled by heating with tank coils or in-line heaters. Liquid-injection incinerators are not as versatile as rotary-kiln incinerators because the former are designed to burn only pumpable liquid wastes and some slurried wastes. These incinerators can be mounted vertically or horizontally. The heart of the liquid-injection system is the waste atomization device or nozzle (burner) which atomizes the waste and mixes it with air into a suspension. Combustion takes place in the combustion chamber. Atomization is usually achieved either mechanically, using a rotary cup, or by pressure atomization nozzles (high-pressure air or steam).

The primary reasons for the wide use of liquid-injection incinerators are the relatively low capital, operating, and maintenance costs due to the simple design. The major disadvantages are high sensitivity to plugging of atomizing nozzles, high sensitivity of the combustion temperature to the heating value of the wastes, and limited ability to handle solids.

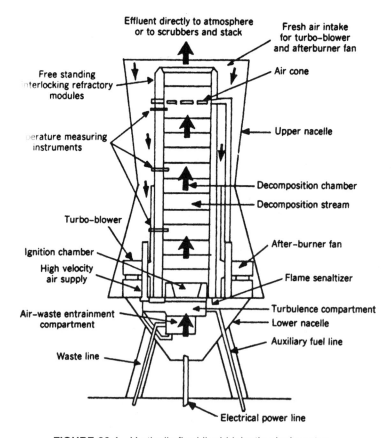

FIGURE 20.1 Vertically fired liquid-injection incinerator.

Rotary-Kiln Incinerators

Rotary-kiln process incinerators (see Figure 20.2) were initially designed for lime processing. The rotary kiln is a cylindrical refractory-lined shell that is mounted at a slight incline from the horizontal plane. Combustion is accomplished in a rotating cylinder inclined at an angle to facilitate mixing the waste materials with circulating air. The kiln accepts all types of solid and liquid waste materials with heat values between 550 and 8300 kcal/kg (between 1000 and 15,000 Btu/lb). Solid wastes and drummed wastes are usually fed by a pack-and-drum feed system, which may consist of a bucket elevator for loose solids and a conveyor system for drummed wastes. Pumpable sludges and slurries are injected into the kiln through another nozzle. Temperatures for burning vary from 810°C to 1600°C.

The primary advantages of the rotary kiln for waste incineration include the following: a wide variety of wastes can be incinerated simultaneously; high operating temperature can be achieved; and a gentle and continuous mixing of incoming wastes can be obtained. Disadvantages of the rotary kiln include the following: there are high capital and operating costs; highly trained personnel must be used to ensure proper operation; and the refractory lining must be replaced frequently if very abrasive and corrosive conditions exist in the kiln.

Fluidized-Bed Incinerators

Fluidized-bed process incinerators have been used mostly in the petroleum and paper industries, as well as for the processing of nuclear wastes, spent cook liquor, wood chips, and sewage sludge. Hazardous wastes in any physical state can be applied to the fluidized-bed process incinerator. Auxiliary equipment includes a fuel burner system, an air supply system, and feed systems for liquid and solid wastes. This process incinerator takes its name from the behavior of a granular bed of nonreactive sand, stirred by passing a gaseous oxidizer (air, oxygen, or nitrous oxide) through the bed at a rate sufficiently high to cause the bed to expand and act as a fluid. Preheating of the bed to start-up temperatures is accomplished by a burner located above and impinging down on the bed. The waste is passed directly into the sand. With the onset of ignition, a continuous exposure of the uniformly distributed waste to the heated sand results in efficient oxidation.

Some advantages of the fluidized incinerator are as follows: (a) It has a high combustion efficiency; (b) the combustion design is simple and does not require moving parts after initial feeding of fuel and waste; therefore, its maintenance cost is low; (c) designs are more compact; (d) comparatively low gas temperatures and excess air requirements minimize the formation of nitric oxide; (e) in some cases, the bed itself neutralizes some of the hazardous products of combustion; (f) the bed mass provides a large surface area for reaction; (g) temperatures throughout the bed are relatively uniform; (h) it can process aqueous waste slurries and can tolerate fluctuation in waste feed rates; (i) if the waste contains sufficient calorific values, the use of auxiliary fuels is unnecessary; moreover, the excess heat may be recycled in some cases; and (j) the bed can function as a heat sink; start-up after weekends may require little or no preheat time.

Disadvantages of fluidized bed incinerators are as follows: (a) Bed diameters and height are limited by design technology; (b) ash removal presents a potential problem; (c) systems requiring low temperatures may have carbon buildup in the bed due to increased residence time; (d) operating costs are high; (e) waste types

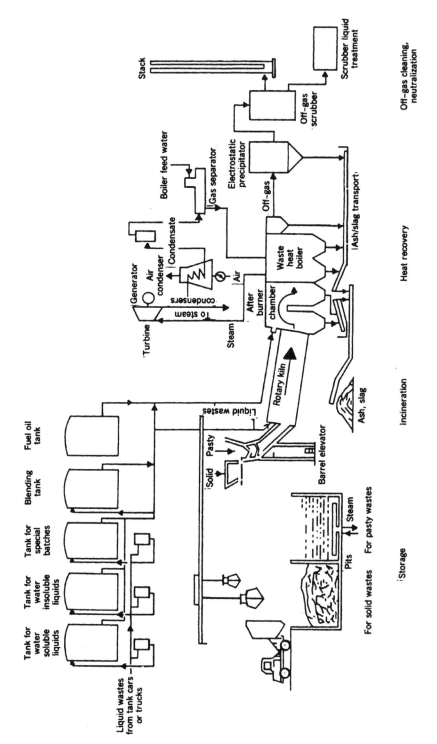

FIGURE 20.2 Rotary-kiln incinerator.

are limited; (f) certain organic wastes will cause the bed to agglomerate; (g) particulate emissions can be a major problem; and (h) it is still a relatively unproven technology.

Multiple-Hearth Incinerators

Multiple-hearth process incinerators were initially developed to dispose of sewage sludge and have limited application for hazardous wastes, such as chemical sludges, oil refinery sludges, and still bottoms. A typical multiple-hearth incinerator consists of a refactory-lined steel shell, a central shaft that rotates, a series of solid flat hearths, a series of rabble arms with teeth for each hearth, an air blower, fuel burners mounted on the walls, an ash removal system, and a waste feeding system. Side ports for tar injection, liquid waste burners, and an afterburner may also be included. Wastes in different physical states are fed to the process incinerator at different levels and are agitated by rotating arms within each chamber. Sludges enter the top and move down successive chambers. Tars and greases are applied along the sides. Gases and liquids are applied through ports in the lower portions. The wastes move steadily downward as effluent gases move steadily upward.

Co-incineration

Co-incineration refers to the joint combustion of hazardous waste with an industrial combustion process to provide additional energy or as a supplemental source of fuel. This is not a unique technology; any existing incineration process can be used for this special case of mixing waste streams to obtain better combustion of particularly intractable waste material. Co-incineration most often occurs in processes such as cement kilns, utility and industrial boilers, and refuse or sludge incinerators. The advantages of co-incineration include utilization of the wastes' heat content to save fuel with usually little or no increase in emissions. The major disadvantage is the limited number of acceptable waste types based on the particular combustion process and the product manufactured.

Other technologies such as cement kilns and molten salt incineration also appear to be satisfactory, but all operational parameters have not been adequately tested. Both processes can be used commercially for burning hazardous wastes if more accurate and reliable data from research and demonstration can be provided to overcome public opposition. Details of these as well as other processes are provided in the literature (3).

20.4 AIR POLLUTION CONCERN AND EMISSION CONTROL

One major concern associated with hazardous waste incineration is the emission of air contaminants. The greatest mass of air contaminants consists primarily of the following criteria pollutants: oxides of nitrogen, oxides of sulfur, and particulate matter. Also, of great concern are trace levels of noncriteria pollutants such as chlorinated by-products, benzene, heavy metals, and acid gases. Available data indicate that criteria pollutant emissions from hazardous waste incinerators are about the same as those from various industrial combustion processes. The emission rates are dependent on the waste incineration rate, chemical composition of the waste, incinerator type, air pollution control equipment, and incinerator system operating parameters. A brief discussion of criteria and noncriteria pollutants is provided below.

Criteria Pollutants

Emissions from hazardous waste incineration consist mostly of the criteria pollutants, not including carbon monoxide and hydrocarbons. Emissions of carbon monoxide and hydrocarbons are normally very low because of the high combustion and destruction efficiency achieved in the incinerators. Therefore, this discussion covers only the oxides of nitrogen, oxides of sulfur, other acid gases, and particulate matter emissions.

The basic operating parameters of a hazardous waste incinerator produce ideal conditions for the formation of significant levels of oxides of nitrogen. There are two mechanisms that generate NO from incineration processes: thermal NO_x and fuel-bound NO_x. Thermal NO_x is formed by the nitrogen in the combustion air combining with available oxygen; fuel-bound NO_x is formed by the chemical conversion of nitrogen in the fuel and waste with available oxygen. At temperatures above 2800°F, thermal NO_x is a major contributor to total NO_x emissions; but fuel-bound NO_x can be a major contributor at temperatures below 2800°F. The available process modifications for NO_x reduction include peak flame temperature reduction, low excess air, low heat release rate, and staged incineration. All of these modifications are in conflict with the principles of efficient incineration and would not be likely choices, except for staged combustion.

Emissions of oxides of sulfur and various other acid gases may also result from hazardous waste incineration. The acid gases emitted may include sulfur dioxide, hydrogen chloride, hydrogen fluoride, and hydrogen bromide. The levels of these emissions are a direct function of the sulfur, chloride, fluoride, and bromide content of the waste feed and the auxiliary fuel. Because of their wide variation in concentrations of the various elements in waste feeds and fuels, uncontrolled concentrations of oxides of sulfur may vary from 10 to 12,000 parts per million (ppm); hydrochloric acid may vary from 50 to 5,000 ppm; and hydrogen fluoride may vary from 2 to 200 ppm. Various scrubbing technologies are used to control SO_x and acid gas emissions. Removal efficiencies exceeding 95% have been reported.

The emissions of particulate matter are strongly influenced by (a) the chemical composition of the waste being incinerated and the auxiliary fuel, (b) the type of incinerator and its operating parameters, and (c) the air pollution control system. Particulate matter primarily consists of inert ash, various salts, and condensed gaseous contaminants. Most of the pollutants of concern other than criteria pollutants such as heavy metals and organic toxic by-products are collected as particulate matter. The metals and toxic by-products will condense as fine particulates or on fine particles as the exhaust gas stream cools. Concentrations of uncontrolled emissions of particulate matter from hazardous waste incinerators can normally be expected to be between 0.2 and 10 grains per dry standard cubic feet of air. Controlled concentrations typically range from 0.015 to 0.05 grains per dry standard cubic feet of air. Incinerators burning waste with high solids or ash content emit higher grain loadings and larger particle sizes. The range of grain loading of particulate matter after control devices is usually less than 0.08 gr/dscf corrected to 7% O_2 to satisfy the regulatory requirement. The technology used to control particulate matter is well established (see Chapter 7). Selection of a control device depends on the inlet grain loading, particle size distribution, acid removal devices, and regulatory requirements. Scrubbers, baghouses, and wet electrostatic precipitators are currently used on hazardous waste incinerators.

Noncriteria Pollutants

Emission of noncriteria pollutants from hazardous waste incinerators may be significantly higher than they would be from more conventional combustion sources which burn fossil fuels. This is because of the chemical composition of the waste materials. For instance, incineration of wastes containing heavy metals may contain vaporized metals in the exhaust. Incineration of chlorinated compounds may release corrosive acidic gases or toxic compounds. Emissions of these compounds may be of greater concern than emissions of criteria pollutants. The following discussion will focus on dioxins, furans, and heavy metals, three of the most frequently discussed types of toxic air pollutants associated with hazardous waste incineration. Also, other noncriteria compounds which may be present in the exhaust stream such as benzene, carbon tetrachloride, chloroform, benzene hexa-chloride, and benzo(*a*)pyrene are of concern because these compounds may be toxic at low exposure levels. Some emission tests of hazardous waste incinerators do not test for the presence of these compounds.

The polychlorinated dibenzodioxins and polychlorinated dibenzofurans are among the most toxic compounds known to man. Based on available emissions tests, it is unclear to what extent dioxins and furans are emitted in the vapor phase versus the condensed phase on fine particles. Available data indicate that the concentrations are in the parts-per-billion (ppb) range with corresponding mass emission rates of well under 1 lb/yr for a regional facility.

Incineration of chlorinated organic compounds produces hydrochloric acid and free chlorine. Chlorine formation results, in part, from an inadequate supply of hydrogen to convert the chlorine in the chlorinated organic compounds to HCl. Free chlorine is more toxic and harder to remove by wet scrubbing than is HCl. Thus, formation of free chlorine in combustion gases is a problem.

In many cases, heavy metals are present in trace quantities in hazardous organic waste streams. Metals such as lead, cadmium, arsenic, and mercury cannot be destroyed by incineration, and therefore the metals will leave the incineration as part of the bottom ash, fly ash, or exhaust gases. The amount of emissions of specific metals is dependent on the metals content of the fuel and waste.

Metal oxides resulting from incineration of metallic waste compounds are mostly submicron in size and may not be effectively removed by conventional air pollution control equipment. Therefore, unless it contains only a small amount of metallic components (such as Hg, As, Se, Pb, or Cd), the waste should not be incinerated. The best control of metal oxide emissions would be removal of such metals from the waste by chemical precipitation or other means prior to inciner-ation (3). The next-best procedure may be mitigated through proper blending of the waste to maintain low levels of metals and proper selection of air pollution control devices.

20.5 INCINERATION OF POLYCHLORINATED BIPHENYLS

In the same year that the RCRA was passed (1976), Congress also passed the Toxic Substances Control Act (TSCA), partly in response to concern about potential health hazards from polychlorinated biphenyl (PCB) contamination. The TSCA imposed a ban on the manufacture of PCBs, and in 1978 and 1979 the EPA issued regulations for proper PCB marking and disposal, which included high-temperature incineration as an approved disposal/treatment method.

The purpose of incinerating PCB-containing wastes is to destroy the PCBs so effectively that any emissions of undestroyed PCBs to the environment will be at such low concentrations that adverse environmental and health impacts are not expected to occur. Thermal destruction tests on PCBs indicate that essentially complete destruction occurs in well-designed incineration systems. The most important cause of incomplete combustion is lack of turbulence or incomplete mixing of fuel, air, and combustion products. The primary concern about compounds such as PCBs was that chronic exposure of a population, even to a low level, may produce body burden levels which will eventually exhibit toxic effects. If incineration is planned, exposure of any population to airborne PCBs, HCl, and Cl_2 from incineration must be minimized and controlled gas; all points of emission must be monitored to detect even the lowest trace amounts of PCBs of intermediate products.

The toxicity of PCBs is well known, and methods to dispose of wastes containing PCBs depend mainly on the concentration of PCBs in the wastes. The EPA regulates the disposal of wastes containing PCBs in concentrations greater than 50 ppm. Wastes with PCB concentrations between 50 and 500 ppm may be burned in high-efficiency boilers or buried in chemical waste landfills, but concentrations greater than 500 ppm must be burned in an approved process incinerator.* Such approval includes an initial report, trial run report, and other requested information. The process incinerator must meet all of the requirements specified in Subpart E and Annex I of 40 CFR Part 761. The EPA stack emission standards require all process incinerators burning PCBs and PCB items to achieve a conversion efficiency or emission removal of more than 99.9999%, and to maintain the introduced PCB wastes for a 2-sec residence time at 1200°C ($+100$°C) and 3% excess oxygen in the stack gas or for a 1.5-sec residence time at 1600°C and 2% excess oxygen in the stack gas. The process incineration system should be monitoring O_2, CO, CO_2, NO_x, HCl, total chlorinated organic content, PCBs, and particulates.

The Rollins incineration facility was the first commercial chemical-waste incinerator in the United States that was approved by the EPA (in 1979) for PCB destruction. The incinerator is a rotary-kiln system. The unit operates at combustion temperatures in excess of 1000°C, with a residence time of 2–3 sec. The unit is unique in that it has two separate burners feeding a common afterburner. Both burners have natural gas ignitors and gas burners to provide initial heating of the refractory lining, flame stability, and supplemental heat if necessary. Number 2 oil is used as auxiliary fuel when wastes have insufficient heat content. All combustion gases are quenched and scrubbed by a venturi scrubber. After being scrubbed, combustion gases pass through absorption trays and a mist eliminator before entering a 100-ft stack. Two 400-hp induced-draft fans in series provide the energy

---

*EPA was considering changes to these regulations at the time of the preparation of this chapter.

for the scrubber and maintain the entire incinerator at negative pressure. Spent scrubber water is neutralized with lime and sent to a settling pond. This unit has since undergone some minor changes.

The ENSCO incineration facility was the second commercial chemical-waste incinerator approved by the EPA (in 1981) for disposal of PCBs. It is also a rotary-kiln system. The unit operates at combustion temperatures between 1000°C and 1200°C, with a residence time of 5 sec. A conveyor moves solids to the kiln. Sludge, semisolids, and slurries are fed into the bottom of the enclosure. A fan pulls air from the enclosure and forces it into the kiln. The outlet gases from the kiln pass into a cyclone for ash removal. Flue gases pass sequentially into a venturi scrubber, then into a cyclone demister, and finally to a stack. HCl is neutralized in the scrubber by lime and caustic, and spent scrubber waste is sent to a sludge lagoon. During trial burns of liquid PCBs and solid electronic capacitors containing PCBs, an extensive series of tests were performed at ENSCO. Samples of all input streams and all output streams to and from the ENSCO incineration facility were collected on a daily basis and analyzed for PCBs. Based on the results of these tests, it was reported that the incineration facility effectively destroyed PCBs. This unit is still active.

Other facilities have successfully destroyed PCBs and been permitted. In addition, the U.S. EPA has developed a mobile incineration system for on-site disposal of various types of hazardous materials. The system is self-contained on three semitrailers, with only water and fuel to be supplied externally. Trailer 1 carries the shredder, stoker, and rotary kiln, which is the primary incineration unit. Trailer 2 carries the secondary combustion chamber (afterburner), which provides temperature, residence time, and oxygen levels suitable for complete destruction of the hazardous contaminants. The unit operates at 1100°C with a 2.2-sec residence time. The gases leaving the chamber are cooled by water quenching it in a wetted throat venturi elbow. The gas stream is then cooled and passed on to Trailer 3, which is equipped with (a) a specialized particulate scrubber for removal of fly ash and (b) a caustic scrubber for removal of SO_2, HCl, and Cl_2. The system can process 75 gal/hr of waste with 99.99% expected efficiency.

Liquid PCB wastes such as mineral oil, and dielectric fluids containing 50–500 ppm of PCB are eligible for incineration in high-efficiency boilers. However, incineration of PCB wastes in high-efficiency boilers is not widely practiced. Concern over potential damage to the boiler and public opposition are two factors that have contributed to its limited use. To burn PCB waste in high-efficiency boilers, design and operational requirements must be specified.

Theory and experiment indicate that essentially complete destruction of PCBs can be achieved in incinerators and high-efficiency boilers. However, inadequate design or operation can lead to incomplete combustion and/or formation of highly toxic compounds such as polychlorinated dibenzofurans (PCDFs) and dibenzo-*p*-dioxins (PCDDs). Theoretical calculations predict that, under oxidizing conditions, formation of PCDFs and PCDDs is not thermodynamically favored.

20.6 SOCIOECONOMIC CONCERNS

Socioeconomic and environmental considerations, together with public acceptance, often dictate the location of a proposed site for a hazardous waste incinerator.

Unfortunately, the controlling factors seem to be socioenvironmental concerns rather than economic or technical factors. Economic concerns include installation and operating costs, costs of transporting and storage of the hazardous wastes, and disposal costs of the residue. If the incineration system is land-based, the costs should include adaptation of the site, which includes access roads, water and sewage problems, topography modifications, soil problems, aesthetic treatment, and compensation for possible losses by neighboring taxpayers from air pollution damage to, and devaluation of, their property. Social and environmental concerns mostly involve potential effects upon public health. The adverse consequences to the public health of living within an environment that receives low levels of various air contaminants, especially toxic contaminants, on a daily basis is not known. However, the inhalation at high levels of many specific metallic or salt particulates, polynuclear organic compounds, and acid gases is known both to have serious respiratory effects and to contribute to development of chronic illnesses such as cancer. These contaminants can be emitted continuously from a hazardous waste incinerator if not properly removed from the stack gas.

The siting of a proposed incinerator, therefore, has been one of the main social problems (3,4). The siting of hazardous waste management facilities is a critical issue. Some of the key comments in an early review paper (that are still applicable today) are cited below (5).

1. The need for properly designed and operated treatment, storage, and disposal facilities exceeds existing capacity. Wastes being generated by ongoing processes must be accommodated as well as those removed from emergency or remedial action sites. Major remedial sites contain tremendous quantities of hazardous wastes and contaminated soil. In some instances, remedial work at a Superfund site may tax the disposal capacity of nearby permitted facilities. As a result, these sites may reach capacity sooner than originally anticipated, creating additional pressure for the siting of new facilities.

2. Siting of such facilities has encountered significant public resistance in the vicinity of the proposed sites—the "not in my back yard" (NIMBY) syndrome. Everyone agrees new sites must be found; however, no one wants one in their community. This is true in spite of the care taken in siting efforts to protect the environment and minimize impacts upon surrounding populations.

3. Siting such facilities poses substantial technical, legal, financial, and political challenges. A further complicating factor is that a large segment of the public is apparently unconvinced that hazardous wastes can be disposed of safely. Furthermore, the siting process is not always perceived by the public as being fair, open, and equitable. Those that live near the selected site feel that they carry too much of the cost, including health concerns, lower property values, and the stigma of living near a site. The challenge to the industries involved and to the federal and state governments is to convince the public that the site is needed, that reasonable steps have been taken to protect public health, that the solution is fair, and that appropriate incentives have been provided for community acceptance.

4. The key to this process is the public misconception that hazardous waste is the result of industry carelessness in disposing of dangerous materials. Programs such as Florida's Amnesty Days enlighten us to the fact that we all live with and handle hazardous materials every day in our kitchens, laundry rooms, garages, and

automobiles. This, in turn, helps us understand these materials and acts to reduce concern by demonstrating that such materials can be managed in a responsible manner. If other creative ways of reducing public emotion related to hazardous wastes can be found, it will allow more rational approaches to solving this problem.

Despite the ever-increasing quantities of hazardous wastes, there are only a limited number of commercially available hazardous waste incinerators in the United States. During the past several years, disposal of oil and hazardous substances resulting from cleanup of accidental spills and abandoned waste sites has become increasingly difficult. As described above, greater public awareness of dangers associated with hazardous waste disposal has led to local opposition to any land disposal sites in their neighborhood. The justifications for new incineration facilities are obvious and critical, but finding a site is extremely difficult.

20.7 SUMMARY

1. Incineration is not a new technology. It has been commonly used for treating organic hazardous waste for many years in Europe and the United States. The major benefits of incineration are that the process actually destroys most of the waste rather than just disposing of or storing it. Incineration can be used for a variety of specific wastes and is reasonably competitive in cost compared to other disposal methods.

2. Incineration of hazardous wastes under the U.S. Resource Conservation and Recovery Act (RCRA) of 1976 is regulated under the *Code of Federal Regulations*, Title 40, Parts 122, 264, and 265. These regulations establish performance standards for hazardous waste incineration regarding:

1. Destruction and removal efficiency

2. Hydrochloric acid gas emissions

3. Particulate emissions

3. There are five major types of process incinerators commercially available for burning combustible hazardous wastes: liquid-injection, rotary-kiln, fluidized-bed, multiple-hearth, and co-incineration. Liquid-injection and rotary-kiln incinerators are at present the most highly developed and most commonly used incinerators for hazardous waste incineration. Both exist throughout the United States in full-scale operation.

4. One major concern associated with hazardous waste incineration is the emission of air contaminants. The greatest mass of air contaminants consists primarily of these criteria pollutants: oxides of nitrogen, oxides of sulfur, and particulate matter. Also, of great concern are trace levels of noncriteria pollutants such as chlorinated by-products, benzene, heavy metals, and acid gases.

5. The purpose of incinerating PCB-containing wastes is to destroy the PCBs so effectively that any emissions of undestroyed PCBs to the environment will be at such low concentrations that adverse environmental and health impacts are not expected to occur. Thermal destruction tests on PCBs indicate that essentially complete destruction occurs in well-designed incineration systems.

6. Socioeconomic and environmental considerations, together with public acceptance, often dictate the location of a proposed site for a hazardous waste incinerator. Unfortunately, the controlling factors seem to be socioenvironmental concerns rather than economic or technical factors.

REFERENCES

1. T. Shen, Y. McGuinn, and L. Theodore, *Hazardous Waste Incineration Student Manual,* U.S. EPA APTI Course 502, 1986.
2. L. Theodore and Y. McGuinn, *Pollution Prevention,* Van Nostrand Reinhold, 1992.
3. L. Theodore and J. Reynolds, *Introduction to Hazardous Waste Incineration,* Wiley-Interscience, New York, 1988.
4. L. Theodore and D. Helfritsch, *Hazardous Waste Incineration,* Theodore Tutorials, East Williston, NY, 1992.
5. U.S. EPA, Office of Policy, Planning and Evaluation, *Assessment of Incineration as a Treatment Method for Liquid Organic Hazardous Wastes: Summary and Conclusions,* 1985.

21

SMALL QUANTITY GENERATORS

21.1 INTRODUCTION

In 1976, Congress passed the Resource Conservation and Recovery Act (RCRA) which directed the U.S. Environmental Protection Agency (EPA) to develop and implement a program to protect human health and the environment from improper hazardous waste management practices. The program is designed to control the management of hazardous waste from its generation to its ultimate disposal — "from cradle to grave."

The EPA first focused on large companies, which generate the greatest portion of hazardous waste. Business establishments producing less than 1000 kg (2200 lbs) of hazardous waste in a calendar month (known as *small quantity generators*) were exempted from most of the hazardous waste management regulations published by EPA in May 1980.

In recent years, however, public attention has been focused on the potential for environmental and health problems that may result from mismanaging even small quantities of hazardous waste. For example, small amounts of hazardous waste dumped on the land may seep into the earth and contaminate underground water that supplies drinking water wells.

In November 1984, the Hazardous and Solid Waste Amendments to RCRA were signed into law. With these amendments, Congress directed EPA to establish new requirements that would bring small quantity generators who generate between 100 and 1000 kg of hazardous waste in a calendar month into the hazardous waste regulatory system. The EPA issued final regulations for these 100- to 1000-kg/mo generators on March 24, 1986. Most of the requirements were effective September 22, 1986 (1).

Compliance with these regulations make it necessary for these generators to determine whether or not they are required by law to meet these requirements and if so how to comply. This chapter will address some of the more pertinent issues related to small quantity generators and will provide specific information which will assist small quantity generators in meeting the requirements.

21.2   DETERMINING THE GENERATOR CATEGORY

Small businesses must first determine whether or not they generate a hazardous waste. It is the responsibility of the generator to determine whether the waste generated is hazardous. The term "hazardous waste" means a solid waste, or combination of solid wastes, which because of its quantity, concentration, or physical, chemical, or infectious characteristics may either (a) cause or significantly contribute to an increase in mortality or an increase in serious, irreversible, or incapacitating reversible illness; or (b) pose a substantial present or potential hazard to human health or the environment when improperly treated, stored, transported, or disposed of, or otherwise managed (2).

There are two ways a waste may be brought into the hazardous waste regulatory system: (a) listing, and (b) identification through characteristics. A waste is considered hazardous if it appears on any one of the four lists of hazardous waste contained in the RCRA regulations. These wastes have been listed because they either exhibit one of the characteristics described below or contain any number of toxic constituents that have been shown to be harmful to health and the environment. The regulations list over 400 hazardous wastes, including wastes derived from manufacturing processes and discarded commercial chemical products. Even if a waste does not appear on one of the EPA lists, it is considered hazardous if it has one or more of the following characteristics:

- It is easily combustible or flammable. This is called an *ignitable* waste. Examples are paint wastes, certain degreasers, or other solvents.

- It dissolves metals or other materials or burns the skin. This is called a *corrosive* waste. Examples are waste rust removers, waste acid or alkaline cleaning fluids, and waste battery acid.

- It is unstable or undergoes rapid or violent chemical reaction with water or other materials. This is called a *reactive* waste. Examples are cyanide plating wastes, waste bleaches, and other waste oxidizers.

Some wastes are considered to be "acutely hazardous." These are wastes that the EPA has determined to be so dangerous in small amounts that they are regulated the same way as are large amounts of other hazardous wastes. Acutely hazardous wastes, for example, may be generated using certain pesticides. They also include dioxin-containing wastes.

Wastes that appear in Appendix C with an asterisk (*) have been designated acutely hazardous. If a business generates more than 1 kg (approximately 2.2 lbs) of acutely hazardous wastes in a calendar month or stores more than that amount for any period of time, it is subject to all of the regulations that apply to generators that generate more than 1000 kg of hazardous waste per calendar month. Contact one of the sources of information listed in Appendix B for more information about acutely hazardous wastes. Table 21.1 contains a list of waste streams generated by small quantity generators.

Table 21.1 Typical waste streams generated by small quantity generators

Type of Business	Type of Hazardous Wastes Generated
Building cleaning and maintenance	Acids/bases Solvents
Chemical manufacturers	Acids/bases Cyanide wastes Heavy metals/inorganics Ignitable wastes Reactives Solvents
Cleaning agents and cosmetics	Acids/bases Heavy metals/inorganics Ignitable wastes Pesticides Solvents
Construction	Acids/bases Ignitable wastes Solvents
Educational and vocational shops	Acids/bases Ignitable wastes Pesticides Reactives Solvents
Equipment repair	Acids/bases Ignitable wastes Solvents
Formulators	Acids/bases Cyanide wastes Heavy metals/inorganics Ignitable wastes Pesticides Reactives Solvents
Funeral services	Solvents Formaldehyde
Furniture/wood manufacturing and refinishing	Ignitable wastes Solvents
Laboratories	Acids/bases Heavy metals/inorganics Ignitable wastes Reactives Solvents
Laundries and dry cleaners	Dry-cleaning filtration Residues Solvents

Continued

Table 21.1 *Continued*

Type of Business	Type of Hazardous Wastes Generated
Metal manufacturing	Acids/bases Cyanide wastes Heavy metals/inorganics Ignitable wastes Reactives Solvents Spent plating wastes
Motor freight terminals and railroad transportation	Acids/bases Heavy metals/inorganics Ignitable wastes Lead-acid batteries Solvents
Other manufacturing: 1. Textiles 2. Plastics 3. Leather	Heavy metals/inorganics Solvents
Pesticide end-users and application services	Heavy metals/inorganics Pesticides Solvents
Printing and allied industries	Acids/bases Heavy metals/inorganics Ink sludges Spent plating wastes Solvents
Wood preserving	Preserving agents

There are three categories of hazardous waste generators: (a) generators of no more than 100 kg/mo (also known as *conditionally exempt small quantity generators*); (b) 100- to 1000-kg/mo small quantity generators; and (c) generators of 1000 kg or more in a month. To determine which category of hazardous waste generator a business falls into—and what requirements must be met—measure or "count" the hazardous wastes the business generates in a calendar month. In general, add up the weight of all hazardous wastes the business generates during a month; the total weight will determine the generator category.

When begining to count the hazardous wastes each month, it may be confusing at first to determine what kinds of hazardous wastes are generated and how much. To have questions answered, call the EPA RCRA/Superfund Hotline or the state hazardous waste management agency listed in Appendix B.

If hazardous waste is accumulated until enough is collected to make transport to a licensed hazardous waste management facility more economical, make sure that:

- No more than 6000 kg of hazardous waste is accumulated *in any* 180 *day period* (270 days are allowed if the waste must be transported over 200 miles to a licensed hazardous waste facility) if the facility is a 100- to 1000-kg/mo generator. Otherwise, a special storage permit must be obtained.

- No more than 1000 kg of hazardous waste is accumulated *at any time* if the facility is a generator of no more than 100 kg/mo.

Generators of No More Than 100 kg/mo

If no more than 100 kg (about 220 lbs or 25 gal) of hazardous waste and no more than 1 kg (about 2 lbs) of acutely hazardous wastes is generated in any calendar month, the generator is a conditionally exempt small quantity generator and is required by the federal hazardous waste laws to:

- Identify all hazardous waste generated.
- Send this waste to a hazardous waste facility, or a landfill or other facility approved by the state for industrial or municipal wastes.
- Never accumulate more than 1000 kg of hazardous waste on the generator's property. Otherwise, the generator becomes subject to all the requirements applicable to 100- to 1000 kg/mo generators.

Generators of 100–1000 kg/mo

If more than 100 and less than 1000 kg (between 220 and 2200 lbs or about 25 to under 300 gal) of hazardous waste and no more than 1 kg of acutely hazardous waste is generated in any month, the generator is classified as a 100- to 1,000-kg/mo generator and is required by the federal hazardous waste laws to:

- Comply with the 1986 rules for managing hazardous waste, including the accumulation, treatment, storage, and disposal requirements.

Generators of 1000 kg/mo or More

If 1000 kg (about 2200 lbs or 300 gal) or more of hazardous waste, or more than 1 kg of acutely hazardous waste is generated in any month, the generator is classified as a generator of 1000 kg/mo or more and is required by the federal hazardous waste laws to:

- Comply with all applicable hazardous waste management rules.

Counting Hazardous Wastes. Do count *all* quantities of "Listed" and "Characteristic" hazardous waste that are:

- Accumulated on-site for any period of time prior to subsequent management.
- Packaged and transported off-site.
- Placed directly in a regulated on-site treatment or disposal unit.
- Generated as still bottoms or sludges and removed from product storage tanks.

You *do not* have to count wastes that:

- Are specifically exempted from counting. Examples of these exempted wastes are:

1. Spent lead-acid batteries that will be sent off-site for reclamation.
2. Used oil that has not been mixed with hazardous waste.

- May be left in the bottom of containers that have been completely emptied through conventional means, for example, by pouring or pumping. Containers that held an acute hazardous waste must be more thoroughly cleaned.

- Are left as residue in the bottom of product storage tanks, if the residue is not removed from the product tank.

- Are reclaimed continuously on-site without storing the waste prior to reclamation, such as dry-cleaning solvents. (Do count any residue removed from the machine, as well as spent cartridge filters.)

- Are managed in an elementary neutralization unit, a totally enclosed treatment unit, or a wastewater treatment unit. An elementary neutralization unit is a regulated tank, container, or transport vehicle (including ships) which is designed to contain and neutralize corrosive wastes.

- Are discharged directly to a publicly owned treatment works (POTW) without being stored or accumulated first. This discharge to a POTW must comply with the Clean Water Act. POTWs are public utilities, usually owned by the city, county, or state, that treat industrial and domestic sewage for disposal.

- Have already been counted once during the calendar month, and treated on-site or reclaimed in some manner, and used again.

21.3   MEETING THE NEW REQUIREMENTS

Once a small quantity generator status has been determined, a facility must (a) obtain a U.S. EPA Identification Number, (b) use the full Uniform Hazardous Waste Manifest system when shipping hazardous waste off-site, (c) offer wastes only to hazardous waste transporters who have U.S. EPA Identification Numbers, and (d) accumulate waste on-site for no more than 180 days, or 270 days if the waste is to be shipped more than 200 miles, unless it obtains a hazardous waste permit. The facility must also ensure that its hazardous waste is managed at a hazardous waste facility with interim status or a permit under RCRA (3).

If a business generates more than 100 kg of hazardous waste in any calendar month, it will need to obtain a U.S. EPA Identification Number. Transporters and facilities that store, treat, or dispose of regulated quantities of hazardous waste must also have U.S. EPA Identification Numbers. These 12-character identification numbers used by EPA and states are part of a national database on hazardous waste activities. To obtain an identification number, call or write the state hazardous waste management agency or EPA regional office (see Appendix B) and ask for a copy of EPA Form 8700-12, "Notification of Hazardous Waste Activity."

The U.S. EPA Identification Number will stay with the business site or location. If the business is moved to another location, the EPA or the state of the new location must be notified and a new form must be submitted. If hazardous waste was previously handled at the new location, and it already has a U.S. EPA Identification Number, the generator will be assigned that number for the site after it has notified the EPA.

21.4 SHIPPING HAZARDOUS WASTE OFF-SITE

The three most important things to remember about shipping hazardous waste off-site are: (a) Choose a hauler and facility that have EPA identification numbers, (b) package and label the wastes for shipping, and (c) prepare a hazardous waste manifest. Under federal regulation, a 100- to 1000-kg/mo generator is allowed to accumulate hazardous wastes on the premises without a permit for up to 180 days (or 270 days if it must be shipped more than 200 miles) as long as no more than 6000 kilograms is ever accumulated. These limits are set so that a small business can accumulate enough waste to make shipping and disposal more economical.

Carefully choosing a hauler and designating a waste management facility is important. The hauler will be handling the wastes beyond the generator's control while the generator is still responsible for its proper management. Similarly, the waste management facility will be the final destination of the hazardous waste for treatment, storage, or disposal. Before choosing a hauler or designating a facility, a generator should check with the following sources:

- Friends and colleagues in business who may have used a specific hazardous waste hauler or designated facility in the past.

- Trade association(s) that may keep files on companies that handle hazardous wastes.

- The Better Business Bureau or Chamber of Commerce to find out if any complaints have been registered against a hauler or facility.

- The state hazardous waste management agency or EPA regional office, which will be able to tell whether or not a company has a U.S. EPA Identification Number and whether or not the company has had any problems.

After checking these sources, the generator should contact the hauler and designated hazardous waste management facility directly to verify that they have U.S. EPA Identification Numbers, and that they can and will handle the waste. Also make sure that they have the necessary permits and insurance, and that the hauler's vehicles are in good condition. Checking sources and choosing a hauler and designated facility may take some time; try to begin checking well ahead of the time you will need to ship the waste. Careful selection is very important.

When preparing hazardous wastes for shipment, the generator must put the wastes in containers acceptable for transportation and make sure the containers are properly labeled. The hauler should be able to assist. If additional information is needed, consult the requirements for packaging and labeling hazardous wastes found in the Department of Transportation (DOT) regulations (49 CFR Part 172). To find out what these requirements are for specific wastes, contact the state transportation agency.

A hazardous waste manifest is a multicopy shipping document that must be filled out and used to accompany hazardous waste shipments. The manifest form is designed so that shipments of hazardous waste can be tracked from their point of generation to their final destination — the so-called "cradle-to-grave" system. The hazardous waste generator, the hauler, and the designated facility must each sign this document and keep a copy. The designated facility operator also must send a copy back to the generator so that the generator can be sure that its shipment arrived. The generator must keep this copy, which will be signed by the hauler and designated facility, on file for three years.

If the generator does not receive a signed copy from the designated hazardous waste management facility within 30 days, it is a good idea to find out why and, if necessary, let the state or EPA know. **REMEMBER**: Just because a generator has shipped the hazardous waste off its site and the waste is no longer in its possession, its liability has *not* ended. The generator is potentially liable under Superfund for *any* mismanagement of its hazardous waste. The manifest will help to track the waste during shipment and make sure it arrives at the proper destination.

States, haulers, recyclers, and designated facilities may require additional information; the generator should check with them before preparing a hazardous waste shipment. The hazardous waste hauler often will be the best source for packaging and shipping information and will help in completing the manifest. The EPA has also prepared some industry-specific information to help in completing the manifest. This industry-specific information is available from EPA Regional Offices and a number of trade associations. If a generator has any trouble obtaining, filling out, or using the manifest, ask the hauler, the designated facility operators, or one of the contacts listed in Appendix B for help.

Federal regulations do allow the generator to haul its hazardous waste to a designated facility. It must, however, obtain an EPA transporter identification number and comply with applicable DOT requirements for packaging, labeling, marking, and placarding its shipment. There are also financial responsibility and liability requirements under the Federal Motor Carrier Act, but the generator may be exempt from these if it uses a vehicle with a Gross Vehicle Weight Rating of less than 10,000 pounds (van or pick-up truck), transports the wastes for commerce within the state in nonbulk shipments (i.e., containers with capacities of less than 3,500 gallons), or transports hazardous wastes which meet the "limited quantity exclusion" requirements of Section 172.101 of the DOT regulations. If a generator decides to transport its own hazardous wastes, it should call the state hazardous waste management agency (see Appendix B) to find out what state regulations apply. Not all states will allow a generator to transport its own hazardous wastes. If there is an accident during transport, the generator is responsible for the clean-up.

21.5 TYPICAL WASTE STREAMS GENERATED BY SELECTED SMALL QUANTITY GENERATORS

The following section provides an in-depth analysis of a select few of specific small quantity generators.

Vehicle Maintenance (4)

If a generator repairs or maintains cars, vans, trucks, heavy equipment, farm equipment or if the generator removes oil or grease, removes rust, dirt or paint, repairs or rebuilds, paints, or replaces lead-acid batteries, then the products used on the vehicles, and on the equipment, tools, hands, or floor may contain hazardous materials, and the wastes generated by using these products may be hazardous wastes. Everyday mechanics, bodymen and others use products containing hazardous materials. Products like rust removers, which contain strong acid or alkaline solutions; carburetor cleaners, which contain flammable or combustible liquids; used rags containing combustible or flammable solvents; paints with flammable or combustible thinners or reducers; and auto and truck batteries contain chemicals or materials which are hazardous to human health and the environment. Table 21.2 lists typical operations/processes which use products that may contain hazardous materials and which probably generate hazardous wastes (also see Table 21.3). If a facility generates 220 lbs (about half of a 55-gal drum) or more of hazardous waste per month, it must fill out a Uniform Hazardous Waste Manifest when this waste is shipped off the property.

Not all vehicle maintenance operations generate hazardous wastes. Under current federal law, you do not have to use a Manifest when you ship used/dead lead acid batteries that are destined for recycling. For used motor oil, see Chapter 18. Environmentally sound methods should be used for the collection, storage and recycling of used motor oil. The individual states may have their own requirements for lead acid batteries or used oil. Check with your state hazardous waste management agency.

Dry-Cleaning and Laundry Plants (5)

The establishments covered under Drycleaning and Laundry Plants include retail Drycleaning stores; industrial and linen supply plants with dry-cleaning operations; leather and fur cleaning plants; self-service laundromats with dry-cleaning equipment; and other establishments with dry-cleaning operations. While not all of these facilities will produce hazardous waste, those facilities using hazardous solvents may be subject to new Resource Conservation and Recovery Act (RCRA) provisions regarding the treatment, storage, disposal, and transportation of small quantities of hazardous waste. These solvents include:

- *Perchloroethylene*, otherwise known as perc, PCE, or tetrachlorethylene
- *Valclene*, also known as fluorocarbon 113 or trichlorotrifluoroethane
- *Petroleum solvents*, such as Stoddard, quick-dry, low-odor and other solvents with a flash point less than 140°F. ("140°F solvent" and other solvents with a flash point equal to or greater than 140°F are not considered hazardous under EPA RCRA designation. If a generator is unsure of the flash point, check with the distributor of the solvent.)

Perchloroethylene and valclene plants potentially produce two types of hazardous wastes: (a) Still residues from solvent distillation (the entire weight), and (b) spent filter cartridges (total weight of the cartridge and remaining solvent after draining). Perchloroethylene plants also may produce cooked powder residue (the total weight of drained powder residues from diatomaceous or other powder filter systems after heating to remove excess solvent), which is a hazardous waste. Petroleum solvent plants potentially produce only one type of hazardous waste, still residues from solvent distillation (the entire weight). However, if 140°F solvent is used, the still residue will not normally be a hazardous waste. Well-drained filter cartridges or drained filter muck (powder residues from diatomaceous filter systems) are solids and do not meet the criteria for classification as an ignitable solid; therefore, they are not hazardous wastes.

If a plant produces 220 lbs or more of hazardous waste per month, it is subject to certain requirements, including the use of a Uniform Waste Manifest when

Table 21.2 Typical operations using materials which may generate hazardous wastes

Typical Process/ Operation	Typical Materials Used	Typical Material Ingredients on Label	General Types of Wastes Generated
Oil and grease removal	Degreasers — gunk, carburetor cleaners, engine cleaners, varsol, solvents, acids/alkalies	Petroleum distillates, aromatic hydrocarbons, mineral spirits	Ignitable wastes, spent solvents, combustible solids, waste acid/ alkaline solutions
Engine, parts, and equipment cleaning	Degreasers — gunk, carburetor cleaners, engine cleaners, solvents, acids/ alkalies, cleaning fluids	Petroleum distillates, aromatic hydrocarbons, mineral spirits, benzene, toluene, petroleum, naphtha	Ignitable wastes, spent solvents, combustible solids, waste acid/ alkaline solutions
Rust removal	Naval jelly, strong acids, strong alkalies	Phosphoric acid hydrochloric acid, hydrofluoric acid, sodium hydroxide	Waste acids, waste alkalies
Paint preparation	Paint thinners, enamel reducers, white spirits	Alcohols, petroleum distillates, oxygenated solvents, mineral spirits, ketones	Spent solvents, ignitable wastes, ignitable paint wastes, paint wastes with heavy metals
Painting	Enamels, lacquers, epoxys, alkyds, acrylics, primers	Acetone, toluene petroleum distillates, epoxy ester resins, methylene chloride xylene, VM&P naphtha, aromatic hydrocarbons, methyl isobutyl ketones	Ignitable paint wastes, spent solvents, paint wastes with heavy metals, ignitable wastes
Spray booth, spray guns, and brush cleaning	Paint thinners, enamel reducers, solvents, white spirits	Ketones, alcohols, toluene, acetone isopropyl alcohol, petroleum distillates, mineral spirits	Ignitable paint wastes, heavy metal paint wastes, spent solvents
Paint removal	Solvents, paint thinners, enamel reducers, white spirits	Acetone, toluene, petroleum distillates, methanol, methylene chloride, isopropyl alcohol, mineral spirits, alcohols, ketones, other oxygenated solvents	Ignitable paint wastes, heavy metal paint wastes, spent solvents
Used lead-acid batteries	Car, truck, boat motorcycle, and other vehicle batteries	Lead dross, less than 3% free acids	Used lead-acid batteries, strong acid/alkaline solutions

shipping hazardous waste off the premises. To determine whether a plant qualifies as a regulated small quantity generator and to complete the Manifest, the weight of the hazardous waste the plant generates will have to be listed. Table 21.4 lists common types and average quantities of hazardous waste produced per 1000 lbs of clothes cleaned (also see Table 21.5).

Generally there are three methods for proper disposal of hazardous wastes that are currently considered acceptable by both EPA and most state hazardous waste management agencies: Disposal in an authorized hazardous waste landfill, disposal at an authorized high-temperature incineration facility, or disposal through an authorized recycler of hazardous wastes. From an environmental perspective, recycling or incineration is generally preferable to land disposal.

Under the new RCRA, EPA was required to issue, by November 8, 1986, new regulations that ban the disposal on or into the land of hazardous waste containing certain solvents, including perchloroethylene and Valclene. Congress set specific prohibition dates for certain high-risk and high-volume waste and established a three-part schedule with specific deadlines for the EPA to develop treatment standards for the remaining listed and characteristic wastes. On November 7, 1986, the EPA published a final rule (51 FR 40572) establishing effective dates and treatment standards for F001-F005 solvent wastes and F020-F023 and F026-F028 dioxin-containing wastes. A second rule on hazardous waste, known as the California list, prohibited the land disposal of liquid hazardous wastes containing certain toxic constituents or exhibiting certain properties unless subjected to prior treatment was published July 8, 1987 (52 FR 25760). The treatment standards for all hazardous wastes identified or listed on or before November 8, 1984, were published on August 17, 1988 (53 FR 31138), June 23, 1989 (54 FR 26594), and June 1, 1990 (55 FR 22520). The November 7, 1986, final rule also established the basic framework for the land disposal restrictions (LDRs) program. The EPA was also required to establish treatment standards for all hazardous wastes listed or identified after November 8, 1984. These rulemakings were published in the Federal Register on August 8, 1992 (57 FR 37194), September 19, 1994 (59 FR 47982), April 8, 1996 (61 FR 15566 and 15660), and May 12, 1997 (62 FR 25998).

21.6 GOOD HOUSEKEEPING AND A SAFE ENVIRONMENT

The four most important things you should remember about managing your waste properly are as follows. Reduce the amount of hazardous waste, conduct a self-inspection, cooperate with state and local inspectors, and call the state hazardous waste management agency or the U.S. EPA with any questions. Good hazardous waste management can be thought of simply as using "good housekeeping" practices such as using and reusing materials as much as possible; recycling or reclaiming waste; treating waste to reduce its hazards; or reducing the amount of waste generated. To reduce the amount of waste generated:

- Do not mix nonhazardous wastes with hazardous ones. For example, do not put nonhazardous cleaning agents or rags in the same container as a hazardous solvent or the entire contents becomes subject to the hazardous waste regulations.

- Avoid mixing several different hazardous wastes. Doing so may make recycling very difficult, if not impossible, or make disposal more expensive.

Table 21.3 Waste descriptions[a]

Waste Type	Designations/Trade Names	DOT Shipping Name	Hazard Class	UN/NA ID Number
Strong Acid/Alkaline Wastes				
Ammonium hydroxide	Ammonium hydroxide, NH$_{40}$H, spirit of Hartshorn, aqua ammonia	Waste ammonium hydroxide (containing not less than 12% but not more than 44% ammonia)	Corrosive material	NA2672
Hydrobromic acid	Hydrobromic acid, HBr	(containing less than 12% ammonia)	ORM-A	NA2672
		Waste hydrobromic acid (not more than 49% strength)	Corrosive material	UN1788
Hydrochloric acid	Hydrochloric acid, HCl, Muriatic acid	Waste hydrochloric acid	Corrosive material	NA1789
Hydrofluoric acid	Hydrofluoric acid, HF, fluorohydric acid	Waste hydrofluoric acid	Corrosive material	UN1790
Nitric acid	Nitric acid, HNO$_2$, Aquafortis	Waste nitric acid (over 40%)	Oxidizer	UN2031
		(40% or less)	Corrosive material	NA1760
Phosphoric acid	Phosphoric acid, H$_3$PO$_4$, Orthophosphoric acid	Waste phosphoric acid	Corrosive material	UN1805
Potassium hydroxide	Potassium hydroxide, KOH, potassium hydrate, caustic potash, potassa	Waste potassium hydroxide solution	Corrosive material	UN1814
Sodium hydroxide	Sodium hydroxide, NaOH, caustic soda, soda lye, sodium hydrate	Dry solid, flake, bead, or granular	Corrosive material	UN1813
		Waste sodium hydroxide solution	Corrosive material	UN1824
Sulfuric acid	Sulfuric acid, H$_2$SO$_4$, oil of vitriol	Dry solid, flake, bead, or granular	Corrosive material	UN1823
		Waste sulfuric acid	Corrosive material	UN1830
Chromic acid	Chromic acid	Waste chromic acid solution	Corrosive material	UN1755
Ignitable Wastes				
Ignitable wastes NOS[b]	Carburetor cleaners, ignitable wastes NOS	Waste flammable liquid NOS	Flammable liquid[c]	UN1993
Aromatic hydrocarbons		Waste combustible liquid NOS	Combustible liquid[d]	NA1993
Petroleum distillates		Waste flammable solid NOS	Flammable solid	UN1325

Ignitable Paint Wastes

Ethylene dichloride	Ethylene dichloride, 1,2-dichloroethane	Waste ethylene dichloride	Flammable liquid	UN1184
Benzene	Benzene	Waste benzene (benzol)	Flammable liquid	UN1114
Toluene	Toluene	Waste toluene (toluol)	Flammable liquid	UN1294
Ethyl benzene	Ethyl benzene	Waste ethyl benzene	Flammable liquid	UN1175
Chlorobenzene	Chlorobenzene, monochlorobenzene, phenylchloride	Waste chlorobenzene	Flammable liquid	UN1134
Methyl ethyl ketone	Methyl ethyl ketone (MEK), methyl acetone, Meetco, butanone, ethyl methyl ketone	Waste methyl ethyl ketone	Flammable liquid	UN1193

Spent Solvents

White spirits, Varson	White spirits, mineral spirits, naphtha	Waste naphtha	Flammable liquid	UN2553
			Combustible liquid	UN2553
1,1,1-Trichloroethane	Aeothane TT, Chlorlen, chloroethene, methyl-chloroform, Alpha T, chlorotene	Waste 1,1,1-trichloroethane	ORM-A	UN2831
Petroleum distillates	Petroleum distillates	Petroleum distillate	Flammable liquid	UN1268
			Combustible liquid	UN1268

Paint Wastes with Heavy Metals

Paints with heavy metals Lead Nickel Chromium	Heavy metals paint	Hazardous waste, liquid or solid, NOS	ORM-E	NA9189

[a] These description may change given variations in waste characteristics or conditions.
[b] NOS — not otherwise specified.
[c] A flammable liquid has a flash point below 100°F.
[d] A combustible liquid has a flash point between 100 and 200°F.

Table 21.4   Typical quantities of hazardous waste from dry cleaning[a]

Waste Type	Cleaning Method		
	Perc	Valclene	Petroleum Solvents
	Average Quantity of Hazardous Waste (pounds)		
Still residues	25	10	20
Spent cartridge filters			
Standard (carbon core)	20	15	[b]
Adsorptive (split)	30	20	[b]
Cooked powder residue	40	NA[c]	NA
Drained filter muck	NA	NA	[b]

[a]Pounds of waste per 1000 lbs of clothes cleaned.
[b]Well-drained filter cartridges or drained filter muck are solids and do not meet the criteria for classification as an ignitable solid; therefore, they are not hazardous wastes.
[c]NA — not available.

- Avoid spills or leaks of hazardous products. (The materials used to clean up such spills or leaks also will become hazardous.)
- Make sure the original containers of hazardous products are completely empty before throwing them away. Use *all* the product.
- Avoid using more of a hazardous product than needed. For example, use no more degreasing solvent or pesticide than needed to do the job. Also, do not throw away a container with unused solvent or pesticide in it.

Reducing hazardous waste means saving money on raw materials and reducing the costs to the business for managing and disposing of hazardous wastes.

Table 21.5   Waste descriptions[a]

Waste	DOT Shipping Name	Hazard Class	UN/NA ID Number
Perc	Waste perchloroethylene or waste tetrachloroethylene	ORM-A	UN 1897
Valclene	Hazardous waste, NOS[b]	ORM-E	UN 9189
Petroleum solvents	Waste petroleum distillate	Combustible liquid[c]	UN 1268
	Waste petroleum naphtha	Combustible liquid[c]	UN 1255

[a]In certain situations, other DOT descriptions may be applicable to the wastes listed.
[b]NOS — not otherwise specified.
[c]If the flash point of the solvent or residue *as disposed of* is less than 100°F, the hazard class would be "flammable liquid." Although the flash point of petroleum dry-cleaning solvents is above 100°F, the presence of contaminants (such as printing inks) could lower the overall flash point to below 100°F.

Another aspect of good housekeeping is cooperating with inspection agencies and using a visit by an inspector as an opportunity to identify and correct problems. Accompanying state or local inspectors on a tour of the facility will enable the generator to ask any questions and receive advice on more effective ways of handling hazardous products and wastes. In addition, guiding the inspectors through the property and explaining the operations may help them to be more sensitive to the particular problems or needs of your business. Inspectors can also serve as a valuable source of information on recordkeeping, manifests, and safety requirements specific to a facility.

The best way to prepare for a visit from an inspector is to conduct self-inspection. The items listed below can serve as a basic guide to developing a self-inspection checklist.

- Do you have some documentation on the amounts and kinds of hazardous waste you generate and how you determined that they are hazardous?
- Do you have a U.S. EPA identification number?
- Do you ship waste off-site? If so, by which hauler and to which designated hazardous waste management facility?
- Do you have copies of manifests used to ship your hazardous waste off-site? Are they filled out correctly? Have they been signed by the designated facility?
- Is your hazardous waste stored in the proper containers?
- Are the containers properly dated and marked?
- Have you designated an emergency coordinator?
- Have you posted emergency telephone numbers and the location of emergency equipment?
- Are your employees thoroughly familiar with proper waste handling and emergency procedures?
- Do you understand when you may need to contact the national response center?

If a generator is still uncertain about how to handle hazardous waste, or has any questions concerning the rules for 100- to 1,000-kg/mo generators, there are several sources listed in Appendix A that can be contacted for answers. Taking responsibility for proper handling of hazardous waste will not only ensure a safer environment and workplace for everyone, but will save the business money.

21.7 SUMMARY

1. Small Quantity Generators (SQGs), as the name implies, address those generators that generate small amounts of hazardous waste and who are required by law to manage these wastes in an environmentally sound manner.

2. Small businesses must first determine whether or not they generate a hazardous waste or not. It is the responsibility of the generator to determine whether the waste generated is hazardous.

3. Once you have determined you are a small quantity generator, you must meet the new requirements.

4. Carefully choosing a hauler and designating a waste management facility is important. The hauler will be handling your wastes beyond your control, while you are still responsible for their proper management.

5. Typical waste streams generated by selected small quantity generators.

6. Good hazardous waste management can be thought of simply as using good housekeeping practices such as using and reusing materials as much as possible recycling or reclaiming waste, treating waste to reduce its hazards, or reducing the amount of waste you generate.

REFERENCES

1. *Understanding the Small Quantity Generator Hazardous Waste Rules*, EPA/530–SW–86–019, September 1986.
2. The Solid Waste Disposal Act as amended by the Hazardous and Solid Waste Amendments of 1984 (Public Law 98-616). Section 1004—Definitions.
3. *Hazardous Waste Requirements for Small Quality Generators of 100 to 1,000 kg/mo.*, EPA/530–SW–86–003.
4. SQG Fact Sheet for Vehicle Maintenance, U.S. EPA V-Series.
5. SQG Fact Sheet for Drycleaning and Laundry Plants, U.S. EPA D-Series.

22

HOUSEHOLD HAZARDOUS WASTE

22.1 INTRODUCTION

Common household products which are no longer used and have components which are classified as "hazardous" constitute a new class of waste called "Household Hazardous Waste." A waste is generally considered hazardous because of its toxicity, corrosivity, ignitability, or reactivity. The term "hazardous waste" is defined by the Resource Conservation and Recovery Act (RCRA)of 1976 to mean a solid waste, or combination of solid wastes, which because of its quantity, concentration, or physical, chemical, or infectious characteristics may (a) cause, or significantly contribute to, an increase in mortality or an increase in serious irreversible or incapacitating reversible illness; or (b) pose a substantial present or potential hazard to human health or the environment when improperly treated, stored, transported, disposed of, or otherwise managed (1).

Many people use and store a variety of different hazardous products in their home, such as furniture polish, wood preservatives, stain removers, paint thinner, drain openers, oven cleaners, and lawn and garden products such as herbicides, pesticides and fungicides, to name a few. They can be found under the kitchen sink, in the bathroom, in the basement or tucked away in the garage. These household hazardous materials may threaten human health and the environment when improperly disposed of. All household products should be used properly and disposed of in a safe manner.

Almost every one of the approximately 82 million households in this nation produces at least some household hazardous wastes. The average individual alone produces approximately one ton of waste a year in the home, some of which is hazardous. The residential waste stream includes everything that is put out in the trash can, wastes that are accumulated and stored in garages or basements, and wastes that are poured down the drain or dumped on the ground.

Because waste generation data are scarce, no one really knows how much of this waste is hazardous. Studies have not focused on the total amount of waste stored in the home, disposed of illegally or put out for collection. Since these hazardous wastes from homes and small commercial operations contribute to hazardous wastes entering sanitary landfills, the Environmental Protection Agency (EPA) is concerned about these wastes making their way into soil, water, or air once they are disposed of. Currently, household hazardous wastes are exempt from the EPA's hazardous waste regulations under the RCRA. These wastes are not exempt, however, from the provisions of the EPA's Superfund program under the Comprehensive Environmental Response, Compensation, and Liability Act (CERCLA).

The overall impact of household hazardous waste disposal is not fully known, but potential concerns include: health problems for homeowners, children, and pets from improper storage and disposal by the homeowner; injuries to refuse collection personnel while waste is being emptied, compacted, or transported; spills and fire hazards at collection and disposal sites; and pollution of air, ground water, and surface water resulting from improper disposal and contamination of septic tanks and wastewater treatment systems from disposal of hazardous wastes down drains (2). This chapter will look at various types of household hazardous wastes and discuss how to properly dispose of them.

22.2 SOLVENTS

An amazing variety of household products contain solvents made from petroleum. A solvent is a substance that can dissolve another substance. Water is the most common solvent; however, many solvents have an organic rather than a water base. Organic solvents are used to clean brushes and rollers, dissolve or thin paints, dilute varnishes and other oil-based products, and to clean up after painting or varnishing. They can cause environmental and health impacts if used improperly. Table 22.1 presents a list of various household solvents and products that contain solvents.

Organic solvents are either chlorinated or nonchlorinated. A chlorinated solvent contains chlorine in its chemical structure. A product is chlorinated if "chloro," 'chlor," or "chloride" is contained in the chemical name of any of the ingredients listed on the label (for example, trichloroethylene). Nonchlorinated solvents are usually labeled "flammable" or "combustible."

Organic solvents are toxic to aquatic life, even in low concentrations. Airborne solvents can heighten smog conditions. Solvents with chlorinated hydrocarbons

Table 22.1 Household solvents and products that contain solvents

Solvents	Products that contain solvents
Paint thinners	Oil-based paints
Furniture strippers	Spot removers
Degreasers	Rug cleaners
Nail polish removers	Furniture oils
Dry-cleaning fluids	Glues, adhesives
Paint brush cleaners	Glues, adhesives

resist ordinary breakdown to less harmful components and are very persistent in the environment.

Always use care when handling solvents. Wear heavy rubber gloves, clothing that covers exposed skin, and eye protection. Do not inhale solvent fumes. After handling solvents, wash your hands and exposed skin before eating or smoking. Keep your work area well-ventilated. Use a fan to draw fumes out of an open window. Don't wash your hands with solvents. Use borax or a hand cleaner marketed for mechanics (3).

Solvents may cause serious health effects if they come into contact with the skin or eyes or are inhaled. Excessive solvent exposure can cause a wide range of symptoms, many quite serious. The most damaging are the halogenated solvents, which are often found in paint strippers, spot removers, and degreasers.

Most solvents are recyclable, although this is not always practical to do at home. Always try to use up the product in its intended manner. Paint thinners that have paint mixed into them can be reused by capping the container tightly and allowing the paint to settle to the bottom of the container (this process may take several months for large volumes). The clean solvent may then be poured off the top and reused, and the sludge that is left can be allowed to dry out (preferably outdoors) and should then be discarded. Paint thinners can also be used up by mixing them into oil-based paints, or they can be reused after filtering them through a coffee filter.

Solvents which contain chlor-, chloro-, or a similar phrase in their chemical name are chlorinated, which is one type of halogenated solvent. Other types may include such phrases as fluoro-, bromo-, or variations of these. These halogenated solvents should be handled carefully.

Solvents should be stored until a household hazardous waste collection program is held in your area. If the solvents must be disposed of immediately, then very small amounts (less than one cup) of nonhalogenated solvents can be evaporated by mixing the solvent with an absorbent and leaving the solvent mixture outdoors. When the absorbent is fully dried, it should be wrapped in a plastic bag and placed with the other trash. This should be done carefully so that children or animals cannot come into contact with the chemical. Always ensure proper ventilation when evaporating solvents (4).

For more than one quart of a chlorinated solvent, store the substance for future use, or until better disposal methods are available, in a box lined with two plastic bags. Keep away from heat sources and out of reach of children and pets.

Special nonchlorinated solvents [i.e., methanol (wood alcohol), isopropronal (rubbing alcohol), ethanol (grain alcohol), acetone (nail polish remover), and methyl ethyl ketone (MEK)], in quantities of less than one quart, may be diluted and washed down the drain or flushed down the toilet if your house is connected to a sewer system. If you have a septic system, do NOT pour the solvents down the drain or flush them. Take the solvents to a friend's house that is connected to a sewer system. Be careful not to breathe fumes as you pour toxics down the drain. If you have larger quantities (more than one quart) of these solvents, there are no good disposal options. They should be used for their intended purpose or stored safety until better disposal methods are available.

Other nonchlorinated solvents, such as paint thinners, mineral spirits, and lacquer thinner, should be handled in the following manner:

- For less than one gallon, evaporate the solvent after pouring it over an absorbent, such as kitty litter, in a box lined with a plastic bag. The box should be kept out of reach of children and pets. When the solvent has evaporated, the box and its contents can be thrown away with your regular garbage.

• For more than one gallon of solvent, there are no good disposal options other than using the substance for its intended use. If you can't use it, give the product to a friend or neighbor (3).

22.3 PAINT

Most paints in use today are either latex or oil-based. While latex paints generally pose no serious environmental concern when properly disposed of, oil-based paints, including enamel, varnish, and lacquer, contain solvents which can damage groundwater supplies unless precautions are taken. Also, some older paints may contain high levels of lead and can cause serious health problems if ingested directly or through contaminating of drinking water supplies.

Paint purchased before 1978 may contain lead as an additive. Lead was banned from paints and similar surface coating materials for use on walls, children's toys, and furniture due to its health and environmental effects. Paints sold today have health effects similar to solvents and may cause similar damage in the environment if improperly handled and disposed of.

When using paint or similar products, such as shellac or varnish, keep your work area well-ventilated. Use a fan to draw fumes out an open window. Use borax or strong soap instead of solvents to wash paint off your skin. If you are painting out of doors, use drop cloths to protect shrubs, plants, and the lawn.

Buy only the amount of paint you need and use up existing paint before buying more. If paint is stored lid side down (close lid tightly first!), the paint will form a seal and prevent hardening or moisture damage. And, of course, store your paints in a dry area where they will not freeze and where children and pets can't get at them (3).

To avoid disposal problems, try to buy only as much paint as you need. When you have leftover paint, try to use it up on a smaller project or give it away to someone else who can use it. Community groups, theater groups, or schools may be able to put your leftover paint to good use.

Paints that are too old or in quantities too small to be reused should be disposed of properly. Latex paints and small quantities (less than half-full cans) of oil-base paint can be allowed to dry out and harden. This should be done outdoors or in an area with very good ventilation, and away from children, animals, or heat. This process may be quickened by stirring the paint frequently, pouring the paint in layers into a cardboard box, or "painting" old pieces of wood or other waste materials. Hardened paint can be disposed with other household trash. The paint may also be solidified by pouring it over clay cat litter and letting it dry out.

Half-full cans or more of oil-based paint or any amounts of paint containing lead should be saved for a household hazardous waste collection program. Different colors of paint may be combined for easier storage, provided that the paints are of the same type (latex or oil-based). Make sure that the label clearly states the type of paint inside.

Artist paints often contain much higher levels of toxic metals than other paints and should not be handled as ordinary paint. These paints should be saved for a

Table 22.2 Commonly used pesticides

Pesticides	Possible Toxic Ingredients
Weed killers	2, 4-D
Slug bait	Metaldehyde, dieldrin, sevin
Flea powder	Carbaryl, dichlorophene
Wood preservatives	Pentachlorophenol, malathion
Disinfectants	Methyl salicylate
Insect repellents	Kerosene, petroleum distillates, DDT, chlordane
Mothballs	Methylene chloride, Mapthalene, DDT
Pest strips	Vapona

collection program. Aerosol paints must also receive special handling, and should also be saved for a collection program (4).

22.4 PESTICIDES

Pesticides are chemicals that are intended to kill unwanted insects, animals, plants, or microorganisms. These products may also be toxic to humans or pets. Many pesticides are not biodegradable; they accumulate in the environment and could eventually contaminate groundwater and food supplies. Pesticides include not only commonly recognized insecticides and herbicides, but also products such as wood preservatives, flea products, and some insect repellents. Table 22.2 provides a list of commonly used pesticides.

Banned pesticides may not be used for any purpose (except in extreme state-declared emergencies). When a pesticide is banned, the manufacturer often accepts unused portions from consumers for proper disposal. Table 22.3 contains a list of banned and restricted-use pesticides.

Pesticides rarely kill or repel strictly one particular organism or plant. They are also poisonous to humans, pets, wildlife, and livestock. Rain and wind carry pesticides across the treated area and into storm sewers or directly into water

Table 22.3 Banned and restricted-use pesticides

Examples of Banned Pesticides	Examples of Restricted-Use Pesticides
DDT	Aldicarb
Aldrin	Lindane
Dieldrin	Vapona[38]
Chlordane	Warfarin
Mercury compounds	2, 4, 5-T
Toxaphene	Paraquat
Heptachlor	

bodies and groundwater. Many pesticides are persistent, that is, they degrade slowly and remain in the environment for long periods of time. They may then move through the food chain as larger organisms eat smaller ones and the pesticides accumulate or concentrate in the larger organisms (this is called bioaccumulation).

When handling pesticides, wear heavy rubber gloves, clothing that covers exposed skin, and eye protection. Avoid wearing contact lenses. After handling pesticides, thoroughly wash your hands and any exposed skin before eating or smoking. When spraying pesticides, treat only the affected area and use a particle mask (available at hardware stores) to avoid breathing the pesticide.

Consider alternatives to pesticides whenever possible. There are many methods of alternative pest control that avoid or reduce the use of strong, persistent chemicals. Call the local EPA office for more information.

Pesticides should never be mixed together, poured on the ground, dumped in water, poured down the drain, or otherwise disposed of anywhere but at a licensed facility.

For disposal of banned and restricted-used pesticides:

1. Call the regional Department of Environmental Conservation office for information on disposal.

2. Continue to store unlabeled or banned pesticides until better disposal methods are available. Place the product, in its original labeled container, inside a large sealed plastic container (such as an ice cream bucket), a box lined with two plastic garbage bags, or a large metal container with a lid. Clearly label the outside container with the name and quantity. Store on a high shelf, in a locked cabinet or closet, or in other locations out of the reach of children and pets. Make sure the storage area is dry.

3. Find a business with similar pesticide wastes that uses a licensed hazardous waste disposal company. The business may be willing to dispose of your wastes for a small fee.

4. Find a community household hazardous waste collection program that will accept the pesticide.

For disposal of usable pesticides:

1. Small amounts (less than one-half gallon or one pound):
   a. For solids and powders, wrap the pesticide (in the original container) securely in newspaper or other absorbent material. Place it inside two plastic bags or a large, sealed plastic container (such as an ice cream bucket). Dispose of the wrapped container with your garbage.

   b. For liquid pesticides, pack the product, in its original container, inside two plastic garbage bags or a sealed plastic container with an equal amount of kitty litter or sand; dispose of it with your garbage.

2. For larger amounts (more than one-half gallon or one pound), if there is no one who could use it properly, follow the instructions given for banned pesticides (3).

22.5 HOUSEHOLD CLEANERS AND OTHER PRODUCTS

Household cleaners and polishes are used to scour sinks, sanitize bathtubs and tiles, disinfect toilets, soften fabric, and remove dirt from clothes. These are all examples of potentially hazardous materials. Discarded pharmaceuticals and some cosmetics can also be household hazardous waste. Table 22.4 provides a list of household cleaners and various other household items that are considered hazardous wastes.

Cleaning products are more likely to enter the environment through everyday use than from improper disposal. These products can accumulate to levels toxic for fish and wildlife if disposed of improperly.

When using any of these items, always follow the directions carefully and store them properly. Avoid breathing fumes or inhaling powders. Wear rubber gloves and protective clothing as necessary (3).

It is always best to use the product up according to directions. If you can't use it, give it so someone who can. For products that must be disposed of, check the label for instructions. Never mix household cleaners. Bleach and ammonia, for example, react to form a deadly gas.

If a product is normally flushed down the drain during use, as most cleaners and detergents are, the product can usually be disposed of by pouring it down the drain slowly, with the water running. Do not dispose of highly toxic or corrosive materials this way. These materials should be saved for a household hazardous waste collection day. Metal polishes, wood polishes and waxes, and other solvent-based cleaners should be used up or safety-stored for a household hazardous waste collection program. Mothballs are flammable and toxic. Unusable mothballs should be safely stored until a household hazardous waste program is held in your area.

Avoid the use of septic tank cleaners or drain openers containing tetrachloroethylene, 1,1,1-trichloroethane, or dichlorobenzene. Any products that contain any halogenated hydrocarbon, aromatic hydrocarbon, or halogenated phenol in an amount greater than one part per hundred by weight are prohibited from sale and/or use in certain areas due to their potential to contaminate groundwater. If you have any of these products, do not use them up. They should be saved for a household hazardous waste collection program.

Most medications (except chemotherapy drugs) may be safely poured down the drain with running water. Large amounts of antibiotics may harm septic systems. If in doubt, call your doctor or pharmacist.

Table 22.4   Examples of household cleaners and other products

Denture cleaners	Astringents
Furniture polish	Oven cleaners
Floor wax	Nail polish
Bleaches	Nail polish remover
Coffee pot cleaners	Makeup remover
Dish and laundry detergents	Antiseptics
Drain cleaners	Outdated drugs
Toilet bowl cleaners	Swimming pool products

22.6 AUTOMOTIVE PRODUCTS

Many people perform routine auto maintenance at home. The wastes this generates pose health and environmental hazards if disposed of improperly. Table 22.5 provides a list of wastes resulting from automotive repairs.

Many do-it-yourselfers dump their auto wastes on the ground, down the storm drain, or in the trash. These wastes are washed into surface and ground water by rain, and more directly, through the storm sewer. Waste oil has the greatest impact because it is insoluble — oil sticks to everything from sand and soil to bird feathers. A small amount can seriously pollute our waterways. Long-term exposure to petroleum pollution can interfere with the feeding and reproduction of aquatic creatures and birds. Edible marine organisms accumulate toxic compounds found in oil in their systems.

Both antifreeze and brake fluid appear to be relatively harmless to the environment if they are disposed of properly. However, antifreeze is poisonous. Many pets and wildlife are attracted by the sweet taste of the antifreeze and may die after drinking from puddles of antifreeze dumped on the ground.

Automotive products rarely present problems to the mechanic when used properly. When using automotive products, be sure that the work area is well-ventilated. If gas or battery acid is splashed onto your skin or in your eyes, wash the affected area continuously for several minutes. Remove affected clothing. Don't wash this clothing with other clothes. Battery acid may not create holes right away, but they may appear later in the wash. Remember to keep automotive products out of reach of children and pets. Never clean your hands with gasoline. Use strong soap, borax, or a hand cleaner advertised for mechanics (3).

Waste oil is easily recyclable. In New York State, full service auto centers are required to accept waste oil brought in to the shop. When you change your oil, collect the waste oil in a plastic milk jug or in containers specially designed for oil collection, storage, and transport. These containers are relatively inexpensive and can be purchased at any automotive supply store.

Never pour antifreeze, contaminated gas, or any other automotive wastes on the ground or in a storm sewer. Antifreeze can only be poured down the drain if your home is connected to a sewer system. If your home is not connected to a sewer system, collect the antifreeze in a waterproof container, seal it tightly, wrap it in newspaper and dispose of it in the garbage.

Wrap used oil filters, empty oil cans, and empty antifreeze containers in newspaper and deposit them in the garbage can. Service stations may be able to accept transmission fluid, brake fluid, diesel fuel, or kerosene. Do not mix these products together, or with your waste oil.

Table 22.5 Wastes resulting from automotive repairs

Automotive Wastes	Toxic Ingredients
Gasoline	Petroleum hydrocarbons, benzene, tetraethyllead
Used motor oil	Lead, polyaromatic hydrocarbons
Antifreeze	Ethylene glycol
Motor oil	Trichloroethylene
Auto battery	Sulfuric acid

Gasoline is toxic and extremely flammable, and should never be used as a cleaner. Leftover gasoline should be used as a fuel. If small amounts of impurities are present, they may be filtered out using a strainer or coffee filter. Water may be eliminated by adding dry gas. If the gasoline cannot be used, bring it to a service station or save it in a proper gasoline container for a houehold hazardous waste collection program.

22.7 GENERAL GUIDELINES FOR SAFE USAGE OF HAZARDOUS HOUSEHOLD PRODUCTS

The following is a listing of general guidelines for safe usage of hazardous household products. These should be adhered to as closely as possible.

1. Always read the label and follow directions and warnings.
2. Keep products out of reach of children and pets. Products should be kept away from table edges and should not be placed on chairs or ladders.
3. Avoid wearing contact lenses, particularly soft or extended wear lenses. Contacts absorb fumes and gases and collect chemicals splashed in the eye.
4. Don't overuse a product. Twice as much does not work twice as well. For example, using twice as much laundry detergent will not get clothes cleaner but will place an additional strain on sewage treatment plants.
5. Never mix products or different brands of the same product. Mixing products can produce explosive reactions or toxic fumes. The most common example is the combination of chlorine bleach and ammonia, which produces highly toxic fumes. Chlorine bleach also creates toxic fumes when mixed with products such as drain and toilet bowl cleaners.
6. Always wear protective equipment when this is recommended on a product label. Follow directions carefully. The use of gloves is not intended to prevent staining on your hands. Gloves offer protection against chemicals entering the bloodstream through skin absorption. While poisoning is associated more with ingestion, it often occurs through skin absorption. Eye goggles and face masks can range from molded paper filters covering the nose and mouth to masks with respirator cartridges. Protective equipment is usually inexpensive ($2–3 for goggles, $1 or less for paper masks) and may be purchased at hardware stores or specialty supply stores.
7. Ensure proper ventilation if you use hazardous materials indoors. One open window is not adequate. Use fans to blow air from your work space to an open window. Make sure the air is escaping outside and is not being recirculated indoors. Take frequent fresh air breaks.
8. If pregnant, avoid exposure to chemicals as much as possible. Many chemicals have not been tested for their effects on unborn children.
9. Keep the numbers for the local poison control center or your family physician near your phone.
10. Keep syrup of ipecac on hand to induce vomiting if necessary. Never use it without the recommendation of a poison control center or physician since vomiting is discouraged in some poison situations.

11. First aid information on product labels can be misleading or incorrect. Always call the poison control center for the most up-to-date information.
12. Clean up after using hazardous products. Carefully seal products. Clean the outside of the containers. Spills should be cleaned up immediately. Rags soaked in flammables should be disposed of in a covered metal container. Always clean paintbrushes right after use to minimize the need for solvents.
13. Buy only the amount of product needed.
14. Avoid aerosols when possible. Aerosol spray products disperse their contents beyond target areas. Small aerosol particles are easily inhaled into the lungs and quickly absorbed into the bloodstream.
15. Avoid using unnecessary hazardous products; use alternative products that are safer. Contact your EPA or state office for more information.

22.8   HOUSEHOLD HAZARDOUS WASTE TABLE*

Table 22.6 lists the major categories of household products which may result in Household Hazardous Waste. The table describes the hazardous ingredients of the products and why they are hazardous. The table also provides possible alternatives (if any), suggests how the products should be used, stored or disposed of. The following definitions and legends are to be used along with the table:

Definitions

Corrosive: A chemical or its vapors that cause destruction or irreversible alteration in body tissues at the site of contact.

Flammable: Can be ignited under almost all temperature conditions.

Irritant: Causes soreness or inflammation of the skin, eye, mucous membranes or respiratory system.

Toxic: May cause injury or death upon ingestion, absorption through skin or inhalation.

Legend

A: When allowed by the State or local wastewater treatment authority, the waste may be disposed of in a sewer. Certain bacteria in the sewage treatment plant can detoxify the chemical. Do not pour on the ground.
B: Keep in tightly closed jar and allow contaminants to settle out. Strain the supernatant through a fine mesh sieve; reuse the liquid. The concentrated contaminants should be stored and taken to a collection day.
C: Air dry latex paints and discard container in the trash if allowed by State or local environmental laws.
R: Recycle your wastes. Carry to a service station or reclamation center.
S: Do not dispose of these substances. These wastes should be safety stored until a hazardous waste program is organized in your community. Fully spent pesticide container may, however, be triple rinsed and the wastewater reused according to the instructions on the label.

*Adopted from Household Hazardous Waste Wheel, Environmental Hazardous Management Institute, Portsmouth, NH, 1987)(5).

U: Fully use these products so that no waste remains except residuals attached to the container. Containers should be rinsed with water. The container may then be disposed with the trash while rinse water may be reused or poured down the drain with great quantities of water. In rare circumstances when these products will not be completely used, the waste should be stored according to S.

22.9 WHAT CAN BE DONE ABOUT HOUSEHOLD HAZARDOUS WASTE?

Because of the very nature of Household Hazardous Waste (HHW) which is either toxic, ignitable, corrosive, reactive, or explosive, various management methods must be considered to minimize its impact on our society. Management methods recommended by the EPA include consumer education, waste exchange, reuse/ recycle, community colllection program, incineration, or disposal in a hazardous waste landfill.

Consumer Education

The purpopse of consumer education is to increase public awareness of the potential problem and to reduce the amount of HHW. The program should focus on:

- Change in buying pattern. Buy only what is needed and buy less toxic products. Use substitutes, such as chlorine bleach to remove mildew. Use of septic tank pumpage, flexible rod, baking soda/vinegar, or hot water to unclog drain. Additional suggestions are available in the literature (6).
- Safe handling and storage of hazardous household products. Keep products in their original containers, completely use all products, avoid mixing products or excessive watering to runoff, predetermine ways to deal with spills (ventilation, protective clothing, use absorbent materials) and publicize nearby poison control center or emergency number.
- Proper disposal method. Inform the public of what HHW can or cannot be drained into the sewers (including storm sewers), of the local or state regulations on the disposal of HHW, and of the potential pollution problem with dumping such wastes on the ground, or in bodies of water.

A network program can be established along with the Consumer Education program. Through cooperation with local or regional business communities, state solid waste authorities and other volunteer groups, a waste exchange, reuse/ recycing program can be established. Such programs could be successful in the recycling of used motor oil, automotive batteries, and paints, household cleaners and other usable and safety-packaged products. Two areas should be addressed in any program: quality control/liability due to misrepresentation and disposal methods for usable products.

Collection and Disposal of HHW

Many communities and States have successfully conducted programs on collection and disposal of HHW. The following suggestions may help to make such a program successful:

- Community support. Solicit support from local and state government, citizen groups, and other service organizations for collection site selection, planning,

Table 22.6 Household hazardous waste table

Product Name	Hazardous Ingredients	Hazard Properties	Alternatives	Waste Management
Household Maintenance Products				
Powder or abrasive cleaners	Trisodiumphosphate, Ammonia, ethanol	Corrosive, irritant, toxic	Rub area with 1/2 lemon dipped in borax-rinse dry	U
Ammonia-based cleaners	Ammonia, ethanol	Irritant, toxic, corrosive	Vinegar, salt & water mixture for surfaces. Baking soda & water for the bathroom	U
Mothballs	Napthalenes, paradichlorobenzene	Toxic	Cedar chips, newspaper, lavender flowers	U
Bleach cleaners	Sodium or potassium hydroxide, hydrogen peroxide, sodium or calcium hypochlorite	Corrosive	For laundry, use 1/2 cup sodium hexametophosphate per 5 gallons water	U
Floor and furniture polish	Diethylene glycol, petroleum distillates, nitrobenzene	Flammable toxic	1 part lemon juice, 2 parts olive or or vegetable oil	S
Rug and upholstery cleaners	Napthalene, perchloroethylene, oxalic acid, diethylene glycol	Irritant, corrosive, toxic	Dry cornstarch sprinkled on rug; vacuum	S
Drain cleaner	Sodium or potassium hydroxide, sodium hypochlorite, hydrochloric acid, petroleum distillates	Corrosive, toxic	Plunger; flush with boiling water, 1/4 cup baking soda; and 2 oz. vinegar	S
Disinfectants	Diethylene or methylene glycol, sodium hypochlorite, phenols	Corrosive, toxic	1/2 cup borax in 1 gallon water	U
Oven cleaner	Potassium hydroxide, sodium hydroxide, ammonia	Corrosive, toxic	Baking soda and water	U
Toilet	Muriatic (hydrochloric) or oxalic acid, paradichlorobenzene calcium hypochlorite	Irritant, corrosive, toxic	Toilet brush and baking soda; mild detergent	S
Silver polish	Acidified thiorea, sulfuric acid	Corrosive, toxic	Place silver in boiling water; add baking soda, salt, and a piece aluminum foil	S

Automotive Products

Product	Components	Hazard	Alternative	Code
Batteries	Sulfuric acid, lead	Corrosive, toxic	Unknown	R
Used oil	Hydrocarbons (e.g., benzene), heavy metals	Flammable, toxic	Unknown	S
Brake fluid	Glycol ethers, heavy metals	Flammable, toxic	Unknown	S
Transmission fluid	Hydrocarbons, mineral oils	Flammable, toxic	Unknown	R
Antifreeze	Ethylene glycol	Toxic	Unknown	A

Paint Products

Product	Components	Hazard	Alternative	Code
Stain/Finish	Mineral spirits, glycol ethers, ketones, halogenated hydrocarbons, naptha	Flammable, toxic	Latex paint or natural earth pigment finishes	S
Wood preservative	Chlorinated phenols, copper/zinc napthenate, creosote, magnesium flurosilicate	Flammable, toxic	Unknown	S
Furniture strippers	Acetone, methyl ethyl ketone, alcohols, xylene, toluene, methylene chloride	Flammable, toxic	Sandpaper	S
Thinners, turpentine	n-Butyl alcohol, acetone, methyl isobutyl ketone, petroleum distillates	Flammable, toxic	Use water-based paints	B
Rust paint	Methylene chloride, petroleum distillates toluene	Flammable, toxic	Unknown	S
Latex or water-based paints	Resins, glycol ethers, esters, pigments, phenyl mercuric acetate	Flammable	Limestone-based whitewash/cassein-based paint	C
Enamel or oil-based paints	Pigments, ethylene, aliphatic hydrocarbons mineral spirits	Flammable, toxic	Latex or water-based paint	S

Lawn and Garden Products

Product	Components	Hazard	Alternative	Code
Herbicides	Triethylamine salt, prometon	Toxic	Strong hosing or hand weeding; keep grass short	S
Organophosphates	Parathion, malathion, diazinon lindane, dichlorvos, chloropyrifos	Toxic	Remove plant debris or wood from garden, insecticidal soap	S

365

Table 22.6 (Continued)

Product Name	Hazardous Ingredients	Hazard Properties	Alternatives	Waste Management
		Lawn and Garden Products		
Chlorinated Hydrocarbons	DDT, aldrin, endrin kepone, dieldrin, heptachlor, chlordane, dicofll, lindane	Toxic	Keep garden clean; import predators, e.g., ladybugs ground beetles, preying mantis; insecticidal soap	S S
Botanicals	Pyrethrins, rotenone, nicotine	Toxic	Insecticidal soap: import predators	S
Carbamates	Carbaryl (Sevin), Temik, carbofuran, propoxur, (Baygon), aldicarb	Toxic	Keep garden weed free, import predators, insecticidal soap	S
Arsenicals	Lead arsenate, calcium arsenate, Paris Green	Toxic	Live traps, remove food supply	S
House plant insecticide	Methoprene, malathion tetramethrin, trichloroethane	Toxic	Mixture of bar soap and water or old dishwater spray on leaves then rinse	S
Fungicides	Captan, folpet, malathion, anilazine	Toxic	Do not overwater, keep areas clean and dry	S
		Miscellaneous Products		
Pool chemicals	Muriatic acid, sodium hypochlorite, algicide	Corrosive, toxic	Unknown	S
Photographic chemicals	Silver, acetic acid, hydroquinone, sodium sulfite, ferrocyanide	Corrosive, toxic, irritant	Unknown	S
Rat and mouse poisons	Lead arsenate, coumarins/warfarin strychnine	Toxic	Live traps, remove food sources	S
Roach and ant killer	Organophosphates, carbamates	Toxic	Roaches: traps or baking soda and powder sugar mixture; Ants: red chili pepper to discourage entry sprinkle boric acid powder on ant trails	S
Flea collars and sprays	Carbamates, pyrethrins, organophosphates	Toxic	Pennyroyal ointment, herbal collar, brewers yeast diet	

arranging for chemists "on loan," obtaining equipment (traffic cones, signs, packaging materials, etc.), and resolving liability issues.

- Allow time for planning and public relations. The goal is to maximize participation and lower the cost per unit of waste disposal; ask for public service announcement spots from TV and radio stations; inform public of what HHW is not acceptable, and the form of packaging required for acceptable waste.

- Fund raising. Look for grants from the state, donations from the business community, or impose a user's fee.

- If possible, ask state or local government to help select/negotiate a contract with the collection and disposal company. There may be potential savings from their participation, especially if such government is also acting on behalf of other communities. Develop a plan to segregate, bulk, recycle, and dispose of unusable HHW.

- Institutionalize the program—do not allow such a program to become a one-time event.

- Get help and advice from others, e.g., the League of Women Voters of Massachusetts at 8 Winter St., Boston, MA 02108, Tel: (617) 357-8380; the Golden Empire Health Planning Center at 2100 21st Street, Sacramento, CA 95818, Tel: (916) 731-5050.

Treatment in a Hazardous Waste Incinerator or Disposal in a Hazadous Waste Landfill

There are several pros and cons of the above treatment/disposal methods.

- When properly operated, an incincerator can destroy combustibles and organic chemicals at a high efficiency. This method is especially cost-effective for segregated household waste. Incinceration is the most effective treatment method to reduce long-term risk because HHW can be removed from the environment, not just relocated.

- Increasing costs are associated with incineration because of the public concerns over siting of incinerators and the cost of transportation to the limited number of incinerators available. The accessibility of state-permitted hazardous waste landfill may increase the costs of disposal; and the land disposal method simply contains HHW and does not treat the waste. Keep in mind that segregation of HHW and the use of substitutes reduce the volume of waste and are the keys to a cost-efective disposal program.

22.10 SUMMARY

1. Common household products which are no longer used and have components which are classified as "hazardous" constitute a new class of waste called "household hazardous waste."

2. An amazing variety of household products contain solvents made from petroleum.

3. Most paints in use today are either latex or oil-based. Whareas latex paints generally pose no serious environmental concern when properly disposed of,

oil-based paints, including enamel, varnish, and lacquer, contain solvents which can damage groundwater supplies unless precautions are taken.

4. Many pesticides are not biodegradable; they accumulate in the environment and could eventually contaminate groundwater and food supplies.

5. Household cleaners and polishes scour sinks, sanitize bathtubs and tiles, disinfect toilets, soften fabric, and remove dirt from clothes. Discarded pharmaceuticals and some cosmetics can also be household hazardous waste.

6. Many people perform routine auto maintenance at home. The wastes this generates pose health and environmental hazards if disposed of improperly.

7. General guidelines which can provide safe usage of hazardous household products.

8. The household hazardous waste table presented describes typical HHW, their hazardous constituents, and possible alternatives.

9. Because of the very nature of HHW which are either toxic, ignitable, corrosive, reactive, or explosive, various management methods must be considered to minimize its impact on our society.

REFERENCES

1. The Solid Waste Disposal Act as amended by the Hazardous and Solid Waste Amendments of 1984 (Public Law 98-616). Section 1004 — Definitions.
2. U.S. EPA, "Managing Household Hazardous Waste," *EPA Journal*, 13(13), 11–12, April 1987.
3. New York State Legislative Commission on Solid Waste Management, *Household Hazardous Waste: An Overview*, October 1988. A background report from the office of the Vice Chairman.
4. New York State Department of Environmental Conservation. *Household Hazardous Waste Fact Sheets*, October 1988.
5. U.S. EPA, *Helpful Hints on Household Hazardous Waste*, June 29, 1988.
6. M. K. Theodore and L. Theodore, Major Environmental Issues Facing the 21st Century, Prentice-Hall, Upper Saddle River, NJ, 1996.

23

SUPERFUND

23.1 INTRODUCTION

In 1986, Congress enacted sweeping amendments to the Comprehensive Environmental Response, Compensation, and Liability Act (CERCLA), the nation's law which implemented Superfund to clean up abandoned hazardous waste sites. Two years later, the U.S. Environmental Protection Agency (EPA) set a course for the Superfund program designed to improve the program's performance and to increase the role of the private sector is paying for the cleanup. As a result of these actions, Superfund has dramatically increased its record of success. Cleanup has been initiated at nearly 65% the sites that are a national priority. Thousands of emergency actions have been taken around the country to make sites safe. The federal effort has been augmented by well over $2 billion in cleanup efforts by responsible parties. Responsible parties are now actively engaged at 60% of the national priority sites. After two decades of work, the program can report substantial environmental progress in cleaning up sites. The following six goals provide the direction for the Superfund program:

1. Make sites safer: Control the imminent threats immediately and address the worst problems at sites first.
2. Make sites cleaner: Accelerate and improve long-term cleanup action at sites.
3. Strengthen enforcement and maximize responsible-party work at sites.
4. Bring innovative technologies to bear when cleaning up Superfund sites.
5. Implement an aggressive program of community involvement.
6. Communicate progress to the public.

This chapter explains the progress and the challenges facing both those who

This chapter was adapted from *Superfund: Focusing on the Nation at Large*, U.S. EPA, September, 1990.

clean up hazardous waste sites and those who live near one. It also describes those sites that are deemed to be of national priority and that have been placed on the National Priorities List (NPL). It provides information on the types of sites on the NPL, and portrays the progress of these sites as they approach construction of long-term cleanup remedies.

23.2 INITIAL LEGISLATION

As the 1970s came to a close, a series of headline stories gave Americans a look at the dangers of dumping industrial and urban wastes on the land. First there was New York's Love Canal. Hazardous waste buried there over a 25-year period contaminated streams and soil, and endangered the health of nearby residents. This resulted in the evacuation of several hundred people. Then the Valley of the Drums in Kentucky attracted public attention. The site of these leaking storage barrels quickly became front page news. The next national hazardous waste headline was Times Beach. Oil contaminated with toxic dioxin tainted the land and water in this eastern Missouri community.

In all these cases, public health and the environment were threatened, lives were disrupted, and property values were depreciated. It was becoming increasingly clear that there were large numbers of serious hazardous waste problems that were falling through the cracks of existing environmental laws. The magnitude of this problem moved Congress to enact CERCLA in 1980. CERCLA was the first federal law dealing with the dangers posed by the nation's abandoned and uncontrolled hazardous waste sites.

The law authorized the federal government to respond directly to releases, or threatened releases, of hazardous substances that may endanger public health and welfare, or the environment. Legal actions could be taken to force parties responsible for causing the contamination to clean up those sites or reimburse the Superfund for the costs of cleanup. If those responsible for site contamination could not be found or were unwilling or enable to clean up a site, the EPA could use monies from the Superfund to clean up a site. The Superfund is actually the trust fund that finances these cleanup actions. CERCLA established a $1.6 billion fund made up of taxes on crude oil and commercial chemicals.

Thus CERCLA, commonly known as the Superfund, was launched as a direct and limited effort to clean up the nation's hazardous waste sites. Congress recognized that the EPA could not address all sites, and therefore directed it to set priorities for federal action under the Superfund. At that time, expectations were high that the $1.6 billion fund created by Congress was sufficient to clean up these priority sites.

The news stories turned out to be just the beginning. Few realized the size of the problem until the EPA began the process of site discovery and site evaluation. Not hundreds, but thousands, of potentially hazardous waste sites existed, and they presented the nation with some of the most complex pollution problems it had ever faced.

Within twelve years of the start of the Superfund program, hazardous waste surfaced as a major environmental concern in every part of the United States. It wasn't just the land that was contaminated by past waste disposal practices. Chemicals in the soil were spreading into the groundwater (a source of drinking water for many) and into streams, lakes, bays, and wetlands. At some sites, toxic vapors were rising into the air. Some pollutants, such as metals and solvents, had damaged vegetation, endangered wildlife, and threatened the health of people who unknowingly worked or played in contaminated soil, drank contaminated water, or ate contaminated plants or animals.

As site discoveries grew, cost estimates rose. Clearly $1.6 billion was not enough to clean up the nation's most serious hazardous waste sites. Realizing the long-term nature of the problem and the enormous job ahead, Congress reauthorized the program in 1986 for another 5 years, adding $8.5 billion to the fund. The monies were made available to the Superfund directly from exise taxes on petroleum and feedstock chemicals, a tax on certain important chemical derivatives, an environmental tax on corporations, appropriations made by Congress from general tax revenues, and any monies recovered or collected from parties responsible for site contamination. The amended law, known as the Superfund Amendments and Reauthorization Act (SARA), was stricter and broader in scope and required that "to the maximum extent practicable" solutions make use of alternative or resource recovery technologies and be permanent. This reauthorization also allowed the EPA the long-needed opportunity to develop a comprehensive management strategy to meet the growing challenges of this technically complex program.

Between 1980 and 1998, the EPA identified 1,396 hazardous waste sites as the most serious in the nation. These sites comprise the "National Priorities List" (NPL), those sites targeted for cleanup under the Superfund. This list is continuously updated. But site discoveries continue, and the EPA estimates that, while some sites will be deleted after lengthy cleanups, this list will continue to grow by approximately 100 sites per year, reaching 2100 sites by the year 2000.

Reauthorization of the Superfund was incorporated into the 1991 Budget legislation passed by Congress and signed by the President. This provided authority to continue funding under the existing program structure through September 30, 1994. As of 1999, the program is still operating under this structure.

From the beginning of the Superfund program, Congress recognized that the federal government could not and should not be responsible for addressing all environmental problems stemming from past disposal practices. Therefore, the EPA was directed to set priorities and establish a list of sites to target. Sites on the NPL are thus a relatively small subset of a larger inventory of potentially hazardous waste sites, but they do compromise the most complex and environmentally compelling cases. The EPA has logged more than 39,783 sites on its national hazardous waste site inventory and assesses each site within 1 year of being logged. Of the assessed sites, 9,245 remain active, while 30,438 have been archived as No Further Remedial Action Planned (NFRAP) sites (2). The NFRAP sites require no further federal action because they do not pose significant human health or environmental risks. The remaining sites are undergoing further assessment to determine if long-term federal cleanup activities are appropriate. Where imminent threats to the public or the environment were evident, the EPA completed or monitored more than 1,800 immediate actions.

23.3 HAZARDOUS WASTE PROBLEMS ARE MULTIFACETED

Today our nation is paying the price for years of abuse. There is no "quick fix." Yesterday's inexpensive and supposedly efficient disposal practices have resulted in the costly and combersome cleanups of today. Improperty disposed hazardous wastes have threatened many environmental resources, and the nature of these toxic "soup" compounds the cleanup problem (see Figure 23.1). Indeed, a national hazardous waste program will probably be necessary for many years.

In 1990, the EPA estimated that the Superfund will spend more than $27 billion on cleanup construction at sites currently on the NPL. Parties responsible for contamination were expected to conduct 65% of the cleanup work, which accounts for billions more in cleanup dollars. It was expected to take about 7 years before all sites currently on the NPL started engineered cleanup activities, and the EPA expected to add sites at the rate of about 100 each year. The average cost of cleanup was $26 million per site, and there was every reason to believe that these costs would climb as some of the more complex sites move into the cleanup phase of the process.

It is virtually impossible to describe the typical hazardous waste site; they are extremely diverse. Many are municipal or industrial landfills. Others are manufacturing plants where operations improperly disposed of wastes. Some are large federal facilities dotted with "hot spots" of contamination from various high-tech or military activities. The chief contributors of these wastes are in our manufacturing sector (see Figure 23.2).

While many hazardous waste sites have been abandoned, a site may still be an active operation, or it may be fully or partially closed down. Sites vary dramatically in size, from a 0.25-acre metal plating shop to a 250-square-mile mining area. The types of wastes they contain vary widely, too. Some of the chief constituents of wastes present in solid, liquid, and sludge forms include heavy metals (a common by-product of many electroplating operations) and solvents or decreasing agents. These are discussed further in the next section.

NPL sites are found in all types of settings. Slightly more are found in rural/suburban areas than in the urban areas, but very few are truly remote from either homes or farms (see Figure 23.3).

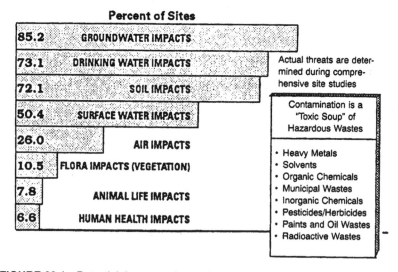

FIGURE 23.1 Potential threats to the environment that led to listing on the NPL.

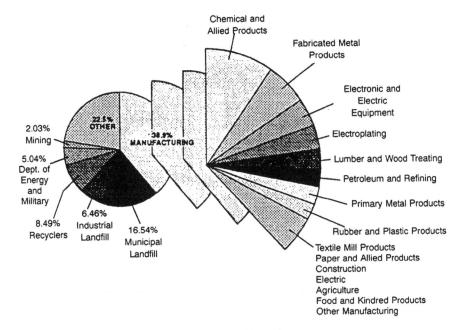

FIGURE 23.2 Wastes at NPL sites — sources.

Yet the idea of a "site", some kind of disposal area or dump, still doesn't portray the entire picture. Transportation spills and other industrial process or storage accidents account for some hazardous waste releases. The result can be fires, explosions, toxic vapors, and contamination of groundwater used for drinking.

Every NPL site is unique, and cleanups must be tailored to the specific needs of each site and the types of wastes that contaminate it. The range of possibilities is

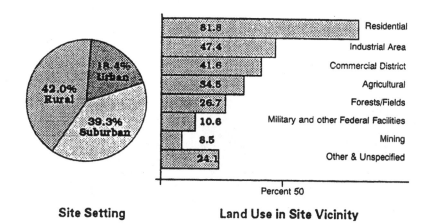

Site Setting   **Land Use in Site Vicinity**

FIGURE 23.3 NPL site locations.

enormous. First, the site's physical characteristics (its hydrology, geology, topography, and climate) determine how contaminants will affect the environment. Then, there is the variation in site type: landfill, manufacturing plant, military base, metal mine, among others. The type of wastes present adds another complex dimension. Information on the health and environmental effects of hazardous wastes comes mainly from laboratory studies of pure chemicals. There still is much to learn about the nature of the complex mixtures of wastes generally found at these sites, how they affect the environment, and how best to control them.

No matter how exhaustive preliminary studies may be, sampling and site observation simply cannot reveal the full extent of the problem at many sites. Uncertainties exist right up until the point where ground is broken for the cleanup work and throughout the final cleanup process. Thus, there is no ready answer to the question, "How long will it take?" On average, and this includes a broad range, 6–8 years will elapse between the start of the cleanup study and remedy completion.

While technological concepts were not fully field-tested in the early 1980s, hazardous waste cleanup efforts have begun to yield the information needed to design permanent site cleanup solutions. Since 1986, the move has been away from "containment" of hazardous wastes. Containment entails segregating the wastes in a particular place, but unfortunately many materials cannot be reliably controlled this way. This is particularly true of liquids, highly mobile substances (like solvents), and high concentrations of toxic compounds. For these wastes, *treatment* is the preferred approach. It reduces the toxicity, mobility, and volume of wastes at Superfund sites.

There has been a progressive increase since 1986 in the frequency with which treatment (rather than containment) has been selected as a remedy for controlling the primary source of contamination at hazardous waste sites. In 1987, some type of waste treatment was being used in about 50% of cleanup remedies that the EPA selected. By 1989, that number had risen to more than 70% (see Figure 23.4). From 1980 to 1996, Superfund cleanup experts applied 1,371 containment technologies and 1,471 treatment technologies at 897 NPL sites. Since the inception of the Superfund program, cleanup crews have handled billions of gallons of groundwater, liquid waste, and surface water. In addition, workers have dealth with millions of cubic yards of soil, solid waste, and sediment (1).

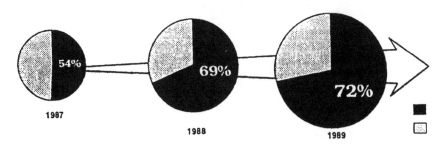

FIGURE 23.4 EPA increased technologies.

23.4 SITE CONTAMINANTS POSE DIVERSE THREATS

Hazardous waste can include products and residues from a variety of industrial, agricultural, and military activities. Some of the hazard lies in the waste itself, its concentration and quantity, and physical or chemical nature. But much of the danger arises from improper handling, storage, and disposal practices. The result is that humans and/or the environment are exposed to contamination.

Wastes were poorly managed in the past because the disposers often failed to understand their toxic effect and realize how strictly they had to be contained. Dangerous chemicals have often migrated from uncontrolled sites. They may percolate from holding ponds and pits into underlying groundwater They may be washed over the ground into lakes, streams, and wetlands. They may evaporate, explode, or blow into the air, spreading hazardous chemicals. They may soak into soil, making land and groundwater unfit for habitat or agriculture. Some hazardous chemicals build up, or bioaccumulate, in plants, animals, and people when they consume contaminated food and water.

Today's EPA-approved hazardous waste disposal facilities and practices require specific safeguards to keep pollutants from entering the environment. But the knowledge of taking preventive precautions was gained at considerable expense, and not before uncontrolled hazardous waste sites had contaminated the environment and threatened human health.

Determining the risks of hazardous waste to human and environmental health is a complex undertaking. The EPA conducts risk assessments at each site, analyzing the possible ways people, animals, and plants could come into contact with contaminants. Risk assessors are concerned about the effects of contact today and potential contact in the future. How long were populations exposed? How serious will the consequences be? Has the nature of waste changed over time? Where various wastes are present, what is their combined effect? Scientists often find the net risk in these situations difficult to quantify.

Risk hinges upon how dangerous the chemical is, how people may come into contact with it, how frequently, and in combination with what other chemicals. The EPA has worked hard to determine the amounts and types of chemicals that can safely exist in water, air, and soil. The Agency for Toxic Substances and Disease Registry also conducts its own independent assessments of the health effects of contamination from Superfund sites. The more sites that are analyzed, however, the longer becomes the list of potentially threatening substances and mixtures. Scientists are working to determine the risks associated with these newfound problems. They are wrestling with the problems posed by the toxic chemical "soups" that have been present in some holding ponds for years. (See Chapters 40 through 43 for more on Risk Assessment.)

Like the sites themselves, possible effects on human and environmental health span a broad spectrum. Adverse effects on people can range from minor physical irritation to serious health disorders. They also can take the form of slowly degenerating health or of sudden serious damage to health. Plants and animals may become contaminated and enter the food chain. A sudden poisoning event like a hazardous waste spill or the breaching of a hazardous waste impoundment can pose serious health risks.

Health and environmental risk is complicated by the fact that if nothing is done, people and ecosystems can suffer a gradual deterioration for years and can show adverse health effects long after the fact. In addition, there is the issue of sensitivity. Certain populations are more sensitive, namely, elderly people, children, and endangered or threatened plants and animals. Some environments are more sensitive in the way they respond to the effects of hazardous chemicals: wetlands, coastal areas, estuaries, and many other water bodies, for example; wildlife refuges; or rare pine or shale barrens. These are fragile and valuable assets that must be protected.

Table 23.1 provides a brief description of specific contaminants that are frequently found in Superfund sites across the nation, along with their effects on human health. This sampling of contaminant groups serves as an illustration of the potential dangers arising from uncontrolled or abandoned hazardous waste sites. It also highlights the dramatic need for the EPA to intercede to protect affected residents and impacted environments.

The goal of EPA's Superfund program is to tackle the immediate dangers first, and then move through the progressive steps necessary to eliminate any long-term risks to public health and the affected environments. But in addition to the Superfund, other major laws help the EPA to control toxic substances in order to avert future hazardous waste sites. Each focuses legislative pressure on reducing contamination at the source, *before* human health and the environment are threatened. These programs were designed to keep toxic substances out of the environment, by either controlling or eliminating them.

- The *Toxic Substances Control Act* strictly regulates the production of substances that pose an unreasonable risk to human health or the environment.

- The *Resource Conservation and Recovery Act* allows the states and the EPA to track hazardous wastes from their production through final disposal to ensure that toxic chemicals and wastes are handled safely and disposed of properly.

- The *Safe Drinking Water Act* allows the EPA to establish maximum safe levels of contaminants in drinking water to protect the public health.

- The *Clean Water Act* controls all forms of water pollution by limiting the concentrations of pollutants discharged or dumped into national waterways. Major oil spills, such as the Exxon Valdez incident, are addressed under this law.

- The *Federal Insecticide, Fungicide, and Rodenticide Act* strictly regulates the manufacture, sale, and uses of pesticides and requires that all pesticide products sold or distributed in the United States be registered with the EPA.

23.5   INDIVIDUAL ROLES IN SUPERFUND SITE CLEANUP

The EPA Manages the Superfund Program

The EPA's Superfund program is responsible for:

- Enforcing Superfund laws and overseeing Superfund cleanup activities.
- Studying sites and evaluating the contamination and its risk to health and the environment.

Table 23.1 Common toxic chemicals at NPL sites

Chemical Contaminants	Sources of Contamination	Environments Affected	Potential Health Threats
Heavy metals	Common by-products of electroplating; batteries and paint pigments; photography; smelting. Mercury is used in thermometers, fluorescent lights, and other products.	Groundwater, surface water, soils, air, animal tissue	Cadmium: Tumors, liver, and kidney damage. Chromium: Hemorrhages and respiratory cancer Mercury: Kidney, brain, and neurological damage. May enter food chain via bioaccumulation. Lead: Brain, bone, and neurological damage. Prolonged exposure may lead to learning disabilities in children.
Volatile organic compounds (VOCs)	Solvents and degreasing agents, gasoline octane enhancers, oils, paints, varnishes, dry-cleaning compounds, chemical manufacture.	Groundwater, soils, air	Cancers, impairment of nervous system resulting in sleepiness, headaches, and possible kidney or liver damage. Chronic exposure to benzene can cause leukemia.
Pesticides and herbicides	Commercial pesticide, and herbicide production; agricultural and industrial applications; defoliants.	Groundwater, surface water, air, animal tissue	Hazardous compounds can accumulate in the food chain or result in diverse health effects ranging from nausea to nervous disorders. Dioxin, a common by-product of pesticide manufacture, is a suspected carcinogen and known to be among the most toxic substances.
Polychlorinated biphenyls (PCBs)	Electric transformers, used in insulators and coolants; adhesives; caulking compounds; other products.	Groundwater, sediments, soils	Stored in the fatty tissues of humans and animals through bioaccumulation. May cause liver damage or cancer.
Creosotes	Wood-preserving operations, combustion by-products.	Sediments, soils, surface water	Polyaromatic hydrocarbons (PAHs) and polynuclear aromatics (PNAs) may cause skin ulcerations and cancers with prolonged exposure.

Sources: Refs. 2 and 3.

- Identifying and responding to hazardous waste emergencies.
- Searching for those who created or contributed to site hazards.
- Negotiating cleanup offers or settlements with cooperating parties or suing uncooperative ones.
- Selecting the best cleanup remedy for each site.
- Monitoring cleanup at all NPL sites, regardless of who does the work.
- Keeping the public informed about progress at each site.
- Helping develop new cleanup technologies and expertise.
- Coordinating cleanup and enforcement activities with the U.S. Army Corps of Engineers and the U.S. Department of Justice.

Over the past decade, the EPA has fostered a group of hazardous waste specialists who can both manage and advise on approaches to site cleanup: remedial program managers (RPMs) and on-scene coordinators (OCSs). The typical RPM oversees long-term site cleanup, and an OSC manages immediate cleanup actions. These technical managers continue to expand their expertise and experience with hazardous waste cleanups.

RPMs and OSCs deal with numerous complexities. They must comply with a variety of federal, state, and local laws and regulations. They must coordinate the activities and interactions of state and local offices, contractors, technical specialists, landowners, and often the private individuals or companies potentially responsible for site contamination. And, as central players in the decisions regarding the cleanup approach, they must balance the technical feasibility of the cleanup strategy with community concern and fiscal realities.

The EPA has also fostered the growth of expertise in the private sector. National environmental engineering firms that perform the design and construction of hazardous waste remedies across the country have gained considerable knowledge about site conditions, contaminants, and technological approaches that work.

States Play an Important Role

The Superfund law authorizes the EPA to transfer funds and management responsibility to states to lead cleanup activities at NPL sites. Over the years since the inception of Superfund, a strengthened EPA/state partnership in the program has developed. States are currently involved with cleanup activities at over a quarter of all Superfund sites. When states take the lead for cleanup activities at a site, their responsibilities usually closely parallel the EPA's.

Citizens Help Shape Decisions

Superfund activities also depend upon local citizen participation. The EPA's job is to analyze the hazards and deploy the experts, but the EPA needs citizen input as it makes choices for affected communities.

Because the people in a community with a Superfund site will be those most directly affected by hazardous waste problems and processes, the EPA encourages citizens to get involved in cleanup decisions. Here are some things citizens can do:

- *Report hazardous waste dumping*, no matter how long ago it occurred. Call the National Response Center toll free at (800) 424-8802.

- *Find out when cleanup investigators will arrive* and share information with them. Citizen's insights have identified polluters, helped the cleanup team decide where to dig and test, and raised specific community concerns that have been factored into cleanup decisions.
- *Get information from the EPA or the state Superfund office.* These offices are responsible for providing information to citizens.
- *Learn about the EPA's Community Involvement Programs.* The EPA keeps citizens informed about site conditions and progress via news releases, free fact sheets, and presentations on environmental and health issues to schools, community groups, and business organizations. Files that contain accurate, up-to-date information on site conditions are usually kept at a school, a library, or the town hall.
- *Engage Experts.* The EPA's Technical Assistance Grants provide up to $50,000 to a community group wishing to hire specialists who can interpret sampling results, technical reports, and other documents. (Call or write the nearest EPA office for specific information.)
- *Write the EPA for information on the status of any site.* Every site or spill ever reported is in the EPA's computer, including the many thousands that turned out not to be hazardous. Citizens can get all the details except for those relating to possible legal actions against owners or possible polluters.

Public comment and involvement have influenced the EPA's plans for cleanups in a number of cases, and citizens have provided the EPA with valuable information about conditions at a site. For example:

- At a site in Illinois, local citizens and businesses expressed concern that the EPA's proposed cleanup alternative would limit the use of a nearby lakeshore and harm the town's economy. In response to these concerns, the EPA developed another cleanup option that preserved the town's use of the lakeshore.
- At a site in Minnesota, local residents expressed a strong preference for treatment of local contaminated wells over connection to the reservoir supply of a nearby city. After careful consideration of information provided by the residents, the EPA proposed a plan to treat the local wells to remove contaminants.
- Local residents are often an excellent source of information. Many have lived in an area for years and can help identify those responsible for contamination and help locate sites where wastes were illegally disposed of in the neighborhood. Many times local residents have called the National Response Center at (800) 424-8802 to report hazardous materials that present an imminent threat.

Although the EPA tries to include the community's preferences in selecting a remedy for the site, requirements of the Superfund law may lead the EPA to select a response action that is not the community's first choice.

Industry Pays for Hazardous Waste Cleanup

Industry pays for hazardous waste cleanup through specific taxes it pays. Over 80% of the fund known as Superfund is supported directly by excise taxes on petroleum and feedstock chemicals and on some imported chemicals, as well as by corporate environmental taxes. Financial settlements from site polluters also are returned to the fund.

Superfund dollars are used to clean up sites when those who cause the contamination can't or won't pay. Companies are unable to pay for a variety of reasons. They may be too small, such as an individual or a small company without sufficient assets. Perhaps they have declared bankruptcy. In other cases, responsible owners can't be identified or found. On the other hand, many companies can and do pay for cleanup at sites they helped to contaminate.

The EPA Is Making Polluters Pay

The EPA spends considerable effort tracking down the potentially responsible parties. A potentially responsible party (PRP) is any individual or company that might have contributed to or caused the contamination problems at a Superfund site. Examples include owners, operators and waste transporters or producers. Many PRPs did not break a law when they disposed of their hazardous wastes. Thus, when the EPA compels a PRP to clean up a site, it is usually imposing *retroactive civil liability* rather than criminal liability. Nonetheless, the PRP can be legally ordered to pay for or conduct the cleanup of its wastes. The EPA begins the search for PRPs as soon as a site is discovered, and it makes a more concentrated effort to find them after a site is added to the NPL. Once a PRP is located and notified of its potential liability, the EPA or the state begins the negotiation process. The negotiations can lead directly to a satisfactory settlement or, if negotiations fail, to a legal order that compels cooperation under the threat of severe financial penalty. Indeed, the Superfund program makes it a *high* priority to find parties who can perform or pay for cleanup, because this helps maximize the use of Superfund dollars.

The EPA uses a variety of enforcement tools (e.g., administrative orders, consent decrees, negotiations) to engage responsible parties in site cleanup. Every successful negotiation of a private-party cleanup means that the money in the Superfund can be directed instead to those sites that represent immediate energies, or that have no hope of every being cleaned up by those responsible.

Even if identifiable PRPs refuse to undertake cleanup, they are likely to pay in the end. The federal government can and does sue them to recover cleanup costs. If a responsible party refuses to comply with an EPA order, and the site is cleaned up under Superfund authority, the EPA may choose to seek "treble damages." That means the uncooperative polluter may pay up to three times the amount of the cleanup costs expended by the government. In cases that require an emergency response, or where legal actions appear too time-consuming given the present danger, the EPA has the authority to perform the cleanup using Superfund dollars and recover costs later.

If a polluter is clearly implicated at a hazardous waste site, it is in the company's best interest to cooperate in cleanup. The company can contain costs if it does the work, rather than getting a bill for up to three times the cost from the EPA in court. The EPA will try to reach settlement with a polluter who is cooperative concerning cleanup actions. Cooperation first, with legal action as necessary, is the process designed to move from the planning stage to field cleanup actions as quickly as possible. The EPA or the state monitors all work and ensures that it meets government-stipulated standards.

Success in making polluters pay is measurable. Participation in cleanups by PRPs increased from 40% in 1987 to more than 60% in 1989. Strictly enforcing laws that enable the EPA to recover cleanup costs has saved the Superfund about $2 billion in work value between 1980 and 1990, with half of that sum recovered since late 1986.

23.6 CLEANUP SUCCESSES: MEASURING PROGRESS*

The Superfund responds immediately to situations posing imminent threats to human health and the environment at both NPL sites and sites not on the NPL. The purpose is to stabilize, prevent, or temper the effects of a hazardous release, or the threats of one. Imminent threats might include tire fires or discarded waste drums leaking hazardous chemicals. Because they reduce the threat a site poses to human health and the environment, immediate cleanup actions are an integral part of the Superfund program.

The EPA has invested considerable resources in identifying sites that present imminent threats and in undertaking the emergency responses required. The EPA also has developed teams of professionals to combat threatening situations. These emergency workers may assist in cleanup of a dangerous spill or advise state and local officials on the need for a temporary water supply, air and water monitoring, removal of contaminated soils, or relocation of residents.

Immediate response to imminent threats was one of the Superfund's most notable achievements. The EPA monitored and completed emergency actions that attacked the most imminent threats of toxic exposure in more than 1800 cases as of September, 1990. These included actions at both NPL sites and sites not on the NPL in communities across the nation. The EPA used its enforcement authority to have responsible parties perform emergency actions in approximately 400 of those cases.

Between 1986 and 1989, the EPA aggressively accelerated its efforts to clean up sites on the NPL. More cleanups were started in 1987, after the Superfund law was amended, than in any previous year. In 1989 clean-up construction was started at more sites than ever before. The start of clean-up construction actions increased by over 200% between late 1986 and 1989. Of the sites of the NPL in September 1990, more than 500 (nearly half) had cleanup construction activity. Success still was, and is measured by "progress through the cleanup pipeline."

As of September 1990, 272 sites had cleanup work underway, and the "pipeline" was full of sites headed for cleanup. Two hundred and sixty four sites had completed remedy selection and were either in the engineering design phase or close to it; in addition, 504 sites were at the "investigation" step, where the nature of the contamination problem was studied.

The steps through the pipeline which a site must undergo before being deleted from the NPL are as follows:

1. *A detailed study at the site.* Analysts observe site conditions and take samples of wastes and any soil, water, and air that may be affected, and then they study the range of possible clean-up strategies.

2. *Remedy selection.* The EPA analyzes findings from the study and formally chooses the best remedy from among the alternatives suggested.

*Cleanup activities were reported as of September 1990.

3. *Engineering design.* The EPA or its designate, often the U.S. Army Corps of Engineers, prepares specifications and drawings for the selected remedy.

4. *Cleanup construction and follow-up.* Although various parties may construct or otherwise carry out the remedy designed, the EPA is always in charge. Cleanup is often followed by a requirment to operate, maintain, or monitor the site for several years. This can extend the official deletion of the site from the NPL by years.

On average, a site spends 6–8 years progressing through these steps. The public has the right and opportunity to comment at every step in the process.

The Superfund pipeline shows stepwise progress in moving sites toward final cleanup. Much of that movement has traditionally been measured in administrative and legal milestones. However, the start of cleanup construction does not necessarily adequately reflect the magnitude of environmental progress that is made. In addition, such real progress often lags behind construction activity. For example, while construction of an incinerator initiates work at a site, actual environmental progress won't take place until hazardous wastes are destroyed. Greater emphasis is now being placed on the *environmental progress* the program is making, the tangible physical evidence that the program is achieving results. To do this, the EPA examines official records and discusses environmental progress with site clean-up managers to measure what has actually been accomplished in the terms of contamination reduction and protecting the public from exposure to hazardous substances. The results of this approach to measuring environmental progress are summarized here and are discussed in detail in the study report entitled *Superfund: Reporting Progress Through Environmental Indicators* (4).

The EPA focused the Superfund Environmental Indicators study on NPL sites where prior to 1990, construction work for site cleanup had actually begun or immediate actions had been completed. While progress made by immediate actions taken at certain sites not on the NPL was also examined, this summary only discusses environmental progress made at NPL sites. The two major categories used to report progress reflect, in part, the way that the EPA approaches cleanup at NPL sites. The two categories are:

1. *Sites where there have been actions taken to address imminent threats.* When a hazardous waste site is discovered the EPA immediately undertakes any necessary emergency actions to make the site safe. These actions may be either temporary or permanent and can range from relocation of affected residents to provision of an alternate water supply or physical removal, treatment, or containment of wastes. These actions are taken in order to reduce imminent danger without delay.

2. *Sites where there have been actions taken to achieve health and environmental goals.* Once sites are safe, the EPA selects a remedy for the site, and it documents the cleanup goals that this remedy must achieve before it may be considered complete. In the study, the EPA investigated how each site is making progress toward meeting these goals. For example, if the groundwater is, or might be, used as drinking water, that goal is the attainment of the national drinking water standards. In some cases, particularly for the land surface, varying goals are established for different areas of a given site, reflecting the presence of different kinds of concentrations of contaminants. Since progress is evaluated for each environmental pathway at every site, it is possible for a single site to show several

levels of achievement for this measure: For example, some land areas may be clean, while work on groundwater or surface water contamination has only just begun.

The EPA had collected information on the quantities of hazardous waste physically handled during the course of clean-up actions. These amounts provide an indicator of the quantities and types of materials removed, contained, or treated by the Superfund program. While the numbers are frequently impressive, this information does not, in itself, demonstrate environmental progress. Total quantities of wastes removed from the environment are best used to supplement the two major categories of progress noted above.

As of September 1990 the EPA documented significant progress at 422 of the nation's highest priority sites, more than one-third of the sites listed on the NPL at that time (see Figure 23.5). Because contamination at NPL sites could require activities both to control imminent threats and provide permanent cleanup, some sites showed progress in more than one of these categories. Specifically, the EPA documented the following progress in reducing risk and achieving human health and environmental goals at NPL sites in September 1990:

- At 356 sites, immediate actions were taken to reduce imminent risk from exposure to site contaminants; at 252 of these sites, clean-up actions were also underway to achieve permanent cleanup goals.

- At 318 sites, actions that would lead to permanent cleanup were initiated or were currently showing progress toward meeting human health and environmental goals for at least one environmental pathway. These environmental pathways, or areas, included contaminated land, groundwater, and surface water. Land contamination included soil and solid and liquid wastes found on or near the ground surface. The groundwater pathway included subsurface waters and aquifers. The surface water included lakes, ponds, slow-moving streams, and marshes.

As a first step, the Superfund program is required to evaluate, stabilize, treat, or otherwise take actions to make dangerous sites safe. At 356 sites, immediate actions to protect nearby populations and to control the imminent threat of exposure to hazardous contaminants had been taken by September 1990. Of those actions, 245

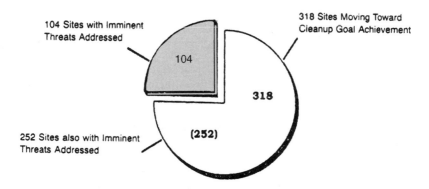

FIGURE 23.5 NPL sites with environmental progress.

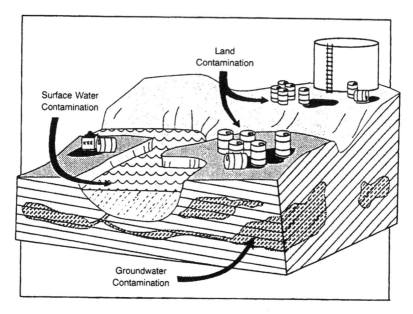

FIGURE 23.6 Hazardous waste effects on environmental pathways.

involved the removal, treatment, or containment of wastes; 293 provided site security to prevent accidental exposure to hazardous substances at the site; and 73 provided an alternate water supply to affected residents. Site numbers add up to greater than 356 sites because more than one risk reduction activity may have occurred at any one site.

Estimates on the magnitude of these actions indicate that almost a quarter of a million people with contaminated household water supplies were provided with an alternate residential water source. At 19 sites, the immediate actions involved the relocation of populations away from contaminated areas. The EPA estimated that more than 3900 people were evacuated or relocated during these site cleanup activities. After cleanup, 39% of these evacuated residents were returned to their homes, while the rest were permanently relocated.

Once all sites are safe, the EPA identifies and addresses the worst contamination problems at individual sites, focusing its efforts on the long-term cleanup of the most threatening areas before addressing any other contaminated pathways. Thus, long-term clean-up activities can be in different stages at a single site.

In the study, the EPA evaluated the status of these long-term clean-up activities in meeting human health and environmental goals. Any one site may have various contaminated pathways, each of which threatens health and the environment in a different way. Figure 23.6 illustrates the land, groundwater, and surface water pathways of contamination that were measured in the September 1990 study.

Today, ongoing cleanup work at a site is described as: *All Clean-up Goals Met, Some Clean-up Goals Met, Cleanup Underway,* or *Developing Clean-up Strategies.* Achievement is evaluated by the clean-up progress made within each specific environmental pathway that was contaminated at a site. Cleanup work is considered to be "underway" when hazardous wastes or contaminated water and soil have actually been moved or treated at a site, but that work has not progressed far enough for observers to be able to claim, with any certainty, a great deal of progress. Work at the site *is* underway, and hazards *may* in fact have been reduced, but the EPA is *not yet claiming credit for even partial success.*

Where several areas of contamination must be cleaned up before goals for a specific environmental pathway are met, "some clean-up goals met" is documented where one or more contaminated areas, such as two out of three lagoons at a site, have been cleaned up to meet permanent health and environmental standards, but not all work for a particular site has been done. In some cases, it is possible to declare full success in cleaning up a part of the environment: The land is clean, the surface water is clean, and so on, but the groundwater may still require clean-up actions. To date, the Superfund program has achieved the greatest degree of success in moving toward cleanup of land contamination.

As of June 1997, at NPL sites, land contamination has been addressed at 804 sites. Some clean-up goals have been met at 202 sites, while all clean-up goals have been met at 321 sites. Cleanup is underway at 160 sites, and another 121 sites are developing clean-up strategies. Land cleanup is often undertaken first because it substantially reduces risk to people, animals, and plants that might otherwise come into direct contact with waste at the site. These actions can also forestall a need for future groundwater cleanup by removing a source that could percolate into the subsurface water (2).

Most Superfund sites have actual or potential groundwater contamination. Many Americans use groundwater as a drinking water source, and its cleanup has proved to be one of the most difficult environmental problems to solve. As of June 1997, groundwater cleanup is occurring at 450 sites. Some clean-up goals have been met at 67 sites, while all clean-up goals have been met at 39 sites. Cleanup is underway at 267 sites, and another 77 sites are developing clean-up strategies. Experience to date suggests that meeting health and environmental goals in this area may take many more years of treatment and monitoring than initially expected even a few years ago (2).

Contaminated surface waters can create substantial hazards for drinking supply, wildlife, and recreational uses. Natural weather conditions such as heavy rainfall may aggravate the situation by spreading contamination via runoff and overflow of contaminants from the site. As of June 1997, clean-up goals for surface-water contamination are being addressed at 271 NPL sites. Some cleanup goals have been met at 30 sites, while all clean-up goals have been met at 42 sites. Cleanup is underway at 40 sites, and another 159 sites are developing cleanup strategies (2).

The sheer volume of hazardous wastes that has been removed in cleaning up NPL sites illustrates an impressive aspect of the environmental progress being made at the nation's Superfund sites. The study cited earlier (4) documented significant work in addressing wastes in each of the contamination pathways at 329 NPL sites of September 1990. Some of the details are as follows:

- Land contamination includes both soils and solid and liquid wastes. Soils and other solid waste removed from the environment thus far total more than 3,909,000 yd^2; this amounts to 2700 football fields covered with wastes to a depth of 1 ft. Liquid wastes total almost a *billion* gallons, or about 4 gal for each resident of the United States.

- Groundwater treated to date totals approximately 3,580,000,000 gal, equivalent to more than 14 gal for each person in the United States.

• Surface water treated to date totals almost 83,000,000 gal, equivalent to about one-third gallon for each person in the United States.

The EPA has gained enough experience in clean-up technologies to understand that environmental protection does not end when the clean-up remedy has been constructed. Many compex technologies, like those designed to clean up ground-water, must operate for several years in order to accomplish their objectives. Therefore, the EPA does not abandon a site after the cleanup work is done. Every 5 years, the EPA reviews each site where residues from hazardous waste cleanup still remain to ensure that human and environmental health are still being safeguarded. The EPA corrects any deficiencies discovered and reports to the public annually on all 5-yr reviews conducted that year.

23.7 SUMMARY

1. The Comprehensive Environmental Response, Compensation, and Liability Act (CERCLA), enacted in 1980, was the first federal law dealing with the dangers posed by the nation's abandoned and uncontrolled hazardous waste sites.

2. Superfund was established under CERCLA as a trust fund to finance site clean-up actions.

3. By 1990, 1236 hazardous waste sites comprised the National Priorities List (NPL), a listing of those sites targeted for cleanup. Approximately 100 sites are added to the NPL each year.

4. Sites vary dramatically in size, type of contaminant(s), and location. Each clean-up action must be specifically tailored to the site.

5. Treatment is the preferred approach to cleanup, rather than containment.

6. Common toxic contaminants include heavy metals, volatile organics, pesti-cides and herbicides, PCBs, and creosotes. Health effects vary from specific anatomical damage to cancer.

7. Local citizens are encouraged to get involved in clean-up activities by reporting hazardous waste dumping, sharing information with investigators, and engaging experts to interpret data.

8. A potentially responsible party (PRP) is any individual or company that might have contributed to or caused the contamination problems at a Superfund site. The PRP can be legally ordered to pay for or conduct the cleanup of its wastes.

9. Of the 1236 sites listed on the NPL in 1990, 272 sites have clean-up work underway, 264 sites have completed remedy selections and are in the engineering design phase, and 504 sites are in the investigation step.

REFERENCES

1. U.S. EPA, *Superfund Environmental Indicators: Cleanup of Hazardous Waste Sites*, December 1998.
2. U.S. EPA, *Toxic Chemicals — What They Are, How They Affect You*, Region 5.
3. U.S. EPA, *Glossary of Environmental Terms*.
4. U.S. EPA, *Superfund: Reporting Progress Through Environmental Indicators*, 1990.

24

UNDERGROUND STORAGE TANKS

24.1 INTRODUCTION

Environmental contamination from leaking underground storage tanks poses a significant threat to human health and the environment. These leaking underground storage tanks (USTs) contaminate our nation's groundwater, which is a major source of drinking water. Nationally, there were over 2 million USTs in use as of December 22, 1988, when the UST regulations became effective. As of early 1999, the number of active USTs had decreased to under one million. However, there were an estimated 5–6 million underground storage tanks containing hazardous substances or petroleum products in use in the United States during the oil boom of the 1950s and 1960s. Originally placed underground as a fire prevention measure, these tanks have substantially reduced the damage from stored flammable liquids. However, an estimated 400,000 underground tanks are thought to be leaking now, and many more will begin to leak in the near future. Products released from these leaking tanks can threaten groundwater supplies, damage sewer lines and buried cables, poison crops, and lead to fires and explosions.

Under the Resource Recovery and Conservation Act (RRCA), underground storage tanks are defined as tanks with 10% or more of their volume, including piping, located underground. Almost half of the tanks to be regulated by EPA are petroleum storage tanks owned by gas stations, and another 47% are petroleum storage tanks owned by a group of other industries that store petroleum products for their own use. Airports, firms with large trucking fleets, farms, golf courses, and manufacturing operations may all own tanks. The remainder of the tanks that will be regulated are used by a variety of industries for chemical storage.

Many of the petroleum tank systems were installed during the oil boom of the 1950s and 1960s. Two 1985 studies of tank age distribution indicate that approximately one-third of the existing motor fuel storage tanks are now over 30 years old or of unknown age. Most of these aging tank systems are constructed of bare steel, not protected against corrosion, and are nearing the end of their useful lives. Many of these old tank systems have already had a leak, or will soon leak unless measures are taken to improve or remove them. When these old tanks are pulled from the ground, many of them have holes where a dip stick was dropped hundreds of times to measure the amount of fuel in the tank.

Exacerbating the problem of old tanks still in use are the thousands of gas stations that closed during the oil crisis of the 1970s. Although these tanks are not the only ones of concern, the abandoned tanks at these stations frequently were not closed properly, and ownership and responsibility for future problems is difficult to determine.

The primary reason for regulating underground storage tanks is to protect water, especially groundwater that is used for drinking water. This is one of the nation's greatest natural resources and one which is extremely difficult to remediate once it is contaminated. Fifty percent of the U.S. population depends on groundwater for drinking water. Rural areas would be seriously affected if their groundwater were contaminated, since it provides 95% of their total water supplies. Groundwater drawn for large-scale agricultural and industrial uses also can be adversely affected by contamination from leaking underground tanks (1).

As we approach the new millennium, UST programs will need to focus resources in addressing the increased number of confirmed releases as a result of the 1998 upgrade requirements which represented the last major regulatory deadline, it took effect on December 22, 1998. The EPA's approach to regulation of the underground storage tank systems was designed to prevent the creation of a new generation of releases and to clean up releases promptly and cost-effectively. The UST technical requirements, which include new tank standards, leak detection, operation and maintenance, and closure, spill, overfill, and corrosion protection, were designed to prevent new releases into the environment. These regulatory requirements were phased-in over a period of ten years from December 22, 1988 to December 22, 1998, to allow owners and operators enough time to comply with the requirements. This meant that as of December 22, 1998, all substandard USTs must be closed and all tanks in operation must be in compliance with all the technical requirements of 40 CFR Part 280, including leak detection and 1998 requirements for spill, overfill, and corrosion requirements.

The flurry of activities resulting from compliance activities to meet the last UST regulatory deadline in December 22, 1998, resulted in numerous closures of substandard tanks. As site assessments and analytical results from sampling activities are conducted in response to permanent closure activities, the number of leaking underground storage tank (LUST) sites is expected to increase significantly. Although the rate of new releases is declining due to the implementation of federal and state UST programs, the presence of methyl tertiary-butyl ether (MTBE) as a new contaminant of concern is posing remediation challenges to both new and old LUST sites. MTBE is a fuel additive that has been used in the United States as an octane-enhancing replacement for lead. More recently, MTBE has been used as a fuel oxygenate added to gasoline. The contamination resulting from MTBE had been elevated to a national level in 1996 when two Santa Monica, California, drinking-water well fields were shut down, at a cost of millions of dollars per year, because of high MTBE concentrations.

Since 1979, MTBE has been used in the United States as an octane-enhancing replacement for lead, primarily in mid- and high-grade gasoline at concentrations as high as 8% by volume. Since the mid-1980s, it has been widely used throughout the country for this purpose. It is also used as a fuel oxygenate at higher concentrations (11–15% by volume) as part of the U.S. EPA's programs to reduce ozone and carbon monoxide levels in the most polluted areas of the country. The Oxygenated Fuel (Oxyfuel) and Reformulated Gasoline (RFG) Programs were initiated by the U.S. EPA in 1992 and 1995, respectively, to meet requirements of the 1990 Clean Air Act Amendments (CAAA). The Oxyfuel Program requires 2.7% oxygen by weight in gasoline during the fall and winter months to reduce carbon monoxide emissions. In order to meet this requirement, gasoline producers must use oxygen-containing compounds termed "fuel oxygenates" (e.g., ethanol, MTBE). When MTBE is used to meet the Oxyfuel requirements, it is added at a concentration of approximately 15% by volume to gasoline. The RFG Program requires 2% oxygen by weight year-round in the most polluted metropolitan areas to reduce ozone and smog. When MTBE is used to meet the RFG requirements, its concentration in gasoline is 11% by volume (2).

24.2 EARLY REGULATIONS

Federal laws were enacted in response to the increasing problems resulting from leaking underground storage tanks. These laws generally are based on many state and industry ongoing efforts. Congress provided for federal regulation of underground storage tanks as part of the 1984 RCRA amendments, but it exempted residential heating oil and small farm tanks. As a first step, Congress required that all owners or operators register their tanks with the appropriate state agency, indicating tank age, location, and contents. Thus, for the first time states had to set up an inventory of tanks in their jurisdictions. The amendments also included interim design requirements to tanks installed after May 1985. Congress further directed the EPA to develop regulations requiring owners to detect leaks from new and existing underground storage tanks and clean up environmentally harmful releases from such tanks. Tank owners must also demonstrate that they are financially capable of cleaning up leaks from tanks and compensating third parties for damages resulting from such leaks.

As required by the Hazardous and Solid Waste Amendments (HSWA), EPA has been developing a comprehensive regulatory program for underground storage tanks. The EPA proposed three sets of regulations pertaining to underground tanks. The first addresses technical requirements for petroleum and hazardous substance tanks, including new tank performance standards, release detection, release reporting and investigation, corrective action, and tank closure. The second proposed regulation addresses financial responsibility requirements for underground petroleum tanks. The third addresses standards for approval of state tank programs.

While amending the Comprehensive Environmental Response, Compensation and Liability Act (CERCLA) in 1986, Congress had amended RCRA to provide $500 million over the ensuing five years for a Leaking Underground Storage Tank Trust Fund. Generated by one-tenth of a cent federal tax on certain products, primarily motor fuels, the trust fund has been made available to the states to help them clean up leaks from underground petroleum storage tanks if certain conditions for use of the fund are met.

Because the number of tanks requiring investigation and attention is too great for EPA to tackle alone, the EPA has developed a program that will be carried out primarily by state and local governments. The national tank program is designed primarily to be a network of state and local programs. The EPA will provide research, regulations, training, technical support, and enforcement backup, as necessary.

New provisions in the Superfund Amendments and Reauthorization Act (SARA) authorize the EPA and states that enter into cooperative agreements with the EPA to issue orders requiring owners and operators of underground storage tanks to undertake corrective action where a leak or spill is suspected or has occurred. This corrective action could include testing tanks to confirm the presence of a leak, excavating the site to determine the exact nature and extent of contamination, and cleaning contaminated soil and water. It also may include providing an alternative water supply to affected residences or temporary or permanent relocation of residents.

EPA has signed cooperative agreements with all states and transferred millions of dollars from the Trust Fund to do this work. With these funds, states have begun cleanups of many of these sites around the country (1).

24.3   THE SCOPE OF THE PROBLEM

The problem of leaking underground storage tanks presents several major challenges not only for the EPA but also for states, local governments, and industry. Perhaps the biggest challenge is to achieve better but less expensive tank design and leak detection technology. Research and development in these areas is strong and the potential for improvement is great. Effective and affordable technology should encourage voluntary compliance among tank owners. It will become less expensive to comply with the law and, at the same time, the environment will be protected from leaks and spills from underground storage tanks.

The large number of underground tanks presents significant regulatory challenges. This is best illustrated by a comparison to the hazardous waste program. In the early 1990s in New England alone, approximately 5,000 handlers of hazardous waste are regulated under RCRA and there are 59 sites on the Superfund National Priorities List. In the same region, the number of underground tanks falling under EPA or state regulations is 150,000, or 30 times larger than the number of regulated RCRA facilities.

Congress directed EPA to require that tank owners demonstrate that they are financially capable of cleaning up leaks from their underground storage tanks and compensating parties for damages resulting from such leaks. These costs could include cleaning up leaked petroleum, supplying drinking water, or compensating individuals for personal injury or property damage. More recently, SARA imposed the requirements that owners or operators of underground storage tanks have a minimum insurance coverage of $1 million per occurrence. A major challenge that EPA, state and local governments, and the regulated community face involves this assurance of tank owner's financial responsibility.

The lack of an adequate pollution liability insurance market makes finding affordable insurance extremely difficult. Insurance programs, state funds, and other assurance mechanisms are sorely needed. The EPA has been working with states and with the insurance industry on ways to develop such mechanisms to help tanks owners.

The cleanup of contaminated sites from leaking underground storage tanks is another major challenge that EPA and the states faces. Often the contamination is in the soil directly above groundwater. The challenge lies in quickly finding the most seriously contaminated sites. These sites need to be addressed before petroleum reaches groundwater supplies.

24.4 THE NEW APPROACH

The EPA is using a new approach to implementing the Congressional mandate to address underground tanks called the "franchise concept." The Agency sees itself as the franchiser, with the responsibility of seeing that the franchises, in this case the states, run their operations successfully. This model permits both uniformity and distinctiveness in individual management styles, allowing states the greatest flexibility in developing a UST program that is tailored to meet their own specific needs and the demands of the regulated communities. The EPA initially will focus on assisting the states to establish comprehensive tank programs, in addition to simultaneously providing a range of services to help them improve their performance. EPA believes that the UST regulatory program will be most effectively carried out by the level of government nearest to the problem. State and local government know their regulated communities and are best able to respond quickly and effectively to their individual problems.

This will be a significant departure from the traditional approach under which the EPA manages all areas of a program until a state demonstrates it can operate independently. The EPA is encouraging states to apply for formal approval of state UST programs to operate "in lieu of" the federal program. Approval of a state program means that the requirements in the state's laws and regulations will be in effect rather than the federal requirements. The EPA plans to approve acceptable state UST programs as quickly as possible and to follow up with activities that provide continual assistance to states and localities for improving their capability and performance. The EPA is to provide necessary assistance for the states to succeed in implementing and enforcing this program. This includes providing special expertise, developing training videotapes (e.g., for tank installers and inspectors), publishing handbooks, and training state and local employees. As EPA works with states and local communities to do a better job of communicating the dangers of leaking tanks to the estimated 750,000 tank owners, more and more owners will replace older tank with protected, safe tanks (1).

24.5 THE FEDERAL REGULATIONS

The federal regulations consist of three major sets of regulations: the technical regulations as set forth in 40 CFR Part 280 addressing technical standards for corrective action requirements for owners and operators of Underground Storage Tanks; the Underground Storage Tank State Program Approval regulations as set forth in 40 CFR Part 281, which addresses regulations for approval of states to run underground storage tank programs in lieu of the federal program; and the financial responsibility requirements as set forth in 40 CFR Part 280, which requires that UST owners and operators demonstrate ability to assume financial responsibility for the costs of corrective action and compensation of third parties arising from the release of petroleum from underground storage tanks. The financial responsibility requirements will help ensure the owners and operators can respond promptly to clean up releases and to compensate third parties for injuries or damages associated with the releases.

Compliance with the following subparts of the Technical Regulation as defined in 40 CFR Part 280 will meet the federal requirements. Owners and operators must contact the UST Program in their individual state for specific state requirements (3). An outline of the key federal regulations is provided below, including the Release Detection requirements (Subpart D, as it appeared in the *Federal Register*) which represents the major thrust of the requirements. The key technical standards are:

SUBPART B — UST SYSTEMS: DESIGN, CONSTRUCTION, INSTALLATION
AND NOTIFICATION

§280.20 Performance standards for new UST systems.
§280.21 Upgrading of existing UST systems.
§280.22 Notification requirements.

SUBPART C — GENERAL OPERATING REQUIREMENTS

§280.30 Spill and overfill control.
§280.31 Operation and maintenance of corrosion protection.
§280.32 Compatibility.
§280.33 Repairs allowed.
§280.34 Reporting and recordkeeping.

SUBPART D — RELEASE DETECTION

§280.40 General Requirements for all UST systems.
§280.41 Requirements for petroleum UST systems.
§280.42 Requirements for hazardous substance UST systems.
§280.43 Methods of release detection for tanks.
§280.44 Methods of release detection for piping.
§280.45 Release detection recordkeeping.

SUBPART E — RELEASE REPORTING, INVESTIGATION,
AND CONFIRMATION

§280.50 Reporting of suspected releases.
§280.51 Investigation due to off-site impacts.
§280.52 Release investigation and confirmation steps.
§280.53 Reporting and cleanup of spills and overfills.

SUBPART F — RELEASE RESPONSE AND CORRECTIVE ACTION
FOR UST SYSTEMS CONTAINING PETROLEUM
OR HAZARDOUS SUBSTANCES

§280.60 General.
§280.61 Initial response.
§280.62 Initial abatement measures and site check.
§280.63 Initial site characterization.
§280.64 Free product removal.
§280.65 Investigations for soil and ground-water cleanup.
§280.66 Corrective action plan.

§280.67   Public participation.

The following entire section from the *Federal Register* is included below because of the profound impact it has meeting the requirements of the regulations.

Subpart D — Release Detection

§280.40   General Requirements for All UST Systems

(a) Owners and operators of new and existing UST system must provide a method, or combination of methods, of release detection that:

    (1) Can detect a release from any portion of the tank and the connected underground piping that routinely contains product;

    (2) Is installed, calibrated, operated, and maintained in accordance with the manufacturer's instructions, including routine maintenance and service checks for operability or running conditions; and

    (3) Meets the performance requirements in §280.43 or §280.44, with any performance claims and their manner of determination described in writing by the equipment manufacturer or installer. In addition, methods used after December 22, 1990, except for methods permanently installed prior to that date, must be capable of detecting the leak rate or quantity specified for that method in §280.43(b), (c), and (d) of detection of 0.95 and a probability of false alarm of 0.05.

(b) When a release detection method operated in accordance with the performance standards in §280.43 and §280.44 indicates a release may have occurred, owners and operators must notify the implementing agency in accordance with Supbart E.

(c) Owners and operators of all UST systems must comply with the release detection requirements of this subpart by December 22 of the year listed in Figure 24.1 (see also Figure 24.2).

(d) Any existing UST system that cannot apply a method of release detection that complies with the requirements of this subpart must complete the closure procedures in Subpart G by the date on which release detection is required for that UST system under paragraph (c) of this section.

§280.41   Requirements for Petroleum UST Systems

Owners and operators of petroleum UST systems must provide release detection for tanks and piping as follows:

(a) *Tanks.* Tanks must be monitored at least every 30 days for releases using one of the methods listed in §280.43(d) through (h) except that:

    (1) UST systems that meet the performance standards in §280.20 or §280.21, and the monthly inventory control requirements in §280.43(a) or (b), may use tank tightness testing (conducted in accordance with §280.43(c)) at

LEAK DETECTION	
NEW TANKS *2 Choices*	• Monthly Monitoring* • Monthly Inventory Control and Tank Tightness Testing Every 5 Years (You can only use this choice for 10 years after installation.)
EXISTING TANKS *3 Choices* *The chart at the bottom of the next page displays these choices.*	• Monthly Monitoring* • Monthly Inventory Control and Annual Tank Tightness Testing (This choice can only be used until December 1998.) • Monthly Inventory Control and Tank Tightness Testing Every 5 Years (This choice can only be used for 10 years after adding corrosion protection and spill/overfill prevention or until December 1998, whichever date is later.)
NEW & EXISTING PRESSURIZED PIPING *Choice of one from each set*	• Automatic Flow Restrictor • Annual Line Testing • Automatic Shutoff Device -and- • Monthly Monitoring* • Continuous Alarm System (except automatic tank gauging)
NEW & EXISTING SUCTION PIPING *3 Choices*	• Monthly Monitoring* (except automatic tank gauging) • Line Testing Every 3 Years • No Requirements

CORROSION PROTECTION	
NEW TANKS *3 Choices*	• Coated and Cathodically Protected Steel • Fiberglass • Steel Tank Clad with Fiberglass
EXISTING TANKS *4 Choices*	• Same Options as for New Tanks • Add Cathodic Protection System • Interior Lining • Interior Lining and Cathodic Protection
NEW PIPING *2 Choices*	• Coated and Cathodically Protected Steel • Fiberglass
EXISTING PIPING *2 Choices*	• Same Options as for New Piping • Cathodically Protected Steel

SPILL / OVERFILL PREVENTION	
ALL TANKS	• Catchment Basins -and- • Automatic Shutoff Devices -or- • Overfill Alarms -or- • Ball Float Valves

* Monthly Monitoring includes: Automatic Tank Gauging, Ground-Water Monitoring, Vapor Monitoring, Other Approved Methods, Interstitial Monitoring

FIGURE 24.1   What do you have to do? Minimum requirements.

least every 5 years until December 22, 1998, or until 10 years after the tank is installed or upgraded under §280.21(b), whichever is later;

(2) UST systems that do not meet the performance standards in §280.20 or §280.21 may use monthly inventory controls (conducted in accordance with §280.43(a) or (b)) and annual tank tightness testing (conducted in accordance with §280.43(c)) until December 22, 1998 when the tank must be upgraded under §280.21 or permanently closed under §280.71; and

TYPE OF TANK & PIPING	LEAK DETECTION	CORROSION PROTECTION	SPILL / OVERFILL PREVENTION
New Tanks and Piping*	At installation	At installation	At installation
Existing Tanks** Installed:	By No Later Than:		
Before 1965 or unknown	December 1989		
1965 - 1969	December 1990		
1970 - 1974	December 1991	December 1998	December 1998
1975 - 1979	December 1992		
1980 - December 1988	December 1993		
Existing Piping**			
Pressurized	December 1990	December 1998	Does not apply
Suction	Same as existing tanks	December 1998	Does not apply

* New tanks and piping are those installed after December 1988
** Existing tanks and piping are those installed before December 1988

IF YOU CHOOSE TANK TIGHTNESS TESTING AT EXISTING USTs ...

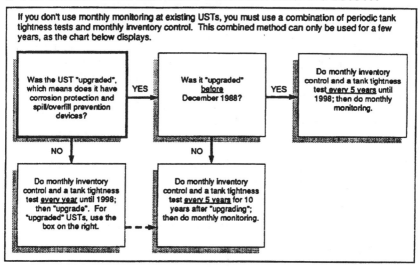

If you don't use monthly monitoring at existing USTs, you must use a combination of periodic tank tightness tests and monthly inventory control. This combined method can only be used for a few years, as the chart below displays.

FIGURE 24.2 When do you have to act? Important deadlines.

(3) Tanks with capacity of 550 gal or less may use weekly tank gauging (conducted in accordance with §280.43(b)).

(b) *Piping.* Underground piping that routinely contains regulated substances must be monitored for releases in a manner that meets one of the following requirements:

(1) Pressurized Piping. Underground piping that conveys regulated substances under pressure must:
  (i) Be equipped with an automatic line leak detector conducted in accordance with §280.44(a); and
  (ii) Have an annual line tightness test conducted in accordance with §280.44(b) or have monthly monitoring conducted in accordance with §280.44(c).
(2) Suction Piping. Underground piping that conveys regulated substances under suction must either have a line tightness test conducted at least every 3 years and in accordance with §280.44(b), or use a monthly monitoring method conduct in accordance with §280.44(c). No release detection is required for suction piping that is designed and constructed to meet the following standards:
  (i) The below-grade piping operates at less than atmospheric pressure;
  (ii) The below-grade piping is sloped so that the contents of the pipe will drain back into the storage tank if the suction is released;
  (iii) Only one check valve is included in each suction line;
  (iv) The check valve is located directly below and as close as practical to the suction pump; and
  (v) A method is provided that allows compliance with paragraphs (b)(2)(ii)–(iv) of this section to be readily determined.

§280.42 Requirements for Hazardous Substance UST Systems

Owners and operators of hazardous substance UST systems must provide release detection that meets the following requirements:

(a) Release detection of existing UST systems must meet the requirements for petroleum UST systems in §280.41. By December 22, 1998, all existing hazardous substance UST systems must meet the release detection requirements for new systems in paragraph (b) of this section.
(b) Release detection at new hazardous substance UST systems must meet the following requirements:
  (1) Secondary containment systems must be designed, constructed and installed to:
    (i) Contain regulated substances released from the tank system until they are detected and removed;
    (ii) Prevent the release of regulated substances to the environment at any time during the operational life of the UST system; and
    (iii) Be checked for evidence of a release at least every 30 days.

Note: The provisions of 40 CFR 265.193, Containment and Detection of Releases, may be used to comply with these requirements.

  (2) Double-walled tanks must be designed, constructed, and installed to:
    (i) Contain a release from any portion of the inner tank within the outer wall; and
    (ii) Detect the failure of the inner wall.
  (3) External liners (including vaults) must be designed, constructed, and installed to:
    (i) Contain 100% of the capacity of the largest tank within its boundary;
    (ii) Prevent the interference of precipitation or groundwater intrusion with the ability to contain or detect a release of regulated substances; and

(iii) Surround the tank completely (i.e., it is capable of preventing lateral as well as vertical migration of regulated substances).

(4) Underground piping must be equipped with secondary containment that satisfies the requirements of paragraph (b)(1) of this section (e.g., trench liners, jacketing of double-walled pipe). In addition, underground piping that conveys a regulated substances under pressure must be equipped with an automatic line leak detector in accordance with §280.44(a).

(5) Other methods of release detection may be used if owners and operators:

    (i) Demonstrate to the implementing agency that an alternate method can detect a release of the stored substance as effectively as any of the methods allowed in §§280.43(b) through (h) can detect a release of petroleum;

    (ii) Provide information to the implementing agency on effective corrective action technologies, health risks, and chemical and physical properties of the stored substance, and the characteristics of the UST site; and,

    (iii) Obtain approval from the implementing agency to use the alternate release detection method before the installation and operation of the new UST system.

§280.43   Methods of Release Detection for Tanks

Each method of release detection for tanks used to meet the requirements of §280.41 must be conducted in accordance with the following:

(a) *Inventory Control.* Product inventory control (or another test of equivalent performance) must be conducted monthly to detect a release of at least 1.0% of flow-through plus 130 gal on a monthly basis in the following manner:

(1) Inventory volume measurements for regulated substance inputs, withdrawals, and the amount still remaining in the tank are recorded on each operating day;

(2) The equipment used is capable of measuring the level of product over the full range of the tank's height to the nearest one-eighth of an inch;

(3) The regulated substance inputs are reconciled with delivery receipts by measurement of the tank inventory volume before and after delivery;

(4) Deliveries are made through a drop tube that extends to within one foot of the tank bottom;

(5) Product dispensing is metered and recorded within the local standards for meter calibration or an accuracy of 6 in.^3 for every 5 gal of product withdrawn; and

(6) The measurement of any water level in the bottom of the tank is made to the nearest one-eighth of an inch at least once a month.

Note: Practices described in the American Petroleum Institute Publication 1621, "Recommended Practice for Bulk Liquid Stock Control at Retail Outlets," may be used where applicable as guidance in meeting the requirements of this paragraph.

(b) *Manual Tank Gauging.* Manual tank gauging must meet the following requirements:

(1) Tank liquid level measurements are taken at the beginning and ending of a period of at least 36 hr during which no liquid is added to or removed from the tank;

(2) Level measurements are based on an average of two consecutive stick readings at both the beginning and end of the period;

(3) The equipment used is capable of measuring the level of product over the full range of the tank's height to the nearest one-eighth of an inch;

(4) A leak is suspected and subject to the requirements of Subpart E if the variation between beginning and ending measurements exceeds the weekly or monthly standards in the following table:

Nominal tank capacity	Weekly standard (one test)	Montly standard (average of four tests)
550 gal or less	10 gal	5 gal
551–1000 gal	13 gal	7 gal
1001–2000 gal	26 gal	13 gal

(5) Only tanks of 550 gal or less nominal capacity may use this as the sole method of release detection. Tanks of 551–2000 gal may use the method in place of manual inventory control in §280.43(a). Tanks of greater than 2,000 gal nominal capacity may not use this method to meet the requirements of this subpart.

(c) *Tank Tightness Testing.* Tank tightness testing (or another test of equivalent performance must be capable of detecting a 0.1-gal/hr leak rate from any portion of the tank that routinely contains product, while accounting for the effects of thermal expansion or contraction of the product, vapor pockets, tank deformation, evaporation or condensation, and the location of the water table.

(d) *Automatic Tank Gauging.* Equipment for automatic tank gauging that tests for the loss of product and conducts inventory control must meet the following requirements:

(1) The automatic product level monitor test can detect a 0.2-gal/hr leak rate from any portion of the tank that routinely contains product; and

(2) Inventory control (or another test of equivalent performance) is conducted in accordance with the requirements of §280.43(a).

(e) *Vapor Monitoring.* Testing or monitoring for vapors within the soil gas of the excavation zone must meet the following requirements:

(1) The materials used as backfill are sufficiently porous (e.g., gravel, sand, crushed rock) to readily allow diffusion of vapors from releases into the excavation area;

(2) The stored regulated substance, or tracer compound placed in the tank system, is sufficiently volatile (e.g., gasoline) to result in a vapor level that is located in the excavation zone in the event of a release from the tank;

(3) The measurement of vapors by the monitoring device is not rendered inoperative by the groundwater, rainfall, or soil moisture or other known interferences so that a release could go undetected for more than 30 days;

(4) The level of background contamination in the excavation zone will not interfere with the method used to detect releases from the tank;

(5) The vapor monitors are designed and operated to detect any significant increase in concentration above background of the regulated substance stored in the tank system, a component or components of that substance, or a tracer compound placed in the tank system;

(6) In the UST excavation zone, the site is assessed to ensure compliance with

the requirements in paragraphs (e)(1) through (e)(4) of this section and to establish the number and positioning of monitoring wells that will detect releases within the excavation zone from any portion of the tank that routinely contains product; and

(7) Monitoring wells are clearly marked and secured to avoid unauthorized access and tampering.

(f) *Groundwater Monitoring.* Testing or monitoring for liquids on the groundwater must meet the following requirements:

(1) The regulated substance stored is immiscible in water and has a specific gravity of less than one;

(2) Groundwater is never more than 20 ft from the ground surface and the hydraulic conductivity of the soil(s) between the UST system and the monitoring wells or devices is not less than 0.01 cm/sec (e.g., the soil should consist of gravels, coarse-to-medium sands, coarse silts, or other permeable materials);

(3) The slotted portion of the monitoring well casing must be designed to prevent migration of natural soils or filter pack into the well and to allow entry of regulated substance on the water table into the well under both high and low groundwater conditions;

(4) Monitoring well shall be sealed from the ground surface to the top of the filter pack;

(5) Monitoring wells or devices intercept the excavation zone or are as close to it as is technically feasible;

(6) The continuous monitoring devices or manual methods used can detect the presence of at least one-eighth of an inch of free product on top of the groundwater in the monitoring wells;

(7) Within and immediately below the UST system excavation zone, the site is assessed to ensure compliance with the requirements in paragraphs (f)(1) through (f)(5) of this section and to establish the number and positioning of monitoring wells or devices that will detect releases from any portion of the tank that routinely contains product; and

(8) Monitoring wells are clearly marked and secured to avoid unauthorized access and tampering.

(g) *Interstitial Monitoring.* Interstitial monitoring between the UST system and a secondary barrier immediately around or beneath it may be used, but only if the system is designed, constructed, and installed to detect a leak from any portion of the tank that routinely contains product and also meets one of the following requirements:

(1) For double-walled UST systems, the sampling or testing method can detect a release through the inner wall in any portion of the tank that routinely contains product;

Note: The provisions outlined in the Steel Tank Institute's "Standard for Dual Wall Underground Storage Tanks" may be used as guidance for aspects of the design and construction of underground steel double-walled tanks.

(2) For UST systems with a secondary barrier within the excavation zone, the sampling or testing method used can detect a release between the UST system and the secondary barrier;

(i) The secondary barrier around or beneath the UST system consists of artificially constructed material that is sufficiently thick and impermeable (at least 10^{-6} cm/sec for the regulated substance stored) to direct a release to the monitoring point and permit its detection;

(ii) The barrier is compatible with the regulated substance stored so that a release from the UST system will not cause a deterioration of the barrier allowing a release to pass through undetected;

(iii) For cathodically protected tanks, the secondary barrier must be installed so that it does not interfere with the proper operation of the cathodic protection system;

(iv) The groundwater, soil moisture, or rainfall will not render the testing or sampling method used inoperative so that a release could go undetected for more than 30 days;

(v) The site is assessed to ensure that the secondary barrier is always above the ground water and not in a 25-year flood plain, unless the barrier and monitoring designs are for use under such conditions; and,

(vi) Monitoring wells are clearly marked and secured to avoid unauthorized access and tampering.

(3) For tanks with an internally fitted liner, an automated device can detect a release between the inner wall of the tank and the liner, and the liner is compatible with the substance stored.

(h) *Other Methods.* Any other type of release detection method, or combination of methods, can be used if:

(1) It can detect a 0.2 gal/hr leak rate or a release of 150 gal within a month with a probability of detection of 0.95 and a probability of false alarm of 0.05; or

(2) The implementing agency may approve another method if the owner and operator can demonstrate that the method can detect a release as effectively as any of the methods allowed in paragraphs (c) through (h) of this section. In comparing methods, the implementing agency shall consider the size of release that the method can detect and the frequency and reliability with which it can be detected. If the method is approved, the owner and operator must comply with any conditions imposed by the implementing agency on its use to ensure the protection of human health and the environment.

§280.44   Methods of Release Detection for Piping

Each method of release detection for piping used to meet the requirements of §280.41 must be conducted in accordance with the following:

(a) *Automatic Line Leak Detectors.* Methods which alert the operator to the presence of a leak by restricting or shutting off the flow of regulated substances through piping or triggering an audible or visual alarm may be used only if they detect leaks of 3 gal/hr at 10 lb/in.^{2} line pressure within 1 hr. An annual test of the operation of the leak detector must be conducted in accordance with the manufacturer's requirements.

(b) *Line Tightness Testing.* A periodic test of piping may be conducted only if it can detect a 0.1 gal/hr leak rate at one and one-half times the operating pressure.

(c) *Applicable Tank Methods.* Any of the methods in §280.43(e) through (h) may be used if they are designed to detect a release from any portion of the underground piping that routinely contains regulated substances.

§280.45   Release Detection Recordkeeping

All UST system owners and operators must maintain records in accordance with §280.34 demonstrating compliance with all applicable requirements of this Subpart. These records must include the following:

(a) All written performance claims pertaining to any release detection system used, and the manner in which these claims have been justified or tested by the equipment manufacturer or installer, must be maintained for 5 years, or for another reasonable period of time determined by the implementing agency, from the date of installation;

(b) The results of any sampling, testing, or monitoring must be maintained for at least 1 year, or for another reasonable period of time determined by the implementing agency, except that the results of tank tightness testing conducted in accordance with §280.43(c) must be retained until the next test is conducted; and

(c) Written documentation of all calibration, maintenance, and repair of release detection equipment permanently located on-site must be maintained for at least one year after the servicing work is completed, or for another reasonable time period determined by the implementing agency. Any schedules of required calibration and maintenance provided by the release detection equipment manufacturer must be retained for 5 years from the date of installation.

24.6   COMPLYING WITH FINANCIAL RESPONSIBILITY REQUIREMENTS

Subtitle I of RCRA requires owners and operators of USTs to meet certain financial responsibility requirements for cleanup and third-party damages resulting from leaks that may occur. The EPA set up a schedule to phase in the financial responsibility requirements for USTs over a two-year period because of concerns about the unavailability of financial assurance mechanisms to large portions of the regulated community. The phase-in was designed to achieve balance between the need to ensure financial capability for UST releases and the necessary time for owners and operators to obtain assurance mechanisms.

The phase-in set different compliance dates for four compliance groups. As of January 1989, the first group, consisting of petroleum marketers owning 1000 or more USTs and nonmarketers with more than $20 million in tangible net worth, were required to comply with the financial responsibility requirements. In October 1989, the second group, consisting of petroleum marketers owning 100–999 USTs, were required to show financial responsibility. By April 26, 1991, the third group, consisting of petroleum marketers owning between 13 and 99 USTs, was required to comply. The fourth group, consisting of marketers with 1–12 USTs at more than one facility or fewer than 100 USTs at a single facility and nonmarketers with net worth of less than $20 million, was expected to comply by October 26, 1991.

On August 5, 1991 the EPA proposed to extend the October 26, 1991, compliance date to December 31, 1992 and was to make a final decision concerning any extension after considering public comments received on the proposed rule.

On December 19, 1991, the U.S. Environmental Protection Agency announced the extension of the compliance deadline for Group IV from October 26, 1991 to December 31, 1993 for petroleum marketers with 1–12 USTs and nonmarketers with a net worth of less than $20 million. The compliance deadline for local government was extended to February 18, 1994. Indian tribes owning USTs on Indian land were extended to December 31, 1998. The Agency believed that the extension would allow states more time to develop and implement financial assistance programs, such as loans, and grants, to assist tank owners in meeting the UST technical and financial responsibility requirements. Financial institutions are requiring owners and operators to meet UST technical requirements before they can purchase a financial mechanism to meet the financial responsibility requirement. This means that many owners under Group IV have to conduct site assessments and upgrade their sites or replace facilities with new UST systems prior to purchasing some sort of financial mechanism.

Owners and operators can comply with the financial responsibility requirements in a number of ways, such as self-insurance (which requires a financial test), guarantee, insurance and risk-retention group coverage, surety bond, letter of credit, use of state-required mechanisms, state funds, or other state assurance, trust fund, and standby trust fund. Owners and operators can use a single means or a combination of methods to satisfy the required coverage of financial requirement.

24.7   CORRECTIVE ACTION RESPONSE PROCESS

Owners and operators of petroleum and hazardous substance UST systems must respond to a leak or spill within 24 hr of release or within another reasonable period of time as determined by the implementing agency. The responses to releases from USTs depend on several different factors, largely site-specific. The recent inclusion of MTBE as a new contaminant of concern will pose new challenges in remediation of LUST sites. The effectiveness of remediation methods is directly linked to the physical and chemical characteristics of the constituent of interest. Because MTBE behaves differently in soil, air, and water than other petroleum constituents, the choice of an effective remediation technology may be different when MTBE is present at a site. Benzene is most often the contaminant of concern in gasoline because of its high solubility and known carcinogenicity. As a result, comparing the characteristics of MTBE with benzene is helpful in showing how remediation technologies may differ when MTBE is added to gasoline. MTBE is 30 times more solubile than benzene in water. Pure MTBE can reach an equilibrium concentration in water of approximately 5% (i.e., 48,000 mg/L). When moving from liquid phase (i.e., free product) to the vapor phase, MTBE is three times more volatile than benzene (i.e., the vapor pressure of MTBE is three times the vapor pressure for benzene). When moving from the dissolved phase (in water) to the vapor phase, MTBE is about ten times less volatile than benzene (i.e., its Henry's law constant is 1/10th benzene). MTBE is much less likely then benzene to adsorb to soil or organic carbon. MTBE is more resistant to biodegradation than benzene. When MTBE is in the soil as the result of a petroleum release, it may separate from the rest of the petroleum, reaching the groundwater first and dissolving rapidly. Once in the groundwater, MTBE travels at about the same rate as the groundwater, whereas benzene and other petroleum constituents tend to biodegrade and adsorb to soil particles.

Because MTBE behaves differently from petroleum hydrocarbons when released into the environment, a remediation investigation may need to be modified to properly characterize the area of MTBE contamination (3). Corrective action usually involves two phases. The first involves the initial corrective actions intended to limit the impact of a sudden or newly discovered release. The second involves long-term, permanent corrective measures. Initial corrective actions (or responses) are those directed at immediately containing and controlling a release. Timing is critical. Therefore, efforts are focused on source control and public welfare. This includes the collection and containment of released material. Permanent corrective measures, on the other hand, involve comprehensive cleanup to protect human health and the environment. Both require an assessment of the situation, which includes the physical and chemical nature of the released substance, the environmental setting of the incident, and the extent of the impact. The options available for addressing both initial and long-term corrective actions are discussed below.

Initial Corrective-Action Options

When a leak in an UST is discovered or occurs suddenly, the initial corrective actions are directed toward collection and containment of the substance released. Initial efforts typically occur within a short time frame, are of brief duration and involve limited resources. This often entails deployment of field personnel and equipment to the scene within hours of the occurrence to minimize the impact of the release (5).

Permanent Corrective-Action Options

After the initial response, the focus of assessment and investigation activities turns toward the need for permanent corrective measures. In situations where the release poses no danger to groundwater resources, human health, and the environment, corrective action may be limited to pumping the tank and removing significantly contaminated soils. The appropriate local, state, and federal agencies should be consulted for more specific corrective-action guidelines.

Two aspects of corrective actions determine their desirability from a performance standpoint: effectiveness and useful life. Effectiveness refers to how well the corrective action technology accomplishes its intended purpose, such as containment, diversion, removal, destruction, or treatment. The effectiveness of available alternatives should be determined either through design specifications or by performance evaluation. Useful life refers to the length of time a corrective action technology can maintain the desired level of effectiveness. Some corrective action technologies deteriorate with time. Such deterioration can sometimes be slowed down through proper system operation and maintenance, but the technology may eventually require replacement. Each corrective action alternative should be evaluated in terms of the projected service lives of its component technologies (6).

Two aspects of UST corrective action technologies that provide information about reliability are their operation and maintenance requirements (i.e., frequency and complexity) and their demonstrated reliability of similar UST sites (6). Demonstrated and expected reliability is a way of measuring the risk and effect of failure. The engineer should evaluate whether the considered technologies have been used effectively at similar sites; whether a combination of technologies has been used effectively; whether the failure of any one technology has an immediate impact on receptors; and whether the alternative has the flexibility to deal with uncontrollable changes at the site.

Another important consideration in the selection process is whether the corrective action can be implemented with relative ease, i.e., whether it can be constructed or installed within the time frame required to achieve the desired results.

24.8 CLEANUP OF RELEASES FROM PETROLEUM USTs: SELECTED TECHNOLOGIES (7)

Only a limited number of technologies to clean soil, air, and water of the contaminants principally associated with gasoline are available that have demonstrated performance records and have progressed to full-scale application. This section reviews these technologies in terms of their removal efficiencies, limitations, and costs.

Recovery of Free Product from Water Table

The two technologies most commonly used to limit the migration of floating gasoline across the water table are the trench method and the pumping well method. A variety of equipment can be used to recover the free product. Typically, skimmers, filter separators, and oil/water separators are used in trench recovery, and single- and dual-pump systems are used with the pumping wells.

The trench method is most effective when the water table is no deeper than 10–15 ft below the ground surface. Excavation of the trench is easy to undertake, and with this method the entire leading edge of the gasoline plume can be captured. However, the trench method does not reverse groundwater flow, so it may not be appropriate when a potable well supply is immediately threatened. The cost of this system is about $100/yd^3 of soil excavated.

A pumping well system is normally used for deep spills, when water table depth exceeds 20 ft below the ground surface. The direction of groundwater flow can be reversed with this system. The cost of this system range from $100 to $200/ft for 4-in. to 10-in. gavel-packed galvanized steel wells. This cost includes engineering and labor.

Skimmers, filter separators, surface-mounted product recovery pumps, aboveground oil/water separators, and dual pump systems can all be used to separate gasoline from groundwaters. Dual pump systems and oil/water separators are typically used for deeper releases. Skimmers can achieve up to 99% recovery of all hydrocarbons floating on the water surface. The cost is about $6,000 to 7,000; the addition of a water table depression pump to expedite gasoline flow can increase capital cost of skimming system to $12,000 to 13,000 (approximately doubles the cost). Filter separators can reduce spill thickness to a sheen.

For additional recovery, the top layer of the gasoline–water mixture must be removed from the well and treated aboveground. Filter separators can only be used to recover spills 20 ft or less below ground surface, and only with surface-mounted pumps. (Submersible pumps would cause the floating separator to sink.) The cost is approximately $6,000 to 7,000; the addition of a water table depression pump to increase gasoline flow can increase capital costs of the separation system to $12,000 to 13,000 (about double).

Aboveground oil–water separators are large tanks into which the recovered gasoline–water mixture is pumped and allowed to separate. Tanks range in size from 1,000-gal units to 10,000-gal units. To achieve the necessary retention time, separators must be sized at least 10 times larger than the groundwater extraction rate. The cost range from about $6,000 (1,000 gal) to $16,000 (10,000 gal).

Dual pump systems can remove up to 99% of free-floating product. The most commonly used gasoline–water separation units consist of a water table depression pump and a product recovery pump. The depression pump creates cones of influence that allow gasoline to accumulate; the product recovery pump, which is equipped with gasoline sensors, brings only the gasoline to the ground surface. The cost is approximately $12,000 to 14,000 for the two types of pumps. Because at least two pumps are required, operation and maintenance (O&M) costs are higher than with other methods.

Case studies of groundwater contamination have led to three conclusions:

1. The cost of recovering free product at a site depends more on the recovery method and equipment required for the cleanup than on the size of the spill.

2. More than one gasoline-recovery option may be feasible at a given spill.

3. The costs of free product recovery are small compared to the cost of restoring hydrocarbon-contaminated groundwater to drinking water standards.

The case studies review involved spills from 2,000 to 100,000 gal. Costs of recovery ranged from $43,000 to 225,000 (including equipment, labor, engineering, and hydrogeologic services). On average, only 29% of the spilled product was recovered, at a cost range of $2 to 93/gal.

Removal of Gasoline From Unsaturated Soils

Soil treatment is an essential component of a corrective action plan. After a spill, hydrocarbons in the unsaturated zone can eventually enter the groundwater, if the soil is not treated. A number of techniques are used, but they vary in cost and effectiveness. Excavation and disposal is the most widely used corrective action for contaminated soil. Other methods include enhanced volatilization, incineration, venting, soil washing/extraction, and microbial degradation. With respect to MTBE, because it has a very high vapor pressure and low affinity for sorption to soil, MTBE can be effectively remediated by soil treatment technologies, typically without any costs beyond those needed for remediating other petroleum constituents. Biomedial methods for soil treatment (i.e., land-farming, bioventing, biopiles) are currently not recommended for removing MTBE because it is considered recalcitrant to biodegradation. This recommendation may change in the future as new research examines the efficacy of specific strains of bacteria and/or improved methods pf biograding MTBE (3).

Excavation and disposal can be 100% effective. Moreover, soil excavation as an adjunct to removal of underground storage tanks (USTs) may help to eliminate the major source of continuing gasoline migration to the subsurface. The following limitations must be kept in mind:

- Standard backhoes (0.5-yd^3 capacity) can reach only a maximum depth of 16 ft. Larger backhoes (3.5 yd^3 capacity) are available that can remove soils at depths of up to 45 ft.
- Excavation is difficult in heavily congested areas or in areas close to or under buildings.
- The more soil brought to the surface, the greater the risk of exposure.
- Although tipping fees at some landfills are a reasonable $12/yd^3, disposal can cost up to $160/yd^3 if the soil is considered hazardous.
- The lack of uniform guidelines among the states for the disposal of contaminated soils means that transport risks may run high, as soil is sent from states with strict guidelines to the more permissive states.

The cost is $200–300/yd^3, which is relatively expensive. The result is that only small quantities of contaminated soil, say, 500 yd^3, are normally excavated and disposed. The trend is toward applying alternatives to land disposal, such as incineration or biodegradation, by which contaminants are destroyed.

Disposal of contaminated soils in batch asphalt plants is a practice not yet reported in the literature, but may be more common than most people realize. Some plants charge $55/yd^3 for accepting gasoline-contaminated soils; other plants refuse such soils because they must then observe the state laws governing hazardous waste treatment facilities.

Theoretically, up to 99.99% of volatile organic compounds (VOCs) can be removed by enhanced volatilization, but this soil treatment method has not been widely applied in the field. Different methods of enhancing volatilization include rototilling, mechanical aeration, pneumatic conveyor systems, and low-temperature thermal stripping. Only thermal stripping has been documented to successfully remove contaminants with vapor pressures comparable to those of gasoline constituents. Limitations to enhanced volatilization include soil characteristics that constrain the movement of gasoline vapors from the soil to the air; contaminant concentrations that may create an explosion hazard; and the need to control dust and organic vapor emissions. The cost is $245–320/yd^3 of soil treated. This technique is most effective with 15,000–18,000 tons of soil.

By complete oxidation, incineration can eliminate 99.99% (or more) of gasoline constituents in soil. This technology is widely practiced and highly reliable. The associated limitations are that the soil must be brought to the surface, which increases the risk of exposure; incineration is usually appropriate only when toxics other than volatiles are present; and the permitting requirements may cause time delays. The cost is $200–640/yd^3 of soil. However, soil volumes of less than 20,000 yd^3 will increase costs considerably.

Venting allows for the removal of gasoline vapors from unsaturated soils without excavation. It has been demonstrated to be effective in recovering as much as 99% of gasoline components in unsaturated soil. The technology has not been widely applied in the field, however, partly because critical design parameters remain undefined. Moreover, its effectiveness is uncertain because soil characteristics may impede free movement of vapors, create an explosion hazard, or cause high levels of organic emissions. Venting is relatively easy to implement and causes minimal disturbance to structures or pavement. The cost is $15–20/yd^3, which is inexpensive. It becomes even more cost-effective when soil volumes exceed 500 yd^3.

With soil washing and extraction, contaminants are leached from the soil into a leaching medium, after which the extracted contaminants are removed by conventional methods. Removal of 99% of volatile organic compounds is possible under ideal conditions, but typical removal rates are less. High percentages of silt and clay in the soil may impede the separation of the solid and liquid after the washing phase. Since the process requires physical separation techniques (e.g., distillation, centrifugation, and evaporation), pilot studies are recommended before final design and implementation. The cost is $150–200/yd^3 of contaminated soil.

Theoretically, gasoline removal efficiencies of 99% or more can be achieved with microbial degradation of contaminants. The technique has not been widely applied in the field, and additional research is required to confirm cost and effectiveness. The advantages of this technique are that the soil is usually treated in situ and volatiles are completely destroyed. Gasolines composed principally of alkenes in the C_5 to C_{10} range would be the quickest to degrade. For its effectiveness, the technique is dependent on oxygen levels, nutrient levels, temperature, and moisture content of the soil. The cost is approximately $66–123/yd^3. The combination of soil venting and microbial degradation is often one of the least costly and most effective corrective actions for treating gasoline-contaminated soils.

Much confusion exists about the hazard posed by gasoline-contaminated soil and how the soil should be treated. An informal survey of several states revealed that none require soil testing during UST excavations. Many states do require a fire marshal to be present to determine explosion hazards at sites where visual inspection show soil to be contaminated. Landfill is the principal mode of disposal of contaminated soils, and time delays are common. None of the states surveyed have regulations preventing open aeration of contaminated soils to reduce the volatile organic compounds. Many excavators admit placing gasoline-contaminated soils on plastic sheets until the volatiles disperse so that the soil can be trucked to the local landfill.

Removing Gasoline Dissolved in Groundwater

Two widely used technologies for removing that portion of the gasoline plume dissolved in groundwater are air stripping and filtration through granular activated carbon (GAC). However, biorestoration is a cost-effective and presently a more widely used alternative. With MTBE a relatively new contaminant of concern in groundwater and because MTBE behaves differently from petroleum hydrocarbons when released into the groundwater, selecting an effective remediation technology for removing MTBE contamination may be more challenging. In contrast with the preferred remediation techniques for petroleum hydrocarbons such as benzene (i.e., bioremediation), pumping contaminated groundwater and treating it above ground (i.e., "pump and treat") may more often be an effective remediation technology for MTBE because MTBE does not adsorb significantly to soil. In addition, because it is highly soluble, most of the MTBE mass may quickly dissolve into groundwater, making pumping an efficient method for removing large quantities of the contaminant. Air sparging is another groundwater remediation technology that has shown some promise. In addition to remediation of the source area, point-of-use treatment with GAC appears to be a common approach to addressing MTBE when contamination is limited to individual homes or private wells. Based on limited research and anecdotal information, the U.S. EPA's Office of Underground Storage Tanks estimates that at approximately 75% of MTBE-contaminated sites, the incremental cost increase of remediation will be less than 50% above the cost of remediating that same petroleum release without MTBE (3).

For most volatile organics found in gasoline, packed towers have maximum removal efficiencies of 99–99.5%. Through air stripping, effluent concentrations of $5\,\mu g/L$ volatile organics can be achieved. Concentrations less than $5\,\mu g/L$ are not usually achievable because the technology is constrained by the size of the tower that would be required to achieve such a high removal efficiency. Critical design parameters include the type of packing material used, the air-to-water ratio, the stripping factor, and the tower height. Plastic packings are the most widely used; they are inexpensive and lightweight. Air-to-water ratios from 20:1 to 100:1 are common for aromatics removal in general and for those in gasoline particular. Stripping factors between 3:1 and 5:1 are best suited for gasoline-related constituents. In designing a packed air tower, the following considerations are important:

- Zoning laws may restrict the maximum height of a tower.
- The tower, blower, and pumps may have to be enclosed, not only for noise reduction but also for aesthetic reasons.
- Influent air must be free of VOCs, so air intake must be situated to avoid "short circuiting" between the influent air and the tower effluent air.
- Gaseous demisters are usually needed to prevent water from leaving the top of the tower.
- Vapor-phase treatment, if required, will significantly increase the cost.

Because more contaminant can be absorbed in an air-to-carbon loading than in a water-to-carbon loading, vapor-phase treatment with GAC may be advantageous. The cost ranges from \$50,000–100,000 (including labor, engineering, and contingencies), which is 50–80% less than comparable costs to treat with GAC. On a volume-treated basis, typical costs at a leaking UST site are \$5–25 per thousand gallons.

GAC adsorption can remove as much as 99.99% of the organic compounds found in gasoline. To achieve effluent concentrations of $5\,\mu g/L$ or less for gasoline constituents, GAC is almost always required. Designing a GAC system is complex, as the following points illustrate:

- Each contaminant competes for carbon pore space.
- The EBCT (empty bed contact time) is directly related to the size of contactor needed; 15 minutes is the usual minimum contact time for gasoline spills.
- Fixed bed columns and pressure filters are normally used in cleaning leaking UST sites; use of pressure filters saves repumping costs because they allow higher surface loading rates and pressure discharge to the distribution system.
- The ability of a compound to be removed with GAC is a function of its solubility. Low-solubility compounds adsorb better than high-solubility compounds. The order in which gasoline components break through (from earliest to latest) is: benzene, ethylbenzene, toluene, xylene, naphthalene, and phenol.

- Some compounds found only in certain gasolines might break through earlier than benzene because of their low adsorption capacities: methyltertiary butyl ether (MTBE), ethylene dibromide (EDB), and ethylene dichloride (EDC). Less than 40% of today's gasolines contain EDB or EDC. Only 10% contain MTBE. It is more expensive to design for the removal of these compounds than it is to design for benzene removal.

- Effectiveness of system may be reduced by excessive iron or manganese, and hardness of the water. If iron concentration exceeds 5 mg/L, removal prior to carbon filtration is recommended.

- Spent carbon from leaking UST sites is usually landfilled. Caution must be exercised in handling gasoline-saturated carbon tanks because they can self-ignite.

- GAC is most effective when used with air stripping. Carbon life can be extended by treating gasoline-contaminated groundwater with packed air towers. A two-phase approach is best. The first phase is to install a packed air tower. Its performance would then be monitored to determine effluent concentrations, and the possible need for a second-phase treatment with GAC.

The cost ranges from $300,000–400,000 for a typical GAC unit. Costs include labor, engineering, and contingencies. O&M costs range from $25,000–30,000/yr.

Under proper condition, bioremediation can reduce by 99% trace concentrations of aromatic hydrocarbons. Its distinct advantage is that the gasoline contaminants are completely destroyed, not merely transferred to another environmental medium. Its applicability depends on dissolved oxygen concentrations, available nutrients, temperature, pH, salinity, concentrations of contaminants, presence of predators, and water content. Through biorestoration, effluent concentrations in the ppm-range (mg/L) can probably be attained; treatment to ppb-levels (μg/L) requires manipulation of the system (encouragement of cometabolism or degradation by an added substrate). Currently, the technology appears to work best as a "polishing" step. Few data exist on costs but it range from $30–40/yd^3 treated to $10,000/acre treated, and from $4–6/lb of contaminant removed.

Restoration of the polluted aquifer can often take months or years, during which time users of the water must find alternative water sources. Two alternatives are point-of-entry treatment systems and extension of the water distribution system.

Systems which treat water at the point of entry into a home are preferable to point-of-use systems that can be placed on individual taps. Research indicates that showering in water that contains volatile gasoline compounds may pose a serious health threat; therefore, only point-of-entry systems are considered appropriate at homes with gasoline-contaminated well water. There are several types of devices: reverse osmosis, ion exchange, distillation, aeration, and carbon adsorption. Carbon adsorption is the most effective in eliminating dissolved gasoline compounds, since it is capable of removing more than 99% of dissolved gasoline compounds, including benzene, toluene, and xylene.

Activated carbon can adsorb dissolved compounds for water, but only up to a point. To eliminate the risk of contaminant breakthrough, two carbon tanks in series are installed, and the effluent water is tested periodically for the presence of VOCs. The most serious limitation associated with carbon adsorption point-of-entry treatment systems is that significant changes in contaminant concentrations may go undetected. If the influent concentrations fluctuate and exceed the design capacity of the system, contaminant breakthrough could occur without the resident knowing it. For this reason, it is recommended only as an interim remedial measure in a home. The cost is broken down as follows: carbon tanks, from $700–900; carbon replacement and disposal, from $100–200 per replacement; testing for VOCs, $250. Additional water quality improvement equipment, such as chemical feed units, softeners, filter, retention tanks, and polishers, are in the $500–950 range for each piece of equipment. Case studies reported in this manual indicate that annual capital and O&M costs are from $4000–5000 per household.

Cost is usually the primary consideration in extending water mains to homes affected by a contaminated well field. This measure is often the appropriate long-term solution. Cost: transmission mains (for long distances) range from $27/linear foot (lf) for 6-in. mains to $40/lf for 12-in. mains, and $84/lf for 24-in. mains. Distribution mains (for short distances between the transmission main and the individual home or building) range from $44/lf for 6-in. mains, to $56/lf or 12-in. mains, to $100/lf for 24-in. mains.

24.9 SUMMARY

1. Environmental contamination from leaking underground storage tanks poses a significant threat to human health and the environment. The discovery of MTBE as a new contaminant of concern will present new clean-up challenges as we enter the new millennium.

2. Federal laws were enacted in response to the increasing problems resulting from leaking underground storage tanks. These laws generally are based on many state and industry ongoing efforts. Congress provided for federal regulation of underground storage tanks as part of the 1984 RCRA amendments but it exempted residential heating oil and small farm tanks.

3. The problem of leaking underground storage tanks presents several major challenges not only for the EPA but also for states, local governments, and industry.

4. The EPA is using a new approach to implementing the Congressional mandate to address underground tanks, the "franchise concept." The EPA sees itself as the franchiser with the responsibility to see that the franchises, in this case the states and counties, run their operations successfully.

5. The federal regulations consist of those major sets of regulations. The technical regulations as set forth in 40 CFR Part 280 addressing technical standards for corrective action requirements for owners and operators of underground storage tanks.

6. Subtitle I of the Resource Conservation and Recovery Act (RCRA) requires owners and operators of underground storage tanks (USTs) to meet certain financial responsibility requirements for cleanup and third-party damages resulting from leaks that may occur.

7. Owners and operators of petroleum and hazardous substance UST systems must respond to a leak or spill within 24 hours of release or within another reasonable period of time as determined by the implementing agency. The re-

sponses to releases from USTs depend on several different factors, largely site-specific.

8. Only a limited number of technologies to clean soil, air, and water of the contaminants principally associated with gasoline are available that (a) have demonstrated performance records and (b) have progressed to full-scale application.

REFERENCES

1. U.S. EPA, Environmental Progress and Challenges: *EPA's Update*, EPA–230–07–88–033, August 1988.
2. U.S. EPA, MTBE Fact Sheet No. 1, EPA 510–F–98–001, January 1998.
3. U.S. EPA, MTBE Fact Sheet No. 2, EPA 510–F–98–002, January 1998.
4. Rules and Regulations, *Federal Register*, 52(185), 27197–37207, September 23, 1988.
5. U.S. EPA, *Underground Storage Tank Corrective Action Technologies*, EPA/625/6–87–015, January 1987.
6. U.S. EPA, *Handbook — Remedial Action at Waste Disposal Sites* (Revised), EPA-625/6–85–006, 1985a.
7. U.S. EPA, *Cleanup of Releases from Petroleum USTs: Selected Technologies*, EPA/530/UST–88/001, April 1988.

25

METALS

25.1 INTRODUCTION

The dangers of human exposure to many metallic chemical elements have long been recognized. Metals such as lead, mercury, cadmium, and arsenic are toxic. Health effects range from retardation and brain damage, especially in children from lead poisoning, to the impairment of the central nervous system as a result of mercury exposure. Since these toxic metals are chemical elements, they cannot be broken down by any chemical or biological process. As a result of this, it is not uncommon to have metallic buildup or bioaccumulation.

Table 25.1 lists four common metals and the health effects associated with each. The following sections will describe these concerns in more detail, and explain how exposure to these substances can be reduced.

25.2 LEAD

Lead was once the most popular metal used in the gasoline and paint industries. In ancient times, it may have been the first metal to be smelted by man. Lead pipes used by Romans are still in use today, and pottery glazed with lead oxide dates back to the bronze age.

Lead is a bluish-white metal with a bright luster. It is soft, malleable, and ductile. It is a poor conductor of electricity and is very resistant to corrosion. Lead ores commonly occur with zinc, copper, and pyrite ores. Galena (lead sulfide) accounts for more than 90% of primary lead production at present. There is also a large secondary lead industry that recycles lead from batteries and other lead scrap.

Lead in Drinking Water

The primary source of lead in drinking water is corrosion of plumbing materials, such as lead service lines and lead solder, in water distribution systems and in houses and larger buildings. Virtually all public water systems serve households

TABLE 25.1 Health effects of common metals

Pollutant	Health Concerns
Lead	Retardation and brain damage, especially in children
Cadmium	Affects the respiratory system, kidneys, prostate, and blood
Mercury	Several areas of the brain as well as kidneys and bowels, are affected
Arsenic	Causes cancer

with lead solders of varying ages, and most faucets are made of materials that can contribute some lead to drinking water. You cannot see, smell, or taste lead, and boiling your water will not get rid of lead. If you think your plumbling might have lead in it, have your water tested for lead. The only way to know if you have lead in your water is to have it tested. Call your local health department or your water supplier to see how to get it tested. Testing your water is easy and cheap ($15 to 25). Household water will contain more lead if it has sat for a long time in the pipes, or is hot or naturally acidic. If your water has not been tested or has high levels of lead, (a) do not drink, cook, or make baby formula with water from the hot water tap: (b) if cold water hasn't been used for more than two hours, run it for 30–60 sec before drinking it or using it for cooking; and (c) consider buying a filter certified for lead removal. Call the EPA's Safe Drinking Water Hotline for more information (1).

The health effects related to the ingestion of too much lead are very serious and can lead to impaired blood formation, brain damage, increased blood pressure, premature birth, low birth weight, and nervous system disorders. Young children are especially at risk from high levels of lead in drinking water (2).

Lead in Gasoline

Lead has long been used in gasoline to increase octane levels to avoid engine knocking. Lead is a heavy metal that can cause serious physical and mental impairment. Children are particularly vulnerable to effects of high lead levels. Two efforts begun 25 years ago are responsible for a 95% reduction in the use of lead in gasoline.

Recognizing the health risks posed by lead, the EPA in the early 1970s required the lead content of all gasoline to be reduced over time. The lead content of leaded gasoline was reduced in 1985 from an average of 1.0 g/gal to 0.5 g/gal, and still further in 1986 to 0.1 g/gal.

In addition to phasing down of lead in gasoline, the EPA's overall automotive emission control program required the use of unleaded gasoline in many cars beginning in 1975. Currently, all the gasoline sold in this country is unleaded (1).

Lead in Paints

Lead compounds were commonly used in the formulation of house paint. As the paint chipped, many young children ate the paint chips, which resulted in brain damage. Since the early 1940s, nontoxic pigments have gradually replaced lead in paints. In 1978, the Consumer Products Safety Commission (CPSC) banned the sale of lead for use in residential paints, and made it illegal to paint children's toys and household furniture with lead-based paints. Most states have outright banned the use of lead in house paints. If a home has been painted with lead-based paint, extreme caution must be taken when sanding or remodeling these surfaces. Lead dust levels are 10–100 times greater in homes where sanding or open-flame burning of lead-based paints has occurred.

If paint being removed is suspected of containing lead, have it tested. Leave lead-based paint undisturbed. Do not sand or burn it off. Cover lead-based paint with wallpaper or other building material. Replace moldings and other woodwork or have them removed and chemically treated off-site(3).

If your home was built before 1978, you should be concerned about lead-based paint hazards. The older your house is, the more likely it is to contain lead-based paint. Even if the original paint has been covered with new paint or another covering, cracked or chipped painted surfaces can expose the older, lead-based paint layers, possibly creating a lead hazard.

If you are removing paint or breaking through painted surfaces, you should be concerned about lead-based paint hazards. If your job involves removing paint, sanding, patching, scraping or tearing down walls, you should be concerned about exposure to lead-based paint hazards. If you are doing other work, such as removing or replacing windows, baseboards, doors, plumbing fixtures, heating and ventilation duct work, or electrical systems, you should be concerned about lead-based paint hazards, since you may be breaking through painted surfaces to do these jobs. Getting the right equipment and knowing how to use it are essential steps in protecting yourself during remodeling or renovating. A high-efficiency particulate air (HEPA) filter-equipped vacuum cleaner is a special type of vacuum cleaner that can remove very small lead particles from floors, window sills, and carpets and keep them inside the vacuum cleaner. Regular household or shop vacuum cleaners are not effective in removing lead dust. They blow the lead dust out through their exhausts and spread the dust throughout the home. HEPA vacuum cleaners are available through laboratory safety and supply catalogs and vendors. They can sometimes be rented at stores that carry remodeling tools. You will also need to use a properly fitted respirator with HEPA filters to filter lead dust particles out of the air you breathe. Make sure you buy specific HEPA filters — they are always purple. Dust filters and dust masks are not effective in preventing you from breathing in lead particles. Protective clothes, such as coveralls, shoe covers, hats, goggles, and gloves should be used to help keep lead dust from being carried into areas outside of the work site. Wet-sanding equipment, wet/dry abrasive paper, and wet-sanding sponges can be purchased at hardware stores; spray bottles for wetting surfaces to keep dust from spreading can also be purchased at general retail and garden-supply stores(4). Most importantly, keep all nonworkers, especially children, pregnant women, and pets, outside of the work areas while doing remodeling or renovation work until cleanup is completed.

Lead in Municipal Solid Waste

Lead is widespread in the municipal solid waste (MSW) stream; it is in both the combustible and noncombustible portions of MSW. Discharges of lead in MSW are overwhelmingly greater than discards of cadmium.

Lead-acid batteries, primarily batteries for automobiles, rank first, by a wide margin, of the products containing lead that enter the waste stream. Trends in

TABLE 25.2   Lead in products discarded in MSW,[a] 1970 to 2000

Products	1970	1986	2000[b]	Tonnage	Percentage
Lead-acid batteries	83,825	138,043	181,546	Increasing	Variable
Consumer electronics	12,233	58,536	85,032	Increasing	Increasing
Glass and ceramics	3,465	7,956	8,910	Increasing	Increasing; stable after 1986
Plastics	1,613	3,577	3,228	Increasing; decreasing after 1986	Fairly stable
Soldered cans	24,117	2,052	787	Decreasing	Decreasing
Pigments	27,020	1,131	682	Decreasing	Decreasing
All others	12,567	2,537	1,701	Decreasing	Decreasing
Totals	*164,840*	*213,652*	*181,887[b]*		

[a]Data is given in short tons.
[b]Estimated.

quantities of lead discarded in products in MSW, ranked by tonnage discarded in 1986, are shown in Table 25.2. The last two columns on the table indicate whether the total tonnage of lead in a product is generally increasing or decreasing, and whether the percentage of total lead contained in MSW contributed by a product is increasing or decreasing.

Lead discards in batteries are shown to be growing steadily, as are discards in consumer electronics. Discards of lead solder in cans and lead in pigments, however, virtually disappeared between 1970 and 1986. Lead discards in other products are known to be relatively small.

Findings about the individual products in MSW that contain lead are as follows:

- *Lead-acid batteries* contributed 65% of the lead in MSW in 1986; this percentage has ranged between 50% and 85% during the 1970 to 1986 period studied. The tonnages in Table 25.2 represent discards after recycling, but of all the products considered, only lead-acid batteries are recycled to a significant extent. Recycling rates, which have ranged from 52 to 80%, have a major effect on the tonnage of lead-acid batteries discarded.

- *Consumer electronics* (television sets, radios, and video cassette recorders) accounted for 27% of lead discards in MSW in 1986. They contributed lead from soldered circuit boards, leaded glass in television sets, and plated steel chassis. Leaded glass accounts for most of the lead in these products.

- *Glass and ceramics*, as reported here, include lead in products such as glass containers, tableware and cookware, and other items such as optical glass. These contributed 4% of lead discards in 1986. (Leaded glass in light bulbs is included in the "All others" category in Table 25.2.)

- *Plastics* use lead in two ways: as a heat stabilizer (primarily in polyvinyl chloride resins) and as a component of pigments in many resins. This category, which includes products such as nonfood packaging, clothing and footwear, housewares, records, furniture, appliances, and other miscellaneous products, accounted for about 2% of lead discards in 1986. Plastics in consumer electronics products are counted under that category.

- *Soldered cans* have experienced a large decline in usage since 1970, when they contributed 14% of the lead in MSW. Leaded solder is currently used in steel food cans, general purpose cans (like aerosols), and shipping containers.

- *Pigments* containing lead compounds have declined greatly since 1970, dropping from 18% of total lead discards to less than 1%. This category includes pigments used in paints, printing inks, textile dyes, etc. Pigments used in plastics, glass and ceramics, and rubber products are accounted for in those categories.

- *All others* include brass and bronze products, light bulbs (which contain lead in solder and in glass), rubber products, used oil, collapsible tubes, and lead foil wine bottle wrappers. Collapsible tubes contributed over 5% of total lead discards in 1970, but their use has declined dramatically since then. None of the other items has exceeded 1% of the total since 1970 (5).

Commonly Used Lead Compounds

Lead is used in many compounds. Some of the more common sources are briefly described below.

Lead monoxide (PbO), commonly called litharge, is the most widely used inorganic lead chemical. It is used in storage battery plates, ceramics and glasses, paint, rubber, and other products. It is also used in the production of other lead chemicals such as lead orthoplumbate.

Lead dioxide (PbO_2) is used as an oxidizing agent in the manufacture of chemicals and dyes and as a curing agent for sulfide polymers. It is the active materials of the positive plate in electric storage batteries.

Lead orthoplumbate (Pb_3O_4) is commonly called red lead. It is a brilliant red pigment, and is used as an inhibitor in surface coatings to prevent corrosion of metals. It is used in storage batteries, leaded glass, lubricants, and rubber. It is also used for making lead dioxide.

Lead sulfide (PbS), or galena, is a common lead mineral. Lead sulfide has semiconducting properties. It is also used for mirror coatings and it is a component of blue lead pigments.

Lead metaborate ($PbO \cdot B_2O_3 \cdot H_2O$) is used in glazes, enamels, and glasses.

Basic lead carbonate [$2PbCO_3 \cdot Pb(OH)_2$] is commonly called white lead. It is the most important basic salt of lead. It is a white pigment and is used in surface coatings, greases, and plastic stabilizers.

Lead silicates. The most common silicate of lead is lead metasilicate ($PbSiO_3$). The silicates are used in ceramics, glasses, paints, rubber, and as stabilizers in plastics.

Basic lead sulfates ($XPbO \cdot PbSO_4$) are used as white or blue pigments in paints and as stabilizers for plastics. They are also used as a filler in rubbers and in inks.

Lead chromate ($PbCrO_4$) is an important pigment and is often formulated in combination with other lead compounds or with inorganic salts of other metals to make a range of colors, including chrome green, chrome yellow, and molybdate chrome orange. Chrome yellows contain lead sulfate; chrome greens contain iron cyanides; and molybdenum chrome oranges contain molybdate and often lead sulfate. These pigments are used in paints, coatings, inks, and leather goods.

Basic lead chromate ($PbCrO_4 \cdot PbO$) is used in pigments commonly called chrome oranges.

Lead chloride ($PbCl_2$) can be prepared by the reaction of lead monoxide or basic lead carbonate with hydrochloric acid. Most of its uses are industrial rather than in products that would commonly enter municipal solid waste. Lead chloride is used as a catalyst, as a cathode for seawater batteries, as a flame retardant in polycarbonates, as a flux for the galvanizing of steel, as a photochemical-sensitizing agent for metal patterns on printed circuit boards, and for other uses.

Lead salts are formed of lead and organic acids. Several lead salts of higher fatty acids (C_{10} and over; commonly called lead soaps) have important uses as paint driers, stabilizers for plastics, additives to lubricating oil, and additives in rubber (6–8).

25.3 CADMIUM

Cadmium is a toxic metal most commonly known for its use in rechargeable nickel–cadmium batteries. It is a relatively rare metal that has some unique characteristics that make it useful in a variety of products. Cadmium is silvery-white in color and is soft, ductile, and easily worked. It has good electrical and thermal conductivity. When exposed to moist air, cadmium oxidizes slowly to form a thin coating of cadmium oxide, which protects the metal from further corrosion.

Cadmium usually occurs as the mineral greenockite (CdS). It is usually mined in association with zinc, but sometimes with lead and copper ores. It is almost never found alone in economical quantities. Secondary (recycled) cadmium production is of minor significance in the United States. Unlike lead, which has been used since ancient times, cadmium has been refined and utilized only relatively recently.

Consumption of cadmium by end use is reported by the Bureau of Mines (7). The end use categories are more limited than those reported for lead, however. The categories reported annually include coating and plating, batteries, pigments, plastics stabilizers, and other (including alloys).

While consumption of lead in the U.S. was over 1.2 million tons in 1986, consumption of cadmium was a relatively small 4800 tons. Overall, the domestic consumption of cadmium in the U.S. declined until 1983, but it has increased since then. In both percentage and tonnage, coating and plating and plastic stabilizers have declined since 1970. Use of cadmium in pigments grew in the early 1970s, but has remained about stable since 1975. Domestic use of cadmium in nickel–cadmium batteries has been significant, although the tonnage has been fairly stable since 1976.

Nickel–Cadmium Batteries

Nickel–cadmium (Ni–Cd) batteries are a major consumer of cadmium in the United States. Ni–Cd batteries were invented in the early years of the twentieth century, but were not used extensively until the mid-1940s when they came into use in the military and industrial sectors. Ni–Cd batteries are secondary batteries (rechargeable).

In the early 1960s Ni–Cd batteries for consumer use were developed, but they did not gain real popularity until the early 1970s. Ni–Cd batteries are now used in many products: pocket calculators; toys; microprocessors; hand tools such as portable drills, flashlights, screwdrivers, hedge trimmers, and soldering irons; and rechargeable appliances such as hand-held vacuums, mixers, can openers, VCRs, portable televisions, cameras, electric shavers, lawn mower engine starters, and alarm systems. Many consumer applications such as appliances have the Ni–Cd battery sealed in; the battery cannot be replaced by the owner.

New consumer uses for Ni–Cd batteries, such as portable laptop computers and cellular telephones, are continually being developed. Ni–Cd batteries are also competing with mercury batteries in hearing aids and pocket calculators, and with carbon–zinc and other primary batteries, because their initial high cost can be offset as they are recharged. Military and industrial uses of nickel–cadmium batteries include railroad signaling, diesel locomotive starting, commercial and jet aircraft starting, satellites, missile guidance systems, naval signaling, television and camera lighting, portable hospital equipment, computer memories, pinball machines, and gasoline pumps (9).

Use of Cadmium Pigments

Yellow pigments composed of cadmium sulfide, barium sulfate, and zinc sulfate are used in textile printing. The quantities used are very small, however, because of high costs and poor tinting capabilities. The same is true for cadmium orange or red pigments (cadmium sulfoselenide or mercadmium compounds).

The plastics industry is the largest end user of cadmium pigments. These pigments disperse well in most polymers and give good color and high opacity and tinting strength. Cadmium pigments also are insoluble in organic solvent, and have good resistance to alkalis.

Cadmium reds and maroons, the most durable of the cadmium pigments, are used in automobile finishes. Cadmium reds are coprecipitated and cocalcined mixtures of cadmium sulfide and cadmium selenide. Mercury cadmium pigments are also used occasionally (9).

Cadmium in Municipal Solid Waste

Like lead, cadmium is widespread in products discarded into MSW, although it occurs in much smaller quantities overall. Since 1980, nickel–cadmium household batteries have been the number one contributor of cadmium in MSW.

Trends in quantities of cadmium discarded in products in MSW (ranked by tonnage discarded in 1986) are shown in Table 25.3. Discards of cadmium in household batteries were small in 1970, but then increased dramatically. Cadmium discards in plastics are relatively stable. Discards of cadmium in consumer electronics are shown to decrease over time, while the other categories are relatively small.

Findings about cadmium discards in individual products in MSW are:

- *Household batteries* (rechargeable nickel–cadmium batteries) have accounted for more than half of cadmium discards in the U.S. since 1980. This growth is projected to continue into the 21st century, unless they are replaced by another type of battery.

TABLE 25.3 Cadmium in products discarded in MSW,[a] 1970 to 2000

Products	1970	1986	2000[b]	Tonnage	Percentage
Household batteries	53	930	2,035	Increasing	Increasing
Plastics	342	520	380	Variable	Variable; decreasing after 1986
Consumer electronics	571	161	67	Decreasing	Decreasing
Appliances	107	88	57	Decreasing	Decreasing
Pigments	79	70	93	Variable	Variable
Glass and ceramics	32	29	37	Variable	Variable
All others	12	8	11	Variable	Variable
Totals	*1,196*	*1,788*	*2,684[b]*		

[a]Data is given in short tons.
[b]Estimated.

- *Plastics* continue to be an important source of cadmium in MSW, contributing 28% of discards in 1986. Cadmium is used in stabilizers in polyvinyl chloride resins and in pigments in a wide variety of plastic resins. Cadmium is found in nonfood packaging, footwear, housewares, records, furniture, and other plastic products.
- *Consumer electronics* (television sets and radios) formerly had cadmium-plated steel chassis in many cases. These chassis have been replaced by circuit boards, so cadmium discards in consumer electronics are declining as the older units are replaced. They contributed 9% of the total in 1986.
- *Appliances* (dishwashers and washing machines) formerly had cadmium-plated parts to resist corrosion. This source of cadmium is declining as cadmium-plated parts are replaced by plastics, which are themselves another source of cadmium discards in appliances. Cadmium discards from appliances accounted for about 5% of total in 1986.
- *Pigments* used in printing inks, textile dyes, and paints may contain cadmium compounds, although this is not a large source of cadmium in MSW (about 4% of the total).
- *Glass and ceramics* may contain cadmium as a pigment, as a glaze, or as a phosphor. This is a relatively small source of cadmium in MSW.
- *All other* sources of cadmium include rubber products, used oil, and electric blankets and heating pads. These contribute very small amounts of cadmium to MSW.

Commonly Used Cadmium Compounds

As with lead, cadmium is commonly used in many compounds. Some of the more common sources are briefly described next.

Cadmium chloride ($CdCl_2$) is used in the manufacture of nickel–cadmium batteries. It is also used as a pigment in dyeing and calico printing and in phosphors.

Cadmium oxide (CdO) has many uses, often in the preparation of cadmium products. It is used in processes in the manufacture of nickel–cadmium batteries, stabilizers for PVC, glass, phosphors, semiconductors, electroplating, and ceramic glass, among other uses.

Cadmium sulfide (CdS) is the most widely used cadmium compound. It is also called cadmium yellow, and is used in red and yellow pigments, in phosphors, as a photoconductor, and for other uses.

Cadmium hydroxide (CdH_2O_2) is used mainly as the active material in the negative electrodes of nickel–cadmium batteries.

Cadmium nitrate (CdN_2O_6) is used in the manufacture of nickel–cadmium batteries.

Cadmium sulfate (CdO_4S) is used in the manufacture of nickel–cadmium batteries.

Cadmium carboxcyclates are incorporated in stabilizers for polyvinyl chloride (PVC).

Cadmium acetate ($C_4H_6CdO_4$) is used for iridescent effects on pottery and porcelain and in electroplating.

Cadmium fluoride (CdF_2) is used in the manufacture of phosphors and glass.

Cadmium selenide (CdSe) is used in photoconductors, semiconductors, and phosphors.

Cadmium telluride (CdTe) is used in phosphors and semiconductors.

Cadmium salts are used as light stabilizers in plastics; cadmium/barium salts are used as heat stabilizers in plastics.

25.4 MERCURY

Mercury and most of its compounds are toxic substances. The dangers of human exposure to mercury have long been recognized. Nineteenth-century hat makers, for example, developed a characteristic shaking and slurring of speech from occupational exposure to large quantities of inorganic mercury during the manufacturing process—symptoms that gave rise to the phrase "mad as a hatter." Such problems result from impairment of the central nervous system. In addition, high levels of mercury can cause kidney damage, birth defects, and in extreme cases, death.

In the United States today the groups most likely to be exposed to unacceptably high levels of mercury are Native American and recreational fishermen, who routinely eat large amounts of their catch. The U.S. Food and Drug Administration considers a mercury level of 1 part per million (ppm) to be its action level—that is, the recommended upper limit for safe consumption. Several states have even stricter standards, which they use to issue health advisories (10).

The most common source of the silvery liquid can be found in the bulbs of thermometers and thermostats. Large quantities of mercury can also be found in one type of barometer or another. Small quantities are also found in fluorescent lights, mercury switches, batteries, hearing aids, smoke detectors, watches and cameras.

Mercury vapor in the atmosphere comes from a variety of natural and artificial sources: releases from the ocean and land surfaces, smelters, power plants, forest fires, mineral deposits, paint volatilization, disposal of batteries and fluorescent lamps, and the application of certain fungicides. In addition, other factors appear to be important in determining how readily mercury enters an aquatic food chain. It has been suggested, for example, that acidity may hasten the process, which has raised a further concern: that acid rain could lower the pH of vulnerable lakes and thus increase mercury contamination of the food chain. The potential importance of this effect, however, remains unclear. Neither the Everglades nor the pristine lakes of the upper Midwest have been substantially impacted by acid deposition.

Once deposited in a lake, mercury undergoes a complex series of physical and chemical changes. Some of it remains in solution, some volatizes back to the atmosphere, and some precipitates into sediments. While acidification has been suspected of accelerating uptake into the food chain, the exact nature of this role has not been clear. In particular, the effect could result either from pH changes or from the addition of sulfate ions in acid rain. The presence of dissolved organic carbon may either foster or inhibit uptake, depending on conditions, but again the reasons have remained unclear.

One of the most important issues is what environmental factors influence the formation of methyl mercury, the element's most toxic chemical compound and the one most easily incorporated into an aquatic food chain. More than 95% of the mercury in fish is in the form of methyl mercury. Most atmospheric mercury occurs simply in its elemental form, which must be oxidized before being carried to the earth in rain. In a lake, transformation of the oxidized form to methyl mercury, a process called methylation, takes place mainly through the action of bacteria. The amount of methyl mercury available in the water at any time depends on a balance between methylation and its opposite reaction, demethylation. Researchers are particularly interested in determining what factors can tip this balance one way or the other.

Fish can absorb the methyl mercury directly through their gills or ingest it by eating smaller organisms. Microscopic plants (phytoplankton) absorb considerable amounts of methyl mercury from the water, so the small fish that eat them receive an already concentrated dose. Large predatory fish then eat the smaller ones, and methyl mercury builds up in their bodies because it is difficult to excrete. Eventually, the magnitude of such bioconcentration becomes staggering. Fish at the top of a food chain may have levels of methyl mercury a million times the level in the surrounding water (10).

25.5 ARSENIC

Arsenic has long been known to cause cancer; however, most people associate arsenic with poisoning rather than with its toxic characteristics. The most common pesticide in farms and gardens was lead arsenate but it has been almost entirely replaced by synthetic pesticides. Around the house, arsenic compounds can be found in medicine cabinets, in rat poison, and in plant killers. However, most of the products containing arsenic are being replaced with products with less toxic effects.

25.6 SUMMARY

1. The dangers of human exposure to many metallic chemical elements have long been recognized. Metals such as lead, mercury, cadmium, and arsenic are toxic. Health effects range from retardation and brain damage, especially in children from lead poisoning, to the impairment of the central nervous system as a result of mercury exposure.

2. Lead was once the most popular metal used in the gasoline and paint industries and as a slodering agent in pipes. The health effects related to the ingestion of too much lead are very serious and can lead to impaired blood formation, brain damage, increased blood pressure, premature birth, low birth weight, and nervous system disorders.

3. Cadmium is a toxic metal most commonly known for its use in rechargeable nickel–cadmium batteries. It is a relatively rare metal that has some unique characteristics that make it useful in a variety of products.

4. Mercury and most of its compounds are toxic substances. The dangers of human exposure to mercury have long been recognized. Nineteenth-century hat makers, for example, developed a characteristic shaking and slurring of speech from occupational exposure. In addition, high levels of mercury can cause kidney damage, birth defects, and in extreme cases, death.

5. Arsenic has long been known to cause cancer; however, most people associate arsenic with poisoning rather than with its toxic characteristics.

REFERENCES

1. U.S. EPA, *Lead Poisoning and Your Children*, EPA–800–B–92–0002, September 1992.
2. U.S. EPA, "Environmental Progress and Challenges", *EPA's Update*, EPA–230–07–88–033, August 1988.
3. U.S. EPA, *The Inside Story, A Guide to Indoor Air Quality*, EPA/400/1–88/004, September 1988.
4. U.S. EPA, *Reducing Lead Hazards When Remodeling Your Home*, EPA–747–R–94–002, April 1994.
5. F. L. Smith, Jr., *A Solid Waste Estimation Procedure: Materials Flows Approach*, U.S. EPA Office of Solid Waste (SW-147), May 1975.
6. U.S. EPA, *Second Report to Congress: Resource Recovery and Source Reduction*, Office of Solid Waste Management Programs (SW-161), 1975.
7. U.S. EPA, *Fourth Report to Congress: Resource Recovery and Source Reduction*, Office of Solid Waste (SW-600), 1977.
8. U.S. EPA, *Characterization of Products Containing Lead and Cadmium in Municipal Solid Waste in the United States, 1970 to 2000, Final Report*, EPA/530–SW–89–015C, January 1989.
9. U.S. EPA, *Characterization of Products Containing Lead and Cadmium in Municipal Solid Waste in the United States, 1970 to 2000, Final Report*, EPA/530–SW–89–015A, January 1989.
10. U.S. EPA, "Mercury in the Environment," *EPA Journal*, 16(8), December 1991.

VI

POLLUTION PREVENTION

26

POLLUTION PREVENTION APPROACHES

26.1 INTRODUCTION

Most environmental protection efforts have traditionally emphasized control of pollution from waste substances after it has been generated. Although it may, in some instances, be effective in protecting the environment, this method of waste management has certain disadvantages. Specifically, this type of pollution control does not always solve the problem of pollution; rather, it alters the problem often by transferring pollution from one medium to another, resulting in no net environmental benefit. This has caused interest to be shifted away from focusing on how to deal with the amounts of waste generated, toward efforts concentrated on reducing the amounts of waste produced. The underlying principle of this new direction, defined by many as *pollution prevention* is based on limiting the amount of waste produced "up front," rather than developing extensive treatment processes "downstream," to ensure that the waste poses no threat to people's health or to the environment.

Pollution prevention clearly has now become a top priority for both industries and regulators. As the commitment to pollution prevention increases, the problems created by the generation of wastes will begin to lessen. For industry, this means that pollution prevention must be an integral part of a company's overall operational strategy. Industry must also view pollution prevention as a continuous process of searching for new areas to assess, investigating new methods of pollution prevention, and analyzing and implementing these methods.

The reader should note that in addition to waste reduction, there are two other areas of pollution prevention that need to be included in any waste management analysis. The first issue is energy conservation (EC). Programs in this area have resulted in cost-saving measures that have directly reduced waste generation. Thus, energy conservation is directly related to pollution prevention since a reduction in energy use often corresponds to less energy production and, consequently, less pollution output. The second additional area of concern is that of health, safety, and accident prevention (HS&AP). Accidents like Chernobyl and Bhopal have increased both public and industry awareness, and helped stimulate regulatory policies concerned with emergency planning, response, etc. This second issue is also directly related to pollution prevention. Further, these two issues are interrelated and are addressed in separate chapters as Energy Conservation (Chapter 29), Health and Safety (Chapter 33), and Accident and Emergency Management (Chapter 34). The traditional main focus of pollution prevention—waste reduction—is treated in Chapter 28.

26.2   REGULATIONS

The concept of pollution prevention was first defined as "waste minimization" and was conceptually introduced when the U.S. Congress specifically stated in the Hazardous and Solid Waste Amendments (1984) to the Resource Conservation and Recovery Act (RCRA),

> The Congress hereby declares it to be the national policy of the United States that, wherever feasible, the generation of hazardous waste is to be *reduced* or *eliminated* as expeditiously as possible. Waste that is nevertheless generated should be treated, stored, or disposed of so as to minimize the present and future threat to human health and the environment.

In its 1986 report to Congress (EPA/530–SW–86–033), the Environmental Protection Agency (EPA) detailed the concept of waste minimization as:

> The *reduction*, to the extent feasible, of hazardous waste that is *generated* or subsequently treated, stored or disposed of. It includes any *source reduction* or *recycling* activity undertaken by a generator that results in either (1) the reduction of total volume or quantity of hazardous waste or (2) the reduction of toxicity of hazardous waste, or both, so long as such reduction is consistent with the goal of minimizing present and future threats to human health and the environment.

However, the most important piece of legislation enacted to date is the Pollution Prevention Act. The Pollution Prevention Act of 1990, signed into law in November 1990, established pollution prevention as a "national objective." The Act notes that:

> There are significant opportunities for industry to reduce or prevent pollution at the source through cost-effective changes in production, operation, and raw materials use. The opportunities for source reduction are often not realized because existing regulations, and the industrial resources they require for compliance, focus upon treatment and disposal, rather than source reduction. Source reduction is fundamentally different and more desirable than waste management and pollution control.

> The Act establishes the pollution prevention hierarchy as national policy, declaring that pollution should be prevented or reduced at the source wherever feasible, while pollution that cannot be prevented should be recycled in an environmentally safe manner. In the absence of feasible prevention or recycling opportunities, pollution should be treated; disposal or other releases into the environment should be used as a last resort.

Source reduction is defined to mean any practice which reduces the amount of any hazardous substance, pollutant or contaminant entering any waste stream or otherwise released into the environment (including fugitive emissions) prior to recycling, treatment or disposal; and which reduces the hazards to public health and the environment associated with the release of such substances, pollutants, or contaminants.

The Pollution Prevention Act also formalized the establishment of the EPA's Office of Pollution Prevention, independent of the single-medium programs, to carry out the functions required by the Act and to develop and implement a strategy to promote source reduction. Among other provisions, the law directs the EPA to

1. Facilitate the adoption of source reduction techniques by businesses and by other federal agencies.
2. Establish standard methods of measurement for source reduction.
3. Review regulations to determine their effect on source reduction.
4. Investigate opportunities to use federal procurement to encourage source reduction.
5. Develop improved methods for providing public access to data collected under federal environmental statutes.
6. Develop a training program on source reduction opportunities, devise model source reduction auditing procedures, establish a source reduction clearinghouse, and establish an annual awards program.

Under the Act, facilities required to report releases to the EPA for the Toxic Release Inventory (TRI) must now also provide information on pollution prevention and recycling for each facility and for each toxic chemical. The information includes (a) the quantities of each toxic chemical entering the waste stream and the percentage change from the previous year, (b) the quantities recycled and the percentage change from the previous year, (c) source reduction practices, and (d) changes in production from the previous year. Finally, the Act requires the EPA to report to Congress within 18 months (and biennially afterward) on the actions needed to implement the strategy to promote source reduction.

States have been at the forefront of the pollution prevention movement, providing a direct link to industry, local governments, and consumers. Through grants to states, the EPA has enhanced the capabilities of states to demonstrate innovative and results-oriented programs and has assisted states in implementing a prevention approach. Local governments can also play a significant role in promoting pollution prevention in the industrial, consumer, transportation, agricultural, and public sectors of the community. Many local governments have already taken the lead in putting successful recycling programs into place. A variety of tools are available to promote prevention. Local governments (including cities, counties, sewer and water agencies, planning departments, and other special districts) can provide

1. Educational programs to raise awareness in business and the community of the need to reduce waste and pollution and conserve resources.
2. Technical assistance programs that provide on-site help to companies and organizations in reducing pollution at the source.

3. Regulatory programs that promote prevention through mechanisms such as codes, licenses, and permits.

4. Procurement policies regarding government purchase of recycled products, reusable products, and products designed to be recycled.

Many local governments have passed resolutions and ordinances relating to waste reduction, energy conservation, automobile use, procurement policies, and so on. Such resolutions and ordinances can be useful steps in signaling a public commitment to operate under environmentally sound principles; they also help to define goals and targets and to delineate the specific responsibilities of different local agencies.

26.3   THE EPA's POLLUTION PREVENTION STRATEGY

The EPA's Pollution Prevention Strategy, released in February 1991, was developed by the Agency in consultation with all program and regional offices. The strategy provides guidance on incorporating pollution prevention into the EPA's ongoing environmental protection efforts and includes a plan for achieving substantial voluntary reductions of targeted high-risk industrial chemicals. The strategy is aimed at maximizing private sector initiatives while challenging industry to achieve ambitious prevention goals. A major component of the strategy is the Industrial Toxics Project. The EPA has identified 17 high-risk industrial chemicals that offer significant opportunities for prevention (see Table 26.1). According to the latest toxic release inventory (TRI), U.S. manufacturers released 2.8 billion pounds of toxic chemicals to the environment in 1993, a decline of 406 million pounds, or 12.6%, over 1992 and a decline of nearly 43% when compared to the base year of 1988.

TABLE 26.1   Target chemicals[a]

Benzene	33.1
Cadmium	2.0
Carbon tetrachloride	5.0
Chloroform	26.9
Chromium	56.9
Cyanide	13.8
Dichloromethane	153.4
Lead	58.7
Mercury	0.3
Methyl ethyl ketone	159.1
Methyl isobutyl ketone	43.7
Nickel	19.4
Tetrachloroethylene	37.5
Toluene	344.6
1,1,1-Trichloroethane	190.5
Trichloroethylene	55.4
Xylene	201.6

[a] Data given is in millions of pounds.

The 17 pollutants identified as targets of the industrial toxics project present both (a) significant risks to human health and the environment and (b) opportunities to reduce such risks through prevention. The list was drawn from recommendations submitted by program offices, taking into account such criteria as health and ecological risk potential for multiple exposures or cross-media contamination, technical or economic opportunities for prevention, and limitations of treatment.   All of the targeted chemicals are included in the EPA's TRI; thus, reductions in their releases can be measured in each year's TRI reports. Several hundred companies who have reported releases of the target chemicals have already been contacted by the EPA. The EPA is seeking their cooperation in making voluntary commitments to reduce releases and in developing pollution prevention plans to carry out these commitments.

The strategy also provides guidance on incorporating pollution prevention into the EPA's existing programs, emphasizing the need for continued strong regulatory and enforcement programs. At the same time, the strategy favors flexible, cost-effective approaches that involve market-based incentives where practical. For example, the strategy calls for the use of "regulatory clusters" through which the EPA will categorize the rules it intends to propose over the next several years for certain chemicals and their sources. The clusters are intended to foster (a) improved evaluation of the cumulative impact of standards, (b) more certainty for industry, and (c) early investment in prevention activities.

The strategy outlines several short-term measures that will address various institutional barriers within the EPA's own organization that limit its ability to develop effective prevention strategies. Such measures include designating special assistants for pollution prevention in each assistant administrator's office, developing incentives and awards to encourage EPA staff to engage in pollution prevention efforts, incorporating prevention into the comprehensive 4-year strategic plans by each program office, and providing pollution prevention training to EPA staff.

The industrial toxics project for the manufacturing sector represents the first focus of a comprehensive EPA strategy. The EPA is seeking to work with the Departments of Agriculture, Energy, and Transportation to develop strategies for preventing pollution from agricultural practices and from energy and transportation use. The EPA has already begun several joint initiatives, including (a) a cooperative grants program for sustainable agriculture research with the Department of Agriculture and (b) a joint program with the Department of Energy to demonstrate energy efficiency and waste reduction in key sectors.

Another important goal of EPA's Pollution Prevention Program is to ensure that pollution prevention training and education are available to government, industry, academic institutions, and the general public. Training and education are needed to help institutionalize prevention as the strategy of choice in all environmental decision-making and protection activities. The EPA's Office of Pollution Prevention has developed a resource guide, *Pollution Prevention Training Opportunities in 1990.* The guide describes the types of training courses, workshops, and seminars being offered in each state and provides contact names and addresses. Other sections of the guide list available instruction manuals, opportunity assessment materials, fact sheets, videos, and state and EPA contacts on pollution prevention.

The Pollution Prevention Information Clearinghouse (PPIC) is a multimedia clearinghouse of technical, policy, programmatic, legislative, and financial information dedicated to promoting pollution prevention through efficient information transfer. The Clearinghouse is operated by the EPA's Office of Research and Development and Office of Pollution Prevention.

The Pollution Prevention Information Exchange System (PIES) is computerized information network of the EPA's Pollution Prevention Information Clearinghouse. The PIES provides on-line interactive access through personal computers to a wide range of pollution prevention information. Open 24 hours a day, requiring no user fees, the PIES represents an entry-port to a worldwide communications network on pollution prevention. The PIES features literature search functions, a national calendar of conferences and workshops relating to pollution prevention, hundreds of case studies of pollution prevention, a message center for interaction and exchange with participants, and direct access to news and documents.

26.4   POLLUTION PREVENTION ASSESSMENT

A pollution prevention assessment (PPA) is defined as a systematic, planned procedure with the objective of identifying ways to reduce or eliminate waste, preferably at the source. Generally, the assessment is preceded by careful planning and organization to set overall pollution prevention goals. Next, the actual assessment procedure begins with a thorough review of a plant's operations and waste streams, along with the selection of specific areas to assess. Once an area is selected as a possible minimization area, various options with the potential for reducing waste can be developed and screened. The technical and economic feasibility of each option is evaluated and, finally, the most promising options are implemented.

Initiating a successful program must begin with a secure commitment from top management, allocation of adequate funding and technical expertise, appropriate organization, and a good understanding of goals and planning. To be successful, pollution prevention must become an integral part of the company's operations. The PPA offers opportunities to reduce operating costs, reduce potential liability, and improve the environment, while also improving regulatory compliance. Much of the material presented in this section has been drawn from EPA's original "Waste Minimization Assessment" procedure, developed primarily for hazardous wastes (1). This procedure has been modified and adapted to apply to what is defined above as a pollution prevention assessment (PPA).

Figure 26.1 depicts the pollution prevention assessment procedure. As shown, the assessment procedure can be divided into four phases, including the following two major phases:

1. Assessment phase, consisting of the collection of data and identification and screening of potential pollution prevention options.
2. Feasibility analysis phase, consisting of the technical and economical evaluation of each option.

Each stage depicted in Figure 26.1 is discussed in more detail below.

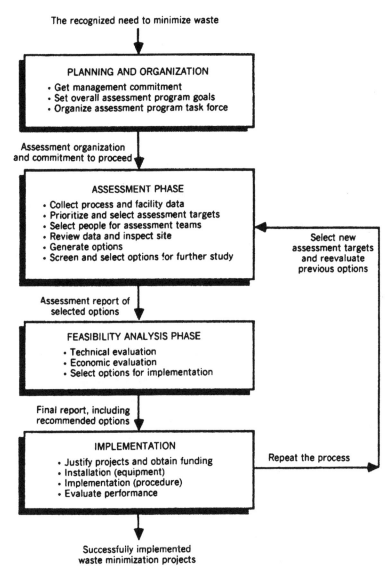

The recognized need to minimize waste

PLANNING AND ORGANIZATION

• Get management commitment
• Set overall assessment program goals
• Organize assessment program task force

Assessment organization
and commitment to proceed

ASSESSMENT PHASE

• Collect process and facility data
• Prioritize and select assessment targets
• Select people for assessment teams
• Review data and inspect site
• Generate options
• Screen and select options for further study

Select new
assessment targets
and reevaluate
previous options

Assessment report of
selected options

FEASIBILITY ANALYSIS PHASE

• Technical evaluation
• Economic evaluation
• Select options for implementation

Final report, including
recommended options

IMPLEMENTATION

• Justify projects and obtain funding
• Installation (equipment)
• Implementation (procedure)
• Evaluate performance

Repeat the process

Successfully implemented
waste minimization projects

FIGURE 26.1 The pollution prevention assessment procedure.

Planning and Organization

This section deals with aspects that are important to the ultimate success of any PPA. First, a commitment from top management to reduce waste generation must be secured. To do this, the benefits, specifically economic benefits, gained from pollution prevention must be clearly shown. Management must be prepared to commit the necessary personnel and financial resources needed for the program. Support at the management level only, however, will not be sufficient in maintaining a successful pollution prevention program. Personnel at every level of the corporate ladder must be dedicated to the common goal of pollution prevention. Initial contact with employees includes solicitation of their view of how the assessment should be performed and which areas are likely candidates for pollution prevention. Gaining further employee support can be sought through incentives, bonuses, and awards. Employees are directly involved in the production of waste and can be instrumental in the overall program if they can be motivated. Recognition and rewards can increase employee cooperation and participation.

Once a company has committed to minimizing the generation of waste, a task force must be established to conduct the actual assessment and convey its efforts to management. A team leader should be selected and should serve as the motivator of the project. This individual needs to be an effective communicator with the rest of the team. This task force should be comprised of individuals from all groups within the company that have an interest in the outcome of the program. The size of the group should reflect the size of the company. Members of different departments within the company can bring varying perspectives and suggestions to the program. The task force should be separated into different teams concentrating on respective areas of the plant with which they each have the most knowledge. An example of a task force, often referred to as *audit team* in industry, is provided below (2) for a metal finishing department in a large aerospace corporation.

Metal finishing department manager.

Process engineer responsible for metal finishing processes.

Facilities engineer responsible for metal finishing department.

Wastewater treatment department supervisor.

Staff environmental engineer.

Assessment Phase

The ultimate purpose of the assessment phase is to develop a comprehensive list of potential pollution prevention options. To accomplish this, a detailed analysis of the plant must be performed.

Collecting data from every process in the plant is the first step in performing the assessment. All waste streams must be identified and characterized. Selecting the principal waste streams or waste-producing operations provides the task force with the main focus for the assessment. The criteria used for selection of principal waste streams should include the composition, quantity, and toxicity of wastes, method and cost of disposal, compliance status, and potential for minimization. Preliminary information should be acquired from hazardous waste manifests, biennial reports, environmental audits, and National Pollution Discharge Elimination System (NPDES) monitoring reports. Routine sampling can also help to provide data on certain chemicals within a particular waste stream. The best means for identifying information on all waste streams is the use of flow diagrams and, subsequently, material balances (3). The material balance includes information on materials entering and leaving a process. Material balances can be used to quantify losses or emissions and provide essential data with which to estimate the size and cost of additional equipment, data to evaluate economic performance, and a baseline for tracking the progress of minimization efforts. Simply, the material balance is represented by the mass conservation principle:

$$\text{Mass in} - \text{Mass out} = \text{Mass accumulated}$$

Material balances should be prepared individually for specific components entering and leaving a process; to be most accurate, balances should be developed for individual units or processes, rather than for a large area of the facility. An overall balance for the entire facility can be constructed from the individual balances. Although characterizing waste streams through material balances can require substantial effort, it is the best way to fully understand the makeup of wastes within a facility. Tracking waste flows and compositions should be performed periodically. In this way, variational flows can be distinguished from constant, continuous flows, thereby providing a more accurate representation.

Feasibility Analysis Phase

Once the assessment phase has identified potential options, the feasibility phase is used to determine whether those options are technically and economically practical, through a detailed evaluation and analysis. The level of analysis is dependent upon the complexity of the pollution prevention project under consideration. Simple options, such as preventive maintenance, normally do not require as detailed an evaluation as would an input substitution option that may result in changes to product specification and process equipment. The detailed evaluation of options should start with source reduction alternatives first.

The technical evaluation determines whether the potential option, once in place, will really work as intended. Performing a technical evaluation requires comprehensive knowledge of pollution prevention techniques, vendors, relevant manufacturing processes, and the resources and limitations of the facility. This evaluation may involve bench scale or pilot scale treatability testing. Some vendors will install equipment on a trial basis. Typical considerations for technical criteria evaluation are as follows:

1. Technical reliability.
2. System safety.
3. Product quality maintenance.
4. Space requirements.
5. Compatibility of proposed equipment with existing system.
6. Downtime necessary for installation.
7. Special expertise requirements.
8. Labor and utility requirements.

All affected groups in the facility should be encouraged to review and comment on the technical evaluation. If an option results in changes in production, effects on product quality must be determined. All options that are shown to be impractical after the technical evaluation are dropped from further evaluation.

Implementation

Once the options for reducing waste have been established in the final report, the project must be justified in order to obtain funding. Here, the initial commitment from management is critical to overcome resistance to change within the company. With funding secured, the selected option is ready to be implemented. The first step, a detailed design of the system, is followed by construction. After the equipment is installed, the personnel involved can be trained and the operation started. Options need to be monitored to ensure their effectiveness and demonstrate that the project has achieved its goals. The extent of pollution prevention needs to be measured, and this can be done most efficiently by recording the quantities of waste generated before and after implementation of the option. The difference, divided by the original volume of waste generated, represents the percentage reduction in waste volume. Because waste generation is directly dependent upon the production rate, the ratio of waste generation to production is also a convenient means of measuring reduction.

It is important to realize that the pollution prevention program is an ongoing effort. Once the highest-priority waste streams have been assessed and dealt with, the assessment task force should continue to identify new opportunities, assess different waste streams, and consider attractive options that were not pursued earlier. The ultimate goal should be to eliminate as much waste generation as possible.

26.5 POLLUTION PREVENTION INCENTIVES

As described earlier, the problems associated with the handling of waste streams have become a major concern for industry as a whole, and for the chemical industry specifically. Overall costs for managing waste, coupled with increasing regulatory and economic consequences, are escalating daily. There are real incentives for avoiding the production of hazardous wastes up front, and they will increase as more regulations are passed and treatment capabilities for waste substances remain limited. These incentives usually result in monetary encouragements and consist of economic benefits, regulatory compliance, reduction in liability, and enhanced public image.

Economic Benefits

Reducing the amount of pollutant/waste produced initially will result in a reduction of the costs associated with the handling of that waste. The obvious costs resulting from the transportation, disposal, and treatment of wastes will all be lowered as the amount of created waste is lowered. Current trends predict that the cost of disposing of waste will continue to increase. In addition, income can be derived through the sale, reuse, or recycling of certain wastes. A number of not-so-obvious gains that are associated with the reduction of waste can be sought, such as a reduction in health and safety costs, reduction in insurance costs, reduction in raw material costs, and a reduction in reporting, manifesting, and permitting costs.

Regulatory agencies can influence companies to investigate pollution prevention techniques through offers of reduced fines and penalties. A pollution prevention program can also be used to bring a firm into compliance with regulations. Permit fees may be structured in a way that promotes pollution prevention activities. For instance, instead of basing a permit fee on the size of a company, an agency might base the fee on the volume and/or toxicity of waste substances produced. Firms actively pursuing pollution prevention would be rewarded by paying a lower fee.

Regulatory Compliance

Federal and state laws require all firms classified as hazardous waste generators or small-quantity generators (SQGs) to implement a pollution prevention program to reduce the quantity of waste to the extent that it is economically feasible. SQGs are facilities that generate more than 10 kg/mo of waste but less than 1000 kg/mo of hazardous waste. A facility generating more than 1000 kg/mo is classified as a generator. Generators are required to sign the following statement on all manifests to certify their pollution prevention efforts:

> Unless I am a small-quantity generator who has been exempted by statute or regulation from the duty to make a waste minimization certification under Section 3002(b) of RCRA, I also certify that I have a program in place to reduce the volume and toxicity of waste generated to the degree I have determined to be economically practicable, and I have selected the method of treatment, storage, or disposal currently available to me which minimizes the present and future threat to human health and the environment.

Firms permitted as hazardous waste generators are generally required under the RCRA to report sampling data on a regular basis. When hazardous wastes are minimized, the frequency of sampling and data reporting is also minimized, thereby saving the firm a considerable amount of money.

In addition, land disposal restrictions and treatment standards provide indirect incentives for reducing (hazardous) waste. Under these regulations, only wastes meeting specified treatment standards will be allowed for disposal in landfills. The level of treatment standards will be based on pollution prevention technologies, and therefore those facilities already employing these reduction techniques will be best suited to meet the standards. With these reduction methods already in place, a company can save on costs associated with waste management.

Reduction in Liability

Both short-term and long-term liabilities can be reduced with pollution prevention programs. Short-term liabilities associated with releases to the environment resulting in noncompliance with permits can be reduced through the overall reduction of hazardous waste. Short-term liabilities connected with personnel exposure and workplace safety will also be lessened. Long-term liabilities resulting from the on-site or off-site disposal of wastes can be reduced as well. Disposing of wastes at a permitted disposal facility does not end a firm's connection to that waste. If a disposal facility is shown to be releasing contaminants into the environment, not only are the owners/operators of that facility liable under CERCLA responsible for undertaking remedial actions to clean up the contamination, but the generators who arranged for the disposal of wastes in that facility are also liable. These generators can be ordered to finance both (a) the investigation of the extent of contamination and (b) the remedial action to address the contamination.

Enhanced Public Image

The public has placed a great deal of emphasis on environmental issues. Recent election campaigns in all levels of government have made the environment a top priority platform issue. Because of this level of concern and awareness, it is becoming increasingly important for companies to share information with the public. Under the Superfund Amendments and Reauthorization Act (SARA), the Emergency Planning and Community Right-to-Know Act (EPCRA) includes mandatory reporting of releases to the environment and optional reporting of pollution prevention activities. Implementation of a pollution prevention program provides a good community relations baseline for improving a firm's public image as a good corporate citizen.

26.6   DETERRENTS TO POLLUTION PREVENTION

The previous section essentially presented the advantages of developing and implementing a pollution prevention program. This section briefly reviews some of the deterrents or impediments to pollution prevention. A "dirty dozen" list is provided below, followed by a short description (and some comments) on each of these deterrents (3,4).

1. Management apathy.
2. Lack of financial commitment.
3. Production concerns.
4. Research, development, and design concerns.
5. Failure to monitor program success.
6. Middle management decisions.
7. Information exchange within an organization.
8. Confusion regarding regulations.
9. Confusion about economic advantages.
10. Bureaucratic resistance to change.
11. Lack of awareness of pollution prevention advantages.
12. Failure to apply multimedia approaches.

1. *Management Apathy.* It is not uncommon for upper-level management in most companies, particularly large ones, to take an indifferent attitude toward pollution prevention. Because administrators are often not technically oriented, most have difficulty realizing the potentially enormous benefits that can be gained from a pollution prevention program.

2. *Lack of Financial Commitment.* It is no secret that many considerations, including resources, individuals, money, economic incentives, and so forth, are critical to the success of a pollution prevention program. These "resources" are often just not available within small companies or within large companies that may be experiencing a financial downturn.

3. *Production Concerns.* The classic song of the production supervisor and/or plant operator is, " I'm meeting deadlines and making money, so don't rock the

boat." Sending a member of the pollution prevention team out to a plant is somewhat akin (at times) to letting a bull loose in a china shop.

4. *Research, Development, and Design Concerns.* The somewhat simplistic mentality described in item 3 can be extended to apply to new projects at the research, development, or design stages. A very common misconception is that any pollution prevention activity is going to delay a (new) project and cost money.

5. *Failure to Monitor Program Success.* Pollution prevention programs are often instituted with high hopes of success but are later abandoned because of a failure to monitor the program properly. Responsible individuals must continuously monitor and record both the successes and failures of the program.

6. *Middle Management Decisions.* In most companies, it is usually middle management that is directly responsible to the top company administrator(s) for earnings. Unfortunately, it is often this group that views this type of program as an added expense that will eat into profits.

7. *Information Exchange Within an Organization.* As indicated in item 1 above, management often has a poor understanding of the advantage of pollution prevention. In addition, the economic, environmental, liability, social, and other advantages that various organizations within a company ascribe to a program of this nature will differ widely.

8. *Confusion Regarding Regulations.* It may be hard to believe, but many companies, particularly small ones, are not aware of the applicable regulations regarding their process or operation. Any permit review process is almost certainly doomed to failure if the responsible individual(s) or group is not cognizant of the pertinent regulations.

9. *Confusion About Economic Advantages.* Most companies are simply not aware of the true costs associated with generating and treating pollutant wastes. However, treatment and/or disposal costs have increased dramatically in recent years, and liability concerns continue to mount. As indicated above, many companies do not realize that pollution prevention opportunities, in many instances, result in increased profits. This situation is prevalent today because, for many companies, particularly large ones, it is difficult to evaluate the economic advantages of pollution prevention. The accounting system is often so complex, and not properly integrated within the various divisions or organizations of a company, that it is nearly impossible to quantify the effects arising from a pollution prevention program.

10. *Bureaucratic Resistance to Change.* It is natural, particularly in large corporations and utilities, to resist any change to an existing process or method of operation. This reluctance to adapt to a changing environmental, regulatory, liability, etc., climate is commonplace in industry.

11. *Lack of Awareness of Pollution Prevention Advantages.* Middle-level managers and upper-level administrators are often unaware of both the pollution problem at their facility and the associated true cost of treatment. Thus, it is understandable why these individuals are not aware of the overall advantages of a pollution prevention program. For more details on this subject the reader is referred to the previous section of this chapter.

12. *Failure to Apply Multimedia Approaches* (2). Unfortunately, an overall approach that examines a system or process from a multimedia point of view is

rarely found in industry today. There is much work to be done in this area, and many environmental improvements can be expected in the years to come.

As has been noted, many companies and individuals are unaware of the advantages of pollution prevention. Until this situation is changed, progress in the environmental arena will — as it has done in the past — come slowly.

26.7   SUMMARY

1. Most environmental protection efforts have traditionally emphasized control of pollution by waste substances after it has been generated. However, this tradition of pollution control does not solve the problem of pollution; rather it alters the problem by transferring pollution from one medium to another, resulting in no net environmental benefit. This has caused interest and focus to shift from ways to deal with the amounts of waste generated toward efforts in reducing the amounts of waste produced.

2. Pollution prevention activities can help a firm to comply with many federal regulations, including:

- Clean Air Act (CAA)
- Clean Water Act (CWA)
- Comprehensive Environmental Response, Compensation, and Liability Act (CERCLA)
- Superfund Amendments and Reauthorization Act (SARA)
- Hazardous and Solid Waste Amendments (HSWA)
- Safe Drinking Water Act (SDWA)
- Toxic Substance Control Act (TSCA)
- The Pollution Prevention Act of 1990

3. One of the key features of the EPA's Pollution Prevention Strategy is the Industrial Toxics Project. The EPA has identified 17 high-risk industrial chemicals that offer significant opportunities for prevention.

4. The EPA's pollution prevention hierarchy includes four major components. In the current working definition used by the EPA, source reduction and recycling are considered the most viable techniques, preceding treatment and disposal.

5. A pollution prevention assessment (PPA) is defined by the EPA as a systematic planned procedure with the objective of identifying ways to reduce or eliminate waste. The assessment procedure includes the collection of data, the identification and screening of potential pollution prevention options, and a feasibility analysis phase, consisting of the technical and economical evaluation of each pollution prevention option.

6. There are numerous advantages to instituting a pollution prevention program. The four key incentives include economic benefits, regulation compliance, reduction in liability, and enhanced public image.

7. Despite the numerous benefits attained with a pollution prevention program, certain misconceptions about possible drawbacks do exist. Concerns have primar-

ily centered on product quality and the large capital expenditures. Generally, product quality should not be compromised as a result of pollution prevention efforts.

REFERENCES

1. U.S. EPA, *Waste Minimization Opportunity Assessment Manual*, EPA/625-7-88/003,1988.
2. Jacob Engineering Group, Inc., *Draft of the EPA Manual for Waste Minimization Audits*, Pasadena, CA, 1987.
3. L. Theodore and Y. C. McGuinn, *Pollution Prevention*, Van Nostrand Reinhold, New York, 1992.
4. L. Theodore: personal notes.

27

DOMESTIC SOLUTIONS

Contributing Author: Mary K. Theodore

27.1 INTRODUCTION

The purpose of this chapter is to offer some suggestions which might be used in the household in the hope of preventing further pollution of the environment. Action by Congress and state legislatures, rulings by courts, pronouncements by important people, or wishing alone cannot clean up the environment or keep it from becoming more polluted (1).

The best place to start and concentrate is, in fact, in the home. It is where the most time is spent and where the individual has the most control over the health, safety, and well-being of those nearest and dearest.

One must be ever vigilant. It begins at home where the emphasis on saving the environment is placed with the individual. From the very beginning, children must be educated. The concept of preserving natural resources must be driven home as conscientiously as the teaching of the ABCs. When children become aware of the cycle — for example, that the sand that one plays in at the beach has a connection with the manufacture of the bottle that one drinks from, and that the disposal of the bottle has an impact on the environment — then society is on the road to recovery. The children of today are the decision-makers, manufacturers, business managers, and the homeowners of tomorrow. Children can become aware of and come to understand the problems associated with environmental management, and they can learn how to initiate real-world solutions to resolve these problems. In addition, individuals of all ages can learn how to work together to preserve the environment for future generations.

In the pages that follow, some areas where the individual can make a difference are addressed. In each of these sections, guidelines are offered, but the emphasis is on the importance of the role of the individual in the battle against pollution. The contribution the individual member of society can make towards the goal of saving the environment for the future cannot be stressed enough. Topic areas that follow include:

Source Reduction and Recycling

Off-Site Collection Techniques

Hazardous Waste

Energy Conservation

Safety Considerations

27.2 SOURCE REDUCTION AND RECYCLING

Preventing pollution at the source should be considered first. Regarding recycling, each individual should reuse, repair, and recycle as often as possible. One use is often not enough. Every time a new product is used instead of a recycled one, some form of natural resource is depleted forever.

The material that is not used — the waste produced in the home and business — is discarded into cans (or the equivalent) and put out for collection. This trash, plus the neighbors' and everyone else's, adds up to a massive pile of waste. According to an Environmental Protection Agency (EPA) statistic, this nation generates enough garbage each year to fill a convoy of trash trucks reaching halfway to the moon. This convoy keeps growing longer every year (2). Eighty percent of the trash goes to landfill, 10% is incinerated, and 10% is recycled. Many of these landfills are almost full, and landfill costs have soared. Not only is it expensive to build new landfill facilities, but finding a suitable location is problematic because of the "not in my back yard" (NIMBY) syndrome.

The national weight percent averages of landfill discards are as follows:

Paper 41%

Yard wastes 18%

Glass 8%

Metals 9%

Plastic 9%

Food waste 8%

Other 7%

Paper waste is a logical target for increased recycling because it represents such a large percentage of municipal solid waste — an estimated 41%. Of this 41%, newsprint and paper packaging (corrugated) are the principal components. Successful recycling of paper would reduce the volume of waste collected and delivered to the landfill. One ton of recovered paper saves 3.3 yd^3 of landfill space. In addition, recycling one ton of paper saves 17 trees. Recycling paper conserves forest lands and other valuable resources. Less energy and water is required to produce paper from wastepaper than from virgin pulp. The paper recycling process reduces air, water, and land pollution when compared to the manufacturing process of paper which uses virgin materials. Individuals could support this endeavor by doing the following:

1. Encourage conservation at home—that is, use both sides of the paper, use junk mail as scrap paper, use crumbled newspaper to clean windows, use newspaper as wrapping paper, etc.

2. Encourage local schools and village offices to follow recycling practices for environmental and economic reasons.

3. Spread the word at fast-food restaurants by commenting to managers about good/bad habits. Let personnel know that you appreciate environmentally conscientious franchises.

4. In any store, do not use a bag unless needed. Have different size cloth bags which can be used and reused for grocery shopping. Some grocery stores reward patrons for using their own bags and/or returning plastic bags to a common bin for cooperative use.

5. When shopping for paper goods, patronize companies which recycle paper and comment on it to storekeepers.

6. Point out wasteful packaging and compliment conscientious practices.

7. Comment on wasteful paper practices used with a computer printer. When buying a new printer, look for paper-saving features.

Yard waste comprises 18% of the landfill. Composting is a viable solution here. Everything that grows yields materials that can be composted and used to supply the garden with organic matter. The simple act of composting food scraps can mean enormous savings. "In September, we composted 217 tons of food waste in 30 different facilities," said Jim Marion, the resource management director for the New York State Department of Correction (3). Multiply that 217 tons from the prisons by the cost of garbage disposal—$115 a ton—and thus the department saved $25,000 a month, or $300,000 a year. On a smaller scale, but as important, is the home compost. To make a small, clean-smelling compost pile from kitchen and yard waste, cut the bottom out of a trash can (or use a wire enclosure) and set on the ground. Pile a day's kitchen waste on the bottom. Shred or break up large pieces; avoid large amounts of any single material. Cover with a layer of dry leaves, grass clippings, and soil. Keep the pile loosely packed and exposed to air. Repeat layering daily. Keep the pile damp. An offensive odor means the pile is packed too tightly or the wrong materials have been used. Earthworms or other small organisms are beneficial. Materials take 6–8 weeks to degrade (break down). Do not use pet feces; large amounts of grease or fat; charcoal; synthetics; floor sweepings; diseased plants; or large bulky materials; or large amounts of meat, cheese, or milk products (4).

Many types of glass can be recycled. Glass food and beverage containers are 100% recyclable and can be reused almost an infinite number of times. However, consideration must be given to the use of energy in cleaning the recycled glass, the detergent runoff, pollution of the water, etc., in this effort. The only glass items that cannot be recycled are light bulbs, ceramic glass, dishes, and plate glass.

Metals are valuable resources that can often be more easily recovered and reused than mined. For this reason, metals recycling is an established practice in many parts of the country. Aluminum and tin (actually tin-coated steel) cans are some metal items that have been successfully recycled over the years.

Plastics make up 9% of the national municipal solid waste stream by weight, and 20% by volume. Approximately half of the plastic municipal waste is packaging; the rest is nondurable consumer goods (such as disposable razor blades) and durable goods (such as appliances). There are still many uncertainties associated with the disposal of plastics. Plastic wastes are very slow to degrade in landfills, but recent data indicate that other wastes, even those considered to be "degradable," such as paper, are also quite slow to degrade. Less than 5% (primarily plastic soft drink bottles and milk jugs) of the plastic waste stream is currently recycled. There are several obstacles to recycling plastics. For example, many objects are made up of different types of plastic material, which makes reprocessing difficult. Progress is being made, however, by plastic manufacturers and the recycling industry to improve the feasibility of plastics recycling (5). Source reduction of plastics can be achieved in a number of ways in the home: Avoid the use of disposable items, try to buy concentrates or other products that require less packaging, and so on.

Additional source reduction and recycling suggestions which are quite simple and can be easily initiated in the home are listed below:

1. Invest in electronic equipment and other durable goods that have good warranties and are easy to repair. Repair rather than throw away equipment, or donate to charities.

2. More than 200 million tires are discarded each year. This number can be reduced by buying high-mileage tires and by maintaining proper air pressure in tires. Also, remember to check the tire pressure every other time the gas tank is filled.

3. Buy products in packages that are recyclable. Avoid buying products packaged in complex, multimaterial packaging.

4. Sign up with a dairy service to have milk products delivered in refillable bottles. If family members drink spring water, use a water service that provides large refillable bottles rather than buying individual bottles at the grocery store.

5. Avoid buying products that have excessive, unnecessary packaging. Buy products in the largest size available to minimize packaging waste (and save money). Buy products in the concentrated form and add water to it (where applicable).

6. If adequate space is available, compost leaves, grass clippings, other yard debris, and food waste to make a natural soil amendment for the yard and garden. If space is not available, donate yard debris to community garden projects.

7. Avoid buying disposable dishes. For picnics or outside parties, invest in reusable, durable plastic plates and cups that can be washed and used again.

8. Do not use disposable razors. Invest in a quality razor and change the blade, or use an electric razor.

9. Invest in resealable containers for storing leftovers; avoid the use of disposable plastic wraps, storage bags, etc.

10. Minimize use of paper towels and napkins. Invest in cloth napkins for everyday use, and use reusable wiping cloths and towels rather than paper towels.

11. Use cloth rather than disposable diapers. Even if one cannot wash diapers at home, diaper services are generally less expensive than buying disposable diapers.

12. Reuse paper and plastic bags. Individuals should always ask themselves if a

bag is needed; do not accept a bag for only one or two items. As described earlier, carry reusable tote bags or bring grocery and produce bags back when shopping.

13. The "brown bag" concept should be replaced by the "reusable bag." More environmentally conscious stores are offering reusable containers for sandwiches, beverages, water, etc., that help eliminate waste such as foil, plastic, etc. Instead of packing a fast-food type of snack, offer a healthy alternative, such as a fruit or vegetable which comes in its own biodegradable wrapping.

14. If you buy pre-prepared microwavable dinners, save the plates for use in outdoor parties or for children.

15. Buy nickel–cadmium batteries and recharge them. When buying recharge-able batteries is not feasible, buy alkaline batteries. They are more expensive than carbon–zinc batteries but are a better value because they last longer (6).

Recycling is just one important remedy for the garbage problem. It is an effective solution because it reduces the amount of waste for disposal. Recycling reduces our reliance on landfills and incinerators. Recycling can cost less than landfilling or incineration. Recycling protects our health and environment when harmful sub-stances are removed from the waste stream. Recycling also conserves our natural resources because it reduces the need for raw materials. With better cooperation, recycling offers great promise for improved management of our trash. A national goal of reducing and recycling 25% of our waste has been set by the EPA. The support and involvement of each individual member of society can make this goal a reality.

27.3 OFF-SITE COLLECTION TECHNIQUES

A variety of methods are used to collect materials away from the home, including (a) single-material and multimaterial curbside collection, (b) drop-off centers, and (c) buy-back centers. In every case, the key to success is convenience for the disposer and for the handler (7). For home use, several environmentally minded stores have come up with containers specifically designed for recycling purposes. These units are "user-friendly" in that they are compact, inexpensive, and neat. They encourage the homemaker to recycle and organize the discards in a convenient and sensible way. Source separated materials can include newspapers, glass, aluminum and bi-metal containers, corrugated cartons, selected plastic containers, and used oil.

Many municipalities provide residents with colorful collection bins or clear and colored plastic bags to separate materials accordingly. Studies have shown that separating trash is easy, taking approximately 15 min per week. A good way to evaluate a program's effectiveness is to measure the amount of recyclables collected. Some type of tally or announcement in the village newspaper or school news would be another way to show progress and encourage continued recycling.

The most effective approach tends to be once-a-week collection on the same day as garbage pickup. Again, convenience to the consumer is the key. Materials can also be collected on a day other than refuse-collection day. If days are not regularly scheduled, however, residents may forget or choose not to store materials for a long period of time.

A multimaterial collection center is a stationary site where residents bring their recyclable materials. In some cases, recyclers are paid for the materials they bring in; in other cases materials are simply dropped off and the funds generated are used to further the program and offset the cost of disposal. According to the EPA, these drop-off centers are the most common form of collection for households. However, the volume of materials collected and the participation rate are considerably lower than those for curbside collection because it requires residents to prepare, store, and transport material. Advantages of this collection method include the ability to collect a wider variety of recyclable materials—that is, they are not limited to two or three, as are many curbside programs. There is also limited expense to this form of collection because less equipment and labor are required to obtain the materials. Recycling centers located at landfills, incinerators, transfer stations, or convenience centers, especially in municipalities without regular refuse collection, experience the best participation rates. They should be convenient to populated areas and be on well-traveled roads. Most consumers who bring materials to a drop-off center live within a 5-mile radius of the site.

Recycling theme centers have proved to be extremely successful in increasing citizen participation. Centers have been designed to replicate a visit to the zoo, circus, riverboat, or old town railroad station. These facilities are designed to make recycling fun for the entire family. Additionally, successful centers must be able to accommodate small vans and pickup trucks for the volume recycler who collects from other people, bars, restaurants, hotels, and hospitals.

Reverse vending machines are popular as a method to recover cans. It is now obvious that the deposit cans have cut down on the amount of cans that would have been discarded in the past.

The role of advertising and publicity cannot be overemphasized. All must be aware of the hours of operation, materials accepted, and the prices paid (if it is a buy-back center). A single-media campaign will not be effective; the publicity must be year-round, employing a variety of methods to educate citizens. Conducting a survey to define operating hours that are the most convenient to the public should also be considered.

All the above collection techniques are central to the individual participating and beginning this process in the home. If the good example of one individual impresses just one other to become environmentally conscientious, progress will be made (7).

27.4 HAZARDOUS WASTE

Common household products which might be found under the kitchen sink, in the bathroom, in the garage, or in the basement probably are hazardous to the health of family members and to the environment. They can be hazardous in various ways. Their use in the house can pollute the air the family breathes. Their improper disposal often pollutes the ambient air (outside the house) and the ground and can leach into the water. The family and the environment can be protected by reducing the number of hazardous products which are bought, substituting a nontoxic solution whenever possible, and by disposing of hazardous waste properly.

Hazardous household products are common products individuals use around the house every day. These include products which are flammable, corrosive, reactive, or poisonous. One should check labels for words like DANGER, WARN-ING, CAUTION, and POISON, which appear on many hazardous substances.

Some of the definitions of hazardous wastes which have appeared in the literature are not as straightforward as they appear. Congress and the EPA have assigned terms to describe wastes (and other substances) that have been enacted under regulation.

1. *Hazardous substance* [Comprehensive Environmental Response, Compensation, and Liability Act (CERCLA)]. Any substance that, when released into the environment, may present substantial danger to public health, welfare, or the environment. Designation as a hazardous substance grows out of the statutory definitions in several environmental laws. Currently there are 717 CERCLA hazardous substances.

2. *Extremely hazardous substances* (CERCLA as amended). Substances which could cause serious, irreversible health effects from a single exposure. For purposes of chemical emergency planning, the EPA has designated 366 substances as extremely hazardous.

3. *Solid waste* [Resource Conservation and Recovery Act (RCRA)]. Any garbage, refuse, sludge, or other discarded material. All "solid waste" is not necessarily solid; it can be liquid or semisolid or can contain gaseous material. Solid waste results from industrial, commercial, mining, and agricultural operations as well as from community activities. Solid waste can be either hazardous or nonhazardous. However, it does not include (a) solid or dissolved material in domestic sewage, (b) certain nuclear material, or (c) certain agricultural wastes.

4. *Hazardous waste* (RCRA). Solid waste (or combinations of solid waste) that, because of its quantity, concentration, or physical, chemical, or infectious characteristics, may pose a hazard to human health or the environment.

5. *Nonhazardous waste* (RCRA). Solid wastes, including municipal wastes, municipal sludge, industrial and commercial wastes, and household waste, that are not hazardous.

It is estimated that the average American generates about 160 lbs of household hazardous waste each year. Typical examples of such discarded materials include pesticides, paints and varnishes, brush cleaners, ammonia, toilet bowl cleaners, bleaches and disinfectants, oven cleaners, furniture polish, swimming pool chemicals, batteries, motor oil, outdated medicines, and many others. Although these substances may be every bit as toxic, corrosive, flammable, or explosive as the industrial wastes regulated under the RCRA, it is unfortunate that federal and state hazardous waste laws do not apply to the comparatively minor household sources. Nevertheless, the cumulative environmental impact of even small amounts of these materials being carelessly discarded by millions of individuals can be significant. For example, groundwater contamination cleanup is expected to cost an additional $100 a year in taxes.

Household hazardous waste disposal practices present a variety of concerns:

1. Stored inside the home, hazardous chemicals pose poisoning risks, particularly for children and pets; some, such as paints and solvents, pose problems or indoor air pollution; others, such as ammonia and chlorine bleach, can result in highly toxic emissions when inadvertently mixed; still others pose serious fire hazards.

2. Home fires involving hazardous chemicals can result in explosions or the generation of toxic fumes which can kill or seriously injure firefighters. Refuse collectors frequently suffer injury when they throw garbage bags into compactor trucks, unaware that they contain corrosive or flammable materials.

3. The environment itself can be seriously degraded when householders pour hazardous liquids into drains and flush them down toilets or into septic systems. People who pour waste motor oil into storm sewers or dump paint cans in the woods can cause long-lasting damage to ground and surface water supplies. Throwing such materials in the trash may ultimately result in the threat of contaminated incinerator ash and air pollution, or in the formation of toxic leachates at municipal landfills.

Listed below are some suggestions to follow which could help reduce the amount of hazardous waste which is produced and protect the family from the risks involved:

1. Avoid purchasing hazardous products whenever possible.

2. Do not buy more of a hazardous product than needed.

3. Follow instructions for use; do not use too much or too often. Use products in well-ventilated areas to avoid inhaling fumes.

4. Store products in original containers; otherwise, label them clearly. Never store hazardous products in food or beverage containers.

5. Give leftovers to those who can use them, but keep in original container.

6. Do not mix products together (e.g., never mix ammonia and chlorine bleach since it emits a lethal gas).

7. Separate hazardous materials from household garbage.

8. Do not dispose of hazardous materials in sewer systems or open bodies of water.

Some nonhazardous remedies for just two domestic applications — cleaning and garden-type activities — are listed below.

Cleaning

1. Baking soda is a nonabrasive scouring powder.

2. Use vinegar and water for windows and smooth surfaces.

3. Use crumbled newspaper to dry windows and mirrors.

4. Mix three parts olive oil with one part vinegar for a furniture polish. Wipe with a clean, soft cloth.

5. Sprinkle vinegar, then a layer of baking soda on oven surfaces. Rub gently with very fine steel wool for tough spots. Wipe with a sponge.

6. Sprinkle baking soda into the toilet bowl, drizzle with vinegar, and scour with a toilet brush.

7. For ant control, wash countertops, cabinets, and floor with equal parts vinegar and water.

8. For roach control, mix equal parts of flour, oatmeal, and plaster of Paris. Keep out of reach of children and pets.

9. Rub toothpaste on wood to remove water stains.

10. Boil cinnamon and cloves in water on the stove for "potpourri" air freshener.

11. Clean upholstery or carpet stains immediately with cold water or club soda.

12. Open drains with a metal snake or plunger.

13. Prevent clogs by pouring boiling water with baking soda and vinegar down drains once a week.

14. Use latex or water-based paints whenever possible. Latex and water-based paints do not require thinners or solvents.

15. Before disposing of oil-based or enamel paint cans, take outdoors, remove the lid, and allow the contents to air dry and harden, but ensure that children and pets are kept away.

In The Garden

1. To control aphids, spray plants with dish suds or soapy water. Rinse off when insects are dead.

2. To control ants, mix equal parts of powdered sugar and powdered borax together and sprinkle on the ant hill or burrow opening. Alternatively, pour a line of cream of tartar, paprika, red chili powder, or dried peppermint leaves at the point of entry.

3. To control snails and slugs, fill a shallow pan with stale beer and place in the garden, or overturn clay pots to attract slugs. Collect and destroy.

4. To control caterpillars, use "stickum" made from $1\frac{1}{2}$ cups rosin, 1 cup linseed oil, 1 tablespoon melted paraffin. Paint around tree trunks.

5. Keep garden clean; debris attracts pests and infected plants will breed them.

6. Instead of chemical fertilizers, use peat moss, manure, or fish meal.

7. Start an organic compost pile.

8. When cutting the lawn, try mulching the grass instead of using lawn mower bag.

9. Use organic gardening techniques.

10. Pull weeds instead of using chemical control.

11. In the fall, cover the garden with plastic to discourage weed germination.

27.5 ENERGY CONSERVATION

The general subject of energy conservation has been treated in rather extensive detail in Chapter 29. Section 29.5, "Domestic Applications," addresses the six major topic areas listed below:

1. Cooling
2. Heating
3. Hot water

4. Cooking

5. Lighting

6. New appliances

The reader is referred to Section 29.5 for details on the above.

Energy conservation measures can also be examined by dividing the home into specific areas. Information on these procedures is given below:

Living Quarters

1. Lighting
   a. Arrange furniture to take advantage of natural light. When purchasing lamps, keep in mind the function of the lamp for a particular situation.
   b. Turn off the lights, television, stereo, and radio when you leave the room.
   c. Use durable fluorescent bulbs. Lighting accounts for about 15% of a home's electric use. By using new screw-in fluorescent bulbs, one can save $60 over the life of each bulb. Fluorescent bulbs are more expensive, but they last ten times longer than incandescents and use 75% less electricity. A twin-tube 48-in., 40-watt fluorescent fixture produces up to four times more light than does one 100-watt incandescent bulb.
2. Weatherization
   a. Use caulking where necessary. One easy and inexpensive way to weatherize a home prior to the cold winter months is to install caulking on windows and doors.
   b. Install storm doors and windows. Storm windows and doors are big energy and money savers. They can reduce heating costs by as much as 15% by preventing warm air from escaping to the outside. Double-glazed and thermopane windows can further minimize heat escape. Temporary plastic storm windows can be installed by using polyethylene or clear vinyl plastic and waterproof tape or wood nailing strips.
3. Heating and cooling
   a. Check the heating system. During the winter, keep the thermostat at 70° during the day, lowering it when the family retires. Installing a timer might be more efficient. Maintain a lower temperature when leaving for the day or when you are away for a more extended period. During the cold winter months, heating is the single biggest energy user in the home. A well-maintained heating system will hold down fuel costs; cold air seeping into the house, along with heated air leaking out through small holes and cracks, can increase cost.
   b. Use fans for cost-effective cooling. During the summer, cooling takes heating's place as the number-one home energy consumer.
   c. Use fans to supplement air conditioning. Fans can help make the air conditioner work at peak efficiency by improving the air circulation in the home.
   d. Make sure the air conditioner is the proper size for the area that is being cooled. The wrong size air conditioner will use more electricity and increase energy bills.

e. Conduct an air-conditioner maintenance check. Regular maintenance will ensure that the air conditioner operates efficiently.

f. Close the intake vent on the air-conditioning unit. On very hot days, one can save energy by closing the fresh air intake vent on room air conditioners.

4. Central air conditioning

a. Buy an energy-efficient model. One cost-effective alternative to having several room air conditioners is to install central air conditioning.

Kitchen

1. Refrigerator

a. Refrigerators are one of the largest energy-users in the home. Clean the refrigerator regularly. Clean and dust the condenser coils and air vents found in the back or bottom of the refrigerator. Use a vacuum cleaner with a brush or nozzle attachment.

b. Set the thermostat at 39° and arrange food so air can circulate around it.

c. If you are away from home for more than a week, use up the food, unplug the refrigerator, and leave the door ajar. However, if the refrigerator door is not being opened often (as when children are home and taking a look, causing the refrigerator to work harder to keep the temperature constant), there is less energy waste.

d. Encourage family members to be energy conscious when going to the refrigerator — knowing which items they want and perhaps having a tray ready to carry the anticipated items — before the door is opened. The refrigerator contents should be arranged efficiently for the purpose of good air circulation, and items should be placed logically with good visibility so that family members can open the door, make a selection, and shut the door with dispatch. Contents should also be placed and replaced in the same position so that food does not go unnoticed and unconsumed, eventually spoiling and thus being wasted.

e. Keep a bottle of drinking water in the refrigerator, instead of running the tap until the water gets cool each time a drink is wanted.

f. Set the freezer thermostat at O°F. A tightly packed freezer keeps food colder with less energy use. When the freezer is not packed, make ice cubes, put them in bags, and fill the freezer with them.

2. Conventional oven

a. Use glass pans to conserve energy. This allows one to set the oven 25° lower because glass retains heat.

b. Do not line drip pans or the oven cavity with aluminum foil. This interferes with heat circulation and can harm the heating components of the oven.

c. When using the oven, do not peek. Every time the door is opened, the temperature falls 25°; the oven then has to work harder to bring the temperature back up.

d. It is not necessary to preheat the oven when cooking dishes that require more than an hour.

e. A toaster-oven-type appliance should be used for small baking needs whenever possible, thus avoiding the bigger energy use of the larger oven.

f. On top of the range, use pots and pans that are properly sized to "fit" the burners. Using a small pan on a large burner wastes energy; the use of a large pan on a small burner is energy-inefficient as well.

3. Microwave oven

   a. Conserve energy with a microwave. A microwave oven is an energy-efficient alternative to a conventional oven. It cooks food more rapidly and uses 70–80% less electricity than a regular oven would use.

4. Dishwasher

   a. Conserve water with a dishwasher. More water is wasted when dishes are handwashed. If dishes must be rinsed or scrubbed beforehand, have a pot filled with water for this purpose.

   b. Run a full dishwasher and save energy. Because 80% of the energy used in automatic dishwashers goes toward heating water, one can save by running the dishwasher only when it is full.

   c. Running the dishwasher in the low peak hours is more cost-efficient.

   d. Open the dishwasher after the rinse cycle and let the dishes air dry if the dishwasher does not have the energy-saver option.

Bathroom

1. Install water flow reduction devices. Fifteen to twenty percent of the energy consumed in the average American home goes to water.

2. Install low-flow showerheads in the shower to limit the flow of water to about 2 gal/min. While in the shower, turn off the water while soaping up or shampooing.

3. Faucet aerators are small "flow control" devices that easily fit in bathroom faucets. Once installed, they reduce hot water use by one-third without affecting water pressure. The running of water while brushing one's teeth can easily use more than 5 gal of water. A running faucet sends 3–5 gal of water down the drain every minute it is on.

4. Toilets are the single greatest water users. Water use in the toilet can be reduced by at least 20% by using a simple water replacement device such as weighted 2-L plastic bottles filled with water. More sophisticated devices are available at hardware stores. The installation of low-flush toilets conserves water by 50%. The plunge ball and flapper valve should be examined periodically to ensure proper seat, and parts should be replaced when needed. Do not use the toilet as a wastepaper basket. Do not flush the toilet every time.

5. Be prudent and prompt about any type of plumbing leak. One leak can waste several thousand gallons of water per year.

Attic

Use insulation to contain heating and cooling costs. Depending on the size and condition of the attic, one could save nearly 25% of the heating and cooling costs by insulating an uninsulated attic.

Basement

1. Use a dehumidifier moderately. Dehumidifiers consume significant amounts of energy. However, if the basement is moist, try running the dehumidifier about 3–5 hr/day. Inspect for any structural flaws or water leaks that may be increasing the moisture level. Use the water collected in the reservoir for ironing or watering house plants.

2. Use hot water sparingly when washing. Because 90% of the energy your washer consumes goes toward heating water, one can save by using hot water only for heavily soiled laundry. Most laundry can be washed in warm or cold water.

3. Dry wash loads consecutively; the heat left over from the previous cycle will increase the efficiency of the dryer. It is important to clean the lint trap on the dryer prior to each use. Separate heavy and light fabrics to keep drying time to a minimum. Mixing different weight fabrics causes the dryer to run longer than necessary.

4. Drain a few gallons of water from the water heater tank every 6 months to remove accumulated sediment and improve the unit's efficiency. As much as 10% can be saved on the water heating costs by simply wrapping a fiberglass blanket around the water heater and fastening it with duct tape or by using a ready-made insulation kit. Most new heaters, however, are already insulated and do not need additional fiberglass blankets (9). In addition, if the house will be unoccupied more than a few days, turn the water heater off to conserve energy.

27.6 SAFETY CONSIDERATIONS

As described in Chapter 34, accidents have occurred since the birth of civilization and were just as damaging in early times as they are today. Anyone who crosses a street or swims in a pool runs the risk of injury through carelessness, poor judgment, ignorance, or other circumstances. This has not changed throughout history.

Accidents at "home" take the lives of more than 20,000 Americans each year. These occurrences are the number one cause of the death of young children; two-thirds of these accidents happen to boys. Accidents claim the lives of more children aged 1 to 14 than do the leading diseases combined.

Often after an accident has occurred, one sadly realizes that if simple safety practices had been followed in a timely manner, the accident could have been prevented. Yet each year, more accidents and injuries take place in the home than anywhere else. Injuries and deaths from fires, burns, and falls lead the list of home accidents. Many accidents are automobile-related. There are an estimated 5.6 million home fires in the United States each year. Building fires claim approximately 5200 lives a year, and most of these victims die in their own homes. Some suggestions to help prevent this tragedy follow.

To Safeguard the Home Against Fire

1. Post emergency numbers for police and fire departments near the phone. Stickers for such purposes are often sent through the mail by the fire department.

2. Use the correct wattage and proper kind of light bulb in an overhead or

ceiling fixture. The wrong type of bulb can lead to overheating and fire. When in doubt, use a bulb no larger than 60 watts.

3. Make sure at least one smoke detector is placed on every floor of the home. Locate the detectors on the ceiling away from air vents and near bedrooms. Test detectors every month to ensure they are working. One might check the unit on the first day of each month or assign the responsibility to a child to encourage fire safety awareness in the family.

4. Develop an emergency exit plan for the home in case of a fire. Practice the plan to make sure everyone can escape quickly and safely. Have a meeting place. Practice at least once in the dark. Most residential fires occur at night and even though more people are on hand, night fires are less likely to be detected because people are sleeping. The location of a smoke detector outside the bedroom can alert the occupant. Focus on these four elements in your fire safety plan: prevention, detection, escape planning and practice, and fire department notification.

5. Clear the chimney and other vents of leaves or debris that can clog them. Have the chimney checked and cleaned by a licensed professional.

6. Do not smoke in bed—smoking is the primary cause of all apartment fires, and nearly a third of them are caused by someone smoking in bed. Mattresses and other bedding materials account for about 10% of all residential fires and 17% of all fire deaths. Other smoking fires start when a cigarette is dropped in upholstered furniture, when smoldering butts are thrown in wastepaper baskets, or by other careless accidents. When shopping for furniture, check materials and methods of construction used for resistance to flame ignitions.

7. More fires start in the kitchen than anywhere else in the house. When something is cooking on the stove, the kitchen should not be left unattended. The handles of pots should always be turned in so that they do not stick out over the edge to tempt a child or to be inadvertently knocked over. Never put anything on the stove that could catch fire. Use protective, fire-retardant potholders. Be careful about igniting clothing when working around stoves, barbecues, or any open flame. Keep sand, baking soda, a fire extinguisher, or a nonflammable cover readily available in case of a fire emergency.

8. In this age of latchkey children and modern technology, microwave safety needs to be addressed. Instruction pertaining to safe use of the microwave is needed. Emergency rooms are seeing more cases of burns involving unsafe use of a microwave. Facial burns have been caused by the steam which escapes when the progress of a food item is being checked. A child's esophagus was damaged after he heated a jelly donut in the microwave and popped it in his mouth, not realizing that the jelly was at a higher temperature than the donut. A similarly tragic accident occurred with the contents of a heated baby bottle. It is important for all to realize that microwaved food must stand the required amount of time to allow it to return to a safe temperature for consumption.

9. Store flammable liquids, such as gasoline, paints, and solvents, away from heating sources. Do not store them in the house or car. Vapors from flammable liquids can ignite even at temperatures below zero. Some chemicals give off deadly gases when combined. Certain combinations are flammable when mixed. All products should be kept in their original containers and should be clearly marked as to their use, possible hazards, and proper disposal (10).

Safety in the Automobile

1. Take a safe-driving course to refresh driving skills. Insurance companies encourage clients to do this by reducing premiums after proof of completion of course. Obey traffic laws and speed limits.

2. The automobile must be kept in top working condition. This includes the engine, tires, lights, doors, etc. An inefficient windshield wiper can be life-threatening in a rain or snow storm. Also to be considered is the inside of the car. Visibility should not be obstructed by clothing, an unsecured dog, etc. Objects should be secured to prevent movement; a sudden stop can make a tissue box or a pair of sunglasses become a disabling projectile.

3. The glove compartment should contain the car manual and a flashlight. Keep a car emergency repair kit. Check the state of the spare tire, and ensure that instructions for assembling and operating the jack are available. Familiarize family members on how to proceed in a car emergency before an emergency occurs. Carry a first-aid kit in the car. Check and replenish the contents of both kits periodically.

4. Eighty percent of all automobile accidents causing injury or death occur less than 25 miles from home or involve cars traveling under 40 miles per hour.

5. Wear seat belts at all times. One is much safer in the car than being thrown from it. Twenty percent of serious injuries from car crashes are the result of unbelted people colliding. A 30-lb child in a 30-mph crash is ripped from one's arms with a force of 900 lb. The chances of being fatally injured are 25 times greater for anyone who is thrown from a car. If one survives being thrown from or through the car, the additional peril of hitting another object or moving vehicles is present.

6. The unexpected action of another vehicle cannot be predicted or controlled. This is particularly problematic when alcohol is present, which figures into more than half of all automobile-related highway fatalities. Alcohol is the largest contributing factor in fatal motor vehicle accidents. There are approximately 2 million alcohol-related accidents annually that produce 23,000 to 26,000 fatalities and leave 300,000 seriously injured victims. The problem more often is the occasional and the social drinkers, who are often unaware of the effects of alcohol on their driving skills. Young drivers between the ages of 16 and 24 are involved in more than one-third of all alcohol-related traffic accidents (11) .

Other specific safety tips that can help make the home safer are provided below.

1. Check the condition of all stairways. Steps, coverings, and handrails should be sturdy and secure. Stairways should be well lighted and free of clutter. Stairway mishaps are among the most dangerous and are more likely to result in serious injury or death.

2. Floors and floor covering should be well maintained. Each year, approximately 84,000 individuals receive hospital emergency treatment for injuries involving floors and flooring materials. Most of these injuries are the result of falls. More than 9,000 people die each year as the result of falls.

3. Keep ladders in good shape. Check for loose rungs, or frayed ropes on extension ladders. Make sure the ladder is on level ground before climbing, and never stand on the top step. The U.S. Consumer Product Safety Commission estimates that in a 1-year period, 65,000 individuals injured in accidents associated with ladders receive hospital emergency room treatment.

4. Make sure clotheslines are above head level. Tree branches that could poke an eye should be trimmed. A root that can cause a fall should be removed. Keep the yard clear of garden tools and litter. Inspect outdoor play equipment and furniture to ensure that they are secure and safe.

5. Do not stand on chairs, boxes, or other makeshift items to get to hard-to-reach places. Use a sturdy step-stool that has a handrail for support.

6. Equip all bathtubs and showers with nonskid mats or strips to help prevent falls. Install grab bars for assistance in getting in and out of tubs or showers. Do not use towel bars as grab-bars.

7. Excessively hot tap water causes over 5000 burns and serious scald injuries annually, mostly to the elderly and the very young. As many as 3200 injuries could be prevented each year if water heaters were set at a safer temperature.

8. Garage doors should be checked regularly to ensure that the springs are not in danger of releasing. An automatic garage door should have an electric eye which reverses the door direction when the eye comes in contact with an object. Similarly, electric car windows should have the same safety feature.

9. Keep the garage door open while running the car engine inside. Be sure to clear snow away from tailpipes and parking lights. Walk around the back of the car before getting in to ensure that the path is clear. A small child or object may not be visible from the driver's seat vantage point. To avoid "backing up" types of accidents, park facing out of driveway or garage.

10. Wear eye protection such as goggles or shields when operating power-tool equipment. Always disconnect and leave these tools in a locked position. Do not store tools or sharp objects overhead; an injury may occur while trying to retrieve it. It has been recorded that 1800 accidents a year have taken place when a child or adult has attempted to remove a tool box from an overhead position. The tool accounting for the most household accidents has been the hammer.

11. Use a special compartment, tray, or utensil block to store sharp knives. Each year, more than 137,000 people are treated in the hospital emergency room for cuts on fingers, hands, and arms which might have been avoided. Teach children the safe way to carry knives and scissors.

12. Every year, over 400 children drown in back-yard pool accidents. Approximately 236 are children under age 5, and 65% are male. Another 4400 children under age 5 are treated in hospital emergency rooms for submersion accidents. Do not rely on fencing to keep young children safe from drowning accidents. Educate children about safety procedures, and never leave them unsupervised by the pool. Parents, guardians, and babysitters should know how to institute cardiopulmonary resuscitation (CPR) or mouth-to-mouth resuscitation immediately and not wait for emergency personnel. A telephone along with emergency numbers and procedures should be poolside. Install ground fault circuit interrupters (GFCIs) near pools (or any high-risk area). Make sure all family members know not to use or touch electrical devices when wet or standing on wet surfaces. In relation to this, all appliances and electric equipment should be located away from the sink and bath.

13. Emergency numbers and basic first-aid instructions should be displayed prominently by the telephone. A first-aid kit should be readily available. Syrup of ipecac should be available in case the Poison Control Center suggests it. Family members should review the steps to be taken in an emergency and have a basic

knowledge of first aid so that they will be prepared to act appropriately when needed. Family members should consider taking a CPR course (12).

Do not be a victim — correct potential accident "sites" now and follow established safety procedures.

27.7 SUMMARY

1. The purpose of this chapter is to offer some suggestions which might be used in the household in the hope of preventing further pollution of the environment. The best place to start and concentrate is, in fact, in the home. It is where the most time is spent and where the individual has the most control over the health, safety, and well-being of those nearest and dearest.

2. Preventing pollution at the source should be considered first. Regarding recycling, each individual should reuse, repair, and recycle as often as possible. One use is often not enough. Every time a new product is used instead of a recycled one, some form of natural resource is depleted forever.

3. A variety of methods are used to collect materials away from the home, including (a) single-material and multimaterial curbside collection, (b) drop-off centers, and (c) buy-back centers.

4. Hazardous household products are common products individuals use around the house every day. These include products which are flammable, corrosive, reactive, or poisonous. One should check labels for words like DANGER, WARNING, CAUTION, and POISON, which appear on many hazardous substances.

5. Energy conservation measures can be applied in the following areas of the home: living quarters, kitchen, bathroom, attic, and basement.

6. Accidents have occurred since the birth of civilization and were just as damaging in early times as they are today.

REFERENCES

1. U.S. EPA, You *Can Make a Difference*, EPA/930-M-90-001, January 1990.
2. U.S. EPA, *The Garbage Problem: An Action Agenda*, EPA/530-SW-89-018, February 1989.
3. A. Raver, "From Kitchen Waste Springs Garden Gold," *New York Times*, The Home Section, November 21, 1991, p. C1.
4. U.S. EPA, *Let's Reduce and Recycle*, Curriculum for Solid Waste Awareness, EPA/530SW-90-005, August 1990.
5. U.S. EPA, *School Recycling Programs, A Handbook for Educators*, EPA/530-SW-90023, August 1990.
6. Author unknown, *Source Reduction Ideas for Businesses*, Rhode Island Solid Waste Management Corporation, undated.
7. Author unknown, *Overview: Solid Waste Disposal Alternatives*, Keep America Beautiful, Inc., undated.
8. Canadian Council of Ministers of the Environment, *Household Hazardous Wastes*, CCME-WM-FS23E, ISBN 0-919074-38-3, September 1990.

9. Author unknown, *A Guide to an Energy Efficient Home*, Long Island Lighting Company, New York, undated.

10. NFPA, *Be a Firesafe Neighbor*, BR-5, 1988.

11. M&M Protection Consultants, *Let's Slip into Something Comfortable...*, SB 4-87-3.0, undated.

12. Author unknown, *79 Tips to Make Your Home Safer,* Long Island Lighting Company, New York, undated.

28

WASTE REDUCTION

28.1 INTRODUCTION

In the current working definition used by the U.S. EPA, source reduction and recycling are considered to be the most viable waste reduction/pollution prevention techniques, preceding treatment and disposal. In its original "Pollution Prevention Policy Statement" published in the January 26, 1989 *Federal Register*, the EPA encouraged organizations, facilities, and individuals to fully utilize source reduction and recycling practices and procedures to reduce risk to public health, safety, and the environment.

Figure 28.1 depicts the EPA hierarchy of preferred approaches to waste reduction/pollution prevention. As one proceeds down the hierarchy, more and more waste is prevented, recycled, or treated, thereby reducing the amount of waste generated and/or released to the environment. As illustrated by Figure 28.1, the hierarchy may be broken down into the following three major components, plus (a fourth) any potential ultimate disposal considerations: source reduction, recycling, treatment, and lastly, ultimate disposal.

The definitions of these four terms that follow are those employed by the authors in this text:

1. Source reduction, consisting of technologies to reduce the volume of wastes initially generated, is the primary approach. The techniques involved are applied to the production process prior to the point of generation. Methods that eliminate or reduce the amount of waste generated by a particular process, either through procedure modifications or through material substitution, are employed.

2. Recycling, the secondary approach, attempts to recover a usable material from a waste stream. The methods involved can take place within the process or at the end of the process and can be implemented either on-site or off-site.

3. The next approach is the use of physical, biological, and chemical treatment methods, including incineration. It should be applied to the wastes remaining after

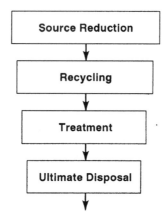

FIGURE 28.1   EPA waste reduction/pollution prevention hierarchy.

all possible source reduction and recycling techniques have been employed. This results in a reduction of the toxicity and volume of waste requiring ultimate disposal.

4. The last approach for managing wastes is ultimate disposal, consisting of landfilling, landfarming, deep-well injection, and ocean dumping. A more detailed flow diagram is provided in Figure 28.2.

Unfortunately, many individuals, particularly those employed by the EPA and other (regulatory) environmental organizations, have interpreted the Pollution Prevention Act and the EPA's subsequent policy statement to mean that pollution prevention refers only to source reduction and, as such, is the preferred method of environmental protection. Recycling, although possessing environmental advantages over other management techniques, is relegated to a secondary position the hierarchy. Treatment and disposal practices are viewed as low-priority options or are simply not considered as part of the hierarchy. However, these four pollution prevention options (source reduction, recycling, treatment, and ultimate disposal) are required in a total systems approach to pollution prevention, thus constituting part of the hierarchy, and are treated as such below.

Although not considered to be a preferable waste reduction technique by the EPA, many in industry contend that treatment, through its intent to protect human health and the environment by reducing or eliminating the quantity and/or toxicity of waste, should be considered as a valid approach. It is not realistic to assume that all waste will be eliminated from the production process. Therefore, for the purposes of this chapter, treatment (as well as disposal) will be considered as an integral approach to waste reduction after source reduction and recycling approaches. Additional details on all components of the hierarchy are provided below.

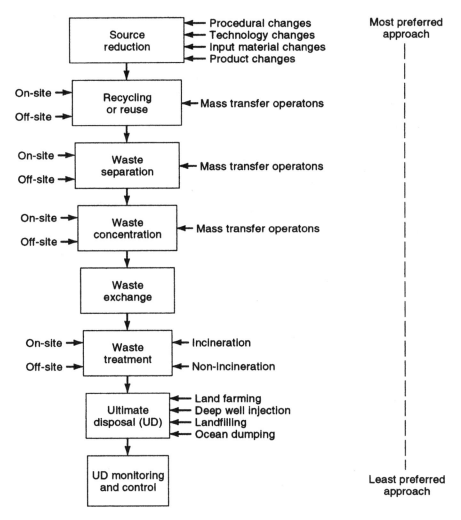

FIGURE 28.2 Waste reduction/pollution prevention hierarchy.

28.2 SOURCE REDUCTION

Once opportunities for waste reduction have been identified by a pollution prevention assessment (see Chapter 26), source reduction techniques should be implemented first. As described above, source reduction involves the reduction of pollutant wastes at their source, usually within a process, and is the most desirable option in the pollution prevention hierarchy. By avoiding the generation of wastes, source reduction eliminates the problems associated with the handling and disposal of wastes.

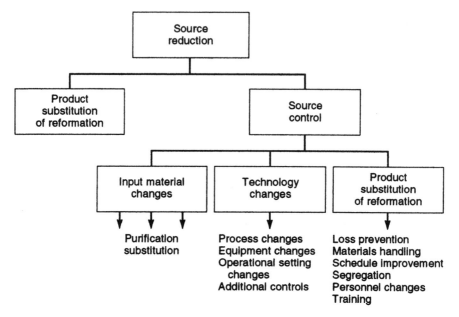

FIGURE 28.3 Source reduction.

A wide variety of facilities can adopt procedures to minimize the quantity of waste generated. Many source reduction options involve a change in procedural or organizational activities, rather than a change in technology. For this reason, these options tend to affect the managerial aspect of production and usually do not demand large capital and time investments. This makes implementation of many source reduction options affordable to companies of any size. Figure 28.3 depicts the source reduction scheme. As is shown, source reduction can be broken down into options involving product substitution or source control (1). These are detailed below.

Procedural Changes

Procedural changes involve the management, organizational, and personnel functions of production. Many of these measures are used in industry largely as efficiency improvements for waste reduction and good management practices. They often require very little capital cost and result in a high return of investment. Companies of any size can implement these practices in all areas of a plant. Evaluating plant procedures can often reveal source reduction opportunities that are relatively inexpensive and easy to implement. As shown in Figure 28.3, procedural changes can include:

1. Loss prevention

2. Materials handling

3. Schedule improvements

4. Segregation

5. Personnel practices

6. Training

Technology Changes

Technology changes involve process and equipment modifications to reduce waste, primarily in a production setting. Technology changes can range from minor changes that can be implemented quickly and at low cost, to major changes involving replacement of processes at a very high cost. Since technology modifications usually require greater personnel and capital cost than procedural changes, they are generally investigated after all possible procedural changes have been instituted. As shown in Figure 28.3, categories of technology modifications include:

1. Process changes

2. Equipment, piping, or layout changes

3. Changes to operational settings

4. Additional automation

Input Material Changes

Input material changes accomplish pollution prevention by reducing or eliminating the waste materials that first enter the process. Changes in materials can also be made to avoid the generation of waste within production processes. Input material changes fall into the categories of material substitution and material purification. The substitute is either less hazardous, produces less pollutant, or results in lower waste generation, but still satisfies end-product specifications. Ideally, the best substitution is the replacement of a hazardous/pollutant material with a nonhazardous one, without damaging the quality of the product.

Examples of industrial applications include:

1. Printing operations—substitution of water-based for solvent-based ink.

2. Furniture manufacture—substitution of water-based for solvent-based paints.

3. Plating operations—replacement of cyanide cadmium plating bath with noncyanide bath.

Product Changes

Product changes are performed by the manufacturer of a product with the intent of reducing waste resulting from a product's use. Product conservation involves the way in which an end-product is used. For example, the manufacture of water-based paints instead of solvent-based paints involves no toxic solvents that make solvent-based paints hazardous. In addition, the use of water-based paints greatly reduces volatile organic emissions to the atmosphere.

Product composition or reformulation involves manufacturing a product with a lower composition of hazardous substances or changing the composition so that no hazardous substances are present. Reformulating a product to contain less hazardous material will reduce the amount of hazardous waste generated during the product's formulation and end use. Using a less hazardous material within a process should reduce the overall amount of hazardous waste produced. For example, a company can use a nonhazardous solvent in place of a chlorinated solvent.

28.3 RECYCLING

Recycling or reuse can take two forms: preconsumer and postconsumer applications. Preconsumer recycling involves raw materials, products, and by-products that have not reached a consumer for an intended end use, but are typically reused within an original process. Postconsumer recycled materials are those that have served their intended end use by a business, consumer, or institutional source and have been separated from municipal solid waste for the purpose of recycling. Regarding preconsumer recycling, techniques allow waste materials to be used for a beneficial purpose. A material is recycled if it is used, reused, or reclaimed. Recycling through use and/or reuse involves returning waste material either to the original process as a substitute for an input material, or to another process as an input material. Recycling through reclamation is the processing of a waste for recovery of a valuable material or for regeneration. Recycling of wastes can provide a very cost-effective waste management alternative. This option can help eliminate waste disposal costs, reduce raw materials costs, and provide income from saleable waste.

Since recycling is the second most preferred option in the waste reduction hierarchy, it should be considered only when all source reduction options have been investigated and implemented. Reducing the amount of waste generated at the source will often be more cost-effective than recycling, since waste primarily is lost raw material or product which requires time and money to recover.

It is important to note that recycling can increase a generator's risk or liability as a result of the associated handling and management of the material involved. The measure of effectiveness in recycling is dependent upon the ability to separate any recoverable waste from other process waste that is not recoverable. Specific information on separation processes is available in the literature (2).

Recycling options can be listed in the following order of preferability:

1. Direct reuse on-site
2. Additional recovery on-site
3. Recovery off-site
4. Sale for reuse off-site
5. Energy recovery

These are detailed below.

Reuse

Reuse involves finding a beneficial purpose for a recovered waste in a different process. Three factors to consider when determining the potential for reuse are:

1. The chemical composition of the waste and its effect on the reuse process.
2. Whether the economic value of the reused waste justifies modifying a process in order to accommodate it.
3. The extent of availability and consistency of the waste to be reused.

Additional On-Site Recovery

Recycling alternatives can be accomplished either on-site or off-site and may depend on a company's staffing or economic constraints. On-site recycling alternatives directly result in less waste leaving a facility and an associated reduction in the reporting and manifesting for that waste. The disadvantages of on-site recycling lie in the capital outlay for recycling equipment, the need for operator training, and additional operating costs. In some cases, the amount of waste generated does not warrant the cost for installation of in-plant recycling systems. In general, however, since on-site alternatives do not involve transportation of waste materials and the liability incurred thereby, they are preferred over off-site alternatives.

Recovery Off-Site

If an insufficient amount of waste is generated on-site to make an in-plant recovery system cost-effective, or if the recovered material cannot be reused on-site, off-site recovery is preferable. Some materials commonly reprocessed off-site are oils, solvents, electroplating sludges and process baths, scrap metal, and lead-acid batteries. The cost of off-site recycling is dependent upon the purity of the waste and the market for the recovered material.

Sale for Reuse Off-Site

As an alternative to both on-site and off-site recycling, the generator may transfer waste to another facility for use as a raw material in its manufacturing operations. Facilities receiving the waste either use it "as is" or subject it to a minimal amount of pretreatment. Supply-and-demand is the key criterion for the success of waste transfer, but there needs to be a method for marketing the waste by determining the existence of a facility capable of utilizing the waste. This need has purportedly been fulfilled by waste exchanges, which serve as brokers of wastes or clearinghouses for information on the availability of wastes (3). Waste exchanges can be either be privately owned or government-funded organizations that facilitate waste transfer by identifying potential users. This exchange often proves to be economically advantageous to both firms involved, since the generator experiences a reduction in waste disposal costs and the purchasing firm experiences a reduction in raw material costs. Liability concerns remain with this option.

Energy Recovery

Recycling can also be achieved in the recovery of energy through the use of waste as a fuel supplement or fuel substitute. Waste may be processed in fossil-fuel-fired plants or in incinerators equipped with an energy recovery system. Note that processes with overall energy efficiencies of less than 60% are generally regarded strictly as incineration and not energy recovery. Usually, a variety of high-Btu wastes with different compositions are blended to produce a fuel with a certain specification.

28.4 OTHER TREATMENT METHODS

Recognizing that not all wastes can be eliminated through source reduction methods or recycling efforts, viable treatment processes must then be looked to for managing remaining wastes. These include incineration (3) and a host of additional treatment alternatives. According to the EPA definition, treatment is "any practice, other than recycling, designed to alter the physical, chemical, or biological character or composition of a hazardous substance, pollutant, or contaminant, so as to neutralize said substance, pollutant, or contaminant or to render it nonhazardous through a process or activity separate from the production of a product or the providing of a service."

Treatment can generally be divided into three categories:

1. Chemical treatment
2. Biological treatment
3. Physical treatment

Chemical treatment is used to convert wastes into other less "hazardous" forms. Biological treatment, like chemical treatment, involves chemical reactions, but here the reactions occur in or around microorganisms (or, occasionally, macroorganisms). Biological treatments are mainly used on organic wastes. Physical treatment is used to concentrate wastes, reduce waste volume, and separate different components for continued treatment or disposal. Additional information is provided below.

Chemical Treatment

Numerous chemical processes are used in the treatment of waste. Most treatment schemes are used in conjunction with other methods to achieve an end result. Some treatment methods are also directly applicable to recycling efforts. Examples of chemical processes are calcination, catalysis, chlorinolysis, electrolysis, hydrolysis, neutralization, oxidation, photolysis, reduction, etc. Specific details on these processes are available in the literature (4).

Biological Treatment

Biological processes also involve chemical reactions, but are differentiated from the chemical category in that these reactions take place in or around microorganisms. The most common use of biological processes in waste treatment is for the decomposition of organic compounds.

The different biological processes include activated sludge, aerated lagoons, anaerobic digestion, composting, enzyme treatment, trickling filter, waste stabilization ponds, etc. All of these, except enzyme treatment, use microorganisms to decompose the waste. Enzyme treatment generally involves extracting the enzyme from the microrganism and using it to catalyze a particular reaction. With proper control, these processes are environmentally sound; additive chemicals are usually not needed and operational expenses are relatively low. Additional details are provided in the literature (4).

Physical Treatment

There are more than 20 types of physical treatment processes known to be used in the handling of wastes; however, very few of these are fully developed or commonly used in industry. Some treatments have been found to have little potential use so that further research in these areas is unlikely. Zone refining, freeze drying, electrophoresis, and dialysis all fall into this category. The difficulty of the operation and/or the high cost of these processes overshadows any possible future use. The most common processes today are sedimentation, filtration, flocculation, and solar evaporation. Most other processes fall in between these two extremes, i.e., they show some potential for future use but are not presently used to any great extent.

Physical treatments may be separated into two categories: phase separation and component separation processes; in the latter, a particular species is separated from a single-phase, multicomponent system. The various physical treatments may fall into one or both of these categories. Sedimentation and centrifugation are used in phase separation, liquid ion exchange and freeze crystallization are used for component separation, and distillation and ultrafiltration are used in both.

Phase separation processes are employed to reduce waste volume and to concentrate the waste into one phase before further treatment and material recovery are performed. Such waste streams as slurries, sludges, and emulsions, which contain more than one phase, are the usual candidates for this category. Filtration, centrifugation, and flotation may be used on slurries that contain larger particles. If the slurry is colloidal, flocculation and ultrafiltration are generally used. If the slurry or sludge is known to contain any volatile components, evaporation or distillation is used to remove them from the waste stream. Because emulsions are so difficult to separate, the type of physical treatment required is usually decided on a case-by-case basis.

Component separation processes remove particular ionic molecular species without the use of chemicals. Most of these are used in wastewater treatment and include such techniques as liquid ion exchange, reverse osmosis, ultrafiltration, air stripping, and carbon adsorption. The first three are used to remove ionic and inorganic components; the last two techniques are used to remove volatile components and gases (4).

28.5 ULTIMATE DISPOSAL

Ultimate disposal is described by many as the final process in the treatment and management of wastes. The term *ultimate disposal methods* was coined by the EPA and originally assigned to the five processes discussed in this section. (The authors have added atmospheric dispersion to the list.) The authors disagree with the nomenclature, preferring that *ultimate disposal* be reserved for a process in which wastes are chemically or biologically rendered innocuous or a process that has been modified so that such wastes are no longer generated. Before disposal, most wastes undergo the various treatments previously described (biologic, chemical, and/or physical) in order to concentrate, detoxify, and reduce the volume of the wastes.

Disposal is defined in Section 10004 of RCRA Subtitle A as

the storage, deposit, injection, dumping, spilling, leaking, or placing of any solid waste or hazardous waste into or in any land or water so that solid waste or hazardous waste or any constituent thereof may not enter the environment or be emitted into the air or discharged into any waters including ground waters.

The ultimate disposition of residuals from treatment processes depends on similar considerations as those connected with the selection of a particular treatment method. Specifically, disposal is based on:

- Federal, state, and local environmental regulations
- Potential environmental hazards
- Liabilities and risks
- Geography
- Demography

The five ultimate disposal methods discussed below are landfarming, deep-well injection, landfilling, and ocean dumping, plus atmospheric dispersion. Landfarming, used for organic wastes, relies on nutrients in the soil to convert wastes into nonhazardous materials that may enrich the soil. The latter four methods concentrate on containing and dispersing, not converting wastes, and may be used on various waste types.

Landfarming

Landfarming is one of several terms used to describe the process of disposing of hazardous and nonhazardous wastes in the upper layer of the soil. The process is not a new one; it has been used for almost 30 years in the disposal of oily petroleum wastes. This application has increased in usage over the years and is now employed to treat up to 50% of these petroleum wastes as well as a number of other biodegradable wastes.

Landfarming has a number of advantages: It can be an effective and low-cost disposal method; it is an environmentally safe and simple process not dependent upon processing equipment; and it is a natural form of waste disposal which can, in some cases, improve the fertility and nature of the soil. There are also some limitations to landfarming. The bulk of the waste disposed of must contain organic components. It is not recommended for use with inorganics, particularly when the pH of the waste is below 7. Wastes containing materials that could pollute the air, groundwater, or the soil itself are not candidates for this method. Some wastes may need pretreatment to make them suitable for landfarming; this additional treatment adds to the total cost of the waste disposal.

Deep-Well Injection

Deep-well injection is an ultimate disposal method that transfers liquid wastes far underground and away from freshwater sources. Like landfarming, this disposal process has been used for many years by the petroleum industry. It is also used to dispose of saltwater in oil fields. When the method first came into use, the injected brine would very often eventually contaminate groundwater and freshwater sands because the site was poorly chosen. The process has since been improved, and laws such as the Safe Drinking Water Act of 1974 ensure that sites for potential wells are better surveyed. Today, injection wells are placed as far away as possible from drinking water sources, usually at least a quarter-mile.

Many factors are considered in the selection of a deep-well injection site. For example, the rock formation surrounding the disposal zone must be strong but permeable enough to absorb the liquid wastes, and the site must be far enough from drinking water sources to prevent contamination. Once a site is selected, it must be tested by drilling a pilot well. The performance data from the pilot well, besides testing permeability and water quality, also aid in the design of the final well and in determining the proper injection rate.

Landfilling

Landfilling is the third ultimate disposal method and is generally used on wastes in the form of sludges. There are two types of landfilling: area fill and trenching. Area fill is essentially accomplished above ground, whereas trenching involves burying the waste. Trenching is the better-established and more popular form of the two. Yet, since trenching requires excavation, area fill has the advantage that it requires less manpower and machinery. Area fill is also less likely to contaminate groundwater since the filling is above ground. Trenching, however, may be used for both stabilized and unstablilized sludges and makes more efficient use of the land. Both techniques require the use of lime and other chemicals to control odors, and cold and wet weather can cause problems with either. Both methods also produce gas, which can cause explosions or harm vegetation, and leachate, which can contaminate ground and surface water.

Most wastes must be subjected to one or more pretreatments such as solidification, degradation, volume reduction, and detoxification before being landfilled. This practice stabilizes the waste and helps decrease the amount of gas and leachate produced from the landfill. Landfilling is similar to landfarming in that both ultimate disposal methods combine wastes and soil. Landfarming, as described above, involves the biochemical reaction between solid nutrients and wastes to degrade and stabilize the waste; as a result, only specific types of wastes can be landfarmed. A larger variety of wastes may be handled by landfilling.

Ocean Dumping

Ocean dumping is the EPA's final "ultimate" disposal method to be considered here. Although it is probably the simplest of the four techniques, its aftereffects and long-term consequences are more complex and less understood than those of other ultimate disposal methods. Regulations passed over the last 20 years have severely limited, and in some cases eliminated, the types and amounts of wastes that may be dumped. Therefore, in some cases, ocean dumping may be no more attractive than any other ultimate disposal method, despite its simplicity.

Currently, only certain types of wastes may legally be ocean-dumped. In the case of some materials, such as industrial wastes, disposal is regulated and may even require some treatment of the waste before disposal. Uncontrolled ocean dumping of untreated wastes occurs in the form of oil spills, rain carrying air pollutants, and runoff carrying land pollutants.

Atmospheric Dispersion

As indicated earlier, this is a recent addition to the "Ultimate Disposal" options. The reader is referred to Chapter 6 for technical details on this method.

28.6   FUTURE TRENDS

The EPA has taken a much stronger position on waste reduction by regarding source reduction as the only "true" pollution activity and treating recycling as an option. Industry's position prior to the Act, and effectively unchanged since, was to attempt to reduce the discharge of pollutant wastes into the environment in the most cost-effective manner. This objective may be achieved in some instances by source reduction, in other cases by recycling, in still other situations by treatment and/or disposal, and usually—as one might expect—with any combination of any of the above.

Traditionally, regulations have been ever-changing, with more stringent controls generally enacted with the passage of time. It is easy to envision that source reduction and perhaps recycling and reuse (instead of treatment or disposal) will become even more economically attractive in the future.

28.7   SUMMARY

1. Once opportunities for waste reduction have been identified by a pollution prevention assessment, source reduction techniques should be implemented first. Source reduction involves the reduction of wastes at their source, usually within a process, and is the most desirable choice in the pollution prevention hierarchy. Source reduction can be broken down into options involving either product changes or source control.

2. Recycling techniques allow waste materials to be used for beneficial purposes. A material is recycled if it is used, reused, or reclaimed. Recycling is the second-most-preferred option in the pollution prevention hierarchy and as such should be considered only when all source reduction options have been investigated and implemented. Recycling options can be listed in the following order of preferability:

- Direct reuse on-site
- Additional recovery on-site
- Recovery off-site
- Sale for reuse off-site
- Energy recovery

3. Recognizing that not all wastes can be eliminated through source reduction methods or recycling efforts, both government and industry accept that other viable treatment processes are available for managing the remaining wastes. Treatment can generally be generated be divided into three categories:

- Chemical treatment
- Biological treatment
- Physical treatment

4. The ultimate disposition of residuals from treatment processes depends on considerations similar to those connected with the selection of a particular treatment method. The selection of a disposal method is based on

- Federal, state, and local environmental regulations

- Potential environmental hazards
- Liabilities and risks
- Geography
- Demography

5. From a waste reduction perspective, it appears certain that greater emphasis will be placed on source reduction and/or recycling.

REFERENCES

1. USEPA, *Waste Minimization, Issues and Options*, Vol. I, EPA 530/-SW-86-041, 1986.
2. L. Theodore and Y.C. McGuinn, *Pollution Prevention*, Van Nostrand Reinhold, New York, 1992.
3. L. Theodore, and J.R. Reynolds, *Introduction to Hazardous Waste Incineration*, Wiley-Interscience, New York, 1988.
4. R. Perry and D.W. Green (eds.), *Perry's Chemical Engineers' Handbook*, 7th edition, McGraw-Hill, New York, 1998.

29

ENERGY CONSERVATION

29.1 INTRODUCTION

The environmental impacts of energy conservation and consumption are far-reaching, affecting air and water (as well as land) quality and public health. Combustion of coal, oil, and natural gas is responsible for air pollution in urban areas, acid rain that is damaging lakes and forests, and some of the nitrogen pollution that is harming estuaries. Although data show that for the period from 1977 to 1989 annual average ambient levels of all criteria air pollutants were down nationwide, 96 major metropolitan areas still exceed the national health-based standard for ozone, and 41 metropolitan areas exceed the standard for carbon monoxide.

Energy consumption also appears to be the primary man-made contribution to global warming, often referred to as the *greenhouse effect* (see Chapter 8 for more details); the U.S. Environmental Protection Agency (EPA) has concluded that energy use—through the formation of carbon dioxide during combustion processes—has contributed approximately 50% to the global warming that has occurred in the last 10 years. Although the scientific community is not unanimous in regard to the causes of global warming, most individuals and groups have indicated that a "reasonable" chance of climatic change exists and have already begun to define the potential implications of such changes, many of which are catastrophic. In light of this situation, the Alliance to Save Energy has challenged Congress to pass meaningful legislation to promote and achieve energy efficiency. Environmental conservation in this area can best be achieved through energy conservation and increased energy efficiency (in combustion, transmission, distribution, etc.).

Following a review of the law of conservation of energy (often referred to as the first law of thermodynamics), the reader is introduced to the second law of thermodynamics. This often misunderstood law provides valuable insight into energy transformation processes and the efficiencies associated with them. This chapter concludes with energy conservation measures that can be implemented in chemical processes and plants, in domestic applications, and by individual actions.

29.2   THE LAW OF CONSERVATION OF ENERGY

From the early recognition of energy, men have studied its effects upon objects, its transfer from object to object, and its transformation from one form to another. This field of study is called *thermodynamics*. Before proceeding to the first law of thermodynamics, often referred to as the law of conservation of energy, certain important terms are defined as follows:

1. A *system* is any portion of the universe which is set aside for study.
2. Once a system has been chosen, the rest of the universe is called the *surroundings*.
3. A system is described by specifying that it is in a certain *state*.
4. The path, or series of values certain variables assume in passing from one state to another, defines a *process*.
5. *Isothermal* means constant temperature.
6. *Isobaric* means constant pressure.
7. *Isochoric* is constant volume.
8. *Adiabatic* specifies no transfer of heat to system from surroundings.
9. A *reversible process* is one where changes occur due to driving forces (e.g., temperature differences that are differentially small); the system may therefore be considered to be in a state of equilibrium during the change.

The first law of thermodynamics specifies that energy is conserved. In effect, this law states that energy is neither created nor destroyed. Thus, the change in energy of a system is exactly equal to the negative of the change in the surroundings. For a system of constant mass (a closed system), the only way the system and surroundings may interchange energy is by work and heat. Work and heat are defined as energy in transit. They are not properties and cannot be stored in a system. Two common forms of work are expansion and electrical. Heat is energy in transit because of a temperature difference. This heat transfer may take place by conduction, convection, or radiation.

The energy balance makes use of the first law to account for all the energy in a chemical process, or in any other process, for that matter. After a system is defined, the energy balance considers the energy entering the system across the boundary, the energy leaving the system across the boundary, and the accumulation of energy within the system. This may be written in equation form as

$$\text{Energy in} - \text{Energy out} = \text{Energy accumulation}$$

This expression has the same form as the law of conservation of mass (see Chapters 6 and 12).

All forms of energy must be included in an energy balance. In many processes certain energy forms remain constant, and changes in them may be neglected. However, these forms should be recognized and understood before their magnitude and constancy can be determined. Some forms of energy are easily recognized in everyday life: the energy of a moving object, the energy given off by a fire, and the energy content of a container of hot water. Other forms of energy are less easily recognized. However, the five key energy terms are kinetic, potential, internal, heat, and work. These are briefly described below.

1. *Kinetic Energy.* The energy of a moving object is called "kinetic energy." A baseball thrown by a pitcher possesses a definite kinetic energy as it travels toward the catcher. A pound of flowing fluid possesses a definite kinetic energy as it travels through a duct.

2. *Potential Energy.* The energy possessed by a mass by virtue of its position in the earth's gravitational field is called "potential energy." A boulder lying at the top of a cliff possesses potential energy with reference to the bottom of the cliff. If the boulder is pushed off the cliff, its potential energy is transformed into kinetic energy as it falls. Similarly, a mass of fluid in a flowing system possesses a potential energy because of its height above an arbitrary reference level.

3. *Internal Energy.* The component molecules of a substance are constantly moving within the substance. This motion imparts internal energy to the material. The molecules may rotate, vibrate, or migrate within the substance. The addition of heat to a material increases its molecular activity and, hence, its internal energy. The temperature of a material is a measure of its internal energy.

4. *Heat.* When energy if transferred between a system and its surroundings, it is transferred either as work or as heat. Thus, heat is energy in transit. This type of energy transfer occurs whenever a hot body is brought into contact with a cold body. Energy flows as heat from the hot body to the cold body until the temperature difference is dissipated — that is, until thermal equilibrium is established. For this reason, heat may be considered as energy being transferred over a temperature difference.

5. *Work.* Work is also energy in transit. Work is done whenever a force acts through a distance.

The first law of thermodynamics may be stated formally — as opposed to equation form — in many ways. One of these is as follows: Although energy assumes many forms, the total quantity of energy is constant, and when energy disappears in one form it must appear simultaneously in other forms.

29.3 THE SECOND LAW OF THERMODYNAMICS

Historically, the study of the second law of thermodynamics was developed by persons such as Carnot (a French engineer), Clausius, and Kelvin in the middle of nineteenth century. This development was made purely on a macroscopic scale and is referred to as the "classical approach to the second law," which does not require the existence of an atomic or molecular theory.

The law of conservation of energy has already been defined in the previous section as the first law of thermodynamics. Its application allows calculations of energy relationships associated with all kinds of processes. The "limiting" law is called the "second law of thermodynamics." Applications involve calculations for maximum power outputs from a power plant and equilibrium yields in chemical reactions. In principle, this law states that water cannot flow uphill and heat cannot flow from a cold to a hot body of its own accord. Other defining statements for this law that have appeared in the literature are provided below.

1. Any process whose sole net result is the transfer of heat from a lower temperature level to a higher one is impossible.

2. No apparatus, equipment, or process can operate in such a way that its only effect (in system and surroundings) is to convert heat taken in completely into work.

3. It is impossible to convert the heat taken into a system completely into work in a cyclical process.

As described earlier, the first law of thermodynamics is a conservation law for energy transformations: Regardless of the types of energy involved in processes — thermal, mechanical, electrical, elastic, magnetic, etc. — the change in the energy of a system is equal to the difference between energy input and energy output. The first law also allows free convertibility from one form of energy to another, as long as the overall quantity is conserved. Thus, this law places no restriction on the conversion of work into heat, or on its counterpart — the conversion of heat into work.

The unrestricted conversion of work into heat is well known to most persons. Frictional effects are frequently associated with mechanical forms of work which result in a temperature rise of the bodies in contact. However, the transformation of heat into work is of greater concern. In nations with a partially developed or developing technological society, the ability to produce energy in the form of work takes on prime importance. Work transformations are necessary to transport people and goods, drive machinery, pump liquids, compress gases, and provide energy input to so many other processes that are taken for granted in highly developed societies. Much of the work output in such societies is available in the basic form of electric energy, which is then converted to rotational mechanical work. Although some of this electric energy (work) is produced by hydroelectric power plants, by far the greatest part of it is obtained from the combustion of fossil fuels or nuclear fuels. These fuels allow the engineer to produce a relatively high-temperature gas or liquid stream that acts as a thermal (heat) source for the production of work. Hence the study of the conversion of heat to work is extremely important — especially in the light of developing shortages of fossil and nuclear fuels, along with the accompanying environmental problems.

The brief discussion of energy-conversion devices above leads to an important second-law consideration — that is, that energy has "quality" as well as quantity. Because work is 100% convertible to heat whereas the reverse situation is not true, work is a more valuable form of energy than heat. Although it is not as obvious, it can also be shown through second-law arguments that heat has "quality" in terms of the temperature at which it is discharged from a system. The higher the temperature at which heat transfer occurs, the greater the possible energy transformation into work. Thus, thermal energy stored at high temperatures generally is more useful to society than that available at lower temperatures. While there is an immense quantity of energy stored in the oceans, for example, its present availability to us for performing useful tasks is quite low. This implies, in turn, that thermal energy loses some of it "quality" or is degraded when it is transferred by means of heat transfer from one temperature to a lower one. Other forms of energy degradation include energy transformations due to frictional effects and electric resistance, among others. Such effects are highly undesirable if the use of energy for practical purposes is to be maximized. The second law provides some means of measuring this energy degradation (1).

In line with the discussion regarding the "quality" of energy, individuals at home and in the workplace are often instructed to "conserve energy." However, this comment, if taken literally, is a misnomer because energy is automatically conserved by the provisions of the first law. (The reader is referred to the presentation in the previous section on energy conservation, particularly the last paragraph.) In reality, the comment "conserve energy" addresses only the environmental concern associated with the "quality" of energy. If the light in a room is not turned off, "quality" energy is degraded although energy is conserved; that is, the electrical energy is converted to internal energy (which heats up the room). Note, however, that this energy transformation will produce a token rise in temperature of the room from which little, if any, "quality" energy can be recovered and used again (for lighting or other useful purposes) (1).

There are a number of other phenomena which cannot be explained by the law of conservation of energy. It is the second law of thermodynamics that provides guidelines to the understanding and analysis of these diverse effects. Among other considerations, the second law can:

1. Provide the means of measuring the quality of energy.
2. Establish the criteria for the "ideal" performance of engineering devices.
3. Determine the direction of change for processes.
4. Establish the final equilibrium state for spontaneous processes.

Finally, the second law (of thermodynamics) also serves to define an important thermodynamic property called *entropy*. It is normally designated as S . Entropy calculations (2), which are beyond the scope of this text, can provide quantitative information on the "quality" of energy and energy degradation. Detailed thermodynamic calculations, in a tutorial form, are also available in the literature (3).

29.4 CHEMICAL PLANT AND PROCESS APPLICATIONS (4)

There are numerous general energy conservation practices that can be instituted at chemical plants. Ten of the simpler ones are detailed below:

1. Lubricate fans.
2. Lubricate pumps.
3. Lubricate compressors.
4. Repair steam and compressed air leaks.
5. Insulate bare steam lines.
6. Inspect and repair steam traps.
7. Increase condensate return.
8. Minimize boiler blowdown.

9. Maintain and inspect temperature-measuring devices.

10. Maintain and inspect pressure-measuring devices.

Some energy conservation practices applicable to specific chemical operations are also provided below:

1. Recover energy from hot gases.

2. Recover energy from hot liquid.

3. Reduce reflux ratio in distillation columns.

4. Reuse hot wash water.

5. Add effects to existing evaporators.

6. Use liquefied gases as refrigerants.

7. Recompress vapor for low-pressure steam.

8. Generate low-pressure steam from flash operations.

9. Use waste heat for absorption refrigeration.

10. Cover tanks of hot liquids to reduce heat loss.

Recycling can also be achieved in the recovery of energy through the use of waste as a fuel supplement or fuel substitute. Waste may be processed in fossil-fuel-fired plants or in incinerators equipped with an energy recovery system. Note that processes with overall energy efficiencies of less than 60% are generally regarded strictly as incineration and not energy recovery. Usually, a variety of high-Btu wastes with different compositions are blended to produce a fuel with a certain specification.

For the purposes of implementing an energy conservation strategy, process changes and/or designs can be divided into four phases, each presenting different opportunities for implementing energy conservation measures:

1. Product conception.

2. Laboratory research.

3. Process development (pilot plant).

4. Mechanical (physical) design.

Energy conservation "training" measures that can be taken in the chemical process industry include the following:

1. Implement a sound operation, maintenance, and inspection (OM&I) program.

2. Implement a pollution prevention program (see Chapter 26).

3. Institute a formal training program for all employees.

It should be obvious to the reader that a multimedia approach (see Chapter 3) that includes energy conservation considerations requires a total systems approach. Much of the environmental engineering work in future years will focus on this area, since it appears to be the most cost-effective way of solving many energy problems.

The following energy conservation practices are recommended at the plant's "office" level:

1. Maintain air-conditioner efficiency, and reduce heated and cooled space.
2. Maintain boiler efficiency.
3. Use natural ventilation whenever and wherever possible, reduce air infiltration, and seal leaks in pipes and ducts.
4. Raise office temperatures in summer.
5. Lower office temperatures in winter.
6. Use shading efficiently.
7. Close windows and doors when and where applicable.
8. Fix broken windows and other air leaks.
9. Do not use lights unnecessarily.
10. Turn off office equipment that is not in use.

Many of these recommendations are (obviously) repeated again in the next section.

29.5 DOMESTIC APPLICATIONS

Domestic applications (repeated to some extent in Chapter 27) involving energy conservation have been divided into six topic areas. These include:

1. Cooling
2. Heating
3. Hot water
4. Cooking
5. Lighting
6. New appliances

Details are provided below.

Cooling

1. Make sure the air conditioner is the proper size for the area to be cooled. The wrong-size air conditioner will use more electricity and increase energy bills. A unit that is too large for a given area will cool the area too quickly, causing the air conditioner to frequently turn itself on and off.

2. The installation of the air conditioner has a lot to do with how efficient it will be. If one has a choice, locate the unit(s) on the north, east, or the best-shaded side of the home. If the unit is exposed to direct sunlight, it has to work much harder and use more energy to cool one's home. Keep shrubbery away from the air conditioner because it blocks vents and reduces the unit's ability to exhaust air.

3. Regular maintenance will ensure that the air conditioner operates efficiently throughout the summer. Check the filter once a month by holding it up to a bright light. Also check the owner's guide to find out how to safely clean the condenser coils and fins on the outside of the unit.

4. One can save on cooling costs by not cooling unoccupied rooms.

5. On hot summer days, the temperature in the attic can reach 150°F. Improving the ventilation in the attic will lower the temperature of the entire house and

make the air conditioner's job a lot easier. Install an attic fan that is controlled by a thermostat to exhaust the hot air.

6. One can save 3% on the cooling costs for every degree the thermostat is raised in the summer. Thus, raising the thermostat from 73°F to 78°F can mean approximate savings of 15% in cooling costs.

7. Fans can make the air conditioner's job easier while saving money. In moderate heat, fans can sometimes completely replace air conditions. Ceiling fans use only about one-tenth the electricity of a typical home air conditioner, and therefore cost only one-tenth as much to operate.

8. To stay most comfortable during the hottest hours of the day, do the cooking, laundry, and bathing in the early morning or late evenings. These activities all increase the level of humidity in the home, making it less comfortable and forcing the air conditioner to work even harder.

9. Drapes, shades, and awnings shield windows from the hot sun, keeping the home cooler. Storm windows also come in handy during the summer since they keep cool air in and hot air out. Weatherstripping and caulking windows and door frames will also keep cool air from leaking out.

10. Like other appliances that heat and cool, refrigerators are big energy users. If the refrigerator door does not shut tightly, check the door seal to see if it needs to be cleaned or replaced. A door leak allows cool air to escape, forcing the refrigerator to use more energy to keep food cold.

Heating

1. Heating is the single biggest energy use in homes. A well-maintained heating system will hold down fuel costs and provide reliable comfort. Have the heating system checked periodically by a professional.

2. Storm windows and doors are big energy and money savers. They can reduce heating costs by as much as 15% by preventing warm air from escaping to the outside. Double-glazed and thermopane windows or even clear plastic across windows can minimize heat escape.

3. Proper insulation in walls, ceilings, and floors also significantly reduces the loss of heat to the outdoors. Insulation will pay for itself in fuel cost savings and home comfort.

4. The many small openings in a home can add up to big heat losses. Caulking and weatherstripping cracks in walls, floors, windows, and doors will save fuel and money. Keeping the fireplace damper closed tightly when not in use will also result in heating cost savings.

5. Letting sunlight in by opening curtains, blinds, and shades over windows facing the sun helps keep the home warm and reduces heating needs. At night or when the sky is overcast, keeping drapes and curtains closed will help keep the warmth indoors.

6. Because dry air makes one feel colder than does moist air at the same temperature, maintaining home humidity will produce personal comfort at a lower thermostat setting and save money. Shallow pans of water on radiator tops or near warm air vents, or a room humidifier, will help raise humidity levels.

7. Keeping the heating thermostat at the lowest temperature comfortable will save on heating costs. For every degree over 70°F one can expect to use 3% more heating fuel.

Hot Water

1. The hot water heater is the second largest energy consumer in the home. Using it efficiently can add up to big savings. For families with an automatic dishwasher, the hot water heater setting can safely be lowered to 130–140°F. Without a dishwasher, the setting can be lowered to 110–120°F.

2. If the house will be vacant for two or more days, lower the temperature of the water heater until you return. If one has a new water heater, drain a few gallons from the tank every 6 months to remove sediment that accumulates and reduces the heater's efficiency.

3. One can save up to 10% on water heating costs by simply wrapping a fiberglass blanket around the water heater and securing it with duct tape, or by installing a ready-made insulation kit.

Cooking

1. A microwave oven is an energy-efficient alternative to a conventional oven. It cooks food more quickly, and it uses 70–80% less electricity than do regular ovens.

2. If cooking on top of the range, use pots and pans that are properly sized to "fit" the burners. Using a small pan on a large burner wastes energy. Cookware with flat bottoms and tight covers is the best choice.

3. In using a conventional oven, try to avoid "peeking" by opening the oven door. Each "peek" can lower the oven temperature by 25°F.

4. Although often recommended, it's not really necessary to preheat the oven for foods with a cooking time of over 1 hr. Also, using glass pans allows one to set the oven 25° lower because glass retains heat.

5. When preparing a meal in the oven, try to use foods that are cooked at about the same temperature. That way, the oven can cook several dishes at the same time and will not waste valuable energy dollars.

Lighting

1. Lighting accounts for about 15% of a home's electric use. New screw-in fluorescent bulbs can replace the incandescent ones most individuals use. Fluorescent bulbs are more expensive, but they last 10 times longer and use 75% less electricity.

2. If one prefers incandescent bulbs, try to use "energy saver" bulbs. These bulbs use halogen gases that allow the filament to burn brighter while consuming less electricity.

3. A lot of energy can be saved by matching, as closely as possible, light bulb wattage to lighting needs. A high-wattage reading light in a hallway or alcove, as an example, is not energy-efficient.

4. One can save energy by turning off incandescent lights when leaving the room. In using fluorescent lighting, however, turn them off only if one will be gone longer than 15 min. Fluorescent lights use as much energy in starting as they use during 15 min of operation, so it's not worthwhile to turn them off for brief periods.

5. Lighting controls or "timers" can help save energy dollars. Timers can be set to turn lights on or off at predetermined times, whereas photocell controls are sensitive to light and turn lamps on and off at sundown and sunrise. Dimmers can vary the level of illumination according to how much light is needed in a given situation.

6. Consider using task lighting (lighting directed at a specific area) instead of overhead or general lighting, which may light unused areas of the room. By limiting lighting only to areas where it is needed, savings in the cost of bulbs and energy can be made.

7. By keeping lights and fixtures clean, lighting efficiency can be improved. As much as 20% of the light generated can be lost to hazing dusts. Also, take advantage of reflected light by keeping portable fixtures as close as possible to light-colored walls or other surfaces. These easy steps may reduce the number and wattage of bulbs.

New Appliances

1. When shopping for a new appliance, check for the yellow Energy Guide label that indicates the unit's energy efficiency. This is particularly important for appliances which use a lot of electricity, such as air conditioners and refrigerators. For air conditioners, the Energy Guide label provides an energy efficiency rating (EER). The higher the EER, the more efficient the air conditioner and the more money saved. Many utilities recommend an EER of 10 or higher.

2. For refrigerators and other appliances, the Energy Guide label provides the estimated yearly energy cost for operating the appliance based on an average national utility rate. With any appliance, it is helpful to compare units in the same size range when trying to determine which model has the lowest annual operating cost.

3. Although very efficient appliances may cost more to buy, they pay for themselves through lower energy bills. For example, by purchasing a very efficient refrigerator, one could save up to $1200 over its life (5).

29.6   INDIVIDUAL EFFORTS

Action by Congress and state legislatures, rulings by courts, pronouncements by important people, or wishing alone cannot conserve energy. Individual efforts by everyone can make things happen and can help to win the battle against wasting energy. Each of us is an important person in that battle. Here are suggestions that can be used to make a difference. Individuals working alone or cooperating with their neighbors, with schools and colleges, with industry, with government, and with nonprofit organizations can make a difference. Here are some suggestions (in addition to those provided in the previous section) that we, as individuals, can act on to help reduce energy "waste."

1. Buy energy-efficient automobiles and other vehicles.
2. A well-tuned internal combustion engine makes a car, boat, lawnmower, or tractor more efficient and safer for the individual and the environment.
3. Carpooling, biking, walking, and using mass transit result in less pollution and in energy savings.
4. Use natural ventilation in the automobile whenever possible.
5. Use natural ventilation in the home whenever possible.
6. Purchase energy-efficient appliances.
7. Avoid travel/trips that are not necessary.
8. Do not waste food.
9. Do not overeat.
10. Make a conscious effort to conduct your life on an energy-efficient basis.

One of the authors of this text is a professor of chemical engineering. It was reported that a university mandated that faculty turn off lights in any room not occupied for over 15 min, close drapes and blinds during the summer, and keep window coverings open during the day in the winter to take advantage of solar gain. Coupled with changes in the building space and hot water temperature, these measures helped to reduce the facility's energy cost by over 25%.

29.7 SUMMARY

1. The environmental impacts of energy conservation and consumption are far-reaching, affecting air, water, and land quality and public health.

2. The law of conservation of energy, often referred to as the first law of thermodynamics, may be stated as follows: Although energy assumes many forms, the total quantity of energy is constant, and when energy disappears in one form, it appears simultaneously in other forms.

3. In principle, the second law of thermodynamics states that water cannot flow uphill and heat cannot flow from a cold to a hot body of its own accord. It also serves to define the "quality" of energy.

4. There are numerous energy conservation practices that can be instituted at chemical plants by the office, maintenance, operation, and production staff.

5. Energy conservation measures at the domestic level include cooling, heating, hot water use, cooking, lighting, and new appliance purchases.

6. The major individual effort at energy conservation involves the purchase and operation of efficient automobiles. Each individual should also make a conscientious effort to conduct one's life on an energy-efficient basis.

REFERENCES

1. L. Theodore, personal notes.
2. J. M. Smith and H. C. Van Ness, *Introduction to Chemical Engineering Thermodynamics*, 4th Ed., McGraw-Hill, New York, 1983.

3. L. Theodore and J. Reynolds, *Thermodynamics*, Theodore Tutorials, McGraw-Hill, New York, 1992.
4. L. Theodore and Y. C. McGuinn, *Pollution Prevention*, Van Nostrand Reinhold, New York, 1992.
5. *Conserving Energy*, A LILCO (Long Island Lighting Co.) publication, Long Island, not dated.

30

ARCHITECTURAL CONSIDERATIONS

Contributing Author: Carmen M. Yon

30.1 INTRODUCTION

The buildings in which we live, work, and play have a strong impact on our environment. Natural resources, such as natural gas, oil, electricity, and water, are consumed in our shopping centers, schools, churches, homes, office buildings, and other workplaces and institutions every day. By-products, in the forms of wastewater, sewage, and smoke, are emitted back into the environment. In fact, the construction, operation, and demolition of buildings in the United States accounts for approximately one-third of the energy consumed in this country.

Release of harmful agents into the air or ground from these buildings may also affect the environment. These releases may be caused by improper design and/or maintenance, or other unforeseen accidents, and may include discharges such as fuel storage leakage, and improper disposal of sewage or other contaminants.

It is beyond the scope of this chapter to detail all the potential environmental problems which could be circumvented by the proper design and construction of buildings. Rather, the intent of this chapter is to identify many of the environmental consequences of construction and utilization of buildings and to show that many solutions are available to help minimize this impact. These discussions are not concrete but are instead geared toward making the reader better understand the issues involved and to stimulate development of new solutions.

30.2   THE BUILDING ENVELOPE

The envelope of a structure is responsible for shielding the occupants from the effects of the elements. This includes protection against the infiltration of water, moisture, or unwanted outside air. At the same time, the envelope must retain the thermal environment within by not allowing either excessive heat loss during periods of heating or heat gain during periods of cooling.

Most buildings are assumed to protect the occupants against water infiltration. If a leak is noted, then the building is defective and requires repair. However, if the envelope is defective against air infiltration, the effects are often less obvious. We may simply pay more money for heating and cooling and learn to adapt to the defects of the house. Most heat loss or gain occurs by either infiltration or conduction. Infiltration occurs when outside air is allowed to enter the building through openings in the envelope. This infiltration most often occurs at doors and windows. While all buildings allow certain levels of infiltration, excessive infiltration may cause certain areas to feel "drafty" and may cause the heating or cooling system to work harder than necessary.

Methods for reduction of infiltration of air are simple. Better-quality windows and doors reduce infiltration through tighter construction joinery. Proper installation of these units to reduce cracks is also beneficial. Proper sealants and weather stripping can be utilized even with existing units and may reduce infiltration by as much as one-third. In larger buildings, the use of vestibules is often very energy-efficient, since these areas act as air locks in areas where doors are used frequently.

The other method by which heat may either enter or exit a structure is through conduction. Conduction is the flow of heat from one material to another. The building envelope's ability to resist this flow of heat is based on the properties of the building materials. The ability of a material to resist conduction is measured by a unit called an R-value; the higher the R-value, the more resistant the material is to heat transmission.

Heat loss due to conduction occurs differently depending on the position of the surface. Because heat rises, most of the heat loss occurs at the ceiling, then at the walls, and the least amount occurs at the floor. Because of this, different recommended R-values have been developed for roof, wall, and floor assemblies. Due to the nature of thermal resistance, minimum values have been derived which maximize the thermal resistance of a building envelope while retaining cost-effectiveness. These values have been derived for ceiling, wall, and floor construction, and vary for different parts of the country.

Recommended R-values are often integrated into the local building codes in many areas. This, however, is a recent phenomenon, and many of the buildings which were built in the past do not meet these standards. Oftentimes, the most efficient method of improving thermal performance in an existing single-family residence is by adding insulation in the attic.

In general, the use of insulation is the most common method of attaining these R-values. In residential structures, fiberglass batt insulation is installed which utilizes air spaces to resist the conduction of heat. In building construction, a requirement for a given R-value correlates to the thickness of the installation and the types of construction materials. For example, to achieve an R-19 insulation value in a wall, which would correspond to the recommended R-value for walls constructed in most states in the United States, 6 in. of insulation would be required.

The proper design of a building envelope can minimize the quantity of energy required to maintain a comfortable interior environment. The proper design depends on both the climate and the knowledge of thermal transmission from structures. Other methods can be employed to minimize power consumption for heating and cooling, which will be discussed in later sections.

30.3 BUILDING SITING

The environment outside of a building's envelope can affect the way the envelope performs. The relationship between the building and site features, such as prevailing winter winds, prevailing summer winds, and the path of the sun, will affect the interior environment.

Diverting chilling winter winds or minimizing their effects on a structure can often reduce the amount of heat loss caused by infiltration and, to a lesser extent, conduction. In order to accomplish this, the direction, velocity, and deviations of the wind, as well as their effects on a structure, must be understood. In some cases, site features may be used to help minimize the impact of the wind. The windward profile and shape of a building may also help.

Of course, while winter winds may not be desirable during periods of heating, summer winds may be desirable during periods of cooling needs. In this case, the building siting may utilize the wind to produce a natural ventilation to remove a buildup of heat and utilize cool night air to cool the building. Again, careful study needs to be taken to ensure that the design will work.

The other major factor in the orientation of a building is its relationship to the sun. Direct sunlight into a building in the summertime can lead to excessive heat buildup in that area. It is important to understand the nature of the path of the sun to understand its effects on a building. For example, the angle of the sun in relation to a particular window can be calculated for different times of the day and different days of the year. Because the sun is higher in the sky during the summer months than it is in winter months, a shading device may be designed which would allow direct sunlight through the window in the winter when the heating effect would be desirable, but would prevent it during the summer when it is hot.

Other methods of shading include the use of vegetation. Deciduous trees, for example, provide a great deal of shade during the summer months. Yet when they loose their leaves during cooler seasons, they allow sunlight to pass through. The proper placement of these trees in relation to a building can therefore be very beneficial.

Other methods of utilizing the site may include partial or even full burial of the building into the site. Because the temperature of the earth remains a near-constant temperature, a structure beneath grade level can be more effectively maintained using less energy.

30.4 SOLAR DESIGN

Several passive solar design principles have been touched on in the previous section. These methods utilize the nature of the sun to heat and, in some cases, even cool buildings. Numerous books are available which propose solar design solutions, analyze the effectiveness of particular solutions, and describe the principles behind solar design (1–7). Many different strategies have been developed to harness the power of the sun. Some of these have been successful, whereas others have not. Many strategies become very expensive and often are noneconomical, whereas others are often very simple and practical.

Solar design often involves utilizing the energy of the sun to heat up a structure, a material, air, or a liquid. Because of the sun's ability to maintain a constant heat output, its energy is often collected and used as a method of preheating another liquid. One such example is a rooftop solar panel. A solar heating panel may produce 100°F water; this may then be used to heat up a quantity of inlet tap water from 50°F to 85°F, after which the hot water heater heats the water the rest of the way. In this example, the heat gained by the solar panel is used to reduce the energy consumption of the hot water heater, while not having to rely on any consistency from the sun.

Other design strategies incorporate solar radiation into the design of the heating system for a structure. The structure is designed to allow sunlight to enter the building and strike an opaque surface. Some of the light energy is reflected, and some energy is converted to heat energy. The heat either will be absorbed by the surface material or will be released into the air. Oftentimes, a large amount of solar energy is admitted into a building and strikes a surface which does not readily absorb heat energy. The heat will build up in the space it entered, and can be distributed to other parts of the building which require heat. In the opposite case, the surface material which the sunlight strikes may absorb the heat energy throughout the course of the day while the sun is out. This stored energy is slowly released overnight when it is needed for heating this area. This can be accomplished by the use of thermal mass or phase change materials. Numerous applications of these strategies exist and have been successful in varying degrees.

The utilization of the energy of the sun, in combination with proper envelope design, siting, and configuration, can help to reduce the requirements for heating and cooling. However, in most cases, it is not able to entirely eliminate the need for a heating system.

30.5   BUILDING MATERIALS

When selecting the materials which go into a building, the designer has an opportunity to be environmentally conscious if the owner shares his/her concern. Many building materials may be produced from waste products of other manufacturing processes or made from recycled products. Recycled products include certain insulations, plastic lumber for use in playground equipment, and the use of automobile tires to build earth-retaining structures.

Consideration must also be given to the potential toxicity of the building materials. A number of building materials have been found to be harmful to the occupants of the buildings. Two such materials are leaded paints and asbestos insulation and flooring. These two materials were widely used in buildings for a number of years. Currently, asbestos removal in older buildings is a major business. Care should, therefore, be taken in the selection of materials which could potentially be hazardous.

30.6 WATER CONSERVATION

Water conservation has become an important topic in today's environmentally conscious society. Numerous urban centers are experiencing water shortages and are building dams to help curb the ever-increasing demand for water. These dams often have an undeniable impact on the natural environment, killing numerous species of animals and plants. Conservation of this resource is very much needed and can be accomplished through personal conservation efforts, as well as conscious building design.

One method of reducing a building's water consumption is through the use of low-volume toilets. The standard toilet uses as much as 5 gal of water per flush, whereas some water-saving models use as little as 2 qt. This can lead to substantial water savings, especially in public and commercial buildings.

A major source of water consumption is the irrigation of landscaped areas. This consumption may be reduced through the careful selection of landscape materials. An extreme example of wasteful water consumption is the common practice of planting Kentucky bluegrass in arid western states. While this grass type thrives in moist climates, it requires a huge amount of irrigation to sustain it in an arid climate. Numerous water-conserving grass types are available which would provide the same green lawn.

Many people are opting to reduce irrigation even more substantially by planting indigenous plant materials, often in lieu of the traditional green lawns. Native grasses, plants, shrubs, and trees are planted and often allowed to grow wildly. In some instances, minor irrigation may be necessary to get plants started or in the event of drought. These landscapes can be exceedingly beautiful as they change with the seasons, and they provide habitats for wildlife and don't require much maintenance.

30.7 LIGHTING SYSTEMS

The majority of the total energy required for lighting is consumed in the buildings which comprise the workplace. Lighting requirements for commercial buildings vary with the type of tasks which are performed within. Many of today's jobs require high levels of illumination because of the nature of the work performed. In order to meet the required lighting levels necessary for various tasks to be successfully completed, large amounts of energy are consumed.

In the process of converting electrical energy into light energy, a great deal of energy is wasted. The efficiency rating of a lighting system is defined as the ratio of the lumens of light energy released per watt of electricity consumed. One type of fluorescent fixture, which has an extremely high efficiency rating, only converts less than one-quarter of the total energy into light. The other two products of this conversion are infrared and dissipated heat. This heat can often become so excessive that energy is required to remove it. In fact, in some instances, buildings have required cooling even during cold winter days, due to excessive heat buildup.

The workplace has one big item in its favor. These buildings are primarily used during the daylight hours. If the light energy of the sun could be controlled, the consumption of energy within the buildings could be reduced dramatically. For example, the required foot-candle (fc) level for reading is 50 fc, and direct sunlight can provide 10,000 fc. A footcandle is a measurement of lighting intensity, measured as lumens per square foot. The sun also provides a much better quality of light.

There are three drawbacks to the full utilization of natural daylight: It is very difficult to control the level of light distribution from the sun to the various parts of the building; the light levels of the sun are constantly changing; and the heat which may be gained as a result of daylighting must be carefully controlled. For these reasons, daylighting design is extremely difficult, but when done properly it has been extremely successful.

The proper design of a lighting system may also reduce energy consumption. By identifying the exact lighting needs for an area, a system can be designed to provide more light where it is needed and less where it is not. Good design may also incorporate a system in which the heat produced from the lights is collected and either used for heating or discharged when heating is not required. In this way, the heating and cooling loads may be reduced.

The use of newer, more energy-efficient light fixtures and lamps can conserve energy, as can selecting the proper size bulb. Today, halogen lights are growing in popularity because of their much higher efficiency, longer life, and whiter light than either conventional fluorescent or incandescent fixtures.

Two other ways to provide more efficient lighting are room color selection and regular maintenance. The colors which are selected for the walls, floor, and ceiling can greatly affect this efficiency. Darker colors absorb a good deal of light while lighter colors will reflect light to other surfaces, thereby creating a lighter space. The frequency with which lamps are cleaned can also contribute to a more efficient system. As an example, a fluorescent light fixture can vary in efficiency anywhere from 10 to 30% based on a 12-mo cleaning cycle versus a 36-mo cycle (8).

The residential consumption of energy for lighting is not as major as that of commercial buildings. The consumption of this energy primarily occurs in the evening when natural daylight is not available. The methods for providing efficient lighting design in commercial lighting systems can also be utilized in residential buildings. These methods include proper design of the lighting system, appropriate choice of bulbs, and the proper use of daylight. The most important consideration in the home, however, is simply to turn off lights which are not in use. After all, it doesn't matter how efficient the system is if it is left on unnecessarily.

30.8 HUMAN COMFORT

Heating Systems

Space heating accounts for approximately 18% of U.S. energy consumption. Numerous types of heating systems are available for use in many different applications. Some systems provide fresh or filtered air which may be both heated and humidified. Other systems may only heat the air that exists in a given area. Various combinations of individual systems are available for heating an entire building. The available space, available utilities, and economics are among the criteria utilized to select a heating system. The proper selection, sizing, and maintenance of a heating system are vital to its efficient use.

The heating systems for most buildings are fueled by either electricity, natural gas, fuel oil, wood, or coal. The selection of the fuel source often is limited by which types of fuel are available in the area. The different fuel systems have different effects on the environment. Most fuels are nonrenewable. The types of pollutants emitted to the atmosphere upon combustion of the fuel varies according to fuel type.

Electricity is the most abundant fuel source for heating. Electric power can be provided from many different generators such as hydroelectric, nuclear, natural-gas-fired, oil-fired, or coal-fired stations. The advantage of an electric system over combustion furnaces is that air pollution control is more efficient at a centralized plant than in individual home furnaces. This system of heating is usually 100% efficient once the electricity enters the building; any losses in efficiency occur at the power plant and in transmission.

Electric heating systems are generally very easy to install and are often used when remodeling or building an addition because they can work independent of the existing system. They are generally more expensive to operate than other systems.

A natural-gas-fired heater is another option. This fuel is not available in all areas, is a nonrenewable resource, and does release pollutants during combustion. It burns cleaner than other fossil fuels and wood. Highly efficient gas furnaces are available with efficiencies up to about 95%.

Fuel oil is another nonrenewable, pollutant-producing fuel. Fuel oil burners are generally about 85% efficient. Oil is usually stored in a tank adjacent to the buildings. This is especially convenient in rural areas which may not have gas service. An additional hazard associated with oil heat is that oil could escape from its holding tank and contaminate the surrounding area.

Coal and wood are two other fuels used for heating. These fuels require space for storage, and do not burn as cleanly as oil or gas. These fuels also require continual maintenance of the combustion process.

As can be seen from the above, no particular heating fuel listed is without its drawbacks. Most of the fuel which is used for space heating comes from fossil fuels, which are becoming less and less abundant as well as more expensive. In the future, alternative fuels which are cleaner-burning and more readily available may be developed, but at this point their development as viable options has not emerged.

Because of this absence of alternative fuels, it is extremely important to reduce consumption of the above-mentioned fuels. Proper building design, including proper envelope, siting, landscaping, and building systems design, can help accomplish this goal (8).

Heating/Cooling Systems

Numerous heating/cooling systems are available for a variety of applications. The proper selection of these systems is vital to their efficient use of energy to provide for the comfort of a building's occupants. It is therefore imperative that the designer of the system understand the dynamics of what factors influence a person's perception of comfort. These factors include humidity, air velocity, air temperature, and the temperature of adjacent surfaces.

The effect of humidity is easily seen and can be quite dramatic. In arid climates, which may have a relative humidity of 20%, a temperature of 90°F can often be comfortable to a person; however, the same person in a more humid climate of, say, 90% relative humidity with a temperature of 90°F would probably be extremely uncomfortable. The direct effects of humidity and temperature on human comfort have been scientifically analyzed and assembled in what is known as a *psychometric chart*. A designer can design a system which may increase the humidity of the air to make a space feel comfortable at a lower temperature. This is often accomplished by the use of a humidifier as a part of the air delivery system of the heating systems. Care must be taken to ensure that excessive humidities are not used during winter months, which could lead to condensation problems. Conversely, humidity can be removed from air to make a space feel cooler without changing the temperature of the air.

We are all aware of the effects of air movement on comfort; the terms "winter wind-chill factor" and "cool summer breeze" are commonly used. Air velocities should be carefully controlled during heating months. Too little air circulation may feel stagnant, whereas excessive velocity can create drafty conditions which require an increase in temperature to maintain the same comfort level. In months requiring building cooling, an increase in air velocity through the mechanical movement of air or natural ventilation can reduce or eliminate the need for cooling.

The temperature of an adjacent surface can often affect the perception of comfort. One example of this is a room which has a "cold" exterior wall. Although the room is comfortable in most areas, as one approaches the wall, the person feels colder. This is often avoided by placing heating resistors and supplying air along exterior walls. The same concept is used in the design of radiant heating systems, such as radiant slabs which may make a room feel warmer by providing a warm floor surface. These systems require careful design to ensure success.

Proper selection of the type of heating/cooling system utilized is very important. Once a system has been selected, the actual mechanical units must be selected and installed. Proper sizing, installation, and maintenance of these systems is essential to their efficiency.

Units should be sized, selected, and installed based on the manufacturer's recommendations to provide maximum efficiency over the life of the unit. A routine maintenance system should be established based on these recommendations to ensure continued efficiency. During installation of the units, insulation should be used as recommended to minimize energy loss from the system itself (8).

30.9 INDOOR AIR QUALITY

With the increase in awareness of the effects of airborne pollutants on humans comes a concern regarding the quality of the air inside the building. This indoor air quality can be controlled through filtration, which also provides fresh outdoor air.

In order to provide a minimum level of air cleanliness, most building codes require that the air in a building be changed at given rates (air changes per hour) based on the types of uses. This is designed to reduce the levels of contaminants which may build up in a space. Often specialized systems such as cooking hoods, paint hoods, and fume hoods are required at the source of these contaminants to expel them directly to the exterior.

While the proper design of air-handling systems to provide adequate removal of contaminants is important, the quality of the air with which it is replaced is of equal concern. The location of the air intake should be positioned to minimize the potential of introducing such elements as automobile exhaust or even site-specific odors. This air, which is then admitted, should be passed through a filter specifically designed to reduce any contaminants present. Indoor air quality is discussed in more detail in Chapter 10.

30.10 SUMMARY

1. The buildings we use every day have a lasting effect on the earth's environment. In order to minimize any detrimental impact caused by these structures, a concerted effort on the part of the designer, owners, managers, and users of these facilities must be made.

2. The dynamic relationships between the design and construction of a structure and its effects on energy consumption must be understood, and strategies must be implemented to minimize this consumption.

3. The effects which buildings directly have on human health also need to be understood and dealt with.

REFERENCES

1. D. Wright and D. A. Andrejko, *Passive Solar Architecture: Logic and Beauty*, Van Nostrand Reinhold, New York, 1982.
2. D. Bainbridge, J. Corbett, and J. Hofacre, *Village Home's Solar House Designs: A Collection of 43 Energy-Conscious House Designs,* Rodale Press, Emmaus, PA, 1979.
3. W. A. Shurcliff, *Solar Heated Buildings of North America: 120 Outstanding Examples*, Brick House Building Publishing Company, Harrisville, NH, 1978.
4. B. Anderson, *Solar Energy: Fundamentals in Building Design*, McGraw-Hill, New York, 1977.
5. J. H. Keyes, *Consumer Handbook of Solar Energy*, Morgan & Morgan, Dobbs Ferry, NY, 1979.
6. E. Mazria, *The Passive Solar Energy Book*, Rodale Press, Emmaus, PA, 1979.
7. D. K. Reif, *Solar Retrofit, Adding Solar to Your Home*, Brick House Building Publishing Company, Andover, MA, 1981.
8. W. J. McGuiness, B. Stein, and J. S. Reynolds, *Mechanical and Electrical Equipment for Buildings*, 6th Ed., John Wiley & Sons, New York, 1980.

VII

ADDITIONAL ENVIRONMENTAL CONCERNS AND MANAGEMENT CONSIDERATIONS

31

ELECTROMAGNETIC FIELDS*

31.1 INTRODUCTION

Most individuals are surrounded by low-level electric and magnetic fields from electric power lines, appliances, and electronic devices. During the 1980s, the public became concerned about such fields because of media reports of cancer clusters in residences and schools near electric substations and transmission lines. In addition, a series of epidemiological studies showed a weak association between exposure to power-frequency electromagnetic fields and childhood leukemia or other forms of cancer.

The high standard of living in the United States is due in large measure to the use of electricity. Technological society developed electric power generation, distribution, and utilization with little expectation that exposure to the resultant electric and magnetic fields (EMF) might possibly be harmful beyond the obvious hazards of electric shocks and burns, for which protective measures were instituted. Today, the widespread use of electric energy is clearly evident by the number of electric power lines and electrically energized devices. Because of the extensive use of electric power, most individuals in the United States are today exposed to a wide range of EMF. It is estimated that at least 100,000 people have been exposed throughout their lives to technology-generated electric and magnetic fields.

Electrical devices act on charged and magnetic objects with electric and magnetic fields in a manner similar to how the moon influences the ocean tides through its gravitational field. Before the advent of man-made electricity, humans were exposed only to the steady magnetic field of the earth and to the sudden occasional increases caused by lightning bolts. But since the advent of commercial electricity within the last century, individuals have been increasingly surrounded by man-made fields generated by power grids and the appliances run by it, as well as by higher-frequency fields from radio and television transmissions.

*Drawn in part (with permission) from *Major Environmental Issues Facing the 21st Century*, M.K. Theodore and L. Theodore, Chapter 40, "Electromagnetic Fields," Prentice-Hall, Upper Saddle River, NJ, 1996.

The most commonly used type of electricity in the home and workplace is AC (alternating current). This type of current does not flow steadily in one direction but moves back and forth. In the United States, it reverses direction 60 times per second. The unit to denote the frequency of alternation is called a Hertz (Hz) in honor of Heinrich Hertz, who discovered radio waves. The presence of electric currents gives rise to both electric and magnetic fields because of the presence of electric charges. Those electric charges with opposite signs attract each other; on the other hand, charges with the same sign repel each other. These charges, if stationary, create what has come to be defined as "electric fields." "Magnetic fields" are created when the charges are nonstationary, that is, are moving. The combination of these two forces is defined as electromagnetic fields, or simply EMF. Since types of fields alternate with alternating electric current, a 60-Hz electric power system will generate 60-Hz electric and magnetic fields.

31.2 EXPOSURE ROUTES AND MEASUREMENT

Electric and magnetic fields at a power frequency of 60 Hz are generated by three main sources: the production, delivery, and use of electric power. However, sources of public exposure include

1. Power generation
2. Transmission
3. Electric circuits in homes and public buildings
4. Electric grounding systems
5. Electric appliances

Electric and magnetic fields have been measured in selected residences and outside environments to help resolve uncertainties in the interpretation of epidemiological results. Although almost all state and local governments, utilities, private firms, and individuals are currently measuring EMF, these measurements are often conducted without adequate supervision and expertise, and lack the standard quality assurance/quality control requirements for research and study projects. Thus, these measurements have not been appropriate for determining public exposure in an absolute sense. In addition, no standardized procedure for EMF measurements exists. Notwithstanding this, instrument development responded to a perceived need for field measurements. In particular, a number of survey instruments that measure electric and magnetic that vary with time are now available, and miniaturized pocket-sized recording instruments have also been developed recently.

Mathematical models to estimate EMF exposure have been developed because measurement of fields at all locations and under all conditions of interest is not practical. Two types of models are available: theoretical and statistical. The application of theoretical models usually involves a detailed computer program. Statistical models are used to develop statistical estimates of average exposure; thus, statistical modeling does not predict individual exposures, but provides estimates for groups of the population.

EMF coupling to biological objects is another area of concern. Interestingly, an electric field immediately adjacent to a body is strongly perturbed and the intensity of the field may differ greatly from that of the unperturbed field. On the other hand, the magnetic field that penetrates the body is essentially unchanged. Both external electric and magnetic fields that vary with time induce electric fields internally and the electric current generated inside the body is proportional to the induced internal electric field.

31.3 HEALTH EFFECTS (1–5)

Since 1980, research into the possible health effects of low-level electromagnetic fields has expanded greatly. Literally thousands of research papers by scientists in both the United States and Europe have been published on the subject. However, the study of potential health effects from these fields is fraught with complexities, contradictions, and what to some observers seem like impossibilities. In many studies, including human epidemiology and laboratory tests on cells and animals, the results obtained with relatively weak electromagnetic fields seem contradictory (6). Nonetheless, there are many epidemiological studies that have reported an association between EMF exposure and health effects. The most frequently reported health effect is cancer. In particular, EMF exposure has been reported to be associated with elevated risks of leukemia, lymphoma, and nervous system cancer in children. Some occupational studies of adults describe an association between EMF exposure and cancer. However, as indicated above, uncertainties remain in the understanding of the potential health effects of EMF.

One should note that EMF analyses are particularly difficult for epidemiologists. The problems that have been attributed by some people to EMFs include several different kinds of cancer, birth defects, behavioral changes, slowed reflexes, and spontaneous abortions. Therefore, the process of deciding which health problems in a community belong to the "cluster" becomes exceedingly difficult. So, the answer to the question, "Can that source be the cause of my problems?" is, "Maybe or maybe not." The source might be the problem, but trying to show that it is can be very difficult, if not impossible.

Despite the complexities and disagreements, scientific opinion has coalesced around a middle ground in recent years. In that middle ground is a great deal of evidence that electromagnetic fields do have some biological effects. It is not yet certain whether such fields can produce adverse health effects, but it is believed there is some nontrivial chance that low-level fields could pose a problem, and a high priority is placed on research aimed at answering that question. However, beyond those holding the middle ground are a few scientists at one extreme who say enough evidence already exists to show that low-level fields, such as those from power lines, do have adverse health effects, in particular cancer, and therefore society should take strong measures to reduce exposures. At the other end of the spectrum is a small group of experts who say that biological effects from such fields would violate the laws of physics, and therefore low-level electromagnetic fields cannot cause cancer or any other disease. They also claim that lab studies on cells or animals that seem to show biological effects with very low-level fields are flawed in some way, for example, that there is some other explanation for the results (6).

For years, scientists assumed that the only harm caused by electromagnetic fields was thermal, that is, their ability to heat up an object. Even so, it was a phenomenon shown to exist only at the higher frequencies of several thousand megahertz, the range in which microwave ovens operate. In the late 1970s, scientists began to question if an association between cancer deaths in Denver children and exposure to extremely low-frequency 60-Hz fields existed. This subject has been debated in the literature since then (7).

Regarding breast cancer, women in electrical jobs are 38% more likely to die of breast cancer than other working women, according to a new study. It found an even higher death rate among female telephone installers, repairers, and line workers. "It's the strongest epidemiological evidence so far that breast cancer may be related to electromagnetic fields in some way, but it's still not very strong evidence," said University of North Carolina researcher Dana Loomis, chief author of the study published in the *Journal of the National Cancer Institute* (8). The new study found that the breast cancer death rate was more than twice as high among female telephone installers, repairers, and line workers, compared with women who worked in nonelectrical occupations. The results were statistically adjusted to factor out income, age, race, and marital status.

The above study also indicated that the risk was 70% higher for female electrical engineers, 28% higher for electrical technicians, and 75% higher for other electrical occupations such as electricians and power line workers. All of those jobs involve sustained-exposure electromagnetic fields, but so do some nonelectrical jobs, such as computer programmers, computer equipment operators, keyboard data enterers, telephone operators, and air traffic controllers. And for each of those five jobs, the study found that female breast cancer mortality was no higher than for the rest of the work force (8).

Environmental agents that cause reproductive and developmental effects are important because they may directly influence health, lifespan, propagation, and functional and productive capacity of children. Some epidemiological studies have reported reproductive and developmental effects from exposure to EMF generated by devices in the workplace and home. Investigations of women and the outcome of their pregnancies have included operators of visual display terminals (VDTs) and users of specific home appliances (electric blankets, heated water beds, and ceiling electric heat). The reports of increased miscarriages and increased malformations suggest that maternal EMF exposure may be associated with adverse effects.

Other studies have reported an increased incidence of nervous system cancer in children whose fathers had occupations with potential EMF exposure. Regarding nervous system effects, neurotransmitters and neurohormones are substances involved in communication both within nervous system and in the transmission of signals from the nervous system to other body organs. Neuroregulator chemicals are released in pulses with a distinct daily or circadian pattern. The studies in which human subjects have been exposed to EMF in controlled laboratory settings describe the following effects: changes in brain activity indicative of possibly slowed information process, slowed reaction time, and altered cardiovascular function, and also including a slowed heart rate and pulse that may indicate direct action on the heart.

The immune system defends against cancer and other diseases. Environmental agents that compromise the effectiveness of the immune system could potentially increase the incidence of cancer and other diseases. Studies on the effect of 60-Hz electric fields on the immune system of laboratory animals found no effect of chronic exposure on rats and mice. Thus, it may be concluded at this time that power frequencies have small or no effects on the immune systems of exposed animals.

While the hazards from these fields may or may not be significant, the fear of them is. In state after state, nervous citizens have delayed or even killed electric utilities' plans to build or expand high-voltage transmission lines. Real estate brokers report that houses next to power lines sell more slowly than others, and for lower prices. Parents with children in schools near power lines are demanding that either the schools or the lines be moved. Meanwhile, lawsuits by cancer victims against power companies are making their way through the courts in many states (7).

It is important to keep the overall EMF health risk in perspective. For example, there are about 2,600 new cases of childhood leukemia every year in the United States. The chance of a given child developing leukemia in any year is about one in 20,000, with the bulk of cases occurring by the age of five. Some epidemiologic studies have suggested that unusually strong magnetic fields may double a child's risk, raising it to one in 10,000. But even in those studies, the vast majority of leukemia cases occurred in houses calculated to have low magnetic fields. In the end, parents must make a personal decision about how much to worry — just as they routinely choose to worry about or ignore other risks in their lives and their children's lives (7).

31.4 MANAGEMENT/CONTROL PROCEDURES (1)

As described earlier, the major source of environmental exposure to EMF is the electric power system, which includes transmission lines, the distribution system (substations, lines, and transformers), and electric circuits in residential and other buildings that provide power to appliances and machinery. Although considerable effort has been focused on the control of EMF from electric utility systems, little work has been done on controlling fields generated by electrically powered appliances, tools, and other devices.

In most circumstances, the strength of low-frequency EMF decreases with distance from the source. One simple mitigation approach is therefore to increase separation distance. There are other methods known to be effective regardless of frequency. They are

1. Shielding
2. Design
3. Location
4. Component choice
5. Filtering

While electric fields can be easily shielded, magnetic fields are much more difficult to shield. Electric fields are shielded to some degree by almost anything, such as trees, bushes, walls, and so forth. Magnetic fields can be reduced by enclosing the source in certain types of metal, such as a material called Mu metal, which is a special alloy. The fields are still present, but the metal has the capability to contain them. This approach to reducing field levels may not be practical for many sources, including power lines. Magnetic field intensity can also be reduced by placing wires close together so that the field from one wire cancels the field from the other. This is now being done in new designs for electric blankets. To some degree the same thing can be done for power lines, but for safety and reliability reasons power lines have minimum required spacing.

Because of the way appliances are made, they have the potential to have very high localized fields, but then the fields decrease rapidly with distance. For example, typical magnetic field strengths not near an appliance are 0.1–4 milligauss (mG), but the field from an electric can opener can be 20,000 mG at 3 cm (approximately 1 in.) from the appliance. At 30 cm (approximately 1 ft), appliance fields are usually around 100 times lower. For the can opener mentioned above, the level would probably be around 20 milligauss. The reader should note that the gauss is a unit for the strength of a magnetic field, also known as magnetic flux density. Magnetic flux density is measured in terms of lines of force per unit area. Remember the patterns that were generated by iron filings on a piece of paper which was placed over a magnet? These patterns are field lines. One normally speaks of magnetic fields in terms of (one) thousandths (1/1000) of a gauss, or milligauss, abbreviated "mG."

When standing under a power line, one is usually at least 20 ft or more away from the line, depending on its height above ground. Under a typical 230-kV transmission line the field is probably less than 120 mG. In contrast, if one moves about 100 ft away from line, the magnetic field is probably about 15 mG, and at 300 ft away from the line, the magnetic field is probably less than 2 mG. From these examples, one can see that distance from the source of the magnetic or electric field can substantially reduce exposure.

Control technology for transmission and distribution lines has been developed and could be applied if warranted. These techniques focus on compaction and shielding of transmission conductors. Compaction is based on the principle that for three-phase, balanced conductor systems, the net field (electric magnetic) of the three phases is zero. A disadvantage of compaction is that it results in an increase in electrical arcing, which affects system reliability. For situations in which compaction was an effective control technology, shielding techniques have been developed that reduce the field at the edge of the right-of-way by approximately tenfold. Compaction techniques that have been developed include gas-insulated transmission lines, superconducting cables, and current cable technologies. In cable or gas-insulated transmission technologies, conductors are inside a metallic sheath in which the electric field exists only between the conductors and the sheath; electric fields external to cable sheaths are essentially zero.

EMF inside the home and schools can be emitted from appliances, wiring systems including the grounding, underground and overhead distribution lines, and transmission lines. A few appliances, especially electric blankets and heated water beds, have been identified as important sources of magnetic field exposure because of their close proximity to the body for long periods of time. Hair dryers and electric shavers, because they too are used close to the body, expose some of the strongest fields but total exposure from these is limited because they are used for minutes per day. Manufacturers have responded by developing low-magnetic-field appliances. Some specific steps one can take to reduce EMF exposure at home, the office, or at school are listed below.

1. Sitting at arms length from a terminal or pulling the keyboard back still further; magnetic fields fall off rapidly with distance.
2. Switching VDTs (but not necessarily the computer) off when not in use.
3. Spacing and locating terminals in the workplace, so that workstations are isolated from the fields from neighboring VDTs. Fields will penetrate partition walls, but do fall quickly with distance.
4. Using electric blankets (or water bed heaters) to warm beds, but unplugging them before sleeping. Magnetic fields disappear when the electric current is switched off. However, electric fields may exist as long as a blanket is plugged in.
5. Not standing close to sources of EMFs such as microwave ovens while in use. Standards are in place to limit microwave emissions. However, the electric power consumption by a microwave oven results in magnetic fields close to the unit that are high. The same is true of other appliances as well.

Existing mass transit systems and emerging technologies such as magnetically elevated trains, electric automobiles, and superconducting magnetic energy storage devices require special consideration. These systems can produce magnetic fields over large areas at different frequencies. Passengers on magnetically levitated trains will be exposed to static fields and to frequencies up to about 1,000 Hz. Existing engineering control technologies may not be sufficient to significantly reduced exposure.

As described earlier, another device that merits special concern is the visual display terminal. In addition to being energized by 60-Hz power, VDTs can produce EMF at frequencies of up to 250,000 Hz. VDT manufacturers, however, have begun to reduce fields by shielding techniques. Metal enclosures are used to shield electric fields, while active magnetic shielding techniques are used to reduce magnetic fields.

The reader should note that there is no simple way to completely block EMFs since the fields are generated by electrical systems and devices in the home, including the wiring and appliances. Electric fields from outside the home (power lines, etc.) are shielded to some extent by natural and building materials, but magnetic fields are not. As noted above, the further a building is from an EMF source, the weaker the fields at the building would be. Keeping fields out of the home would mean keeping any electricity from coming into or being used in the home. Often the fields from sources inside the home (e.g., appliances, wiring, etc.) will result in higher fields than from sources outside the home.

At this point, enough evidence suggests a possible health hazard to justify taking simple steps to reduce exposure to electromagnetic fields. The large dilemma is whether the risks justify making major changes in huge, complex electric power systems that could disrupt the reliable, relatively inexpensive electric service Americans have come to take for granted (7).

Some questions are too large to be answered by individuals or families at this time. How much should a community spend to route transmission lines away from a school? Should a high-voltage line be put on taller towers to minimize fields at ground level even though the expense will result in higher electricity rates (7)?.

It seems sensible to focus on simple ways of reducing exposure to electromagnetic fields rather than to make radical changes. M. Granger Morgan, a public policy expert at Carnegie Mellon University, has proposed a strategy he calls "prudent avoidance," involving simple low-cost or no-cost measures. Although prudent avoidance can be as easy as leaning back from a computer screen, other methods of avoiding EMFs are more difficult such as moving out of a house near a power line. Whether such a move is prudent or paranoid depends largely on one's own feelings about the nature of the risk (7).

31.5 FUTURE TRENDS

If scientists eventually reach a consensus that low-level electromagnetic fields do cause cancer or some other adverse health effect, then regulations defining some safe exposure level will have to be written at some later date. But so far the data are not complete enough for regulators.

Future research is also questionable as to direction and extent at this time because much of the research into the effects of electromagnetic fields, especially that on mechanisms that could cause health effects, is disciplinary, highly complicated, and has raised more questions than it has answered. It may take more than a decade to elucidate the mechanisms. It appears that the technical profession presently does not know if EMF exposure is harmful (aside from the concern for electric shocks and for extreme exposure). It does not know if certain levels of EMFs are safer or less safe than other levels. With respect to most chemicals, one assumes exposure at higher levels is worse than less exposure at lower levels. This may or may not be true for EMFs, however. More research is required to identify response relationships. There is some evidence from laboratory studies that suggest that there could be "windows" for various effects. This means that biological effects are observed at some frequencies' intensities, but not at others. Also, it is not known if continuous exposure to a given field causes a biological effect, or if repeatedly entering and exiting of the field causes effects. There is no number to which one can point and say, "That is a safe or hazardous level of EMF." Many years may pass before scientists have clear answers on cancer or on any other possible health problems that could be caused by electromagnetic fields. But over the long run, avoiding research probably will not be acceptable. It appears that the public will continue to demand research funding and answers to these questions.

Finally, the tendency to sensationalize electric and magnetic fields reporting was recently discussed by McElfresh (9). Information on the latest 53 publications (mainly newspapers) that reviewed 5 major EMF studies was presented. It is hoped that the future will provide more objective reporting by the media on EMF issues.

31.6 SUMMARY

1. International and national organizations, industrial associations, federal and state agencies, Congress, and the public have expressed concern about the potential health effects of exposure to EMF.

2. Exposure assessment research is a high priority research area because it is essential to the successful interpretation of the biological response and is critically important for risk assessment studies.

3. Research on human reproductive effects should emphasize the need to attempt replication of isolated reports of increased miscarriages and increased malformations, and reports of increased incidence of nervous system cancer in children whose fathers had occupations with potential EMF exposure.

4. The potential need for future controls to reduce risks from exposure to EMF is the rationale for control technology research. This research presently is a low priority because no firm cause-and-effect relationship between human health risk and EMF exposure has been established.

5. If scientists eventually reach a consensus that low-level electromagnetic fields do cause cancer or some other adverse health effect, then regulations defining some safe exposure level will have to be written at some later date.

REFERENCES

1. S. Recupero, "The Danger Associated with Electric and Magnetic Fields," drawn, in part, from a term paper submitted to L. Theodore, Manhattan College, 1994.
2. U.S. EPA, *EMF: An EPA Perspective*, Washington, DC, December 1992.
3. A. Leonard, R. Neutra, M. Yost, and G. Lee, *EMF Measurements and Possible Effects*, California Department of Health Services, 1990.
4. Department of Engineering and Public Policy, *EMF from 60-Hertz Electric Power*, Pittsburgh, PA, 1989.
5. R. Wilson, "Currents of Concern." *NY Newsday*, May 1, 1990.
6. B. Hileman, "Health Effects of Electromagnetic Fields Remain Unresolved," *Chemical and Engineering News* (C&EN), November 8, 1993.
7. "Electromagnetic Fields," *Consumer Reports*, May 1, 1994.
8. D. Fagin, *New York Newsday*, June 15, 1994.
9. R. McElfresh, "Responsible Reporting of Environmental Issues by the Media," panel discussion, AWMA Annual Meeting, Cincinnati, June 1994.

32

MONITORING METHODS

32.1 INTRODUCTION

The primary focus of this book has been to look at environmental problems and the various solutions to these problems. In many instances, we are unaware that a problem exists until harm has been done. The damage may be in the form of disease to the surrounding population or destruction of the surrounding ecosystem. Monitoring problem areas or potential problem areas can help to limit future damage.

This chapter focuses on methods used to obtain data regarding contamination of soil, air, and water. Information is provided on how to set up a monitoring program and how to ensure that the data obtained are reliable. Much of this information is based on the American Society for Testing and Materials (ASTM) standards and the U.S. Environmental Protection Agency's Contract Laboratory Program which was established under Superfund. A number of additional references are available for collection and preservation of samples.

32.2 SELECTING A METHODOLOGY

The method used to sample a given area will directly affect the accuracy of the analysis. It is imperative that an appropriate methodology be selected in order to obtain the most reliable results possible. Several factors should be considered when selecting a method, including

- The program objective (documenting exposures, determining regulatory compliance, locating a source).
- The type of material to be sampled (soil, vegetation, air, water, sludge, etc.).
- The physical and chemical properties of the contaminant.
- Other contaminants that affect the results.

- Regulatory requirements.
- Safety requirements.
- Difficulty of utilizing a particular method.
- Cost.
- Reliability.
- Scale of sample area (small-scale site related to individual persons versus a large-scale site).
- Short versus long-term sampling requirement.

In addition to the above, basic professional judgment also plays a role in selecting a sampling method.

32.3   STANDARD PRACTICES FOR SAMPLING OF WATER

This section includes three procedures for water sample collection. The first procedure is for the collection of a grab sample of water at a specific site representing conditions only at the time of sampling. It is the only procedure suitable for bacteriological analyses and some radiological test procedures. The second practice is for collection of a composite sample at a specific site, portions of which are collected at varied time intervals. Alternatively, the composite may consist of portions collected at various sites or may consist of a combination of both site and time variables. The third procedure provides a continuous flowing sample, from one or more sampling sites, suitable for on-stream analyzers (1).

The goal of sampling is to obtain for analysis a portion of the main body of water that is truly representative. The most critical factors necessary to achieve this are points of sampling, time of sampling, frequency of sampling, and maintenance of integrity of the sample prior to analysis. Homogeneity is frequently lacking, necessitating multiple-point sampling. If it is impractical to utilize a most-representative sampling point, it may be practical to determine and understand interrelationships so that results obtained at a minimum number of points may be extrapolated. A totally representative sample should not be an absolute prerequisite to the selection of a sampling point. With adequate interpretation, a nonrepresentative sample can yield valuable data about trends and can indicate areas where more representative data would be available. Most samples collected from a single point in a system must be recognized as being nonrepresentative to some degree. Therefore, it becomes important to recognize the degree of representation in the sample and to make it a part of the permanent record. Otherwise, an artificial degree of precision is assigned to the data when it is recorded.

The following general rules are applicable to all sampling procedures. First, the samples must represent the conditions existing at the point taken. Second, the samples must be of sufficient volume and must be taken frequently enough to permit reproducibility of testing requisite for the desired objective, as conditioned by the method of analysis to be employed. Third, the samples must be collected, packed, shipped, and manipulated prior to analysis in a manner that safeguards against change in the particular constituents or properties to be examined.

Grab Samples

The procedure for collecting grab samples is applicable to sampling water from sources such as wells, rivers, streams, lakes, oceans, and reservoirs for chemical, physical, bacteriological, or radiological analysis. A grab sample represents the conditions existing only at the point and time of sampling.

A reasonably accurate estimate of the composition of a raw water piped from a large body of water (such as the Great Lakes) far enough from the shoreline to avoid variation from inflowing tributaries and waste discharges may be made by taking individual samples at infrequent intervals (such as biweekly or monthly) sufficient to cover seasonal changes. If samples are taken from near the shoreline of such a body of water or from a river, take them at shorter intervals (for instance daily) to provide more exact knowledge of the variations in composition where these are of importance in the use to which the water is to be put. If greater variations or cycles of pollution occur or closer surveillance of plant intake water is required, collect more frequent samples (for example, at hourly intervals).

Water undergoing continuous or intermittent treatment must be sampled with such frequency that adequate control is ensured. The interval between samples is directly related to the rate at which critical characteristics can reach intolerable limits.

Normally, samples are taken without separation of particulate matter. If constituents are present in colloidal or flocculent suspension, take the sample so that they are present in representative proportion.

Consult the specific method of analysis for any given constituent to determine the volume of sample required. Frequently, the required volume will vary with concentration level of any given constituent. The mininum volume collected should be three to four times the amount required. When sampling highly radioactive water, smaller sample sizes may be desirable to reduce the radiation hazard.

Choose the point of sampling for open bodies of water with extreme care so that a representative sample of the water to be tested is obtained. Avoid surface scum. Because of a wide variety of conditions found in streams, lakes, reservoirs, and other bodies of water, it is not possible to prescribe the exact point of sampling. Where the water in a stream is mixed so as to approach uniformity, a sample taken at any point in the cross-section is satisfactory. For large rivers or for streams not likely to be uniformly mixed, more samples are desirable and are usually taken at a number of points at the surface across the entire width and at a number of depths at each point. When boats are used, take care to avoid collecting samples where the turbulence caused by a propeller or by oars has disturbed the characteristics of the water. Ordinarily, samples are taken at these points and then combined to obtain an integrated sample of such a stream of water. Alternatively, the single grab samples may be tested—for example, to determine the point of highest bacterial density.

Choose the location of the sampling point with respect to the information desired and in conformity to local conditions. Allow sufficient distance downstream (with respect to stream flow at the time of sampling) from a tributary or source of pollution to permit thorough mixing. If this is not possible, it is better to sample the stream above the tributary or source of pollution and, in addition, to sample the tributary source of pollution. In general, a distance of 1–3 miles below the tributary is sufficient. Collect samples at least one-half mile below dams or waterfalls to allow time for the escape of entrained air. When lakes, reservoirs, or other bodies of water are sampled, it is necessary to avoid nonrepresentative areas such as those created by inlet streams, more stagnant areas, or abrupt changes in shorelines, unless determining the effect of such local conditions is a part of the sampling program. It is desirable to take a series of samples from any source of water to determine whether differences in composition are likely to exist before final selection of the sampling point.

For sampling to determine chemical and physical analyses of unconfined water at any specific depth in ponds, lagoons, reservoirs, and so on, during which contact with air or agitation of the water would cause a change in concentration of characteristics of a constituent to be determined, use a sampling apparatus so constructed that the solution at the depth to be sampled flows through a tube to the bottom of the container, and that a volume of sample equal to four to ten times the volume of the receiving container passes through it. When no determinations of dissolved gases are to be made, any less complicated apparatus may be used that will permit the collection of a sample at a desired depth, or of an integrated sample containing water from all points in a vertical section.

When samples are to be shipped, do not fill the bottle entirely, in order to allow some room for expansion when the contents are subjected to a change in temperature. An air space of 10% usually suffices for this purpose, although this does not protect against bursting of the container due to freezing. However, if contact with air would cause a change in the concentration or characteristics of a constituent to be determined, secure the sample without contact with air and completely fill the container.

Add chemical preservatives to samples for chemical or physical examination only as specified in specific test methods. Quick freezing has been found to be beneficial in preserving some organic constituents. Note any preservatives added on the label.

Refrigerate or ice the samples collected for biological examination immediately after collection. Hold or transport at a temperature of not more than 4°C. Do not hold samples for microbiological analyses longer than 6 hr from time of collection to analysis. Consider field examination if these time limits cannot be met.

Analyze drinking water samples delivered to the laboratory by the collector on the same day. A specific exception is made for drinking water samples mailed or sent by public transportation to control laboratories. They are permitted to be held up to 30 hours.

In general, allow as short a time as possible to elapse between the collection of a sample and its analysis. Under some conditions, analysis in the field is necessary to secure reliable results. The actual time that may be allowed to intervene between the collection and analysis of a sample varies with the type of examination to be conducted, the character of the sample, and the time interval allowable for applying corrective treatment.

Composite Samples

Composite sampling is applicable for subsequent chemical and physical analyses. It may not be suitable for sample collection for radiological examination, particularly for short-lived radionuclides. Composite samples are not suited for bacteriological examination, for constituents which change with contact to air, or for purgable organics.

Composite samples may be made by mutual agreement of the interested parties by combining individual (grab) samples taken at frequent intervals or by means of automatic samplers. Consult individual test methods for the effect of time interval and temperature prior to analysis (2–4). Indicate whether or not the volume of sample is proportional to the rate of flow. At the end of a definite period, mix the composite sample thoroughly so that determinations on a portion of the composite sample will represent the average for the stable constituents. Variations of unstable constituents may be determined by analysis on the individual samples.

In sampling process waters, collect composite samples in at least one 24-hr period. If the process is cyclic in nature, collect samples during at least one complete process cycle. Collect increments for composite samples at regular intervals from 15 min to 1 hr, and in proportion to the rate of flow of the water. This may be conveniently done by taking a simple multiple in milliliters per minute, in gallons per minute, or in some other unit of flow. Choose a suitable factor to give the proper volume (about 4 liters) for the composite sample. When samples are taken from a stream, composite samples for analysis normally consist of equal quantities of daily samples for a suitable number of consecutive days — for example, 7 days. The point of sampling should be the same as described for collection of grab samples.

Continual Sampling

Continual sampling is applicable to sampling water from sources such as wells, rivers, streams, lakes, oceans, and reservoirs on a continual basis for use in chemical, physical, or radiological analyses. The apparatus employed consists of the following:

- Delivery valve or pump.
- Piping system.
- Flow regulation system.
- Waste disposal system.

Sampling is essentially on a continuous basis. Intermittent operation is possible through use of sample bypass equipment, although this is seldom used except in measuring variables with a time relationship, such as rate of oxygen uptake. In these cases, deviations from this method are handled under descriptions of the specific measurement involved.

The size, quantity, and, in some cases, type of particulate matter often account for one or more of the variables to be measured and, in other cases, introduce errors in the analysis if the variables are disturbed. The water delivery system should flow fast enough to keep the heavier particles in suspension, and the system volume should be large enough to prevent undesirable filter action through restrictions.

When simultaneous samples from several locations are required, water is drawn continually from each individual source proportionate to flow and is then mixed into a single sample. The selection of sampling points is determined in a manner similar to that of selection of grab sample sites.

Because pumps employing suction principles disturb the gas–liquid balance, use a submersible-type pump for pumping samples from open bodies of water whenever the measurements to be made concern dissolved gases such as oxygen or carbon dioxide. Pumps, screens, valves, and piping must be selected of corrosion-resistant material to prevent sample contamination from corrosion products and to prevent need for undue maintenance. The debris screens employed around the pump intake should be of sufficient size to preclude a significant pressure drop developing across the screen in the event of partial clogging.

Manufacturers of continual analyzers and samplers will generally specify minimum volume and pressure requirements for proper operation. Sample pump selection must be based on these minimums and the configuration of sample piping. The piping system between the pump and the sample container should be designed so that the pump is operating against the lowest practical head. The piping system should be constructed with a continual rise in elevation from the pump to the point of delivery without reverse bends in which sediment and algae can accumulate. To prevent freezing in outdoor installations, remove the check valve from the pump in order that the piping will drain in the event of power failure.

In a continually operating sample system, the time lag between the system intake and the point of sample delivery is a function of the flow rate of the water and the dimensions of the intervening pipe. Usually, the system dimensions make this time so short that its effect on the accuracy of the determination is negligible. Wherever special precautions should be observed, they will be described in the particular method covering that analysis.

Groundwater Monitoring Wells

The quality of groundwater has become an issue of national concern. Groundwater monitoring wells are one of the more important tools for evaluating the quality of groundwater, delineating contamination plumes, and establishing the integrity of hazardous material management facilities. The goal in sampling groundwater monitoring wells is to obtain samples that are truly representative of the aquifer or groundwater in question.

Water that stands within a monitoring well for a long period of time may become unrepresentative of formation water because chemical or biochemical change may cause water-quality alterations; and even if it is unchanged from the time it entered the well, the stored water may not be representative of formation water at the time of sampling. Because the representativeness of stored water is questionable, it should be excluded from samples collected from a monitoring well.

There is a fairly large choice of equipment presently available for groundwater sampling from single-screened wells and well clusters. The sampling devices can be categorized into eight basic types, as described in detail in ASTM D4448. The type of equipment selected will depend upon the sampling methodology (5).

32.4 STANDARD PRACTICES FOR SAMPLING OF SOILS

There are two portions of the soil that are important to the environmental scientist. The surface layer (0–15 cm) reflects the deposition of airborne pollutants, especially those recently deposited pollutants. Pollutants that have been deposited by liquid spills or by long-term deposition of water-soluble materials may be found at depths ranging up to several meters. Plumes emanating from hazardous waste dumps or from leaking storage tanks may be found at considerable depths. The methods of sampling each of these are slightly different, but all make use of one of two basic techniques. Samples can be collected with some form of core sampling or auger device, or they may be collected by use of excavations or trenches. In the latter case, the samples are cut from the soil mass with spades or short punches. The ASTM has developed a number of methods that have direct application to soil sampling. These often need to be modified slightly to meet the needs of the environmental scientist who requires samples for chemical analyses because the ASTM methods are designed primarily for engineering tests. The techniques that are utilized should be closely coordinated with the analytical laboratory in order to meet the specific requirements of the analytical methods used (6).

The methods outlined below are for the collection of soil samples alone. At times, it is desirable to collect samples of soil water. In these cases, use can be made of some form of suction collector. The statistical designs would be the same no matter which of the soil water collectors was used. In those cases where suction devices are used, the sampling media is water and not soil even though the samples are a good reflection of soluble chemicals that may be moving through the soil matrix. These methods are not discussed below.

Surface Sampling

Surface soil sampling can be divided into two categories, namely, the upper 15 cm and the upper meter. The very shallow pollution such as that found downwind from a new source or at sites of recent spills of relatively insoluble chemicals can be sampled by use of one of the methods described below. The deeper pollutants found in the top meter are the soluble, recent pollutants or those that were deposited on the surface a number of years ago. These have begun to move downward into the deeper soil layers. One of the methods discussed in the following subsection, "Shallow Subsurface Sampling," should be used in those cases.

A number of studies of surface soils have made use of a punch or thin-walled steel tube that is 15–20 cm long to extract short cores from the soil. The tube is driven into the soil with a wooden mallet; the core and the tube are extracted; and the soil is pushed out of the tube into a stainless steel mixing bowl and composited with other cores. The soil punch is fast and can be adapted to a number of analytical schemes, provided that precautions are taken to avoid contamination during shipping and in the laboratory. An example of how this method can be adapted would be to use the system to collect samples for volatile organic chemical analysis. The tubes could be sealed with a Teflon plug and coated with a vapor sealant such as paraffin, or better, some nonreactive sealant. These tubes could then be decontaminated on the outside and shipped to the laboratory for analyses.

Surface samples may also be collected using a seamless steel ring, approximately 15–30 cm in diameter. The ring is driven into the soil to a depth of 15–20 cm. The ring is extracted as a soil-ring unit, and the soil is removed for analysis. These large cores should be used where the results are going to be expressed on a per-unit area basis. This allows a constant area of soil to be collected each time. Removal of these cores is often difficult in very loose sandy soil and in very tight clayey coils. The loose will not stay in the ring. The clayey soil is often difficult to break loose from the underlying soil layers, and thus the ring must be removed with a shovel.

Perhaps the most undesirable sample collection device is the shovel or scoop. This technique is often used in agriculture, but where samples are being taken for chemical pollutants, the inconsistencies are too great. Samples can be collected using a shovel or trowel if area and/or volume are not critical. Usually the shovel is used to mark out a boundary of soil to be sampled. The soil scientist attempts to take a constant depth of soil, but the reproducibility of sample sizes is poor; thus the variation is often considerably greater than with one of the methods listed above.

Shallow Subsurface Sampling

Precipitation may move surface pollutants into the lower soil horizons or move them away from the point of deposition by surface runoff. Sampling pollutants that have moved into the lower soil horizons requires the use of a device that will extract a longer core than can be obtained with the short probes or punches. Three basic methods are used for sampling these deeper soils:

- Soil probes or soil augers.
- Power-driven corers.
- Trenching.

The soil probe collects 30–45 cm of soil in intact, relatively undisturbed soil cores, whereas the auger collects a "disturbed" sample in approximately the same increments as the probe. Power augers can extract cores up to 60 cm long. With special attachments, longer cores can be obtained with the power auger if this is necessary. The requirements for detail often desired in research studies or in cases where the movement of the pollutants is suspected to be through very narrow layers cannot be met effectively with the augers. In these cases, some form of core sampling or trenching should be used.

Two standard tools used in soil sampling are the soil probe (often called a King tube) and the soil auger. These tools are designated to acquire samples from the upper 2 m of the soil profile. The soil probe is nothing more than a stainless steel or brass tube that is sharpened on one end and fitted with a long, T-shaped handle. These tubes usually have an approximately 2.5-cm inside diameter, although larger tubes can be obtained. The cores collected by the tube sampler or soil probe are considered to be "undisturbed" samples, although in reality this is probably not the case. The tube is pushed into the soil in approximately 20- to 30-cm increments. The soil core is then removed from the probe and placed either in the sample container or in a mixing bowl for compositing.

The auger is approximately 3 cm in diameter and is used to take samples when the soil probe will not work. The samples are "disturbed"; therefore, this method should not be used when it is necessary to have a core to examine or when very fine detail is of interest. The auger is twisted or screwed into the soil and then extracted. Because of the length of the auger and the force required to pull the soil free, only about a 20- to 30-cm maximum length can be extracted at one time. In very tight clays, it may be necessary to limit the length of each pull to about 10 cm. Consecutive samples are taken from the same hole, and thus cross-contamination is a real possibility. The soil is compacted into the threads of the auger and must be extracted with a stainless steel spatula. Larger-diameter augers such as the bucket auger, the Fenn auger, and the blade augers can also be used if larger samples are needed. These range in size from 8 to 20 cm in diameter.

If distribution of pollutant with depth is of interest, the augers and the probes are not recommended because they tend to contaminate the lower samples with material from the surface. Also, many workers have sustained back injuries trying to extract a hand auger or soil probe that has been inserted too far into the soil. A foot jack is a necessary accessory if these tools are to be used. The foot jack allows the tube to be removed from the soil without use of the back muscles.

Trenching is used to carefully remove sections of soil during studies where a detailed examination of pollutant migration patterns and detailed soil structure are required. It is perhaps the least cost-effective sampling method because of the relatively high cost of excavating the trench from which the samples are collected. It should therefore be used only in those cases where detailed information is desired.

A trench approximately one meter wide is dug to a depth approximately one foot below the desired sampling depth. The maximum effective depth for this method is about two meters, unless done in some stepwise fashion. Where a number of trenches are to be dug, a backhoe can greatly facilitate sampling. The samples are taken from the side of the pit using the soil punch or a trowel.

The sampler takes the surface 15-cm sample using the soil punch or by carefully excavating a 10-cm slice of soil that is 10 square centimeters on the surface. The soil can be treated as an individual sample or can be composited with other samples collected from each face of the pit. After this initial sample is taken, the first layer is completely cut back, thereby exposing clean soil at the top of the second layer to be sampled. Care must be exercised to ensure that the sampling area is clear of all material from the layers above. The punch or trowel is then used to take samples from the shelf created by the excavation from the side of the trench. This process is repeated until all samples are taken. The resulting hole appears as a set of steps cut into the side of the trench.

An alternate procedure that is also effective results from using the punch to remove soil cores from the side of the trench at each depth to be sampled. Care must be taken to guard against soil sloughing down the side of the hole. A shovel should be used to carefully clean the soil sampling area prior to driving the punch into the trench side.

Sampling for Underground Plumes

Sampling for underground plumes is perhaps the most difficult of all of the soil sampling methods. Often it is conducted along with groundwater and hydrological sampling. The equipment required usually consists of large, vehicle-mounted augers and coring devices, although there are some small tripod-mounted coring units available that can be carried by several people using backpacks.

The procedure listed here closely follows ASTM method D1586-67 in many respects. The object of the sampling is to take a series of 45.7-cm (18-in.) or 61-cm (24-in.) undisturbed cores with a split spoon sampler. (Longer cores can be obtained by combining several of the shorter tubes into one long split spoon). A 15.2-cm (6-in.) auger is used to drill down to the desired depth for sampling. The split spoon is then driven to its sampling depth through the bottom of the augered hole and the core extracted.

The ASTM manual calls for the use of a 63.5-kg (140-lb) hammer to drive the split spoon. The hammer is allowed to free fall 76 cm (30 in.) for each blow to the spoon. The number of blows required to drive the spoon 15.2 cm (6 in.) is counted and recorded. The blow counts are a direct reflection of the density of the soil, and they can be used to obtain some information on the soil structure below surface. Unless this density information is needed for interpretive purposes, it may not be necessary to record the blow counts. In soft soils, the split spoon can often be forced into the ground by the hydraulic drawdown on the drill rig. This is faster than the hammer method and does not require the record-keeping necessary to record the blow counts.

Samples should be collected at least every 1.5 m (5 ft) or in each distinct stratum. Additional samples should be collected where sand lenses or thin silt and sand layers appear in the profile. This sampling is particularly important when information on pollution migration is critical. Soluble chemicals are likely to move through permeable layers such as sand lenses. This appears to be especially important in tight clay layers where the main avenue of water movement is through the porous sandy layers.

Detailed core logs should be prepared by the technical staff present at the site during the sampling operation. These logs should note the depth of sample, the length of the core, and the depth of any features of the soil such as changes in physical properties, color changes, the presence of roots, rodent channels, and so on. If chemical odors are noted or unusual color patterns are detected, these should be noted also. Blow counts from the hammer should be recorded on the log along with the data mentioned above.

The procedure using samples collected every 1.5 m (5 ft) is most effective in relatively homogeneous soils. A variation in the method is to collect samples of every distinct layer in the soil profile. Large layers may be sampled at several points if they are unusually thick. A disadvantage of this approach is the cost for the analyses of the additional samples acquired at a more frequent interval. The soil horizons or strata are the avenues through which chemical pollutants are likely to migrate. Some are more permeable than others and thus are more likely to contain traces of the chemicals if they are moving through the soil. Generally speaking, the sands and gravels are more prone to contamination than are the clays because of increased permeability. This is especially true out on the leading edges of the plume and shortly after a pollutant begins to move. Low levels found in these layers can often serve as a warning of a potential problem at a later date.

A disadvantage of this type of sampling is the impact of the vehicle on yards and croplands. Special care must be taken to protect yards, shrubs, fences, and crops. The yards must be repaired, all holes backfilled, and all waste removed. Plastic sheeting should be used under all soil handling operations such as subsampling, compositing, and mixing.

Miscellaneous Tools

Hand tools such as shovels, trowels, spatulas, scoops, and pry bars are helpful for handling a number of the sampling situations. Many of these can be obtained in stainless steel for use in sampling hazardous pollutants. A set of tools should be available for each sampling site where cross-contamination is a potential problem.

These tool sets can be decontaminated on some type of schedule in order to avoid having to purchase an excessive number of these items.

A hammer, screwdriver, and wire brushes are helpful when working with the split spoon samplers. The threads on the connectors often get jammed because of soil in them. This soil can be removed with the wire brush. Pipe wrenches are also a necessity, as is a pipe vise or a plumber's vise.

32.5   STANDARD PRACTICES FOR SAMPLING OF AMBIENT AIR

This section presents the broad concepts of sampling the ambient air for the concentrations of contaminants. Detailed procedures are not discussed. General principles in planning a sampling program are given, including guidelines for the selection of sites and the location of the air sampling inlet. The reader is referred to the reference materials for details, including background information, air-quality modeling techniques, and special-purpose air-sampling programs (7, 8).

Investigations of atmospheric contaminants involve the study of a heterogeneous mass under uncontrolled conditions. Interpretation of the data derived from the air-sampling program must often be based on the statistical theory of probability. Extreme care must be observed to obtain measurements over a sufficient length of time to obtain results that may be considered representative.

The variables that may affect the contaminant concentrations are the atmospheric stability (temperature–height profile), turbulence, wind speed and direction, solar radiation, precipitation, topography, emission rates, chemical reaction rates for the formation and decomposition of contaminants, and the physical and chemical properties of the contaminant. To obtain concentrations of gaseous contaminants in terms of weight per unit volume, the ambient temperature and atmospheric pressure at the location sampled must be known.

Because the analysis of the atmosphere is influenced by phenomena in which all factors except the method of sampling and composition are beyond the control of the investigator, statistical consideration must be given to determine the adequacy of the number of samples obtained, the length of time that the sampling program is carried out, and the number of sites sampled. The purpose of the sampling and the characteristics of the contaminant to be measured will have an influence in determining this adequacy. Regular or, if possible, continuous measurements of the contaminant with simultaneous pertinent meteorological observations should be obtained during all seasons of the year. Statistical techniques may then be applied to determine the influence of the meteorological variables on the concentrations measured (9).

The choice of sampling techniques and measurement methodology, the characteristics of the sites, the number of sampling stations, and the amount of data collected all depend on the objectives of the monitoring program. These objectives may be one or more of the following:

• Health and vegetation effects studies.
• Trend analysis.
• Evaluation of pollution abatement programs.
• Establishment of air-quality criteria and standards relating to effects.
• Enforcement of control regulations.

- Development of air pollution control strategies.
- Activation of alert or emergency procedures.
- Land use, transportation, and energy systems planning.
- Background evaluations.
- Atmospheric chemistry studies.

In order to cover all the variable meteorological conditions that may greatly affect the air quality in an area, air monitoring for lengthy periods of time may be necessary to meet most of the above objectives. The topography, demography, and micrometeorology of the area, as well as the contaminant measured, must be considered in determining the number of monitoring stations required in the area. Photographs and a map of the locations of the sampling stations is desirable in describing the sampling station.

Unless the purpose of the sampling program is site-specific, the sites monitored should, in general, be selected as to avoid undue influence by any local source that may cause local elevated concentrations that are not representative of the region to be characterized by the data. Monitoring sites for determining the impact on air quality by individual sources should be selected, if possible, so as to isolate the effect of the source being considered. When there are many sources of the contaminant in the area, the sites sampled should be strategically located so that with wind direction data obtained simultaneously near the sites, the monitoring results will provide evidence of the contributions of the individual sources. Multiple samplers or monitors operating simultaneously upwind and downwind from the source are often very valuable and efficient.

The meteorological parameters that are most important in an atmospheric sampling program are: wind direction and speed, the degree of persistence in direction, and gustiness; temperature and its changes with height above ground; the mixing height, that is, the height above ground that the pollutants will diffuse to during the afternoon; and solar radiation and hours of sunshine, humidity, precipitation, and barometric pressure. These parameters are important in assessing the pollution potential of an area and should be considered in the planning of a monitoring program and in the interpretation of the data. Pertinent meteorological and climatological information may be obtained from the local weather department. In many localities, however, the micrometeorology may be unique and meteorological investigations to provide data specific to the area may be needed (10–12).

Topography can influence the contaminant concentrations in the atmosphere. For example, a valley will cause persistence in wind directions and intensify low-level nocturnal inversions that will limit the dispersion of pollutants emitted into it. Mountains or plateaus may act as barriers affecting the flow of air as well as the contaminant concentrations in their vicinity. Consideration should be given to the influence of these features as well as that of large lakes, the sea, and oceans (9, 10).

The choice of procedure for the air sampling is dependent on the contaminant to be measured. The reader is referred to ASTM Practice D3249 for recommendations for general ambient air analyzer procedures. ASTM-recommended methods have been published for most of the common contaminants that are sampled. Automatic instruments providing a continuous record of the concentrations of the contaminant should be utilized whenever possible to save manpower and increase efficiency. Very often factors such as temperature, humidity, and vibrations, as well as the power line voltage, can influence the output of the air-monitoring instrument, and these should be controlled.

The monitors must be supplied with sample air that represents the ambient air under investigation. Careful consideration should be given to the sample conveying system. A duct system is often utilized for this purpose. There should be as few abrupt enlargements and elbows as possible, because these may affect the uniformity and hence the concentration of the contaminants measured. The material for the duct should be such that there will be little or no interaction between it and the contaminants in the air sampled. Employ temperature control to limit the condensation forming in the sampling lines. Take the samples from straight sections of the duct; also, the inlet lead to the monitoring instrument should be as short as possible.

The following guidelines are recommended for sampling locations unless site-specific measurements are desired. The height of the inlet to the sampling duct should normally be from 2.5 to 5 m above ground, whenever possible. The height of the inlet above the sampling station structure or vegetation adjacent to the station should be greater than 1 m. Sampling should preferably be through a vertical inlet with an inverted cone over the opening. For a horizontal inlet, there should be a minimum of 2 m from the face of the structure. For access to representative ambient air in the area sampled, the elevation angle from the inlet to the top of nearby buildings should be less than 30°. To be representative of the area in which a large segment of the population is exposed to contaminants emitted by automobiles, the inlet should be at a distance greater than 15 m from the nearest high-volume traffic artery. Photochemical oxidants or ozone samplers should be located at distances greater than 50 m from high-volume traffic locations. Particulate matter samplers should be sited at locations that are greater than 200 m from unpaved streets or roads (13–15).

Air sampling can be conducted for long or very short periods, depending on what type of information is needed. Instantaneous or grab sampling is the collection of an air sample over a short period whereas longer period sampling is called integrated air sampling. There is no sharp dividing line between the two sampling methods; however, grab samples are usually taken in a period of less than 5 min.

Grab samples represent the environmental concentration at a particular point in time. It is ideal for following several phases of a cyclic process and for determining airborne concentrations of brief duration but is seldom used to estimate 8-hr average concentrations. Grab samplers consist of various devices. An evacuated flask or plastic bag can be useful in gas and vapor concentration analysis. After introducing the sample of air into the container, it is sealed to prevent loss or further contamination and is then sent to a laboratory for analysis. There, trace analysis procedures such as gas chromotography, infrared spectrophotometry, or other methods are used to determine concentrations of gaseous contaminants.

In integrated air sampling, a known volume of air is passed through a collection medium to remove the contaminant from the sampled airstream. It is the preferred method of determining time-weighted average exposures. Integrated sampling consists of one or a series of samples taken for the full or partial duration of the time-averaging period. The time-averaging period can be from 15 min to 8 hr, depending upon whether a ceiling, short-term, or 8-hr exposure limit is being evaluated. Details of the apparatus or instruments employed in sampling the air or in carrying out associated meteorological investigations are discussed in the references.

The procedure for sampling should be undertaken in the following steps. First, conduct a general exploratory survey of the area including the topography, an inventory of sources for the contaminants, the height of their emissions, traffic, and land-use data. Next, complete a preliminary meteorology analysis to identify wind direction frequencies, wind velocity, and temperature–height profiles. Exploratory short-term temporary sampling requires a number of temporary sampling stations, to determine the need and the best sites for extensive long-term monitoring. Using air-quality models, as well as the input of emission inventory and meteorological information for the area, an estimate for the levels of air quality over the area may be calculated. The model results will provide guidance in determining the locations for monitors that will measure the maximum levels and the number of monitors required to characterize the air quality in the area of concern (16, 17).

32.6 SAMPLE PACKAGING AND SHIPMENT

This section provides general requirements for shipping soil and water samples. Similar methods are required for shipping air samples. These procedures are based on requirements for participation in the Federal Contract Laboratory Program (18). The laboratory for analyzing a sample should be consulted regarding packaging requirements before the initiation of a sampling program. Samples must be packaged for shipment in compliance with current U.S. Department of Transportation (DOT) and commercial carrier regulations. All required government and commercial carrier shipping papers must be filled out and shipment classifications made according to current DOT regulations.

Traffic reports, dioxin shipment records, packing lists, chain-of-custody records, and any other shipping/sample documentation accompanying the shipment must be enclosed in a waterproof plastic bag and taped to the underside of the shipping cooler lid. Coolers must be sealed with custody seals in such a manner that the custody seal would be broken if the cooler were opened. Shipping coolers must have clearly visible return address labels on the outside.

Inside the cooler, sample containers must be enclosed in clear plastic bags through which sample tags and labels are visible. Table 32.1 describes the containers which are used for collection of various samples. Dioxin samples (as well as water and soil should samples be suspected of having medium or high concentration, or of containing dioxin) must be enclosed in a metal can with a clipped or sealable lid (paint cans are normally used for this purpose) and surrounded by packing material such as vermiculite. The outer metal can must be labeled with the number of the sample contained inside.

Water samples for low- or medium-level organics analysis and low-level inorganics analysis must be shipped cooled to VC with ice. No ice is to be used in shipping inorganic low-level soil samples or medium/high-level water samples, organic high-level water or soil samples, or dioxin samples. Ice is not required in shipping soil samples, but may be utilized at the option of the sampler. All cyanide

Table 32.1 Types of sample collection bottles

Description of Bottle Used	Sample Type[a]
80-oz amber glass bottle with Teflon-lined black phenolic cap	Extractable organics; low-concentration water samples
40-mL glass vial with teflon-lined silicon septum and black phenolic cap	Volatile organics; low- and medium-concentration water samples
1-liter high-density polyethylene bottle with white poly cap	Metals, cyanide; low-concentration water samples
120-mL wide-mouth glass vial with white poly cap	Volatile organics; low- and medium-concentration soil samples
16-oz wide-mouth glass jar with Teflon-lined black phenolic cap	Metals, cyanide; medium-concentration water samples
8-oz wide-mouth glass jar with Teflon-line black phenolic cap	Extractable organics; low- and medium-concentration soil samples
	Metals, Cyanide; low- and medium-concentration soil samples
	Dioxin; soil samples
	Organic and inorganics; high-concentration liquid and solid samples
4-oz wide-mouth glass jar with Teflon-lined black phenolic cap	Extractable organics; low- and medium-concentration soil samples
	Metals, cyanide; low- and medium-concentration soil samples
	Dioxin; soil samples
	Organics and Inorganics; high-concentration liquid and solid samples
1-liter amber glass bottle with Teflon-lined black phenolic cap	Extractable organics; low-concentration water samples
32-oz wide-mouth glass jar with Teflon-lined black phenolic cap	Extractable organics; medium-concentration water samples
4-liter amber glass bottle with Teflon-lined black phenolic cap	Extractable organics; low-concentration water samples

[a]This column specifies the *only* type(s) of samples that should be collected in each container.

samples, however, must be shipped cooled to 4°C. Low- and medium-level water samples for inorganic analysis require chemical preservation.

Waterproof, metal ice chests or coolers are the only acceptable type of sample shipping container. Cardboard and styrofoam containers should not be used. Shipping containers should be packed with noncombustible, absorbent packing material (vermiculite is recommended) surrounding the plastic-enclosed, labeled sample bottles (or labeled metal cans containing samples) to avoid sample breakage in transport. Sufficient packing material should be used so that sample containers will not make contact during shipment. Earth or ice should never be used to pack samples. Ice should be in sealed plastic bags to prevent melting ice from soaking packing material which, when soaked, makes handling of samples difficult in the lab.

Samples for organics analysis must be shipped "Priority One/Overnight." If shipment requires more than a 24-hr period, sample holding times can be exceeded, thereby compromising the integrity of the sample analyses. Samples for inorganics analysis should be held until sampling for the entire area is complete and shipped for delivery within 2 days. Three days is the recommended period for collection of the samples. All samples should be shipped through a reliable commercial carrier.

32.7 SAMPLE DOCUMENTATION

Each sample must be properly documented to ensure timely, correct, and complete analysis for all parameters requested, and, most importantly, to support use of sample data in potential enforcement actions concerning a site. The documentation system provides the means to individually identify, track, and monitor each sample from the point of collection through final data reporting. As used herein, a sample is defined as a representative specimen which is collected at a specific location of a site at a particular point in time for a specific analysis and which may refer to field samples, duplicates, replicates, splits, spikes, or blanks that are shipped from the field to a laboratory.

Sample Traffic Report

The sample documentation system is usually based on the use of the EPA Sample Traffic Report, a four-part carbonless form printed with a unique sample identification number. One Traffic Report (TR) and its preprinted identification number is assigned by the sampler to each sample collected. The two types of TRs currently in use are Organic and Inorganic.

To provide a permanent record for each sample collected, the sampler completes the appropriate TR, recording the case number, site name or code, location and site/spill ID, analysis laboratory, sampling office, dates of sample collection and shipment, and sample concentration and matrix. Numbers of sample containers and volumes are entered by the sampler beside the analytical parameter(s) requested for particular sample portions.

A strip of adhesive sample labels printed with the TR sample number comes attached to the TR for the sampler's use in labeling sample bottles. The sampler affixes one of these numbered labels to each container making up the sample. In order to protect the label from water and solvent attack, each label must be covered with clear waterproof tape. The sample labels, which bear the TR identification number, permanently identify each sample collected and link each sample component throughout the analytical process.

Sample Tag

To render sample data valid for enforcement uses, individual samples must be traceable continuously from the time of collection until the time of introduction as evidence during litigation. One mechanism utilized is the use of the "sample tag." Each sample removed from a waste site and transferred to a laboratory for analysis is identified by a sample tag containing specific information regarding the sample, as defined by the EPA National Enforcement Investigations Center (NEIC). Following sample analysis, sample tags are retained by the laboratory as physical evidence of sample receipt and analysis, and may later be introduced as evidence in Agency litigation proceedings.

The information recorded on an EPA sample tag includes:

- Case number(s) — the unique number(s) assigned to identify the sampling event.
- Sample number — the unique sample identification number used to document that sample.
- Project code — the number assigned by EPA to the sampling project.
- Station number — a two-digit number assigned by the sampling team coordinator.
- Date — a six-digit number indicating the month, day, and year of collection.
- Time — a four-digit number indicating the military time of collection.
- Station location — the sampling station description as specified in the project plan.
- Samplers — signatures of samplers on the project team.
- Tag number — a unique serial number preprinted or stamped on the tag.
- Lab sample number — reserved for laboratory use.

Additionally, the sample tag contains appropriate spaces for noting that the sample has been preserved and indicating the analytical parameter(s) for which the sample will be analyzed. Each sample tag is completed and securely attached to the sample container. Samples are then shipped under chain-of-custody procedures as described below.

Chain-of-Custody Record

Official custody of samples must be maintained and documented from the time of sample collection up to introduction as evidence in court, in accordance with EPA enforcement requirements. The following custody documentation procedure was developed by NEIC and is used in conjunction with sample documentation.

A sample is considered to be in an individual's custody if the following criteria are met: It is in your possession or it is in your view after being in your possession; or it was in your possession and then was locked up or sealed to prevent tampering; or it is in a secured area. Under this definition the team member actually performing the sampling is personally responsible for the care and custody of the samples collected until they are transferred or dispatched properly. In follow-up, the sampling team leader reviews all field activities to confirm that proper custody procedures were followed during the field work.

The Chain-of-Custody Record is employed as physical evidence of sample custody. The sampler completes a Chain-of-Custody Record to accompany each cooler shipped from the field to the laboratory.

Information similar to that entered on the sample tag is recorded on the Chain-of-Custody Record. Information includes the project number, sampler's signatures, and the case number. For each station number, the sampler indicates the following: date; time; whether the sample is a composite or grab; station location; number of containers; analytical parameters; sample number(s); and sample tag number(s). When relinquishing the samples for shipment, the sampler signs in the space indicated on the form, entering the date and time the samples are relinquished.

The top, original signature copy of the Chain-of-Custody Record is enclosed in plastic (with the sample documentation) and secured to the inside of the cooler lid. A copy of the custody record is retained for the sampler's files.

Shipping coolers are secured, and custody seals are placed across cooler openings. As long as custody forms are sealed inside the sample cooler and custody seals remain intact, commercial carriers are not required to sign off on the custody form.

The laboratory representative who accepts the incoming sample shipment signs and dates the Chain-of-Custody Record to acknowledge receipt of the samples, completing the sample transfer process. It is then the laboratory's responsibility to maintain internal log books and records that provide a custody record throughout sample preparation and analysis.

32.8 SUMMARY

1. It should be noted that the methods provided in this chapter are for purposes of general overview only. Before beginning any sampling program, background research must be conducted to determine (a) proper equipment for both sampling and personal protection, (b) proper sampling methodology and analytical methods, and (c) appropriate health and safety practices to be employed. This is especially important when handling materials which may be hazardous or radioactive.

2. Three types of water sampling are grab samples, composite samples, and continuous flowing samples. Under site-specific condition, these types of sampling can be used for almost any body of water.

3. Two portions of the soil that are important in sampling are: (a) the surface layer, which reflects the deposition of airborne pollutants; and (b) the soil at various depths, which represents soluble pollutants or traveling plumes of pollutants.

4. Variables that affect air contaminant concentrations include atmospheric stability, turbulence, wind speed and direction, solar radiation, precipitation, topography, emission rates, chemical reaction rates for their formation and decomposition, and the physical and chemical properties of the contaminant. All of these factors need to be accounted for when developing and analyzing an air-sampling system.

5. Special requirements for sample packaging and shipment should be determined by contacting the laboratory before starting the sample collection.

6. Critical sample documentation includes the sample traffic report, the sample tag, and the Chain-of-Custody Report.

REFERENCES

1. ASTM, 1990 *Annual Book of ASTM Standards*, Volume 11.04, ASTM D3370.
2. G. Friedlander, J. W. Kennedy, and J. Miller, *Nuclear and Radiochemistry*, second edition, John Wiley & Sons, New York, 1964.

3. R. T. Overman and H. M. Clark, *Radioisotope Techniques*, McGraw-Hill, New York, 1960.

4. *Hazardous Material Regulations of the D.O.T.*, Title 49, Parts 170 to 190. Consult current edition. Contact the U.S. Government Printing Office, Superintendent of Documents, Washington, DC 20401.

5. ASTM, 1990 *Annual Book of ASTM Standards*, Volume 11.04, ASTM D4448.

6. U.S. EPA, *Preparation of Soil Sampling Protocol*, EPA 600-483-020, August 1983.

7. ASTM, 1991 *Annual Book of ASTM Standards*, Volume 11.03, ASTM D1357.

8. S. Calvert and H. M. Englund, *Handbook of Air Pollution Technology*, New York, John Wiley & Sons, 1984.

9. L. D. Kornreich, *Proceedings of the Symposium on Statistical Aspects of Air Quality Data*, EPA-650/4-74-038, October 1974.

10. R. E. Munn, *Descriptive Micrometeorology*, Academic Press, New York, 1966.

11. D. H. Slade, *Meteorology and Atomic Energy*, TID-24180, U.S. Atomic Energy Commission, National Bureau of Standards, U.S. Department of Commerce, Springfield, VA, 1968.

12. A. L. Morris and R. C. Barras, *Air Quality Meteorology and Ozone*, ASTM Special Technical Publication, STP 653, Philadelphia, 1978.

13. U.S. EPA, *Guidance for Air Quality Monitoring Network Design and Instrument Siting*, QAQPS No. 1.2-012, September 1975.

14. U.S. EPA, *Selecting Sites for Carbon Monoxide Monitoring*. EPA-450/3-75-077, September 1975.

15. US. EPA, *Optimum Site Exposure Criteria for SO_2 Monitoring*, EPA-450/3-77-0113, April 1977.

16. J. H. Seinfeld, "Optimal Locations of Pollutant Monitoring Stations in an Airshed," *Atmospheric Environment*, Vol. 6, Pergamon Press, Elmsford, NY, 1972, pp. 847–858.

17. M. M. Benairic, *Urban Air Pollution Modeling*, MIT Press, Cambridge, MA, 1980.

18. U.S. EPA, *User's Guide to the Contract Laboratory Program*, December 1986.

33

HEALTH AND SAFETY

33.1 INTRODUCTION

The U.S. Environmental Protection Agency (EPA) estimates that approximately 57 million tons of hazardous waste are produced each year in the United States. These wastes must be treated and stored or disposed of in a manner that protects the environment from adverse affects of the various constituents of those wastes. In response to the need to protect human health and the environment from improper disposal of these hazardous wastes, anyone working in a hazardous or unsafe environment must receive health and safety training. More specifically, the Occupational Safety and Health Administration (OSHA) and the EPA regulations require employers to provide health and safety training to all employees involved in (a) the clean-up of uncontrolled hazardous waste sites, (b) corrective actions under the Resource Conservation and Recovery Act, (c) emergency response actions involving the release of hazardous substances and (d) the clean-ups at sites recognized by any government agency as uncontrolled hazardous waste sites. In determining who should receive health and safety training, employers should also consider the liability that they or their agency may incur in the event of an accident. It is imperative that anyone working in a hazardous environment receive the appropriate health and safety training as required by law (1).
Key regulations are addressed in the next section.

Personnel responding to environmental incidents involving hazardous substances may encounter a wide range of health and safety problems. Besides hazards associated with the physical, chemical, and toxicological properties of the materials involved, other safety concerns (see Chapter 34, "Accident and Emergency Management"), such as electrical hazards, heat stress, cold exposure, faulty equipment, and construction dangers, can also have adverse effects on personnel.

To ensure the safety of response personnel, an effective, comprehensive health and safety program must be established and followed (2). The sections to follow in this chapter will primarily address health and safety concerns associated with hazardous and/or toxic substances and hazardous waste sites.

33.2 HEALTH AND SAFETY REGULATIONS

The United States Occupational Safety and Health Administration (OSHA) has regulations governing employee health and safety at hazardous waste operations and during emergency responses to hazardous substance releases. These regulations (29 CFR 1910.120) contain general requirements for safety and health programs, site characterization and analysis, site control, training, medical surveillance, engineering controls, work practices along with personal protective equipment, exposure monitoring, informational programs, material handling, decontamination, emergency procedures, illumination, sanitation, and site excavation.

The EPA's Standard Operating Safety Guides supplement and complement these regulations, but for specific legal requirements, the OSHA's regulations must be used. Other OSHA regulations may pertain to employees working with hazardous materials or working at hazardous waste sites. These, as well as state and local regulations, must also be considered when developing worker health and safety programs (2).

The OSHA Hazard Communication Standard was first promulgated on November 25, 1983, and is printed in 29 CFR Part 1910.120. The standard was developed to inform workers who are exposed to hazardous chemicals of the risks associated with those chemicals. The purpose of the standard is to ensure that

- The hazards of all chemicals produced or imported are evaluated.
- Information concerning chemical hazards is transmitted to employers and employees.

Hazard information must be transmitted from manufacturers to employers via material safety data sheets (MSDS)and containers labels. This information must be transmitted from employers to employees by means of comprehensive hazard communication programs, which include MSDSs, container labels, employee information and training programs. A subsection on MSDS is provided in this section.

The basic requirements of the Hazardous Communication Standard are as follows:

- Have an MSDS on file for every hazardous chemical present or used in the workplace.

- Keep MSDSs readily accessible during each work shift, and inform all employees how to obtain MSDSs. If employees travel between workplace during a work shift, the MSDSs may be kept at a central location at the primary workplace, as long as the necessary information is immediately available in the event of an emergency.

- Ensure that every container that holds hazardous chemicals in the workplace is clearly labeled with the identity of the hazardous chemical and appropriate hazard warnings.

- Do not remove or deface labels on incoming containers of hazardous chemicals.

- Before their initial assignment or whenever a new hazard is introduced into their work area, inform employees of both (a) the requirements of the standard operations in their work area where hazardous chemicals are present, and (b) the location and availability of the written hazard communication program.

- Train employees how to identify and protect themselves from chemical hazards in the work area, how to recognize the physical and health hazards of chemicals in their work area, and how to obtain and use the employer's written hazard communication program and appropriate hazard information.

- Develop, implement, and maintain a written communication program for each workplace which describes how material safety data sheets, labeling, and employee information and training requirements will be met. The written program must also include a list of hazardous chemicals present in the workplace, the methods which will be used to inform employees of the hazards associated with performing nonroutine tasks, and chemicals present in unlabeled pipes.

Employers who have multiemployer workplaces must include methods the employer will use to provide contractors at his facility with material safety data sheets. These employers must also describe how they will inform subcontractor's employees about (a) the precautions which must be followed and (b) the labeling system used in the workplace.

Labeling Hazardous Chemicals

Accurate, complete, and effective labels are critical to the success of a hazardous communication program. Hazardous chemicals labels tell employees if a container is dangerous and what the dangers are. Even if employees never take the time to review material safety data sheets for chemicals in their work areas, essential hazard information must be apparent to employees each time they look at labels while handling hazardous chemicals.

In order to comply with the labeling requirements of hazardous communication standard, labeling programs must meet the following minimum requirements:

- Containers holding hazardous chemicals which are subject to an OSHA substance-specific health standard are labeled in accordance with that standard.

- Each container of hazardous chemicals in the workplace is labeled, tagged, or marked with the identity of the hazardous chemical and appropriate hazard warnings. (Exception: Warnings in the form of signs, placards, process sheets, or other written materials may be used instead of labels for individuals stationary process containers.)

- Existing labels on incoming containers of hazardous chemicals are not to be removed or defaced. (Exception: Employers can remove the labels as long as the container is immediately marked with the required information.)

- Labels are in English, but can be available in other languages in addition to English. Labels are prominently displayed on containers or available in the work area throughout each work shift.

- New labels are not required if existing labels already convey the required information.

Hazardous Chemical Labeling Systems

Different hazardous labeling systems have been developed by organizations in order to help standardize the information presented on labels. Three common systems presently in use include

- National Fire Protection Association (NFPA) 704 Standard.
- American National Standards Institute (ANSI) Standard Z129.1.
- National Paints and Coating Association Hazardous Materials Information Systems (HMIS).

The NFPA system uses hazard rankings and pictographs within a distinctive three-color diamond to convey health, reactivity, and flammability hazard information.

Material Safety Data Sheets

Material Safety Data Sheet (MSDSs) form the backbone of hazard communication program. The MSDS is the primary mechanism used to (a) document chemical hazard information discovered during the hazard determination process, and (b) to pass that information to end users of the chemical.

In order to comply with the requirements of the Hazard Communication Standard, an MSDS program must meet the following requirements:

- Obtain or develops an MSDS for each hazardous chemical.
- Ensures each MSDS is complete.
- Maintains copies of MSDSs and keeps them readily accessible during each work shift.
- Makes MSDSs available to OSHA upon request.

Material Safety Data Sheets can have many different formats, and may be kept in any form. Some manufacturers use completely different formats, such as operating procedures for a specific process or marketing brochures which include safety data. Regardless of which format is selected, employers must ensure that in all cases the required information is provided for each hazardous chemical in the facility. In addition, the MSDS must be in English.

Because this system does not include chemical identity information, it cannot be used alone for compliance with the hazard communication standard. The ANSI standard recommends the use of signal words such as CAUTION, WARNING, and DANGER to convey the degree of hazard. Standard language is also used to convey chemical hazards, precautions for chemical storage and use, and first-aid procedures.

The HMIS is similar to the NFPA system in that the health, flammability, and reactivity hazards are rated by severity on a 4 to 0 scale with the same color scheme. However, the HMIS label is rectangular rather than diamond-shaped, and contains the identity of the hazardous material as listed on the material safety data sheet.

33.3 EXPOSURE TO TOXIC SUBSTANCES

Toxic substances (including radioactive material and biological agents) or chemically active substances present a special concern because they can be inhaled, ingested, or be absorbed through (or destructive to) the skin. They may exist in the air or, due to site activities, become airborne. Liquids or sludges can splash on the skin. The effects of these substances can vary significantly. Ingested or inhaled, the substances may cause no apparent illness, or they can be fatal. On the skin they may cause no demonstrable effects. Other substances, however, may damage the skin or be absorbed through it, leading to systemic toxic effects.

Two types of potential exposures exist:

1. *Acute:* Exposures occur for relatively short periods of time, generally minutes to 1–2 days. Concentrations of toxic air contaminants are high relative to their protection criteria. In addition to inhalation, airborne substances might directly contact the skin, or liquids and sludges may be splashed on the skin or into the eyes, leading to toxic effects.

2. *Chronic:* Continuous exposure occurs over longer periods of time, generally months to years. Concentrations of inhaled toxic contaminants are relatively low. Direct skin contact by immersion, splash, or by contaminated air involves contact with substances exhibiting low dermal activity.

In general, acute exposures to chemicals in air are more typical in (a) transportation accidents and fires, or (b) releases at chemical manufacturing or storage facilities. High concentrations of contaminants in air usually do not persist for long periods of time. Acute skin exposures may occur when workers must be in close contact with the substances in order to control the release, for example, while patching a tank car, while offloading a corrosive material, while uprighting a drum, or while containing and and treating the spilled material.

Chronic exposures are usually associated with longer-term removal and remedial operations. Contaminated soil and debris from emergency operations may be involved, soil and ground water may be polluted, or temporary impoundment systems may contain diluted chemicals. Abandoned waste sites typically represent chronic exposure problems. As activities start at these sites, personnel engaged in certain activities (sampling, handling containers, or bulking compatible liquids) face an increased risk of acute exposures from splashes, or from vapors, gases, or particulates that might be generated.

At any specific incident, the hazardous properties of the materials may only represent a potential risk. For example, if a tank car of liquified natural gas is involved in an accident but remains intact, the risk from fire and explosion is low. In other incidents, the risks to response personnel are high. For instance, when toxic or flammable vapors are being released from a ruptured tank truck. The continued health and safety of response personnel requires that the risks (real or potential) at an episode be assessed and appropriate measures instituted to reduce or eliminate the threat to response personnel (2).

Chemicals and chemical groups affect different parts of the body. One chemical, such as an acid or base, may affect the skin whereas another, such as carbon tetrachloride, attacks the liver. Some chemicals will affect more than one organ or system. The organ or system being attacked is referred to as the target organ. The damage done to an organ can differ in severity depending on chemical composition, length of exposure, and the concentration of the chemical. When two chemicals enter the body and one intensifies the damage done by the other, it is called a synergistic effect. Synergism complicates almost any exposure due to a lack of toxicological information. For just one chemical, it may typically take a toxicological research facility approximately two years of studies to generate valid data. The data produced in that two-year time frame applies only to that one chemical. With the addition of another chemical, the original may become a totally different actor. Without drawing this out, it becomes apparent that there are many unknowns. Effects can vary in severity from minimal and temporary to serious permanent damage.

The National Institute of Occupational Safety and Health (NIOSH) recommends standards for industrial exposure which OSHA uses in its regulations. The NIOSH packet guide to chemical hazards contains a wealth of information on specific chemicals, their chemical name, structure/formula, synonyms, trade names, exposure limits, time-weighted average, short-term exposure limit, "immediately dangerous to life and health" concentrations (IDLHs), chemical and physical properties such as molecular weight, boiling point, lower explosive limit (LEL), measurement methods, personal protection and sanitation, recommendations for respirators selection, and health hazards information.

33.4   HEALTH AND SAFETY REQUIREMENTS

Any incident represents a potentially hostile situation. Chemicals that are combustible, explosive, corrosive, toxic, or reactive, along with biological and radioactive materials, can affect the general public or the environment as well as response personnel. Physical hazards may also be encountered. Workers may fall, trip, be struck by objects, or be subjected to danger from electricity and heavy equipment. Injury and illness may also occur due to the physical stress of response personnel. While the response activities needed at each individual incident are unique, there are many similarities. One similarity is that all responses require protecting the health and ensuring the safety of response personnel.

To reduce the risks to personnel responding to hazardous substance incidents, an effective health and safety program must be developed and followed. At a minimum, a comprehensive worker health and safety program should address

- Safe work practices
- Engineered safeguards
- Medical surveillance
- Environmental and personnel monitoring
- Personnel protective equipment
- Education and training
- Standard operating safety procedures

As part of a comprehensive program, standard operating safety procedures provide instructions on how to accomplish specific tasks in a safe manner. In concept and principle, standard operating safety procedures are independent of the type of incident.

Medical Program

To safeguard the health of response personnel, a medical program must be developed, established, and maintained. This program has two essential components: routine health care and emergency treatment.

Routine Health Care. At a minimum, routine health care and maintenance should consist of:

- Pre-employment medical examinations to establish the individual's state of health, baseline physiological data, and ability to wear personnel protective equipment.
- Annual examinations, of which, the frequency and content will be determined by the examining physician. The examination may vary depending on the length and type of work assignment, the frequency of exposure, and the individual's physical condition.
- More frequent examinations (determined by the physician) due to the workers's assignment and potential exposure levels.
- Special medical examinations, care, and counseling in case of known or suspected exposures to toxic substances. Any special tests needed depend on the chemical substance to which the individual has been exposed.
- Termination examinations conducted at the end of employment or upon reassignment. The content of the examination should be similar to the baseline examination.

Emergency Medical Care and Treatment. The Medical Program must address emergency medical care and treatment of response personnel, including possible exposures to toxic substances and injuries resulting from accidents or physical hazards. The following items should be included in emergency care provisions:

- Name, address, and telephone number of the nearest medical treatment facility. This should be conspicuously posted. A map and directions for locating the facility, plus the travel time, should be readily available.
- The facilities to provide care and treatment of personnel exposed or suspected of being exposed to toxic (or otherwise hazardous) substances. If the facility lacks toxicological capability, arrangements should be made for consultant services.
- Administration arrangements for accepting patients.
- Arrangements to quickly obtain ambulance, emergency, fire, and police services. Telephone numbers and procedures for obtaining these services should be conspicuously posted.
- Emergency showers, eyewash fountains, and first-aid equipment readily available on-site. Personnel should have advanced first aid and emergency lifesaving training.
- Provisions for the rapid identification of the substance to which the worker has been exposed (if this has not previously been done). This information must be given to medical personnel.
- Procedures for decontamination of injured workers and preventing contamination of medical personnel, equipment, and facilities.

- Protocols for heat stress and cold exposure monitoring, and working in adverse weather conditions.
- Medical evacuation requirements.

The EPA's Environmental Response Team's "Occupational Medical Monitoring Program Guidelines for SARA Field Activity Personnel," June 2, 1988, addresses specific medical monitoring concerns and procedures.

Because of the nature and risk of the work associated with hazardous material incidents and because the potential exposure to harmful substances may have an adverse effects on an employee, it is essential that proper records be maintained and retained.

Medical records should contain the following information:

- Any occupational exposure.
- Employees' use of respirators and personnel protective clothing.
- Any work-related injuries.
- Physician's written opinion of medical problems and treatment.
- Record of all medical examinations.

As part of the medical program, response personnel should be instructed in the signs and symptoms that might indicate potential exposure to toxic substances. Some of these are:

Observable by Others

- Changes in complexion, skin discoloration.
- Lack of coordination.
- Changes in demeanor.
- Excessive salivation.
- Pupillary response.
- Changes in speech pattern.
- Breathing difficulties.
- Difficulties with coordination.
- Coughing.

Nonobservable by Others

- Headaches.
- Dizziness.
- Blurred vision.
- Cramps.
- Irritation of eyes, skin, or respiratory tract.
- Behavior changes.

Health and Safety Training

Safety and health training must be an integral part of the total response health and safety program. Safety training must be continuous and frequent for response personnel to maintain their proficiency in the use of equipment and their knowledge of safety requirements.

All personnel involved in responding to environmental incidents and who could be exposed to hazardous substances, health hazards, or safety hazards must receive safety training prior to carrying out their response functions. Health and safety training must, at a minimum, include:

- Use of personnel protective equipment, for example, respiratory protective apparatus and protective clothing.
- Safe work practices, engineering controls, and standard operating safety procedures.
- Hazard recognition and evaluation.
- Medical surveillance requirements, symptoms and signs which might indicate medical problems, and first aid.
- Site safety plans and plan development.
- Site control and decontamination.
- Use of monitoring equipment, if applicable.

Training must be as practical as possible and include hands-on use of equipment and exercises designed to demonstrate and practice classroom instruction. Formal training should be followed by at least three days of on-the-job experience working under the guidance of an experienced, trained supervisor. All employees should, at a minimum, complete an 8-hr safety refresher training course annually. Health and safety training must comply with OSHA's training requirements as defined in 29 CFR 1910.120.

Qualified Safety Personnel

Personnel responding to chemical incidents must make many complex decisions regarding safety. Making these decisions correctly requires more than elementary knowledge. For example, selecting the most effective personal protective equipment requires not only expertise in the technical areas of respirators, protective clothing, air monitoring, physical stress, etc., but also experience and professional judgment.

Only a competent, qualified person (safety specialist) has the technical judgment to evaluate a particular incident and determine the appropriate safety requirements. It's through a combination of professional education, on-the-job experience, specialized training, and continual study, that the safety professional acquires the expertise to make sound decisions.

Standard Operating Safety Practices

Standard operating safety procedures should include safety precautions and operating practices, that all responding personnel should follow. These would include:

Personal Precautions

- Eating, drinking, chewing gum or tobacco, smoking, or any practice that increases the probability of hand-to-mouth transfer and ingestion of material is prohibited in any area designated contaminated.
- Hands and face must be thoroughly washed upon leaving the work area.
- Whenever decontamination procedures for outer garments are in effect, the

entire body should be thoroughly washed as soon as possible after the protective garment is removed.

- No facial hair which interferes with a satisfactory fit of the mask-to-face-seal is allowed on personnel required to wear respirators.
- Contact with contaminated or suspected contaminated surfaces should be avoided. Whenever possible, do not walk through puddles, leachate, discolored surfaces, kneel on ground, lean, sit, or place equipment on drums, containers, or the ground.
- Medicine and alcohol can potentiate the effects from exposure to toxic chemicals. Prescribed drugs should not be taken by personnel on response operations where the potential for absorption, inhalation, or ingestion of toxic substances exists unless specifically approved by a qualified physician. Alcoholic beverages should be avoided, in the off-duty hours, during response operations.

Operations

- All personnel going on-site must be adequately trained and thoroughly briefed on anticipated hazards, equipment to be worn, safety practices to be followed, emergency procedures, and communications.
- Any required respiratory protection and chemical protective clothing must be worn by all personnel going into areas designated for wearing protective equipment.
- Personnel on-site must use the buddy system when wearing respiratory protection. At a minimum, two other persons, suitably equipped, are required as safety backup during initial entries.
- Visual contact must be maintained between pairs on-site and safety personnel. Entry team members should remain close together to assist each other during emergencies.
- During continual operations, on-site workers act as safety backup to each other. Off-site personnel provide emergency assistance.
- Personnel should practice unfamiliar operations prior to doing the actual procedure.
- Entrance and exit locations must be designated and emergency escape routes delineated. Warning signals for site evacuation must be established.
- Communications using radios, hand signals, signs,or other means must be maintained between initial entry members at all times. Emergency communications should be prearranged in case of radio failure, necessity for evacuation of site, or other reasons.
- Wind indicators visible to all personnel should be strategically located throughout the site.
- Personnel and equipment in the contaminated area should be minimized, consistent with effective site operations.
- Work areas for various operational activities must be established.
- Procedures for leaving a contaminated area must be planned and implemented prior to going on-site. Work areas and decontamination procedures must be established based on expected site conditions.

33.5   SITE SAFETY PLAN

The purpose of the site safety plan is to establish policies and procedures for protecting the health and safety of response personnel during all operations conducted at an incident. It contains information about the known or suspected hazards, routine and special safety procedures that must be followed, and other instructions for safeguarding the health of the responders.

A site safety plan shall be prepared and reviewed by qualified personnel for each hazardous substance response. Before operations at an incident commence, all safety aspects of site operations should be thoroughly examined. A safety plan is then written based on the anticipated hazards and expected work conditions. The plan should be conspicuously posted or distributed to all response personnel and discussed with them. The safety plan must be periodically reviewed to keep it current and technically correct.

In nonemergency situations, for example, long-term remedial action at abandoned hazardous waste sites, safety plans are developed simultaneously with the general work plan. Workers can become familiar with the plan before site activities begin. Emergency responses generally require the use of a generic safety plan, standing standard operating procedures, and special verbal instructions until (if time permits) a plan can be written.

The plan must contain safety requirements for routine (but hazardous) response activities and also for unexpected site emergencies. The major distinction between routine and emergency site safety planning is the ability to predict, monitor, and evaluate routine activities. A site emergency is unpredictable and may occur at any time.

Categories of Hazardous Materials Responses

Three general categories of response exist: emergencies, hazardous waste site investigations, and remedial actions. Although considerations for personnel safety are generic and independent of the response category, in scope, detail, and length, safety requirements and plans vary considerably. These variations are generally due to the reason for responding (or category of response), information available, and the severity of the incident with its concomitant dangers to the responder.

Emergencies

1. *Situation.* Emergencies generally require prompt action to prevent or reduce undesirable effects. Immediate hazards of fire, explosion, and release of toxic vapors or gases are of prime concern. Emergencies vary greatly in respect to types and quantities of material, hazards, numbers of responders involved, type of work required, population affected, and other factors. Emergencies usually last from a few hours to a few days.

- *Information Available.* varies from none to much. Usually, information about the materials involved and their associated hazards, is quickly obtained in transportation related incidents, or incidents involving fixed facilities. Determining the substances involved in other incidents, such as mysterious spills or illegal dumping requires considerable time and effort.

- *Time Available.* Little time. Generally requires prompt action to bring the incident under control.

- *Reason for Response.* To implement prompt and immediate actions to control dangerous or potentially dangerous situations.

2. *Effects on Plan.* In emergencies, time is not available to write lengthy and detailed safety plans. Therefore, general safety plans for emergency response (generic plans) are developed prior to responding and are implemented when an emergency occurs.

Responding organizations must rely on (a) their existing generic safety plan and written standard operating safety procedures adapted to meet incident-specific conditions, and (b) the use of verbal safety instructions.

Since there is a heavy reliance on verbal communications, an effective system to keep all responders informed must be established. Whenever possible, these incident-specific instructions should be written and posted.

Site Investigations

1. *Situation.* In nonemergency responses, for example, preliminary inspections at abandoned wastes sites or more comprehensive waste site investigations, the objective is to determine and characterize (a) the chemicals and hazards involved, (b) the extent of contamination, and (c) risks to people and the environment. In general, initial inspections, detailed investigations, and extent of contamination surveys are limited in the activities that are required and number of people involved. Initial or preliminary inspections generally require 1–5 days. Complete investigations may last over a longer period of time (months).

- *Information Available.* Much background information is often available, but may not be specific enough for making initial safety decision. On-site information more fully developed through additional surveys and investigations.

- *Time Available.* In most cases adequate time is available to make a preliminary evaluation of the site's characteristics and to develop a written site-specific safety plan.

- *Reason for Response.* To gather data to verify or refute existing information, to gather information to determine scope of subsequent investigations, or to collect data for planning remedial action.

2. *Effects on Plan.* Sufficient time is available to determine, on a preliminary basis, the hazards anticipated and other conditions associated with the site and to write initial safety plans. In scope and detail, these plans tend to be brief and contain safety requirements for specific on-site work relevant to collecting data. As information is developed through additional investigations, the safety plan is modified and, if necessary, more detailed and specific requirements are added.

Remedial Actions

1. *Situation.* Remedial actions are cleanups which may take many years to complete. They commence after more immediate problems at an emergency have been controlled, or they involve the mitigation of hazards and restoration of abandoned hazardous waste sites. Numerous activities are required involving the efforts of many people, a detailed logistics and support base, extensive equipment, and more involved work activities.

- *Information Available.* Much known about on-site hazards.
- *Time Available.* Ample time for work planning.
- *Reason for Response.* Systematic and complete control, cleanup, and restoration.

2. *Effects on Plan.* Since ample time is available before work commences, site safety plans tend to be comprehensive and detailed. From prior investigations much detail may be known about the materials or hazards at the site and extent of contamination.

Preliminary Site Evaluation and Safety Plan

A preliminary evaluation of a hazardous waste site's characteristics must be performed by a qualified person, prior to anyone going on the site. The information obtained is used to determine the appropriate health and safety control procedures needed to protect initial entry-team personnel from identified or suspected hazards. After initial site entry, a more detailed evaluation of site characteristics is made based upon information collected by the entry team. The preliminary site safety plan is then modified and refined.

Of immediate concern are known or expected substances that are immediately dangerous to life and health (IDLH) through skin absorption or inhalation, or other conditions that may cause death or serious injury. Some examples of these conditions are fire or explosive potential, visible vapor clouds, radioactive labeled-material, and confined space entry.

A preliminary evaluation of the site's characteristics shall include:

- Incident location and name.
- Site description, topography, and size.
- Descriptions of the activities or tasks to be done.
- Duration of planned of planned activities.
- Site accessibility.
- Hazardous substances and health hazards involved or expected.
- Chemical, physical, and toxicological properties of the hazardous substances involved.
- Behavior and dispersion of material involved.
- Availability and capabilities of emergency assistance.

Additional information that might be useful is:

- Types of containers, storage, or transportation methods.
- Prevailing weather condition and forecast.
- Surrounding populations and land use.
- Ecologically sensitive areas.
- Facility records.
- Preliminary assessment reports.
- Off-site survey results.

The information initially available, collected during a preliminary inspection, or obtained through subsequent investigations provides a basis for developing a detailed, site-specific safety plan. This type of information is then used along with the reason for responding to develop a comprehensive safety plan.

The safety plan is tailored to the conditions imposed by the incident and to its environmental setting. As work progresses and as additional information becomes available, the safety plan is reviewed, modified, and kept current.

General Requirements for Routine Operations

Routine operations are all those activities that may be required in responding to an emergency or a remedial action at a hazardous waste site in order to identify, evaluate, and control (including clean up) the incident. These activities may involve a high degree of risk, but are standard operations generally involved in responding to that type of incident.

Safety practices for routine operations closely parallel accepted procedures used in industrial hygiene and industrial safety. Whenever a hazardous incident progresses to the point where operations become more routine, the associated site safety plan becomes a more refined document. As a minimum, the following must be included as part of the site safety plan for routine operations.

1. *Key Personnel and Alternates.* The plan must identify the incident manager as well as the site safety and health officer (and alternates) and any other personnel responsible for site safety. It should also identify key personnel associated with other site operations. The names, telephone numbers, addresses, and organizations of these people must be listed in the plan and posted in a conspicuous place.

2. *Known Hazards and Risks.* All known or suspected physical, biological, radiological, or chemical hazards must be described. It is important that all health-related data be kept up-to-date. As air, water, soil, or hazardous substance monitoring and sampling data become available, they must be evaluated, significant risk or exposure to workers noted, potential impact on public assessed, and changes made in the plan. These evaluations need to be repeated frequently since much of the plan is based on this information.

3. *Routine or Special Training Requirements.* Personnel must be trained not only in general safety procedures and use of safety equipment, but in any specialized work they may be expected to do.

4. *Levels of Protection.* The levels of protection to be worn at locations on-site or by work functions must be designated. This includes the specific types of respirators and type of chemical protective clothing to be worn for each level. No one shall be permitted in areas requiring personnel protective equipment unless they have been trained in its use and are wearing it.

5. *Site-Specific Medical Requirements.* Specialized medical requirements should be determined when unusual hazards are expected to be encountered.

6. *Environmental Surveillance Program.* A program to monitor site hazards must be implemented. This would include air monitoring and sampling and other kinds of media sampling at or around the site that would identify chemicals present, their hazards, possible routes of migration off-site, and associated safety requirements.

7. *Work Areas.* Work areas (exclusion zone, contamination reduction zone, and support zone) need to be designated on the site map and the map posted. The size

of zones, zone boundaries, and access control points into each zone must be marked and made known to all site workers.

8. *Site Control Procedures.* Control procedures must be implemented to prevent unauthorized access. Site security procedures — fences, signs, security patrols, and check-in procedures — must be established. Procedures must also be established to control authorized personnel into work zones where personnel protection is required.

9. *Decontamination.* Decontamination procedures for personnel and equipment must be established. Arrangements must also be made for the proper disposal of contaminated material, solutions, and equipment.

10. *Emergency Response Plan.* A plan for responding safely and effectively to emergency situations that might develop at the site must be developed and included as part of the overall site safety plan.

11. *Confined Space Entry.* Procedures to assure the safety of personnel who may have to make confined space entry must be established.

12. *Weather-Related Problems.* Weather conditions can affect site work. Temperature extremes, high winds, precipitation, and storms can impact personnel safety. Work practices must be established to protect workers from the effects of weather and shelters provided, when necessary. Temperature extremes (especially heat) and their effect on people wearing protective clothing, must be considered and procedures established to monitor for and minimize heat stress.

On-Site Emergencies

The plan must address site emergencies — occurrences that require immediate actions to prevent additional problems or harm to responders, the public, property, or the environment. In general, all responses present a degree of risk to the workers. During routine operations risk is minimized by establishing good work practices and using personnel protective equipment. Unpredictable events such as fire, chemical exposure, or physical injury may occur and must be anticipated. The plan must contain detailed information for managing these contingencies.

To accomplish this, the contingency plan must

1. Establish site emergency procedures
   a. List the names and emergency functions of on-site personnel responsible for emergency actions along with the special training required.
   b. Post the location of the nearest telephone (if none is at the site).
   c. Provide alternative means for emergency communications.
   d. Provide a list of emergency services organizations that may be needed. Names, telephone numbers, and locations must be posted. Arrangements for using emergency organizations should be made beforehand. Organizations that might be needed are

      • Fire and rescue agency
      • Police department
      • Health department
      • Explosive experts
      • Local hazardous material response units

- Emergency services offices
- Radiation experts

  e. Address and define procedures for the rapid evacuation of workers. Clear, audible warning signals should be established. Well-marked emergency exits must be located throughout the site, as well as internal and external communications plans developed.

  f. A complete list of emergency equipment should be attached to the safety plan. This list should include emergency equipment available on-site, as well as all available medical, rescue, transport, fire-fighting, and mitigative equipment available off-site.

2. Address emergency medical care.

   a. Determine the location of the nearest medical or emergency care facility and determine its capability to handle chemical exposure cases.

   b. Arrange for, in advance, treating, admitting, and transporting of injured or exposed workers.

   c. Post the location of medical or emergency care facilities, the required travel time, directions, and telephone number.

   d. Determine the location of a local physician's office, along with travel directions and hours of availability, and post the physician's telephone number if other medical care is not available.

   e. Determine the nearest ambulance service and post its telephone number.

   f. List the names of the responding organization's physicians, safety officers, or toxicologists and the telephone numbers of each. Also include the nearest poison control center, if applicable.

   g. Maintain accurate records on any exposure or potential exposure or injuries to site workers during an emergency (or routine operations).

3. Advise workers of their duties during an emergency. In particular, it is imperative that the site safety officers, standby rescue personnel, decontamination workers, and emergency medical technicians practice emergency procedures.

4. Incorporate into the plan adequate procedures for the decontamination of injured workers and for their transport to medical care facilities. Contamination of transport vehicles, medical care facilities, or medical personnel may occur and should be addressed in the plan. Whenever feasible, these procedures should be discussed with appropriate medical personnel in advance of operations.

5. Establish procedures in cooperation with local and state officials for evacuating residents who live near the site.

Implementation of the Site Safety Plan

The site safety plan, (standard operating safety procedure or a generic safety plan for emergency response) must be written to avoid misinterpretation, ambiguity, and mistakes that can result from verbal orders. The plan must be reviewed and approved by qualified personnel. Once the safety plan is implemented, it needs periodic examination and modification, if necessary, to reflect any changes in site work and conditions.

When there is more than one organization involved at the incident, the development of a safety plan should be a coordinated effort among the various agencies. Once the plan has been reviewed and approved by a qualified safety professional, lead personnel from each organization should sign the plan to document that they are in agreement with the provisions as well as to verify that their organization will follow it accordingly.

A safety and health officer must be appointed to ensure that the requirements of the safety plan are implemented. The safety officer has the authority to halt all operations if conditions become unsafe. In addition, the safety officer is responsible for instructing personnel on the provisions of the safety plan. Frequent safety meetings should be held to keep personnel informed about site hazards, changes in operating plans, modifications of safety requirements, and for any additional exchanges of information. All those on-site must comply with the provisions set forth in the safety plan.

Frequent audits by the incident manager or the safety officer should be made to determine compliance with the plan's requirements. Any deviations should be brought to the attention of the incident manager and any deficiencies corrected. Modifications in the plan should be reviewed and approved by appropriate personnel (2).

33.6 LEVELS OF PROTECTION

Response personnel must wear protective equipment when there is a probability of contact with hazardous substances that could affect their health. This includes contact with vapors, gases, or particulates that may be generated by site activities, and direct contact with skin-affecting substances. Full-facepiece respirators protect the lungs, gastrointestinal tract, and eyes against airborne toxicants. Chemical-resistant clothing protects the skin from contact with skin destructive and absorbable chemicals. Good personal hygiene limits or helps prevent ingestion of material.

Equipment to protect the body against contact with known or anticipated toxic chemicals has been divided into four categories according to the degree of protection afforded:

Level A. Should be worn when the highest level of respiratory, skin, and eye protection is needed.

Level B. Should be worn when the highest level of respiratory protection is needed, but a lesser degree of skin protection is needed.

Level C. Should be worn when a lesser level of respiratory protection is needed than Level B. Skin protection criteria are similar to Level B.

Level D. Should be worn only as a work uniform and not on any site with respiratory or skin hazards. It provides no protection against chemical hazards.

The Level of Protection selected should be based on the hazard and risk of exposure.

Hazard: Type and measured concentration of the chemical substance in the ambient atmosphere and its toxicity.

Risk: Potential for exposure to substances in air, splashes of liquids, or other direct contact with material due to work being done.

In situations where the type of chemical, concentration, and possibilities of contact are not known, the appropriate level of protection must be selected based on professional experience and judgment until the hazards can be better characterized.

Personnel protective equipment reduces the potential for contact with toxic substances. Additionally, safe work practices, decontamination, site entry protocols, and other safety procedures further ensure the health and safety of responders. Together, these provide an integrated approach for reducing harm to response personnel.

Levels of Protection

Level A Protection

1. Personnel protective equipment
   a. Pressure-demand, supplied-air respirator approved by the Mine Safety and Health Administration (MSHA) and National Institute for Occupational Safety and Health (NIOSH). Respirators may be:
      - Pressure-demand, self-contained breathing apparatus (SCBA), or
      - Pressure-demand, airline respirator (with an escape bottle for atmospheres with, or having the potential for, Immediately Dangerous to Life and Health (IDLH) contaminant concentrations).
   b. Fully encapsulating chemical-resistant suit
   c. Coveralls (optional)
   d. Long cotton underwear (optional)
   e. Chemical-resistant gloves (inner)
   f. Chemical-resistant boots with steel toe and shank. (Depending on suit construction, worn over or under suit boot)
   g. Hard hat (optional) [under suit (optional)]
   h. Disposable gloves and boot covers (optional) (worn over fully encapsulating suit)
   i. Cooling unit (optional)
   j. Two-way radio communications (inherently safe)
2. Criteria for selection. Meeting any of these criteria warrants use of Level A Protection:
   a. The chemical substance has been identified and requires the highest level of protection for skin, eyes, and the respiratory system.
   b. Substances with a high degree of hazard to the skin are suspected to be present, and skin contact is possible. Skin contact includes: splash, immersion, or contamination from atmospheric vapors, gases, or particulates.
   c. Operations must be conducted in confined, poorly ventilated areas until the absence of substances requiring Level A protection is determined.
   d. Direct readings on field Flame Ionization Detectors (FID) or Photoionization Detectors (PID) and similar instruments indicate high levels of unidentified vapors and gases in the air.
3. Guidance on selection
   a. Fully encapsulating suits are primarily designed to provide a gas- or

vapor-tight barrier between the wearer and atmospheric contaminants. Therefore, Level A is generally worn when high concentrations of airborne substances that could severely effect the skin are known or presumed to be present. Since Level A requires the use of a self-contained breathing apparatus, more protection is afforded to the eyes and respiratory system. Until air surveillance data are available to assist in the selection of the appropriate level of protection, the use of Level A may have to be based on indirect evidence of the potential for atmospheric contamination or other means of skin contact with substances having severe skin-affecting properties. Conditions that may require Level A protection include

- *Confined Spaces.* Enclosed, confined, or poorly ventilated areas are conducive to build-up of toxic vapors, gases, or particulates. An entry into an enclosed space does not automatically warrant Level A protection, but should serve as a cue to carefully consider the justification for a lower level of protection.

- *Suspected or Known Highly Toxic Substances.* Various substances that are highly toxic, especially through skin absorption, require Level A. Technical grade pesticides, concentrated phenolic compounds, poison "A" compounds, fuming corrosives, and a wide variety of organic solvents are of this type. Carcinogens, and infectious substances known or suspected to be involved may require Level A protection. Field instruments may not be available to detect or quantify air concentrations of these materials. Until these substances are identified and their concentrations determined, maximum protection is necessary.

- *Visible indicators.* Visible air emissions from leaking containers or railroad or truck tank cars, as well as smoke from chemical fires and others, indicate high potential for concentrations of substances that would be extreme respiratory or skin hazards.

- *Job functions.* Initial site entries are generally walk-throughs in which instruments and visual observations are used to make a preliminary evaluation of the hazards.

- In initial site entries, Level A should be worn when
  - There is a probability for exposure to high concentrations of vapors, gases, or particulates.
  - Substances are known or suspected of being extremely toxic directly to the skin or by being absorbed.

Subsequent entries are to conduct the many activities needed to reduce the environmental impact of the incident. Levels of Protection for later operations are based not only on data obtained from the initial and subsequent environmental monitoring, but also on the protective properties of suit material as well. The probability of contamination and ease of decontamination must also be considered. Examples of situations where Level A has been worn are:

- Excavating soil to sample buried drums suspected of containing high concentrations of dioxin.
- Entering a cloud of chlorine to repair a valve broken in a railroad accident.

- Handling and moving drums known to contain oleum.
- Responding to accidents involving cyanide, arsenic, and undiluted pesticides.

b. The fully encapsulating suit provides the highest degree of protection to skin, eyes, and respiratory system given that the suit material resists chemicals during the time the suit is worn. While Level A provides maximum protection, all suit materials may be rapidly permeated and degraded by certain chemicals. These limitations should be recognized when specifying the type of fully encapsulating suit. Whenever possible, the suit material should be matched with the substance it is used to protect against.

Level B Protection

1. Personnel protective equipment

   a. Pressure-demand, supplied-air respirator (MSHA/NIOSH approved). Respirators may be

   - Pressure-demand, self-contained breathing apparatus, or
   - Pressure-demand, airline respirator (with escape bottle for IDLH or potential for IDLH atmosphere)

   b. Chemical-resistant clothing (includes overalls and long-sleeved jacket or hooded, one- or two-piece chemical-splash suit or disposable chemical-resistant, one-piece suits)

   c. Long cotton underwear (optional), or

   d. Coveralls (optional)

   e. Chemical-resistant gloves (outer)

   f. Chemical-resistant gloves (inner)

   g. Chemical-resistant boots (outer) with steel toe and shank

   h. Chemical-resistant boot covers (outer, disposable) (optional)

   i. Hard hat (face shield optional)

   j. Two-way radio communications (inherently safe)

2. Criteria for selection. Meeting any one of these criteria warrants use of Level B protection:

   a. The type and atmospheric concentration of toxic substances has been identified and requires a high level of respiratory protection, but less skin protection than Level A. These would be:

   - Atmospheres with IDLH concentrations, but the substance or its concentration in air does not represent a severe skin hazard, or
   - Chemicals or concentrations involved do not meet the selection criteria permitting the use of air-purifying respirators.

   b. The atmosphere contains less than 19.5% oxygen.

   c. It is highly unlikely that the work being done will either

   i. generate high concentrations of vapors, gases or particulates, or

   ii. involve splashes of material that will affect the skin.

   d. Atmospheric concentrations of unidentified vapors or gases are indicated by direct readings on instruments such as the FID or PID or similar instruments, but vapors and gases are not suspected of containing concentrations of skin toxicants.

3. Guidance on selection

a. Level B does not afford the maximum skin (and eye) protection as does a fully encapsulating suit, since the chemical-resistant clothing is not considered gas-, vapor-, or particulate-tight. However, a good quality hooded, chemical-resistant, one-piece garment, with taped wrist, ankles, and hood does provides a reasonable degree of protection against splashes of liquids and lower concentrations of chemicals in the ambient air. At most abandoned, outdoor hazardous waste sites, ambient atmospheric gas or vapor levels usually do approach concentrations sufficiently high to warrant Level A protection. In all but a few circumstances, Level B should provide the protection needed for initial reconnaissance. Subsequent operations require a reevaluation of Level B protection based on the probability of being splashed by chemicals, their effect on the skin, or the presence of hard-to-detect air contaminants. The generation of highly toxic gases, vapors, or particulates due to the work being done must also be considered.

b. The chemical-resistant clothing required in Level B is available in a wide variety of styles, materials, construction detail, and permeability. One- or two-piece garments are available with or without hoods. Disposable suits with a variety of fabrics and design characteristics are also available. Taping joints between the gloves, boots, and suit, and between hood and respirator reduces the possibility for splash and vapor or gas penetration, but is not a gas-tight barrier. These factors and other selection criteria all affect the degree of protection afforded. Therefore, a specialist should select the most effective chemical-resistant clothing based on the known or anticipated hazards and job function. Level B equipment does provide a high level of protection to the respiratory tract. Generally, if a self-contained breathing apparatus is required, selecting chemicalresistant clothing (Level B) rather than a fully encapsulating suit (Level A) is based on the need for less protection against known or anticipated substances affecting the skin. Level B skin protection is selected by:

- Comparing the concentrations of known or identified substances in air with skin toxicity data.

- Determining the presence of substances that are destructive to or readily absorbed through the skin by liquid splashes, unexpectedly high levels of gases, vapor, or particulates, or by other means of direct contact.

- Assessing the effect of the substance (at its measured air concentrations or potential for splashing) on the small areas left unprotected by chemical-resistant clothing. A hooded garment, taped to the mask with boots and gloves taped to the suit, further reduces the area for potential skin exposure.

c. For initial site entry and reconnaissance at an open site, approaching whenever possible from upwind, Level B protection (with good quality hooded, chemical-resistant clothing) should protect response personnel, provided the conditions described in selecting Level A are known or judged to be absent.

Level C Protection

1. Personnel protective equipment
   a. Air-purifying respirator, full-face, canister-equipped (MSHA/NIOSH approved)
   b. Chemical-resistant clothing (includes: coveralls or hooded one-piece or two-piece chemical splash suit or chemical-resistant hood and apron; disposable chemical-resistant coveralls)
   c. Coveralls (optional)
   d. Long cotton underwear (optional)
   e. Gloves (outer), chemical-resistant
   f. Gloves (inner), chemical-resistant
   g. Boots (outer), chemical-resistant with steel toe and shank
   h. Chemical-resistant boot covers (outer, disposable) (Optional)
   i. Hard hat (face shield optional)
   j. Escape mask (optional)
   k. Two-way radio communications (inherently safe)

2. Criteria for selection. Meeting all of these criteria permits use of Level C protection:
   a. Oxygen concentrations are not less than 19.5% by volume.
   b. Measured air concentrations of identified substances will be reduced by the respirator below the substance's threshold limit value (TLV) and the concentration is within the service limit of the canister.
   c. Atmospheric contaminant concentrations do not exceed IDLH levels.
   d. Atmospheric contaminants, liquid splashes, or other direct contact will not adversely affect any body area left unprotected by chemical resistant clothing.
   e. Job functions do not require self-contained breathing apparatus.
   f. Direct readings are a few ppms above background on instruments such as the FID or PID.

3. Guidance on selection
   a. Level C protection is distinguished from Level B by the equipment used to protect the respiratory system, assuming the same type of chemical-resistant clothing is used. The main selection criterion for Level C is that atmospheric concentrations and other selection criteria permit wearing air-purifying respirators. The air-purifying device must be a full-face respirator (MSHA/NIOSH approved) equipped with a canister suspended from the chin or on a harness. Canisters must be able to remove the substances encountered. Half-masks or cheek cartridge equipped, full-face masks should be used only with the approval of a qualified health and safety professional. In addition, a full-face, air-purifying mask can be used only if:
      • Substance has adequate warning properties.
      • Individual passes a qualitative fit-test for the mask.

- Appropriate cartridge/canister is used, and its service limit concentration is not exceeded.
- Site operations are not likely to generate unknown compounds or excessive concentrations of already identified substances.

b. An air surveillance program is part of all response operations when atmospheric contamination is known or suspected. It is particularly important that the air be thoroughly monitored when personnel are wearing air-purifying respirators. Periodic surveillance using direct-reading instruments and air sampling is needed to detect any changes in air quality necessitating a higher level of respiratory protection.

c. Level C protection with a full-face, air-purifying respirator should be worn routinely in an atmosphere only after the type of air contaminant is identified, concentrations measured and the criteria for wearing air-purifying respirator met. A decision on continuous wearing of Level C must be made after assessing all safety considerations, including

- The presence of (or potential for) organic or inorganic vapors or gases against which a canister is ineffective or has a short service life.
- The known (or suspected) presence in air of substances with low TLVs or IDLH levels.
- The presence of particulates in air.
- The errors associated with both the instruments and monitoring procedures used.
- The presence of (or potential for) substances in air which do not elicit a response on the instrument used.
- The potential for higher concentrations in the ambient atmosphere or in the air adjacent to specific site operations.

d. The continuous use of air-purifying respirators (Level C) must be based on the identification of the substances contributing to the total vapor or gas concentration and the application of published criteria for the routine use of air-purifying devices. Unidentified ambient concentrations of organic vapors or gases in air approaching or exceeding a few ppm above background require, at a minimum, Level B protection.

Level D Protection

1. Personnel protective equipment
   a. Coveralls
   b. Gloves (optional)
   c. Leather or chemical-resistant boots/shoes with steel toe and shank
   d. Safety glasses or chemical splash goggles (optional)
   e. Hard hat (face shield optional)
   f. Escape mask (optional)
2. Criteria for selection. Meeting any of these criteria allows use of Level D protection:
   a. No contaminants are present.
   b. Work functions preclude splashes, immersion, or potential for unexpected inhalation of any chemicals.

Level D protection is primarily a work uniform. It can be worn only in areas where there is no possibility of contact with contamination.

Protection in Unknown Environments

In all incident response, selecting the appropriate personnel protective equipment is one of the first steps in reducing health effects from toxic substances. Until the toxic hazards at an incident can be identified and personnel safety measures commensurate with the hazards instituted, preliminary safety requirements must be based on experience, judgment, and professional knowledge.

Of primary concern in evaluating unknown situations are atmospheric hazards. Toxic concentrations (or potential concentrations) of vapors, gases, and particulates, low oxygen content, explosive potential, and the possibility of radiation exposure all represent immediate atmospheric hazards. In addition to making air measurements to determine these hazards, visual observation and review of existing data can help determine the potential risks from other materials.

Once immediate hazards other than toxic substances have been eliminated, the initial on-site survey and reconnaissance continues. Its purpose is to further characterize toxic hazards and, based on these findings, refine preliminary safety requirements. As data is obtained from the initial survey, the Level of Protection and other safety procedures are adjusted. Initial data also provide information upon which to base further monitoring and sampling requirements. No one method can determine a level of protection in all unknown environments. Each situation must be examined individually.

Additional Considerations for Selecting Levels of Protection

Other factors which should be considered in selecting the appropriate Level of Protection are:

1. *Heat and Physical Stress.* The use of protective clothing and respirators increases physical stress, in particular, heat stress on the wearer. Chemical protective clothing greatly reduces natural ventilation and diminishes the body's ability to regulate its temperature. Even in moderate ambient temperatures, the diminished capacity of the body to dissipate heat can result in one or more heat-related problems.

All chemical-protective garments can be a contributing factor to heat stress. Greater susceptibility to heat stress occurs when protective clothing requires the use of a tightly fitted hood against the respirator face piece, or when gloves or boots are taped to the suit. As more body area is covered, less cooling takes place, increasing the probability of heat stress. Whenever any chemical-protective clothing is worn, a heat stress recovery monitoring program must occur.

Wearing protective equipment also increases the risk of accidents. It is heavy and cumbersome, decreases dexterity and agility, interferes with vision, and is fatiguing to wear. These factors all increase physical stress and the potential for accidents. In particular, the necessity of selecting Level A protection should be balanced against the increased probability of heat stress and accidents. Level B and C protection somewhat reduces accident probability because the equipment is lighter and less cumbersome and vision problems are less serious.

2. *Air Surveillance.* A program must be established for routine, periodic air surveillance. Without an air surveillance program, any atmospheric changes could

go undetected and jeopardize response personnel. Surveillance can be accomplished with (a) various types of air pumps and filtering devices followed by analysis of the filtering media, (b) portable real-time monitoring instruments located strategically on-site, (c) personal dosimeters, and (d) periodic walk-through by personnel carrying direct-reading instruments.

3. *Decision-Logic for Selecting Protective Clothing.* No adequate criteria, similar to the respiratory protection decision-logic are available for selecting protective clothing. A concentration of a known substance in the air approaching a TLV or permissible exposure requires an encapsulating suit. A hooded, high quality, chemical-resistant suit may provide adequate protection. The selection of Level A over Level B is a judgment that should be made by a qualified individual considering the hazards and risk.

Hazards: The physical form of the potential contaminant must be considered. Airborne substances are more likely to contact personnel wearing nonencapsulating suits, which are not considered gas- or vapor-tight. Liquids contacting the skin are generally considered more hazardous than contact with vapors, gases, and particulates. Effect of the contaminant on skin:

- Highly hazardous substances are those that are easily absorbed through the skin, causing systemic effects, or those that cause severe skin destruction.

- Less hazardous substances are those that are not easily absorbed through the skin (causing systemic effects) or that do not cause severe skin destruction

Risk: Concentration of the contaminant, the higher the concentration, the higher the probability of injury.

Work function: Site work activities dictate the probability of direct and indirect skin contact.

Instability of the situation: A higher level of protection should be considered when there is a probability of a release involving vapor or gases, splashes, or immersion in liquids, or loss of container integrity.

4. *Atmospheric Conditions.* Atmospheric conditions such as stability, temperature, wind direction and wind velocity, as well as barometric pressure determine the behavior of contaminants in air or the potential for volatile material being released into the air. These parameters should be considered when determining the need for and level of protection required.

5. *Work in the Exclusion Zone.* For operations in the Exclusion Zone (area of potential contamination), different levels of protection may be selected and various types of chemical-resistant clothing worn. The selection would be based on measured air concentrations, the job function, the potential for skin contact or inhalation of the materials present, and ability to decontaminate the protective equipment used.

6. *Escape Masks.* Carrying an escape, self-contained breathing apparatus of at least five minute duration is optional while wearing Level C or Level D protection. For initial site entry, a specialist should determine, on a case-by-case basis, whether they should be carried or be strategically located in areas that have higher possibilities for harmful exposure.

Vapor or Gas Concentrations as Indicated by Direct-Reading Instruments
Instruments such as the FID and PID can be used to detect the presence of many organic vapors or gases either as single compounds or mixtures. Dial readings are frequently referred to, especially with unidentified substances, as total vapor and gas concentrations (in ppm). More correctly, they are deflections of the needle on the dial indicating an instrument response and do not directly relate to the total concentration in the air. They must not be used as the sole criteria for selecting levels of protection (2).

33.7 SUMMARY

1. In response to the need to protect human health and the environment from improper disposal of these hazardous wastes, anyone working in a hazardous or unsafe environment must receive health and safety training.

2. The U.S. Occupational Safety and Health Administration (OSHA) has regulations governing employee health and safety at hazardous waste operations and during emergency responses to hazardous substance releases.

3. Toxic (including radioactive material and biological agents) or chemically active substances present a special concern because they can be inhaled, ingested, or be absorbed through or destructive to the skin. Ingested or inhaled, the substances may cause no apparent illness or they can be fatal. Other substances, however, may damage the skin or be absorbed through it, leading to systemic toxic effects.

4. To reduce the risks to personnel responding to hazardous substance incidents, an effective health and safety program must be developed and followed.

5. The purpose of the site safety plan is to establish policies and procedures for protecting the health and safety of response personnel during all operations conducted at an incident. It contains information about the known orsuspected hazards, routine and special safety procedures that must be followed, and other instructions for safeguarding thehealth of the responders.

6. Response personnel must wear protective equipment when there is a probability of contact with hazardous substances that could affect their health. This includes vapors, gases, or particulates that may be generated by site activities, and direct contact with skin-affecting substances. Full facepiece respirators protect lungs, gastrointestinal tract, and eyes against airborne toxicants. Chemical-resistant clothing protects the skin from contact with skin-destructive and absorbable chemicals. Good personal hygiene limits or helps prevent ingestion of material.

REFERENCES

1. *Hazardous Waste Operations and Emergency Response*, 29 CFR Part 1910, Final Rule Section 1910.120 (2) Applications.
2. U.S. EFA, Office of Emergency and Remedial Response Division, *Standard Operating Safety Guides*, July 1988.
3. 29 CFR Part 1900 to 1910 (SS 1901.1 to 1910.999) Revised as of July 1, 1991, Section 1910.116.
4. *Respiratory Protection*, U.S. Department of Labor, Occupational Safety and Health Administration, OSHA 3079.

34

ACCIDENT AND EMERGENCY MANAGEMENT

34.1 INTRODUCTION

Accidents are a fact of life, whether they be a careless mishap at home, an unavoidable collision on the freeway, or a miscalculation at a chemical plant. Even in prehistoric times, long before the advent of technology, a club-wielding caveman might have swung at his prey and inadvertently toppled his friend in what can only be classified as an "accident." As Man progressed, so did the severity of his misfortunes. The "Modern Era" has brought about assembly lines, chemical manufacturers, nuclear power plants, and so on, all carrying the capability of disaster. To keep pace with the changing times, safety precautions must constantly be upgraded. It is no longer sufficient, as with the caveman, to shout the warning, "Watch out with that thing!" Today's problems require more elaborate systems of warnings and controls to minimize the changes of serious accidents.

Industrial accidents occur in many ways—a chemical spill, an explosion, a nuclear plant out of control, and so on. There are often problems in transport, with trucks overturning, trains derailing, or ships capsizing. There are "acts of God," such as earthquakes and storms. The one common thread through all of these situations is that they are rarely expected and frequently mismanaged.

Early Accidents (1)
Accidents have occurred since the birth of civilization, and were just as damaging then as they are today. Anyone who crosses a street, rides in a car, or swims in a pool runs the risk of injury through carelessness, poor judgment, ignorance, or other circumstances. This has not changed throughout history. In the following pages, a number of accidents and disasters which took place before the advances of modern technology will be examined.

Catastrophic explosions have been reported as early as 1769, when one-sixth of the city of Frescia, Italy was destroyed by the explosion of 100 tons of gunpowder stored in the state arsenal. More than 3000 people were killed in this, the second deadliest explosion in history.

The worst explosion in history occurred in Rhodes, Greece, in 1856. A church on this island, which had gunpowder stored in its vaults, was struck by lightning. The resulting blast killed an estimated 4000 people. This remains the highest death toll for a single explosion (2).

One of the most legendary disasters occurred in Chicago, Illinois, in October 1871. The "Great Chicago Fire," as it is now known, is alleged to have started in a barn owned by Patrick O'Leary, when one of his cows overturned a lantern. Whether or not this is true, the O'Leary house escaped unharmed because it was upwind of the blaze, but it left the barn destroyed along with 2124 acres of Chicago real estate.

Catastrophic accidents have occurred on the sea as well as on land. In 1947 an unusual incident involved both. The French freighter *Grandcamp* arrived at Texas City, Texas, to be loaded with 1400 tons of ammonium nitrate fertilizer. Sometime that night, a fire broke out in the hold of the *Grandcamp*. The ship's crew made only limited attempts to fight the flames, apparently fearing water would damage the rest of the cargo. Because the *Grandcamp* was docked only 700 ft from a Monsanto chemical plant which produced styrene, a highly combustible ingredient in synthetic rubber, the *Grandcamp* was ordered towed from the harbor.

As tugboats prepared to hook up their lines, the *Grandcamp* exploded in a flash of fire and steel fragments. The blast rattled windows 150 miles away, registered on a seismograph in Denver, and killed many people standing on the dock. The Monsanto chemical plant exploded minutes later, killing many survivors of the first blast, shattering most of the Texas City business district, and setting fires throughout the rest of the city. As the fires burned out of control, the freighter *High Flyer*, also loaded with nitrates, exploded in the harbor. This third explosion proved too much for the people of Texas City, who had responded so efficiently to the initial two blasts. Hundreds of people were forced to leave the city, letting the fire burn itself out. The series of explosions had killed 468 people and seriously injured 1000 others. The final death toll may have been as high as 1000, because the dock area contained a large population of migrant workers without permanent address or known relatives. This disaster was probably caused by careless smoking aboard the *Grandcamp* (2,3).

Recent Major Accidents (1)

The advances of modern technology have brought about new problems. Perhaps the most serious of these is the threat of a nuclear power plant accident, known as a *meltdown*, as discussed in Chapter 19. In this section, several of this era's most infamous (non-nuclear) accidents will be examined.

An explosion at the Nypro Ltd. caprolactum factory at Flixborough, England, on June 1, 1974, was one of the most serious in the history of the chemical industry and the most serious that has occurred in the history of the United Kingdom. Of those working on the site at the time, 28 were killed and 36 others injured. Outside of the plant, 53 people were reported injured, while 1821 houses and 167 shops suffered damage. The estimated cost of the damage was well over $100 million.

The worst disaster in the recent history of the chemical industry occurred in Bhopal, in central India, on December 3, 1984. A leak of methyl isocyanate from a chemical plant, where it was used as an intermediate in the manufacture of a pesticide, spread into the adjacent city and caused the poisoning death of over 2500 people and the injuring of approximately 20,000 others. The owner of the plant, Union Carbide Corporation, reported that the accident was "the result of a unique combination of unusual events," although there has also been talk of possible sabotage. No data have been presented to support this latter claim. The cause of the accident was attributed to faults in the design of the plant's safety system — which was the responsibility of Union Carbide.

An accident occurred recently which, although it did not involve the loss of human life, can still be considered a disaster. On January 2, 1988, a 48-ft-high fuel tank at the Ashland Oil terminal in Pennsylvania ruptured. Nearly 3.9 million gallons of No. 2 distilled fuel poured out. The force caused the tank to jump backward 100 ft and sent a wave 35 ft high crashing into another tank, 100 ft away. Much of the spilled fuel was trapped by a containment dike; however, 600,000 gallons escaped into the Monongahela River at Floreffe, about 25 miles upstream from Pittsburgh. Soon after the spill, a rumor began circulating that there was a possible gasoline leak as well. This raised concerns about a fire, leading to the evacuation of 250 homes. The fuel that had spilled into the frigid water began to emulsify and sink. The extremely cold weather caused ice to form on the river. It is nearly impossible to recover oil that sinks below the skirts of the recovery booms or that becomes trapped in the ice. Various methods were used to remove the oil from the river. Chemists developed a method which mixed the contaminated water with powdered carbon and bentonite, which gives the slurry higher absorbency. The mixture is then pumped to a treatment plant where other chemicals are added to balance acidity and make the oil coagulate in a settling tank. This treatment is not new, but the chemists had to come up with the right combinations of chemicals to handle the oil. At one point, the Environmental Protection Agency (EPA) allowed the use of a substance called Elastol for the first time. Elastol congeals spilled hydrocarbons into a mass that can be easily recovered.

Advances in Safety Features (4)

Today's sophisticated equipment and technologies require equally sophisticated means of accident prevention. Unfortunately, the existing methods of detection and prevention are often assumed to be adequate until proven otherwise. This latter means of determining a technology's effectiveness can sometimes be costly and can often lead to loss of life. Chemical manufacturers and power plants are businesses, and thus are not as likely to "unnecessarily" update their present controls.

Prior to the advent of technology, there was still a need for safety features and warnings; yet these did not exist. Many accidents occurred because of a lack of knowledge of the system, process, or substance being dealt with. Many of the pioneers of modern science were sent to an early grave by their experiments. Karl Wilhelm Scheele, a Swedish chemist who discovered many chemical elements and compounds, often sniffed or tasted his finds. He eventually died of mercury poisoning. Likewise, Madame Curie died from leukemia contracted from overexposure to radioactive elements. Had either of these brilliant scientists any idea of the danger of their work, their methods would certainly have changed significantly. In those days, a safety precaution was often devised by trial and error; if inhaling a certain gas was found to make someone sick, the prescribed precaution was not to smell it. Today, the physical properties of most known compounds are readily found in handbooks, so that proper care can be exercised when working with these chemicals. Labs are equipped with exhaust hoods and fans to minimize a buildup of gases; in addition, safety glasses and eyewash stations are required, as are gloves and smocks.

Natural disasters are now often accurately predicted, buying precious time in which warnings can be made and possible evacuation plans implemented. Radar equipment commonly tracks storms, and seismographs detect slight rumblings in the earth which can provide early warning of potential earthquakes. Volcanic eruptions can be predicted by using seismic event counters and aerial scanning of infrared anomalies. Where natural disasters would often occur unexpectedly in the past, their occurrences today are more predictable. This allows more time for preparation and less likelihood of loss of life.

The use of computers and modern instrumentation has greatly enhanced plant safety. System overloads, uncontrollable reactions, and unusual changes in heat or pressure can be detected, with the information being relayed to a computer. The computer can then shut down the system, or take the steps necessary to minimize the danger. Industry has come a long way from sniffing and tasting its way to safety.

34.2 SUPERFUND AMENDMENTS AND REAUTHORIZATION ACT OF 1986 (5)

The Superfund Amendments and Reauthorization Act (SARA) of 1986 renewed the national commitment to correcting problems arising from previous mismanagement of hazardous wastes. While SARA was similar in many respects to the original law, it also contained new approaches to the program's operation. The 1986 Superfund legislation did the following (6):

1. It reauthorized the program for five more years and increased the size of the cleanup fund from $1.6 to 8.5 billion.
2. It set specific cleanup goals and standards, and stressed the achievement of permanent remedies.
3. It expanded the involvement of states and citizens in decision-making.
4. It provided for new enforcement authorities and responsibilities.
5. It increased the focus on human health problems caused by hazardous waste sites.

The law is more specific than the original statute with regard to such things as remedies to be used at Superfund sites, public participation, and accomplishment of cleanup activities. The most important part of SARA with respect to public participation is Title III. Title III addresses the important issues regarding community awareness and participation in the event of a chemical release and is an important part of SARA which addresses hazardous materials releases and is subtitled the Emergency Planning and Community Right-to-Know Act of 1986. Title III establishes requirements for emergency planning, hazardous emissions reporting, emergency notification, and "community right to know." The objectives of Title III are to improve local chemical emergency response capabilities, primarily through improved emergency planning and notification, and to provide citizens and local governments with access to information about chemicals in their localities. Title III has four major sections which aid in the development of contingency plans. They are as follows:

1. Emergency Planning (Sections 301–303)
2. Emergency Notification (Section 304)
3. Community Right-to-Know Reporting Requirements (Sections 311 and 312)
4. Toxic Chemicals Release Reporting — Emissions Inventory (Section 313).

Title III has also developed time frames for the implementation of the Emergency Planning and Community Right to Know Act of 1986. Details on these are provided in the sections that follow.

34.3   THE NEED FOR EMERGENCY RESPONSE PLANNING

Emergencies have occurred in the past and will continue to occur in the future. A few of the many common sense reasons to plan ahead for emergencies are as follows (7):

1. Emergencies will happen; it is only a question of time.
2. When emergencies occur, the minimization of loss and the protection of people, property, and the environment can be achieved through the proper implementation of an appropriate emergency response plan.
3. Minimizing the losses caused by an emergency requires planned procedures, understood responsibility, designated authority, accepted accountability, and trained, experienced people. A fully implemented plan can do this.
4. If an emergency occurs, it may be too late to plan. Lack of preplanning can turn an emergency into a disaster.

A particularly timely reason to plan ahead is to ease the "chemophobia" or fear of chemicals, which is so prevalent in society today. So much of the recent attention to emergency planning and newly promulgated laws are a reaction to the tragedy at Bhopal. Either a total lack of information or misinformation is the probable cause of "chemophobia." Fire is hazardous, and yet it is used regularly at home. Most adults have understood the hazard associated with fire since the time of the caveman. By the same token, hazardous chemicals, necessary and useful in our technological society, are not something to be afraid of. Chemicals need to be carefully used and their hazards understood by the general public. An emergency plan which is well designed, understood by the individuals responsible for action, and understood by the public can ease the concern over emergencies and reduce chemophobia. People will react during an emergency; how they react can be somewhat controlled through education. The likely behavior during an emergency when ignorance is pervasive is panic.

An emergency plan can minimize loss by helping to ensure the proper response in an emergency. "Accidents become crises when subsequent events and the actions of people and organizations with a stake in the outcome combine in unpredictable ways to threaten the social structures involved" (8). The wrong response can turn an accident into a disaster as easily as no response. One example is a chemical fire which is doused with water, causing the fire to emit toxic fumes; the same fire would be better left to burn itself out. Another example is the evacuation of people from a building into the path of a toxic vapor cloud; they might well be safer staying indoors with closed windows. Still another example is the members of a rescue team becoming victims because they were not wearing proper breathing protection. The proper response to an emergency requires an understanding of the hazards. A plan can provide the right people with the information needed to respond properly during an emergency.

Other than the above-mentioned common-sense reasons to plan, there are legal reasons. Recognizing the need for better preparation to deal with chemical emergencies, Congress enacted the Superfund Amendments and Reauthorization Act (SARA) of 1986; this act was discussed in detail in the last section. One part of SARA is a free-standing act called Title III, the Emergency Planning and Community Right-to-Know Act of 1986. This act requires federal, state, and local governments to work together with industry in developing emergency plans and "community right-to-know" reporting on hazardous chemicals. These new requirements build on the EPA's Chemical Emergency Preparedness Program and numerous state and local programs which are aimed at helping communities deal with potential chemical emergencies (9).

Most larger industries have long had emergency plans designed for on-site personnel. The protection of people, property, and thus profits has made emergency plans and prevention methods common in industry. On-site emergency plans are often a requirement by insurance companies. Expansion of these existing industry plans to include all significant hazards and all people in the community is a way to minimize the effort required for emergency planning.

34.4   THE PLANNING COMMITTEE

Emergency planning should grow out of a team process coordinated by a team leader. The team may be the best vehicle for gathering people with various kinds of expertise into the planning process, thus producing a more accurate and complete plan. The team approach also encourages a planning process that will reflect a consensus of the entire community. Some individual communities and/or areas that include several communities had already formed advisory councils before the SARA requirements. These councils can serve as an excellent resource for the planning team (10).

When selecting the members of a team that will bear overall responsibility for emergency planning, the following considerations are important:

1. The members of the group must have the ability, commitment, authority, and resources to get the job done.
2. The group must possess, or have ready access to, a wide range of expertise relating to the community, its industrial facilities, transportation systems, and the mechanics of emergency response and response planning.
3. The members of the group must agree on their purpose and be able to work cooperatively with one another.
4. The group must be representative of all elements of the community, with a substantial interest in reducing the risks posed by emergencies.

While many individuals have an interest in reducing the risks posed by hazards, their differing economic, political, and social perspectives may cause them to favor different means of promoting safety. For example, people who live near an industrial facility with hazardous materials are likely to be greatly concerned about avoiding any threat to their lives. They are likely to be less concerned about the costs of developing accident prevention and response measures than some of the other team members. Others in the community, those representing industry or the budgeting group, for example, are likely to be more sensitive to the costs involved. They may be more anxious to avoid expenditures for unnecessary, elaborate prevention and response measures. Also, industry facility managers, although concerned with reducing risks posed by hazards, may be reluctant, for proprietary reasons, to disclose materials and process information beyond what is required by law. These differences can be balanced by a well-coordinated team, which is responsive to the needs of its community.

Among the agencies and organizations with emergency response responsibility, there may be differing views about the role they should play in case of an incident. The local fire department, an emergency management agency, and public health agency are all likely to have some responsibilities during an emergency. However, each of these organizations might envision a very different set of actions for their respective agencies at the emergency site. The plan will serve to detail the actions of each response group during an emergency.

In organizing the community to address the problems associated with emergency planning, it is important to bear in mind that all affected parties have a legitimate interest in the choices among planning alternatives. Therefore, strong efforts should be made to ensure that all groups with an interest in the plan are included in the planning process. The need for control of the committee during the planning process, as well as for control during the plan implementation, is amplified by the number of different groups involved in the community. Each of these groups has a right to participate in the planning, and a well-structured, organized planning committee should serve all the community groups.

By law, the planning committee should include the following (11):

1. Elected and state officials.
2. Civil defense personnel.
3. First-aid personnel.
4. Local environmental personnel.
5. Transportation personnel.
6. Owners and operators of facilities subject to the SARA.
7. Law enforcement personnel.
8. Firefighting personnel.
9. Public health personnel.
10. Hospital personnel.
11. Broadcast and print media.
12. Community groups.

There are other individuals who could also serve the community well and should be a part of the committee, such as (12)

1. Technical professionals.
2. City planners.
3. Academic and university researchers.
4. Local volunteer help organizations.

The local government has a great share of the responsibility for emergency response within its community. The official who has the power to order evacuation, fund fire and emergency units, and educate the public is a key person to emergency planning and the response effort. For example, the entire plan will fail if a timely evacuation is necessary and not ordered on time. Although politics should be disassociated from technical decisions, the emergency planning is a place where such linkage is inevitable. Distasteful options which require political courage are often necessary (e.g., having to evacuate a section of town where there is some doubt about the necessity of evacuation), but the consequence of not evacuating might be deadly. A public official can build public support for future candidacy by using the issue of chemical safety as a bandwagon, but mistakes in handling emergencies are measured by a strong instrument—the election—where a failed emergency plan can be fatal to a political career. Politics is a social feedback device which, when used properly, can aid the government leader in making correct decisions. A political career can also be destroyed by an error in reading the social feedback. Developing an effective plan can save the elected official hours of media criticism after a crisis, because the plan can provide the details of events organized by someone on the team as they occur. Because of the power an elected official has locally, that person is likely to take the leadership place on the committee.

The management or control of the committee during its planning, and especially during implementation of the plan, is essential. As discussed earlier, the committee will be made up of different individuals with different priorities and the emergency plan will be generated by them. The different groups will have their own legitimate interests. Each interest will have to be weighed against its value to the plan. To have a respect for the interests of each of the individuals, as well as a respect for their contribution, is an essential attribute of the committee leader. The committee leader is likely to be chosen for several reasons, but among these should be the following:

1. The degree of respect held for the person by groups and individuals with an interest in the emergency plan.
2. The availability of time and resources this person will have to serve the committee.
3. The person's history of working relationships with concerned community agencies and organizations.
4. The person's management skills and communications skills.
5. The person's existing responsibility and background experience related to emergency planning, prevention, and response.

Personal considerations, as well as institutional ones, should be weighed when selecting a committee leader. If a person being evaluated for the position of committee leader has all the right resources to address the emergency planning and implementation, but is unable to interact with local officials, then someone else may be a better choice. Since the committee leader must coordinate this large group of people with different priorities and expertise, the choice of the leader is critical to the success of the committee (13).

34.5 HAZARDS SURVEY

In order to characterize the types and extents of potential disasters, a survey of hazards or foreseeable threats in the community must be performed and evaluated. Without information on the types of events which are possible or on the potentially affected areas, an appropriate plan cannot be developed. A plan for a city with a river, for example, may not be applicable to a desert city on a seismic fault. An inventory of the community assets, hazard sources, and risks must be done prior to the actual plan writing.

Duplication can be an enemy of cost-efficiency. Some emergency plans may already exist in the community, and these should be used wherever possible. Community groups that may have developed such plans are

1. Civil Defense.
2. Fire Department.
3. Red Cross.
4. Public Health.
5. Local Industry Council.

Such existing plans should be studied, and their applicability to the proposed community plan should be evaluated.

The resources of the community should be listed and then compared to the needs of the plan. Local government departments, such as transportation, water, power, and sewer, may have such resources. Some examples of these are

1. Trucks.
2. Equipment (e.g., backhoes, flatbeds).
3. Laboratory services (e.g., water department).
4. Fire vehicles.
5. Police vehicles.
6. Emergency suits.
7. Breathing apparatus.
8. Gas masks.
9. Number of trained emergency people.
10. Number of volunteer personnel (e.g., Red Cross).
11. Buses or cars.
12. Communication equipment (e.g., hand radios).
13. Local TV and radio stations.
14. Ambulances.
15. Trained medical technicians and first-aid personnel.
16. Stocks of medicines.

17. Burn treatment equipment.
18. Fallout shelters.

The potential sources of hazards should be listed for risk assessment. The SARA requires certain industries to provide information to the planning committee. The committee should gather the information about small as well as large industries in order to evaluate the significant risks. The information regarding chemicals required by the SARA to be provided includes:

1. The chemical name.
2. The quantity stored over a period of time.
3. The type of chemical hazard (e.g., toxic, flammable, ignitable, corrosive).
4. Chemical properties and characteristics (e.g., liquid at certain temperatures, gas at certain pressures, reacts violently with water).
5. Storage description and storage location on the site.
6. Safeguards or prevention measures associated with the hazardous chemical storage or handling design, such as dikes, isolation of incompatible substances, and fire-resistance equipment.
7. Control features for accident prevention such as temperature and pressure controllers and fail-safe design devices, if included in the handling design.
8. Recycle control loops intended for accident prevention.
9. Emergency shutdown features.

The planning committee should designate hazard sources on a community map. This information probably already exists and can be obtained from local groups such as the transportation department, environmental agency, city planning department, community groups, and industry. Some of the data to locate on the community map are

1. Industry and other chemical locations.
2. Wastewater and water treatment plants which have chlorine stored.
3. Potable and surface water.
4. Drainage and runoff.
5. Population location and density in different areas.
6. Transportation routes for children.
7. Commuter routes.
8. Truck transport roads.
9. Railroad lines, yards, and crossings.
10. Major highways, noting merges and downhill curves.
11. Hospitals and nursing homes.
12. Fallout shelters.

The potential for natural disaster based on the history and knowledge of the region and earth structure should be indicated in the plan. Items such as seismic fault zones, flood plains, hurricane potentials, and winter storm potentials should be noted.

The risk inventory or risk evaluation is the next part of the hazard survey. It is not practical to expect the plan to cover every potential accident. The hazards should be evaluated, and then the plan should be focused on the most significant ones. This risk assessment stage requires the technical expertise of many people to compare the pieces of data and determine the relevance of each.

In performing the risk evaluation, much data have to be evaluated. Among the important factors to be considered are the following:

1. The routes of transport of hazardous substances should be reviewed to determine where a release could occur.

2. Industry sites are not the only sources of hazards. The proximity of hazards to people and other sensitive environmental receptors should be examined.

3. The toxicology of different exposure levels should be reviewed.

Once the significant risks are listed, the hazard survey is complete and the plan can then be developed.

34.6 PLAN FOR EMERGENCIES

Successful emergency planning begins with a thorough understanding of the event or potential disaster being planned for. The impacts on public health and the environment must also be estimated. Some of the types of emergencies that should be included in the plan are (14)

1. Earthquakes.
2. Explosions.
3. Fires.
4. Floods.
5. Hazardous chemical leaks — gas or liquid.
6. Power or utility failures.
7. Radiation incidents.
8. Tornadoes or hurricanes.
9. Transportation accidents.

In order to estimate the impact on the public or the environment, the affected area or emergency zone must be studied. A hazardous gas leak, fire, or explosion may cause a toxic cloud to spread over a great distance. An estimate of the minimum affected area and thus the area to be evacuated should be performed, based on an atmospheric dispersion model. There are various models which can be used. While the more difficult models produce more realistic results, the models which are simpler and faster to use may provide adequate data for planning purposes (15). A more thorough discussion of atmospheric dispersion is presented in Chapter 6 of this book.

In formulating the plan, some general assumptions may be made:

1. Organizations do a good job when they have specific assignments.
2. The various resources will need coordination.

3. Most resources that are necessary are likely to be already available in the community (in plants or city departments).
4. People react more rationally when they have been apprised of the situation.
5. Coordination is basically a social process, not a legal one.
6. Disorganization and reorganization are common in a large group.
7. Flexibility and adaptability are basic requirements for a coordinated team.

The objective of the plan should be to prepare a procedure to make maximum use of the combined resources of the community in order to accomplish the following:

1. Safeguard people during emergencies.
2. Minimize damage to property and the environment.
3. Initially contain and ultimately bring the incident under control.
4. Effect the rescue and treatment of casualties.
5. Provide authoritative information to the news media (who will transmit it to the public).
6. Secure the safe rehabilitation of the affected area.
7. Preserve relevant records and equipment for the subsequent inquiry into the cause and circumstances.

The key components that should be contained in the emergency action plan include (9)

1. Emergency actions other than evacuation.
2. Escape procedures when necessary.
3. Escape routes clearly marked on a site map, and perhaps also on the roads.
4. A method for accounting for people after evacuation.
5. Rescue and medical duties.
6. Reporting emergencies to the proper regulatory agencies.
7. Notification of the public by an alarm system.
8. Contact and coordination person responsibilities.

34.7 TRAINING OF PERSONNEL

The education of the public regarding the real hazards in their community and how to respond to an emergency is critical to the public support of the emergency plan. Public opinion surveys show, however, that Americans are not yet prepared to deal either with the information provided by SARA Title III or with the reality of hazards in their community. Most people understand that hazards exist but do not realize that the potential hazards are all around them in their own communities. The common perception is that hazards exist elsewhere, as do the resulting emergencies (16). The education of the populace about the true hazards associated with routine discharges from plants in the neighborhood and preparing that populace for emergencies is a real challenge to the community committee. People must be taught how to react to an emergency. This includes how to recognize and report an incident, how to react to alarms, and what other action to take. A possible result of SARA Title III may be, initially, a fear on the part of the public of industrial discharges. News stories, based on hazardous chemical inventory, accidental releases, or annual emissions reports of questionable accuracy or taken out of context, can be misleading. It is hoped that this can be put into perspective through training programs.

The personnel at an industrial plant, particularly the operators, are trained in the operation of the plant. These people are critical to proper emergency response. They must be taught to recognize abnormalities in operations and to report them immediately. Plant operators should also be taught how to respond to various types of accidents. Emergency squads at plants can also be trained to contain the emergency until outside help arrives, or, if possible, to terminate the emergency. Shutdown and evacuation procedures are especially important when training plant personnel.

Training is important for the emergency teams to ensure that their roles are clearly understood and that accidents can be reacted to safely and properly without delay. The emergency teams include the police, firefighters, medical people, and volunteers who will be required to take action during an emergency. These people must be knowledgeable about the potential hazards. For example, specific antidotes for different types of medical problems must be known by medical personnel. The whole emergency team must also be taught the use of personal protective equipment.

Local government officials also need training. Because these officials have the power to order an evacuation, they must be aware of under what circumstances such action is necessary. The timing of an evacuation is critical; this must be understood by these people prior to the emergency itself. Local officials also control the use of city equipment and therefore must be knowledgeable as to what is needed for an appropriate response to a given emergency.

Media personnel must also be involved in the training program. The public gets information through the media; it is important that the information the public receives is accurate. If the information is incorrect or distorted, an emergency can easily cause panic. For this reason, it is important that the media people be somewhat knowledgeable about the potential hazards and the details of emergency responses.

Training for emergencies should be routinely done under the following circumstances:

1. When a new member is added to the group.
2. When someone has a new responsibility within the community.
3. When new equipment or materials are to be used in emergency response.
4. When emergency procedures are revised.
5. When a practice drill shows inadequacies in performance of duties.
6. At least once annually.

Any training program should address the following questions:

1. How are potential hazards recognized? (This can be done by periodic review of hazards and accident prevention measures.)
2. What are the necessary precautions to be taken when responding to an emergency (e.g., personal protective equipment)?

3. Where are the evacuation routes?

4. To whom should a hazard be reported?

5. What actions should be taken in order to respond properly to special alarms or signals?

It is important for emergency procedures to be performed as planned. This requires training on a regular basis so that people understand and remember how to react. The best plan on paper is likely to fail if the persons involved are reading it for the first time as the emergency is occurring. People must be trained before an emergency happens.

34.8   NOTIFICATION FOR PUBLIC AND REGULATORY OFFICIALS

Notifying the public of an emergency is a task which must be accomplished with caution. People will react in different ways. Many will simply not know what to do, others will not take the warning seriously, and still others will panic. Proper training in each community can help minimize any panic and condition the public to make the right response.

Methods of communicating the emergency will differ for each community, depending upon its size and resources. Some techniques for notifying the public are

1. The sounding of fire department alarms in different ways to indicate certain kinds of emergencies.

2. Chain phone calls (this method usually works well in small towns).

3. Police cars or volunteer teams with loudspeakers.

Once the emergency is communicated, an appropriate response by the public must be evoked. For this response to occur, an accepted plan that people know and understand must be put into effect. Panic may occur, and this plan should be flexible enough to include the appropriate countermeasures. An emergency can quickly become a total disaster if there is a panicked public.

The reporting of information to the emergency coordinator must be carefully screened. A suspected "crank call" should be checked out before an alarm is sounded. An obvious risk in not taking immediate action is that the plan will not be implemented in time. Therefore, if a crank call cannot be verified as such, a response must begin. In this case, local police should be dispatched to the scene of a reputed emergency quickly to verify the report firsthand.

The media (e.g., news, radio, and television) can be a major resource for communication. One job of the emergency coordinator is to prepare the information that is to be reported to the media. The emergency plan should include a procedure to pass along information to the media promptly and accurately.

Besides notifying the response team and the community about what procedures to follow in an emergency, there are also requirements to report certain types of emergencies to government agencies. For example, state and federal laws require the reporting of hazardous releases and nuclear power plant problems. There are also more specific requirements under SARA Title III for reporting chemical releases. Facilities where a listed hazardous substance is produced, stored, or used must immediately notify the local emergency planning committee and the State

Emergency Response Commission if there is a release of a substance specifically listed in the SARA. These substances include (a) approximately 400 extremely hazardous chemicals on the list prepared by the Chemical Emergency Preparedness Program and (b) chemicals subject to reportable quantities requirements of the original Superfund (9). The initial notification can be made by telephone, by radio, or in person. Emergency notification requirements involving transportation incidents can be satisfied by dialing 911 or calling the operator. The emergency planning committee should provide a means to get information on transportation accidents reported quickly to the coordinator.

The SARA requires that the notification of an industrial emergency include

1. The name of the chemical released.
2. Whether it is known to be acutely toxic.
3. An estimate of the quantity of the chemical released into the environment.
4. The time and duration of the release.
5. Where the chemical was released (e.g., air, water, land).
6. Known health risks and necessary medical attention.
7. Proper precautions, such as evacuation.
8. The name and telephone number of the contact person at the plant or facility where the release occurred.

A written follow-up emergency notice is required as soon as is practical after the release. The notice should include:

1. An update of the initial information.
2. Additional information on response actions already taken; known or anticipated health risks; and advice on medical attention.

As of October 1986, the reporting and written notice are required by law.

34.9 PLAN IMPLEMENTATION

Once an emergency plan has been developed, its successful implementation can be ensured only through constant review and revision. A number of ongoing steps which should be taken are

1. Inventory checks on a routine basis of equipment, personnel, hazards, and population densities.
2. Auditing of the emergency procedure.
3. Training on routine basis.
4. Practice drills.

The coordinator must ensure that the emergency equipment is always in a state of readiness. The siting of the control center and locating its equipment is also the coordinator's responsibility. A main control center and an alternate one should be provided for. The location should be chosen carefully. The following items should be present at the control center:

1. Copies of the current emergency plan.
2. Maps and diagrams of the area.
3. Names and addresses of key functional personnel involved in the plan.
4. Means to initiate alarm signals in the event of a power outage.
5. Communication equipment (e.g., phones, radio, TV, and two-way radios).
6. Emergency lights.
7. Evacuation routes mapped out on the area map.
8. Self-contained breathing equipment for possible use by the control center crew.
9. Cots plus other miscellaneous furniture.

Inspection of emergency equipment such as fire trucks, police cars, medical vehicles, personal safety equipment, and alarms should be routinely done.

The plan should be audited on a regular basis, at least annually, to ensure that it is current. Items to be updated include the list of potential hazards and emergency procedures (adapted to any newly developed technology). A guideline for auditing the emergency response plan is available from the Chemical Manufacturers Association (17).

More extensive information regarding the subject matter of this chapter (4), as well as calculational procedures (18), can be found in the literature.

34.10 SUMMARY

1. Accidents are a fact of life, whether they be a careless mishap at home, an unavoidable collision on the freeway, or a miscalculation at a chemical plant.

2. The Superfund Amendments and Reauthorization Act (SARA) of 1986 renewed the national commitment to correcting problems arising from previous mismanagement of hazardous wastes.

3. An emergency plan can minimize loss by helping to ensure the proper response in an emergency.

4. Emergency planning should grow out of a team process coordinated by a team leader. The team may be the best vehicle for gathering people with various kinds of expertise into the planning process, thus producing a more accurate and complete plan.

5. In order to characterize the types and extents of potential disasters, a survey of hazards or foreseeable threats in the community must be performed and evaluated.

6. Successful emergency planning begins with a thorough understanding of the event or potential disaster being planned for. The impacts on public health and the environment must also be estimated.

7. The education of the public regarding the real hazards in their community and how to respond to an emergency is critical to the public support of the emergency plan. Public opinion surveys show, however, that Americans are not yet prepared to deal either with the information provided by SARA Title III or with the reality of hazards in their community.

8. Notifying the public of an emergency is a task which must be accomplished with caution. People will react in different ways. Many will simply not know what to do, others will not take the warning seriously, and still others will panic. Proper training in each community can help minimize any panic and condition the public to make the right response.

9. Once an emergency plan has been developed, its successful implementation can be ensured only through constant review and revision.

REFERENCES

1. L. Theodore et al., *Accident and Emergency Management Student Manual*, U.S. EPA APTI, Contributing Author: J. O'Byrne, 1988.

2. J. Cornell, *The Great International Disaster Book*, 3rd ed., Charles Scribner's Sons, New York, 1976.

3. *Catastrophe! When Man Loses Control*, prepared by the editors of the *Encyclopaedia Britannica*, Bantam Books, NY, 1979

4. L. Theodore et al., *Accident and Emergency Management*, John Wiley & Sons, New York, 1989.

5. L. Theodore et al., *Accident and Emergency Management Student Manual*, U.S. EPA APTI, Contributing Author: L. Girardi Schoen, 1988.

6. "SARA: A First Year in Review," *Waste Age*, February 1988.

7. M. Krikorian, *Disaster and Emergency Planning*, Institute Press, Loganville, GA, 1982.

8. W. Beranek et al., "Getting Involved in Community Right-to-Know," *C & E News*, 62, October 26, 1987.

9. "Other Statutory Authorities, Title III: Emergency Planning and Community Right-to-Know", *EPA J.*, 13(1), 000-000, 1987.

10. National Response Team, *Hazardous Materials Emergency Planning Guide*, March 1987.

11. R. H. Schulze, *Superfund Amendments and Reauthorization Act of 1986 (SARA Title III)*, Trinity Consultants Inc., May 1987.

12. J. T. O'Reilly, *Emergency Response to Chemical Accidents: Planning and Coordinating Solutions*, McGraw-Hill, New York, 1987.

13. P. Shrivastava, *Bhopal: Anatomy of a Crisis*, Ballinger, Cambridge, MA, 1987.

14. E. J. Michael, O. W. Bell, and J. W. Wilson, *Emergency Planning Considerations for Specialty Chemical Plants*, Stone and Webster Engineering Corporation, Boston, MA, 1986.

15. U.S. EPA, *Title III Fact Sheet: Emergency Planning Community Right-to-Know*, 1987.

16. "CMA, Title III: The Right to Know, The Need to Plan," *Chemocology*, March 1987.

17. CMA, *Community Awareness & Emergency Response*, Program Handbook, Washington, D.C., April 1985.

18. L. Theodore and Y. C. McGuinn, *Health, Safety and Accident Management: Industrial Applications*, an ETS Theodore Tutorial, ETS International, Roanoke, VA, 1992.

35

ENVIRONMENTAL TRAINING

Contributing Authors: Leo H. Stander, Jr. and Charles D. Pratt

35.1 INTRODUCTION

Employee development is an investment employers make in the future of their organizations. Like any other investment, the "return" depends upon the quality of the investment and its applicability to the organization. The goal of training is the development of the human resource — that is, to help employees reach their full potential as professionals. Employee development, sometimes referred to as *human resource development*, comprises a variety of mechanisms such as on-the-job training, internships, exchanges, and full- and part-time study at educational institutions, and self-study. These activities are used to attract and retain the best employees, to build expertise in a field, to improve employee morale, and to maintain professional credibility.

Training can be viewed as a means of removing some of the obstacles to the successful and efficient accomplishment of organizational goals. There are never enough resources to provide all the training that employees would request, if the opportunity were presented. However, essential training must take place for an organization to be able to perform all its activities. It is clear that once critical needs are identified, training resources are usually found. Good management requires proper preparation of human resources to meet the ever-changing environment in which the organization exists. No better example exists than the situation being faced by companies which produce or use chemicals now considered hazardous. The training and retraining workload requirements produced by the Superfund Amendments and Reauthorization Act (SARA) of 1986, the Clean Air Act of 1990, the Pollution Prevention Act of 1990, and the proposed (at the time of preparation of this book) water pollution legislation, are staggering (1). Some of the options presented in this chapter may be useful as a means of providing additional training.

Environmental training can also lead to more effective work systems if they are then based on the fundamental principles of instructional technology. The most important step in the process is assessing actual training needed to effectively perform the work that needs to be done, or to comply with the many governmental regulations that require training.

The purpose of this chapter is to provide general information concerning a few of these developmental activities. Long- and short-term academic and professional training will be described, as will employee exchange programs and summer and cooperative student employment. The chapter concludes with a section on training myths.

35.2 ACADEMIC TRAINING

Academic training involves going back to school for additional education and may include both full-time and part-time attendance at colleges and universities. Many employers encourage employees to participate in such activities because it shows a willingness to obtain and maintain expertise in a field of endeavor and to improve technical and managerial capabilities. They encourage participation by allowing time off from work and by directly paying or reimbursing the employee for the tuition, fees, books, and other incidental expenses which may be incurred. Individuals can obtain training towards a baccalaureate or advanced degree either through part-time study while working on the job or through long-term training.

Long-term training can be defined as away-from-the-job training on a full-time basis for a period covering 90 or more consecutive days. Usually, this type of training involves a concentrated academic regimen of study and coursework towards a postgraduate degree. Long-term training in the fields of environmental management can be obtained from colleges or universities which provide courses in environmental science and engineering.

Many employers allow for such training and provide incentives, especially where the training is job-related. However, since the number of requests for these training opportunities far exceeds the resources available, employees must compete for approval. Usually, the employer has established eligibility requirements for candidates and has established criteria for evaluating the requests for financial assistance. Such criteria include relevancy of the proposed training to the organization; ability and motivation of the candidates; quality- and cost-effectiveness of the proposed training; and the candidate's career potential. Though the individual's efforts of obtaining such approval may be arduous, the benefits are even higher. For example, in the U.S. Environmental Protection Agency (EPA), an approved candidate can receive full salary and benefits plus tuition, fees, and book allowance. Some state agencies, such as the Tennessee Department of Health and Environment, offer comparable long-term training incentives.

Because of the "investment" made by the employer in the applicant's training, assurances of a "return" are usually required. Such assurances may include either (a) an applicant's agreement to remain with the employer for a prescribed time period after completion of study or (b) a reimbursement of all the employer's training expenses. This time period ranges from 2 to 3 years.

For those individuals who are unemployed or who may be employed where long-term training incentives are not authorized, other sources of financial assistance are available. Assistance in the form of scholarships, fellowships, or traineeships is available from colleges and universities and from professional associations and societies. As with the long-term training resources that are provided by employers, this financial assistance is limited. Eligibility requirements are prescribed, and established ranking criteria are used to select the most qualified applicants. Interested applicants should "shop around" in technical publications, such as the *Journal of the Air and Waste Management Association*, and in college or university financial assistance offices to find grant-in-aid programs which match their specific interests, needs, skills, and employment situations.

Fellowships are also available from some state and federal environmental agencies. For example, the New Jersey Department of Environmental Protection, Office of Environmental Health Assessment, offers four types of fellowships for a variety of areas. At the federal level, the EPA offers several types of financial assistance for academic training in environmental protection. Though most of the EPA's assistance is reserved for employees of state and local environmental agencies, some assistance may be available for individuals as well.

The EPA's air pollution traineeships are available to help individuals attend the schools identified as Area Training Centers (i.e., Rutgers University, the University of Florida, the University of Cincinnati, the University of Illinois at Chicago, the University of Texas at Arlington, the University of Washington, and the California Polytechnic State University). The EPA traineeships are limited to full-time employees of state and local agencies with responsibilities in air pollution control. The criteria for receiving these awards are the same as the criteria for admittance to the school's graduate programs. This assistance is in the form of a forgivable loan administered by the school's financial assistance offices. To receive these traineeships, applicants must agree to repay their loans by returning to work for, or by being employed by, state or local air pollution control agencies for a period of two years or by reimbursing the school for the training expenses.

Another type of resource available to individuals interested in securing additional academic instruction in their profession is part-time study. Part-time study involves attending one or more classes during an academic term while being employed on a full-time basis. These classes may be job-related or of personal interest. As with long-term training, they can be part of a curriculum designed to assist the individual in securing a degree. Many institutions have academic programs specifically designed to accommodate the student who is also a professional. Such accommodations include scheduling classes in the evenings and on weekends so there is minimal interference with the student's work schedules. As with long-term training, resources are available and employees are encouraged to take advantage of them, particularly if the courses are job-related. Because there are more requests for these training opportunities than there are available resources, competition among applicants is encountered. However, because resources can be spread out to cover more opportunities, the competition may not be as intense.

Financial assistance for part-time study is available from employers in both the private and public sectors, particularly where the courses are job-related. For example, Union Carbide Corporation will reimburse up to 75% of their employees' costs of tuition, fees, necessary supplies, and required textbooks for approved courses. Approved courses include those needed to secure a formal college degree and those considered by management to be directly related to the employee's duties and responsibilities. The remaining amount will be reimbursed by the company after all the requirements for an advanced degree are met. Governmental agencies, such as the EPA, Tennessee Department of Health and Environment, and the Toledo Environmental Services Agency, also have training costs reimbursement incentives for their employees.

The EPA also has a fellowship grant program designed to enhance the capabilities of career personnel in state and local pollution control agencies. These grants provide funding for tuition, fees, books, and supplies. In addition, a small stipend covering other expenses is provided. This grant program allows students to attend academic institutions of their choice on a part-time basis and to receive training in a specific field or credits toward a baccalaureate or advanced graduate degree.

Another training opportunity available from some universities is the transmittal of courses by low-power television broadcast to an employer's facilities. This is particularly useful where several individuals may need to take a course at the same time. These courses are offered for academic credit and can be used to help fulfill requirements toward an advanced degree.

35.3 EMPLOYEE EXCHANGE PROGRAMS

Though not a specific training resource for all individuals involved in environmental management, it is an available technique for employees to use in securing additional training and experience. Employee exchange programs are opportunities for individuals in public sector occupations to work in other public sector occupations or in academia or industry, and vice versa. In theory, such an exchange can result in better cooperation and understanding and can improve delivery of government services. Two examples of such exchange programs are Intergovernmental Personnel Act (IPA) assignments and assignments provided by the President's Commission on Executive Exchange.

IPA assignments (2) involve employee exchanges between the U.S. Government and a state or local governmental agency, an educational institution, or certain nonprofit organizations such as professional societies and research organizations. IPA assignments are made under the authority of the IPA of 1970. An assignment occurs when a federal or nonfederal employee temporarily leaves his/her organization to work for another organization in another sector. The assignments must be designed so that they are beneficial to both organizations and can be made on a full-time, part-time, or intermittent basis. Initially the assignments last for a period of up to two years but may be extended for a period not to exceed four years. Costs are shared by the participating organizations. IPA appointments can be used to acquire hard-to-find skills for the solutions of common problems. They often provide training and firsthand experience in the application of environmental programs and regulations at the grass-roots level. One of the co-authors of this book, Dr. Louis Theodore, a chemical engineering professor at Manhattan College, participated in the IPA program during the 1973–1974 academic year.

Appointments under the President's Commission on Executive Exchange provide senior government executives with opportunities to be placed in a corporate assignment to gain fresh insights from the private sector. The program provides senior corporate executives with opportunities to take assignments in senior government positions and to become acquainted with decision-making processes in the federal government. These assignments can last no longer than one year. Although these assignments may not specifically be related to environmental management activities, the knowledge acquired from such an exchange can be of assistance in managing various program operations.

35.4 SUMMER EMPLOYMENT AND COOPERATIVE TRAINING

Summer employment and cooperative (co-op) training are two techniques sometimes used by employers to evaluate potential new hires before they complete their studies. It also gives students an opportunity to improve their qualifications with professional work experience before graduation. These techniques allow prospective employees to develop job skills while providing employers with low-cost ideas and fresh perspectives.

Co-op education allows a student to alternate terms of study with terms of part-time or full-time employment in business, industry, or government. Students can blend classroom study with supervised employment in an area relevant to their vocational and educational goals. Co-op students usually alternate full-time employment with full-time study during their final year before receiving their baccalaureate degree. However, at most universities, students are eligible to start their co-op program following the freshman year. During the periods of employment, co-op students receive salaries and sometimes earn credits toward a degree. Once they receive their baccalaureate degree, many of the co-op students return as permanent employees of the company where they were employed during college. Through this program, an employer can evaluate a prospective employee's educational background and ability to perform job-related tasks effectively. Companies that have employed co-op students include Merck Company, Inc., Shering Corp., Ciba Geigy, AT&T, Veritech, and Texaco USA.

Summer employment with industry or with a government agency is also a way a college student can gain knowledge in the field of environmental management. For example, the U.S. Public Health Service's Commissioned Officer Student Training and Extern Program (COSTEP) is a recruiting device which offers opportunities for students in health-related fields, such as engineering or environmental science, to gain knowledge and experience during summers after their sophomore year. These individuals are commissioned as officers in the Commissioned Corps of the Public Health Service and receive the pay, benefits, and allowances of commissioned officers. Similar opportunities exist in state and local environmental agencies, and in industry.

35.5 SHORT-TERM TRAINING

Short-term training is intense instruction on a given technical or administrative topic. Offerings can last from a few hours to 90 days, but usually last less than a week. Short-term training utilizes a variety of group learning formats, including workshops, seminars, classes, and, to a certain extent, conferences. This type of training is a staple of professional development and is the principal technique used by employers to upgrade their employees from novices to journeymen to experts in their fields.

Short-term training is, by far, the most prevalent training resource available to professionals in environmental management. Literally hundreds of offerings are available each year, covering a variety of topics. They are sponsored and taught by consultants, university faculty, government agency personnel, industrial associations, professional societies and organizations, and even trade publications. Information on such offerings can be obtained from direct-mail advertising, professional journals, and newsletters.

Some employers conduct short-term training for their own employees, whereas others offer training to interested participants at a nominal fee. Charges for such courses offered and conducted by the U.S. EPA's Air Pollution Training Institute (APTI) cost approximately $100 per day for lecture courses and $150 per day for laboratory courses, with typical courses lasting from 3 to 5 days. There is no charge for employees of state and local agencies with responsibilities in air pollution control. Courses offered through the Air and Waste Management Association as part of a Specialty Conference or the Annual Meeting usually cost $60 per 2-hr session or $200 or $500 for a 1- or 2-day session. Courses or workshops offered by consultants or trade groups typically cost between $300 and $750 for 1- to 3-day day sessions. However, these figures can be misleading because the total cost to the company, agency, or organization will be significantly higher. For example, the inclusion of travel, room and board, car rental, and so on, can increase the cost for a 4-day course to approximately $2500 (3). This does not include the "cost" associated with the individual being absent from work during the course; this cost estimate ranges from $500 to $1000 per day (3).

Short course offerings are conducted in a variety of locations, such as an employer's own premises and remote resort settings. The locations are usually determined by available resources and attendee needs. The topics covered during these short courses range from a general overview of environmental technical and management activities to more specialized courses. The course content is usually designed with the intended student in mind—that is, either (a) general overview information for managers or for people newly involved in the area or (b) specific technical information designed to help individuals perform their duties more effectively, as in the case of laboratory or inspection personnel.

35.6   SELF-STUDY TRAINING

During the past few years, self-study training has become an important part of the Federal Training program. Self-study allows students to participate without incurring travel expenses or time away from the office. On the negative side, there is a lack of interaction with other students and no direct guidance is provided by the instructor. Some individuals consider self-study training as an imposition on their private time, and they feel that it takes away from their productive work time. Despite these possible shortcomings, the majority of individuals find that a great deal of information and knowledge can be obtained through self-study training.

Self-study programs meet three objectives: (a) They are designed to reach a wide audience, (b) they are relatively inexpensive, and (c) they encourage "learning by doing." Over the past few years, the Air Pollution Training Institute of the EPA has enrolled more students in the self-study program than in the short course training program. This is partly because of the increased demand for the new training packages and a general cutback in short courses. There is evidence of a general increase in acceptance of the self-study programs and their value as a supplement to on-the-job and short course training.

Economically speaking, there is little doubt of the cost-effectiveness of such a program. In many cases, the travel costs alone prevent prospective attendees from participating in other types of training. Currently, the cost of many self-study programs in environmental protection are underwritten by the federal government or public organizations. There is little or no cost to an employer to initiate an internal training program based on the self-study courses available from the EPA and other sources. For example, Westinghouse Electric Corp. has recently initiated such a program for some of its employees, and many agencies have made some use of these readily materials.

An important characteristic of the self-study format is learning-by-doing. There are two levels of learning-by-doing: The first involves the mechanics of writing answers into blocks and receiving feedback by comparing the answers. The second level enters when the student uses a course manual in a work situation and begins to apply the principles contained in the course. Also, many of the self-study materials contain worksheets and trouble-shooting guides to assist the student in his work.

More recently, a set of over 20 unique environmental tutorials, entitled *Theodore Tutorials* (4), has been developed (by one of the authors of this book) that can be used for self-study purposes. One of the key features of these tutorials is that a *Theodore Tutorial* is a self-instructional problem workbook. It attempts to meet the challenge of effective instruction for "students" with diverse backgrounds, including nontechnical education, with an approach that lies between programmed instruction and the conventional textbook. The general format for these *Theodore Tutorials* is as follows. The material is divided into three parts: Basic Operations, Problems, and Solutions. The first part of each tutorial workbook provides a series of Basic Operations that are required when solving most engineering problems. Each basic operation refers to a particular calculation from chemistry, physics, thermodynamics, and so forth, that many individuals refer to as Basic Concepts. Each Basic Operation is presented with a title, problem statement, pertinent data, and solution in programmed instructional format. The second part of each tutorial contains problems. This section deals with individual calculations associated with the title topic in question. These problems are laid out in such a way as to develop the reader's technical understanding of the subject in question. Each problem contains a title, introduction, problem statement and data, solution in pro-grammed-instructional format, and a concluding comment statement. The more difficult problems can be found at or near the end of each set. Various options are available to the reader in attempting to solve each problem. These include

1. Employing the problem statement only.

2. Employing the problem statement and the introduction.

3. Employing the problem statement and the introduction, plus the right side of the solution outline in programmed-instructional format.

4. Employing the problem statement and the introduction, plus the right- and left-hand sides of the solution outline in programmed-instructional format.

If difficulty is encountered in solving the problems, the reader can refer to the solution section (in Part 3) that immediately follows the Problem section. An Appendix, consisting of a discussion of SI units, conversion constants, physical and chemical properties, glossary, and so on, is also included.

From the perspective of an organization, the self-study approach has much to offer in terms of meeting its goals by controlling training costs and by providing an ongoing information exchange and training program. From the students' perspective, self-study offers a means to learn at their own pace and to build a reference library for future use. A self-study program can provide both (a) a valuable introduction into a new area or discipline and (b) a synthesis of historical information upon which decisions can be made. The time required to assemble the documents contained in a course is far greater than the time it takes to complete the self-study material. In air pollution control, the major source of self-study materials is EPA's APTI at Research Triangle Park, North Carolina. Currently, they list approximately 30 course titles in their "Chronological Schedule of Courses," a copy of which may be obtained by writing to the Institute at Mail Drop 17, Research Triangle Park, North Carolina 27711. The Institute currently provides these courses at no cost to the students; however, where audio-visual materials have been provided, students are requested to return them when finished. Other sources of self-study training materials include universities; the Air and Waste Management Association (AWMA), which offers a videotape program; and the National Environmental Health Association, which offers a limited number of similar courses. Costs vary among these providers; however, the costs for such courses are significantly lower than those for short courses or for long-term academic training.

35.7   TRAINING MYTHS

For a training program to be successful over the long run, management support to its training staff is essential. Managers must take time out to understand training standards, resource requirements, and organizational impacts. While internal experts or consultants can develop specific information for the training programs, support for the program should come from the top down, be felt by staff and employees, and result in the desired organizational change. However, management should be made aware of some of the common misconceptions or training myths making the rounds in industry and government. Some of these are provided below.*

1. **Formal training? It's a luxury — we're doing fine without it.** In an atmosphere of budget-cutting, training often is eyed as easy fat to trim. Oftentimes, managers become tempted to look at a cost that does not appear to reap immediate, measurable dollar benefits and begin a process of rationalization to defend the cut. This is dangerous logic.

Before you follow this path, make sure someone informs you of the true costs of floor-drained chemicals, gas releases to the community, employee injuries due to weak personal protective equipment programs, and waste increase due to the "I didn't know, I wasn't sure, and nobody told us" syndrome.

*Source: M. J. Cherniak, Changing Attitudes Toward Training's Role, *Environmental Protection*, September, 1991. Adapted with permission.

In short, training can save money through improved work efficiency, fewer accidents and errors, and improved employee knowledge and morale. Anyone in business for the long haul is smart to look at training as a tool, not a burden.

2. **I read about a great video program and manual. Let's have everyone see it.** Video training is a quick way to dig a morale hole. Videos are a valuable training aid — emphasis on the word *aid*. While the idea of a one-time cost to purchase what may be an expensive resource is attractive, you should consider that there are problems inherent in this type of "instruction."

Unless it is interactive (more expensive), it offers no hands-on practice, represents a limited opportunity to get questions answered, and is proven to be less likely to be remembered. In a world of passive, television-influenced learners, video training loses its impact quickly when it has no support from other teaching methods. Besides, I have not run across many managers who have been willing to sit through such sessions themselves.

Make up your mind: "Train 'em or entertain 'em." Let your competent in-house instructors select and purchase videos as an adjunct to meet *their* objectives and program needs. Otherwise, your employees will be frustrated because you made them sit through a session in which they could not participate.

3. **John already knows the subject. Let him teach it.** The hole gets deeper. Believe me when I say that this is pretty common and usually disastrous thinking. All of us have had enough good and bad teachers to know that it is not something anybody can just pick up and do well. Training your trainers must be an integral part of any training program.

I recently sat through an ammonia safety training session presented by an ammonia specialist. He had no objectives, could not explain his points in understandable terms, and had difficulty relating to the audience. At a critical point early in his presentation, he answered a question with "Let me try to explain this in a way you can understand it." The trainees were no longer just lost; they were angry. I thought the session was, for all practical purposes, over. The organization's money was wasted.

Would-be trainers need guidance, tips, and confidence. By the way, did anyone bother to ask John if he gets nervous in front of groups?

4. **Just send Jane and John to a course. Then they can come back and teach it to the rest of us.** This point is closely related to the previous one and is a back-door rationalization to keep training budgets slim and trim. I think it was either Albert Einstein or my Uncle Sal who said, "It just ain't that easy." Simply put, attending a workshop or seminar does not give one the expertise, background, or resources to perform that training. The preparation, planning, and teaching expertise do not just transfer magically to someone; they come in time.

Attending the course(s) is part of the education process, an important part. But look at it this way: I've watched people catch salmon for years. I still haven't caught one myself. Successful trainers need training and confidence, not just a few days sitting in on a course. Managers must accept this truth.

35.8 SUMMARY

1. Employee development is an investment employers make in the future of their organizations. Like any other investment, the "return" depends upon the quality of

the investment and its applicability to the organization. The goal of training is the development of the human resource—that is, to help employees reach their full potential as professionals.

2. Academic training involves going back to school for additional education and may include both full-time and part-time attendance at colleges and universities. Many employers encourage employees to participate in such activities because it shows a willingness to obtain and maintain expertise in a field of endeavor and to improve technical and managerial capabilities.

3. Though not a specific training resource for all individuals involved in environmental management, an employee exchange program is an available technique for employees to use in securing additional training and experience. Employee exchange programs are opportunities for individuals in public sector occupations to work in other public sector occupations or in academia or industry, and vice versa.

4. Summer employment and co-op training are two techniques sometimes used by employers to evaluate potential new hires before they complete their studies. It also gives students an opportunity to improve their qualifications with professional work experience before graduation.

5. Short-term training is intense instruction on a given technical or administrative topic. Offerings can last from a few hours to 90 days, but usually last less than a week.

6. Self-study programs meet three objectives: (a) They are designed to reach a wide audience, (b) they are relatively inexpensive, and (c) they encourage "learning by doing."

7. For a training program to be successful over the long run, management support to its training staff is essential. Managers must take time out to understand training standards, resource requirements, and organizational impacts.

REFERENCES

1. W. Bunner, Regulations Dictate Training Assessment, *Environmental Protection*, January/February, 1991.
2. *EPA Intergovernmental Personnel Act (IPA) Handbook*, Special Resources Program, Personnel Management Division, Office of Administration, U.S. EPA, 1987.
3. L.Theodore, personal notes, 1998.
4. *Theodore Tutorials,* International Air & Waste Management Association Bookstore, Pittsburgh, PA, 1999.

36

ECONOMIC CONSIDERATIONS

36.1 INTRODUCTION

Although the technical parameters influencing the selection and design of a given type of an environmental control system may be unique, cost is the only parameter relevant to all systems. This is especially so where different types of systems can control a source to achieve an emission limit. Here, cost is often the parameter used to select the optimum system from the alternatives available.

Economics plays an important role in setting many state and federal environmental control regulations. The extent of this role varies with the type of regulation. For some types of regulations, cost is explicitly used in determining their stringency. This use may involve a balancing of costs and environmental impacts, costs and dollar valuation of benefits, or environmental impacts and economic consequences of control costs. For other types of regulations, cost analysis is used to choose among alternative regulations with the same level of stringency. For these regulations, the environmental goal is determined by some set of criteria which do not include costs. However, cost-effectiveness analysis is employed to determine the minimum economic way of achieving the goal. For some regulations, cost influences enforcement procedures or requirements for demonstration of progress toward compliance with an environmental quality standard. For example, the size of any monetary penalty assessed for noncompliance as part of an enforcement action needs to be set with awareness of the magnitude of the control costs being postponed or bypassed by the noncomplying facility. For regulations without a fixed compliance schedule, demonstration of reasonable progress toward the goal is sometimes tied to the cost of attaining the goal on different schedules.

Economic considerations are a vital input into two other types of analyses that also sometimes have a role in standard-setting. Cost is needed for a benefit–cost analysis that addresses the economic efficiency of alternative regulations. Cost is also an input into any analysis of the economic impact of each regulatory alternative. An economic impact analysis usually deals with the consequences of the regulation for small businesses, employment, prices, and market and industry structure (1).

An understanding of the economics involved in environmental management control is important in making decisions at both the engineering and management levels. Every engineer or scientist should be able to execute an economic evaluation of a proposed project. If the project is not profitable, it should obviously not be pursued; and the earlier such a project can be identified, the fewer are the resources that will be wasted.

Before the cost of an environment project can be evaluated, the factors contributing to the cost must be recognized. There are two major contributing factors, namely, capital costs and operating costs; these are discussed in Sections 36.2–36.4. Once the total cost of the project has been estimated, the engineer must determine whether or not the process (or change) will be profitable. This involves converting all cost contributions to an annualized basis. Section 36.5 provides a basis for comparing alternate proposals and for choosing the best proposal. Project optimization is the subject of Section 36.6, where a brief description of a perturbation analysis is presented. Several examples, based on a hazardous waste incineration facility, are also provided in this chapter.

Detailed cost estimates are beyond the scope of this chapter. Such procedures are capable of producing accuracies in the neighborhood of $\pm 10\%$; however, such estimates generally require many months of engineering work. This chapter is designed to give the reader a basis for a preliminary cost analysis only, with an expected accuracy of approximately $\pm 20\%$.

36.2   CAPITAL EQUIPMENT COSTS

Equipment cost is a function of many variables, one of the most significant of which is *capacity*. Other important variables include equipment type and location, operating temperature, and degree of equipment sophistication. Preliminary estimates are often made from simple cost–capacity relationships that are valid when the other variables are confined to narrow ranges of values; these relationships can be represented by approximate linear (on log–log coordinates) cost equations of the form (2)

$$C = \alpha Q^{\beta} \qquad (36.2.1)$$

where C represents cost; Q represents some measure of equipment capacity; and α and β represent empirical constants that depend mainly on equipment type. It should be emphasized that this procedure is suitable for rough estimation only; actual estimates from vendors are more preferable. Only major pieces of equipment are usually included in this analysis; smaller peripheral equipment such as pumps and compressors are usually not included.

The development below examines capital equipment cost data and information for a hazardous waste incineration (HWI) facility that uses the above approach (3). The most common types of HWI systems in use are the liquid-injection, rotary-kiln, and, to lesser extent, the multiple-hearth (among others). Cost–capacity relationships for two types of liquid-injection systems are presented here. Both include a waste heat boiler and a flue gas scrubbing system, in addition to the incinerator.

The first type, designated as Case A in Table 36.1, is designed to burn relatively clean liquids that contain an insignificant amount of chlorine or salts. Normal operating temperatures seldom exceed 1800°F, and residence times are usually short (0.30.5 sec). The incinerator consists of a horizontally aligned cylindrical steel shell, combustion air blower, flame safeguard, and combustion control instrumentation. Case B is a liquid injection system designed to handle problematic liquid wastes such as highly chlorinated materials, salty aqueous wastes, and liquefied tars. Units in this category are obviously more costly and are designed to accommodate operating temperatures of up to 2200–2400°F; residence times range from 1.5 to 25 sec. These units can be either vertical or horizontal cylindrical steel shells and include combustion air blowers, flame safeguards, and combustion control instrumentation. The applicable capacity range for both Case A and Case B systems is 1–100 million Btu/hr.

The rotary-kiln system in Table 36.1 includes the kiln-afterburner and auxiliaries plus waste heat boiler (optional) and a scrubbing-flue gas handling system. The cost of this system is considerably higher than that of the liquid-injection systems. The applicable capacity range for the constants in Table 36.1 is 5–100 million Btu/hr. The kiln temperatures range from 1600 to 1800°F, whereas the afterburner

Table 36.1   Values of α and β for use with Equation 36.2.1[a]

Incinerator[b]	α [dollar/(Btu/hr)]	β
Liquid injector (Case A)	30.1	0.503
Liquid injector (Case B)	41.6	0.512
Rotary kiln	220	0.478
Multiple hearth (Case A)	517	0.351
Multiple hearth (Case B)	586	0.630

Waste Heat Boiler[c]	α [dollar/(lb/hr)]	β
Case A	462	0.515
Case B	1944	0.401
Case C	9231	0.278

Scrubber[d]	α (dollar/acfm)	β
Case A	145	0.717
Case B	46	0.816

[a]Costs calculated using these constants are based on the value of the dollar in 1982.
[b]Units of Q are Btu/hr.
[c]Units of Q are lb/hr.
[d]Units of Q are acfm.

temperatures range from 1800 to 2200°F with a residence time of approximately 2 sec. [*Note:* A (venturi) scrubber is usually required for particulate control.]

Hearth incineration systems include the multiple-chamber combustion unit, a waste heat boiler (optional), and a scrubbing-flue gas handling system. Case A is valid for capacities ranging from 1 to 10 million Btu/hr, and Case B is valid for capacities ranging from 10 to 50 million Btu/hr.

Waste heat boilers are primarily used in facilities that can accommodate 10–50 million Btu/hr, where the steam can be used elsewhere. Their use is less likely in smaller facilities and in commercial facilities where there is usually no market for the steam (2). In Table 36.1, waste heat boilers are divided into three categories. Case A is for smaller facilities using a fire-tube boiler with a flue gas flow rate ranging from 2,000 to 100,000 lb/hr. Case B is for water-tube boilers of standard design with a gas flow rate ranging from 15,000 to 150,000 lb/hr. Water-tube boilers operating with extremely high particulate loadings and high acid gas environments make up Case C, where the applicable range is 15,000–150,000 lb/hr. The constants α and β for all three boiler types are based on inlet temperatures of 1,800–2,200°F and outlet temperatures of approximately 500°F. For the fire-tube boiler, the steam pressure is in the neighborhood of 150 psig. Water-tube steam pressures range from 150 to 200 psig.

Scrubbing system designs and costs also vary considerably from one system to the next. Typical scrubbing and flue gas handling systems include quenchers, venturi scrubbers, acid gas absorbers, caustic recycle systems, ID fans, stacks, and auxiliaries. System costs calculated from Table 36.1 are based on a venturi pressure drop of 30 in H_2O.

In Table 36.1, scrubbers are divided into two categories based on whether or not a waste heat boiler is involved. Case A is for systems receiving hot combustion gas directly from the incinerator (no waste heat boiler) at a temperature of 1,800–2,200°F. The applicable range of capacities for constants listed in Table 36.1 for this case is 1,000–200,000 acfm. Case B involves scrubbing systems receiving gas from waste heat boilers at a temperature of approximately 550°F. Because of the much lower operating temperatures, less costly materials of construction and smaller-sized equipment can be employed. The applicable capacity range is 1,000–50,000 acfm. Because scrubbing systems vary so much in type, *cost adjustment factors* (CAFs) are used to account for this variation (1); these are listed in Table 36.2. To use the CAFs for any of the four cases listed, the cost resulting from Equation 36.2.1 is multiplied by the appropriate CAF.

Similar methods for estimating the costs of storage tanks, waste conveyors, pumps, compressors, site development, and so on, are available in the literature (2). If greater accuracy is needed, however, actual quotes from vendors should be used.

Again, the equipment cost estimation model just described is useful for a very preliminary estimation. If present-day cost values are needed and if old price data are available, the use of an indexing method may be employed to update the data. The method consists of adjusting the earlier cost data to present values using factors that correct for inflation. A number of such indices are available; one of the most commonly used is the *chemical engineering fabricated equipment cost index* (FECI) (2,4), past values of which are listed in Table 36.3. Other indices for construction, labor, buildings, engineering, and so on, are also available in the literature (2,4).

Table 36.2 Cost adjustment factors for scrubbing systems[a]

Adjustment from Baseline Scenario	CAF
Extremely high venturi ΔP (100 in. H_2O)	2.0
Same, but no acid gas absorption	1.2
No venturi required	0.85
No absorption system required	0.6

[a]See Ref. 2.

Generally, it is not wise to use past cost data older than 5–10 years, even with the use of the cost indices. Within that time span, the technologies used in the processes may have changed drastically. The use of the indices could cause the estimates to be much greater than the actual costs. Such an error might lead to the choice of an alternative proposal other than the least costly.

36.3 CAPITAL COSTS

The usual technique for determining the *capital costs* (i.e., *total* capital costs, which include equipment design, purchase, and installation) for a project and/or process can be based on the *factored method* of establishing direct and indirect installation costs as a function of the known equipment costs. This is basically a *modified Lang method*, whereby cost factors are applied to known equipment costs (6, 7).

Table 36.3 Fabricated equipment cost index

Year	Index
1992	???
1991	363.0 (estimate)
1990	357.6
1989	355.4
1988	342.5
1987	323.8
1986	318.4
1985	325.3
1984	334.1
1983	327.4
1982	326.0
1981	321.8
1980	291.6
1979	261.7
1978	238.6
1977	216.6
1976	200.8
1975	192.2

Source: Ref. 5.

The first step is to obtain from vendors (or, if less accuracy is acceptable, from one of the estimation techniques previously discussed) the purchase prices of the primary and auxiliary equipment. The total base price, designated by X, which should include instrumentation, controls, taxes, freight costs, and so on, serves as the basis for estimating the direct and indirect installation costs. The installation costs are obtained by multiplying X by the cost factors, which are available in the literature (6–11). For more refined estimates, the cost factors can be adjusted to more closely model the proposed system by using adjustment factors that take into account the complexity and sensitivity of the system (6, 7).

The second step is to estimate the direct installation costs by summing all the cost factors involved in the direct installation costs, which include piping, insulation, foundation and supports, and so on. The sum of these factors is designated as the *direct installation cost factor* (DCF). The direct installation costs are then the product of the DCF and X.

The third step consists of estimating the indirect installation costs. The procedure here is the same as that for the direct installation costs — that is, all the cost factors for the indirect installation costs (engineering and supervision, startup, construction fees, etc.) are added; the sum is designated the *indirect installation cost factor* (ICF). The indirect installation costs are then the product of ICF and X.

Once the direct and indirect installation costs have been calculated, the total capital cost (TCC) may be evaluated as

$$\text{TCC} = X + (\text{DCF})(X) + (\text{ICF})(X) \tag{36.3.1}$$

The conversion of the TCC to an annualized basis involves an economic parameter known as the *capital recovery factor* (CRF). It may be calculated from

$$\text{CRF} = \frac{(i)(1 + i)^n}{(1 + i)^n - 1} \tag{36.3.2}$$

where $n = $ *projected* lifetime of the system (yr)
    $i = $ annual interest rate (expressed as a fraction)

The CRF is a positive, fractional number. The *annualized capital cost* (ACC) is the product of the CRF and TCC and represents the total installed equipment cost distributed over the lifetime of the facility.

Some guidelines in purchasing equipment are listed below:

- Do not buy or sign any documents unless provided with certified independent test data.

- Previous clients of the vendor company should be contacted and their facilities visited.

- Prior approval from local regulatory officials should be obtained.

- A guarantee from the vendors involved should be required. Startup assistance is usually needed, and assurance of prompt technical assistance should be obtained in writing. A complete and coordinated operating manual should be provided.

- Vendors should provide key replacement parts if necessary.
- Finally, 10–15% of the cost should be withheld until the installation is completed.

36.4 OPERATING COSTS

Operating costs can vary from site to site because these costs partly reflect local conditions — for example, staffing practices, labor, and utility costs. Operating costs, like capital costs, may be separated into two categories: direct and indirect costs. *Direct* costs are those that cover material and labor and are directly involved in operating the facility. These include labor, materials, maintenance and maintenance supplies, replacement parts, waste (e.g., residues after incineration) disposal fees, utilities, and laboratory costs. *Indirect* costs are those operating costs associated with, but not directly involved in, operating the facility; costs such as overhead (e.g., building–land leasing and office supplies), administrative fees, local property taxes, and insurance fees fall into this category.

The major direct operating costs are usually those associated with labor and materials. *Materials* costs for the project involve the cost of chemicals needed for the operation of the process (10). Labor costs differ greatly, but are a strong function of the degree of controls and/or instrumentation. Typically, there are three working shifts per day with one supervisor per shift. On the other hand, the plants may be manned by a single operator for only one-third or one-half of each shift; that is, usually only operator, supervisor, and site manager are necessary to run the facility. Salary costs vary from state to state and depend significantly on the location of the facility. The cost of *utilities* generally consists of that for electricity, water, fuel, and steam. The annual costs are estimated with the use of material and energy balances.

Costs for *waste disposal* [e.g., for the incinerator (ash) or scrubber (water) residues] can be estimated on a per-ton-capital basis. Costs of landfilling ash can run significantly upwards of $100/ton if the material is hazardous, and can be as high as $10/ton if it is nonhazardous. The cost of handling a scrubber effluent can vary, depending on the method of disposal. For example, if conventional sewer disposal is used, the effluent probably has to be cooled and neutralized before disposal; the cost for this depends on the solids concentration. Annual *maintenance* costs can be estimated as a percentage of the capital equipment cost. The annual costs of replacement parts can be computed by dividing the cost of the individual part by its expected lifetime. The life expectancies can be found in the literature (7). *Laboratory* costs depend on the number of samples tested and the extent of these tests; these costs can be estimated as 10–20% of the operating labor costs.

The *indirect* operating costs consist of overhead, local property tax, insurance, and administration, less any credits. The *overhead* comprises payroll, fringe benefits, social security, unemployment insurance, and other compensation that is indirectly paid to the plant personnel. This cost can be estimated as 50–80% of the operating labor, supervision, and maintenance costs (11, 12). Local *property taxes* and *insurance* can be estimated as 1–2% of the total capital cost (TCC), while administration costs can be estimated as 2% of the TCC.

The total operating cost is the sum of the direct operating costs and the indirect operating costs, less any credits that may be recovered (e.g., the value of recovered steam). Unlike capital costs, operating costs are always calculated on an annual basis.

36.5   PROJECT EVALUATION

In comparing alternate processes or different options of a particular process from an economic point of view, it is recommended that the total capital cost be converted to an annual basis by distributing it over the projected lifetime of the facility. The sum of both the *annualized capital costs* (ACCs) and the *annual operating costs* (AOCs) is known as the *total annualized cost* (TAC) for the facility. The economic merit of the proposed facility, process, or scheme can be examined once the total annual cost is available. Alternate facilities or options (e.g., a baghouse versus an electrostatic precipitator for particulate control, or two different processes for accomplishing the same degree of waste destruction) may also be compared. Note that a small flaw in this procedure is the assumption that the operating costs remain constant throughout the lifetime of the facility.

Once again, the development to follow examines the economics associated with the construction of an HWI facility (4). If an on-site (internal) incineration system is under consideration for construction, the total annualized cost should be sufficient to determine whether or not the proposal is economically attractive as compared to other proposals. If, however, a commercial incineration process is being considered, the profitability of the proposed operation becomes an additional factor. The method presented here assumes a facility lifetime of 10 years and also assumes that the land is already available.

One difficulty in this analysis is estimating the revenue generated from the facility, because both technology and costs can change from year to year. Also affecting the revenue generated is the amount of waste handled by the facility. Naturally, the more waste handled, the greater the revenues. If a reasonable estimate as to the revenue that will be generated from the facility can be made, a rate of return can be calculated.

This method of analysis is known as the *discounted cash flow method using an end-of-year convention;* that is, the cash flows are assumed to be generated at the end of the year, rather than throughout the year (the latter obviously being the real case). An expanded explanation of this method can be found in any engineering economics text (13). The data required for the analysis are the TCC, the annual after-tax cash flow (A), and the working capital (WC). Generally, for HWI facilities, WC includes the on-site fuel inventory, caustic soda solution, maintenance materials (spare parts, etc.), and wages for approximately 30 days. The WC is expended at the startup of the plant (time = 0 years) and is assumed to be recoverable after the life of the facility (time = 10 years). For simplicity, it is assumed that the WC is 10% of the TCC and that the TCC is spread evenly over the number of years used to construct the facility. (For this example, this construction period is assumed to be 2 years).

Usually, an after-tax rate of return on the initial investment of at least 30% is desirable. The method used to arrive at a rate of return will be discussed briefly. An annual after-tax cash flow can be computed as the annual revenues (R) minus the annual operating costs (AOC) and minus income taxes (IT). Income taxes can be estimated at 50% (this number may be lower with the passage of new tax laws) of taxable income (TI):

$$IT = 0.5(TI) \tag{36.5.1}$$

The taxable income is obtained by subtracting the AOC and the depreciation of the plant (D) from the revenues generated (R):

$$TI = R - AOC - D \tag{36.5.2}$$

For simplicity, straight-line depreciation is assumed; that is, the plant will depreciate uniformly over the life of the plant. For a 10-year lifetime, the facility will depreciate 10% each year:

$$D = 0.1(TCC) \tag{36.5.3}$$

The annual after-tax cash flow (A) is then

$$A = R - AOC - IT \tag{36.5.4}$$

This procedure involves a trial-and-error solution. There are both positive and negative cash flows. The positive cash flows consist of A and the recoverable working capital in year 10. Both should be discounted *backward* to time $= 0$, the year the facility begins operation. The negative cash flows consist of the TCC and the initial WC. In actuality, the TCC is assumed to be spent evenly over the 2-year construction period. Therefore, one-half of this flow is adjusted *forward* from after the first construction year (time $= -1$ year) to the year the facility begins operating (time $= 0$). Forward adjustment of the 50% TCC is accomplished by multiplying by an economic parameter known as the *single-payment compound amount factor* (F/P), given by

$$F/P = (1 + i)^m \tag{34.5.5}$$

where i is the rate of return (fraction) and m is the number of years (in this case, 1 year).

For the *positive* cash flows, the annual after-tax cash flow (A) is discounted backward by using a parameter known as the *uniform series present worth factor* (P/A). This factor is dependent on both interest rate (rate of return) and the lifetime of the facility and is defined by

$$P/A = \frac{(1 + i)^n - 1}{i(1 + i)^n} \tag{34.5.6}$$

where n is the lifetime of the facility (in this case, 10 years). [*Note:* The P/A is the inverse of the CRF (*capital recovery factor*).] The recoverable working capital at year 10 is discounted backward by multiplying WC by the *single present worth factor* (P/F) which is given by

$$P/F = \frac{1}{(1 + i)^n} \tag{34.5.7}$$

where n is the lifetime of the facility (in this case, 10 years). The positive and negative cash flows are now equated and the value of i, the rate of return, may be determined by trial and error from Equation 36.5.8:

$$\text{Term } 1 + \text{Term } 2 = \text{Term } 3 + \text{Term } 4 \qquad (36.5.8)$$

where Term 1 $[((1 + i)^{10} - 1)/i(1 + i)^m]A$ (worth at year = 0 of annual after-tax cash flows), Term 2$(1/(1 + i)^{10})$WC (worth at year = 0 of recoverable WC after 10 years), Term 3 = (WC + 1/2 TCC) (assumed expenditures at year = 0), and Term 4 = 1/2 (TCC)$(1 + i)^1$ (worth at year = 0 of assumed expenditures at year = -1).

36.6   PROJECT OPTIMIZATION

Once a particular process scheme has been selected, it is common practice to optimize the process from a capital cost and O&M (operation and maintenance) standpoint. There are many optimization procedures available, most of them too detailed for meaningful application in this chapter. These sophisticated optimization techniques, some of which are routinely used in the design of conventional chemical and petrochemical plants, invariably involve computer calculations. Use of these techniques in environmental management analysis is usually not warranted, however.

One simple optimization procedure that is recommended is the *perturbation study*. This involves a systematic change (or *perturbation*) of variables, one by one, in an attempt to locate the optimum design from a cost and operation viewpoint. To be practical, this often means that the engineer must limit the number of variables by assigning constant values to those process variables that are known beforehand to play an insignificant role. Reasonable guesses and simple or short-cut mathematical methods can further simplify the procedure. Much information can be gathered from this type of study because it usually identifies those variables that significantly impact on the overall performance of the process and also helps identify the major contributors to the total annualized cost.

36.7   SUMMARY

1. Although the technical parameters influencing the selection and design of a given type of an environmental control system may be unique, cost is the only parameter relevant to all systems. This is especially so where different types of systems can control a source to achieve an emission limit. Here, cost is often the parameter used to select the optimum system from the alternatives available.

2. Equipment cost is a function of many variables, one of the most significant of which is capacity. Other important variables include equipment type and location, operating temperature, and degree of equipment sophistication.

3. The usual technique for determining the capital costs (i.e., total capital costs, which include equipment design, purchase, and installation) for a project and/or process can be based on the *factored method* of establishing direct and indirect installation costs as a function of the known equipment costs.

4. Operating costs can vary from site to site because these costs partly reflect local conditions — for example, staffing practices, labor, and utility costs. Operating costs, like capital costs, may be separated into two categories: direct and indirect costs.

5. In comparing alternate processes or different options of a particular process from an economic point of view, it is recommended that the total capital cost be converted to an annual basis by distributing it over the projected lifetime of the facility. The sum of both the annualized capital costs (ACCs) and the annual operating costs (AOCs) is known as the *total annualized cost* (TAC) for the facility. The economic merit of the proposed facility, process, or scheme can be examined once the total annual cost is available.

6. Once a particular process scheme has been selected, it is common practice to optimize the process from a capital cost and O&M (operation and maintenance) standpoint.

REFERENCES

1. U.S. EPA, *OAQPS Control Cost Manual*, 4th ed., EPA 450/3-90-006, 1990.

2. R. J. McCormick and R. J. DeRosier, *Capital and O&M Cost Relationships for Hazardous Waste Incineration*, Acurex Corp., Cincinnati, OH, EPA Report 600/2-87-175, October 1984.

3. L. Theodore and J. Reynolds, *Introduction to Hazardous Waste Incineration*, Wiley-Interscience, New York, 1988.

4. J. Matley, "CE Cost Indexes Set Slower Pace," *Chemical Engineering*, 75–76, April 29, 1985.

5. *Chemical Engineering*, McGraw-Hill, New York, 1992.

6. R. B. Neveril, *Capital and Operating Costs of Selected Air Pollution Control Systems*, Gard, Inc., Niles, IL, EPA Report 450/5-80-002, December 1978.

7. W. M. Vatavuk and R. B. Neveril, "Factors for Estimating Capital and Operating Costs," *Chemical Engineering*, 157–162, November 3, 1980.

8. G. A. Vogel and E. J. Martin, "Hazardous Waste Incineration, Part 1 — Equipment Sizes and Integrated-Facility Costs," *Chemical Engineering*, 143–146, September 5, 1983.

9. G. A. Vogel and E. J. Martin, "Hazardous Waste Incineration, Part 3 — Estimating Capital Costs of Facility Components," *Chemical Engineering*, 75–78, October 17, 1983.

10. G. A. Vogel and E. J. Martin, "Hazardous Waste Incineration, Part 3 — Estimating Capital Costs of Facility Components," *Chemical Engineering*, 87–90, November 28, 1983.

11. G. D. Ulrich, *A Guide to Chemical Engineering Process Design and Economics*, John Wiley & Sons, New York, 1984.

12. G. A. Vogel and E. J. Martin, "Hazardous Waste Incineration, Part 4 — Estimating Operating Costs," *Chemical Engineering*, 97–100, January 9, 1984.

13. E. P. DeGarmo, J. R. Canada, and W. G. Sullivan, *Engineering Economy*, 6th edn., Macmillan, New York, 1979.

14. C. Hodgman, S. Selby, and R. Weast, eds., CRC *Standard Mathematical Tables*, 12th ed., Chemical Rubber Company, Cleveland, OH (presently CRC Press, Boca Raton, FL), 1961.

VIII

NEW TECHNOLOGIES
AND APPROACHES

37

BIOREMEDIATION

37.1 INTRODUCTION

Bioremediation is by far the most commonly applied remediation technology. Bioremediation is a process that utilizes microorganisms to transform harmful substances into nontoxic compounds. It is one of the most promising new technologies for treating chemical spills and hazardous wastes. For over two decades it has been used to degrade petroleum products and hydrocarbons. It is a potentially effective treatment technique for many of the 10,000 to 15,000 oil spills that occur each year.

This process uses naturally occurring microorganisms such as bacteria, fungi, or yeast to degrade harmful chemicals into less toxic or nontoxic compounds. Bioremediation can be used as an *in situ* remediation technology that uses indigenous microorganisms to treat contaminated soil and groundwater in place without removing the soil or groundwater or it can be used to treat soil and groundwater in an *ex situ* mode by applying the bioremediation technique in either composite piles or bioreactors.

Composite piles takes many different forms such as biopiles, biocells, bioheaps, and biomounds, to name a few. This technology involves heaping contaminated soils into piles or cells and stimulating aerobic microbial activity within the soils through aeration and/or addition of minerals, nutrients, and moisture. Composite piles have proved to be effective in reducing concentrations of nearly all the constituents of petroleum products typically found at underground storage tank sites. Land farming (see Chapter 28), also known as land treatment or land application, is another above-ground bioremediation technology for soils that reduces the concentrations of petroleum constituents. If the contaminated soils are shallow (less than 3 ft below the ground surface), it may be possible to effectively stimulate microbial activity without excavating the soils. If the petroleum-contaminated soils is deeper than 5 ft, the soils should be excavated and reapplied on the ground surface.

In situ bioremediation can be used to treat both soil and groundwater. *In situ* bioremediation can be implemented in a number of treatment modes, including aerobic (oxygen respiration); anixic (nitrate respiration) and anaerobic (non-oxygen respiration) and co-metabolic. The aerobic mode has proved most effective in reducing contaminant levels of aliphatic and aromatic petroleum hydrocarbons typically found in gasoline and diesel fuels. In the aerobic application, groundwater is oxygenated by one of three methods: (a) direct sparging of air or oxygen through an injection well; (b) saturation of water with air or oxygen prior to injection; and (c) the addition of oxygen-releasing compounds such as hydrogen peroxide. Whichever method of oxygenation is used, it is important to ensure that oxygen is distributed throughout the area of contamination.

Microorganisms, like all living organisms, need nutrients (such as nitrogen, phosphate, and trace metals), carbon, and energy to survive. Microorganisms break down a wide variety of organic (carbon-containing) compounds found in nature to obtain energy for their growth. Many species of soil bacteria, for example, use petroleum hydrocarbons as a food and energy source, transforming them into harmless substances consisting mainly of carbon dioxide, water, and fatty acids. Bioremediation harnesses this natural process by promoting the growth of microorganisms that can degrade contaminants and convert them into nontoxic byproducts. In many instances, environmental conditions can be altered to enhance the bioremediation process. By altering the types of microorganisms present, nutrients, and climatic conditions (i.e., pH, moisture, temperature, and oxygen levels) microbial degradation can be enhanced.

Bioremediation can be an attractive option for many reasons. First, it is an ecologically sound "natural" process. New strains of bacteria which most efficiently break down organic wastes often appear naturally. As a result, the population of the strains explodes, propelling the "breaking down of hazardous wastes," or bioremediation process, forward. When soil bacteria are exposed to organic contaminants, they tend to develop an increased ability to degrade those substances. These bacteria can increase in numbers when a food source (the waste) is present. When the contaminant is degraded, the microbial population naturally declines. Second, instead of merely transferring the contaminants from one place to another, e.g., to a hazardous waste landfill, bioremediation destroys the target chemicals — residues from the biological treatment are usually harmless products. Third, it is usually less expensive than other technologies. Finally, bioremediation can often be accomplished *where* the problem is located. This eliminates the need to transport large quantities of contaminated waste off-site and the potential threats to health and the environment that can arise during such transport (1).

37.2 CHARACTERISTICS OF BIOREMEDIATION SYSTEMS

Bioremediation systems utilize biological processes for the destruction of contaminants in the soil and the groundwater. Biological systems incorporate natural microorganisms into the treatment scheme through modifications of the existing soil and groundwater environment to encourage their growth and reproduction. In the process of growth, these microorganisms require a source of energy (electron donor), and a means of extracting this energy from the electron donor via an appropriate electron acceptor, as follows:

Microbes + Electron Donor (Energy & Carbon Source) + Nutrients + Electron Acceptor → More Microbes + Oxidized End Products

The application of biological systems for waste treatment encourages the use of waste contaminants of concern as the electron donor, while supplying indigenous microorganisms with the electron acceptors and nutrients that they require.

In the application of biological treatment for waste destruction, the limiting factor in full-scale engineered systems is the rate of transfer of the electron acceptor to the reaction site. Electron acceptors are those chemicals that can be utilized by biological systems to extract energy from electron donors so that energy is available for cell replication and growth. The main electron acceptors of interest include oxygen, nitrate, sulfate, carbon dioxide, and organic carbon. Oxygen is the preferred electron acceptor, as its use results in a maximum energy yield to the microorganism, thereby yielding the maximum possible amount of cell production and organism growth per unit amount of electron donor utilized. The subsequent utilization of electron acceptor by the microorganisms follows the sequence nitrate, sulfate, carbon dioxide, and finally organic carbon based on the relative energy yield resulting from their use. The approximate relative energy release from each of these electron acceptors is summarized in Table 37.1.

In addition to yielding higher energy releases as compared to the other terminal electron acceptors, a wide variety of chemicals are degraded. Using it as an electron acceptor, it generally yields a more rapid rate of contaminant degradation and yields oxidized end products that can be safely released into the environment. Oxygen-based biological systems are also the most preferred engineered systems because of their inherent stability and process performance.

Biological treatment processes applicable to fuel-contaminated sites are directed towards aerobic conversion of organic contaminants of concern to carbon dioxide and water (termed mineralization), or the oxidation of complex parent compounds to smaller, more oxidized constituents. With compounds of gasoline being straight, branched-chain, and aromatic hydrocarbons, few if any intermediate products can be expected, and bioremediation, if feasible, will result in complete contaminant mineralization. A variety of engineered biological systems for soil and groundwater treatment have been developed. For contaminated soils, *ex situ* and *in situ* treatments include the slurry phase reactor, land treatment and soil piles, bioventing and biosparging. For contaminated groundwater, *ex situ* and *in situ* treatments include the suspended growth reactor, fixed film reactor, hybrid reactor, and biofilters. Biological systems have advantages over other types of treatment technologies in that if they are applicable to a given site/soil/waste situation they:

- Lead to complete destruction of the contaminants of concern under ambient pressure and temperature conditions.

- Due to low temperature and pressure reaction conditions, they are highly cost-effective systems.

Table 37.1 Relative energy yield from various terminal electron acceptors

Electron Acceptor	Relative Energy Yield
Oxygen	50
Nitrate	45
Sulfate	2
Carbon dioxide	1
Organic carbon	1

- They generate no chemical sludges and generally require minimal chemical addition to maintain operating conditions at optimal levels.
- If the processes are conducted *in situ*, no solid residues are produced at all.
- Advances in process design have extended the applicability of biological reactors to soils, slurries, and water and gas phase treatment.

Because they are biological in nature, however, their applicability is limited to:

- Those sites where the contaminants of concern are biological, i.e., the contaminants must be able to serve as a source of energy and carbon for the cells to grow. Specialized applications to chlorinated solvents and some polycyclic aromatic hydrocarbons (PAHs) have been demonstrated utilizing co-metabolic reactions, but these techniques will generally not be applicable to most underground storage tank (UST) sites.
- Those sites where toxicants or chemicals that can inhibit microbial activity are not present. These inhibitary materials include high concentrations of some organic compounds, and particularly moderate to high concentrations of heavy metals.

Biological processes are applicable for a broad spectrum of organic chemicals of environmental concern. Table 37.2 presents common constituents of fuels that have been shown susceptible to biodegradation. It should be noted that high concentrations ($> \approx 1\%$) of otherwise biodegradable compounds may not be amenable to biodegradation because of toxicity and/or inhibition that develops at these high contaminant levels. When free products exist, some inhibition can be expected, and product recovery is highly recommended to enhance the potential performance of bioremediation schemes.

A number of microbial and environmental factors affect soil and groundwater bioremediation.

- *Microbial populations:* An acclimated indigenous population of microorganisms capable of degrading the compounds of interest must exist at the site if bioremediation is to be successful. A wide variety of organisms have been isolated from the environment with the capability to degrade petroleum hydrocarbons commonly found in gasoline, diesel, jet fuels, etc. It is highly likely, then, that the addition of specially cultured organisms will be necessary at an UST site. In addition, delivery and control over microbial amendments is difficult at best, and microbial augmentation at a site should not be

Table 37.2   Biodegradable UST-related compounds

Compound	Aerobic	Anaerobic
Straight chain alkanes	X	X
Branched alkanes	X	X
Cyclic alkanes	X	
Aromatic hydrocarbons	X	X
Two- and three-ringed polycyclic hydrocarbons	X	

Source: U.S. EPA, 1985.

considered. If an active population of indigenous microorganisms does not exist at a site, inhibitory and/or toxic conditions should be suspected, and alternatives to bioremediation should be considered.

- *Oxygen:* As indicated above, oxygen is the preferred electron acceptor, and is necessary for aerobic biodegradation of organic contaminants. Residual oxygen concentrations of >1.0 mg/L in the aqueous phase, and $>2\%$ vol should be maintained in the gas phase for vapor systems to ensure that oxygen is not limiting the overall reaction rate.

- *Soil water:* This parameter is important in soil-based systems as the microorganisms rely on the soil water to supply a habitat for their growth and survival. The soil water also provides a media for transfer of contaminants from the product or solid phases to the microorganisms. Soil water should be in the range of 25% to 85% field capacity (the water content of the soil after it freely drains by gravity) to sustain microbial activity. Optimal soil water content is generally found in the range of 75% field capacity.

- *pH:* pH is a measure of the hydrogen ion concentration, and should be in the range of 7 (neutral pH) for optimal biological treatment performance. pH within the range of 5.5 to 8.5 will support biological activity, but should be adjusted within the range to 7 if possible to improve process performance. Because soil pH is difficult to impossible to modify, it can be used as a primary indicator of the feasibility of soil bioremediation based on site assessment data.

- *Nutrients:* Nutrients can be classified into major, minor and trace element groupings. The common major nutrients of concern for bioremediation include nitrogen and phosphorus, while the minor nutrients include sodium, potassium, calcium, magnesium, iron, chloride, and sulfur. The major nutrients are required in order-of-magnitude higher levels than the minor and the trace nutrients, and subsequently are the nutrients of concern in terms of managed bioremediation. A typical C:N:P weight ratio of 100:10:1 is often utilized to ensure that adequate levels of N and P exist for unhindered bioremediation. These values are approximately half of that found in cell material (generally estimated to be $C_5H_7O_2N$, where the C:N:P is $\approx 50{:}10{:}1$) due to the assumption that half of the carbon in the contaminant is used to produce cell material, while the balance is used for energy production by the cells. One additional note is important. For *in situ* remediation systems (i.e., bioventing), nutrient addition has not been shown to improve bioactivity, and it is generally thought that nutrient addition for these systems is unnecessary.

- *Temperature:* Biological systems can operate over a wide range of temperature conditions from 5 to 60°C. Three temperature ranges have been identified based on the growth of distinct groups of microorganisms:
  - Psychrophilic, $< 15°C$
  - Mesophilic, 15 to 45°C
  - Thermophilic, $> 45°C$

In general, most UST site conditions would result in mesophilic temperatures, and should be adequate to support active microbial growth. If systems are exposed to temperatures below 10°C for extended periods of time during winter months, their performance should be expected to deteriorate until temperatures are raised again. A rule of thumb suggests that reaction rates will increase or decrease by a factor of 2 for each 10°C rise or fall of temperature. In addition, soil-based treatment systems in northern climates have not shown a significant lag period in system performance when temperatures are raised, indicating that summertime performance should rapidly resume as soil/groundwater temperatures are raised within the mesophilic range.

· *Toxicants in Waste:* Because of the biological nature of bioremediation systems any material that disrupts the biochemicals processes taking place within the microorganisms employed in the treatment system will cause a disruption and eventual failure of that system. A variety of organic and inorganic toxicants can adversely affect the biological treatment system. The microbial consortium existing within the biological treatment system can acclimate to some of the materials, and by design (i.e., blending of contaminated soil with adjacent uncontaminated soil in a soil pile or land farm system) concentrations of the toxicants can be reduced below inhibitory levels to allow their eventual degradation over time (2).

37.3 PRELIMINARY SCREENING OF *IN SITU* GROUNDWATER BIOREMEDIATION

An initial screening should be conducted to determine if bioremediation is a feasible technology for remediating the contaminated site. This preliminary assessment will allow you to quickly determine whether or not bioremediation should be option in the overall corrective action process without investing significant resources.

1. *Nature and Extent of the Contamination:* Determine the types of contaminants and the extent of the contamination from both historical data and the site assessment. The chemical characteristics of the contaminants will dictate their biodegradability. Are the chemical constituents potentially biodegradable? If yes, the biodegradation is a likely candidate for remediation. For example, heavy metals are not degraded by bioremediation. The biodegradability of organic constituents depends on their chemical structures and physical/chemical properties (e.g., water solubility, water/octanol partition coefficient). Highly soluble organic compounds with low molecular weights will tend to be more rapidly degraded than slightly soluble compounds with high molecular weights. The low water solubilities of the more complex compounds render them less bioavailable to petroleum-degrading organisms. Consequently the larger, more complex chemical compounds may be slow to degrade or even be recalcitrant to biological degradation (e.g., asphaltenes in No. 6 fuel oil).

2. *Nature of Soil:* Determine the types of soil within the contaminated area. This data is obtained from several sources — primarily from the site assessment. The site assessment will include sampling and analysis for soil, soil gas, and groundwater. It will also provide crucial information on the geology/hydrogeology, the nature and the extent of the contamination, migration pathways, and points of exposure. The soil analysis will determine whether or not bioremediatiom would be an effective remedial option. Bioremediation is generally effective in permeable media (i.e., sandy, gravelly acquifer media). In general, an aquifer medium of lower permeability will require a longer time to clean up than a more permeable medium. The differences in soil structures and textures will impact the flow and distribution

of the contaminants and reactants through the subsurface. The location, distribution, and disposition of petroleum contamination in the subsurface can significantly influence the likelihood of success for bioremediation.

37.4 EVALUATION OF *IN SITU* GROUNDWATER BIOREMEDIATION EFFECTIVENESS

In order to conduct a detailed evaluation of an *in situ* bioremediation option, you will need to identify specific soil and constituent characteristics and properties, compare them to ranges where *in situ* groundwater bioremediation is potentially effective and decide whether treatability studies are necessary to determine effectiveness. Figure 37.1 details some of the decisions necessary during the evaluation.

While the preliminary screening will focus on hydraulic conductivity and the constituent biodegradability, the detailed evaluation should consider a broader range of site and constituent characteristics.

For Soil Characteristics
- Hydraulic conductivity
- Soil Structure and stratification
- Groundwater mineral content
- Groundwater pH
- Groundwater temperature
- Microbial presence
- Terminal electron acceptors
- Nutrient concentrations

For Constituent Characteristics
- Chemical structure
- Concentration and toxicity
- Solubility

Soil Characteristics that Affect *In Situ* Groundwater Bioremediation

Hydraulic Conductivity. Hydraulic conductivity is a measure of water's ability to move through the aquifer medium and is one of the most important factors in determining the potential effectiveness of *in situ* groundwater bioremediation. Hydraulic conductivity can be determined from aquifer tests such as slug tests and pumping tests. For aquifers with hydraulic conductivity greater than 10^{-4} cm/sec, *in situ* groundwater bioremediation is effective. For sites with lower hydraulic conductivities (e.g., 10^{-4} to 10^{-6} cm/sec), the technology also could be effective but it must be carefully evaluated, designed, and controlled.

Soil Structure and Stratification

Soil structure and stratification are important to *in situ* groundwater bioremediation because they affect groundwater flowrates and patterns when water is extracted and injected. Structural characteristics such as microfracturing can result in higher permeabilities than expected for certain soils (e.g., clays). In this case, flow will increase in the fractured media but not in the unfractured media. The

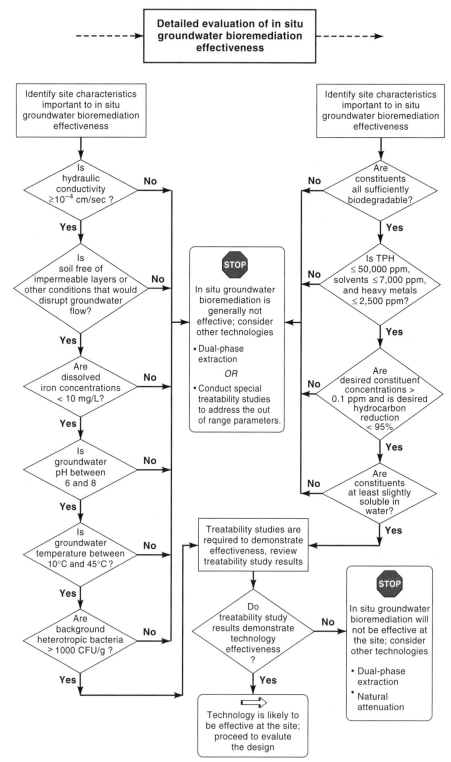

FIGURE 37.1 *In situ* groundwater remediation evaluation process flow chart.

stratification of soils with different permeabilities can dramatically increase the lateral flow of groundwater in the more permeable strata while reducing the flow through less permeable strata. This preferential flow behavior can lead to reduced effectiveness and extended remedial times for less-permeable strata.

The intergranular structure and stratification of aquifer media can be determined by reviewing soil logs from wells or borings and examining geologic cross-sections. It will be necessary to verify that soil types have been properly identified, that visual observations of soil structure have been documented, and that boring logs are of sufficient detail to define soil stratification. Stratified soils may require special design consideration (e.g., special injection wells) to ensure that these less permeable strata are adequately handled.

Fluctuations in the groundwater table should also be determined. Significant seasonal or daily (e.g., tidal, precipitation-related) fluctuations will submerge some of the soil in the unsaturated zone which should be considered during design of the system.

Groundwater Mineral Content

Excessive calcium, magnesium, or iron in groundwater can react with phosphate, which is typically supplied as a nutrient in the form of tripolyphosphate, or carbon dioxide, which is produced by microorganisms as a by-product of aerobic respiration. The products of these reactions can adversely affect the operation of an *in situ* bioremediation system. When calcium, magnesium or iron reacts with phosphate or carbon dioxide, crystalline precipitates or "scale" is formed. Scale can constrict flow channels and can also damage equipment, such as injection wells and sparge points. In addition, the precipitation of calcium or magnesium phosphates ties up phosphorous compounds, making them unavailable to microorganisms for use as nutrients. This effect can be minimized by using tripolyphosphates in a mole ratio of greater than 1:1 tripolyphosphates to total minerals (i.e., magnesium and calcium). At these concentrations, the tripolyphosphates act as a sequestering agent to keep the magnesium and calcium in solution (i.e., prevent the metal ions from precipitating and forming scale).

When oxygen is introduced into the subsurface as a terminal electron acceptor, it can react with dissolved iron (Fe^{+2}) to form an insoluable iron precipitate, ferric oxide. This precipitate can be deposited in aquifer flow channels, reducing permeability. The effects of iron precipitation tend to be most noticeable around injection wells, where oxygen concentration in groundwater is highest and can render injection wells inoperable. Table 37.3 provides a guide to assessing the potential impact of dissolved iron in groundwater.

Other parameters that could be good indicators of potential groundwater scaling are hardness, alkalinity, and pH. In particular, very hard water (i.e., $\geqslant 400\text{--}500\,\text{mg/L}$ carbonate hardness) tends to promote scaling. The potential adverse effects caused by excessive mineral content (e.g., calcium, magnesium, iron, total carbonates) in the groundwater warrants careful attention during site characterization activities.

Groundwater pH

Extreme pH values (i.e, less than 5 or greater than 10) are generally unfavorable for microbial activity. Typically, optimal microbial activity occurs under neutral pH conditions (i.e., in the range of 6–8). The optimal pH is site-specific. For

Table 37.3   Dissolved iron and *in situ* bioremediation effectiveness

Dissolved Iron Concentration (mg/L)	Effectiveness
$Fe^{+2} < 10$	Probably effective
$10 \leqslant Fe^{+2} \leqslant 20$	Injection wells require periodic testing and may need periodic cleaning or replacement
$Fe^{+2} > 20$	Not recommended

example, aggressive microbial activity has been observed at lower pH conditions outside of this range (e.g., 4.5 to 5) in natural systems. Because indigenous microorganisms have adapted to the natural conditions where they are found, pH adjustment, even toward neutral, can inhibit microbial activity. If man-made conditions (e.g., releases of petroleum) have altered the pH outside the neutral range, pH adjustment may be needed. If the pH of the groundwater is too low (too acid), lime or sodium hydroxide can be added to increase the pH. If the pH is too high (too alkaline), then a suitable acid (e.g., hydrochloric, muriatic) can be added to reduce the pH. Changes to the pH should be closely monitored because rapid changes of more than 1 or 2 units can inhibit microbial activity and may require an extended acclimation period before the microbes resume their activity.

Groundwater Temperature

Bacterial growth rate is a function of temperature. Subsurface microbial activity has been shown to decrease significantly at temperatures below 10°C and essentially to cease below 5°C. Microbial activity of most bacterial species important to petroleum hydrocarbon biodegradation also diminishes at temperatures greater than 45°C. Within the range of 10°C to 45°C, the rate of microbial activity typically doubles for every 10°C rise in temperature. In most cases, for *in situ* groundwater bioremediation, the bacteria living in an aquifer system are likely to experience relatively stable temperature with only slight seasonal variations. In most areas of the United States, the average groundwater temperature is about 13°C, but groundwater temperatures may be somewhat lower or higher in the extreme northern and southern states.

Microbial Presence

Soil normally contains large numbers of diverse microorganisms, including bacteria, algae, fungi, protozoa, and actinomycetes. Of these organisms, the bacteria are the most numerous and biochemically active group, particularly at low oxygen levels, and they contribute significantly to *in situ* groundwater bioremediation.

At a contained site, the natural microbial population undergoes a selection process. First, there is an acclimation period, during which microbes adjust to their new environment and new source of food. Second, those organisms that adapt most quickly tend to grow fastest and can use up nutrients that other microbes would need. Third, as the environmental conditions change and the nature of the food supply changes, the microorganism populations change as well. Organisms capable of withstanding the stress of their changing environment will generally be those that will contribute to the bioremediation of the site.

To determine the presence and population density of naturally occurring bacteria that will contribute to degradation of petroleum constituents, laboratory analysis of soil samples from the site should be completed. These analyses, at a minimum, should include plate counts for total heterotrophic bacteria (i.e., bacteria that can use organic compounds as an energy source) and hydrocarbon-degrading bacteria. Although heterotrophic bacteria are normally present in all soil environments, plate counts of less 1000 colony-forming units (CFU)/g of soil could indicate depletion of oxygen or other essential nutrients or the presence of toxic constitutents. However, concentrations as low as 100 CFU/g of soil can be stimulated to acceptable levels, assuming toxic conditions (e.g., exceptionally high concentrations of heavy metals) are not present. These conditions are summarized in Table 37.4.

Some Corrective Action Plans (CAPs) propose the addition of microorganisms (bioaugmentation) into the aquifer environment when colony plate counts are low. However, research has shown that most *in situ* bioremediation projects have been successfully completed without microbial augmentation. Experience with microbial augmentation shows that it varies in effectiveness. Except in coarse-grained, highly permeable material, microbes tend not to move very far past the point of injection; therefore, their effectiveness is limited in extent. In general, microbial augmentation does not adversely affect bioremediation, but it could represent an unnecessary cost.

Terminal Electron Acceptors

Microorganisms require carbon as an energy source to sustain their metabolic functions, which include growth and reproduction. The metabolic process used by bacteria to produce energy requires a terminal electron acceptor (TEA) to enzymatically oxidize the carbon source (organic matter) to carbon dioxide.

$$\text{Organic Matter} + O_2 + \text{Biomass} \rightarrow CO_2 + H_2O + \Delta H_f$$

where ΔH_f is energy generated by the reaction to fuel other metabolic processes including growth and reproduction. In this example, oxygen serves as the TEA.

Microorganisms are classified by the carbon and TEA sources they use to carry out metabolic processes. Bacteria that use organic compounds as their source of carbon are called heterotrophs; those that use inorganic carbon compounds such as carbon dioxide are called autotrophs. Bacteria that use oxygen as their TEA are

Table 37.4 Heterotrophic bacteria and *in situ* groundwater bioremediation effectiveness

Total Heterotrophic Bacteria (CFU/g dry soil)	Effectiveness
> 1000	Generally effective
100–1000	May be effective; needs further evaluation to determine whether toxic conditions are present and/or whether population responds to stimulation (e.g., increased supply of electron acceptor and/or nutrients)
< 100	Not generally effective

called aerobes; those that use a compound other than oxygen (e.g., nitrate, sulfate) are called anaerobes; and those that can utilize both oxygen and other compounds as TEAs are called facultative. For *in situ* groundwater bioremediation applications directed at petroleum products, bacteria that are both aerobic (or facultative) and heterotrophic are most important in the degradation process.

Nutrient Concentrations

Microorganisms require inorganic nutrients such as nitrogen and phosphate to support cell growth and sustain biodegradation processes. Nutrients may be available in sufficient quantities in the aquifer but, more frequently, nutrients need to be added to maintain adequate bacterial populations.

A rough approximation of maximum nutrient requirements can be based on the stoichiometry of the overall biodegradation process:

$$\text{C-source} + \text{N-source} + O_2 + \text{Minerals} + \text{Nutrients} \rightarrow$$
$$\text{Cell mass} + CO_2 + H_2O + \text{other metabolic by-products}$$

Different empirical formulas of bacterial cell mass have been proposed; the most widely accepted are $C_5H_7NO_2$ and $C_{60}H_{87}O_{32}N_{12}P$. Using the empirical formulas for cell biomass and other assumptions, the carbon:nitrogen:phosphorus ratios necessary to enhance biodegradation fall in the range of 100:10:1 to 100:1:0.5, depending on the constituents and bacteria involved in the biodegradation process.

Chemical analyses of soil samples (collected from below the water table) and groundwater samples should be completed to determine the available concentrations of nitrogen (expressed as ammonia) and phosphate. Soil analyses are routinely conducted in agronomic laboratories that test soil fertility for farmers. These concentrations can be compared to the nitrogen and phosphorus requirements calculated from the stoichiometric ratios of the biodegradation process. Some microbes can use nitrate as a nitrogen source. The drinking water standard for nitrate is 40 mg/L and there may be regulatory prohibitions against injecting nitrate into groundwater. If nitrogen addition is necessary, slow release sources should be used and addition of these materials should be monitored throughout the project to prevent degradation of water quality. In addition, excessive nitrogen additions can lower soil pH, depending on the amount and type of nitrogen added.

Because of water quality and soil chemistry considerations, *in situ* groundwater bioremediation should be operated at near nutrient-limited conditions.

Constituent Characteristics That Affect *In Situ* Groundwater Bioremediation

Chemical Structure. The chemical structures of the constituents to be treated by *in situ* groundwater bioremediation are important for determining the rate at which biodegradation will occur. Although nearly all constituents in petroleum products typically found at UST sites are biodegradable, the more complex the molecular structure of the constituent, the more difficult the product is to treat and the greater the time required for treatment. Most low-molecular-weight (nine carbon atoms or less) aliphatic and monoaromatic constituents are more easily biodegraded than higher-molecular-weight aliphatic or polyaromatic organ constituents. Straight-chain aliphatic (i.e, alkanes, alkenes, and alkynes) hydrocarbon compounds are more readily degraded than their branched isomers, and monoaromatic compounds (e.g., benzene, ethyl benzene, toluene, xylenes) are more rapidly degraded than the two-ring compounds (e.g., naphthalene), which in turn are more rapidly degraded than the larger multiringed compounds (i.e., polyaromatic hydrocarbons or polynuclear aromatic hydrocarbons). The larger, more complex chemical structures may be slow to degrade or be essentially resistant to biological degradation (e.g., asphaltenes in No. 6 fuel oil). Table 37.5 lists, in order of decreasing rate of potential biodegradability, some common constituents found at petroleum UST sites.

Petroleum hydrocarbon contamination is sometimes accompanied by other organic contaminants, including both nonchlorinated solvents (e.g., alcohols, ketones, esters, acids) and chlorinated compounds (e.g., trichloroethane, chlorinated phenols, polychlorinated biphenyls (PCBs)). The nonchlorinated solvents tend to be readily biodegradable but can exert toxic effects at high concentrations. Lightly chlorinated compounds (e.g., chlorobenzene, dichlorobenzene, chlorinated phenols, lightly chlorinated PCBs) are typically degradable under aerobic conditions. The more highly chlorinated compounds tend to be more resistant to aerobic degradation, but they can be degraded by dechlorination under anaerobic conditions. Several common chlorinated solvents (e.g., chlorinated ethanes, ethenes) can be degraded under aerobic conditions if they exist in the presence of another contaminant that can behave as a co-metabolite (e.g., methane, toluene, phenol).

Evaluation of the chemical structure of the constituents proposed for reduction by *in situ* groundwater bioremediation at the site will allow you to determine which constituents will be the most difficult to degrade. You should verify that remedial time estimates, treatability studies, and operation and monitoring plans are based on the constituents that are the most difficult to degrade in the biodegradation process.

Concentration and Toxicity

High concentrations of petroleum organics or heavy metals in site soils can be toxic to or inhibit the growth and reproduction of bacteria responsible for biodegradation. In addition, very low concentrations of organic material will result in diminished levels of bacterial activity.

In general, concentrations of petroleum hydrocarbons (measured as total petroleum hydrocarbons) in excess of 50,000 ppm, organic solvent concentrations in excess of 7,000 ppm, or heavy metals in excess of 2,500 ppm in the groundwater or aquifer medium are considered inhibitory and/or toxic to aerobic bacteria. Review of the CAP to verify that the average concentrations of petroleum hydrocarbons and heavy metals in the soils and groundwater to be treated are below these levels is necessary.

In addition to maximum concentrations, you should consider the clean-up concentrations proposed for the treated soils. Below a certain "threshold" constituent concentration, the bacteria cannot obtain sufficient carbon from degradation of the constituents to maintain adequate biological activity. The threshold level determined from treatability studies conducted in the laboratory is likely to be much lower than what is achievable in the field under less than optimal conditions. Although the threshold limit varies greatly depending on bacteria-specific and

Table 37.5 Chemical structure and biodegradability

Biodegradability	Example Constituents	Products in Which Constituents is Typically Found
More degradable	n-Butane, 1-pentane n-Octane	Gasoline
	Nonane	Diesel fuel
	Methyl butane, Dimethylpentenes, Methyloctanes	Gasoline
	Benzene, toluene, ethylbenzene, xylenes	Gasoline
	Propylbenzenes	Diesel, kerosene
	Decanes	Diesel
	Dodecanes	Kerosene
	Tridecanes	Heating fuels
	Tetradecanes	Lubricating oils
Less degradable	Naphthalenes	Diesel
	Fluoranthenes	Kerosene
	Pyrenes	Heating oil
	Acenaphthenes	Lubricating oils
Resistant	Asphaltenes	Fuel oil no. 6
	MTBE	Gasoline

constituent-specific features, constituent concentrations below 0.1 ppm in the total aquifer matrix may be difficult to achieve. However, concentrations in the groundwater for these specific constituents may be below detection levels.

Experience has shown that reductions in petroleum hydrocarbon concentrations greater than 95% can be very difficult to achieve because the presence of "resistant" or nondegradable petroleum constituents. Identify the average starting concentrations and the desired clean-up concentrations in the CAP. If a clean-up level lower than 0.1 ppm is required for any individual constituent or a reduction in petroleum hydrocarbon concentration of greater than 95% is required to reach the clean-up level, either a treatability study should be required to demonstrate the ability of bioremediation to achieve these reductions at the site, or another technology should be considered. Another option is to combine one or more technologies to achieve clean-up goals.

Solubility

Solubility is the amount of a substance (e.g., hydrocarbon) that will dissolve in a given amount of another substance (e.g., water). Therefore, a constituent's solubility provides insight as to its fate and transport in the aqueous phase. Constituents that are highly soluble have a tendency to dissolve into the groundwater and are more available for biodegradation. Conversely, chemicals that have low water solubilities tend to remain in the adsorbed phase and will biodegrade more slowly. In general, lower molecular weight constituents tend to be more soluble and biodegrade more readily than do higher molecular weight or heavier constituents. In the field, aqueous concentrations rarely approach the solubility of a substance because dissolved concentrations tend to be reduced through competitive dissolution of other constituents and degradation processes such as biodegradation, dilution, and adsorption.

37.5 EVALUATIONS OF *IN SITU* GROUNDWATER BIOREMEDIATION SYSTEM DESIGN

Once you have verified that *in situ* groundwater bioremediation has the potential to be effective, you can evaluate the design of the proposed remedial system. The CAP should include a discussion of the rationale for the design and present the conceptual engineering design. Detailed engineering design documents might also be included, depending on state requirements.

The following design elements are typically required:

- *Volume and area of aquifer* to be treated is generally determined by site characterization combined with regulatory action levels or a site specific risk assessment.

- *Initial concentration of constitutents of concern* can be measured during initial site characterization and during treatability studies. These concentrations will be used to predict likely toxic effects of the contaminants on indigenous microorganisms and to estimate electron acceptor and nutrient requirements, and the extent of treatment required.

- *Required final constituent concentrations* are generally defined by your state as remediation action levels or determined on a site-specific basis using transport models and/or risk assessment calculations. These limits will define the areal extent of the aquifer to be remediated.

- *Estimates of electron acceptor and nutrient requirements.* As a rule of thumb, 3 lb of oxygen are added per pound of hydrocarbon as an electron acceptor. For nutrients, a maximum ratio of 100:10:1 for C:N:P is typically used (assume 1 lb of hydrocarbon is equal 1 lb of carbon). Often systems require substantially less, on the order of 100:1:0.5, especially if plugging of injection wells/galleries is a problem.

- *Layout of injection and extraction wells.* Probably the most critical factor is ensuring that the contaminant plume is hydraulically controlled. This will prevent it from spreading and concentrate bioremediation efforts on the contaminants. For large complex sites, designing this layout can be facilitated by groundwater modeling. Injection wells and infiltration galleries can be located upgradient of the contaminant source, with extraction wells located downgradient of the source. Alternatively, injection points can be located along the centerline of the plume. The latter arrangement can typically achieve shorter remediation times, but at greater expense.

- *Design Area of Influence (A.I.).* The A.I. is an estimate of the volume/area of aquifer to which an adequate amount of electron acceptor and nutrient can be supplied to sustain microbial activity. Establishing the design A.I. is not a trivial task because it depends on many factors including intrinsic permeability of the soil, soil chemistry, moisture content, and desired remediation time.

Although the A.I. should usually be determined through field pilot studies, it can be estimated from groundwater modeling or other empirical methods. For sites with stratified geology, the area of influence should be defined for each soil type. The A.I. is important in determining the appropriate number and spacing of extraction or injection wells or infiltration galleries.

- *Groundwater extraction and injection flow rates* can vary from a few to a few hundred gallons per minute, depending primarily on the hydraulic conductivity of the aquifer. Although flow rates can be estimated by groundwater modeling, they are best determined by pilot studies. In general, only about 75% of extracted water can be readily reinjected using either injection wells or infiltration galleries.

- *Site Construction Limitations.* Locations of buildings, utilities, buried objects, etc., must be identified and considered in the design process.

- *Electron Acceptor System.* For aerobic processing, air, oxygen, or hydrogen peroxide can be used; for anaerobic processing, alternative electron acceptors (e.g., nitrate, sulfate, or ferric iron) can be used. The electron acceptors may be introduced using a direct air/oxygen sparge system into the injection well (air sparging) or a water injection system.

- *Nutrient Formulation and Delivery System.* Site characterization and bench-scale treatability studies will determine if nutrients are required. The nutrients selected should be compatible with aquifer chemistry to minimize precipitation and flow-channel fouling.

- *Bioaugmentation.* Microorganisms can be added to the injected or infiltrated water to increase microbial activity.

- *Extracted Groundwater Treatment and Disposition.* The above ground treatment system for extracted groundwater should be of sufficient size to process the volume of water extracted. Disposition of treated groundwater will depend on specific state policies. Some states discourage reinjection, although in most instances, reinjection makes good technical sense without causing adverse impact on the receiving groundwater. Groundwater treatment systems could entail biological, chemical, and/or physical treatment. The selection of the appropriate extracted groundwater treatment technology will depend on the proposed duration of the operation, size of the treatment system, and cost.

- *Remedial Cleanup Time.* Imposed remedial clean-up time could affect the design of the remedial system. Ultimately, the duration of the cleanup will depend on the rate of biological activity attainable, the bioavailability of the contaminants of concern, and the locations and spacings of the injection/extraction wells.

- *Ratio of Injection/Infiltration to Extraction.* The percentage of the treated water that is reinjected or reinfiltrated should be based on hydraulic control. Because dispersion and diffusion at the boundary of the A.I. is likely to allow some migration of contaminated groundwater, less groundwater is generally injected or recharged to the aquifer than is extracted. This provides for better hydraulic containment of the contamination.

- *Free Product Recovery System.* A system designed to recover free product should be used to reduce source input effects to the groundwater and generally optimize saturated zone remediation.

37.6 COMPONENTS OF AN *IN SITU* GROUNDWATER BIOREMEDIATION SYSTEM

Once the design rationale is defined, the design of the *in situ* groundwater bioremediation system can be developed. Figure 37.2 is a schematic diagram of a typical *in situ* groundwater bioremediation system using injection wells. A typical *in situ* groundwater bioremediation system design includes the following components and information:

* Extraction well(s) orientation, placement, and construction details
* Injection well(s) or infiltration gallery(ies) orientation, placement, and construction details
* Filtration system to remove biomass and particulates that could promote clogging of injection wells or galleries
* Extraction groundwater treatment system (e.g., biological, chemical oxidation, granular carbon adsorption) and methods for disposal or reuse of treated groundwater (surface discharge, discharge to a sewer, reinjection)
* Nutrient solution preparation system and storage
* Microorganism addition system (if required)
* Electron acceptor system (e.g., air, oxygen, hydrogen peroxide)
* Monitoring well(s) orientation, placement, and construction details
* System controls and alarms

Extraction wells are generally necessary to achieve hydraulic control over the plume to ensure that it does not spread contaminants into areas where contamination does not exist or accelerate the movement toward receptors. Placement of extraction wells is critical, especially in systems that also use nutrient injection wells or infiltration galleries. These additional sources of water can alter the natural groundwater flow patterns, which can cause the contaminant plume to move in an unintended direction or rate. Without adequate hydraulic control, this situation can lead to worsening of the original condition and complicate the cleanup or extend it.

Nutrient injection systems may not be necessary at all, if the groundwater contains adequate amounts of nutrients, such as nitrogen and phosphorus (3).

37.7 SUMMARY

1. Bioremediation is by far the most commonly applied remediation technology. Bioremediation is a process that utilizes microorganisms to transform harmful substances into nontoxic compounds.

2. The application of biological systems for waste treatment encourages the use of waste contaminants of concern as the electron donor, while supplying indigenous microorganisms with the electron acceptors and nutrients that they require.

3. An initial screening should be conducted to determine if bioremediation is a feasible technology for remediating the contaminated site. This preliminary assessment will allow you to quickly determine whether or not bioremediation should be an option in the overall corrective action process without investing significant resources.

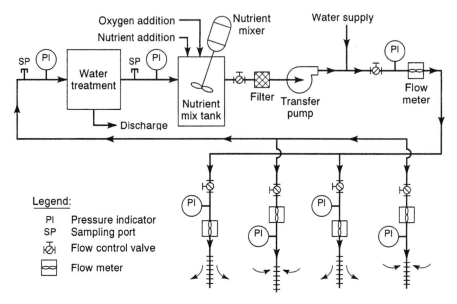

FIGURE 37.2   Schematic diagram of a typical groundwater bioremediation system using injection wells.

4. In order to conduct a detailed evaluation of an *in situ* bioremediation option, you will need to identify specific soil and constituent characteristics and properties, compare them to ranges where *in situ* groundwater bioremediation is potentially effective, and decide whether treatability studies are necessary to determine effectiveness.

5. Once you have verified that *in situ* groundwater bioremediation has the potential to be effective, you can evaluate the design of the proposed remedial system. The CAP should include a discussion of the rationale for the design and present the conceptual engineering design.

6. A typical *in situ* groundwater bioremediation system design includes extraction wells, injection wells or infiltration gallery(ies), the filtration system, the extraction groundwater treatment system, the nutrient solution preparation system and storage, a microorganism addition system (if required), the electron acceptor system, monitoring wells, system controls, and alarms.

REFERENCES

1. U.S. EPA, *Common Cleanup Methods At Superfund Sites*, Office of Emergency and Remedial Responses, EPA 540/R-94/043, August 1994.

2. U.S. EPA, *Bioremediation UST Corrective Workshop*, February 1993.

3. U.S. EPA, *How To Evaluate Alternative Cleanup Technologies for Underground Storage Tank Site*, EPA 510-B-94-003, October 1994.

38

SOIL VAPOR EXTRACTION*

38.1 INTRODUCTION

Soil vapor extraction (SVE) is rapidly becoming the technology of choice for remediation of soils contaminated with volatile organic compounds (VOCs). This technology is relatively low-cost and very effective in reducing concentrations of volatile constituents in petroleum products adsorbed to soils in the unsaturated (valdose) zone. Soil vapor extraction (SVE) is also referred to as soil venting, vacuum extraction, *in situ* volatilization or enhanced volatilization. The growing interest in this technology for removing volatile compounds is due primarily to the ease with which SVE systems can be installed and operated to achieve removal of volatile organic contaminants from the subsurface. SVE embraces two fundamental concepts; first, the contaminants must be transformed into a vapor form; and second, the vapors must be able to move through the subsurface and be extracted. In this technology, a vacuum is applied to the soil matrix to create a negative pressure gradient that causes movement of vapors toward extraction wells. Volatile constituents are readily removed from the subsurface through the extraction wells. The extracted vapors are then treated, as necessary, and discharged to the atmosphere or reinjected to the subsurface (where permissible).

Vapor extraction systems have many advantages that make this technology applicable to a broad spectrum of contaminated sites (1):

- SVE is an *in situ* technology that can be implemented with a minimum of site disturbance. In many cases, normal business operations may continue throughout the clean-up period.

- SVE has potential for treating large volumes of soil at reasonable costs, in comparison to other available technologies.

- SVE systems are relatively easy to install and use standard, readily available equipment. This allows for rapid mobilization and implementation of remedial activities.

*This chapter was adapted from *How To Evaluate Cleanup Technologies for Underground Storage Tanks Sites,* EPA 510-B-94-003, October 1994.

- SVE effectively reduces the concentration of volatile organic contaminants in the vadose zone, which in turn reduces the potential for further transport of contaminants due to vapor migration and infiltrating precipitation.

- SVE can serve as an integral component of a complete remedial program.

- Discharge vapor treatment options allow design flexibility required to satisfy site-specific air discharge regulations.

This technology has proved to be effective in reducing concentrations of volatile organic compounds (VOCs) and certain semivolatile organic compounds (SVOCs) found in petroleum products at underground storage tank (UST) sites. SVE is generally more successful when applied to the lighter (more volatile) petroleum products, such as gasoline. Diesel fuel, heating oils, and kerosene, which are less volatile than gasoline, are not readily treated by SVE but may be suitable for removal by bioventing. SVE is generally not successful when applied to lubricating oils, which are nonvolatile, but these oils may be suitable for removal by bioventing.

38.2 PRELIMINARY EVALUATION FOR SVE EFFECTIVENESS

Although SVE has become a widely used technology, it is necessary to conduct a preliminary evaluation to determine the plausibility of this technology at a particular site. This preliminary screening is the first step in determining whether or not SVE should be proposed as a component of a CAP. A typical SVE system is shown in Figure 38.1.

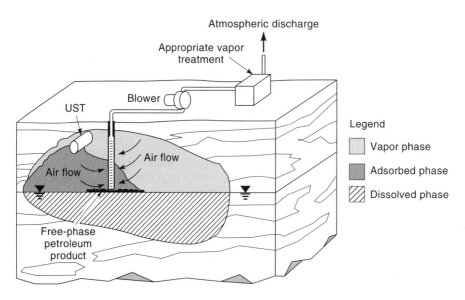

FIGURE 38.1   Typical soil vapor extraction system.

The evaluation process can be divided into the following steps.

- *Step 1:* An initial screening of SVE effectiveness, which will allow you to quickly gauge whether SVE is likely to be effective, moderately effective, or ineffective.
- *Step 2:* A detailed evaluation of SVE effectiveness, which provides further screening criteria to confirm whether SVE is likely to be effective. To complete the detailed evaluation, you will need to find specific soil and constituent characteristics and properties, compare them to ranges where SVE is effective, decide whether pilot studies are necessary to determine effectiveness, and conclude whether SVE is likely to work at a site.
- *Step 3:* An evaluation of the SVE system design, which will allow you to determine if the rationale for the design has been appropriately defined based on pilot study data or other studies, whether the necessary design components have been specified, and whether the construction process flow designs are consistent with standard practice.
- *Step 4:* An evaluation of the operation and monitoring plans, which will allow you to determine whether start-up and long-term system operation monitoring is of sufficient scope and frequency and whether remedial progress monitoring plans are appropriate.

The evaluation process is summarized in the SVE evaluation process flow chart in Figure 38.2. This process flow chart will serve as a road map to help decide whether or not SVE is an appropriate technology.

The following lists compare some of the favorable and unfavorable aspects of the use of soil vapor extraction.

Advantages of Using Soil Vapor Extraction
- Proven performance; readily available equipment; easy installation.
- Minimal disturbance to site operation.
- Short treatment times; usually 6 months to 2 years under optimal conditions.
- Cost-competitive: $20–50/ton of contaminated soil.
- Easily combined with other technologies (e.g., air sparging, bioremediation and vacuum-enhanced, dual-phase extraction).
- Can be used under buildings and other locations that cannot be excavated.

Disadvantages of Using Soil Vapor Extraction
- Concentration reductions greater than about 90% are difficult to achieve.
- Effectiveness is less certain when applied to sites with low-permeability soil or stratified soils.
- May require costly treatment for atmospheric discharge of extracted vapors.
- Air emission permits generally required.
- Only treats unsaturated-zone soils; other methods may also be needed to treat saturated-zone soils and groundwater.

Although the theories that explain how SVE works are well understood, determining whether SVE will work at a given site is not simple. Experience and judgement are needed to determine whether SVE will work effectively. The key

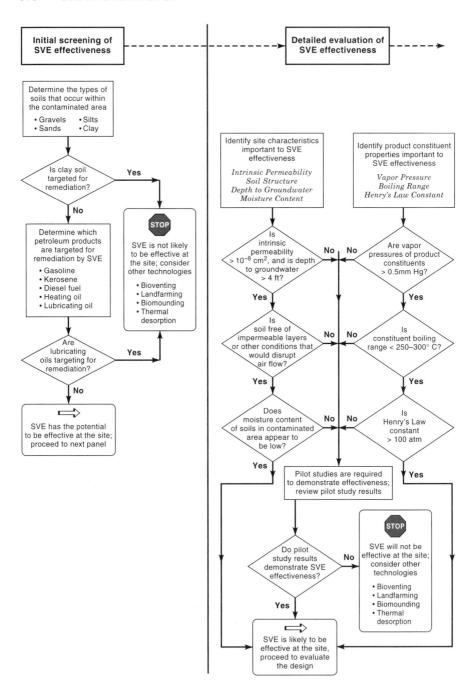

FIGURE 38.2 SVE evaluation process flow chart.

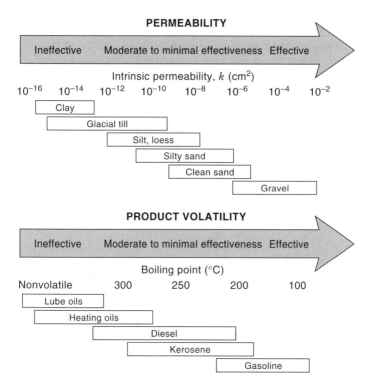

FIGURE 38.3 Initial screening for SVE effectiveness.

parameters that should be used to decide whether SVE will be a viable remedy for a particular site are the following:

- *Permeability* of the petroleum-contaminated soils. Permeability of the soil determines the rate at which soil vapors can be extracted.
- *Volatility* of the petroleum constituents. Volatility determines the rate (and degree) at which petroleum constituents will vaporize from the soil-adsorbed state to the soil vapor state.

In general, the type of soil (e.g., clay, silt, sand) will determine its permeability. Fine-grained soils (e.g., clays and silts) have lower permeability than coarse-grained soils (e.g., sands and gravels). The volatility of a petroleum product or its constituents is a measure of its ability to vaporize. Because petroleum products are highly complex mixtures of chemical constituents, the volatility of the product can be roughly approximated by its boiling point range. Figure 38.3 provides a range of soil permeablilties for typical soil types as well as ranges of volatility (based on boiling point range) for typical petroleum products. Use this screening tool to make an initial assessment of the potential effectiveness of SVE. To use this tool, you should scan the proposed corrective action plan to determine the soil type present and the type of petroleum product released at the site.

38.3   DETAILED EVALUATION OF SVE EFFECTIVENESS

The key parameters used to evaluate permeability of the soil and the volatility of the specific constituent will allow a more through evaluation of the effectiveness of SVE and will identify areas that could require special considerations.

For permeability of soil, the following key parameters should be addressed:

* Intrinsic permeability
* Soil structure and stratification
* Depth to groundwater
* Moisture content

For constituent volatility, the following key parameters should be addressed:

* Vapor pressure
* Product composition and boiling point
* Henry's law constant

Factors That Contribute To Permeability of Soil

Intrinsic Permeability.   Intrinsic permeability is a measure of the ability of soils to transmit fluids and is the single most important factor in determining the effectiveness of SVE. Intrinsic permeability ranges over 12 orders of magnitude (from 10^{-16} to 10^{-3} cm^2) for the wide variety of earth materials, although a more limited range applies for common soil types (10^{-13} to 10^{-5} cm^2). Intrinsic permeability is best determined from field tests, but can be estimated within one or two orders of magnitude from soil boring logs and laboratory tests. Coarse-grained soils (e.g., sands) have greater intrinsic permeability than fine-grained soils (e.g., clays or silts). Note that the ability of a soil to transmit air, which is of prime importance to SVE, is reduced by the presence of soil water, which can block the soil pores and reduce air flow. This is especially important in fine-grained soil, which tend to retain water.

Intrinsic permeability can be determined in the field by conducting permeability tests or SVE pilot studies, or in the laboratory using soil core samples from the site. Use the values presented in Table 38.1 to determine if intrinsic permeability is within the effectiveness range for SVE.

At sites where the soils in the saturated zone are similar to those within the unsaturated zone, hydraulic conductivity of the soils may be used to estimate the permeability of the soils. Hydraulic conductivity is a measure of the ability of soils

Table 38.1   Intrinsic permeability and SVE effectiveness

Intrinsic Permeability (k)	SVE Effectiveness
$k \geqslant 10^{-8}$ cm^2	Generally effective
$10^{-8} \geqslant k \geqslant 10^{-10}$ cm^2	May be effective; needs further evaluation
$k < 10^{-10}$ cm^2	Marginal effectiveness to ineffective

to transmit water. Hydraulic conductivity can be determined from aquifer tests, including slug tests and pumping tests. Hydraulic conductivity can be converted to intrinsic permeability using the following equation:

$$k = K(\mu/\rho g)$$

where

$$k = \text{intrinsic permeability (cm}^2)$$
$$K = \text{hydraulic conductivity (cm/s)}$$
$$\mu = \text{water viscosity (g/cm} \cdot \text{s)}$$
$$\rho = \text{water density(g/cm}^3)$$
$$g = \text{acceleration due to gravity (cm/s}^2)$$

For water at 20°C, $\mu/\rho g = 1.02 \times 10^{-5}$ cm/s.

Intrinsic permeability is also typically stated in units of darcy. To convert k from cm^2 to darcy, multiply by 10^8.

Soil Structure And Stratification.

Soil structure and stratification are important to SVE effectiveness because they can affect how and where soil vapors will flow within the soil matrix under extraction conditions. Structural characteristics such as microfracturing can result in higher permeabilities than expected for certain soil components (e.g. clays). However, the increased flow availability will be confined within the fractures but not in the unfractured media. This preferential flow behavior can lead to ineffective extraction or significantly extended remedial times. Stratification of soils with different permeabilities can increase the lateral flow of soil vapors in the more permeable stratum while dramatically reducing the soil vapor flow through the less permeable stratum.

You can determine the intergranular structure and stratification of the soil by reviewing soil boring logs for wells or borings and by geologic cross-sections. You should verify that soil types have been identified, that visual observations of soil structure have been documented, and that sampling intervals are of sufficient frequency to define any soil stratification. Stratified soils may require special consideration in design to ensure that less-permeable strata are addressed.

Depth To Groundwater.

Fluctuations in the groundwater table should also be considered when reviewing a CAP. Significant seasonal or daily (tidal or precipitation-related) fluctuations may, at times, submerge some of the contaminated soil or a portion of the extraction well screen, making it unavailable for air flow. This is most important for horizontal extraction wells, where the screen is parallel to the water table surface.

SVE is generally not appropriate for sites with a groundwater table located less than 3 ft below the land surface. Special considerations must be taken for sites with a groundwater table located less than 10 ft below the land surface because groundwater upwelling can occur within SVE wells under vacuum pressures, potentially occluding well screens and reducing or eliminating vacuum-induced soil vapor flow. Use Table 38.2 to determine whether the water-table depth is of potential concern for SVE effectiveness.

Table 38.2 Depth to groundwater and SVE effectiveness

Depth to Groundwater (ft)	SVE Effectiveness
> 10	Effective
3–10	Need special controls (e.g., horizontal wells or groundwater pumping)
< 3	Not generally effective

Moisture Content. High moisture content in soils can reduce soil permeability and thereafter the effectiveness of SVE by restricting the flow of air through soil pores. Airflow is particularly important for soils within the capillary fringe where oftentimes a significant portion of the constituents can accumulate. Fine-grained soils create a thicker capillary fringe than coarse-grained soils. The thickness of the capillary fringe can usually be determined from soil boring logs (i.e., in the capillary fringe, soils are usually described as moist or wet). The capillary fringe usually extends from inches to several feet above the groundwater table elevation. SVE is not generally effective in removing contaminants from the capillary fringe. When combined with other technologies (e.g., pump-and-treat to lower the water table or air sparging to strip contaminants from the capillary fringe) the performance of SVE-based systems is considerably increased.

Moist soils can also occur as a result of stormwater infiltration in unpaved areas without sufficient drainage. This moisture may be a persistent problem for fine-grained soils with slow infiltration rates. SVE does dehydrate moist soils to some extent, but the dehydration process may hinder SVE performance and extend operational time.

Factors That Contribute To Constituent Volatility

Vapor Pressure. Vapor pressure is the most important constituent characteristic in evaluating the applicability and potential effectiveness of an SVE system. The vapor pressure of a constituent is a measure of its tendency to evaporate. More precisely, it is the pressure that a vapor exerts when in equilibrium with its pure liquid or solid form. Constituents with higher vapor pressures are more easily extracted by SVE systems. Those with vapor pressures higher than O.5 mm Hg are generally considered amenable for extraction by SVE.

Gasoline, diesel fuel, and kerosene are each composed of over a hundred different chemical constituents. Each constituent will be extracted at a different rate by an SVE system, generally according to its vapor pressure. Table 38.3 lists vapor pressures of selected petroleum constituents.

Product Composition And Boiling Point. The most commonly encountered petroleum products from underground storage tank releases are gasoline, diesel fuel, kerosene, heating oils, and lubricating oils. Because of their complex constituent composition, petroleum products are often classified by their boiling point range. Because the boiling point of a compound is a measure of its volatility, the applicability of SVE to a petroleum product can be estimated from its boiling point

Table 38.3 Vapor pressures of common petroleum constituents

Constituent	Vapor Pressure (mm Hg at 20°C)
Methyl *t*-butyl ether	245
Benzene	76
Toluene	22
Ethylene dibromide	11
Ethylbenzene	7
Xylenes	6
Naphthalene	0.5
Tetraetyllead	0.2

range. The boiling point ranges for common petroleum products are shown in Table 38.4.

In general, constituents in petroleum products with boiling points less than 250 to 300°C are sufficiently volatile to be amenable to removal by SVE. Therefore, SVE can remove nearly all gasoline constituents, a portion of kerosene and diesel fuel constituents, and a lesser portion of heating oil constituents. SVE cannot remove lubricating oils. Most petroleum constituents are biodegradable, however, and might be amenable to bioventing. Injection of heated air also can be used to enhance the volatility of these products because vapor pressure generally increases with temperature. However, energy requirements for volatility enhancement are so large as to be economically prohibitive.

Henry's Law Constant. Another indicator of the volatility of a constituent is its Henry's law constant. Henry's law constant is the partitioning coefficient that relates the concentration of a constituent dissolved in water to its partial pressure in the vapor phase under equilibrium conditions. In other words, it describes the relative tendency for a dissolved constituent to partition between the vapor phase and the dissolved phase. Therefore, the Henry's law constant is a measure of the degree to which constituents that are dissolved in soil moisture (or groundwater) will volatilize for removal by the SVE system. Henry's law constants for several common constituents found in petroleum products are shown in Table 38.5. Constituents with Henry's law constants of greater than 100 atmospheres are generally considered amenable to removal by SVE.

Table 38.4 Petroleum Product Boiling Point Ranges

Product	Boiling Point Range (°C)
Gasoline	40 to 225
Kerosene	180 to 300
Diesel fuel	200 to 338
Heating oil	>275
Lubricating oils	Nonvolatile

Table 38.5 Henry's law constant of common petroleum constituents

Constituent	Henry's Law Constant (atm)
Tetraethyllead	4700
Ethylbenzene	359
Xylenes	266
Benzene	230
Toluene	217
Naphthalene	72
Ethylene dibromide	34
Methyl t-butyl ether	27

Other Considerations

There are other site-specific aspects to consider when evaluating the potential effectiveness of an SVE system. For example, it may be anticipated that SVE would be only marginally effective at a site as the result of low permeability of the soil or low vapor pressure of the constituents. In this case, bioventing may be the best available alternative for locations such as under a building or other inaccessible areas. SVE may also be appropriate near a building foundation to prevent vapor migration into the building. Here, the primary goal may be to control vapor migration and not necessarily to remediate soil.

Pilot Scale Studies

At this stage, as a result of the considerations described above, you will be in a position to decide if SVE is likely to be highly effective, somewhat effective, or ineffective. If it appears that SVE will be only marginally to moderately effective at a particular site, make sure that SVE pilot studies have been completed at the site and that they demonstrate SVE effectiveness. Pilot studies are an extremely important part of the design phase. Data provided by pilot studies are necessary to properly design the full-scale SVE system. Pilot studies also provide information on the concentration of volatile organic compounds that are likely to be extracted during the early stages of operation of the SVE system.

While pilot studies are important and recommended for evaluating SVE effectiveness and design parameters for any site, they are particularly useful at sites where SVE is expected to be only marginally to moderately effective. Pilot studies typically include short-term (1 to 30 days) extraction of soil vapors from a single extraction well, which may be an existing monitoring well at the site. However, longer pilot studies (up to 6 months) which utilize more than one extraction well may be appropriate for larger sites. Different extraction rates and wellhead vacuums are applied to the extraction wells to determine the optimal operating conditions. The vacuum influence at increasing distances from the vapor extraction well is measured using vapor probes or existing wells to establish the pressure field induced in the subsurface by operation of the vapor extraction system. The pressure field measurements can be used to define the design radius of influence for SVE. Vapor concentrations are also measured at two or more intervals during the pilot study to estimate initial vapor concentrations of a full-scale system. The vapor concentration, vapor extraction rate, and vacuum data are also used in the design process to select extraction and treatment equipment.

In some instances, it may be appropriate to evaluate the potential of SVE effectiveness using a screening model such as HyperVentilate 1993. HyperVentilate can be used to identify required site data, decide if SVE is appropriate at a site, evaluate air peameability tests, and estimate the minimum number of wells needed. It is not intended to be a detailed SVE predictive modeling or design tool.

38.4 EVALUATION OF THE SVE SYSTEM DESIGN

Once it has been verified that SVE is applicable, scrutinize the design of the system. A pilot study that provides data used to design the full-scale SVE system is highly recommended. The CAP should include a discussion of the rationale for the design and presentation of the conceptual engineering design. Detailed engineering design documents might also be included, depending on state requirements.

Rationale For The Design

Consider the following factors in evaluating the design of the SVE system in the CAP.

- Design *radius of influence* (ROI) is the most important parameter to be considered in the design of an SVE system. The ROI is defined as the greatest distance from an extraction well at which a sufficient vacuum and vapor flow can be induced to adequately enhance volatilization and extraction of the contaminants in the soil. As a rule-of-thumb, the ROI is often considered to be the distance from the extraction well at which a vacuum of at least 0.1 inches of water is observed.

    The ROI depends on many factors, including lateral and vertical permeability, depth to the groundwater table, the presence or absence of a surface seal, the use of injection wells, and the extent of soil heterogeneity. Generally, the design ROI can range from 5 ft (for fine-grained soils) to 100 ft (for coarse-grained soils). For sites with stratified geology, design ROI should be defined for each soil type. The ROI is important for determining the appropriate number and spacing of extraction wells. The ROI should be determined based on the results of pilot study testing; however, at sites where pilot tests cannot be performed, the ROI can be estimated using air-flow modeling or other empirical methods.

- *Wellhead vacuum* is the vacuum pressure that is required at the top of the extraction well to produce the desired vapor extraction flow rate from the extraction well. Although wellhead vacuum is usually determined through pilot studies, it can be estimated and typically ranges from 3 to 100 in. of water vacuum. Less permeable soils generally require higher wellhead vacuum pressures to produce a reasonable radius of influence. It should be noted, however, that high vacuum pressures (e.g., greater than 100 in. of water) can cause upwelling of the water table and occlusion of the extraction well screens.

- *Vapor extraction flow rate* is the volumetric flow rate of soil vapor that will be extracted from each vapor extraction well. Vapor extraction flow rate, radius of influence, and wellhead vacuum are interdependent (e.g., a change in the extraction rate will cause a change in the wellhead vacuum and radius of influence). Vapor extraction flow rate should be determined from pilot studies but may be calculated using mathematical or physical models. The flow rate will contribute to the operational time requirements of the SVE system. Typical extraction rates can range from 10 to 100 cubic feet per minute (cfm) per well.

- *Initial constituent vapor concentrations* can be measured during pilot studies or estimated from soil gas samples or soil samples. They are used to estimate constituent mass removal rate and SVE operational time requirements and to determine whether treatment of extracted vapors will be required prior to atmospheric discharge or reinjection.

    The initial vapor concentration is typically orders of magnitude higher than the *sustained vapor extraction concentration*, and the former can be expected to last only a few hours to a day before dropping off significantly. Vapor treatment is especially important during this early phase of remediation.

- Required *final constituent concentrations* in soils or vapors are either defined by state regulations as "remedial action levels," or determined on a site-specific basis using fate and transport modeling and risk assessment. They will determine what areas of the site require treatment and when SVE operation can be terminated.

- Required *remedial clean-up time* may also influence the design of the system. The designer may reduce the spacing of the extraction wells to increase the rate of remediation to meet clean-up deadlines or client preferences, as required.

- *Soil volume to be treated* is determined by state action levels or a site-specific risk assessment using site characterization data for the soils.

- *Pore volume calculations* are used along with extraction flow rate to determine the pore volume exchange rate. The exchange rate is calculated by dividing the soil pore space within the treatment zone by the design vapor extraction rate. The pore space within the treatment zone is calculated by multiplying the soil porosity by the volume of soil to be treated. Some literature suggests that one pore volume of soil vapor should be extracted at least daily for effective remedial design progress.

    The time required to exchange one pore volume of soil vapor can be calculated using the following equation:

$$E = \varepsilon V/Q$$

$$\text{with units of, } E = \frac{(\text{m}^3 \text{ vapor/m}^3 \text{ soil}) \cdot (\text{m}^3 \text{ soil})}{(\text{m}^3 \text{ vapor/hr})} = \text{hr}$$

where,

E = pore volume
ε = soil porosity (m^3 vapor/m^3 soil)
V = volume of soil to be treated (m^3 soil)
Q = total vapor extraction flowrate (m^3 vapor/hr),

- *Discharge limitations and monitoring requirements* are usually established by state regulations but must be considered by designers of an SVE system to ensure that monitoring ports are included in the system hardware. Discharge limitations imposed by state air quality regulations will determine whether off-gas treatment is required.
- *Site construction limitations* such as building locations, utilities, buried objects, residences, and the like must be identified and considered in the design process.

38.5 COMPONENTS OF AN SVE SYSTEM

Once the rationale for the design is defined, the actual design of the SVE system can be developed. A typical SVE system design will include the following components and information:

- Extaction wells
- Well orientation, placement, and construction details
- Manifold piping
- Vapor pretreatment design
- Blower selection
- Instrumentation and control design
- Optional SVE components
  - Injection wells
  - Surface seals
  - Groundwater depression pumps
  - Vapor treatment systems

A schematic of a soil vapor extraction system is shown in Figure 38.4.

Extraction Wells

Well Orientation. An SVE system can use either vertical or horizontal extraction wells. Orientation of the wells should based on site-specific needs and conditions. Table 38.6 lists site conditions and the corresponding appropriate well orientation.

Well Placement And Number of Wells. The number and location of extraction wells can be determined by using several methods. In the first method, divide the area of the site requiring treatment by the area of influence for a single well to obtain the total number of wells needed. Then, space the wells evenly within the treatment area to provide areal coverage so that the areas of influence cover the entire area of contamination.

$$\text{Area of Influence for a single well} = \Pi \cdot (\text{ROI})^2$$

$$\text{Number of wells needed} = \frac{\text{Treatment area } (m^2)}{\text{Area of influence for single extraction well } (m^2/\text{well})}$$

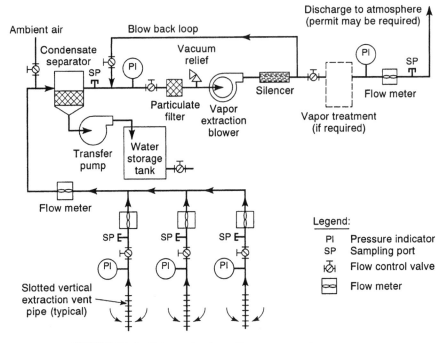

FIGURE 38.4   Schematic of a soil vapor extraction system.

In the second method, determine the total extraction flow rate needed to exchange the soil pore volume within the treatment area in a reasonable amount of time (8 to 24 hr). Determine the number of wells required by dividing the total extraction flow rate needed by the flow rate achievable with a single well.

$$\text{Number of wells needed} = (\varepsilon V/t)/q$$

where

> ε = soil porosity (m^3 vapor/m^3 soil)
> V = volume of soil in treatment area (m^3 soil)
> q = vapor extraction rate from single extraction well (m^3 vapor/hr).
> t = pore volume exchange time (hr)

In the example below, an 8-hr exchange time is used.

$$\text{Number of wells needed} = \frac{\left(\dfrac{\text{m}^3 \text{ vapor}}{\text{m}^3 \text{ soil}}\right) \cdot \left(\dfrac{\text{m}^3 \text{ soil}}{8 \text{ hr}}\right)}{\dfrac{\text{m}^3 \text{ vapor}}{\text{hr}}}$$

Table 38.6   Well orientation and site conditions

Well Orientation	Site Conditions
Vertical extraction well	Shallow to deep contamination (5 to 100+ ft Depth to groundwater > 10 ft)
Horizontal extraction well	Shallow contamination (< 25 ft). More effective than vertical wells at depths < 10 ft. Construction difficult at depths > 25 ft
	Zone of contamination confined to a specific stratigraphic unit

Consider the following additiotial factors in determining well spacing.

* Use closer well spacing in areas of high contaminant concentrations to increase mass removal rates.
* If a surface seal exists or is planned for the design, space the wells slightly farther apart because air is drawn from a greater lateral distance and not directly from the surface. However, be aware that this increases the need for air injection wells.
* At sites with stratified soils, wells that are screened in strata with low intrinsic permeabilities should be spaced more closely than wells that are screened in strata with higher intrinsic permeabilities.

Well Construction — Vertical Well Construction

Vertical extraction wells are similar in construction to groundwater monitoring wells and are installed using the same techniques. Extraction wells are usually constructed of polyvinyl chloride (PVC) casing and screening. Extraction well diameters typicallly range from 2 to 12 in., depending on flow rates and depth, a 4-in. diameter is most common. In general, 4-in. diameter wells are favored over 2-in. diameter wells because 4-in. diameter wells are capable of higher extraction flow rates and generate less frictional loss of vacuum pressure.

Figure 38.5 depicts a typical vertical extraction well. Vertical extraction wells are constructed by placing the casing and screen in the center of a borehole. Filter pack material is placed in the annular space between the casing/screen and the walls of the borehole. The filter pack material extends 1 to 2 ft above the top of the well screen and is followed by a 1- to 2-ft-thick bentonite seal. Cement–bentonite grout seals the remaining space up to the surface. Filter pack material and screen slot size must be consistent with the grain size of the surrounding soils.

The location and length of the well screen in vertical extraction wells can vary and should be based on the depth to groundwater, the stratification of the soil, and the location and distribution of contaminants. In general, the length of the screen has little effect on the ROI of an extraction well. However, because the ROI is affected by the intrinsic permeability of the soils in the screened interval (lower intrinsic permeability will result in a smaller ROI, other parameters being equal), the placement of the screen can affect the ROI.

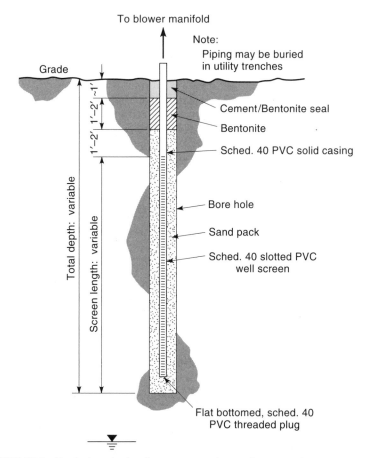

To blower manifold

Note:
Piping may be buried
in utility trenches

Grade

Cement/Bentonite seal

Bentonite

Sched. 40 PVC solid casing

Bore hole

Sand pack

Sched. 40 slotted PVC
well screen

Total depth: variable

Screen length: variable

Flat bottomed, sched. 40
PVC threaded plug

FIGURE 38.5 Typical vertical soil vapor extraction well construction.

- At a site with homogeneous soil conditions, ensure that the well is screened throughout the containated zone. The well screen may be placed as deep as the seasonal low water table. A deeper well helps to ensure remediation of the greatest amount of soil during seasonal low groundwater conditions.

- At a site with stratified soils or lithology, check to see that the screened interval is within the zone of lower permeability because preferred flow will occur in the zones of higher permeability.

Horizontal Well Construction. Look for horizontal extraction wells or trench systems in shallow groundwater conditions. Figure 38.6 shows a typical shallow horizontal well construction detail. Horizontal extraction wells are constructed by placing slotted (PVC) piping near the bottom of an excavated trench. Gravel backfill surrounds the piping. A bentonite seal or impermeable liner is added to

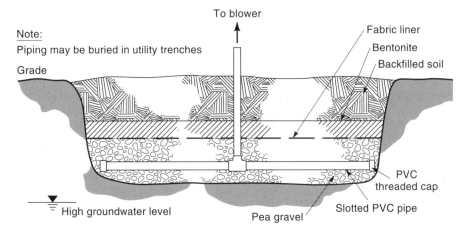

FIGURE 38.6   Typical horizontal soil vapor extraction well construction.

prevent air leakage from the surface. When horizontal wells are used, the screen must be high enough above the groundwater table that normal groundwater table fluctuations do not submerge the screen. Additionally, vacuum pressures should be monitored such that they do not cause upwelling of the groundwater table that could occlude the well screen(s).

Manifold Piping

Manifold piping connects the extraction wells to the extraction blower. Piping can either be placed above or below grade depending on site operations, ambient temperature, and local building codes. Below-grade piping is most common and is installed in shallow utility trenches that lead from the extraction wellhead vault(s) to a central equipment location. The piping can either be manifolded in the equipment area or connected to a common vacuum main that supplies the wells in series, in which case flow control valves are sited at the wellhead. Piping to the well locations should be sloped toward the well so that condensate or entrained groundwater will flow back toward the well.

Vapor Pretreatment

Extracted vapor can contain condensate, entrained groundwater, and particulates that can damage blower parts and inhibit the effectiveness of downstream treatment systems. In order to minimize the potential for damage to blowers, vapors are usually passed through a moisture separator and a particulate filter prior to entering the blower. Check the CAP to verify that both a moisture separator and a particulate filter have been included in the design.

Blower Selection

The type and size of blower selected should be based on both the vacuum required to achieve design vacuum pressure at the extraction wellheads (including upstream and downstream piping losses) and the total flow rate. The flow rate requirement should be based on the sum of the flow rates from the contributing vapor extraction wells. In applications where explosions might occur, blowers must have explosion-proof motors, starters, and electrical systems. There are three basic types of blowers that can be used in an SVE system.

- Centrifugal blowers (such as squirrel-cage fans) should be used for high-flow (up to 280 standard ft^3/min), low-vacuum (less than 30 in. of water) applications.
- Regenerative and turbine blowers should be used when a higher (up to 80 in. of water) vacuum is needed.
- Rotary lobe and other positive displacement blowers should be used when a very high (greater than 80 in. of water) vacuum and moderate air flow are needed.

Monitoring And Controls

The parameters typically monitored in an SVE system include:

- Pressure (or vacuum)
- Air/vapor flow rate
- Contaminant mass removal rates
- Temperature of blower exhaust vapors

The equipment in an SVE system used to monitor these parameters provides the information necessary to make appropriate system adjustments and track remedial progress. The control equipment in an SVE system allow the flow and vacuum pressure to be adjusted at each extraction well of the system, as necessary. Control equipment typically includes flow control valves. Figure 38.7 lists typical monitoring and control equipment for an SVE system, where each of these pieces of monitoring equipment should be placed, and the types of equipment that are available.

Additional SVE system components might also be used when certain site conditions exist or pilot studies dictate they are necessary. These components include:

- Injection and passive inlet wells
- Surface seals
- Groundwater depression pumps
- Vapor treatment systems

Injection and Passive Inlet Wells. Air injection and inlet wells are designed to help tune air flow distribution and may enhance air flow rates from the extraction wells by providing an active or passive air source to the subsurface. These wells are often used at sites where a deeper zone (i.e., > 25 ft) is targeted for SVE or where the targeted zone for remediation is isolated from the atmosphere by low permeability materials. They are used also to help prevent short-circuiting of air flow from the atmosphere at sites with shallower target zones. Passive wells have little effect unless they are placed close to the extraction well. In addition, air injection is used

Monitoring Equipment	Location in System	Example of Equipment
Flow meter	• At each wellhead • Manifold to blower • Blower discharge	• Pitot tube • In-line rotameter • Orifice plate • Venturi or flow tube

Vacuum gauge	• At each wellhead or
	  manifold branch
	• Before and after filters
	  upstream of blower
	• Before and after vapor
	  treatment
		• Manometer
		• Magnehelic gauge
		• Vacuum gauge

Vapor treatment sensor	• Manifold to blower
	• Blower discharge (prior
	  to vapor treatment)
		• Bi-metal dial type
		  thermometer

Sampling port	• At each wellhead or manifold branch • Manifold to blower • Blower discharge	• Hose barb • Septa fitting

Vapor sample collection
equipment (used through a
sampling port)
	• At each wellhead or
	  manifold branch
	• Manifold to blower
	• Blower discharge
		• Tedlar bags
		• Sorbent tubes
		• Sorbent canisters
		• Polypropylene tubing for
		  direct GC injection

Control Equipment		

Flow control valves	• At each wellhead or
	  manifold branch
	• Dilution or bleed valve at
	  manifold to blower
		• Ball valve
		• Gate/globe valve
		• Butterfly valve

FIGURE 38.7 Monitoring and control equipment.

to eliminate potential stagnation zones (areas of no flow) that sometimes exist between extraction wells.

Air injection wells are similar in construction to extraction wells but can be designed with a longer screened interval in order to ensure uniform air flow. Active injection wells force compressed air into soils. Passive air inlet wells, or inlets, simply provide a pathway that helps extraction wells draw ambient air to the subsurface. Air injection wells should be placed to eliminate stagnation zones, if present, but should not be placed such that the injected air will force contaminants to an area where they will not be recovered (i.e., off-site).

Surface Seals. Surface seals might be included in an SVE system design to prevent surface water infiltration that can reduce air flow rates, reduce emissions of filgitive vapors, prevent vertical short-circuiting of air flow, or increase the design ROI. These results are accomplished because surface seals force fresh air to be drawn from a greater distance from the extraction well. If a surface seal is used, the lower pressure gradients result in decreased flow velocities. This condition may require a higher vacuum to be applied to the extraction well.

Surface seals or caps should be selected to match the site conditions and regular business activities at the site. Options include high density polyethylene (HDPE) liners (similar to landfill liners), clay or bentonite seals (with cover vegetation or other protection), or concrete or asphalt paving. Existing covers (e.g., pavement or concrete slab) might not provide sufficient air confinement if they are constructed with a porous subgrade material.

Groundwater Depression Pumps. Groundwater depression pumping might be necessary at a site with a shallow groundwater table. Groundwater pumps can reduce the upwelling of water into the extraction wells and lower the water table and allow a greater volume of soil to be remediated. Because groundwater depression is affected by pumping wells, these wells must be placed so that the surface of the groundwater is depressed in all areas where SVE is occurring. Groundwater pumping, however, can create two additional waste streams requiring appropriate disposal: (a) groundwater contaminated with dissolved hyrocarbons; and (b) liquid hydrocarbons (i.e., free product, if present).

Vapor Treatment System. Look for vapor treatment systems in the SVE design if pilot studies data indicate that extracted vapors will contain VOC concentrations in excess of state or local air emission limits. Available vapor treatment options include granular activated carbon (GAC), catalytic oxidation, and thermal oxidation.

GAC is a popular choice for vapor treatment because it is readily available, simple to operate, and can be cost-competitive. Catalytic oxidation, however, is generally more economical than GAC when the contaminant mass loading is high. However, catalytic oxidation is not recommended when concentrations of chemical constituents are expected to be sustained at levels greater than 20% of their lower explosive limit (LEL). In these cases, a thermal oxidizer is typically employed because the vapor concentration is high enough for the constituents to burn. Biofilters, an emerging vapor-phase biological treatment technique, can be used for vapors with less than 10% LEL, appear to be cost-effective, and may also be considered (see next chapter for more details.)

38.6 EVALUATION OF OPERATION AND MONITORING PLANS

Make sure that a system operation and monitoring plan has been developed for both the system start-up phase and for long-term operations. Operations and monitoring are necessary to ensure that system performance is optimized and contaminant mass removal is tracked.

Start-Up Operations

The start-up phase should include 7 to 10 days of manifold valving adjustments. These adjustments should optimize contaminant mass removal by concentrating vacuum pressure on the extraction wells that are producing vapors with higher contaminant concentrations, thereby balancing flow and optimizing contaminant

Phase	Monitoring Frequency	What to Monitor	Where to Monitor
Start-up (7–10 days)	Daily	• Flow • Vacuum • Vapor concentrations	• Extraction vents • Manifold • Effluent stack
Remedial (ongoing)	Biweekly or monthly	• Flow • Vacuum • Vapor concentrations	• Extraction vents • Manifold • Effluent stack

FIGURE 38.8   System monitoring recommendations.

mass removal. Flow measurements, vacuum readings, and vapor concentrations should be recorded daily from each extraction vent, from the manifold, and from the effluent stack.

Long-Term Operations

Long-term monitoring should consist of flow-balancing, flow and pressure measurements, and vapor concentration readings. Measurements should take place at biweekly to monthly intervals for the duration of the system operational period. Figure 38.8 provides a brief synopsis of system monitoring recommendations.

Remedial Progress Monitoring

Monitoring the performance of the SVE system in reducing contaminant concentrations in soils is necessary to determine if remedial progress is proceeding at a reasonable pace.

The mass removed during long-term monitoring intervals can be calculated using vapor concentration and flow rate measurements taken at the manifold. The instantaneous and cumulative mass removal is then plotted versus time. The contaminant mass removed during an operating period can be calculated using the equation provided below. This relationship can be used for each extraction well (and then totalled) or for the system as a whole, depending on the monitoring data that is available.

$$M = C \cdot Q \cdot t$$

where

$$M = \text{cumulative mass removed (kg)}$$
$$C = \text{vapor concentration (kg/m}^3\text{)}$$
$$Q = \text{extraction flow rate (m}^3/\text{hr)}$$
$$t = \text{operational period (hr)}$$

$$\text{Dimensionally, Mass removed (kg)} = \frac{\text{kg}}{\text{m}^3} \cdot \frac{\text{m}^3}{\text{hr}} \cdot \text{hr}$$

Remedial progress of SVE systems typically exhibits asymptotic behavior with

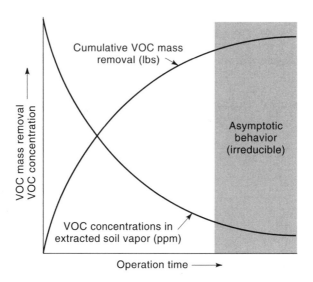

FIGURE 38.9   Relationship between concentration reduction and mass removal.

respect to both vapor concentration reduction and cumulative mass removal (Figure 38.9). At this point, the composition of the vapor should be determined and compared with soil vapor samples. This comparison will enable confirmation that there has been a shift in composition toward less volatile components. Soil vapor samples may indicate the composition and extent of the residual contamination. When asymptotic behavior begins to occur, the operator should closely evaluate alternatives that increase mass removal rate such as increasing flow to extraction wells with higher vapor concentrations by terminating vapor extraction from extraction wells with low vapor concentrations or pulsing. Pulsing involves the periodic shutdown and startup operation of extraction wells to allow the subsurface environment to come to equilibrium (shutdown) and then begin extracting vapors again (startup). Other, more aggressive steps to curb asymptotic behavior can include installation of additional injection wells or extraction wells.

If asymptotic behavior is persistent for periods greater than about six months and the concentration rebound is sufficiently small following periods of temporary system shutdown, termination of operations may be appropriate if residual levels are at or below regulatory limits. If not, operation of the system as a bioventing system with reduced vacuum and air flow may be an effective remedial alternative.

38.7 SUMMARY

1. Soil vapor extraction (SVE)is rapidly becoming the technology of choice for remediation of soils contaminated with volatile organic compounds (VOCs). This technology is relatively low cost and very effective in reducing concentrations of volatile constituents in petroleum products adsorbed to soils in the unsaturated (valdose) zone.

2. Although SVE has become a widely used technology, it is necessary to conduct a preliminary evaluation to determine the plausability of this technology at a particular site.

3. The key parameters used to evaluate permeability of the soil and the volatility of the specific constituent will allow a more through effectiveness evaluation and will identify areas that could require special considerations.

4. Once you have verified that SVE is applicable, you can scrutinize the design of the system. A pilot study that provides data used to design the full-scale SVE system is highly recommended. The CAP should include a discussion of the rationale for the design and presentation of the conceptual engineering design.

5. A typical SVE system design will includes extaction wells, well orientation, placement, and construction details, manifold piping, vapor pretreatment design, blower selection, instrumentation, and control design and optional SVE components.

6. Make sure that a system operation and monitoring plan has been developed for both the system start-up phase and for long-term operations. Operations and monitoring are necessary to ensure that system performance is optimized and contaminant mass removal is tracked

REFERENCES

1. U.S. EPA, *Soil Vapor Extraction Technology*, Office of Research and Development, EPA/540/2-91/003.
2. U.S. EPA, *How To Evaluate Alternative Cleanup Technologies For Underground Storage Tank Site*, EPA 510-B-94-003, October 1994.

39

BIOFILTRATION

39.1 INTRODUCTION

Air streams containing volatile organic compounds (VOCs) can be biologically treated with a technology known as biofiltration. Although the term "filtration" is more appropriate for physical separation, the term *biofiltration* refers to the biochemical destruction of VOCs through the actions of microorganisms, usually on immobilized solids.

In a typical system, contaminated air streams are passed through a bed of solids (see section 39.3 for more details). VOCs and oxygen are transferred to liquid biofilms attached to solid surfaces where the pollutants are usually converted to harmless end products. The process takes place in equipment defined as biofilters. Biofilters often use porous media that are usually of an organic base (peat moss, wood, compost, etc.); they do not involve a liquid stream circulating through the bed of solids. Nonporous media (ceramics and plastics) can also be used in what are known as trickling biofilters or biotrickling filters. These involve a nutrient-rich liquid stream continuously circulating through the system. However, biofilters are usually the preferred choice (1).

Biofiltration is a biologically driven process that exploits the above-mentioned ability of microorganisms to remove biodegradable substances in gas streams. Although the method has been widely used to eliminate odorous gaseous effluents from wastewater treatment plants, stock farms, compost factories, and various other process gas streams, its application in the control of volatile organic compound (VOC) emissions has begun to attract attention and interest in recent years.

In a general sense, the biofiltration process is similar to the conventional activated sludge treatment (see Chapters 13 and 14) for wastewater purification since in both instances microorganisms often oxidize VOC chemicals to carbon dioxide and water in a moist, oxygen-rich environment. The principal difference is that in biofiltration, the microorganisms are usually immobilized on the solid filter material, while in wastewater treatment they are dispersed as a suspension in the liquid (water) phase.

Biofiltration is occasionally an efficient and economic alternative to other more conventional, air pollution control (APC) technologies. Included in these options are catalytic and thermal incineration, condensation, adsorption, and absorption. Until recently, biofiltration was rarely considered appropriate for air pollution control applications; the control of gaseous pollutants was accomplished by one of the conventional methods mentioned above. Details on these control technologies can be found in Chapter 7.

Biofiltration originated in Europe for use in connection with odor abatement. Over the past three decades, biofilters have developed from what were systems for abating odors into technically sophisticated and controlled units for removing specific chemicals from industrial sources. In 1990, the Clean Air Act Amendments further motivated interest in this area. For these reasons, along with some of the economic and environmental advantages of biofiltration (to be discussed later), the biofilter market has grown in the United States.

As indicated previously, the first applications of biofiltration technology were for odor control, particularly at municipal wastewater treatment and composting facilities. The first known application in the U.S. was the installation of a soil bed to remove odors at a sewage pump station in Long Beach, California, in 1953. The initial applications of soil beds in Europe were made around 1959 for odor control at a municipal wastewater treatment facility in Nuremberg, Germany, and in 1964 at a compost facility in Genève-Villette, Switzerland.

In the late 1970s and 1980s, interest in biofiltration as an innovative, environmentally friendly technology was stimulated in Europe by more restrictive air pollution regulations. Federal and local governments in Germany and Holland, in particular, provided financial support for biofiltration research, development, and application. This work included determination of process kinetics, improved bioreactor designs, and application of the technology to a much broader variety of air emissions.

39.2 PRINCIPLES

The principles governing biofiltration are essentially similar to those of any biofilm processes. It consists of a multistep process that occurs within the bed of a biofilter. Initially, a chemical (usually a VOC) in the gas phase passes the interface between gas flowing in the pore space of the bed and the aqueous biofilm surrounding the solid medium. The chemical then diffuses through the biofilm to the microorganism(s). Finally, the microorganism(s) obtain energy from the oxidation of the chemical as a primary substrate, or they cometabolize the chemical via nonspecific enzymes. Simultaneously, there is diffusion and uptake of nutrients, such as nitrogen and phosphorus in available forms, and oxygen within the biofilm. The end result is the conversion of these chemicals to harmless end products such as carbon dioxide (CO_2), water (H_2O), inorganic salts and biomass, which ultimately leave the system (2).

The capability and efficiency of an operating biofilter to eliminate pollutants in a waste gas is a result of both physical and biological factors. The physical factors include mass transfer processes, flow behavior and profile of the gaseous phase, and average residence time. The microbiological phenomena involve the rate of pollutant elimination by oxidation, a chemical or biochemical reaction.

Engineers and scientists have been trying to develop theoretical models that enhance the fundamental knowledge of the transport and biological processes, and at the same time provide quantitative information on equipment design. Since the interaction between both the physical and microbiological phenomena is extremely complex (3,4), simplifying assumptions are necessary to model the reacting system. These assumptions bring into question the validity and applicability of the numerous models appearing in the literature. Although these kinetic models can provide some basis for sizing a biofilter in the case of a single-component off-gas, they have limited utility for off-gases with a mixture of pollutants. Complications due to interactions between compounds can have either a positive or negative impact on the biodegradation rate. These interactions include co-metabolism, which can increase the degradation rate of recalcitrant compounds (5); cross-inhibition, which can diminish the degradation rate; and vertical stratification, where the most readily degradable compounds are metabolized at the inlet portion of the biofilter and less degradable compounds pass through to be metabolized at higher levels (6). Pilot-scale testing is highly recommended to correctly design a biofilter for a multicomponent waste gas stream (7).

39.3 PROCESS SYSTEM

The main components and design considerations of a biofiltration system typically include

1. Raw gas collection and transportation
2. Preconditioning
3. Humidification
4. Influent gas distribution
5. Filter material

Michelsen (4) has provided a description of each of these five components of the system. This information is given below. A schematic flowsheet of a typical system is given in Figure 39.1.

1. *Raw Gas Collection and Transportation.* The gas to be treated by biofiltration is collected from within an enclosure, from emission sources with local exhaust ducts, or by other accumulation systems such as canopies. The collected off-gas is transported with ductwork and a fan through the preconditioning and humidification equipment, gas distribution system, and organic filter material.

2. *Raw Gas Preconditioning.* Pretreatment of the raw gas is generally required to adjust and/or convert the influent airstream into a range of conditions suitable for biofiltration. These are detailed below in subsections a–c.

    a. *Particulates.* Particulates such as dust, grease, oils, acid mist, or other aerosols can be detrimental to the biofilter by obstructing and clogging both the pores of the filter media or vents in the air distribution system. A fabric filter, venturi scrubber, or other separator is typically installed to remove particulates to avoid depositing these nonvolatile contaminants in the biofilter; the reader is again referred to Chapter 7 for information on these control devices.

    b. *Temperature.* The operating temperature of a biofilter is mainly a function of

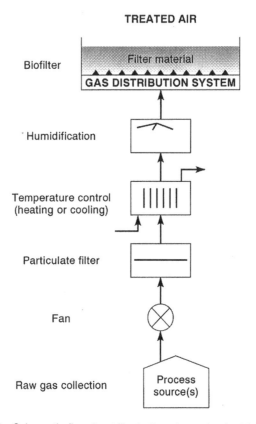

FIGURE 39.1 Schematic flowsheet illustrating elements of a biofilter system.

the influent gas temperature. Heat losses due to ambient conditions or moisture evaporation from either the filter or gas stream and heat generation from biological degradation can also influence operating temperature. In many applications, cooling or heating of the influent gas stream is necessary to regulate the temperature, moving it into a favorable range for biofiltration.

The principal microorganisms active in a biofilter are mesophilic bacteria. Mesophilic bacteria function over a wide temperature range (between 5 and 50°C) with an optimum at about 37°C. Lower temperatures tend to suppress microbial growth while higher temperatures inhibit activity and can be fatal to the bacteria. Other microorganisms also active in biodegradation reactions utilizing the pollutants can function over a broader temperature range. Fungi are generally more adaptable to temperature extremes. Thermophilic bacteria are also active at higher temperatures. Cryophilic bacteria are active at lower temperatures.

The optimum biofilter operating temperature is typically around 35°C. While a higher temperature increases microbial activity, it also decreases the water solubility of the pollutants, so destruction efficiency (fortunately) tends to be relatively flat over a fairly wide range. The recommended operating temperature range for high efficiency is between 20 and 40°C. Precise control is generally not required. In cases where the off-gas temperature is equal to ambient conditions, lower removal efficiency is typical during cold winter months if the inlet air is unheated (5).

In situations where the off-gas is cooled, temperature adjustment is typically combined with humidification, as discussed in detail below. An important operating cost factor is the capability to utilize inexpensive adiabatic vs. more costly alternatives to cool the inlet temperature to the recommended temperature range.

c. *Other Factors.* In some situations the raw gas may contain substances which are toxic to the microorganisms, either due to their high concentration or chemical nature. In these cases the raw gas may be made amenable to biofiltration by either modifying the gas collection system to avoid gathering the toxic component, pretreatment to eliminate the toxic component, and/or dilution of the gas stream.

3. *Humidification.* The microorganisms inhabiting the biofilter require moisture for survival and metabolism. Maintaining a sufficient moisture content in the filter medium is therefore a critical operating condition to control in order to sustain high biofiltration efficiency. Insufficient moisture will diminish the biological activity. Inadequate inlet gas humidity can also induce drying, shrinking, and compaction of the filter medium, leading to cracks/fissures in the filter and resulting in low efficiency due to incompletely treated raw gas short circuiting the biofilter.

Humidification of the influent air to approach saturation is recommended to keep the filter material moist. Gas moisture is generally adjusted using a water spray humidifier (typically a single water spray tower) which can also serve to remove particulates and adjust the temperature. Relying solely on spraying water onto the surface of the biofilter for moisture control, without humidification of the influent air, is not advised since localized drying of the filter can result (6).

Microbial reactions which destroy the pollutants are exothermic and can cause the gas temperature to increase across the filter. When the off-gas has a high pollutant concentration, a temperature increase of several degrees Celsius is possible. Since the inlet and outlet gases both approach moisture saturation, the temperature increase will tend to dry out the filter material. For these cases, automatic water spray nozzles are installed to supplement the influent humidity and maintain a favorable moisture content.

4. *Influent Gas Distribution System.* The primary function of the gas distribution system is to homogeneously disperse the influent gas into the biofilter for consistent loading of the filter medium. In upflow biofilters, the gas distribution system generally performs the additional functions of

a. Providing a system for drainage, collection, and diversion of excess water in the filter material

b. Providing a containment seal at the base of the biofilter to avoid the potential concern of filter leachate contaminating the soil or groundwater

c. Forming a structural support floor upon which the filter packing is loaded.

The first distribution system was the soil bed design which utilizes a ductwork header feeding waste gas via a horizontal network of perforated pipes. The perforated pipe network, approximately 2 to 3 ft below ground level, is covered by the filter bed which is typically soil and perhaps compost. Newer biofilter air distribution systems utilize a slotted or vented plate design of interlocking concrete (or plastic) blocks, concrete slabs, or metal grating. In some enclosed biofilters, gas distribution is downflow. However, the reader should note that the media bed itself serves as an excellent gas distributor.

5. *Filter Material.* The filter material is an extremely crucial component of a biofilter as it is the microbial population's habitat. The material must provide a favorable environment for metabolic activity in order to realize and sustain high biofiltration performance. A wide variety of biofilter materials have been utilized, with the most common being soil or compost produced from leaves, bark, wood chips, activated sludge, paper, or other materials of organic origin. Peat and heather have been used in some applications, as well as inert materials. A strong adsorbent such as activated carbon is added in some situations to increase biofiltration efficiency.

   In selecting the filter material, the following considerations are important (7):

   a. The filter material's particle size, distribution, and pore structure must provide a high surface area for microbial attachment.
   b. Since the elimination reactions occur in the biofilm, a large surface-to-volume ratio is desirable for rapid transfer of the pollutant to the aqueous phase.
   c. The filter material should have good water retention since the microorganisms require high moisture for growth.

   Generally, the filter material is the source of inorganic nutrients required for microbial growth. Filter materials with a high organic content generally contain a high concentration of nutrients. In some situations, periodic addition of nutrients is required to maintain optimum biofilter performance, depending on the specific filter material and the pollutants and their concentrations.

   Since the filter bed is effectively a packed column through which the influent gas passes, a porous structural matrix with low pressure drop is desired for homogeneous air distribution. Compaction of the filter material should be minimized to avoid reducing air permeability and the formation of cracks through which gas can channel. In order to maintain a proper porous structure and avoid compaction, large particles such as wood and bark chips, plastic spheres, ceramics, or lava are added to the filter medium mixture to provide strength and create voids for low pressure drop.

   The requirement for a low pressure differential across the filter material limits the maximum height. The height of the compost filter bed is typically 1.0 to 1.5 m (3–5 ft). The maximum height is 2.0 m for an optimized low pressure drop filter material. The minimum height is limited by the need for a pressure differential across the filter for a uniform gas distribution, and at least 0.5 m is recommended.

   In addition to high air permeability, the filter material should have good drainage characteristics to avoid a higher than optimum moisture content. Leachate from a biofilter is often due to rainwater in an open system or excess irrigation where water sprays are used for moisture control. Drainage from the filter bed may contain undegraded or only partially degraded gas pollutants, acidic end products from biodegradation reactions, or soluble components from the filter material. Thus, biofilter leachate is a potential source of wastewater contamination. Discharge treatment if often required by local environmental regulations. Leachate is typically recirculated to the humidifier to minimize water discharge. When the drainage is recirculated, it is periodically sewered and the humidification system flushed to avoid plugging with solids suspended in the leachate.

The optimum condition for microbial growth is a neutral pH range. Most microorganisms are inhibited below a pH of about 3 although some, such as fungi, still function. Acidic products are often formed by oxidation of some pollutants, particularly volatile inorganics such as H_2S, SO_2, and NO_x, and halogenated organics. In these situations the filter material must have buffering capacity to neutralize the acidic products formed by oxidation reactions to avoid low pH and the deterioration of bed performance. Neutralizing compounds such as lime, limestone, or dolomite are typical additives used to enhance buffering capacity.

Since the filter particles are utilized by the microbes as a source of nutrients, they decompose over time and become smaller in size, which can lead to high gas flow resistance, compaction, and formation of cracks. Selection of a filter material with a low rate of attrition and decomposition is desired for both lower maintenance (and cost) and longer life.

39.4 DESIGN CONSIDERATIONS (4)

For biofiltration to be selected as a viable technical and cost-effective option for a specific application requires demonstration of three key design considerations: destruction efficiency, economic evaluation, and a positive comparison with alternative air pollution control (APC) methods. These three consideratons lead to the review of the following factors.

Application Characterization

In considering biofiltration as an appropriate APC technology, the key attributes are the concentration and biodegradability of the pollutants in the raw gas. Low concentrations of highly degradable components favor biofiltration.

Technology Evaluation

If biofiltration is the appropriate option, the second step of the evaluation is determination of the destruction efficiency (DE). For some single-component raw gases the destruction efficiency and biofilter sizing can be determined from either kinetic modeling and/or (usually) vendor experience with similar emissions. Many industrial emissions contain a mixture of compounds at varying concentrations. For such situations, pilot-scale testing is recommended. This key variable must be sufficiently high to achieve the required DE of the chemical in question. If biofiltration can achieve the destruction efficiency requirements, economic evaluation "Economics" is necessary to determine whether it is the most cost-effective alternative.

The following six conclusions can be drawn regarding the capability of biofiltration for the treatment of different VOCs.

1. Aliphatic compounds with less than seven carbon atoms are very degradable.
2. Aliphatic compounds with more than six carbon atoms and some aromatic compounds are moderately degradable. A longer retention time, however, is required.
3. Some compounds, such as benzene, degrade poorly.
4. Chlorinated hydrocarbon compounds with one or two chlorine atoms can be treated by biofiltration but degrade slowly.
5. Higher-concentration emissions with large variability in gas volumetric flow, chemical composition, or concentration tend to be less suitable for biofiltration.
6. Biofiltration has been applied most successfully in the treatment of a relatively continuous and uniform off-gas which contains a dilute concentration of readily biodegraded pollutants.

Sizing

The unit's size is perhaps the most important design consideration. The size is obtained directly from the average residence time. This key variable must be sufficiently high to achieve the required DE of the chemical in question. Required biofilter gas residence time information usually includes

1. Volumetric flow rate
2. Volumetric flow rate variation
3. Concentration of the pollutant compounds
4. Concentration variation of the pollutant compounds
5. Biodegradability (ease of) of the compounds
6. The required destruction efficiency
7. Filter-bed packing material
8. The biofilter physical design, including the type and height of the bed

As indicated earlier, based on current experience, biofiltration has been applied most successfully in the treatment of waste gases with a dilute concentration of easily biodegraded pollutants. Destruction efficiency above 90% can be achieved for many common VOCs. Efficiency can exceed 99% for odors and gases with low concentrations of readily degraded compounds.

Economics

Because of the wide variety of potential applications and different peripheral requirements, the cost of a biofilter system is very site-specific. The factors which influence capital cost of a biofilter project are size, type of design, space availability, and preconditioning requirements. Capital costs vary from a low of $5 per cubic foot per minute ($5/cfm) to a high of about $100/cfm. The low cost range of $5–10/cfm is for the installation of open soil bed systems to treat low concentrations of rapidly biodegrading compounds. A typical annual operating cost of about °7/cfm has been reported in the literature (8). The operating cost of biofiltration for appropriate applications is almost always lower than other APC technologies.

39.5 COMPARISON WITH OTHER ALTERNATIVES

In the treatment of low-concentration off-gases, biofiltration tends to compete with scrubbers, activated carbon adsorption, and thermal and catalytic oxidation (with heat recovery). These conventional APC methods can often achieve the required efficiency but can also have cost and operational drawbacks in comparison to biofiltration.

Several capital and operating cost comparisons between biofiltration and conventional APC techniques have been published (9–13). For appropriate applications, biofiltration is usually lower in capital and operating costs. For applications where more expensive biofilter designs (enclosed or multilevel) are installed, the capital cost is generally competitive with other options.

Greater acceptance of biofiltration as a viable APC option is anticipated as its application becomes more widespread. The successful operation of the several recently installed full-scale biofilters in the U.S. is expected to expand acceptance of the technology.

39.6 ADVANTAGES AND DISADVANTAGES

Michelsen (4) has also reported on the advantages and disadvantages of biofiltration. These are detailed below.

Advantages

1. Biofiltration is a natural process and as such is considered a safe, environmentally friendly technology. It does not utilize a combustion source or require fuel to destroy the pollutants. It does not generate carbon dioxide in excess of that formed by the oxidation of the pollutants. It does not transfer the pollutants cross-media.

2. Biofilters are generally lower in capital and operating cost for the treatment of large-volume waste gases containing low concentrations of biodegradable pollutants.

3. Biofiltration can achieve high destruction efficiency in the treatment of many common air pollutants. It can effectively treat mixtures of volatile organic and inorganic compounds.

4. Biofiltration is a relatively simple technology which has low operational and maintenance requirements. It does not require large quantities of chemical additives. Maintaining favorable conditions for the microorganisms, such as filter moisture content and pH, is essential to sustain high destruction efficiency.

Disadvantages

1. While biofiltration is a versatile APC technology for many odor, VOC, and other air toxic applications, it is not the most appropriate option for treating some air emission sources. Biofiltration is less successful for the treatment of waste gases with the following characteristics:

    a. High concentrations of pollutants

    b. Contains chemicals which are difficult to biodegrade or toxic to the microor-

ganisms
c. Emissions with widely fluctuating volumes (intermittent), chemical components, or concentrations

2. The microorganisms must acclimate to the influent gas, requiring the passage of time before degrading pollutants at maximum efficiency. Acclimation time can vary from a few days to as long as several months, depending on the source of filter material and the specific pollutant chemicals. Microbial acclimation is a concern when treating fluctuating emissions.

3. Biofilters have long residence times which require a substantial volume of filter material. Since bed height is limited by pressure drop and strength of the porous packing matrix, biofilter area requirements can be very large. Multilevel bed designs are often constructed to overcome space limitations.

4. Biofiltration is an emerging APC technology which is relatively unfamiliar, and therefore unproven to many industrial engineers and environmental authorities. Regulatory and business management approval of nonconventional technology is oftentimes difficult.

39.7 SUMMARY

1. Air streams containing volatile organic compounds (VOCs) can be biologically treated with a technology known as biofiltration. In a typical system, contaminated air streams are passed through a bed of solids.

2. Biofiltration is a biologically driven process that exploits the ability of microorganisms to remove biodegradable substances in gas streams. Although the method has been widely used to eliminate odorous gaseous effluents from wastewater treatment plants, stock farms, compost factories, and various other process gas streams, its application in the control of volatile organic compound (VOC) emissions has begun to attract attention and interest in recent years.

3. The main components of a biofiltration system typically include raw gas collection and transportation, preconditioning, humidifcation, influent gas distribution, and filter material.

4. For biofiltration to be selected as a viable technical and cost-effective option for a specific application requires demonstration of three key design considerations destruction efficiency, economic evaluation, and a positive comparison with alternative air pollution control (APC) methods.

5. Because of the wide variety of potential applications and different peripheral requirements, the cost of a biofilter system is very site specific. The factors which influence capital cost of a biofilter project are size, type of design, space availability, and preconditioning requirements.

6. Greater acceptance of biofiltration as a viable APC option is anticipated as its application becomes more widespread. The successful operation of the several recently installed full-scale biofilters in the U.S. is expected to expand acceptance of the technology.

REFERENCES

1. B. Baltzis, S. Wozdyla and S. Zarooh, "Modeling Biofiltration of VOC Mixtures Under

Steady-State Conditions," *Journal of Environmental Engineering,* 599–609, June 1997.

2. W. Swanson and R. Loehr, "Biofiltration: Fundamentals, Design and Operations Principles, and Applications," *Journal of Environmental Engineering,* 538–546, June 1997.

3. W. Perry and D. Green, "Waste Management," in *Handbook of Chemical Engineering, 7th ed.,* McGraw-Hill Book Co., New York, 375–394, 1998.

4. J. Michelsen, "Biofiltration," *Handbook of Air Pollution Control Engineering and Technology,* J. McKenna, J. Mycock, and L. Theodore, eds., CRC Press, Boca Raton, FL., 1994.

5. D. Kampbell, J. T. Wilson, H. W. Read, and T. T. Stocksdale, "Removal of Volatile Aliphatic Hydrocarbons in a Soil Bioreactor," *JAPCA,* 37(10), 1236–1240 (1987).

6. K. Kosky and C. R. Neff, "Innovative Biological Degradation System for Petroleum Hydrocarbons Treatment," presented at the NWWA and API Conf. Exp. Petroleum Hydrocarbons and Organic Chemicals in Ground Water, Houston, TX, Nov. 9–11, 1988.

7. G. Leson and A. M. Winer, "Biofiltration: An Innovative Air Pollution Control Technology for VOC Emissions," *Journal of Air Waste Management Association,* 41(8), 1045–1054, 1991.

8. *McIlvaine Scrubber-Adsorber Newsletter,* November 1991.

9. H. Bohn, "Consider Biofiltration for Decontaminating Gases," *Chemical Engineering Progress,* 34–40, April 1992.

10. J. Don and L. Feenstra, "Odour Abatement Through Biofiltration," paper presented at Symposium Louvain La Neuve, Belgium, April 1984.

11. W. Prokop and H. L. Bohn, "Soil Bed System for Control of Rendering Plant Odors," *Journal of the Air Pollution Control Association,* December 1985.

12. R. Zurlinden, M. T. Turner, and G. W. Gruwell, "Treatment of Volatile Organic Compounds in a Pilot Scale Biofilter," paper presented at the Air and Waste Management Association 86th Meeting and Exhibition, Denver, CO., June 13–18, 1993.

13. D. Barshter, S. W. Paff, and A. B. King, "Biofiltration — Room Temperature Incineration," paper presented at the Air and Waste Management Association 86th Meeting and Exhibition, Denver, CO, June 13–18, 1993.

IX

RISK-RELATED TOPICS

40

RISK-BASED
DECISION MAKING

40.1 INTRODUCTION

In the 1980s, to satisfy the need to start corrective action programs quickly, many regulatory agencies decided to uniformly apply, at underground storage tank (UST) clean-up sites, regulatory clean-up standards developed for other purposes. It became increasingly apparent that applying such standards without consideration of the extent of actual or potential human and environmental exposure was an inefficient means of providing adequate protection against the risks associated with UST releases. The EPA now believes that risk-based corrective-action processes are tools that can facilitate efforts to clean up sites expeditiously, as necessary, while still assuring protection of human health and the environment (1).

Risk-based decision making and risk-based corrective action are decision-making processes for assessing and responding to a chemical release. The processes take into account effects on human health and the environment, inasmuch as chemical releases vary greatly in terms of complexity, physical and chemical characteristics, and in the risk that they may pose. Risk-based corrective action (RBCA) was initially designed by the American Society for Testing and Materials (ASTM) to assess petroleum releases, but the process may be tailored for use with any chemical release.

The United States Environmental Protection Agency (the EPA) and several state environmental agencies have developed similar decision-making tools. The EPA refers to the process as "risk-based decision making." While the ASTM RBCA standard deals exclusively with human health risk, the EPA advises that, in some cases, ecological goals must also be considered in establishing clean-up goals.

Risk-based decision making and the RBCA process integrate risk and exposure assessment practices, as suggested by the EPA. The processes help to identify which assessment and remediation activities protect both human health and the environment. If a chemical release occurs, or is even suspected, risk-based decision making may be implemented. When utilizing these processes, it is important to establish appropriate safety and health practices and to determine any regulatory limitations prior to their use.

40.2   RISK ASSESSMENT

Chapters 41, 42, and 43 treat in greater detail *how* to evaluate risks to health and the environment. For the purposes of this chapter, a few definitions of common terms will suffice. Risk is the probability that persons or the environment will suffer adverse consequences as a result of an exposure to a substance. The amount of risk is determined by a combination of the concentration of the chemical the person or the environment is exposed to, the rate of intake or dose of the substance, and the toxicity of the substance. Risk assessment is the procedure used to attempt to quantify or estimate risk. Risk-based decision making distinguishes between the "point of exposure" and the "point of compliance." The point of exposure is the point at which the environment or the individual comes into contact with the chemical release. A person may be exposed by methods such as inhalation of vapors, as well as physical contact with the substance. The point of compliance is a point in between the point of release of the chemical (i.e., its source), and the point of exposure. The point of compliance is selected to provide a safety buffer for effected individuals and/or environments.

40.3   WHY USE RISK-BASED DECISION MAKING?

The use of the risk-based decision making process allows for efficient allocation of limited resources, such as time, money, regulatory oversight, and qualified professionals. Advantages of using this process include

- Decisions are based on reducing the risk of adverse human or environmental impacts.
- Site assessment activities are focused on collecting only that information that is necessary to make risk-based corrective action decisions.
- Limited resources are focused on those sites that pose the greatest risk to human health and the environment at any time.
- Compliance can be evaluated relative to site-specific standards applied at site-specific point(s) of compliance.
- Higher quality, and in some cases faster, clean-ups may be achieved than are currently possible.
- Documentation is developed that can demonstrate that the remedial action is protective of human health, safety, and the environment.

By using risk-based decision making, decisions are made in a consistent manner. Protection of both human health and the environment is accounted for.

A variety of EPA programs involved in the protection of groundwater and cleanup of environmental contamination utilize the risk-based decision making approach. Under the EPA's regulations dealing with cleanup of underground storage tank (UST) sites, regulators are expected to establish goals for clean-up of UST releases based on consideration of factors that could influence human and environmental exposure to contamination. Where UST releases affect groundwater being used as public or private drinking water sources, EPA generally recommends that clean-up goals be based on health-based drinking water standards; even in such cases, however, risk-based decision making can be employed to focus corrective action (1). (For more on USTs, refer to Chapter 24).

In the Superfund program, risk-based decision making plays an integral role in determining whether a hazardous waste site belongs on the National Priorities List. Once a site is listed, qualitative and quantitative risk assessments are used as the basis for establishing the need for action and determining remedial alternatives. To simplify and accelerate baseline risk assessments at Superfund sites, EPA has developed generic soil screening guidance that can be used to help distinguish between contamination levels that generally present no health concerns and those that generally require further evaluation. (For more on Superfund, refer to Chapter 23). The Resource Conservation and Recovery Act (RCRA) Corrective Action program also uses risk-based decision making to set priorities for clean-up so that high-risk sites receive attention as quickly as possible, to assist in the determination of clean-up standards, and to prescribe management requirements for remediation of wastes.

40.4 THE RISK-BASED CORRECTIVE ACTION APPROACH (2,3)

The risk-based corrective action (RBCA) process is implemented in a tiered approach, with each level involving increasingly sophisticated methods of data collection and analysis. As the analysis progresses, the assumptions of earlier tiers are replaced with site-specific data and information. Upon evaluation of each tier, the results and recommendations are reviewed, and it is determined whether more site-specific analysis is required. Generally, as the tier level increases, so do the costs of continuing the analysis.

The first step is the site assessment, which is the identification of the sources of the chemical(s) of concern, any obvious environmental impacts, any potentially impacted human and environmental receptors (e.g., workers, residents, lakes, streams, etc.), and potentially significant chemical transport pathways (e.g., groundwater flow, atmospheric dispersion, etc.). The site assessment also includes information collected from historical records and a visual inspection of the site. An example of criteria used for a site classification in outline form follows.

Example of Site Classification—Criteria and Prescribed Scenarios (2)
1. Immediate threat to human health, safety, or sensitive environmental receptors.

    a. Explosive levels, or concentrations of vapors that could cause acute health effects, are present in a residence or other building.
    b. Explosive levels of vapors are present in subsurface utility system(s), but no building or residences are impacted.
    c. Free-product is present in significant quantities at ground surface, on surface-water bodies, in utilities other than water supply lines, or in surface water runoff.
    d. An active public water-supply well, public water-supply line, or public surface-water intake is impacted or immediately threatened.
    e. Ambient vapor/particulate concentrations exceed concentrations of concern from an acute exposure or safety viewpoint.
    f. A sensitive habitat or sensitive resources (sport fish, economically important species, threatened and endangered species, and so forth) are impacted and affected.

2. Short-term (0 to 2 years) threat to human health, safety, or sensitive environmental receptors.

   a. There is potential for explosive levels, or concentrations of vapors that could cause acute effects, to accumulate in a residence or other building.
   b. Shallow contaminated surface soils are open to public access, and dwellings, parks, playgrounds, daycare centers, schools, or similar use facilities are within 500 ft (152 m) of those soils.
   c. A nonpotable water-supply well is impacted or immediately threatened.
   d. Groundwater is impacted, and a public or domestic water-supply well producing from the impacted aquifer is located within 2-years' projected groundwater travel distance down-gradient of the known extent of chemical(s) of concern.
   e. Groundwater is impacted, and a public or domestic water supply well producing from a different interval is located within the known extent of chemicals of concern.
   f. Impacted surface-water, storm-water, or groundwater discharges within 500 ft (152 m) of a sensitive habitat or surface-water body used for human drinking water or contact recreation.

3. Long-term (>2 years) threat to human health, safety, or sensitive environmental receptors.

   a. Subsurface soils [>3 ft (0.9 m) below ground surface] are significantly impacted, and the depth between impacted soils and the first potable aquifer is less than 50 ft (15 m).
   b. Groundwater is impacted, and potable water-supply wells producing from the impacted interval are located >2 years' groundwater travel time from the dissolved plume.
   c. Groundwater is impacted, and nonpotable water-supply wells producing from the impacted interval are located >2 years groundwater travel time from the dissolved plume.
   d. Groundwater is impacted, and nonpotable water-supply wells that do not produce from the impacted interval are located within the known extent of chemical(s) of concern.
   e. Impacted surface-water, storm-water, or groundwater discharges within 1500 ft (457 m) of a sensitive habitat or surface-water body used for human drinking water or contact recreation.
   f. Shallow contaminated surface soils are open to public access, and dwellings, parks, playgrounds, daycare centers, schools, or similar use facilities are more than 500 ft (152 m) from those soils.

4. No demonstrable long-term threat to human health or safety or sensitive environmental receptors. Priority 4 scenarios encompass all other conditions not described in Priorities 1, 2, and 3 and that are consistent with the priority description given above. Some examples are as follows:

   a. Nonpotable aquifer with no existing local use impacted.
   b. Impacted soils located more than 3 ft (0.9 m) below ground surface and greater than 50 ft (15 m) above nearest aquifer.
   c. Groundwater is impacted, and nonpotable wells are located down-gradient and outside the known extent of the chemical(s) of concern, and they produce from a nonimpacted zone.

Once the applicable criteria are met, the site is then classified according to the urgency of need for initial response action, based on information collected during the site assessment. Associated with site classifications are initial response actions that are to be implemented simultaneously with the RBCA process. Sites should be reclassified as actions are taken to resolve concerns or as better information becomes available.

A Tier 1 Evaluation is then conducted using a "look-up table." The look-up table contains screening level concentrations for the various chemicals of concern. The *look-up table* is defined as a tabulation for potential exposure pathways (e.g., inhalation, ingestion, etc.), media (e.g., soil, water, and air), a range of incremental carcinogenic risk levels which are used as target levels for determining remediation requirements, and potential exposure scenarios (e.g., residential, commercial, industrial, and agricultural). If a look-up table is not provided by the regulatory agency or available from a previous evaluation, the person conducting the RBCA analysis must develop the look-up table. If a look-up table is available, the user is responsible for determining that the risk-based screening levels (RBSLs) in the table are based on currently acceptable methodologies and parameters.

The RBSLs are determined using typical, non-site-specific values for exposure parameters and physical parameters for media. The RBSLs are calculated according to methodology suggested by the EPA (4,5). The value of creating a look-up table is that users do not have to repeat the exposure calculations for each site encountered. The look-up table is only altered when reasonable maximum exposure parameters, toxicological information, or recommended methodologies are updated. Some states have compiled such tables that, for the most part, contain identical values (as they are based on the same assumptions). The look-up table is used to determine whether site conditions satisfy the criteria for a quick regulatory closure or warrant a more site-specific evaluation.

If further evaluation is required, a Tier 2 Evaluation provides the user with an option to determine site-specific target levels (SSTLs) and point(s) of compliance. It is important to note that both Tier 1 RBSL and Tier 2 SSTLs are based on achieving similar levels of protection of human health and the environment. However, in Tier 2 the non-site-specific assumptions and point(s) of exposure used in Tier 1 are replaced with site-specific data and information. Additional site-assessment data may be needed. For example, the Tier 2 SSTL can be derived from the same equations used to calculate the Tier 1 RBSL, except that site-specific parameters are used in the calculations. The additional site-specific data may support alternate fate and transport analysis. At other sites, the Tier 2 analysis may involve applying Tier 1 RBSLs at more probable point(s) of exposure.

At the end of Tier 2, if it is determined that more detailed evaluation is again warranted, a Tier 3 evaluation is then conducted. A Tier 3 Evaluation provides the user with an option to determine SSTLs for both direct and indirect pathways using site-specific parameters and point(s) of exposure and compliance when it is judged that Tier 2 SSTLs should not be used as target levels. Tier 3, in general, can be a substantial incremental effort relative to Tiers 1 and 2, as the evaluation is much more complex and may include additional site assessment, probabilistic evaluations, and sophisticated chemical fate/transport models.

With the RBCA process, the user compares the target levels (RBSLs or SSTLs) to the concentrations of the chemical(s) of concern at the point(s) of compliance at the conclusion of each tier evaluation. If the concentrations of the chemical(s) of concern exceed the target levels at the point(s) of compliance, then either remedial action, interim remedial action, or further tier evaluation should be conducted. When it is judged that no further assessment is necessary or practicable, a remedial alternatives evaluation should be conducted to confirm the most cost-effective option for achieving the final remedial action target levels (RBSLs or SSTLs as appropriate).

Detailed design specifications may then be developed for installation and operation of the selected measures. The selected measures may include some combination of source removal, treatment, and containment technologies, as well as engineering and institutional controls. Examples of these include the following: soil venting, bioventing, air sparging, "pump-and-treat," and natural attenuation/ passive remediation. The remedial action must continue until such time as monitoring indicates that concentrations of the chemical(s) of concern are not above the RBSL or SSTL, as appropriate, at the points of compliance or source area(s), or both. When concentrations of chemical(s) of concern no longer exceed the target levels at the point of compliance, then the user may elect to deem the RBCA process complete. If achieving the desired risk reduction is impracticable due to technology or resource limitations, an interim remedial action, such as removal or treatment of "hot spots," may he conducted to address the most significant concerns, change the site classification, and facilitate reassessment of the tier evaluation.

After completion of the RBCA activities, most regulatory agencies require the submission of a RBCA report. The RBCA report typically contains a variety of site characterization items, including a site description; summaries of the site ownership and use, past releases or potential source areas, and the current and completed site activities; a description of regional hydrogeologic conditions and site-specific hydrogeologic conditions; and a summary of beneficial use. A site map of the location should be provided, which includes designations for local land use and groundwater supply wells, as well as the location of structures, above-ground storage tanks, buried utilities and conduits, and suspected/confirmed sources. Site photos should also be provided. The report also typically provides a discussion of the RBCA process, including a summary and discussion of the risk assessment (hazard identification, dose-response assessment, exposure assessment, and risk characterization), and the methods and assumptions used to calculate the RBSL and/or SSTL; a summary of the tier evaluation; a summary of the analytical data and the appropriate RBSL or SSTL used; and a summary of the ecological assessment. Additional data used needed for the analysis, such as groundwater elevation map; geologic cross section(s); and dissolved plume map(s) of the chemical(s) of concern, should also be included in the report.

40.5 SUMMARY

1. Risk-based decision making and risk-based corrective action are decision-making processes for assessing and responding to a chemical release which take into account effects on human health and the environment inasmuch as chemical releases vary greatly in terms of complexity, physical and chemical characteristics, and in the risk that they may pose.

2. Risk is the probability that individuals or the environment will suffer adverse consequences as a result of an exposure to a substance. Risk assessment is the procedure used to attempt to quantify or estimate risk.

3. The use of the risk-based decision making process allows for efficient allocation of limited resources, such as time, money, regulatory oversight, and qualified professionals.

4. The Risk Based Corrective Action (RBCA) process is implemented in a tiered approach, with each level involving increasingly sophisticated methods of data collection and analysis. As the analysis progresses, the assumptions of earlier tiers are replaced with site-specific data and information.

REFERENCES

1. *Use of Risk-Based Decision Making*, OSWER Directive 9610.17, U.S. Environmental Protection Agency, Washington, D.C., March 1995.

2. *Standard Guide for Risk-Based Corrective Action Applied to Petroleum Release Sites*, ASTM E1739-95, American Society for Testing and Materials, Philadelphia, PA.

3. *Ecological Assessment of Hazardous Waste Sites: A Field and Laboratory Reference Document*, EPA/600/3-89/013, NTIS No. PB-89205967, Environmental Protection Agency, Washington, D.C., March 1989.

4. *Integrated Risk Information System (IRIS)*, U.S. Environmental Protection Agency, Washington, D.C., October 1993.

5. *Health Effects, Assessment Summary Tables (HEAST)*, OSWER OS-230, U.S. Environmental Protection Agency, Washington, D.C., March 1992.

ENVIRONMENTAL RISK ASSESSMENT

41.1 INTRODUCTION

How is it possible to make decisions dealing with environmental risks from pesticides to air pollution in a complex society with competing interests and viewpoints, limited financial resources, and a lay public that is deeply concerned about the risks of cancer and other illness? Risk assessment and risk management, taken together constitute a decision-making approach that can help the different parties involved avoid stalemate and thus enable the larger society to work out its environmental problems rationally and with good results.

This chapter explores the theory and practice of risk assessment/risk management and attempts to put this problem-solving approach in the context of today's environmental challenges. It describes the challenges to environmental decision making today, including a national tradition of focusing narrowly on separate environmental problems, the danger of stalemate on crucial environmental issues, and a divergence of public and scientific views on what are the most risky environmental problems. The risk assessment/risk management decision-making approach is explained in terms of its relevance to today's environmental needs; how it works, generally and in specific situations; and how it can be improved. Included in this chapter are sections specifically on risk assessments, on risk management, and on risk communication, a term widely used to refer to public discussion about environmental risks. The risk communication section includes an explanation of why this tool is important and how it works.

41.2 DEFINITIONS

Risk assessment and risk management give a framework for setting regulatory priorities and for making decisions that cut across different environmental areas. This kind of framework has become increasingly important in recent years for several reasons, one of which is the considerable progress made in pollution control in this country. Twenty years ago, it wasn't hard to figure out where the first priorities should be. The worst pollution problems were all too obvious. Now that a number of clean-up areas are moving toward final control stages, the real priority problems and their solutions are not so obvious.

As a practical matter, it often comes down to the question of whether the final increment of a control program is cost-effective, given the resources available, or whether those same resources would be better spent on other, more pressing environmental problems. For example, it is known that the last 5% of pollution control is usually the most difficult and the most costly on a percentage basis. Is it worth it? Risk assessment and risk management help answer such pragmatic questions, and also enable evaluation of regulatory efforts to ensure that the environment is being made safer and that pollution is not just moving from one place to another.

Environmental risk assessment may be broadly defined as a scientific enterprise in which facts and assumptions are used to estimate the potential for adverse effects on human health or the environment that may result from exposures to specific pollutants or other toxic agents. Risk management, as the term is used by the U.S. Environmental Protection Agency (EPA) and other regulatory agencies, refers to a decision-making process which involves such considerations as risk assessment, technological feasibility, economic information about costs and benefits, statutory requirements, public concerns, and other factors. Risk communication is the exchange of information about risk.

Risk assessment may be also defined as the characterization of potential adverse effects to humans or to an ecosystem resulting from exposure to environmental hazards. Risk assessment supports risk management, the set of choices centering on whether and how much to control future exposure to the suspected hazards. Risk managers face the necessity of making difficult decisions involving uncertain science, potentially grave consequences to health or the environment, and large economic effects on industry and consumers. What risk assessment provides is an orderly, explicit, and consistent way to deal with scientific issues in evaluating whether a hazard exists and what the magnitude of the hazard may be. This evaluation typically involves large uncertainties, because the available scientific data are limited, and the mechanisms for adverse health impacts or environmental damage are only imperfectly understood.

When you look at risk, how do you decide how safe is safe, or how clean is clean? To begin with, the technician must look at both sides of the risk equation — that is, both the toxicity of a pollutant and the extent of public exposure. Look at both *current* and *potential* exposure, considering *all* possible exposure pathways. In addition to human health risks, look at potential *ecological* or other *environmental* effects. Even in conducting the most comprehensive risk assessment, there are always uncertainties, and one must make assumptions.

From a risk management standpoint, whether dealing with a site-specific situation or a national standard, the deciding question is ultimately, "What degree of risk is acceptable?" In general, this does not mean a "zero risk" standard, but rather a concept of *negligible* risk: At what point is there really no significant health or environmental risk, and at what point is there an adequate safety margin to protect public health and the environment? In addition, some environmental statutes require consideration of benefits together with risks in making risk management decisions.

It is possible to promote the goal of zero risk by emphasizing preventive policies so that pollution does not occur in the first place. For example, regulators can strive to ensure that pesticides do not have the potential to leach into groundwater. However, it is simply not possible to develop "zero risk" environmental programs across the board. Public health and environmental risks can be minimized, but it is not likely that we can eliminate all such risks. Ours is not a risk-free society.

41.3 PUBLIC PERCEPTION OF RISK

In making an effort to understand the significance of risk analyses, it is helpful to place the estimated risks in the same perspective as other everyday risks that have been determined by a similar methodology. Table 41.1 lists a number of risks for comparison. These have been derived from actual statistics and reasonable estimates (1,2). People often overestimate the frequency and seriousness of dramatic, sensational, dreaded, well-publicized causes of death and underestimate the risks

Table 41.1   Lifetime risks to life commonly faced by individuals

Cause of Risk	Lifetime (70-Yr) Risk, per Million Persons
Cigarette smoking	252,000
All cancers	196,000
Construction	42,700
Agriculture	42,000
Police killed in line of duty	15,400
Air pollution (Eastern United States)	14,000
Motor vehicle accidents (traveling)	14,000
Home accidents	7,700
Frequent airplane traveler	3,500
Pedestrian hit by motor vehicle	2,900
Alcohol, light drinker	1,400
Background radiation at sea level	1,400
Peanut butter, 4 tablespoons per day	560
Electrocution	370
Tornado	42
Drinking water containing chloroform at maximum allowable EPA Limit	42
Lightning	35
Living 70 years in zone of maximum impact from modern municipal	1
Smoking 1.4 cigarettes	1
Drinking 0.5 L of wine	1
Traveling 10 mi by bicycle	1
Traveling 30 mi by car	1
Traveling 1000 mi by jet plane (air crash)	1
Traveling 6000 mi by jet plane (cosmic rays)	1
Drinking water containing trichloroethylene at maximum allowable EPA limit	0.1

from more familiar, accepted causes that claim lives one by one. Indeed, risk estimates by "experts" and lay people (or "the public") on many key environmental problems differ significantly. This problem and the reasons for it are extremely important, because in our society the public generally does not trust experts to make important risk decisions alone.

This situation was illustrated by a 1987 EPA study. In commissioning this study nearly a year earlier, Administrator Lee M. Thomas had sought to "compare the risks currently associated with major environmental problems, given existing levels of control." Thomas's explicit premise was that "in a world of limited resources, it may be wise to give priority attention to those pollutants and problems that pose the greatest risks to our society."

To assess and compare these problems, the EPA created a special task force of about 75 career managers and experts from all EPA programs. The task force compared four different types of risks existing now for each of 31 environmental problem areas: cancer risk, noncancer health risks, ecological effects, and welfare effect (e.g., materials damage). While the task force did not try to "weight" or "add" the different types of risks for problem areas, they did develop rough rankings of problems within risk types. Beyond these rankings, the task force made no assertions about what EPA's priorities ought to be, noting that policy-makers must consider many other factors besides risk when they set priorities, such as legislation, economics, technology, and public mandate.

The examination of public mandate raised some interesting issues. A rough analysis of public polling data by the Roper Organization, Inc. on environmental problems made it clear that the EPA's actual priorities and legislative authorities correspond more closely with public opinion than they do with the EPA task force's estimates of the relative risk. The most significant differences concern hazardous waste and chemical plant accidents (high public concern, medium/low risk ranking by the task force), pesticides, indoor air pollution, consumer product exposure, worker exposure to chemicals, and global warming (medium/low public concern, relatively high risk ranking by the task force).

The most obvious reason for the differences is that the general public simply does not have all the information that was available to the task force experts. The subject is vast, and it is hard for anyone to have full knowledge of the information. Indeed, the experts themselves had to expend considerable effort to develop their rankings, and all of them were surprised by at least some of the findings.

Beyond this fact, it is interesting to observe that the judgments expressed in the polling data are consistent with an important finding by various researchers: People often overestimate the frequency and seriousness of dramatic, sensational, dreaded, well-publicized causes of death and underestimate the risks from more familiar, accepted causes that claim lives one by one. The EPA report should help people gain a better knowledge of the information and help close the gap between the experts and the public.

It is also important to note that the experts and the public were answering somewhat different questions. The EPA task force purposely dealt with a limited number of dimensions of risks, ignoring most of the intangible aspects that are of great value to the public: the degree to which risks are familiar, generally accepted, voluntary, controllable by the individual, and so on. These differences reflect a more general pattern of experts taking a societal (macro) perspective, whereas the lay public usually takes a more individual or personal (micro) perspective.

These factors provide important additional insights in explaining the differences between the task force's relative rankings and the public's. Hazardous waste disposal provides the most dramatic illustration. Recognizing the degree of public concern on this issue, the task force double-checked its rankings of active and inactive hazardous waste sites. The task force noted that in certain locations hazardous waste does pose a very serious risk, but relatively few people live near enough to be directly affected. Thus the total national impacts on public health and welfare and environment do not match the national concern. The intrusive, involuntary nature of the risk, the fact that slow-moving groundwater can stay polluted for a very long time, the presence of any identifiable "scapegoat," and the difficulty many people have in seeing any overriding benefit to having a hazardous waste site nearby are also important factors.

Interestingly enough, the fact that hazardous waste is only a problem in some locations has not been lost on the public. While 76% of the people interviewed by Roper called chemical waste disposal a "most serious" environmental problem and the same percentage said there is not enough regulation of industrial toxic waste, only 36% were aware of toxic waste problems in their own communities and only 16% considered toxic wastes to be near enough to their homes to be a threat to their personal health.

In contrast, indoor air pollution, consumer product exposure, and, to some extent, pesticide problems and worker exposure to chemicals are risks to which nearly everyone is exposed. The task force ranked these risks relatively high, yet the public ranked them medium/low. These risks are not dramatic and come from familiar, diffuse, generally accepted sources; it is usually difficult, if not impossible, in these cases to finger a "scapegoat"; and the benefits from the substances causing each of these problems are clear.

Global warming was also ranked low by the public and relatively high by the task force. However, this appears to be a somewhat special case. The task force ranked it high because of the massive potential implications for the entire world. The most probable explanation of the low public ranking is threefold: (a) The consequences are very much in the future and are hard for many to imagine because they extend beyond ordinary experience, (b) the problem is diffuse and there are many causes (the "scapegoat" problem), and (c) there is simply a general lack of public familiarity with the issue. If more people knew about global warming, its implications would probably cause them to rank it much higher. This is a "new" issue and although polling data are not yet available to confirm it, the level of concern appears to be rising.

The most obvious message for those involved in environmental problems — namely, representatives of government, representatives of industry, public interest groups, and the science community — is to recognize how people may react to the risks, to understand why the risks have been assessed technically as high or low, and to tailor policies and communications to accommodate differing perspectives.

41.4 RISK COMMUNICATION

Frustrations abound when it comes to the way public opinion regards environmental risks and drives environmental protection. Some risks are large, sometimes frighteningly large, whereas others are small, sometimes vanishingly small. This is so whether they are placed on the measuring rod of total population life expectancy or on that of the probability of premature death for small numbers of exposed people. The same holds true of nonfatal disease and ecological harm — at least this is what available scientific evidence suggests.

On the surface, it appears practical to remedy the most severe risks first, leaving the others until later — or maybe, if they are small enough, never remedying the others at all. But the behavior of individuals in everyday life often does not conform with this view. For example, there are toxic waste dumps where all evidence indicates that risks are minimal. Yet, the presence of such dumps can lead to numbing anxiety on the part of some, to loss of property values, and to disruption of communities. Elsewhere, facilities to dispose safely of similar wastes may be resisted by all means possible, including threats of civil disobedience. And at the same time, individuals may show little concern for hazardous products in ordinary commerce, resist efforts to protect wetlands vital to ecological integrity, not choose to test their homes for naturally occurring radon, and ignore safe-use labels for pesticides in home use. Examples of hysteria in the face of apparently trivial risks, along with examples of apathy in the face of apparently serious ones, form an unsettling litany to risk managers.

To make judgments wisely requires that individuals know what experts' estimates of the risks are, what it would cost in terms of their other values to reduce them, and how certain and free of bias all of this is.

Scientific precision is not needed, but a sense of whether a risk is "big," "medium," "small," or "infinitesimal" is.

The challenge of risk communication is to provide this information in ways so that it can be incorporated in the views of people who have little time or patience for arcane scientific discourse. Success in risk communication is not to be measured by whether the public chooses the set of outcomes that minimizes risk as estimated by the experts; it is achieved instead when those outcomes are knowingly chosen by a well-informed public.

If you make a list of environmental risks in order of how many people they kill each year, then list them again in order of how alarming they are to the general public, the two lists will be very different. The first list will also be very debatable, of course; it is not really known how many deaths are attributable to, say, geological radon or toxic wastes. But enough is known to be nearly certain that radon kills more Americans each year then all Superfund sites combined. Yet, those who choose not to test their homes for radon are deeply worried about toxic wastes. The conclusion is inescapable: The risks that kill are not necessarily the risks that anger and frighten.

To bridge the gap between the two, risk managers in government and industry have started turning to risk communication. They want help convincing the public that one part per million of dimethylmeatloaf in the air or water may not be such a serious hazard after all. Sometimes they want this help even when one part per million of dimethylmeatloaf is a serious hazard, hoping that clever risk communication can somehow replace effective risk management. But often the best evidence suggests that the dimethylmeatloaf really does endanger our health less than, say, eating peanut butter (not to mention the really big hazards, like cigarette smoking). Can risk communication get people to ease off on the dimethylmeatloaf and worry instead about their peanut butter consumption?

No. What risk communication can do is help risk managers understand why the public properly takes dimethylmeatloaf more seriously than peanut butter. This understanding, in turn, can lead to changes in dimethylmeatloaf policy that will help bring the public and expert assessments of the risk closer together.

The core problem is a definition. To the experts, risk means expected annual mortality. But to the public (and even the experts when they go home at night), risk means much more than that. Let's redefine terms. Call the death rate (what the experts mean by risk) "hazard." Call all other factors, collectively, "outrage." Risk, then, is the sum of hazard and outrage. The public pays too little attention to hazard; the experts pay absolutely no attention to outrage. Not surprisingly, they each rank risks differently.

Risk perception scholars have identified more than 20 "outrage factors." Here are a few of the main ones:

- *Voluntariness.* A voluntary risk is much more acceptable to people than a coerced risk, because it generates no outrage. Consider the difference between getting pushed down a mountain on slippery sticks and deciding to go skiing.

- *Control.* Almost everybody feels safer driving than "riding shotgun." When prevention and mitigation are in the individual's hands, the risk (though not the hazard) is much lower than when they are in the hands of a government agency.

- *Fairness.* People who must endure greater risks than their neighbors, without access to greater benefits, are naturally outraged, especially if the rationale for so burdening them looks more like politics than science. Greater outrage, of course, means greater risk.

- *Process.* Does the agency come across as trustworthy or dishonest, concerned or arrogant? Does it tell the community what's going on before the real decisions are made? Does it listen and respond to community concerns?

- *Morality.* American society has decided over the last two decades that pollution isn't just harmful, it's evil. But talking about cost–risk tradeoffs sounds very callous when the risk is morally relevant. Imagine a police chief insisting that an occasional child-molester is an "acceptable risk."

- *Familiarity.* Exotic, high-tech facilities provoke more outrage than do familiar risks (your home, your car, your jar of peanut butter).

- *Memorability.* A memorable accident — for example, Love Canal, Bhopal, or Times Beach — makes the risk easier to imagine and thus, as we have defined the term, more risky. A potent symbol, the 55-gal drum, can do the same things.

- *Dread.* Some illnesses are more dreaded than others; compare AIDS and cancer with, say, emphysema. The long latency of most cancers and the undetectability of most carcinogens add to the dread.

- *Diffusion in Time and Space.* Hazard A kills 50 anonymous people a year somewhere across the country. Hazard B has one chance in 10 of wiping out its neighborhood of 5000 people sometime in the next decade. Risk assessment tells us the two have the same expected annual mortality: 50. "Outrage assessment" tells us that A is probably acceptable and that B is certainly not.

These "outrage factors" are not distortions in the public's perception of risk. Rather, they are intrinsic parts of what is meant by risk. They explain why people worry more about Superfund sites than about geological radon, and more about industrial emissions of dimethylmeatloaf than about aflatoxin in peanut butter.

There is peculiar paradox here. Many risk experts resist the pressure to consider outrage in making risk management decisions; they insist that "the data" alone, not the "irrational" public, should determine policy. But two decades of data indicate that voluntariness, control, fairness, and the rest are important components of society's definition of risk. When a risk manager continues to be surprised by the public's response of outrage, it is worth asking just whose behavior is irrational.

The solution is implicit in this reframing of the problem. Because the public responds more to outrage than to hazard, risk managers must work to make serious hazards more outrageous. Recent campaigns against drunk driving and sidestream cigarette smoke provide two models of successful efforts to increase public concern about serious hazards by feeding the outrage.

Similarly, to decrease public concern about modest hazards, risk managers must work to diminish the outrage. When people are treated with fairness, honesty, and respect for their right to make their own decisions, they are a lot less likely to overestimate small hazards. At that point, risk communication can help explain the hazard. But when people are not treated with fairness, honesty, and respect for their right to make their own decisions, there is little that risk communication can do to keep them from raising hell, regardless of the extent of the hazard. Most of us wouldn't have it any other way.

41.5 SEVEN RULES OF RISK COMMUNICATION (4,5)

There are no easy prescriptions for successful risk communication. However, those who have studied and participated in recent debates about risk generally agree on the following seven rules. These rules apply equally well to the public and private sectors.

Although many of the rules may seem obvious, they are continually and consistently violated in practice. Thus, a useful way to read these rules is to focus on why they are frequently not followed.

1. Accept and involve the public as a legitimate partner. A basic tenet of risk communication in a democracy is that people and communities have a right to participate in decisions that affect their lives, their property, and the things they value.

> **Guidelines:** Demonstrate your respect for the public and underscore the sincerity of your effort by involving the community early, before important decisions are made. Involve all parties that have an interest or a stake in the issue under consideration. If you are a government employee, remember that you work for the public. If you do not work for the government, the public still holds you accountable.

Point to Consider

• The goal of risk communication in a democracy should be to produce an informed public that is involved, interested, reasonable, thoughtful, solution-oriented, and collaborative; it should not be to defuse public concerns or replace action.

2. Plan carefully and evaluate your efforts. Risk communication will be successful only if carefully planned.

Guidelines: Begin with clear, explicit risk communication objectives, such as providing information to the public, motivating individuals to act, stimulating response to emergencies, or contributing to the resolution of conflict. Evaluate the information you have about the risks, and know its strengths and weaknesses. Classify and segment the various groups in your audience. Aim your communications at specific subgroups in your audience. Recruit spokespeople who are good at presentation and interaction. Train your staff, including technical staff, in communication skills; reward outstanding performance. Whenever possible, pretest your messages. Carefully evaluate your efforts and learn from your mistakes.

Points to Consider

- There is no such entity as "the public"; instead, there are many publics, each with its own interests, needs, concerns, priorities, preferences, and organizations.
- Different risk communication goals, audiences, and media require different risk communication strategies.

3. Listen to the public's specific concerns. If you do not listen to people, you cannot expect them to listen to you. Communication is a two-way activity.

Guidelines: Do not make assumptions about what people know, think, or want done about risks. Take the time to find out what people are thinking: Use techniques such as interviews, focus groups, and surveys. Let all parties that have an interest or a stake in the issue to be heard. Identify with your audience and try to put yourself in their place. Recognize people's emotions. Let people know that you understand what they said, addressing their concerns as well as yours. Recognize the "hidden agenda," symbolic meanings, and broader economic or political considerations that often underlie and complicate the task of risk communication.

Point to Consider

- People in the community are often more concerned about such issues as trust, credibility, competence, control, voluntariness, fairness, caring, and compassion than about mortality statistics and the details of quantitative risk assessment.

4. Be honest, frank, and open. In communicating risk information, trust and credibility are your most precious assets.

Guidelines: State your credentials, but do not ask or expect to be trusted by the public. If you do not know an answer or are uncertain, say so. Get back to people with answers. Admit mistakes. Disclose risk information as soon as possible (emphasizing any reservations about reliability). Do not minimize or exaggerate the level of risk. Speculate only with great caution. If in doubt, lean toward sharing more information, not less, or people may think you are hiding something. Discuss data uncertainties, strengths, and weaknesses, including the ones identified by other credible sources. Identify worst-case estimates as such, and cite ranges of risk estimates when appropriate.

Point to Consider

- Trust and credibility are difficult to obtain. Once lost, they are almost impossible to regain completely.

5. Coordinate and collaborate with other credible sources. Allies can be effective in helping you communicate risk information.

> **Guidelines:** Take time to coordinate all intraorganizational communications. Devote effort and resources to the slow, hard work of building bridges with other organizations. Use credible and authoritative intermediaries. Consult with others to determine who is best able to answer questions about risk. Try to issue communications jointly with other trustworthy sources (for example, credible university scientists, physicians, or trusted local officials).

Point to Consider

- Few things make risk communication more difficult than do conflicts or public disagreements with other credible sources.

6. Meet the needs of the media. The media are a prime transmitter of information on risks; they play a critical role in setting agendas and in determining outcomes.

> **Guidelines:** Be open with and accessible to reporters. Respect their deadlines. Provide risk information tailored to the needs of each type of media (for example, graphics and other visual aids for television). Prepare in advance and provide background material on complex risk issues. Do not hesitate to follow up on stories with praise or criticism, as warranted. Try to establish long-term relationships of trust with specific editors and reporters.

Point to Consider

- The media are frequently more interested in politics than in risk, more interested in simplicity than in complexity, and more interested in danger than in safety.

7. Speak clearly and with compassion. Technical language and jargon are useful as professional shorthand, but they are barriers to successful communication with the public.

> **Guidelines:** Use simple, nontechnical language. Be sensitive to local norms, such as speech and dress. Use vivid, concrete images that communicate on a personal level. Use examples and anecdotes that make technical risk data come alive. Avoid distant, abstract, unfeeling language about deaths, injuries, and illnesses. Acknowledge and respond (both in words and with actions) to emotions that people express: anxiety, fear, anger, outrage, and helplessness. Acknowledge and respond to the distinctions that the public views as important in evaluating risks: voluntariness, controllability, familiarity, dread, origin (natural or human), benefits, fairness, and catastrophic potential. Use risk comparisons to help put risks in perspective, but avoid comparisons that ignore distinctions that people consider important. Always try to include a discussion of actions that are underway or that can be taken. Tell people what you cannot do. Promise only what you can do, and be sure to do what you promise.

Points to Consider

- Regardless of how well you communicate risk information, some people will not be satisfied.

- Never let your efforts to inform people about risks prevent you from acknowledging, and saying, that any illness, injury, or death is a tragedy.

- If people are sufficiently motivated, they are quite capable of understanding complex risk information, even if they may not agree with you.

41.6 RISK MANAGEMENT CASE STUDY: FIFRA AND THE DINOSEB CASE

The Federal Insecticide, Fungicide, and Rodenticide Act (FIFRA), which governs the EPA's regulation of pesticides, is often called a "balancing" statute because it requires the EPA to weigh the risks of pesticides against their economic and social benefits when making regulatory decisions. Under the FIFRA, all pesticides intended for use in the United States must be registered (licensed) by the EPA to ensure that they do not cause "unreasonable adverse effects on the environment." In the context of the FIFRA, unreasonable adverse effects are defined to mean: "any unreasonable risk to man or the environment, taking into account the economic, social, and environmental costs and benefits of the use of any pesticide."

The risk/benefit mandate of the FIFRA makes pragmatic sense when you consider that pesticides, almost by definition, yield risks as well as agricultural and other pest-control benefits. Because pesticides typically perform their intended function because they are toxic to something, there is generally no such thing as a "zero risk" pesticide. Reflecting Congress's recognition that pesticide uses involve trade-offs between benefits and risks, the FIFRA calls upon the EPA to make administrative judgments as to how much risk is reasonable in light of the specific benefits to be obtained from pesticide uses.

Registration under the FIFRA is a license for the sale of a pesticide for use on a specific crop or other site under the circumstances prescribed by its approved labeling. Pesticide registration is not an "either/or" proposition whereby the EPA either gives blanket approval to the sale and use of a pesticide, or else disapproves its registration. On the contrary, in cases where proposed or continued uses of a pesticide raise risk concerns, the FIFRA affords the EPA a spectrum of risk management options to bring down risks, wherever possible, with limited impacts on benefits. Depending on the nature of the EPA's concerns, such options might include: requiring protective apparel and or equipment to minimize risks to pesticide applicators; reducing the rate or frequency of application or otherwise modifying application practices to lower pesticide residue levels on harvested crops; or imposing regional restrictions against using a pesticide in areas where it could leach into groundwater.

On a graduated scale of risk management options available under the FIFRA, regulatory action by the EPA to remove some or all uses of a pesticide from commerce by initiating cancellation proceedings is an option of last resort. Yet there are cases where the EPA does opt to cancel a pesticide, or even call an immediate halt to its use for the duration of formal cancellation proceedings, such as in the case of dinoseb, a chemical with herbicidal, fungicidal, insecticidal, and desiccant properties that has been widely used in recent decades, primarily in agriculture.

On October 7, 1986, the EPA issued a formal notice of intent to cancel and deny all registrations for pesticide products containing dinoseb, citing evidence that it may cause birth defects in children born to women exposed to dinoseb during pregnancy, sterility or decreased fertility in males, acute toxic poisoning, and other potential adverse effects on health and the environment. On the same day, the EPA issued an emergency suspension order effecting an immediate stop to dinoseb use during the time required to complete cancellation proceedings on the pesticide. (Under the FIFRA, an "emergency suspension" takes effect immediately, whereas under an "ordinary suspension," pesticide registrants may request an expedited hearing before the suspension takes effect.) The dinoseb order was the third such emergency suspension order the EPA had issued under the FIFRA.

Consider the case of dinoseb in the context of the EPA's pesticide risk assessment and risk/benefit "balancing" process.

Dinoseb Risk Assessment

What were the studies that led to the EPA's emergency suspension and cancellation initiatives on dinoseb, and how did the EPA use these studies for risk assessment?

- *Birth Defects.* In recent laboratory studies, dinoseb had caused birth defects in the offspring of three test animal species (rabbits, rats, and mice). Based on this multi-test evidence from studies using several different routes of exposure, EPA scientists concluded that dinoseb causes birth defects in laboratory animals and has the potential to cause birth defects in humans. Based on statistical data from the rabbit study (an oral feeding study), a "no observed effects level" (NOEL) was provisionally set at 3 milligrams per kilogram of body weight per day (mg/kg/day), meaning that adverse effects in test animal offspring were apparent at all oral exposure levels higher than 3 mg/kg/day.

- *Male Reproductive Effects.* In rodent feeding studies, dinoseb had caused adverse reproductive effects in males. Based on the evidence in mice and rats, EPA scientists concluded that dinoseb causes adverse reproductive effects in laboratory animals and should be considered a potential cause of human male reproductive disorders.

- *Acute Toxicity.* The LD_{50} of a pesticide (the dose at which 50% of test animals succumb to the toxicity of the chemical) is typically used as a measure of its acute toxicity. Test data cited by the EPA in its dinoseb suspension and cancellation notices showed the dinoseb LD_{50} by dermal exposure to be approximately 75 mg/kg, an LD_{50} low enough to be considered indicative of very high toxicity. There was also direct evidence of the acute toxicity of dinoseb in humans, including at least one human fatality attributed to accidental exposure to dinoseb during spray application.

In addition to the effects just described, dinoseb belongs to a class of chemicals (dinitrophenols) known to induce cataracts in humans, and cataracts have been observed in the eyes of three species of laboratory animals following dinoseb exposure. Dinoseb has also induced tumors in female mice and may have the potential to affect the immunological system. Apart from its potential human health effects, dinoseb also has the potential to adversely affect wildlife.

The toxicity profile just outlined raised very significant concerns regarding the teratogenicity (birth defects) and other hazards of dinoseb. On the other hand, from the standpoint of pesticide risk assessment, the toxicological characteristics of a pesticide chemical are only half the picture. The second basic component of risk is the extent to which people and the environment are actually exposed to the pesticide when it is used in accordance with widespread and commonly recognized practice.

In the case of dinoseb, three basic exposure scenarios were identified:

- Possible dietary exposure to the public through consumption of food or drinking water containing residues of dinoseb.

- Occupational exposures to workers who mix, load, or apply dinoseb.

- Secondary or "coincidental" exposures to bystanders, farmworkers, and others who could be exposed to dinoseb through spray drift, contact with residues in treated fields, or even contact with contaminated clothing or farm equipment immediately after dinoseb application.

In conducting pesticide risk assessments, as in the case of dinoseb, the EPA makes a practice of evaluating all potential toxic effects, but generally focuses its quantitative risk assessment and risk/benefit balancing process on the effect observed at the lowest dose level. For dinoseb, this was the 3 mg/kg/day NOEL cited earlier for dinoseb-induced birth defects in rabbit offspring. In quantitative calculations, the EPA scientists compare this NOEL from laboratory studies with expected human exposure levels to obtain numerical "margins of safety" (NOEL divided by exposure equals margin of safety, or MOS). To protect people from significant health risks, the EPA generally considers an MOS greater than 100 to be acceptable when calculated from animal data. Where an MOS is less than 100, the EPA typically considers the comparative impacts of possible risk management measures.

- *Risks from Dietary Exposure.* EPA scientists calculated MOS values for risks of birth defects from potential dietary exposure to dinoseb residues in food and drinking water. Even when certain "worst-case" assumptions regarding dietary exposure levels were factored into these calculations, the MOS for the risk of birth defects occurring from consumption of food from crops treated with dinoseb was found to be ample, i.e., over 2700. Similarly, from consumption of drinking water in areas where dinoseb may have leached to underground aquifers, the MOS was roughly 2450.

- *Risks from Occupational Exposures to Dinoseb.* Based on experimental data from field studies performed with dinoseb and other agricultural pesticides, exposure levels were estimated for the various kinds of workers involved in the use of dinoseb on various crops sites: mixer/loaders, pilots, airplane flaggers, "ground boom" applicators, and hand-sprayers. For these various kinds of workers, exposure levels were estimated for a range of plausible exposure conditions. In many instances, estimated worker exposure levels were equal to or greater than the NOEL of 3 mg/kg/day for birth defects in test animals treated with dinoseb. If a worker is exposed to a pesticide at a level that is equal to its NOEL in laboratory animals, he or she is said to have an MOS of 1. Thus, in the case of dinoseb, the EPA found virtually no MOS against the occurrence of birth defects in pregnant workers handling the pesticide.

- *Risks from Secondary Exposure to Dinoseb.* The EPA did not have adequate exposure data to calculate MOS values for secondary exposures to dinoseb. However, there are grounds for inferring that significant secondary exposures do occur, including data from the State of California revealing that acute poisonings from spray drift of dinitrophenol pesticides occur annually.

Dinoseb Benefit Assessment

Based on data from the U.S. Department of Agriculture and other sources, the EPA conducted an assessment of the benefits of dinoseb by calculating the short-term and long-term economic impacts expected to occur if dinoseb were unavailable for registered uses. Dinoseb use sites included soybeans, peanuts, cotton, snap beans, potatoes, green peas, grapes, alfalfa, almonds, walnuts, berries, hops, noncrop areas, and a variety of "minor use" crops and sites.

For both short- and long-term scenarios, estimated economic losses were due mainly to increased pest control costs and expected yield losses for some crop sites. For both scenarios, the largest user impacts were projected for potato and peanut growers, whereas the extent of impacts on the production of green peas, snap beans, berries, and hops was uncertain. Apart from these uncertainties, the overall annual impacts of removing dinoseb from the marketplace were estimated at the user level to be in the range of $80 to $90 million. The information available to the EPA did not point to significant market and consumer impacts, except for possible short-term peanut price increases.

Regulatory Options Considered

In the case of dinoseb, the EPA was satisfied that there were adequate margins of safety to protect public health from any risks due to dietary and drinking-water exposures to the pesticide. On the other hand, the EPA's MOS calculations pointed to extraordinarily high risks of birth defects from occupational exposures to dinoseb, and there was reason to believe that secondary exposures to dinoseb also presented significant risks to unborn children. The evidence available to the EPA also indicated that occupational and secondary exposures to dinoseb posed additional risks of adverse reproductive effects in males and acute toxic poisoning. Focusing on these exposure routes, the EPA considered a number of possible risk management options to determine whether such measures could reduce the risks of birth defects and other potential adverse effects to acceptable levels in view of the known benefits of dinoseb.

- *Additional Protective Clothing.* The risks of birth defects in children born to workers involved in the use of dinoseb were found to be unacceptable even with the protection afforded by the requisite apparel specified by dinoseb product labels: goggles or a face shield, impermeable gloves, and an apron when handling dinoseb concentrate; and long-sleeved shirts, long-legged pants, and shoes and socks when handling the concentrate or spraying the prepared formula. To further minimize worker exposure, the EPA considered the possible additional requirement of Tyvek® suits (synthetic, disposable coveralls) for workers who handle dinoseb. However, the EPA decided against this special requirement, due in part to practicality and enforcement problems. In addition, the EPA had concerns about the hazards of heat stress that may result when this type of synthetic clothing is worn in temperatures above 80°F.

- *Protective Farm Equipment.* As part of the exposure and risk assessment of dinoseb, EPA scientists calculated MOS values for workers with and without the use of such protective farm equipment as closed loading systems and enclosed tractor cabs. Although MOS values were higher with the use of this equipment, they were still below 100. Consequently, this option was deemed ineffective to mitigate the risks of dinoseb use.

- *Lower Application Rates.* The EPA also calculated comparative MOS values for low dinoseb application rates (0.625 pounds active ingredient per acre, as directed by the label for some fungicidal uses) versus high application rates (9–12 pounds per acre, as recommended by labels for certain herbicidal uses). MOS values were comparatively higher for the lower application rates, but still well below 100 and therefore unacceptable.

- *Gender-Based Restrictions.* The EPA considered a number of gender-based restrictions to reduce the risks of birth defects associated with exposure to dinoseb. Among other things, the EPA considered label changes to prohibit women of childbearing age from mixing, applying, or handling dinoseb in any way, or to alert pregnant women to the risks of dinoseb exposure. For the purposes of risk management, the impact of such restrictions is limited to direct occupational exposures to dinoseb. As a practical matter, gender-based restrictions were considered inadequate to control secondary exposures to female bystanders, farmworkers, and others. Moreover, such restrictions could not mitigate dinoseb-related risks of male reproductive effects or acute toxicity.

- *Reformulation.* Through comparative MOS calculations, the EPA considered the risk management impacts of reformulating dinoseb to reduce worker exposure. None of the available technologies was found to reduce dinoseb risks to acceptable levels for workers performing the various tasks involved in dinoseb application.

Suspension and Cancellation Initiatives
In the case of dinoseb, the EPA's risk assessment and risk/benefit balancing processes led the EPA to conclude that the risks associated with registered uses of the pesticide could not be reduced to reasonable levels by any means short of immediately removing the pesticide from the marketplace. For all crops and use sites, based on all available risk data and benefits information, the risks of continued use of dinoseb were deemed to outweigh the benefits, not only in the long term, but also during the interval of time required to conduct "ordinary" suspension and cancellation hearings under the FIFRA. Thus, the EPA opted for the most drastic remedial option available under the FIFRA: emergency suspension, calling an immediate halt to the sale and use of a pesticide while cancellation proceedings were being conducted.

41.7 SUMMARY

1. Environmental risk assessment may be broadly defined as a scientific enterprise in which facts and assumptions are used to estimate the potential for adverse effects on human health or the environment that may result from exposures to specific pollutants or other toxic agents.

2. Risk management refers to a decision-making process which involves such considerations as risk assessment, technological feasibility, economic information about costs and benefits, statutory requirements, public concerns, and other factors.

3. Risk communication is the exchange of information about risk.

4. People often overestimate the frequency and seriousness of dramatic, sensational, dreaded, well-publicized causes of death and underestimate the risks from more familiar, accepted causes that claim lives one by one.

5. The Seven Rules of Risk Communication are as follows: Accept and involve the public as a legitimate partner; plan and evaluate your efforts carefully; listen to the public's specific concerns; be honest, frank, and open; coordinate and collaborate with other credible sources; meet the needs of the media; and speak clearly and with compassion.

6. The Federal Insecticide, Fungicide, and Rodenticide Act (FIFRA), which governs the EPA's regulation of pesticides, is often called a "balancing" statute because it requires the EPA to weigh the risks of pesticides against their economic and social benefits when making regulatory decisions.

REFERENCES

1. T. Main, Inc., *Health Risk Assessment for Air Emissions of Metals and Organic Compounds from the Perc Municipal Waste to Energy Facility,* prepared for Penobscot Energy Recovery Company (PERC), Boston, MA, December 1985.

2. R. Wilson and E. A. Crouch, "Risk Assessment and Comparisons: An Introduction," *Science*, April 1987.

3. U.S. EPA, *Unfinished Business: A Comparative Assessment of Environmental Problems,* 1987.

4. V. T. Covello, D. von Winterfeldt, and P. Slovic, "Risk Communication: A Review of the Literature," *Risk Abstracts*, 3(4), 171–182, 1986.

5. C. J. Davies, V. T. Covello, and F. W. Allen, *Risk Communication, Proceedings of the National Conference on Risk Communication, January 29–31, 1986*, Conservation Foundation, Washington, D.C., 1987.

42

HEALTH RISK ASSESSMENT

Contributing Author: Elizabeth Butler

42.1 INTRODUCTION

There are many definitions for the word risk. It is a combination of uncertainty and damage; a ratio of hazards to safeguards; a triple combination of events, probability, and consequences; or even a measure of economic loss or human injury in terms of both the incident, likelihood, and the magnitude of the loss or injury (1). People face all kinds of risks every day, some voluntarily and others involuntarily. Therefore, risk plays a very important role in today's world. Studies on cancer caused a turning point in the field of risk studies because it opened the eyes of risk scientists and health professionals to the broad area of risk assessments.

Since 1970 the field of risk assessment has received widespread attention within both the scientific and regulatory committees. It has also attracted the attention of the public. Properly conducted risk assessments have received fairly broad acceptance, in part because they put into perspective the terms toxic, hazard, and risk.

Toxicity is an inherent property of all substances. It states that all chemical and physical agents can produce adverse health effects at some dose or under specific exposure conditions. In contrast, exposure to a chemical that has the capacity to produce a particular type of adverse effect represents a hazard. Risk, however, is the probability or likelihood that an adverse outcome will occur in a person or a group that is exposed to a particular concentration or dose of the hazardous agent. Therefore, risk is generally a function of exposure and/or dose. Consequently, health risk assessment is defined as the process or procedure used to estimate the likelihood that humans or ecological systems will be adversely affected by a chemical or physical agent under a specific set of conditions (2).

Drawn in part (with permission) from *Major Environmental Issues Facing the 21st Century,* M. K. Theodore and L. Theodore, Chapter 35, "Health Risk Assessment," originally published by Prentice-Hall, Upper Saddle River, NJ, 1996.

The term risk assessment is not only used to describe the likelihood of an adverse response to a chemical or physical agent, but it has also been used to describe the likelihood of any unwanted event. The subject is treated in more detail in Chapter 43. These include risks such as: explosions or injuries in the workplace; natural catastrophes; injury or death due to various voluntary activities such as skiing, sky diving, flying, and bungee jumping; diseases; death due to natural causes; and many others.

The general subject of environmental risk assessment was addressed in the two previous chapters. Before proceeding to applying risk assessment to both health and hazard concerns, one needs to once again define risk management and relate it to risk assessment.

Risk assessment and risk management are two different processes, but they are intertwined. Risk assessment and risk management give a framework not only for setting regulatory priorities but also for making decisions that cut across different environmental areas. *Risk management* refers to a decision-making process that involves such considerations as risk assessment, technology feasibility, economic information about costs and benefits, statutory requirements, public concerns, and other factors. Therefore *risk assessment* supports risk management in that the choices on whether and how much to control future exposure to the suspected hazards may be determined. Regarding both risk assessment and risk management, this chapter addresses these subjects primarily from a health perspective; Chapter 43 addresses these subjects primarily from a safety and accident hazard perspective.

42.2 THE HEALTH RISK EVALUATION PROCESS

Health risk assessments provide an orderly, explicit way to deal with scientific issues in evaluating whether a hazard exists and what the magnitude of the hazard may be. This evaluation typically involves large uncertainties because the available scientific data are limited, and the mechanisms for adverse health impacts or environmental damage are only imperfectly understood. When examining risk, how does one decide how safe is "safe," or how clean is "clean"? To begin with, one has to look at both sides of the risk equation, i.e., both the toxicity of a pollutant and the extent of public exposure. Information is required at both the current and the potential exposure, considering all possible exposure pathways. In addition to human health risks, one needs to look at potential ecological or other environmental effects. In conducting a comprehensive risk assessment, it should be remembered that there are always uncertainties, and these assumptions must be included in the analysis (1).

In recent years, several guidelines and handbooks have been produced to help explain approaches for doing health risk assessments. As discussed by a special National Academy of Sciences committee convened in 1983, most human or environmental health hazards can be evaluated by dissecting the analysis into four parts; hazard identification, dose-response assessment or hazard assessment, exposure assessment, and risk characterization (see Figure 42.1). For some perceived hazards, the risk assessment might stop with the first step, hazard identification, if no adverse effect is identified or if an agency elects to take regulatory action

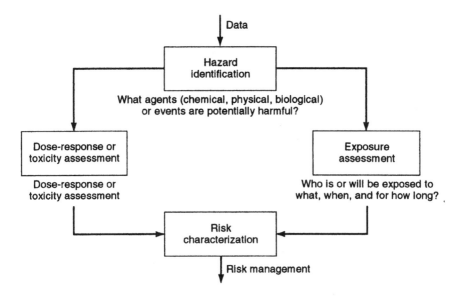

FIGURE 42.1 The health risk evaluation process (1).

without further analysis (2). Regarding hazard identification, a hazard is defined as a toxic agent or a set of conditions that has the potential to cause adverse effects to human health or the environment. Hazard identification involves an evaluation of various forms of information in order to identify the different hazards. Dose-response or toxicity assessment is required in an overall assessment; responses/effects can vary widely since all chemicals and contaminants vary in their capacity to cause adverse effects. This step frequently requires that assumptions be made to relate experimental data from animals and humans. Exposure assessment is the determination of the magnitude, frequency, duration, and routes of exposure of human populations and ecosystems. Finally, in risk characterization, toxicology, and exposure data/information are combined to obtain qualitative or quantitative expression of risk.

Risk assessment also involves the integration of the information and analysis associated with the above four steps to provide a complete characterization of the nature and magnitude of risk and the degree of confidence associated with this characterization. A critical component of the assessment is a full elucidation of the uncertainties associated with each of the major steps. Under this broad concept of risk assessment are encompassed all of the essential problems of toxicology. Risk assessment takes into account all of the available dose-response data. It should treat uncertainty not by the application of arbitrary safety factors, but by stating them in quantitatively and qualitatively explicit terms, so that they are not hidden from decision makers. Risk assessment, defined in this broad way, forces an assessor to confront all the scientific uncertainties and to set forth in explicit terms the means used in specific cases to deal with these uncertainties (3). An expanded presentation on each of the four health risk assessment steps is provided below.

42.3 HAZARD IDENTIFICATION

Hazard identification is the most easily recognized of the actions taken by regulatory agencies. It is defined as the process of determining whether human exposure to an agent could cause an increase in the incidence of a health condition (cancer, birth defect, etc.) or whether exposure by a nonhuman receptor, for example, fish, birds, or other wildlife, might adversely be affected. It involves characterizing the nature and strength of the evidence of causation. Although the question of whether a substance causes cancer or other adverse health effects in humans is theoretically a yes–no question, there are few chemicals or physical agents for which the human data are definitive. Therefore, the question is often restated in terms of effects in laboratory animals or other test systems: "Does the agent induce cancer in test animals?" Positive answers to such questions are typically taken as evidence that an agent may pose a cancer risk for any exposed human. Information for short-term in vitro tests and structural similarity to known chemical hazards may, in certain circumstances, also be considered as adequate information for identifying a hazard (2).

A hazards identification for a chemical plant or industrial application can include information about

1. Chemical identities.
2. The location of facilities that use, produce, process, or store hazardous materials.
3. The type and design of chemical containers or vessels.
4. The quantity of material that could be involved in airborne release.
5. The nature of the hazard (e.g., airborne toxic vapors or mist, fire, explosion, large quantities stored or processed, most likely to accompany hazardous materials spills or releases (4).

An important aspect of hazards identification is a description of the pervasiveness of the hazard. For example, most environmental assessments require knowledge of the concentration of the material in the environment, weighted in some way to account for the geographical magnitude of the site affected; that is, a 1-acre or 300-acre site, a 1,000 gal/min or 1,000,000 gal/min stream. All too often environmental incidents regarding chemical emission have been described by statements like "concentrations as high as 150 ppm" of a chemical were measured at a 1,000-acre waste site. However, following closer examination, one may find that only 1 of 200 samples collected on a 20-acre portion of a 1,000-acre site showed this concentration, and that 2ppm was the geometric mean level of contamination in the 200 samples.

An appropriate sampling program is critical in the conduct of a health risk assessment. This topic could arguably be part of the exposure assessment, but it has been placed within hazard identification because if the degree of contamination is small, no further work may be necessary. Not only is it important that samples be collected in a random or representative manner, but the number of samples must be sufficient to conduct a statistically valid analysis. The number needed to ensure statistical validity will be dictated by the variability between the results. The larger the variance, the greater the number of samples needed to define the problem (2).

The means of identifying hazards is complex. Different methods are used to collect and evaluate toxic properties (those properties that indicate the potential to cause biological injury, disease or death under certain exposure conditions). One method is the use of epidemiological studies that deal with the incidence of disease among groups of people. Epidemiological studies attempt to correlate the incidence of cancer from an emission by an evaluation of people with a particular disease and people without the disease. Long-term animal bioassays are the most common method of hazard determination. (A bioassay as referred to here is an evaluation of disease in a laboratory animal.) Increased tumor incidence in laboratory animals is the primary health effect considered in animal bioassays. Exposure testing for a major portion of an animal's lifetime (2 to 3 years for rats and mice) provides information on disease and susceptibility, primarily for carcinogenicity (the development of cancer).

The understanding of how a substance is handled in the body, transported, changed and excreted, and of the response of both animals and humans, has advanced remarkably. There are many questions concerning these animal tests as to what information they provide, which kinds of studies are the best, and how the animal data compares with human data. In an attempt to answer these questions, epidemiological studies and animal bioassays are compared to each other to determine if a particular chemical is likely to pose a health hazard to humans. Many assumptions are made in hazard assessments. Fox example, it is assumed that the chemical administered in a bioassay is in a form similar to that present in the environment. Another assumption is that animal carcinogens are also human carcinogens. An example is that there is a similarity between animal and human metabolism, and so on. With these and other assumptions, and by analyzing hazard identification procedures, lists of hazardous chemicals have been developed.

42.4 DOSE-RESPONSE

Dose-response assessment is the process of characterizing the relation between the dose of an agent administered or received and the incidence of an adverse health effect in exposed populations, and estimating the incidence of the effect as a function of exposure to the agent. This process considers such important factors as intensity of exposure, age pattern of exposure, and possibly other variables that might affect response, such as sex, lifestyle, and other modifying factors. A dose-response assessment usually requires extrapolation from high to low doses and extrapolation from animals to humans, or one laboratory animal species to a wildlife species. A dose-response assessment should describe and justify the methods of extrapolation used to predict incidence, and it should characterize the statistical and biological uncertainties in these methods. When possible, the uncertainties should be described numerically rather than qualitatively.

Toxicologists tend to focus their attention primarily on extrapolations from cancer bioassays. However, there is also a need to evaluate the risks of lower doses to see how they affect the various organs and systems in the body. Many scientific papers focus on the use of a safety factor or uncertainty factor approach, since all adverse effects other than cancer and mutation-based developmental effects are believed to have a threshold — a dose below which no adverse effect should occur. Several researchers have discussed various approaches to setting acceptable daily intakes or exposure limits for developmental and reproductive toxicants. It is thought that an acceptable limit of exposure could be determined using cancer models, but today they are considered inappropriate because of thresholds (2).

For a variety of reasons, it is difficult to precisely evaluate toxic responses caused by acute exposures to hazardous materials. First, humans experience a wide range of acute adverse health effects, including irritation, narcosis, asphyxiation, sensitization, blindness, organ system damage, and death. In addition, the severity of many of these effects varies with intensity and duration of exposure. Second, there is a high degree of variation in response among individuals in a typical population. Third, for the overwhelming majority of substances encountered in industry, there are not enough data on toxic responses of humans to permit an accurate or precise assessment of the substance's hazard potential. Fourth, many releases involve multiple components. There are presently no rules on how these types of releases should be evaluated. Fifth, there are no toxicology testing protocols that exist for studying episodic releases on animals. In general, this has been a neglected area of toxicology research. There are many useful measures available to employ as benchmarks for predicting the likelihood that a release event will result in serious injury or death. Several works in the literature (5,6) review various toxic effects and discuss the use of various established toxicological criteria.

Dangers are not necessarily defined by the presence of a particular chemical, but rather by the amount of that substance one is exposed to, also known as the dose. A dose is usually expressed in milligrams of chemical received per kilogram of body weight per day. For toxic substances other than carcinogens, a threshold dose must be exceeded before a health effect will occur, and for many substances, there is a dosage below which there is no harm, i.e., a health effect will occur or at least be detected at the threshold. For carcinogens, it is assumed that there is no threshold, and, therefore, any substance that produces cancer is assumed to produce cancer at any concentration. It is vital to establish the link to cancer and to determine if that risk is acceptable. Analyses of cancer risks are much more complex than those for noncancer risks.

Not all contaminants or chemicals are equal in their capacity to cause adverse effects. Thus, clean-up standards or action levels are based in part on the compounds' toxicological properties. Toxicity data employed are derived largely from animal experiments in which the animals (primarily mice and rats) are exposed to increasingly higher concentrations or doses. As described above, responses or effects can vary widely from no observable effect to temporary and reversible effects, to permanent injury to organs, to chronic functional impairment, to, ultimately, death.

42.5   EXPOSURE ASSESSMENT

Exposure assessment is the process of measuring or estimating the intensity, frequency, and duration of human or animal exposure to an agent currently present in the environment or of estimating hypothetical exposures that might arise from the release of new chemicals into the environment. In its most complete form, an exposure assessment should describe the magnitude, duration, schedule, and route of exposure, size, nature, and classes of the human, animal, aquatic, or wildlife populations exposed, and the uncertainties in all estimates. The exposure assessment can often be used to identify feasible prospective control options and to predict the effects of available control technologies for controlling or limiting exposure (2).

Much of the attention focused on exposure assessment has come recently. This is because many of the risk assessments performed in the past used too many conservative assumptions, which caused an overestimation of the actual exposure. Obviously, without exposure(s) there are no risks. To experience adverse effects, one must first come into contact with the toxic agent(s). Exposures to chemicals can occur via inhalation of air (breathing), ingestion of water and food (eating and drinking), or absorption through the skin. These are all pathways to the human body.

Generally, the main pathways of exposure considered in this step are atmospheric transport, surface and groundwater transport, ingestion of toxic materials that have passed through the aquatic and terrestrial food chain, and dermal absorption. Once an exposure assessment determines the quantity of a chemical with which human populations may come in contact, the information can be combined with toxicity data (from the hazard identification process) to estimate potential health risks (1). Thus, the primary purpose of an exposure assessment is to determine the concentration levels over time and space in each environmental media where human and other environmental receptors may come into contact with the chemicals of concern. In addition, there are four components of an exposure assessment: potential sources, significant exposure pathways, populations potentially at risk, and exposure estimates (2).

The two primary methods of determining the concentration of a pollutant to which target populations are exposed are direct measurement and computer analysis, also known as computer dispersion modeling. Measurement of the pollutant concentration in the environment is used for determining the risk associated with an exiting discharge source. Receptors are placed at regular intervals from the source, and the concentration of the pollutant is measured over a certain period of time (usually several months or a year). The results are then related to the size of the local population. This kind of monitoring, however, is expensive and time-consuming. Many measurements must be taken because exposure levels can vary under different atmospheric conditions or at different times of the year. Computer dispersion modeling predicts environmental concentrations of pollutants (see Chapters 6 and 12 for more information on dispersion modeling). In the prediction of exposure, computer dispersion modeling focuses on the discharge of a pollutant and the dispersion of that discharge by the time it reaches the receptor. This method is primarily used for assessing risk from a proposed facility or discharge. Sophisticated techniques are employed to relate reported or measured emissions to atmospheric, climatological, demographic, geographic, and other data in order to predict a population's potential exposure to a given chemical.

42.6 RISK CHARACTERIZATION

Risk characterization is the process of estimating the incidence of a health effect under the various conditions of human or animal exposure described in the exposure assessment. It is performed by combining the exposure and dose-response assessments. The summary effects of the uncertainties in the preceding steps should also be described in this step.

The quantitative estimate of the risk is the principal interest to the regulatory agency or risk manager making the decision. The risk manager must consider the results of the risk characterization when evaluating the economics, societal aspects, and various benefits of the assessment. Factors such as societal pressure, technical uncertainties, and severity of the potential hazard influence how the decision makers respond to the risk assessment. There is room for improvement in this step of the risk assessment (2).

A risk estimate indicates the likelihood of occurrence of the different types of health or environmental effects in exposed populations. Risk assessment should include both human health and environmental evaluations (i.e., impacts on ecosystems). Ecological impacts include actual or potential effects on plants and animals (other than domesticated species). The number produced from the risk characterization, representing the probability of adverse health effects being caused, must be evaluated. This is performed because certain agencies will only look at risks of specific numbers and act on them.

There are two major types of risk: maximum individual risk and population risk. Maximum individual risk is defined exactly as it implies, that is, the maximum risk to an individual person. This person is considered to have a 70-yr lifetime of exposure to a process or a chemical. Population risk is basically the risk to a population. It is expressed as a certain number of deaths per thousand or per million people. These risks are often based on very conservative assumptions which may yield too high a risk.

42.7 SUMMARY

1. Health risk assessment is defined as the process or procedure used to estimate the likelihood that humans or ecological systems will be adversely effected by a chemical or physical agent under a specific set of conditions.

2. The health risk evaluation process consists of four steps: hazard identification, dose-response assessment or hazard assessment, exposure assessment, and risk characterization.

3. In hazard identification, a hazard is a toxic agent or a set of conditions that has the potential to cause adverse effects to human health or the environment.

4. In dose-response assessment, effects are evaluated and these effects vary widely because their capacities to cause adverse effects differ.

5. Exposure assessments is the determination of the magnitude, frequency, duration, and routes of exposure to human populations and ecosystems.

6. In risk characterization, the toxicology and exposure data are combined to obtain a quantitative or qualitative expression of risk.

7. A major avenue for reducing risk will involve source reduction of hazardous materials.

REFERENCES

1. L. Theodore, J. Reynolds, and K. Morris, *Health Safety and Accident Prevention: Industrial Applications*, A Theodore Tutorial, East Williston, NY, 1996.

2. D. Paustenbach, *The Risk Assessment of Environmental and Human Health Hazards; A Textbook of Case Studies,* John Wiley & Sons, New York, 1989.

3. Rodricks, J., and R. Tardiff, *Assessment and Management of Chemical Risks,* American Chemical Society, Washington D.C., 1984.

4. US EPA, *Technical Guidance for Hazards Analysis,* December 1987.

5. D. B. Clayson, D. Krewski, and I. Munro, *Toxicological Risk Assessment,* CRC Press, Inc., Boca Raton, FL, 1985.

6. V. Foa, E. A. Emmett, M. Maron, and A. Colombi, *Occupational and Environmental Chemical Hazards,* Ellis Horwood Limited, Chichester, England, 1987.

43

HAZARD RISK ASSESSMENT

Contributing Author: Elizabeth Capasso

43.1 INTRODUCTION

Risk evaluation of accidents serves a dual purpose. It estimates the probability that an accident will occur and also assesses the severity of the consequences of an accident. Consequences may include damage to the surrounding environment, financial loss, or injury to life. This chapter is primarily concerned with the methods used to identify hazards and the causes and consequences of accidents. Issues dealing with health risks have been explored in the previous chapter. Risk assessment of accidents provides an effective way to help ensure either that a mishap does not occur or reduces the likelihood of an accident. The result of the risk assessment also allows concerned parties to take precautions to prevent an accident before it happens.

Regarding definitions, the first thing an individual needs to know is what exactly is an accident. An accident is an unexpected event that has undesirable consequences (1). The causes of accidents have to be identified in order to help prevent accidents from occurring. Any situation or characteristic of a system, plant, or process that has the potential to cause damage to life, property, or the environment is considered a hazard. A hazard can also be defined as any characteristic that has the potential to cause an accident. The severity of a hazard plays a large part in the potential amount of damage a hazard can cause if it occurs.

Risk is the probability that human injury, damage to property, damage to the environment, or financial loss will occur. An acceptable risk is a risk whose probability is unlikely to occur during the lifetime of the plant or process. An acceptable risk can also be defined as an accident that has a high probability of occurring, but with negligible consequences. Risks can be ranked qualitatively in categories of high, medium, and low. Risk can also be ranked quantitatively as annual number of fatalities per million affected individuals. This is normally denoted as a number times one millionth that is, for example 3×10^{-6}; this representation indicates that on the average, 3 (workers) will die every year for every million individuals.

Another quantitative approach that has become popular in industry is the Fatal Accident Rate (FAR) concept. This determines or estimates the number of fatalities over the lifetime of 1000 workers. The lifetime of a worker is defined as 10^5 hours, which is based on a 40-hr work week for 50 years. A reasonable FAR for a chemical plant is 3.0, with 4.0 usually taken as a maximum. A FAR of 3.0 means that there are 3 deaths for every 1000 workers over a 50-year period (2). The FAR for an individual at home is approximately 3.0.

Risk Evaluation Process for Accidents

There are several steps in evaluating the risk of an accident (see Figure 43.1). These are detailed below if the system in question is a chemical plant.

1. A brief description of the equipment and chemicals used in the plant is needed.

2. Any hazard in the system has to be identified. Hazards that may occur in a chemical plant include fire, toxic vapor release, slippage, corrosion, explosions, rupture of a pressurized vessel, and runaway reactions.

3. The event or series of events that will initiate an accident has to be identified. An event could be a failure to follow correct safety procedures, improperly repaired equipment, or failure of a safety mechanism.

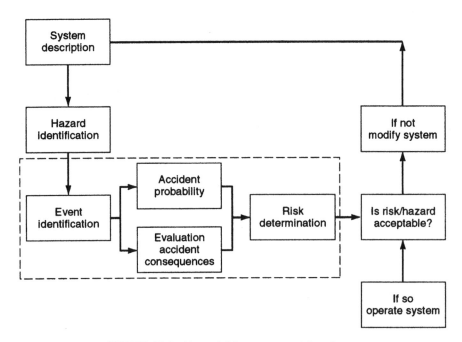

FIGURE 43.1   Hazard risk assessment flowchart.

4. The probability that the accident will occur has to be determined. For example, if a chemical plant has a given life, what is the probability that the temperature in a reactor will exceed the specified temperature range? The probability can be ranked from low to high. A low probability means that it is unlikely for the event to occur in the life of the plant. A medium probability suggests that there is a possibility that the event will occur. A high probability means that the event will probably occur during the life of the plant.

5. The severity of the consequences of the accident must be determined. This will be described later in detail.

6. If the probability of the accident and the severity of its consequences are low, then the risk is usually deemed acceptable and the plant should be allowed to operate. If the probability of occurrence is too high or the damage to the surroundings is too great, then the risk is usually unacceptable and the system needs to be modified to minimize these effects.

The heart of the hazard risk assessment algorithm provided is enclosed in the dashed box (see Figure 43.1). The algorithm allows for reevaluation of the process if the risk is deemed unacceptable (the process is repeated starting with either step one or two).

43.2 HAZARD/EVENTS IDENTIFICATION

Hazard or event identification provides information on situations or chemicals and their releases that can potentially harm the environment, life, or property. Information that is required to identify hazards includes chemical identities, quantities and location of chemicals in question, chemical properties such as boiling points, ignition temperatures, and toxicity to humans. There are several methods used to identify hazards. The methods that will be discussed are the process checklist and the hazard and operability study (HAZOP).

A process checklist evaluates equipment, materials, and safety procedures (1). A checklist is composed of a series of questions prepared by an engineer or scientist who knows the procedure being evaluated. It compares what is in the actual plant to a set of safety and company standards.

Some questions that may be on a typical checklist are:

1. Was the equipment designed with a safety factor?

2. Does the spacing of the equipment allow for ease of maintenance?

3. Are there pressure relief valves on the equipment in question?

4. How toxic are the materials that are being used in the process, and is there adequate ventilation?

5. Will any of the materials cause corrosion to the pipe(s)/reactor(s)/system?

6. What precautions are necessary for flammable materials?

7. Is there an alternate exit in case of fire?

8. If there is a power failure, what fail-safe procedure(s) does the process contain?

9. What hazard is created if any piece of equipment malfunctions?

10. Who is first contacted in the event of an accident?

These questions and others are answered and analyzed. Changes are then made to reduce the risk of an accident. Process checklists are updated and audited at regular intervals.

A hazard and operability study is a systematic approach to recognizing and identifying possible hazards that may cause failure of a piece of equipment (2,3). This method utilizes a team of diverse professional backgrounds to detect and minimize hazards in a plant. The process in question is divided into smaller processes (subprocesses). Guide words are used to relay the degree of deviation from the subprocesses' intended operation. The guide words can be found in Table 43.1. The causes and consequences of the deviation from the process are determined. If there are any recommendations for revision they are recorded and a report is made.

A summary of the basic steps of a HAZOP study is (2,3):

1. Define objectives.

2. Define plant limits.

3. Appoint and train a team.

4. Obtain complete preparative work (i.e., flow diagrams, sequence of events, etc.).

5. Conduct examination meetings that select subprocesses, agree on intentions of subprocesses, state and record intentions, use guide words to find deviations from the intended purpose, determine the cause and consequences of deviation, and recommend revisions.

6. Issue meeting reports.

7. Follow up on revisions.

There are other methods of hazard identification. A "what-if" analysis presents certain questions about a particular hazard and then tries to find the possible consequences of that hazard. The human error analysis identifies potential human errors that will lead to an accident. They can be used in conjunction with the two previously described methods.

Table 43.1   Guide words used to relay the degree of deviation from intended subprocess operation

Word	Subprocess Operation
No	No part of intended function is accomplished
Less	Quantitative decrease in intended activity
More	Quantitative increase in intended activity
Part of	The intention is achieved to a certain percent
As well as	The intention is accomplished along with side effects
Reverse	The opposite of the intention is achieved
Other than	A different activity replaces the intended activity

43.3 CAUSES OF ACCIDENTS

The primary causes of accidents are mechanical failure, operational failure (human error), unknown or miscellaneous, process upset, and design error. Figure 43.2 gives the relative number of accidents that have occurred in the petrochemical field (4).

There are three steps that normally lead to an accident.

1. Initiation
2. Propagation
3. Termination

The path than an accident takes through the above three steps can be determined by means of a fault tree analysis (1). Generally, a fault tree may be viewed as a diagram that shows the path that a specific accident takes. The first thing needed to construct a fault tree is the definition of the initial event. The initial event is a hazard or action that will cause the process to deviate from normal operation. The next step is to define the existing conditions needed to be present in order for the accident to occur. The propagation event (e.g., the mechanical failure of equipment related to the accident) is discussed. Any other equipment or components that needs to be studied have to be defined. This includes safety equipment that will bring about the termination of the accident. Finally, the normal state of the system in question is determined. The termination of an accident is the event that brings the system back to its normal operation. An example of an accident would be the failure of a thermometer in a reactor. The temperature in the reactor could rise and a runaway reaction might take place. Stopping the flow to the reactor and/or cooling the contents of the reactor could terminate the accident.

The reader should view a fault tree as a technique of graphical notation used to analyze complex systems. Its objective is to spotlight conditions that cause a system to fail. Fault tree analysis also attempts to describe how and why an accident or any other undesirable event has occurred. It is also used to describe how and why a potential accident or other undesirable event could take place. Thus, fault tree analysis finds wide application in hazard analysis and risk assessment of process and plant systems.

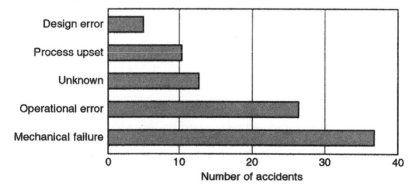

FIGURE 43.2 Causes of accidents in the petrochemical field.

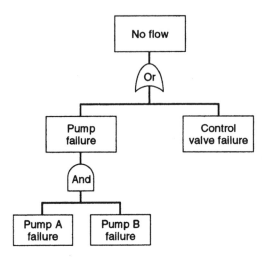

FIGURE 43.3 Fault tree diagram for a water pumping system consisting of two pumps (A and B).

Fault tree analysis seeks to relate the occurrence of an undesired event, the "top event," to one or more antecedent events, called "basic events." The top event may be, and usually is, related to the basic events via certain intermediate events. A fault tree diagram exhibits the causal chain linking the basic events to the intermediate events and the latter to the top event. In this chain, the logical connection between events is indicated by so-called "logic gates." The principal logic gates are the AND gate, symbolized on the fault tree by AND, and the OR gate symbolized by OR.

As a simple example of a fault tree, consider a water pumping system consisting of two pumps, A and B, where A is the pump ordinarily operating and B is a standby unit that automatically takes over if A fails. Flow of water through the pump is regulated by a control valve in both cases. Suppose that the top event is no water flow, resulting from the following basic events: failure of a pump A *and* failure of pump B, *or* failure of the control valve. The fault tree diagram for this system is shown in Fig. 43.3.

43.4 CONSEQUENCES OF ACCIDENTS

Consequences of accidents can be classified qualitatively by the degree of severity. A quantitative assessment is beyond the scope of the text; however information is available in the literature (5). Factors that help to determine the degree of severity are the concentration in which the hazard is released, length of time that a person within the environment is exposed to a hazard, and the toxicity of the hazard. The worst-case consequence or scenario is defined as a conservatively high estimate of the most severe accident identified (1). On this basis one can rank the consequences of accidents into low, medium, and high degrees of severity (6). A low degree of severity means that the hazard is nearly negligible, and the injury to person, property, or the environment is observed only after an extended period of time. The degree of severity is considered to be medium when the accident is serious, but not catastrophic, the toxicity of the chemical released is great, or the concentration of a less toxic chemical is large enough to cause injury or death to persons and damage to the environment unless immediate action is taken. There is a high degree of severity when the accident is catastrophic or the concentrations and toxicity of a hazard are large enough to cause injury or death to many persons, and there is long-term damage to the surrounding environment. Figure 43.4 provides a graphical qualitative representation of the severity of consequences (6).

Event trees are diagrams that evaluate the consequences of a specific hazard. The safety measures and the procedures designed to deal with the event are presented. The consequences of each specific event that led to the accident are also presented. An event tree is drawn (sequence of events that led up to the accident). The accident is described. This allows the path of the accident to be traced. It shows what could be done along the way to prevent the accident. It also shows other possible outcomes that could have arisen had a single event in the sequence been changed (2,3). Thus, an event tree provides a diagrammatic representation of event sequences that begin with a so-called initiating event and terminate in one or more undesirable consequences. In contrast to a fault tree, which works backward from an undesirable consequence to possible causes, an event tree works forward from the initiating event to possible undesirable consequences. The initiating event may be equipment failure, human error, power failure, or some other event that has the potential for adversely affecting an ongoing process.

The following illustration of event tree analysis is based on one reported by Lees (7). Consider a situation in which the probability of an external power outage in any given year is 0.1. A backup generator is available, and the probability that it will fail to start on any given occasion is 0.02. If the generator starts, the probability that it will not supply sufficient power for the duration of the external power outage is 0.001. An emergency battery power supply is available; the probability that it will be inadequate is 0.01.

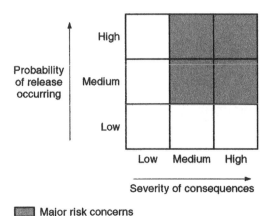

FIGURE 43.4 Qualitative probability–consequence analysis.

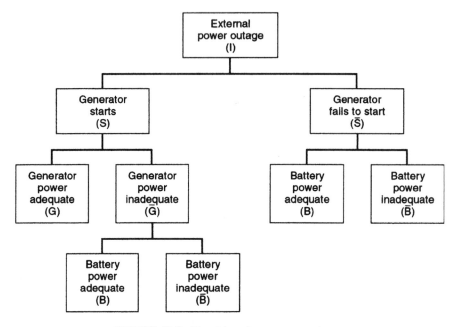

FIGURE 43.5   Event tree for a power outage.

Figure 43.5 shows the event tree with the initiating event, external power outage, denoted by I. Labels for the other events on the event tree are also indicated. The event sequences I S $\bar{G}$ $\bar{B}$ and I $\bar{S}$ $\bar{B}$ terminate in the failure of emergency power supply. Applying the applicable multiplication theorem, one obtains

$$P(I \ S \ \bar{G} \ \bar{B}) = (0.1)(0.98)(0.001)(0.01) = 0.098 \times 10^{-5}$$
$$P(I \ \bar{S} \ \bar{B}) = (0.02)(0.01) = 2.0 \times 10^{-5}$$

Therefore the probability of emergency power supply failure in any given year is 2.098×10^{-5}, the sum of these two probabilities.

43.5   CAUSE–CONSEQUENCE ANALYSIS

Cause–consequence risk evaluation combines event tree and fault tree analysis to relate specific accident consequences to causes (2). The process of cause conse- quence evaluation usually proceeds as follows:

1. Select an event to be evaluated.
2. Describe the safety systems(s)/procedure(s)/factors(s) that interfere with the path of the accident.
3. Perform an event tree analysis to find the path(s) an accident may follow.

4. Perform a fault tree analysis to determine the safety function that failed.

5. Rank the results on a basis of severity of consequences.

As its name implies, cause–consequence analysis allows one to see how the possible causes of an accident and the possible consequences that result from that event interact with each other.

43.6 FUTURE TRENDS

For the most part, future trends will be found in hazard accident prevention, not hazard analysis. To help promote hazard accident prevention, companies should start employee training programs. These programs should be designed to alert the technical staff and employees about the hazards they are exposed to on the job. Training should also cover company safety policies and the proper procedures to follow in case an accident does occur. A major avenue to reducing risk will involve source reduction of hazardous materials. Risk education and communication are two other areas that will need improvement.

43.7 SUMMARY

1. Risk assessment of accidents estimates the probability that hazardous materials will be released and also assesses the severity of the consequences of an accident.

2. The risk evaluation process defines the equipment, hazards, and events leading to an accident. It determines the probability that an accident will occur. The severity and acceptability of the risk are also evaluated.

3. Hazard identification provides information on situations or chemicals that can potentially harm the environment, life, or property. The processes described are process checklist, event tree, and hazard and operability study.

4. Accidents occur in three steps: initiation, propagation, and termination. The primary causes of accidents are mechanical failure, operation failure (human error), unknown or miscellaneous, process upset, and design error.

5. Consequences of accidents are classified by degree of severity into low, medium, and high.

6. Cause–consequence analysis allows one to see the possible causes of an accident and the possible accident that results from a certain event.

7. For the most part, future trends can be found in hazard prevention.

REFERENCES

1. *AIChE, Guidelines for Hazard Evaluation Procedures*, Prepared by Batelle Columbus Division for the Center for Chemical Process Safety of the American Institute of Chemical Engineers, 1985.

2. L. Theodore, J. Reynolds, and F. Taylor, *Accident and Emergency Management*, John Wiley & Sons, New York, 1989.

3. L. Theodore, J. Reynolds, and K. Morris, *Health, Safety and Accident Prevention, Industrial Applications*, A Theodore Tutorial, East Williston, NY, 1996.

4. J. Crowl and J. Louvar, *Chemical Safety Fundamentals With Applications*, Prentice-Hall, Englewood Cliffs, NJ, 1990.

5. R. Ballandi, *Hazardous Waste Site Remediation, The Engineer's Perspective*, Van Nostrand Reinhold, New York, 1988.

6. Author Unknown, *Technical Guidance For Hazard Analysis*, EPA, FEMA, USDOT, Washington, DC, 1978.

7. P. Lees, *Loss Prevention in the Process Industries*, Vol. I, Butterworths, Boston, MA, 1980.

44

ENVIRONMENTAL COMMUNICATION

44.1 INTRODUCTION

Environmental communication is playing a pivotal role in environmental manage-ment. As Americans experience a new wave of environmentalism, how environ-mental issues and concerns are communicated will shape and impact the direction of environmental policies. Since 1987, public concern about the environment has grown faster than concern about virtually any other national problem. There is a growing public demand for cleaner water, air, and land. How federal, state, and local environmental agencies communicate with industries, interest groups, con-cerned citizens, and the public in general will play a critical role in developing and implementing innovative waste management practices.

Protecting the environment and the public health is a huge task in which every one plays an important role. People rely on local governments or organizations to pick up trash, operate sewer systems, deliver drinking water, maintain roads, and regulate land use—all of which directly affect the environment. State governments, in turn, support such activities through their public health, agriculture, transporta-tion, and environmental departments. America's Indian tribes have what is known as "tribal sovereignty" in dealing with environmental and other problems. There are also many private groups that focus on public health, safety, and the environment. Increasingly, private businesses are providing products and services that support environmental protection.

As a result, the Environmental Protection Agency (EPA) shares the job of maintaining environmental quality with a host of varied organizations and individ-uals. The list includes other federal agencies, states, tribes, localities, businesses, interest groups, and individual citizens. Without the help of all of these groups, environmental protection would not be possible (1).

Environmental communications takes many forms, depending on the audience it is intended to serve. At times, it involves testifying in court, testifying before the Congress of the United States or appearing before committees representing the

Congress, delivering speeches at educational facilities ranging from pre-K level to "ivy-league" universities, addressing seminars, conferences, or workshops, responding to press releases or making appearances on television, preparing technical documents, handbooks, guidance documents or pamphlets on environmental issues for audiences in both the public and private sectors. These are just a few of the activities involving environmental communications.

The Environmental Protection Agency (EPA) established the new "Office of Communications" to address this complex and critical element in environmental protection. Environmental communications plays a key role in keeping with the EPA's mission to inform the regulated communities and citizens on the environment. As we approach the new millenium, websites and electronic communications will add a new dimension to environmental communication. The availability of environmental laws, regulations, policies, technicals and guidance documents and information in an electronic form available to all stakeholders provide a fast and effective means of environmental communication. Websites such as "www.epa.gov" assure the public, the regulated communities, and virtually anyone else access to environmental information. The EPA offers a great deal of compliance assistance information and materials through its websites. This form of environmental communication is no doubt also more cost-effective and "environmentally friendly." The printing and distribution of hundred of thousands of documents will no longer be necesssary.

44.2   STATUTORY REQUIREMENTS INVOLVING ENVIRONMENTAL COMMUNICATIONS

There are several major statues and regulations which mandate communication activities with the general public. Since its founding in December 1970, EPA has been given the task of implementing a wide range of environmental statutes. Many of these statutes owe their very existence to years of citizen lobbying, so it is no surprise that some should specifically mandate "public participation." The Federal Water Pollution Control Act of 1972 (later known as the Clean Water Act), the Safe Drinking Water Act of 1974, and the Resources Conservation and Recovery Act of 1976 all contain language requiring EPA to involve the public in their implementation.

The philosophy behind public participation found expression in Section 101(e) of the Federal Water Pollution Control Act of 1972: "A high degree of informed public participation in the control process is essential to the accomplishment of the objectives we seek — a restored and protected natural environment" (2). In what way essential? Legislators — inspired by the dedication of clean air activists in the so-called "Breathers' Lobby" — envisioned the public as the conscience of EPA. Their hope was that concerned citizens, both individually and in groups, would monitor EPA and ensure that the agency actually did its job (3).

The Comprehensive Environmental Response, Compensation, and Liability Act and the Superfund Amendments and Reauthorization Act of 1986 encourage community residents to participate in the process of determining the best way to clean up an abandoned site. To ensure effective and substantive two-way communications from the outset at each remedial response site, a community relations program is tailored to local circumstances. Often, EPA or state staff will interview residents, local officials, and civic leaders to learn all they can about the site and about the community's concerns. These interviews are conducted before and during field work on the remedial investigation. The new Superfund thus formalized existing EPA community relations policy and public participation requirements outlined in the National Contingency Plan. It also required EPA to

- Publish a notice and brief analysis of the proposed remedial action plan.
- Provide an opportunity for the public to comment on that plan.
- Provide an opportunity for a public meeting to allow for two-way communication on the remedial action plan.
- Make a copy of the transcript of the public meeting available to the public.
- Prepare a response to each significant comment made on the proposed remedial action plan.

Community relations activities are somewhat different during a removal action, where human health and the environment must be protected from an immediate threat. During the initial phase of these response actions, the EPA's primary responsibility is to inform the community about actions being taken and the possible effect on the community.

The new Superfund also required the EPA to develop a grant program to make funding for technical assistance available to those who may be affected by a release. The purpose of these grants was to help concerned citizens understand and interpret technical information on the nature of the hazard and recommended alternatives for cleanup. Grants are limited by law to one grant of no more than $50,000 per NPL site. In addition, the grant recipient must contribute at least 20% of the total cost of the grant (4). The Emergency Planning and Community Right-to-Know Act (see Chapter 34 "Accident and Emergency Management," for additional details) created a new relationship among government at all levels, business and community leaders, environmental and other public-interest organizations, and individual citizens. For the first time, the law made citizens full partners in preparing for emergencies and managing chemical risks. Each of these groups and individuals has an important role in making the program work (5):

- **Local communities and states** have the basic responsibility for understanding risks posed by chemicals at the local level, for managing those risks, for reducing those risks, and for dealing with emergencies. By developing emergency planning and chemical risk management at the levels of government closest to the community, the law helps to ensure the broadest possible public representation in the decision-making process.
- **Citizens, health professionals, public-interest and labor organizations, the media, and others** are working with government and industry to use the information for planning and response at the community level. The law gives everyone involved access to more of the facts they need to determine what chemicals mean to the public health and safety.
- **Industry** is responsible for (a) operating as safely as possible using the most appropriate techniques and technologies; (b) gathering information on the chemical it uses, stores, and releases into the environment and providing it to government agencies and local communities; and (c) helping set up procedures to handle chemical emergencies. Beyond meeting the letter of the law,

some industry groups and individual companies are reaching out to their communities by explaining the health hazards involved in using chemicals, by opening communications channels with community groups, and by considering changes in their practices to reduce any potential risks to human health or the environment.

- **The federal government** is responsible for providing national leadership and assistance to states and communities so they will have the tools and expertise they need to (a) receive, assimilate, and analyze all Title III data; and (b) take appropriate measures in accidental risk and emission reduction at the local level. The EPA is also working to ensure that (a) industry complies with the law's requirements; (b) the public has access to information on annual toxic chemical releases; and (c) the information is used in various EPA programs to protect the nation's air, water, and soil from pollution. The EPA is also working with industry to encourage voluntary reductions in the use and release of hazardous chemicals wherever possible (5).

The Emergency Planning and Community Right-to-Know Act has forged a closer, more equal relationship among citizens, health professionals, industry, public-interest organizations, and the local, state, and federal government agencies responsible for emergency planning and response, public health, and environmental protection.

In the past, most of the responsibility for these activities fell upon experts in government and industry. To the extent that citizens or their representatives participated, it was generally "from the outside looking in," as they did what they could to influence decisions that were, for the most part, out of their hands. But under the provisions of the Emergency Planning and Community Right-to-Know Act, all of these groups, organizations and individuals have vital roles to play in making the law work for the benefit of everyone. The law requires facilities to provide information on the of hazardous chemicals in communities directly to the people who are most affected, both in terms of exposure to potential risks and the effects of those risks on public health and safety, the environment, jobs, the local economy, property values, and other factors.

These "stakeholders" are also the people who are best able to do something about assessing and managing risks — through inspections, enforcement of local codes, reviews of facility performance, and, when appropriate, political and economic pressures.

This relationship between the Title III data and community action can best occur at the local level, through the work of the Local Emergency Planning Committee (LEPC). For example, if a local firm has reported the presence of extremely hazardous substances at its facility, several accidents, substantial quantities of chemicals, and continuing releases of toxic chemicals, a community has the data it needs to seek appropriate corrective action. In short, the law opens the door to community-based decision making on chemical hazards for citizens and communities throughout the nation.

LEPCs are crucial to the success of the Emergency Planning and Community Right-to-Know Act. Appointed by State Emergency Response Commissions (SERCs), local planning committees must consist of representatives of all of the following groups and organizations: elected state and local officials; law enforcement, civil defense, firefighting, first aid, health, local environmental and transportation agencies; hospitals; broadcast and print media; community groups; and representatives of facilities subject to the emergency planning and community right-to-know requirements (5).

It was clear from the outset that the public could not put persistent and informed pressure on EPA without a steady flow of information and guidance from the EPA. Meeting that need has been the purpose of EPA's public participation programs. Their mission is threefold:

- To keep the public informed of important developments in EPA's program areas.

- To provide technical information and, if necessary, translate that information into plain English.

- To ensure that the agency takes community viewpoints into account in implementing these programs.

44.3 METHODS OF ENVIRONMENTAL COMMUNICATION

Environmental communications take many forms, the most common being printed materials in the form of technical documents, pamphlets, handouts, brochures, magazines, journals, issue papers, newspaper articles, editorials, and the like. In addition to the more common and traditional means of communication, a move towards the use of radio, television, and websites is playing a pivotal role in environmental communication as we move into the new millenium.

The 1990 National Environmental Education Act contains a provision that specifically calls on EPA to work with "noncommercial educational broadcasting entities" to educate Americans on environmental problems. Educational broadcasting is uniquely positioned to do just that and over the last two decades has, in fact, provided many programs on environmental issues to the public.

Educational, or public, broadcasting reaches vast numbers of Americans. Nationwide there are more than 400 public radio stations and more than 330 public television stations. Many of these stations are local, and independently operated. As such, they are acutely aware of the needs and concerns of their local communities. What's more, public broadcasting gets good marks from the public on educational value. In a recent survey, 91 percent of respondents gave public broadcasting a B on how well the industry is doing in increasing people's understanding of news events and public affairs. Public broadcasting got an A-performance grade on helping to educate children informally.

Ultimately, public broadcasting considers all of its programming — both television and radio — to be educational: children's programming, drama, music and dance, science and nature, skill-building how-to's, as well as its much-acclaimed public affairs and documentary programming. Public broadcasting reaches not only into homes but directly into schools as well. More than twice as many teachers use public broadcasting programming and other instructional materials than use other broadcast and cable services, according to survey results announced by the National Education Association.

Public television has been a pioneer in the uses of noncommercial programming for close to 40 years. It has presented high-quality, noncommercial educational preschool programming, instructional programming for students from kindergarten through grade 12, and college credit and postgraduate courses; it has also provided literacy training, continuing professional education, and job training and retraining programs.

One of public television's more unique innovations has been "interactive" educational programming, which involves viewer participation. The Satellite Educational Resources Consortium (SERC), composed of public television stations and state departments of education, delivers each school day live, interactive, for-credit high school courses to students in 23 states. SERC also provides live, interactive teacher in-service and staff developments training seminars and workshops. In addition, public broadcasting has not been content to educate and inform solely through over-the-air broadcasting. Almost all public television stations provide outreach activities supplement and support for their programming.

Public radio and television both continue to feature programs on environmental subjects. Public radio offers several such programs: Examples include American Energy Update, The Environment Show, and Pollution Solutions; in addition, Terra Firma consists of three- to five-minute radio modules that provide provocative probes into the dysfunctional thinking behind the environmental crisis. Other, more general radio programs, such as All Things Considered and Morning Edition, regularly touch on environmental topics.

Where does public broadcasting get all this programming? From just about everywhere. Public broadcasting stations produce much of their own programming. Other programming is produced by independent producers, both here and abroad. The reservoir of producing entities is vast. And most programs rely on funding from many sources: public broadcasting stations, national public broadcasting organizations like the Public Broadcasting System (PBS), National Public Radio (NPR), and the Corporation for Public Broadcasting; government agencies; private corporations and foundations, and the viewing public.

Once a program or series is produced, it can make its way to the viewing public in one of several ways. Since each station is independent, the management of an individual station may decide to air a program acquired directly from a producer. Stations also receive programs via satellite from the Public Broadcasting Service as well as the four public broadcasting regional organizations. The Southern Educational Communications Association (SECA), for instance distributes via satellite nearly 800 hours of general audience programming to public television stations every year, nearly 70% produced by its membership. As do other regional organizations, SECA provides stations with materials they can use to promote these programs to their viewing public. Noncommercial educational broadcasting has the resources, the technology, and the grassroots relationships with the education community to help EPA provide information to the public and school children on vital environmental issues and problems (6).

Spurred by growing concerns about global environmental problems, the entertainment industry is in the midst of a massive consciousness-raising effort on a variety of environmental issues. The environment is not the first social issue to be adopted by show business, but it may well be the catalyst for the most far-reaching public interest campaign yet launched by the industry.

Show business has had a long history of involvement in public affairs, dating back to World Wars I and II, when Hollywood actively promoted home-front activities. More recently, especially since the advent of television, the industry has fought illiteracy and drunk driving and taken on other social causes. TV images have aroused widespread concern for the starving in Africa, called attention to the homeless and hungry here at home, and helped the Red Cross raise $100 million for aid to the victims of Hurricane Hugo and the California earthquake. The entertainment business has a proud record of supporting civil liberties.

Until the early 1990s, entertainment industry environmentalism was associated largely with a small group of stars such as Robert Redford, Paul Newman, Joanne Woodward, Meryl Streep, and Judy Collins, the Ted Turner broadcasting interests, and occasional news or educational TV specials. But now Hollywood has gone green in a big way. Andy Spahn, president of one of the two major Hollywood organizations focusing on environmental issues: "We're in it for as long as it takes. They tell us we may have as little as 10 to 12 years to correct or reverse some of the most serious threats. You might say that length of time is our minimum commitment."

Two Hollywood groups, the Environmental Media Association and the Earth Communications Office, are spearheading the entertainment industry's approach to creating national and international environmental awareness. In general, their goal is to create a steady stream of environmental messages written into plot lines of regular programs and motion pictures, entertainment specials, and other outlets such as special events and new music and songs. These messages are intended to complement ongoing public service announcements, occasional news specials, and science programs on cable or public television. Stars and other industry leaders are also being asked to take the kinds of environmental leadership roles that Redford and Streep have assumed in former years (7).

44.4 THE CITIZEN'S ROLE IN ENVIRONMENTAL ENFORCEMENT*

The two most important things to do when you see a potential pollution problem are as follows: (a) Make careful observations of the problem, and (b) report it to the proper authorities.

You should fully record your observations. Write down when you observed the problem (both date and time), where you observed the pollution, and how you came to notice the pollution. If the pollution problem has occurred more than once or is continuing, write that down. If possible, try to identify the person or source responsible. If it is a truck dumping wastewater or garbage, write down the license plate of the truck. If you have noticed a particular type of smell, write down your best description of the odor of the smell. If the pollution is visible and you have a camera, take a picture. If possible, you may want a friend, neighbor, or family member to confirm your observations.

Once you have carefully observed the problem and written down your observations, you should call the appropriate local or state authorities to inform them of your observations. Look in your local telephone book in the government pages for the county or city office that might handle the problem. Typically, such offices will be listed as environmental, public health, public works, water pollution, air

*This section and Sections 44.5 and 44.6 have been adopted from U.S. EPA literature (8). The original version was written in the first person, and in order to maintain this passage's continuity, the first-person mode of presentation has been for the most part retained.

pollution, or hazardous waste agencies. If you cannot find a county or city office, look for a state government environmental office. It may require a few calls to find the correct office, but hang in there!

Once you reach the appropriate office, give the official all the information on what you observed and ask him or her to look into the problem. You should ask the official whether the problem you identified is likely to be illegal, how common it is, and how and when the office will investigate. Make sure you get the person's name and telephone number. If the person does not call you back or respond promptly, call the person back and ask what is going on.

If the city or county environmental agency does not respond adequately to your telephone call, you may call back and ask to speak to the official's supervisor or boss. If the supervisor is not available, get his or her name and address. You may then write this person a letter describing the problem you have observed and explaining your dissatisfaction with the office's response to it. Or you could contact the appropriate state environmental office directly, by telephone or letter.

If you cannot get an adequate response from local or state environmental offices, or you cannot find a local or state office to call, you may call the U.S. EPA regional office that covers your area.

If the pollution problem persists and the local, state, and regional U.S. EPA offices appear unwilling or unable to help, you may contact U.S. EPA headquarters in Washington, D.C. If you do not believe the government agencies have adequately responded to the pollution problem, and you believe the pollution is illegal and the problem appears to be continuing, you may have certain individual rights under the citizen suit provisions of the various federal environmental laws that you can assert to remedy the pollution problem yourself. You may wish to contact your own attorney or a public interest environmental group. A listing of national and state environmental groups is contained in the *Conservation Directory*, published annually by the National Wildlife Federation, Washington, D.C., and available in many public libraries. if you win such a lawsuit, the polluter will likely be required to correct the problem causing the pollution, pay penalties to the United States for violating the law, and pay your attorney's fees.

Finally, if you are told that the pollution problem you have observed is legal, but you believe it should not be legal, you are free to suggest changes in the law by writing to your U.S. Senator or Representative in Washington, D.C., or to your state governor or state legislators to inform them of the problem. Local libraries should have the names and addresses of these elected officials.

44.5   WATER AND AIR POLLUTION VIOLATIONS

Periodically, people may become concerned that pollution of a river, stream, lake, or ocean is occurring. This concern may be caused by the sight of an oil sheen on the surface of a river, stream, or lake. It might be caused by their observing a discoloration of the water in a stream or a pipe discharging apparently noxious liquids into a water body. Concern might also arise because an unusual odor is emanating from a body of water, or a bulldozer is seen filling in a marsh or wetland.

While some water pollution is an unfortunate consequence of modern industrial life, there are national, state, and local laws that limit the amount and kinds of water pollution allowed, and in some cases these laws completely prohibit certain types of water pollution. Sometimes it will be easy for a citizen to identify water pollution that is a violation of the law, and sometimes it will be difficult to identify the water pollution problem without sophisticated equipment.

The following subsections describe a few general types of water and air pollution problems a citizen might observe.

Rivers and Lakes

A citizen might observe wastewater flowing out of a pipe directly into a stream, river, lake, or even an ocean. Persons are only allowed to discharge wastewater into a water body if they have received a National Pollutant Discharge Elimination System (NPDES) permit and they are complying with the requirements of that permit. NPDES permits limit the amount of pollutants which persons are allowed to discharge. Unfortunately, it is often difficult to tell with the naked eye if a person is complying with the terms of a NPDES permit. However, some reliable indicators of violations are a discharge that leaves visible oil or grease on the water, a discharge that has a distinct color or odor, or one that contains a lot of foam and solids. Further, if there are dead fish in the vicinity of the discharge, this is a strong indicator of a water pollution violation.

Citizens should be aware that all persons who discharge wastewater to U.S. waters must report their discharges. These monthly reports, commonly called Discharge Monitoring Reports (DMRs), indicate the amount of pollutants being discharged and whether the discharger has complied with its permit during the course of the month. These reports (DMRs) are available to the public through state environmental offices or EPA regional offices.

Wetlands or Marshes

Under the Federal Clean Water Act, persons are only allowed to fill wetlands (commonly known as marshes or swamps) pursuant to the terms of a special discharge permit. "Filling a wetland" generally means that a person is placing fill or dredge material (like dirt or concrete) into the wetland in order to dry it out so that something can be built on the wetland. The Section 404 wetlands program is jointly administered by the U.S. Army Corps of Engineers and EPA. In general, the United States is committed to preserving its wetlands (sometimes called the "no net loss" program) because of the valuable role wetlands play in our environment. In brief, wetlands provide a habitat for many forms of fish, wildlife, and migratory birds; they help control flooding and erosion; and they filter out harmful chemicals that might otherwise enter nearby water bodies.

In general, there is usually no way to know if a wetland is being filled legally or illegally without knowing whether the person has a Section 404 permit and knowing the terms of that permit. However, if you notice fill activity going on in a suspicious manner, e.g., late at night, this may suggest that the wetland is being filled illegally. If you see a wetland being filled and are curious whether there is a permit authorizing such filling, you may call the local Army Corps of Engineers' office or the EPA regional office in your state. If possible, you should tell the Army Corps or EPA the location of the wetland being filled, what kind of filling activity you noticed, and who is doing the filling.

Drinking Water

The nation's drinking water is protected through the Federal Safe Drinking Water Act. Under this law, suppliers of drinking water are required to ensure that their water complies with federal standards, known at maximum contaminant levels (MCLs), for various pollutants and chemicals, such as coliform bacteria. If drinking water suppliers exceed a federal standard, they are required to immediately notify their users and implement measures to correct the problem. While you may not be able to tell if your drinking water is meeting all federal standards without testing equipment, if you notice any unusual smell, taste, or color in your water, you should immediately notify the person who supplies your water and the appropriate state agency.

In many of the circumstances when citizens become aware of a water pollution problem, there are actions that they can take to begin the process of correcting the problem and forcing the violator to comply with the law. The first step is always to make careful observations of the pollution event that you are observing. It is best to make a written record of the time and place of the sighting. As many details as possible should be recorded concerning the nature of pollution, for instance its color, smell, location, and its "oiliness." It is extremely important, if possible, that the source of the pollution be identified, including the name and address of the perpetrator. If the pollution is visible and you have a camera, take a picture. If possible, you may want a friend, neighbor, or family member to confirm your observations.

Once you have carefully observed the problem and written down your observations, you should call the appropriate local or state authorities to inform them of your observations. Look in your local telephone book in the government pages for the county or city office that might handle the problem. Typically, such offices will be listed as environmental, public health, public works, or water pollution agencies. If you cannot find a county or city office, look for a state government environmental office.

As the next step, a determination must be made as to the legality of the discharge. If the discharge is, in fact, illegal, the perpetrator must be confronted, the discharging of pollutants or the filling of the wetland must be halted, and, if feasible, the environmental damage caused by the perpetrator's actions must be corrected. Confrontation of the polluter is most practically achieved by contacting the local, state, or federal environmental protection agency. In general, the state environmental agency is responsible for making a preliminary assessment of the legality of the pollution event observed, for investigating the event, and, if necessary, for initiating an enforcement action to bring the polluter into compliance with the law. The citizen may also contact the U.S. EPA regional office that covers your state for assistance.

If the pollution problem persists and the local, state, and regional U.S. EPA offices appear unwilling or unable to help, you may contact U.S. EPA headquarters in Washington, D.C. If you do not believe the federal, state, or local governments have adequately responded to the pollution problem, and you believe the pollution is illegal and appears to be continuing, you may have certain individual rights under the citizen suit provisions of the various federal environmental laws that you can assert to remedy the pollution problem yourself. The Federal Clean Water Act provides that a citizen adversely affected by water pollution may bring a lawsuit on behalf of the United States to correct the problem. If you want to do this, you will probably need a lawyer to make an assessment of the illegality of the pollution event and your chances of succeeding in a lawsuit. There are a number of public interest organizations who can also be contacted that are in the business of bringing this kind of lawsuit. (A listing of national and state environmental groups is contained in the *Conservation Directory*, 1987, 32nd edition, published by the National Wildlife Federation, Washington, D.C.) If you win such a lawsuit, the polluter will likely be required to correct the problem causing the pollution, pay penalties to the United States for violating the law, and pay your attorney's fees.

Finally, if you have obtained "insider" information that water pollution is occurring, the Clean Water Act protects you from recrimination if the polluter is your employer. Your employer may not fire you or otherwise discriminate against you based on your "blowing the whistle."

Smoke or Odor

There are several air pollution situations a citizen might observe. You might observe visible emissions of air pollutants, such as black clouds of smoke, coming from a source such as a factory or power plant. You might also notice a discharge of air pollution because you can smell a strong odor. In either of these situations, these discharges may or may not be a violation of the Clean Air Act.

The Clean Air Act does allow some pollution discharges. The goal of the Clean Air Act is to keep the overall concentration of the major air pollutants at a level that will protect the public health. States then decide how they are going to meet these air pollution goals. A state may decide not to regulate a particular category of air pollution sources at all and to concentrate its efforts elsewhere in meeting its goals. Regulated sources may have permits from the state allowing them to discharge a certain level of pollution.

The best course of action for a citizen to take in these two situations is first to try to determine the exact source of the pollution. If it is a visible discharge, take a photograph. Also, note the exact time, day, and location you observe the pollution. Then notify your local or state air pollution or environmental agency of your observations. They should be able to determine if the source you observed is regulated, and if so, whether the discharge of pollution you observed is legal. The EPA usually defers to the state for enforcement. Only in limited, appropriate circumstances does the EPA intervene to take enforcement action. However, if you have difficulty in getting a response from your or local agency, contact the nearest regional office of EPA and report your observations.

To repeat, there are two ways to proceed if you suspect that an air or water pollution is occurring: Either contact your state EPA or the U.S. EPA to disclose your information, and/or initiate your own citizen's lawsuit.

44.6 OTHER ENVIRONMENTAL VIOLATIONS

Asbestos

Another situation a citizen might encounter involves construction work. Many old buildings contain the hazardous material, asbestos. Asbestos is extremely harmful to human health if inhaled or ingested. When buildings containing asbestos are renovated or demolished, the asbestos is broken up and can become airborne and, therefore, a health hazard.

EPA regulations require all parties associated with renovations and demolitions involving asbestos to notify EPA of the work and follow certain work practice requirements aimed at eliminating or at least minimizing the amount of airborne asbestos. These requirements largely consist of wetting the asbestos at all stages of the process so that it does not become airborne. The regulations also require the asbestos to be stored and disposed of in a particular manner.

There are several ways a citizen might help identify a violation of the asbestos regulations. If you pass a construction site, you may notice large amounts of white dust coming from the site or scattered around the site. These could be violations if the debris in question contains asbestos. One way a citizen could verify that asbestos is involved is looking for a brand-name label stamped on insulation that is still intact. Otherwise, trained inspectors will have to take samples and laboratory analysis of the debris must be done to verify that it contains asbestos.

The most effective action to take is to notify the nearest EPA regional office about the site. EPA personnel can then check their records to see if they have received notice of the demolition or renovation, and can do an inspection if it seems likely that asbestos is involved.

Auto Warranties

The Clean Air Act requires that motor vehicles sold in the United States meet prescribed emissions standards. In order to ensure that vehicle emissions remain low for the useful life of the vehicle, manufacturers are required to provide broad emission warranty coverage for vehicles that are less than five years old and have been operated for less than 50,000 miles. This warranty applies to defects in any part whose primary purpose is to control emissions, such as the catalytic converter, and in any part that has an effect on emissions, such as the carburetor (except parts that have normal replacement intervals, such as spark plugs).

Manufacturers must make emissions warranty repairs free of charge for any labor or parts. If you believe you are entitled to an emissions warranty repair, contact the person identified by the manufacturer in your owner's manual or warranty booklet.

If you are not satisfied with the manufacturer's response to your emissions warranty claim, you may contact the EPA for assistance by writing to Field Operations and Support Division (EN-397F), U.S. Environmental Protection Agency, Washington, D.C. 20460.

Removing Emission Control Devices

The Clean Air Act also seeks to prevent automotive pollution by prohibiting the removal or rendering inoperative emission-control devices by new and used car dealers, repair shops and fleet operators. In addition, gasoline retailers are prohibited from introducing leaded gasoline into motor vehicles which require unleaded gasoline, and gasoline that is sold as unleaded must not contain excess lead or alcohol. If you know of a violation of the antitampering or motor vehicle fuel rules, please contact EPA by writing to the address listed above.

The Clean Air Act also has a provision allowing citizens to sue any person alleged to be in violation of an emission standard under the Clean Air Act (42 U.S.C. Section 7604).

Abandoned Sites, Barrels, etc.

When citizens see leaking barrels (or barrels that look like they might leak), pits or lagoons on abandoned property, they should avoid contact with the materials, but note as thoroughly as possible their number, size,and condition (e.g., corroded, open, cracked) and the material leaking (e.g., color, texture, odor), and report these to the local fire department or the hazardous waste hotline (800-424-8802 or 202-267-2675). If possible, take a photograph of the area, but do not get too close to the materials. If the substances are hazardous, the statute most likely involved is the Comprehensive Environmental Response, Compensation and Liability Act (CERCLA or the Superfund law), and EPA or the state should take the lead. Under CERCLA, citizens have the opportunity to, and are encouraged to, involve themselves in the community relations program which includes citizen participation in the selection of a remedial action.

A citizen may file suit against any person, including the United States, who is alleged to be in violation of any standard, regulation, condition, requirement or order that has become effective under CERCLA provided that the citizen gives the violator, EPA, and the state sixty days' notice of the intent to sue. A citizen suit cannot be brought, however, if the United States is diligently prosecuting an action under CERCLA.

Hazardous Waste Facilities

When citizens encounter leaks, discharges or other suspect emissions from a hazardous waste treatment, storage or disposal (TSD) facility, they should contact their state hazardous waste office or the local EPA Regional office to determine if the facility has a Resource Conservation and Recovery Act (RCRA) permit or has been granted interim status to operate while it applied for a RCRA permit. Any citizen may obtain copies of a TSD facility's permit and monitoring reports, which would document any violations, from the state agency or EPA Regional office.

A citizen may bring a civil judicial enforcement action against a RCRA violator provided he gives the violator, EPA, and the state sixty days notice of the intent to sue, during which time the state or EPA may pursue an enforcement action. With certain limitations, a citizen may also bring an action against any person who has contributed to or who is contributing to the past or present handling of any solid waste, including hazardous waste, that may present an imminent and substantial endangerment to human health or to the environment.

Transportation Spills

If you see a spill from a truck, train, barge or other vehicle, you should report it immediately to the local fire or police. If it is possible to read any labels on the vehicle, without getting too close, then you should report this information as well.

If you see a spill from a barge, ship, or other vessel into navigable waters or the ocean, such as an oil spill from tanker, you should report the spill and location to the United States Coast Guard, or call the hazardous waste hotline (800-424-8802 or 202-267-2675).

Citizens who provide information leading to the arrest and conviction of persons who commit certain criminal violations under CERCLA may be eligible for a reward of up to $10,000. These awards are often offered in connection with either

(a) a violator's failure to make a required report on a release of a hazardous substance or (b) the destruction or concealment of required records. When citizens encounter instances of pollution involving pesticides or toxic substances, the law that was actually violated will most often be the Clean Water Act, the Clean Air Act, or the Resource Conservation and Recovery Act. Most violations of the Toxic Substance Control Act or the Federal Insecticide, Fungicide, and Rodenticide Act will be discovered only by persons with special training or with access to information that is not generally available to the public.

Toxic Substance Control Act

Violations of the Toxic Substances Control Act (TSCA) that the public might observe include:

- Demolition of a building containing asbestos without proper measures to keep the asbestos contained.
- Improper storage or disposal of transformers containing PCBs (poly-chlorinated biphenyls).
- Improper storage of asbestos.

If you think you are seeing such a violation, you should contact Office Compliance Monitoring (EN-342), U.S. Environmental Protection Agency, Washington, D.C. 20460, or call the National Response Center for Oil and Hazardous Material Spills at 800-424-8802.

Citizen suits are authorized under TSCA (15 U.S.C. Section 2619). Citizens may sue violators of provisions concerning PCBs, asbestos, required testing of chemical substances, notification to EPA before manufacturing or importing new chemicals, or beginning a significant new use of chemicals.

Federal Insecticide, Fungicide, and Rodenticide Act

Citizens may encounter violations of the provisions of the Federal Insecticide, Fungicide, and Rodenticide Act (FIFRA) that govern the use of pesticides. FIFRA requires that pesticides be used by the public only as specified on the label. Many pesticides are labeled for use only by specially licensed applicators. Others have been banned from almost all uses, except for particular uses where no other pesticide is effective.

Violations of FIFRA that citizens may observe include

- Sale or use of banned pesticides that are not registered with EPA. These would lack the EPA registration number that must appear on every pesticide label.
- Use of pesticides in a manner inconsistent with the directions on the label.
- Application of restricted-use pesticides by unlicensed applicators.
- False or misleading labeling or advertisement of pesticides.

If you think you are seeing such a violation, you should contact Office of Compliance Monitoring (EN-342), U.S. Environmental Protection Agency, Washington, D.C. 20460, or call the Hazardous Material Spills at 800-424-8802. There is no citizen suit authority under FIFRA.

Emergency Planning and Community Right-to-Know Act

The Emergency Planning and Community Right-to-Know Law (EPCRA) described earlier requires a wide range of businesses that manufacture, import, process, use, or store chemicals to report certain information to federal, state, and local governments. For example, these businesses are required to report annual estimates of the amounts and types of toxic chemicals they released or disposed of during each calendar year. The data must be reported to EPA and to state agencies, and they are available to the public through an EPA compilation called the Toxics Release Inventory. The data in this inventory may be used by the public to examine the practices of particular manufacturers.

If you believe that a business that was subject to the EPCRA requirement failed to report to the Toxics Release Inventory, you should contact Office of Compliance Monitoring (EN-342), U.S. Environmental Protection Agency, Washington, D.C. 20460. A business' failure to report toxic releases may also be challenged through a citizen suit under 42 U.S.C. Section 11046(a)(1).

44.7 ACCESSIBILITY IN ENVIRONMENTAL COMMUNICATION

The availability and accessibility of means to ensure environmental communication is crucial in establishing effective communications. Hotlines, toll-free numbers, and information lines provide the consumer a vital link with EPA's environmental programs, technical capabilities, and services. EPA is among several environmental agencies currently using these state-of-the-art means of information dissemination and service to the public.

Websites

The EPA offers a great deal of information through its state of the art websites. Once you access the main homepage, you can then access a multitude of related websites. A sample of just a few are identified below:

www.epa.gov	EPA's Homepage
www.epa.gov/oeca/oc	Compliance Assistance Home Page
www.epa.gov/ocea/smbusi.html	Small Business policy

Toll-Free Numbers Offered by EPA Headquarters

- **RCRA/Superfund Hotline — National Tool-Free 800-424-9346 Washington, D.C. Metro 202-382-3000.** The EPA's largest and busiest toll-free number, the RCRA/Superfund Hotline, answers nearly 100,000 questions and document requests each year. Since 1980 when it began, the hotline has expanded significantly and continues to grow, with 21 information specialists now covering 14 incoming lines, eight hours each day. Hotline specialists answer regulatory and technical questions and provide documents on virtually all aspects of the RCRA and Superfund programs. Because of the complexity and changing nature of these programs, the hotline is used widely by the regulated community, people involved in managing and cleaning up hazardous waste, federal, state, and local governments, and the general public. The RCRA/Superfund Hotline can be reached Monday through Friday from 8:30 a.m. to

4:30 p.m. Eastern Standard Time (EST). If the lines are busy, either wait for the next available operator or call back.

- **National Response Center Hotline — National Tool-free 800-424-8802, Washington, D.C. Metro 202-426-2675.** Operated by the U.S. Coast Guard, the National Response Center Hotline responds to all kinds of accidental releases of oil and Hazardous substances. Callers should contact this hotline to report chemical spills. The National Response Center Hotline is available 24 hours a day, seven days a week, every day of the year.

- **Chemical Emergency Preparedness Program (CEPP) Hotline National Toll-Free 800-535-0202 Washington, D.C. Metro and Alaska 202-479-2449.** A relatively new and increasingly popular service, CEPP Hotline has been in operation since late 1985, responding to questions concerning community preparedness for chemical accidents. The Superfund Amendments and Reauthorization Act (SARA) has increased the CEPP Hotline's responsibilities, which include Emergency Planning and Community Right-to-Know, SARA Title III, questions and requests. The CEPP Hotline, which complements the RCRA/Superfund Hotline, is maintained as an information rather than an emergency number. Calls are answered Monday through Friday from 8:30 a.m. to 4:30 p.m. EST.

- **National Pesticides Telecommunications Network (NPTN) National Toll-Free 800-858-7378 (858-P-E-S-T), Texas 806-743-3091.** Operating 24 hours a day, seven days a week, every day of the year, the NPTN provides information about pesticides to the medical, veterinary, and professional communities as well as to the general public. Originally a service for physicians wanting information on pesticide toxicology and on recognition and management of pesticide poisonings, the NPTN has expanded to serve the public by providing impartial information on pesticide products, basic safety practices, health and environmental effects, and clean-up and disposal procedures. Staffed by pesticide specialists at Texas Tech University's Health Sciences Center School of Medicine, this Hotline handles about 18,000 calls each year. Call any time, day or night.

- **Asbestos Hotline — National Toll-Free 800-334-8571, extension 6741.** Formerly the Asbestos Technical Information Service, the Asbestos Hotline at Research Triangle Institute, NC, has evolved from an information number for laboratories doing asbestos analyses to a broader service, providing technical information concerning asbestos abatement problems. The Asbestos Hotline is now available to meet the asbestos information needs of private individuals, government agencies, and the regulated industry. The hotline handles about 10,000 calls each year, and operates Monday through Friday from 8:15 to 5:00 p.m. EST.

- **Small Business Hotline — National Toll-Free 800-368-5888 Washington, D.C. Metro 703-557-1938.** Sponsored by EPA Small Business Ombudsman's Program, this hotline assists small businesses in complying with environmental laws and EPA regulations. The Small Business Hotline gives companies easy access to the Agency, and investigates and resolves problems and disputes with EPA. Acting as a liaison with Agency program offices, the hotlines ensures that EPA considers small business issues during its normal regulatory activities. The Small Business Hotline operates Monday through Friday from 8:30 a.m. to 5:00 p.m. EST., handling over 7,000 inquires each year.

- **EPA National Recruitment Program Number — National Toll-Free 800-338-1350 Washington, D.C. Metro 202-382-3305.** An integral part of EPA's National Recruitment Program, this toll-free service enables the potential hirees to contact the Agency for employment information, and assists EPA managers in locating and hiring qualified employees to fill vacant positions. Recruitment for many Superfund positions currently is a priority of this service. Operating Monday through Friday from 8:30 a.m. to 4:30 p.m. EST., the Recruitment Program Number handles about 6,000 calls each year.

- **Inspector General's Whistle-Blower Hotline — National Toll-Free 800-424-4000 Washington, D.C. Metro 202-382-4977.** The EPA. Inspector General's Office maintains the Whistle-Blower Hotline to receive reports of Agency-related waste, fraud, abuse, or mismanagement from the public and from EPA and other government employees. EPA employees may make complaints or give information to the Inspector General's office confidentially and without fear of reprisal. The Whistle-Blower Hotline is staffed to answer calls in person from 10:00 a.m. to 3:00 p.m. EST., Monday through Friday; at other times, callers may leave a message to be answered during the next work day. This hotline handles about 1,500 calls each year.

Commercial Numbers Offered by EPA Headquarters

- **TSCA Assistance Information Service 202-554-1404.** The TSCA Assistance Information Service provides information on TSCA regulations to the chemical industry, labor and trade organizations, environmental groups, and the general public. Technical as well as general information is available. To help businesses comply with TSCA, a variety of services are offered, including regulatory advice and aid, publications, and audiovisual materials. The TSCA Assistance Information Service handles about 2,500 calls a month and can be reached from 8:30 a.m. to 5:00 p.m. EST, Monday through Friday.

- **Control Technology Center Hotline 919-541-0800.** A component of EPA's Air Toxics Strategy, the newly established Control Technology Center Hotlines provides information to state and local pollution control agencies on sources of emissions of air toxics. Sponsored by EPA's Office of Air Quality Planning and Standards in Research Triangle Park, NC, this hotline takes about 100 calls a month, and can be reached from 8:00 a.m. to 4:30 p.m., Monday through Friday.

- **Public Information Center (PIC) 202-829-3535.** EPA's Public Information Center (PIC) answers inquires from the public about EPA, its programs, and activities, and offers a variety of general, nontechnical information materials. The public is encouraged to reach the PIC through its commercial telephones line or by writing to PIC (PM-211B), U.S. EPA, 401 M Street, S.W., Washington, D.C. 20460 (9).

- **Center for Environmental Research Information — Central point of distribution for EPA results and reports: 513-569-7391.**

- **Hazardous Waste Ombudsman.** Assists citizens and the regulated community who have had problems voicing a complaint or getting a problem resolved about hazardous waste issues. There is a Hazardous Waste Ombudsman at

EPA Headquarters and in each of the regional offices. In Washington the number is 202-260-9361.

- **National Small Flows Clearinghouse.** Provides information on wastewater treatment technologies for small communities: 800-624-8301.

- **Pollution Prevention Information Clearinghouse.** Provides information on reducing waste through source reduction and recycling: 703-821-4800.

- **Radon Information.** For information about radon, you should call the Radon Office in your individual state. The Radon Office at EPA Headquarters also responds to requests for information: 202-260-9605.

- **Safe Drinking Water Hotline.** Provides information on EPA's Drinking water regulations. Operates Monday through Friday from 8:30 a.m. to 4:30 p.m. Eastern Time: 800-426-4791; in the Washington D.C. area, 202-260-5534 (1).

44.6 SUMMARY

1. Environmental communication is playing a pivotal role in environmental management. As Americans experience a new wave of environmentalism, how we communicate environmental issues and concerns will shape and impact the direction of environmental policies.

2. There are several major statutes and regulations which mandate communication activities with the general public. Since its founding in December 1970, EPA has been given the task of implementing a wide range of environmental statutes, many of which contain language requiring EPA to involve the public in their implementation.

3. Environmental communications take many forms, the most common being printed materials in the form of technical documents, pamphlets, handouts, brochures, magazines, journals, issue papers, newspaper articles, editorials and the like. In addition to the more common and traditional means of communication, we have seen a move towards the use of radio and television as a means of environmental communication.

4. The two most important things to do when you see a potential pollution problem are (a) make careful observations of the problem, and (b) report it to the proper authorities.

5. There are two ways to proceed if one suspects that air or water pollution is occurring: Contact the local regulatory body or the U.S. EPA, and/or initiate a citizen's lawsuit.

6. There are a host of environmental violations that occur for which a citizen can take action. These include asbestos, auto warranties, removing emission control devices, abandon sites, hazardous waste facilities, transportation spills and deviations from environmental regulations.

7. The availability and accessibility of means to ensure environmental communication is critical in establishing effective communications. Hotlines, toll-free numbers, and information lines provide the consumer a vital link with EPA's environmental programs, technical capabilities, and services.

REFERENCES

1. U.S. EPA, *Preserving Our Future Today*, Communications and Public Affairs (A-107) 21K-1012, October 1991.
2. *Federal Water Pollution Control Act*, Sect. 101(e), 1972.
3. U.S. EPA, Office of Public Affairs, *EPA Journal*, 11(10), December 1985.
4. U.S. EPA, *The New Superfund, What it is, How it Works,* August 1987.
5. U.S. EPA, *Chemicals in Your Community, A Guide to Emergency Planning and Community Right-to-Know Act,* September 1988.
6. U.S. EPA, *EPA Journal*, 17(4), September/October 1991, 22K-1000.
7. U.S. EPA, *EPA Journal*, 16(1), January/February 1990, 20K-9001.
8. U.S. EPA, Office of Enforcement, *The Public Role in Environmental Enforcement*, March 1990.
9. U.S. EPA Office of Public Affairs, *EPA Journal*, 13(3), August 1987.

X

RECENT DEVELOPMENTS

45

ISO 14000

Contributing Author: Abdool Jabar

45.1 INTRODUCTION

Participation in ISO 14000 is becoming one of the most sought-after statuses in the more general move towards globalization of environmental management. In recent years, there has been heightened international interest in and commitment to improved environmental management practices by both the public and private sectors. This interest is reflected in the success of collaborative international efforts to address environmental problems and in the global recognition of trade-related environmental issues. The Montreal Protocol (the environmental side-agreements of The North American Free Trade Agreement) and the mandates resulting from the 1992 Earth Summit of the United Nations Conference on Environment and Development in Rio de Janeiro are all indications that industry and government all over the world are prepared to put together a plan to effectively manage the environment.

Another indication of interest in improved environmental practices is the emergence of voluntary environmental management standards developed by national standards bodies throughout the world. To address the growing need for an international consensus approach, ISO, the International Organization for Standardization, has undertaken the development of international voluntary environmental standards (1).

Standards are documented agreements containing technical specifications or precise criteria to be used consistently as rules, guidelines, or definitions of characteristics, to ensure that materials, products, processes, and services are fit to be used for their intended purposes. For example, the format of credit cards, phone cards, and "smart cards" that have become commonplace is derived from an ISO International Standard. Adhering to the standard, which defines such features as an optimal thickness (0.76 mm), means that the cards can be used worldwide. International Standards thus contribute to making life simpler and to increasing the reliability and effectiveness of the goods and services we use (2).

The development of International Standards began with the creation of the International Electrochemical Commission (IEC) in 1906 to make standards for the electrochemical industry. Standards in other areas were developed by the International Federation of National Standardization (ISA) which was established in 1926. ISA's emphasis was on standards pertaining to the mechanical engineering industry. Due to the advent of the Second World War, ISA's activities ceased.

In 1946, a meeting was held in London with delegates from 25 countries to discuss the development of international standards. After the meeting, it was decided to establish an international organization to coordinate and unify industrial standards. From this international organization, the ISO was formed and started to function in 1947. The first standard published by this body was "Standard Reference Temperatures for Industrial Length Measurement."

45.2   WHAT IS THE ISO?

The International Organization for Standardization is a private, nongovernmental, international standards body based in Geneva, Switzerland. ISO promotes international harmonization and development of manufacturing, product, and communications standards. ISO has promulgated over 8,000 internationally accepted standards for everything from paper sizes to film speeds. More than 130 countries participate in the ISO as "Participating" members or as "Observer" members. The United States of America is a full voting, Participating member and is officially represented by the American National Standards Institute (ANSI).

Many people will have noticed the seeming lack of correspondence between the official title when used in full, International Organization for Standardization, and the short form, ISO. Shouldn't the acronym be IOS? That would have been the case if it were an acronym. However, ISO is a word derived from the Greek word isos, meaning equal. From "equal" to "standard," the line of thinking that led to the choice of "ISO" as a name of the organization is easy to follow. In addition, the name ISO is used around the world to denote the organization, thus avoiding the plethora of acronyms resulting from the translation of "International Organization for Standardization" into the different national languages of members, e.g., IOS in English or OIN in French. Whatever the country, the short form of the organization's name is always ISO (2).

45.3   NEED FOR INTERNATIONAL STANDARDS

The impetus toward international standards is deeply rooted in economic rewards and an expansion into a global economy. The standardization of goods and services will not only increase potential market share but allow goods and services to be available to more consumers. Different countries producing articles using the same technologies but using different sets of standards limit the amount of trade that can be done among countries. Industries that depend on exports realized that there is need for consistent standards in order to trade freely and extensively. International standards have been established for many technologies in different industries such as textiles, packaging, communication, energy production and utilization, distribution of goods, banking, and financial services. With the expansion of global trade, the importance of international standards will continue to grow. The advent of International Standards makes it possible for one to purchase a video cassette recorder made in South Korea and use a videotape made in Mexico or buy a computer made in the United States and use disks made in China. International certification programs such as those developed by National Association of Corrosion Engineers (NACE) set standards acceptable all over the world. Industries can be assured of qualified services in a timely manner. However, there are still some areas that need standardization, e.g., a videotape which is taped in the United Kingdom cannot be played on a VCR in the United States because the systems are not standardized.

45.4 HOW THE STANDARDS ARE DEVELOPED

The technical work of ISO is highly decentralized, carried out in a hierarchy of some 2850 technical committees and working groups. In these committees, qualified representatives of industry, research institutes, government authorities, consumer bodies, and international organizations from all over the world come together as equal partners in the resolution of global standardization problems. Some 30,000 experts participate in meetings every year.

ISO standards are developed according to the following principles:

- *Consensus:* The views of all interests are taken into account: manufacturers, vendors and users, consumer groups, testing laboratories, governments, and research organizations.
- *Industry-wide:* Global solutions to satisfy industries and customers worldwide
- *Voluntary:* International standardization is market-driven and therefore based on voluntary involvement of all interests in the market-place.

There are three main phases in ISO standards development process. The first phase begins with "the need" for a standard, which is usually expressed by an industry sector. The sector then communicates this need to a national member body. The latter proposes the new work item to the ISO as a whole. Once the need for an International Standard has been recognized and formally agreed upon the first phase of development involves definition of the technical scope of the future standard. This phase is usually carried out in workgroups, which are comprised of technical experts from countries interested in the subject matter. Once agreement has been reached on which particular technical aspects are to be covered in the standard, the second phase begins, during which countries negotiate the detailed specifications within the standard. This is the consensus-building phase.

The final phase comprises the formal approval of the resulting draft International Standard (the acceptance criteria stipulate approval by two-thirds of ISO members that have actively participated in the standards development process, and approval by 75% of all members that vote), following which the agreed text is published as an International Standard. It is now possible to publish interim documents at different stages in the standardization process.

Most standards require periodic revision. Several factors combine to render a standard out of date: technological evolution, new methods and materials, and new quality and safety requirements. To take account of these factors, ISO has established the general rule that all ISO standards should be reviewed at intervals of not more than five years. On occasion, it is necessary to revise a standard earlier. To date, ISO's work has resulted in some 12,000 International Standards, representing more than 300,000 pages in English and French (terminology is often provided in other languages as well) (2).

45.5 DEVELOPMENT OF ENVIRONMENTAL STANDARDS

At the Rio De Janiero conference, it was recognized that there was need for international environmental management standards. Due to the success of the ISO 9000 series of standards, the United Nations Conference on Environment and Development (UNCED) asked ISO to develop international environmental standards. ISO established a Strategic Advisory Group on the Environment (SAGE) to look into the global standardization of environmental management practices. This body was made up of representatives of governments, national standardization organizations, and business and environmental professionals.

The SAGE considered whether such standards would

- Promote a common approach to environmental management similar to quality management.

- Enhance an organization's ability to attain and measure environmental performance.

- Facilitate lower trade barriers.

At the conclusion of their study, SAGE recommended that an ISO technical committee formally consider and produce final "consensus" standards. Thus, in January, 1993, Technical Committee 207 (TC 207) was established. The number 207 was chosen because it fell in the sequence of numbers for technical committees. Canada was awarded the secretariat for TC 207 and the inaugural plenary session was held in June 1993; over 200 delegates from over 30 countries and organizations attended. The largest block of countries with voting rights is from Europe. The TC 207 framework also includes coordination and cooperation with international and regional governmental organizations including the European Union (EU), the General Agreement of Tariffs and Trade (GATT), and the Organization for Economic Cooperation and Development (OECD). TC 207 meets annually to review the progress of its subcommittees (3). The structure of TC 207 with its subcommittees and working groups is shown in Figure 45.1.

The United States has established a Technical Advisory Group (TAG) consisting of academia, industry, government and environmental groups to participate in TC 207. The structure of U.S. TAG is shown in Figure 45.2.

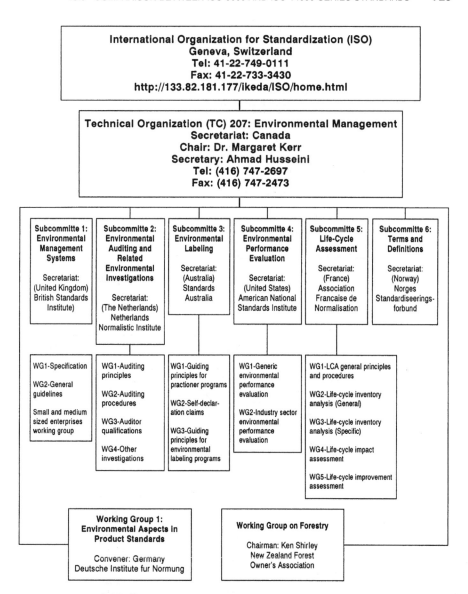

FIGURE 45.1   Structure of ISO Technical Committee 207.

45.6   COMPARISON BETWEEN ISO 9000 AND ISO 14000 SERIES STANDARDS

As shown in Figure 45.3, the ISO 14000 series of standards is made up of one standard ISO 14001, which organizations have to comply with and others that are guidance to assist organizations comply with ISO 14001. ISO 14001 outlines the basis for establishing an Environmental Management System (EMS). The core

ISO 14001	*Environmental management systems — Specifications with guidance for use
ISO 14004	*Environmental management systems — General guidelines on principles, systems, and supporting techniques
ISO 14010	*Guidelines for environmental auditing — General principles on environmental management systems
ISO 14011/1	*Guidelines for environmental auditing – Audit procedures — Audit of environmental management systems
ISO 14012	*Guidelines for environmental auditing – Qualification criteria for environmental auditors
ISO 14015	Environmental site assessments
ISO 14020	Goals and principles of all environmental labeling
ISO 14021	Environmental labels and declarations — Self declaration environmental claims — Terms and definitions
ISO 14022	Environmental labels and declarations — Self declaration environmental claims — Symbols
ISO 14023	Environmental labels and declarations — Self declaration environmental claims — Testing and verification
ISO 14024	Environmental labels and declarations — Environmental labeling Type I — Guiding principles and procedures
ISO 14025	Environmental labels and declarations — Environmental information profiles — Type III guiding principles and procedures
ISO 14031	Evaluation of environmental performance
ISO 14040	Environmental management — Life cycle analysis — Principles and Framework
ISO 14041	Environmental management — Life cycle analysis — Life cycle inventory analysis
ISO 14042	Environmental management — Life cycle analysis — Impact assessment
ISO 14043	Environmental management — Life cycle analysis — Interpretation
ISO 14050	Terms and Definitions — Guide on the Principles for ISO/TC 207/SC6 terminology work
ISO Guide 64	Guide for inclusion of environmental aspects in product standards

*Completed and published.

FIGURE 45.2 Structure of the U.S. Technical Advisory Group (TAG).

American National Standards Institute (ANSI) (212) 642-4900	
US TAG to ISO/TC 207	
Chair: Joe Cascio (703) 750-6401 jcascio@gnet.org.	Vice-chair: Mary McKiel (202) 260-3584 mckiel.mary@epamail.epa.gov
ST1 — Environmental Management Systems	
Chair: Joel Charm (201) 455-4057 joel.charm@alliedsignal.com	EPA Rep: Jim Horne (202) 260-5802 horne.james@epamail.epa.gov
ST2 — Environmental Auditing	
Chair: Cornelius (Bud) Smith (203) 778-6114 (203) 778-6487 Fax	EPA Rep: Cheryl Wasserman (202) 260-8797 wasserman.cheryl@epamail.epa.gov
ST3 — Environmental Labeling	
Chair: Jim Connaughton (202) 736-8364 jconnaugh@sidley.com	EPA Rep: Julie Winters (202) 260-4000 winters.julie@epamail.epa.gov
ST4 — Environmental Performance Evaluation	
Chair: John Master (610) 359-4810 (610) 359-4862 Fax	EPA Rep: John Harman (202) 260-6395 harman.john@epamail.epa.gov
ST5 — Life Cycle Assessment	
Chair: James Fava (610) 701-3636 favaj@wcpost2.rfweston.com	EPA Rep: John Harman (202) 260-6395 harman.john@epamail.epa.gov
ST6 — Terms and Definitions	
Chair: Christopher Bell (202) 736-8118 cbell@sidley.com	EPA Rep: Mary McKiel (202) 260-3584 mckiel.mary@epamail.epa.gov
SWG — Environmental Aspects in Product Standards	
Chair: Stanley Rhodes (510) 832-1415	EPA Rep: John Shoaff (202) 260-1831 shoaff.john@epamail.epa.gov

FIGURE 45.3 Standards in the ISO 14000 series.

sections of the EMS consist primarily of the five subsections highlighted below:

4.1 Environmental Policy

4.2 Planning

 4.2.1 Environmental aspects

 4.2.2 Legal and other requirements

 4.2.3 Objectives and targets

 4.2.4 Environmental management programs

4.3 Implementation and operation

 4.3.1 Structure and responsibility

Organizations that meet the ISO 14001 requirements can seek registration in a process similar to ISO 9000 registration. The ISO 14000 series is complementary to the ISO 9000 series. Whereas the ISO 9000 series deals with Quality Management Standards, the ISO 14000 series deals with Environmental Management Standards. Like the ISO 9000 series, the ISO 14000 series is voluntary and does not replace regulations, legislations, and other codes of practice that an organization has to comply with. Rather it provides a system for monitoring, controlling, and improving performance regarding those requirements. ISO 14000 is a package that ties the mandatory requirements into a management system which is made up of objectives and targets focusing on meeting and exceeding the mandatory requirements, with an additional focus on prevention and continuous improvements. The ISO 14000 uses the same fundamental systems as ISO 9000, such as documentation control, management system auditing, operational control, control of records management policies, audits, training statistical techniques, and corrective and preventive actions (4).

Although there are numerous similarities between ISO 9000 and ISO 14000, there are some definite difference, e.g., ISO 14000 has clearer statements about communication, competence, and economics than those that are currently found in ISO 9000. Also, ISO 14000 incorporates the setting of objectives and quantified targets, emergency preparedness, considering the views of interested parties, and public disclosure of the organization's environmental policy.

For instance, an organization with an ISO 9000 registration will find that it is far along toward gaining ISO 14000 registration right from the outset. Even though there are differences, the management systems are generally consistent within both standards. The ISO approach to management serves as a model which needs to be adapted to meet the needs of the organization and integrated into existing management systems. The standards have been designed to be applied by any organization in any country regardless of the organization's size, process, economic situation, and regulatory requirements (4).

Why Consider Using the ISO 14000 Environmental Management System Model?

An organization may elect to comply with the ISO 14000 as a

- Model for an Environmental Management System.
- Format against which to audit the Environmental Management System.

- Method of demonstrating the Environmental Management System compliance
- Process for third party and/or customer recognition.
- Public declaration of their Environmental Management System

Benefits of implementing the ISO 14000 Environmental Management System may include the following:

- Enhanced compliance to legislation.
- Facilitated financial and real estate transactions, where environmental performance is a factor.
- Reduced costs associated with consumer audits.
- Ability to bid for contracts (protection or increase of market share).
- Market forces (a real or perceived "greening" of the marketplace).
- Economic return from increased efficiency of resource use.
- Increased ability to adapt to changing circumstances.

Some organizations may choose to implement an ISO 14000 program but not seek registration.

Registration to ISO 14000 should be considered if it

- Is a customer or industry requirement
- Complements market strategy
- Is perceived as a valuable motivational factor

Registration to an ISO 9000 standard generally requires twelve to eighteen months of effort depending on the complexity of the organization and the existing systems. It is anticipated that it will take at least the same length of time to develop and implement an Environmental Management System which compiles with the ISO 14000 requirements (4).

How Do You Implement ISO 14000?

As in an ISO 9000 program, senior management commitment is required before embarking on an ISO 14000 program. Once senior management is committed to implementing an ISO 14001 program, the project planning begins. This planning includes scheduling; budgeting; assigning personnel, responsibilities, and resources; and, if required, retaining specialized external assistance.

Senior management needs to provide a focus for the Environmental Management System by defining the organization's environmental policy. The policy must include, among other things, a commitment to continuous improvement, prevention of pollution and compliance with legislation and regulations. It must be specific enough to form the basis for concrete actions; "parenthood"-type feel-good statements are not adequate. When documented by management, this environmental policy must be implemented, maintained and communicated within the organization, and made available to the public.

Next, an initial review the organization's existing environmental program is needed. This review includes the consideration of all applicable environmental regulations, existing processes, documentation, work practices, and effects of current operations. Once the initial review is completed, a strategic or implementation plan can be developed. Implementation planning is similar to project management and the steps, scope, time-frame, costs and responsibilities need to be defined in order to develop and implement an Environmental Management System that meets the organization's targets and objectives, and promotes continuous improvement. The strategic plan sets the framework for participation of the responsible and affected parties within the organization.

Both in the initial review and on an ongoing basis, the organization's activities, products, and services require evaluation to determine their interaction with the environment. Environmental issues such as noise, emissions, environmental impact, waste reduction, and energy must be identified. The organization then needs to identify the aspects which can interact with the environment and which ones it can control or influence. The identified impacts are then used as a basis for setting environmental objectives within the organization. Objectives also need to take into account relevant legal and regulatory requirements; financial, operational, and business requirements; and the views of interested parties. Interested parties may be people or groups, such as neighbors or interest groups, concerned with the organization's environmental performance.

Objectives of the organization need to be determined and specific targets set. An objective is an overall goal which may be as simple as "meeting or exceeding regulations" or "reduction in energy consumption" and the targets provide quantified measurements. Objectives and targets are set by the organization, not by the ISO 14000 standard. Identifying the impacts, judging their significance, and setting reasonable objectives and targets are some of the major "environmental" challenges presented by ISO 14000.

Once the targets and objectives are set, the organizations need to implement the strategic plan. Beyond the environmental challenges, management functions will have to be adapted to meet the requirements of the Environmental Management System Standard. The level of conceptual challenge this will present to ISO 9000 firms, where the corporate culture will already be changing, will be less than for non-ISO 9000 firms, but there will be some new areas that require attention (4).

Maintaining an ISO 14000 Environmental Management System
Once the Environmental Management System is implemented, its progress needs to be continually measured and monitored. Routine measurement and monitoring must be undertaken of the activities which have been identified as having the potential for a significant impact on the environment.

Routine auditing and review are the keys to continuous improvement. Environmental as well as management components will be required in the audit program. Audits of an organization's Environmental Management System do not replace, but rather complement, the issue-specific environmental audits that may be conducted externally by regulators and consultants or internally by environmental engineers or other qualified personnel. Where issue specific audits address regulatory compliance, site assessment, or emissions, the Environmental Management System audits address effectiveness of the management system. Periodic Environmental Management System audits are needed to determine if the Environmental Management System conforms to the requirements of ISO 14001, and that the program is implemented, proportional to the nonconformance, to eliminate recurrence.

To ensure the continuing effectiveness of the Environmental Management System, management needs to regularly review and evaluate information such as the results of audits, corrective action, current and proposed legislation, results of monitoring, and complaints. This review allows management to look at the system and ensure that it is, and will remain, suitable and effective. The management review may result in changes to policies or systems as the organization evolves and as technology advances. An organization's Environmental Management System is not a stagnant system but must continually evolve to meet the organization's ever-changing needs.

45.7 RELATIONSHIP BETWEEN MANAGEMENT SYSTEMS INTEGRATING THE ISO 14000 ENVIRONMENTAL MANAGEMENT SYSTEM

The Environmental Management System must be integrated with the organization's other activities. If it is seen as a separate program, it will be difficult or impossible to maintain. The objectives, targets, procedures, and systems must be part of routine operations related to the ongoing activities of the organization.

It is important to remember that ISO 14000 is not an add-on program. Nor is it about "environmentalism" or being "green." An effective Environmental Management System is the consistent and systematic control of procedures or operations, products, or services which can have a significant impact on the environment and on the overall cost on business and industry. It is obviously concerned with environmental performance, but what it is about is effective corporate management. An organization which has effectively integrated an ISO 14000 Environmental Management System with its other business management systems is well on its way towards managing its processes with a view towards compliance, consistency and continuous improvement, and can accrue the accompanying benefits (4).

45.8 SUMMARY

1. Participation in ISO 14000 is becoming one of the most sought-after statuses in a move toward globalization of environmental management. In recent years, there has been heightened international interest in and commitment to improved environmental management practices by both the public and private sectors. The interest is reflected in the success of collaborative international efforts to address environmental problems and in the global recognition of trade-related environmental issues.

2. The International Organization for Standardization is a private, nongovernmental international standards body based in Geneva, Switzerland. ISO promotes international harmonization and development of manufacturing, product, and communications standards. ISO has promulgated over 8,000 internationally accepted standards from everything from paper sizes to film speeds. To address the growing need for an international consensus approach, ISO, the International Organization for Standardization, has undertaken the development of international voluntary environmental standards

3. The impetus toward international standards is deeply rooted in economic rewards and an expansion into a global economy. The standardization of goods and services will not only enhance potential market share but allow goods and services to be available to more consumers.

4. The technical work of ISO is highly decentralized, carried out in a hierarchy of some 2850 technical committees and working groups. In these committees, qualified representatives of industry, research institutes, government authorities, consumer bodies, and international organizations from all over the world come together as equal partners in the resolution of global standardization problems. Some 30,000 experts participate in meetings every year.

5. Due to the success of the ISO 9000 series of standards, United Nations Conference on Environment and Development (UNCED) asked ISO to develop international environmental standards. ISO established a Strategic Advisory Group on the Environment (SAGE) to look into the global standardization of environmental management practices. This body was made up of representatives of governments, national standardization organizations, and business and environmental professionals.

6. The ISO 14000 series of standards is made up of one standard, ISO 14001, with which organizations have to comply and others which are used as guidance to assist organizations comply with ISO 14001.

7. The Environmental Management System must be integrated with the organization's other activities. If it is seen as a separate program, it will be difficult or impossible to maintain. The objectives, targets, procedures, and systems must be part of routine operations related to the ongoing activities of the organization. It is important to remember that ISO 14000 is not an add-on program.

REFERENCES

1. U.S. EPA, *EPA Standards Network Factsheet; ISO 14000: International Environmental Management Standards,* Office of Research and Development, EPA/626/F-97/004, April 1998.
2. Introduction to ISO, WWW.ISO.CH/INFOE/INTRO.
3. W. M. Von Zharen, *ISO 14000 — Understanding Environmental Standards.* Government Institutes, Rockville, MD, 1996.
4. I. Fredericks and D. McCullum, *International Standards for Environmental Management Systems,* WWW.MGMT14K.COM/EMS.

46

ENVIRONMENTAL AUDITS

46.1 INTRODUCTION

Environmental auditing is fast becoming an integral component of a facility's management plan not only promoting compliance with regulatory requirements but also limiting environmental liabilities in the form of costly penalties and third party lawsuits. Corporations have come to realize the significant benefits resulting from conducting environmental audits. These benefits range from drastic reduction of fines from federal and state environmental protection agencies through implementation of their audit policies, to participation in the flow of lucrative "green" dollars through businesses that promote and reward other environmentally conscious entities. Consumers often seek out and patronize these businesses for their environmental policies.

Effective environmental auditing can lead to higher levels of overall compliance and reduced risk to human health and the environment. The U.S. Environmental Protection Agency (EPA) endorses the practice of environmental auditing and supports its accelerated use by regulated entities to help meet the goals of federal, state, and local environmental requirements. Auditing serves as a quality assurance check to help improve the effectiveness of basic environmental management by verifying that management practices are in place, functioning, and adequate.

Although there are numerous benefits one can derived from an environmental audit, penalties assessed from noncompliance with environmental laws and pollution liability in the form of remediation cost seem to be the two most convincing reasons for conducting environmental audits. In addition, federal and state agencies responsible for enforcing environmental laws offer strong incentives to facilities which voluntarily conduct environmental audits, self-disclose, and promptly correct violations. These include not seeking gravity-based civil penalties or reducing them by 75%, declining to recommend criminal prosecution for regulated entities that self-police, and refraining from routine request for audit reports from those entities.

In the biggest settlement reached under the EPA's audit policy, the EPA and GTE Corporation have come to an agreement that recognizes GTE's cooperation in disclosing and resolving 600 violations at 314 GTE facilities in 21 states. The violations occurred under the Emergency Planning and Community Right to Know Act (EPCRA) and the Spill Prevention Countermeasure and Control (SPCC) requirements of the Clean Water Act (CWA). The EPA has opened the door for responsible companies working to stay in compliance with environmental laws. The EPA's Audit Policy slashes, and in some cases eliminates penalties for those who voluntarily discover, promptly disclose, and expeditiously correct violations, offering powerful incentives for the regulated community to comply. As of March 1, 1998, 247 companies have disclosed environmental violations under the EPA's Audit Policy at more than 700 facilities nationally (1).

46.2   DEFINITION OF ENVIRONMENTAL AUDITING

Environmental auditing is a systematic, documented, periodic, and objective review by regulated entities of facility operations and practices related to meeting environmental requirements. Audits can be designed to accomplish any or all of the following: verify compliance with environmental requirements; evaluate the effectiveness of environmental management systems already in place; or assess risks from regulated and unregulated materials and practices.

Environmental audits evaluate, and are not a substitute for, direct compliance activities such as obtaining permits, installing controls, monitoring compliance, reporting violations, and keeping records. Environmental auditing may verify but does not include activities required by law, regulation, or permit (e.g., continuous emissions monitoring, composite correction plans at wastewater treatment plants, etc.). Audits do not in any way replace regulatory agency inspections. However, environmental audits can improve compliance by complementing conventional federal, state, and local oversight.

The EPA clearly supports auditing to help ensure the adequacy of internal systems to achieve, maintain, and monitor compliance. By voluntarily implementing environmental management and auditing programs, regulated entities can identify, resolve, and avoid environmental problems.

The EPA does not intend to dictate or interfere with the environmental management practices of private or public organizations. Nor does EPA intend to mandate auditing (though in certain instances EPA may seek to include provisions for environmental auditing as part of settlement agreements, as noted below). Because environmental auditing systems have been widely adopted on a voluntary basis in the past, and because audit quality depends to a large degree upon genuine management commitment to the program and its objectives, auditing should remain a voluntary activity.

An organization's auditing program will evolve according to its unique structures and circumstances. Effective environmental auditing programs appear to have certain discernible elements in common with other kinds of audit programs. These elements are important to ensure project effectiveness (2).

46.3   WHY CONDUCT AN ENVIRONMENTAL AUDIT?

Environmental auditing has developed for sound business reasons, particularly as a means of helping regulated entities manage pollution control affirmatively over time instead of reacting to crises. Auditing can result in improved facility environmental performance, help communicate and effect solutions to common environmental problems, focus facility managers' attention on current and upcoming regulatory requirements, and generate protocols and checklists which help facilities better manage themselves. Auditing also can result in better-integrated management of environmental hazards, since auditors frequently identify environmental liabilities which go beyond regulatory compliance.

One of the most compelling reasons to voluntarily conduct an environmental audit should be to avoid criminal prosecution. Because senior managers of regulated entities are ultimately responsible for taking all necessary steps to ensure compliance with environmental requirements. The EPA has never recommended criminal prosecution of a regulated entity based on voluntary disclosure of violations discovered through audits and disclosed to the government before an investigation was already under way. Thus, EPA will not recommend criminal prosecution for a regulated entity that uncovers violations through environmental audits or due diligence, promptly discloses and expeditiously corrects those violations, and meets all other conditions of Section D of the policy.

There are fundamentally two types of environmental audits:

1. *Compliance Audit:* An independent assessment of the current status of a party's compliance with applicable statutory and regulatory requirements. This approach always entails a requirement that effective measures be taken to remedy uncovered compliance problems, and is most effective when coupled with a requirement that the root causes of noncompliance also be remedied.

2. *Management Audit:* An independent evaluation of a party's environmental compliance policies, practices, and controls. Such evaluation may encompass the need for: (a) a formal corporate environmental compliance policy, and procedures for implementation of that policy; (b) educational and training programs for employees; (c) equipment purchase, operation, and maintenance programs; (d) environmental compliance officer programs (or other organizational structures relevant to compliance); (e) budgeting and planning systems for environmental compliance; (f) monitoring, record-keeping, and reporting systems; (g) in-plant and community emergency plans; (h) internal communications and control systems; and (i) hazard identification and risk assessment.

46.4 ELEMENTS OF AN EFFECTIVE AUDITING PROGRAM

An effective environmental auditing system will likely include the following general elements:

1. **Explicit top management support for environmental auditing and commitment to follow up on audit findings.** Management support may be demonstrated by a written policy articulating upper management support for the auditing program and for compliance with all pertinent requirements, including corporate policies and permit requirements as well as federal, state, and local statutes and regulations. Management support auditing program also should be demonstrated by an explicit written commitment to follow up on audit findings in order to correct identified problems and prevent their recurrence.

2. **An environmental auditing function independent of audited activities.** The status or organizational locus of environmental auditors should be sufficient to ensure objective and unobstructed inquiry, observation, and testing. Auditor objectivity should not be impaired by personal relationships, financial or other conflicts of interest, interference with free inquiry or judgment, or fear of potential retribution.

3. **Adequate team staffing and auditor training.** Environmental auditors should possess or have ready access to knowledge, skills, and disciplines needed to accomplish audit objectives. Each individual auditor should comply with the company's professional standards of conduct. Auditors, whether full-time or part-time, should maintain their technical and analytical competence through continuing education and training and certification.

4. **Explicit audit program objectives, scope, resources, and frequency.** At a minimum, audit objectives should include assessing compliance with applicable environmental laws and evaluating the adequacy of internal compliance policies, procedures, and personal training programs to ensure continued compliance.

Audits should be based on a process which provides auditors all corporate policies, permits, and federal, state, and local regulations pertinent to the facility; and checklists or protocols addressing specific features that should be evaluated by auditors.

Explicit written audit procedures generally should be used for planning audits, establishing audit scope, examining and evaluating audit findings, communicating audit results, and following up on findings.

5. **A process that collects, analyzes, interprets, and documents information sufficient to achieve audit objectives.** Information should be collected before and during an on-site visit regarding environmental compliance: (1), environmental management effectiveness (2), and other matters (3), related to audit objectives and scope. This information should be sufficient, reliable, relevant, and useful to provide a sound basis for audit finds and recommendations.

- Sufficient information is factual, adequate, and convincing so that a prudent, informed person would be likely to reach the same conclusions as the auditor.
- Reliable information is the best attainable through use of appropriate audit techniques.
- Relevant information supports audit findings and recommendations and is consistent with the objectives for the audit.
- Useful information helps the organization meet its goals.

The audit process should include a periodic review of the reliability and integrity of this information and the means used to identify, measure, classify, and report it. Audit procedures, including the testing and sampling techniques employed, should be selected in advance to the extent practical and expanded or altered if circumstances warrant. The process of collecting, analyzing, interpreting, and documenting information should provide reasonable assurance that audit objectivity is maintained and audit goals are met.

6. **A process that includes specific procedures to promptly prepare candid, clear, and appropriate written reports on audit findings corrective actions, and schedules for implementation.** Procedures should be in place to ensure such information is communicated to managers, including facility and corporate management, who can evaluate the information and ensure correction of identified problems. Procedures also should be in place for determining what internal findings are reportable to state or federal agencies.

7. **A process that includes quality assurance procedures to assure the accuracy and thoroughness of environmental audits.** Quality assurance may be accomplished through supervision, independent internal reviews, external reviews, or a combination of these approaches (2).

46.5 EPA'S AUDIT POLICY: INCENTIVES FOR SELF-POLICING

The EPA recognized that environmental auditing and sound environmental management generally can provide potentially powerful tools toward greater protection of public health and the environment. The EPA published the *Audit Policy: Incentives for Self-Policing: Discovery, Disclosure, Correction and Prevention of Violations* on December 22, 1995 (60 FR 66706), as excerpted below (3):

The Environmental Protection Agency (EPA) today issues its final policy to enhance protection of human health and the environment by encouraging regulated entities to voluntarily discover, and disclose and correct violations of environmental requirements. Incentives include eliminating or substantially reducing the gravity component of civil penalties and not recommending cases for criminal prosecution where specified conditions are met, to those who voluntarily self-disclose and promptly correct violations. The policy also restates EPA's long-standing practice of not requesting voluntary audit reports to trigger enforcement investigations. This policy was developed in close consultation with the U.S. Department of Justice, states, public interest groups and the regulated community, and will be applied uniformly by the Agency's enforcement programs.

This policy was effective 30 days after publication on December 22, 1995.

Additional documentation relating to the development of this policy is contained in the environmental auditing public docket. Documents from the docket may be obtained by calling (202) 260-7548, requesting an index to docket #C-94-01, and faxing document requests to (202) 260-4400.

I. Explanation of Policy

  A. Introduction

    The Environmental Protection Agency today issues its final policy to enhance protection of human health and the environment by encouraging regulated entities to discover voluntarily, disclose, correct and prevent violations of federal environmental law. Effective 30 days from today, where violations are found through voluntary environmental audits or efforts that reflect a regulated entity's due diligence, and are promptly disclosed and expeditiously corrected, EPA will not seek gravity-based (i.e., non-economic benefit) penalties and will generally not recommend criminal prosecution against the regulated entity. EPA will reduce gravity-based penalties by 75% for violations that are voluntarily discovered, and are promptly disclosed and corrected, even if not found through a formal audit or due diligence. Finally, the policy restates EPA's long-held policy and practice to refrain from routine requests for environmental audit reports.

    The policy includes important safeguards to deter irresponsible behavior and protect the public and environment. For example, in addition to prompt disclosure and expeditious correction, the policy requires companies to act to prevent recurrence of the violation and to remedy any environmental harm which may have occurred. Repeated violations or those which result in actual harm or may present imminent and substantial endangerment are not eligible for relief under this policy, and companies will not be allowed to gain an economic advantage over their competitors by delaying their investment in compliance. Corporations remain criminally liable for violations that result from conscious disregard of their obligations under the law, and individuals are liable for criminal misconduct.

The issuance of this policy concludes EPA's eighteen-month public evaluation of the optimum way to encourage voluntary self-policing while preserving fair and effective enforcement. The incentives, conditions and exceptions announced today reflect thoughtful suggestions from the Department of Justice, state attorneys general and local prosecutors, state environmental agencies, the regulated community, and public interest organizations. EPA believes that it has found a balanced and responsible approach, and will conduct a study within three years to determine the effectiveness of this policy.

B. Public Process

One of the Environmental Protection Agency's most important responsibilities is ensuring compliance with federal laws that protect public health and safeguard the environment. Effective deterrence requires inspecting, bringing penalty actions and securing compliance and remediation of harm. But EPA realizes that achieving compliance also requires the cooperation of thousands of businesses and other regulated entities subject to these requirements. Accordingly, in May of 1994, the Administrator asked the Office of Enforcement and Compliance Assurance (OCA) to determine whether additional incentives were needed to encourage voluntary disclosure and correction of violations uncovered during environmental audits.

EPA began its evaluation with a two-day public meeting in July of 1994, in Washington, D.C., followed by a two-day meeting in San Francisco on January 19, 1995 with stakeholders from industry, trade groups, state environmental commissioners and attorneys general, district attorneys, public interest organizations and professional environmental auditors. The Agency also established and maintained a public docket of testimony presented at these meetings and all comment and correspondence submitted to EPA by outside parties on this issue.

In addition to considering opinion and information from stakeholders, the Agency examined other federal and state policies related to self-policing, self-disclosure and correction. The Agency also considered relevant surveys on auditing practices in the private sector. EPA completed the first stage of this effort with the announcement of an interim policy on April 3 of this year, which defined conditions under which EPA would reduce civil penalties and not recommend criminal prosecution for companies that audited, disclosed, and corrected violations.

Interested parties were asked to submit comment on the interim policy by June 30 of this year (60 Fed.Reg 16875), and EPA received over 300 responses from a wide variety of private and public organizations. (Comments on the interim audit policy are contained in the Auditing Policy Docket, hereinafter, "Docket".) Further, the American Bar Association SONREEL Subcommittee hosted five days of dialogue with representatives from the regulated industry, states and public interest organizations in June and September of this year, which identified options for strengthening the interim policy. The changes to the interim policy announced today reflect insight gained through comments submitted to EPA, the ABA dialogue, and the Agency's practical experience implementing the interim policy.

C. Purpose

This policy is designed to encourage greater compliance with laws and regulations that protect human health and the environment. It promotes a higher standard of self-policing by waiving gravity-based penalties for violations that are promptly disclosed and corrected, and which were discovered through voluntary audits or compliance management systems that demonstrate due diligence. To further promote compliance, the policy reduces gravity-based penalties by 75% for any violation voluntarily discovered and promptly disclosed and corrected, even if not found through an audit or compliance management system.

EPA's enforcement program provides a strong incentive for responsible behavior by imposing stiff sanctions for noncompliance. Enforcement has contributed to the dramatic expansion of environmental auditing measured in numerous recent surveys. For example, more than 90% of the corporate respondents to a 1995 Price-Waterhouse survey who conduct audits said that one of the reasons they did so was to find and correct violations before they were found by government inspectors. (A copy of the Price-Waterhouse survey is contained in the Docket as document VIII-A-76.)

At the same time, because government resources are limited, maximum compliance cannot be achieved without active efforts by the regulated community to police themselves. More than half of the respondents to the same 1995 Price-Waterhouse survey said that they would expand environmental auditing in exchange for reduced penalties for violations discovered and corrected. While many companies already audit or have compliance management programs, EPA believes that the incentives offered in this policy will improve the frequency and quality of these self-monitoring efforts.

D. Incentives for Self-policing

Section C of EPA's policy identifies the major incentives that EPA will provide to encourage self-policing, self-disclosure, and prompt self-correction. These include not seeking gravity-based civil penalties or reducing them by 75%, declining to recommend criminal prosecution for regulated entities that self-police, and refraining from routine requests for audits. (As noted in Section C of the policy, EPA has refrained from making routine requests for audit reports since issuance of its 1986 policy on environmental auditing.)

D.1. Eliminating Gravity-Based Penalties

Under Section C(1) of the policy, EPA will not seek gravity-based penalties for violations found through auditing that are promptly disclosed and corrected. Gravity-based penalties will also be waived for violations found through any documented procedure for self-policing, where the company can show that it has a compliance management program that meets the criteria for due diligence in Section B of the policy.

Gravity-based penalties (defined in Section B of the policy) generally reflect the seriousness of the violator's behavior. EPA has elected to waive such penalties for violations discovered through due diligence or environmental audits, recognizing that these voluntary efforts play a critical role in protecting human health and the environment by identifying, correcting and ultimately preventing violations. All of the conditions set forth in Section D, which include prompt disclosure and expeditious correction, must be satisfied for gravity-based penalties to be waived.

As in the interim policy, EPA reserves the right to collect any economic benefit that may have been realized as a result of noncompliance, even where companies meet all other conditions of the policy. Economic benefit may be waived, however, where the Agency determines that it is insignificant.

After considering public comment, EPA has decided to retain the discretion to recover economic benefit for two reasons. First, it provides an incentive to comply on time. Taxpayers expect to pay interest or a penalty fee if their tax payments are late; the same principle should apply to corporations that have delayed their investment in compliance. Second, it is fair because it protects responsible companies from being undercut by their noncomplying competitors, thereby preserving a level playing field. The concept of recovering economic benefit was supported in public comments by many stakeholders, including industry representatives (see, e.g., Docket, II-F-39, II-F-28, and II-F-18).

D.2. 75% Reduction of Gravity

The policy appropriately limits the complete waiver of gravity-based civil penalties to companies that meet the higher standard of environmental auditing or systematic compliance management. However, to provide additional encouragement for the kind of self-policing that benefits the public, gravity-based penalties will be reduced by 75% for a violation that is voluntarily discovered, promptly disclosed and expeditiously corrected, even if it was not found through an environmental audit and the company cannot document due diligence. EPA expects that this will encourage companies to come forward and work with the Agency to resolve environmental problems and begin to develop an effective compliance management program.

Gravity-based penalties will be reduced 75% only where the company meets all conditions in Sections D(2) through D(9). EPA has eliminated language from the interim policy indicating that penalties may be reduced "up to" 75% where "most" conditions are met, because the Agency believes that all of the conditions in D(2) through D(9) are reasonable and essential to achieving compliance. This change also responds to requests for greater clarity and predictability.

D.3. No Recommendations for Criminal Prosecution

EPA has never recommended criminal prosecution of a regulated entity based on voluntary disclosure of violations discovered through audits and disclosed to the government before an investigation was already under way. Thus, EPA will not recommend criminal prosecution for a regulated entity that uncovers violations through environmental audits or due diligence, promptly discloses and expeditiously corrects those violations, and meets all other conditions of Section D of the policy.

This policy is limited to good actors, and therefore has important limitations. It will not apply, for example, where corporate officials are consciously involved in or willfully blind to violations, or conceal or condone noncompliance. Since the regulated entity must satisfy all of the conditions of Section D of the policy, violations that caused serious harm or which may pose imminent and substantial endangerment to human health or the environment are not covered by this policy. Finally, EPA reserves the right to recommend prosecution for the criminal conduct of any culpable individual.

Even where all of the conditions of this policy are not met, however, it is important to remember that EPA may decline to recommend prosecution of a company or individual for many other reasons under other Agency enforcement policies. For example, the Agency may decline to recommend prosecution where there is no significant harm or culpability and the individual or corporate defendant has cooperated fully.

Where a company has met the conditions for avoiding a recommendation for criminal prosecution under this policy, it will not face any civil liability for gravity-based penalties. That is because the same conditions for discovery, disclosure, and correction apply in both cases. This represents a clarification of the interim policy, not a substantive change.

D.4. *No Routine Requests for Audits*

EPA is reaffirming its policy, in effect since 1986, to refrain from routine requests for audits. Eighteen months of public testimony and debate have produced no evidence that the Agency has deviated, or should deviate, from this policy.

If the Agency has independent evidence of a violation, it may seek information needed to establish the extent and nature of the problem and the degree of culpability. In general, however, an audit which results in prompt correction clearly will reduce liability, not expand it. Furthermore, a review of the criminal docket did not reveal a single criminal prosecution for violations discovered as a result of an audit self-disclosed to the government.

E. Conditions

Section D describes the nine conditions that a regulated entity must meet in order for the Agency not to seek (or to reduce) gravity-based penalties under the policy. As explained in the Summary above, regulated entities that meet all nine conditions will not face gravity-based civil penalties, and will generally not have to fear criminal prosecution. Where the regulated entity meets all of the conditions except the first (D(1)), EPA will reduce gravity-based penalties by 75%.

E.1. *Discovery of the Violation Through an Environmental Audit or Due Diligence*

Under Section D(1), the violation must have been discovered through either a) an environmental audit that is systematic, objective, and periodic as defined in the 1986 audit policy, or b) a documented, systematic procedure or practice which reflects the regulated entity's due diligence in preventing, detecting, and correcting violations. The interim policy provided full credit for any violation found through "voluntary self-evaluation," even if the evaluation did not constitute an audit. In order to receive full credit under the final policy, any self-evaluation that is not an audit must be part of a "due diligence" program. Both "environmental audit" and "due diligence" are defined in Section B of the policy.

Where the violation is discovered through a "systematic procedure or practice" which is not an audit, the regulated entity will be asked to document how its program reflects the criteria for due diligence as defined in Section B of the policy. These criteria, which are adapted from existing codes of practice such as the 1991 Criminal Sentencing Guidelines, were fully discussed during the ABA dialogue. The criteria are flexible enough to accommodate different types and sizes of businesses. The Agency recognizes that a variety of compliance management programs may develop under the due diligence criteria, and will use its review under this policy to determine whether basic criteria have been met.

Compliance management programs which train and motivate production staff to prevent, detect and correct violations on a daily basis are a valuable complement to periodic auditing. The policy is responsive to recommendations received during public comment and from the ABA dialogue to give compliance management efforts which meet the criteria for due diligence the same penalty reduction offered for environmental audits. (See, e.g., II-F-39, II-E-18, and II-G-18 in the Docket.)

EPA may require as a condition of penalty mitigation that a description of the regulated entity's due diligence efforts be made publicly available. The Agency added this provision in response to suggestions from environmental groups, and believes that the availability of such information will allow the public to judge the adequacy of compliance management systems, lead to enhanced compliance, and foster greater public trust in the integrity of compliance management systems.

E.2. Voluntary Discovery and Prompt Disclosure

Under Section D(2) of the final policy, the violation must have been identified voluntarily, and not through a monitoring, sampling, or auditing procedure that is required by statute, regulation, permit, judicial or administrative order, or consent agreement. Section D(4) requires that disclosure of the violation be prompt and in writing. To avoid confusion and respond to state requests for greater clarity, disclosures under this policy should be made to EPA. The Agency will work closely with states in implementing the policy.

The requirement that discovery of the violation be voluntary is consistent with proposed federal and state bills which would reward those discoveries that the regulated entity can legitimately attribute to its own voluntary efforts.

The policy gives three specific examples of discovery that would not be voluntary, and therefore would not be eligible for penalty mitigation: emissions violations detected through a required continuous emissions monitor, violations of NPDES discharge limits found through prescribed monitoring, and violations discovered through a compliance audit required to be performed by the terms of a consent order or settlement agreement.

The final policy generally applies to any violation that is voluntarily discovered, regardless of whether the violation is required to be reported. This definition responds to comments pointing out that reporting requirements are extensive, and that excluding them from the policy's scope would severely limit the incentive for self-policing (see, e.g., II-C-48 in the Docket).

The Agency wishes to emphasize that the integrity of federal environmental law depends upon timely and accurate reporting. The public relies on timely and accurate reports from the regulated community, not only to measure compliance but to evaluate health or environmental risk and gauge progress in reducing pollutant loadings. EPA expects the policy to encourage the kind of vigorous self-policing that will serve these objectives, and not to provide an excuse for delayed reporting. Where violations of reporting requirements are voluntarily discovered, they must be promptly reported (as discussed below). Where a failure to report results in imminent and substantial endangerment or serious harm, that violation is not covered under this policy (see Condition D(8)). The policy also requires the regulated entity to prevent recurrence of the violation, to ensure that noncompliance with reporting requirements is not repeated. EPA will closely scrutinize the effect of the policy in furthering the public interest in timely and accurate reports from the regulated community.

Under Section D(4), disclosure of the violation should be made within 10 days of its discovery, and in writing to EPA. Where a statute or regulation requires reporting be made in less than 10 days, disclosure should be made within the time limit established by law. Where reporting within ten days is not practical because the violation is complex and compliance cannot be determined within that period, the Agency may accept later disclosures if the circumstances do not present a serious threat and the regulated entity meets its burden of showing that the additional time was needed to determine compliance status.

This condition recognizes that it is critical for EPA to get timely reporting of violations in order that it might have clear notice of the violations and the opportunity to respond if necessary, as well as an accurate picture of a given facility's compliance record. Prompt disclosure is also evidence of the regulated entity's good faith in wanting to achieve or return to compliance as soon as possible.

In the final policy, the Agency has added the words, "or may have occurred," to the sentence, "The regulated entity fully discloses that a specific violation has occurred, or may have occurred ..." This change, which was made in response to comments received, clarifies that where an entity has some doubt about the existence of a violation, the recommended course is for it to disclose and allow the regulatory authorities to make a definitive determination.

In general, the Freedom of Information Act will govern the Agency's release of disclosures made pursuant to this policy. EPA will, independently of FOIA, make publicly available any compliance agreements reached under the policy (see Section H of the policy), as well as descriptions of due diligence programs submitted under Section D.1 of the Policy. Any material claimed to be Confidential Business Information will be treated in accordance with EPA regulations at 40 C.F.R. Part 2.

E.3. Discovery and Disclosure Independent of Government or Third Party Plaintiff

Under Section D(3), in order to be "voluntary," the violation must be identified and disclosed by the regulated entity prior to: the commencement of a federal state or local agency inspection, investigation, or information request; notice of a citizen suit; legal complaint by a third party; the reporting of the violation to EPA by a "whistle blower" employee; and imminent discovery of the violation by a regulatory agency.

This condition means that regulated entities must have taken the initiative to find violations and promptly report them, rather than reacting to knowledge of a pending enforcement action or third-party complaint. This concept was reflected in the interim policy and in federal and state penalty immunity laws and did not prove controversial in the public comment process.

E.4. Correction and Remediation

Section D(5) ensures that, in order to receive the penalty mitigation benefits available under the policy, the regulated entity not only voluntarily discovers and promptly discloses a violation, but expeditiously corrects it, remedies any harm caused by that violation (including responding to any spill and carrying out any removal or remedial action required by law), and expeditiously certifies in writing to appropriate state, local and EPA authorities that violations have been corrected. It also enables EPA to ensure that the regulated entity will be publicly accountable for its commitments through binding written agreements, orders or consent decrees where necessary.

The final policy requires the violation to be corrected within 60 days, or that the regulated entity provide written notice where violations may take longer to correct. EPA recognizes that some violations can and should be corrected immediately, while others (e.g., where capital expenditures are involved), may take longer than 60 days to correct. In all cases, the regulated entity will be expected to do its utmost to achieve or return to compliance as expeditiously as possible.

Where correction of the violation depends upon issuance of a permit which has been applied for but not issued by federal or state authorities, the Agency will, where appropriate, make reasonable efforts to secure timely review of the permit.

E.5. Prevent Recurrence

Under Section D(6), the regulated entity must agree to take steps to prevent a recurrence of the violation, including but not limited to improvements to its environmental auditing or due diligence efforts. The final policy makes clear that the preventive steps may include improvements to a regulated entity's environmental auditing or due diligence efforts to prevent recurrence of the violation.

In the interim policy, the Agency required that the entity implement appropriate measures to prevent a recurrence of the violation, a requirement that operates prospectively. However, a separate condition in the interim policy also required that the violation not indicate "a failure to take appropriate steps to avoid repeat or recurring violations"—a requirement that operates retrospectively. In the interest of both clarity and fairness, the Agency has decided for purposes of this condition to keep the focus prospective and thus to require only that steps be taken to prevent recurrence of the violation after it has been disclosed.

E.6. No Repeat Violations

In response to requests from commenters (see, e.g., II-F-39 and II-G-18 in the Docket), EPA has established "bright lines" to determine when previous violations will bar a regulated entity from obtaining relief under this policy. These will help protect the public and responsible companies by ensuring that penalties are not waived for repeat offenders. Under condition D(7), the same or closely-related violation must not have occurred previously within the past three years at the same facility, or be part of a pattern of violations on the regulated entity's part over the past five years. This provides companies with a continuing incentive to prevent violations, without being unfair to regulated entities responsible for managing hundreds of facilities. It would be unreasonable to provide unlimited amnesty for repeated violations of the same requirement.

The term "violation" includes any violation subject to a federal or state civil judicial or administrative order, consent agreement, conviction or plea agreement. Recognizing that minor violations are sometimes settled without a formal action in court, the term also covers any act or omission for which the regulated entity has received a penalty reduction in the past. Together, these conditions identify situations in which the regulated community has had clear notice of its noncompliance and an opportunity to correct.

E.7. Other Violations Excluded

Section D(8) makes clear that penalty reductions are not available under this policy for violations that resulted in serious actual harm or which may have presented an imminent and substantial endangerment to public health or the environment. Such events indicate a serious failure (or absence) of a self-policing program, which should be designed to prevent such risks, and it would seriously undermine deterrence to waive penalties for such violations. These exceptions are responsive to suggestions from public interest organizations, as well as other commenters. (See, e.g., II-F-39 and II-G-18 in the Docket.)

The final policy also excludes penalty reductions for violations of the specific terms of any order, consent agreement, or plea agreement. (See, II-E-60 in the Docket.) Once a consent agreement has been negotiated, there is little incentive to comply if there are no sanctions for violating its specific requirements. The exclusion in this section applies to violations of the terms of any response, removal or remedial action covered by a written agreement.

E.8. Cooperation

Under Section D(9), the regulated entity must cooperate as required by EPA and provide information necessary to determine the applicability of the policy. This condition is largely unchanged from the interim policy. In the final policy, however, the Agency has added that "cooperation" includes assistance in determining the facts of any related violations suggested by the disclosure, as well as of the disclosed violation itself. This was added to allow the agency to obtain information about any violations indicated by the disclosure, even where the violation is not initially identified by the regulated entity.

F. Opposition to Privilege

The Agency remains firmly opposed to the establishment of a statutory evidentiary privilege for environmental audits for the following reasons:

F.1. Privilege, by definition, invites secrecy, instead of the openness needed to build public trust in industry's ability to self-police. American law reflects the high value that the public places on fair access to the facts. The Supreme Court, for example, has said of privileges that, "[w]hatever their origins, these exceptions to the demand for every man's evidence are not lightly created nor expansively construed, for they are in derogation of the search for truth." United States v. Nixon, 418 U.S. 683 (1974). Federal courts have unanimously refused to recognize a privilege for environmental audits in the context of government investigations. See, e.g., United States v. Dexter, 132 F.R.D. 8, 9-10 (D.Conn. 1990) (application of a privilege "would effectively impede [EPA's] ability to enforce the Clean Water Act, and would be contrary to stated public policy.")

F.2. Eighteen months have failed to produce any evidence that a privilege is needed. Public testimony on the interim policy confirmed that EPA rarely uses audit reports as evidence. Furthermore, surveys demonstrate that environmental auditing has expanded rapidly over the past decade without the stimulus of a privilege. Most recently, the 1995 Price

Waterhouse survey found that those few large or mid-sized companies that do not audit generally do not perceive any need to; concern about confidentiality ranked as one of the least important factors in their decisions.

F.3. A privilege would invite defendants to claim as "audit" material almost any evidence the government needed to establish a violation or determine who was responsible. For example, most audit privilege bills under consideration in federal and state legislatures would arguably protect factual information — such as health studies or contaminated sediment data — and not just the conclusions of the auditors. While the government might have access to required monitoring data under the law, as some industry commenters have suggested, a privilege of that nature would cloak underlying facts needed to determine whether such data were accurate.

F.4. An audit privilege would breed litigation, as both parties struggled to determine what material fell within its scope. The problem is compounded by the lack of any clear national standard for audits. The "in camera" (i.e., non-public) proceedings used to resolve these disputes under some statutory schemes would result in a series of time-consuming, expensive mini-trials.

F.5. The Agency's policy eliminates the need for any privilege as against the government, by reducing civil penalties and criminal liability for those companies that audit, disclose and correct violations. The 1995 Price Waterhouse survey indicated that companies would expand their auditing programs in exchange for the kind of incentives that EPA provides in its policy.

F.6. Finally, audit privileges are strongly opposed by the law enforcement community, including the National District Attorneys Association, as well as by public interest groups. See, e.g., Docket, II-C-21, II-C-28, II-C-52, IV-G-10, II-C-25, II-C-33, II-C-52, II-C-48, and II-G-13 through II-G-24.)

G. Effect on States

The final policy reflects EPA's desire to develop fair and effective incentives for self-policing that will have practical value to states that share responsibility for enforcing federal environmental laws. To that end, the Agency has consulted closely with state officials in developing this policy, through a series of special meetings and conference calls in addition to the extensive opportunity for public comment. As a result, EPA believes its final policy is grounded in common-sense principles that should prove useful in the development of state programs and policies.

As always, states are encouraged to experiment with different approaches that do not jeopardize the fundamental national interest in assuring that violations of federal law do not threaten the public health or the environment, or make it profitable not to comply. The Agency remains opposed to state legislation that does not include these basic protections, and reserves its right to bring independent action against regulated entities for violations of federal law that threaten human health or the environment, reflect criminal conduct or repeated noncompliance, or allow one company to make a substantial profit at the expense of its law-abiding competitors. Where a state has obtained appropriate sanctions needed to deter such misconduct, there is no need for EPA action.

H. Scope of Policy

EPA has developed this document as a policy to guide settlement actions. EPA employees will be expected to follow this policy, and the Agency will take steps to assure national consistency in application. For example, the Agency will make public any compliance agreements reached under this policy, in order to provide the regulated community with fair notice of decisions and greater accountability to affected communities. Many in the regulated community recommended that the Agency convert the policy into a regulation because they felt it might ensure greater consistency and predictability. While EPA is taking steps to ensure consistency and predictability and believes that it will be successful; the Agency will consider this issue and will provide notice if it determines that a rulemaking is appropriate.

Statement of Policy: Incentives for Self-Policing

Discovery, Disclosure, Correction and Prevention

A. Purpose

This policy is designed to enhance protection of human health and the environment by encouraging regulated entities to voluntarily discover, disclose, correct and prevent violations of federal environmental requirements.

B. Definitions

For purposes of this policy, the following definitions apply:

"Environmental Audit" has the definition given to it in EPA's 1986 audit policy on environmental auditing, i.e., "a systematic, documented, periodic and objective review by regulated entities of facility operations and practices related to meeting environmental requirements."

"Due Diligence" encompasses the regulated entity's systematic efforts, appropriate to the size and nature of its business, to prevent, detect and correct violations through all of the following:

> a) Compliance policies, standards and procedures that identify how employees and agents are to meet the requirements of laws, regulations, permits and other sources of authority for environmental requirements;
> b) Assignment of overall responsibility for overseeing compliance with policies, standards, and procedures, and assignment of specific responsibility for assuring compliance at each facility or operation;
> c) Mechanisms for systematically assuring that compliance policies, standards and procedures are being carried out, including monitoring and auditing systems reasonably designed to detect and correct violations, periodic evaluation of the overall performance of the compliance management system, and a means for employees or agents to report violations of environmental requirements without fear of retaliation;
> d) Efforts to communicate effectively the regulated entity's standards and procedures to all employees and other agents;

e) Appropriate incentives to managers and employees to perform in accordance with the compliance policies, standards and procedures, including consistent enforcement through appropriate disciplinary mechanisms; and

f) Procedures for the prompt and appropriate correction of any violations, and any necessary modifications to the regulated entity's program to prevent future violations.

"Environmental audit report" means the analysis, conclusions, and recommendations resulting from an environmental audit, but does not include data obtained in, or testimonial evidence concerning, the environmental audit.

"Gravity-based penalties" are that portion of a penalty over and above the economic benefit., i.e., the punitive portion of the penalty, rather than that portion representing a defendant's economic gain from non-compliance. (For further discussion of this concept, see "A Framework for Statute-Specific Approaches to Penalty Assessments", #GM-22, 1980, U.S. EPA General Enforcement Policy Compendium).

"Regulated entity" means any entity, including a federal, state or municipal agency or facility, regulated under federal environmental laws.

C. Incentives for Self-Policing

C.1. No Gravity-Based Penalties: Where the regulated entity establishes that it satisfies all of the conditions of Section D of the policy, EPA will not seek gravity-based penalties for violations of federal environmental requirements.

C.2. Reduction of Gravity-Based Penalties by 75%: EPA will reduce gravity-based penalties for violations of federal environmental requirements by 75% so long as the regulated entity satisfies all of the conditions of Section D(2) through D(9) below.

C.3. No Criminal Recommendations:

(a) EPA will not recommend to the Department of Justice or other prosecuting authority that criminal charges be brought against a regulated entity where EPA determines that all of the conditions in Section D are satisfied, so long as the violation does not demonstrate or involve:

i) a prevalent management philosophy or practice that concealed or condoned environmental violations; or
ii) high-level corporate officials' or managers' conscious involvement in, or willful blindness to, the violations.

(b) Whether or not EPA refers the regulated entity for criminal prosecution under this section, the Agency reserves the right to recommend prosecution for the criminal acts of individual managers or employees under existing policies guiding the exercise of enforcement discretion.

C.4. No Routine Request for Audits: EPA will not request or use an environmental audit report to initiate a civil or criminal investigation of

the entity. For example, EPA will not request an environmental audit report in routine inspections. If the Agency has independent
reason to believe that a violation has occurred, however, EPA may seek any information relevant to identifying violations or determining liability or extent of harm.

D. Conditions

D.1. Systematic Discovery: The violation was discovered through:
(a) an environmental audit; or
(b) an objective, documented, systematic procedure or practice reflecting the regulated entity's due diligence in preventing, detecting, and correcting violations. The regulated entity must provide accurate and complete documentation to the Agency as to how it exercises due diligence to prevent, detect and correct violations according to the criteria for due diligence outlined in Section B. EPA may require as a condition of penalty mitigation that a description of the regulated entity's due diligence efforts be made publicly available.

D.2. Voluntary Discovery: The violation was identified voluntarily, and not through a legally mandated monitoring or sampling requirement prescribed by statute, regulation, permit, judicial or administrative order, or consent agreement. For example, the policy does not apply to:

(a) emissions violations detected through a continuous emissions monitor (or alternative monitor established in a permit) where any such monitoring is required;
(b) violations of National Pollutant Discharge Elimination System (NPDES) discharge limits detected through required sampling or monitoring;
(c) violations discovered through a compliance audit required to be performed by the terms of a consent order or settlement agreement.

D.3. Prompt Disclosure: The regulated entity fully discloses a specific violation within 10 days (or such shorter period provided by law) after it has discovered that the violation has occurred, or may have occurred, in writing to EPA;

D.4. Discovery and Disclosure Independent of Government or Third Party Plaintiff: The violation must also be identified and disclosed by the regulated entity prior to:

(a) the commencement of a federal, state or local agency inspection or investigation, or the issuance by such agency of an information request to the regulated entity;
(b) notice of a citizen suit;
(c) the filing of a complaint by a third party;
(d) the reporting of the violation to EPA (or other government agency) by a "whistle blower" employee, rather than by one authorized to speak on behalf of the regulated entity; or
(e) imminent discovery of the violation by a regulatory agency;

D.5. Correction and Remediation: The regulated entity corrects the violation within 60 days, certifies in writing that violations have been corrected, and takes appropriate measures as determined by EPA to remedy any environmental or human harm due to the violation. If more than 60 days will be needed to correct the violation(s), the regulated entity must so notify EPA in writing before the 60-day period has passed. Where appropriate, EPA may require that to satisfy conditions 5 and 6, a regulated entity enter into a publicly available written agreement, administrative consent order or judicial consent decree, particularly where compliance or remedial measures are complex or a lengthy schedule for attaining and maintaining compliance or remediating harm is required;

D.6. Prevent Recurrence: The regulated entity agrees in writing to take steps to prevent a recurrence of the violation, which may include improvements to its environmental auditing or due diligence efforts;

D.7. No Repeat Violations: The specific violation (or closely related violation) has not occurred previously within the past three years at the same facility, or is not part of a pattern of federal, state or local violations by the facility's parent organization (if any), which have occurred within the past five years. For the purposes of this section, a violation is:

(a) any violation of federal, state or local environmental law identified in a judicial or administrative order, consent agreement or order, complaint, or notice of violation, conviction or plea agreement; or
(b) any act or omission for which the regulated entity has previously received penalty mitigation from EPA or a state or local agency.

D.8. Other Violations Excluded: The violation is not one which (i)resulted in serious actual harm, or may have presented an imminent and substantial endangerment to, human health or the environment, or (ii) violates the specific terms of any judicial or administrative order, or consent agreement.

D.9. Cooperation: The regulated entity cooperates as requested by EPA and provides such information as is necessary and requested by EPA to determine applicability of this policy. Cooperation includes, at a minimum, providing all requested documents and access to employees and assistance in investigating the violation, any noncompliance problems related to the disclosure, and any environmental consequences related to the violations.

E. Economic Benefit

EPA will retain its full discretion to recover any economic benefit gained as a result of noncompliance to preserve a "level playing field" in which violators do not gain a competitive advantage over regulated entities that do comply. EPA may forgive the entire penalty for violations which meet conditions 1 through 9 in section D and, in the Agency's opinion, do not merit any penalty due to the insignificant amount of any economic benefit.

F. Effect on State Law, Regulation or Policy

EPA will work closely with states to encourage their adoption of policies that reflect the incentives and conditions outlined in this policy. EPA remains firmly opposed to statutory environmental audit privileges that shield evidence of environmental violations and undermine the public's right to know, as well as to blanket immunities for violations that reflect criminal conduct, present serious threats or actual harm to health and the environment, allow noncomplying companies to gain an economic advantage over their competitors, or reflect a repeated failure to comply with federal law. EPA will work with states to address any provisions of state audit privilege or immunity laws that are inconsistent with this policy, and which may prevent a timely and appropriate response to significant environmental violations. The Agency reserves its right to take necessary actions to protect public health or the environment by enforcing against any violations of federal law.

G. Applicability

G.1. This policy applies to the assessment of penalties for any violations under all of the federal environmental statutes that EPA administers, and supersedes any inconsistent provisions in media-specific penalty or enforcement policies and EPA's 1986 Environmental Auditing Policy Statement.

G.2. To the extent that existing EPA enforcement policies are not inconsistent, they will continue to apply in conjunction with this policy. However, a regulated entity that has received penalty mitigation for satisfying specific conditions under this policy may not receive additional penalty mitigation for satisfying the same or similar conditions under other policies for the same violation(s), nor will this policy apply to violations which have received penalty mitigation under other policies.

G.3. This policy sets forth factors for consideration that will guide the Agency in the exercise of its prosecutorial discretion. It states the Agency's views as to the proper allocation of its enforcement resources. The policy is not final agency action, and is intended as guidance. It does not create any rights, duties, obligations, or defenses, implied or otherwise, in any third parties.

G.4. This policy should be used whenever applicable in settlement negotiations for both administrative and civil judicial enforcement actions. It is not intended for use in pleading, at hearing or at trial. The policy may be applied at EPA's discretion to the settlement of administrative and judicial enforcement actions instituted prior to, but not yet resolved, as of the effective date of this policy.

H. Public Accountability

H.1. Within 3 years of the effective date of this policy, EPA will complete a study of the effectiveness of the policy in encouraging:

(a) changes in compliance behavior within the regulated community, including improved compliance rates;

(b) prompt disclosure and correction of violations, including timely and accurate compliance with reporting requirements;

(c) corporate compliance programs that are successful in preventing violations, improving environmental performance, and promoting public disclosure;

(d) consistency among state programs that provide incentives for voluntary compliance.

EPA will make the study available to the public.

H.2. EPA will make publicly available the terms and conditions of any compliance agreement reached under this policy, including the nature of the violation, the remedy, and the schedule for returning to compliance.

46.6   SUMMARY

1. Environmental auditing is fast becoming an integral component of a facility's management plan, not only promoting compliance with regulatory requirements but also limiting environmental liabilities in the form of costly penalties and third-party lawsuits. Corporations have come to realize the significant benefits resulting from conducting environmental audits.

2. Environmental auditing is a systematic, documented, periodic, and objective review by regulated entities of facility operations and practices related to meeting environmental requirements.

3. Environmental auditing has developed for sound business reasons, particularly as a means of helping regulated entities manage pollution control affirmatively over time instead of reacting to crises.

4. An effective environmental auditing system will likely include explicit top-management support for environmental auditing and commitment to follow-up on audit findings; an environmental auditing function independent of audited activities; explicit audit program objectives, scope, resources, and frequency; a process that collects, analyzes, interprets, and documents information sufficient to achieve audit objectives; and a process that includes quality assurance procedures to assure the accuracy and thoroughness of environmental audits.

5. The EPA recognized that environmental auditing and sound environmental management generally can provide potentially powerful tools for greater protection of public health and the environment. The EPA published the *Audit Policy: Incentives for Self-Policing: Discovery, Disclosure, Correction, and Prevention of Violations.*

REFERENCES

1. U.S. EPA, *Audit Policy Update*, Office of Enforcement and Compliance Assurance, EPA 300-N-98-003, March 1998.

2. U.S. EPA, *Restatement of Policies Related to Environmental Auditing*, FRL-5021-5, July 28, 1994.

3. U.S. EPA, *Audit Policy: Incentives for Self-Policing*, 60 FR 66706, December 22, 1995.

47

ENVIRONMENTAL ETHICS*

Contributing Author: Ruth Richardson

47.1 INTRODUCTION

In 1854, President Franklin Pierce petitioned Chief Seattle — the leader of the Coastal Salish Indians of the Pacific Northwest — to sell his tribe's land to the United States. In his response to President Pierce and the white Europeans immigrants' pursuit to own and "subdue" the earth, Chief Seattle penned thoughts as environmentally pensive and poignant as any uttered in the more than 140 years since: "Continue to contaminate your bed and you will one day lay in your own waste" (1).

His message fell on the deaf ears of the U.S. government and public. Cries for respect for the earth such as his remained few and far between for the next century. In the wake of events such as the Industrial Revolution, the First and Second World Wars, and the Cold War, a concern for the environment played little if any part in influencing either public policy or private endeavors.

More than a hundred and forty years later, however, Chief Seattle's words echo in every Superfund site, landfill, and oil spill. Public opinion has swung to the green side and a new ethic has evolved: an environmental ethic. As can readily be seen, however, the recent movement toward environmentalism has not created new moral codes. Instead, it has changed the emphasis and expanded the concept of the "common good" that lies at the heart of determining if an action is ethical.

This chapter will first present the variety of moral theories and philosophies that have governed ethics historically. The progressive movement of environmentalism into an influential ethical force is then sketched. Once these historical developments have been presented, today's dilemma of coordinating technology with environmental responsibility will be explored. Finally, the future trends evidenced by present and past activities will be discussed.

*Drawn in part (with permission) from *Major Environmental Issues Facing the 21st Century*, M. K. Theodore and L. Theodore, Chapter 45, "Environmental Ethics," Prentice Hall, Upper Saddle River, NJ, 1996.

47.2 MORAL ISSUES

The conflict of interest between Chief Seattle (and Native Americans in general) and President Pierce (and the European-American expansion) provides a perfect example of how ethics and the resulting codes of behavior they engender can differ drastically from culture to culture, from religion to religion, and even from person to person. This enigma, too, is noted again and again by Seattle (2):

> I do not know. Our ways are different from your ways.... But perhaps it is because the red man is a savage and does not understand.... The air is precious to the red man, for all things share the same breath... the white man does not seem to notice the air he breathes.... I am a savage and do not understand any other way. I have seen a thousand rotting buffaloes on the prairie, left by the white man who shot them from a passing train. I am a savage and I do not understand how the smoking iron horse can be more important than the buffalo we kill only to stay alive.

Chief Seattle sarcastically uses the European word "savage" and all its connotations throughout his address. When one finishes reading the work it becomes obvious which viewpoint (President Pierce's or his own) Chief Seattle feels is the savage one. What his culture holds dearest (the wilderness) the whites see as untamed, dangerous, and savage. What the whites hold in highest regard (utilization of the earth and technological advancement) the Native Americans see as irreverent of all other living things. Each culture maintains a distinct and conflicting standard for the welfare of the world. Opposing viewpoints and moralities such as these are prevalent throughout the world and have never ceased to present a challenge to international, national, state, community, and interpersonal peace.

It is generally accepted, however, that any historical ethic can be found to focus on one of four different underlying moral concepts:

1. *Utilitarianism* focuses on good consequences for all.
2. *Duties Ethics* focus on one's duties.
3. *Rights Ethics* focus on human rights.
4. *Virtue Ethics* focus on virtuous behavior.

Utilitarians hold that the most basic reason why actions are morally right is that they lead to the greatest good for the greatest number. "Good and bad consequences are the only relevant considerations, and, hence all moral principles reduce to one: 'We ought to maximize utility'" (2).

Duties Ethicists concentrate on an action itself rather than the consequences of that action. To these ethicists there are certain principles of duty such as "Do not deceive" and "Protect innocent life" that should be fulfilled even if the most good does not result. The list and hierarchy of duties differs from culture to culture, religion to religion. For Judeo-Christians, the Ten Commandments provide an ordered list of duties imposed by their religion (2).

Often considered to be linked with Duties Ethics, Rights Ethics also assesses the act itself rather than its consequences. Rights Ethicists emphasize the rights of the people affected by an act rather than the duty of the person(s) performing the act. For example, because a person has a *right* to life, murder is morally wrong. Rights Ethicists propose that duties actually stem from a corresponding right. Since each person has a *right* to life, it is everyone's *duty* to not kill. It is because of this link and their common emphasis on the actions themselves that Rights Ethics and Duty Ethics are often grouped under the common heading Deontological Ethics (3).

The display of virtuous behavior is the central principle governing Virtue Ethics. An action would be wrong if it expressed or developed vices — for example, bad character traits. *Virtue Ethicists*, therefore, focus upon becoming a morally good person.

To display the different ways that these moral theories view the same situation one can ex-plore their approach to the following scenario that Martin and Schinzinger present (2):

> On a midnight shift, a botched solution of sodium cyanide, a reactant in organic synthesis, is temporarily stored in drums for reprocessing. Two weeks later, the day shift foreperson cannot find the drums. Roy, the plant manager, finds out that the batch has been illegally dumped into the sanitary sewer. He severely disciplines the night shift foreperson. Upon making discreet inquiries, he finds out that no apparent harm has resulted from the dumping. Should Roy inform government authorities, as is required by law in this kind of situation?

If a representative of each of the four different theories on ethics just mentioned were presented with this dilemma, their decision-making process would focus on different principles.

The Utilitarian Roy would assess the consequences of his options. If he told the government, his company might suffer immediately under any fines administered and later (perhaps more seriously) due to exposure of the incident by the media. If he chose not to inform authorities, he risks heavier fines (and perhaps even worse press) in the event that someone discovers the cover-up. Consequences are the utilitarian Roy's only consideration in his decision-making process.

The Duties Ethicist Roy would weigh his duties and his decision would probably be more clear-cut than his utilitarian counterpart. He is obliged foremost by his duty to obey the law and must inform the government.

The Rights Ethicist mindframe would lead Roy to the same course of action as the duties ethicist — not necessarily because he has a duty to obey the law but because the people in the community have the right to informed consent. Even though Roy's inquiries informed him that no harm resulted from the spill, he knows that the public around the plant has the right to be informed of how the plant is operating.

Vices and virtues would be weighed by the Virtue Ethicist Roy. The course of his thought process would be determined by his own subjective definition of what things are virtuous, what things would make him a morally good person. Most likely, he would consider both honesty and obeying the law virtuous, and withholding information from the government and public as virtueless and would, therefore, tell the authorities.

The scenario used here will be revisited later in this chapter through the eyes of environmentalism to illustrate how this movement is changing the focus of old theories about morality.

47.3 MODERN-DAY MAINSTREAM ENVIRONMENTALISM

Minds like John Muir and Rachel Carson were unique in their respective generations. Their ideas of respect for all flora and fauna were far from predominant in the American mainstream. John Muir was a naturalist and an activist, and on of the founders of the Sierra Club in 1892. Muir believed that the Creator gave all life an equal right to exist, and to destroy plants and animals was ungodly (4). Rachel Carson was a marine biologist. Her 1962 benchmark book *Silent Spring* took environmentalism from pure naturalism into the scientific realm. The evidenced claims she made about the harm caused to wildlife by a range of pesticides (most notably DDT) were as controversial as they were ground-breaking (4). Over the next decade the younger generation embraced a new concern for the environment. The older generation, however, generally dealt with this young movement with opposition rather than cooperation. This was due in large part to the confrontational attitude of many of the youths as well as the perceived threat that the industry-restricting movement itself caused to their economic well-being. As the younger generation grew into positions of power and learned more cooperative tactics, their environmentalist ideas moved from the fringes to the mainstream. En route, the conversion was carried out in the form of both personal growth and government legislation. However, two factions of environmentalism still seem to exist; pure environmentalism (environmentalism for its own sake) and environmentalism for humanity's sake. While they share a common concern for the well-being of the natural world, fundamental differences exist between the two.

One of the most common arguments against the destruction of rainforest land is that any one of the plant or insect species destroyed in the process could contain the elusive cure for cancer or AIDS. With this argument, the ultimate concern is for humanity: We should preserve the natural world because it is best for the human race to do so. This could be considered environmentalism for humanity's sake, and there are a number of other manifestations of it in today's world. The war against the destruction of ozone in the earth's stratosphere is waged largely in the interest of human welfare. While the greenhouse effect has the potential to harm wildlife also, this effect is secondary to that on humanity — both today and in the lives of future generations. This type of environmentalism displays the inherent egocentric attitude of humankind. This faction maintains "an ethic that is second-arily ecological" (5). Here the natural world should be protected because of humanity's dependence on its homeostasis.

The second, more "extremist" form of environmental morality is "primarily ecological" (5). As Aldo Leopold proclaimed, "A thing is right when it tends to preserve the integrity, stability, and beauty of the biotic community. It is wrong if it tends otherwise" (6). Here, humanity has a binding responsibility to protect the homeostasis of the natural world. In this view humanity is considered a part of the interdependent environment rather than something above it. The Native American's adoration of the Great Spirit — which favored the human species no more than any other — is the religious embodiment of such a viewpoint.

The renewed awareness of the environment and awakened concern for its well-being has influenced the ethical world to the point that it has uprooted the focus of the moral correctness of an action. This effect on ethical theories was predicted by John Passmore in 1974: "What it needs for the most part is not so much a 'new ethic' as a more general adherence to a perfectly familiar ethic. For the major sources of our ecological disasters — apart from ignorance — are greed and short-sightedness" (3).

Aldo Leopold made the following observation on personal ethics in his 1949 *A Sand County Almanac*: The scope of one's ethics is determined by the inclusiveness of the community with which one identifies oneself. Leopold parallels the mistreatment of the earth to the mistreatment of slaves that were handled as property. The slave owners were not ethically obliged to the slaves because they considered them outside rather than part of their community. Just as the realm of community grew to include the ex-slaves, it must once again expand to incorporate the whole land community (6). The incorporation of environmentalism into everyday ethics, therefore, does not require a redefinition of one's ethics but, rather, a redefinition of one's "community." This can be applied to each of the ethical theories presented above.

For the utilitarian it requires counting the natural world among those affected by bad and good consequences. The focus of utilitarianism is broadened to include effects on future generations and the welfare of living things other than humans. For the deontological ethicists, the recognition of the environment as part of the community gives it inherent rights and, in turn, imposes on humans the duty to respect those rights. For the virtue ethicists, the virtue of respecting all members of the community would bind them to consider the environment when making decisions.

In the scenario presented earlier, Roy's moral thought process would be affected by the inclusion of the environment into his community regardless of the ethical school of thought he associated himself with. Although his discreet inquiries informed him that no apparent harm resulted from the chemical spill, an environmental impact analysis would have to be made for the utilitarian Roy to fully assess good and bad consequences. If future harm were likely, it may be essential to let the government know so that remediation techniques may be employed at the dumping site. The decision of the rights and duties ethicist Roy would be influenced by their obligation to the environment as well as the surrounding human community. During the virtue ethicist Roy's decision-making process, he would consider which option was the most virtuous with respect to the environment. In each of these cases, an ecologically ethical Roy would have to obtain a reasonable estimate, with the help of the government if necessary, of the environmental effects — immediate and long-term-of the dumping.

In each of these new twists upon old theories on ethics, there exists the fundamentals of a "land ethic." The ethical umbrella is expanding to include under its cover all living beings. Fields of conduct such as disposal and treatment of owned property and land are now becoming judged ethically rather than on the grounds of economic feasibility and personal whimsy.

The mainstreaming of environmentalism is by no means worldwide. The countries in which the greatest impact is seen are the same countries where extensive industrialization exists. Industrialization itself has been crucial to the development of the environmental movement. Not only do its environmental problems and pollution generate concern, citizens of industrialized nations enjoy lives with the luxury of free time and options necessary to be able to devote themselves to such a concern. In poorer countries and communities, the struggle of everyday survival far outweighs any aesthetic concern for the environment. Abraham Maslow's concept of a "hierarchy of needs" can be applied in explaining the difficulty of establishing the environmental movement in impoverished communities and third-world countries.

Maslow maintains that there exist the following "hierarchy of needs" for every human being. He finds five *levels of need:*

1. Survival (physiological needs): food, shelter, health.
2. Security (safety needs): protection from danger and threat.
3. Belonging (social needs): friendship, acceptance, love.
4. Self-esteem (ego needs): self-respect, recognition, status.
5. Self-actualization (fulfillment needs): creativity, realization of individual potentialities.

Maslow maintains that these levels form a hierarchy; lower levels must be satisfied before the individual can give attention to higher levels (3). Until the lower levels of need are at least partially satisfied, a person cannot commit him or herself to the pursuit of higher-leveled needs. For example, a person who is struggling to find any source of food will not be preoccupied with how environmentally conscious the farmer was in the use of fertilizers or pesticides while cultivating the food.

Consider, for example, a town such as many in the mountains of Appalachia where one industry—coal mining—provides all of the town's employment and generates most of the taxes used by the town in running schools and other municipal operations. When the coal-mining company turns to strip mining—a process that essentially rips the mountains to shreds and contaminates groundwater with the heavy metals released—can the miners be expected to jeopardize the welfare of their entire families by protesting because the methods of their employer are environmentally negligent? Their survival needs for food and shelter supersede any idealistic desire they have to preserve the environment. Abuse of this natural hierarchy has been defined as environmental racism and is epitomized by the disproportionately large number of landfills, chemical plants, and toxic dumps in the poorer communities and countries.

47.4 TECHNOLOGY AND ENVIRONMENTALISM

In the ethical theories presented here, established hierarchies of duties, rights, virtues, and desired consequences exist, so that situations can be resolved even when no single course of action satisfies all of the maxims. The entry of environmentalism into the realm of ethics raises questions concerning where it falls in this hierarchy. Much debate continues over these questions of how much weight the natural environment should be given in ethical dilemmas, particularly in those where ecological responsibility seems to oppose economic profitability and technological advances. Those wrapped up in this technology/economy/ecology debate can generally be divided into three groups:

1. Environmental extremists.
2. Technologists to whom ecology is acceptable provided it does not inhibit technological or economic growth.
3. Those who feel technology should be checked with ecological responsibility.

Each is briefly discussed below.

After his year-and-a-half of simple living on the shores of Walden Pond, Henry David Thoreau professed "in wildness is the preservation of the world" (3). He rejected the pursuit of technology and industrialization. While most would agree with his vision of nature as being inspirational, few would choose his way of life. Even so, the movement rejecting technological advances in favor of simple, sustainable, and self-sufficient living is being embraced by more and more people who see technology as nothing but a threat to the purity and balance of nature. Often called environmental extremists by other groups, they even disregard "environmental technologies" that attempt to correct pollution and irresponsibilities, past and present. They see all technology as manipulative and uncontrollable and choose to separate themselves from it. To them, the environment is at the top of the hierarchy.

On the other extreme are the pure technologists. They view the natural world as a thing to be subdued and manipulated in the interest of progress—technological and economic. This is not to say one won't find technologists wandering in a national park admiring the scenery. They do not necessarily deny the beauty of the natural environment, but they see themselves as separate from it. They believe that technology is the key to freedom, liberation, and a higher standard of living. It is viewed, therefore, as inherently good. They see the environmental extremists as unreasonable and hold that even the undeniably negative side effects of certain technologies are best handled by more technological advance. The technologists place environmental responsibility at the bottom of their ethical hierarchy.

Somewhere in the middle of the road travels the third group. While they reap the benefits of technology, they are concerned much more deeply than the technologists with the environmental costs associated with industrialization. It is in this group that most environmental engineers find themselves. They are unlike the environmental extremists since, as engineers, they inherently study and design technological devices and have faith in the ability of such devices to have a positive effect on the condition of the environment. They also differ from the technologists. They scrutinize the effects of technologies much more closely and critically. While they may see a brief, dilute leak of a barely toxic chemical as an unacceptable side effect of the production of a consumer product, the technologists may have to observe destruction—the magnitude of that caused by Chernobyl—before they consider rethinking a technology they view as economically and socially beneficial. In general, this group sees the good in technology but stresses that it cannot be reaped if technological growth goes on unchecked.

47.5 ENGINEERING ETHICS

The ethical behavior of engineers is more important today than at any time in the history of the profession. The engineers' ability to direct and control the technologies they master has never been stronger. In the wrong hands, the scientific advances and technologies of today's engineer could become the worst form of corruption, manipulation, and exploitation. Engineers, however, *are* bound by a code of ethics that carry certain obligations associated with the profession. Some of these obligations include those that lead them to

1. Support ones professional society.
2. Guard privileged information.
3. Accept responsibility for one's actions.
4. Employ proper use of authority.

5. Maintain one's expertise in a state of the art world.

6. Build and maintain public confidence.

7. Avoid improper gift exchange.

8. Practice conservation of resources and pollution prevention.

9. Avoid conflict of interest.

10. Apply equal opportunity employment.

11. Practice health, safety, and accident prevention.

12. Maintain honesty in dealing with employers and clients.

There are many codes of ethics that have appeared in the literature. The preamble for one of these codes is provided below (2):

> Engineers in general, in the pursuit of their profession, affect the quality of life for all people in our society. Therefore, an Engineer, in humility and with the need for Divine guidance, shall participate in none but honest enterprises. When needed, skill and knowledge shall be given without reservation for the public good. In the performance of duty and in fidelity to the profession, Engineers shall give utmost.

47.6   FUTURE TRENDS

Although the environmental movement has grown and matured in recent years, its development is far from stagnant. To the contrary, change in individual behavior, corporate policy, and governmental regulations are occurring at a dizzying pace.

Because of the Federal Sentencing Guidelines, the Defense Industry Initiative, as well as a move from compliance to a values-based approach in the marketplace, corporations have inaugurated company-wide ethics programs, hot lines, and senior line positions responsible for ethics training and development. The Sentencing Guidelines allow for mitigation of penalties if a company has taken the initiative in developing ethics training programs and codes of conduct.

In the near future, these same Guidelines will apply to infractions of environmental law (7). As a result, the corporate community will undoubtedly welcome ethics integration in engineering and science programs generally, but more so in those that emphasize environmental issues. Newly hired employees, particularly those in the environmental arena, who have a strong background in ethics education will allay fears concerning integrity and responsibility. Particular attention will be given to the role of public policy in the environmental arena as well as in the formation of an environmental ethic.

Regulations instituted by federal, state, and local agencies continue to become more and more stringent. The deadlines and fines associated with these regulations encourage corporate and industrial compliance of companies (the letter of the law), but it is in the personal convictions of corporate individuals that the spirit of the law lies, and therein also the heart of a true ecological ethic.

To bolster this conviction in the heart, there must be the emergence of a new *dominant paradigm*. This is defined as "the collection of norms, beliefs, values, habits, and survival rules that provide a framework of reference for members of a society. It is a mental image of social reality that guides behavior and expectations" (3). The general trend in personal ethics is steadily "greener" and is being achieved at a sustainable pace with realistic goals.

A modern-day author suggests the following: The flap of one butterfly's wings can drastically affect the weather (8). While this statement sounds much like what might come from a romantic ecologist, it is actually part of a mathematical theory explored by the contemporary mathematician James Gleick in his book *Chaos, Making a New Science*. The "butterfly" theory illustrates that the concept of interdependence, as Chief Seattle professed it, is emerging as more than just a purely environmental one. This embracing of the connectedness of all things joins the new respect for simplified living and the emphasis on global justice, renewable resources, and sustainable development (as opposed to unchecked technological advancement) as the new, emerging social paradigm. The concept of environmentalism is now *widely* held; its future is becoming *deeply* held.

47.7 SUMMARY

1. In 1854, Chief Seattle penned warnings as environmentally pensive and poignant as any uttered in the 140 years since: "Continue to contaminate your bed and you will one day lay in your own waste."

2. It is generally accepted that any historical ethic can be grouped into one of the following: (a) utilitarianism, (b) duties ethics (c) rights ethics, and (d) virtue ethics.

3. The incorporation of environmentalism into everyday ethics does not require a redefinition of one's ethics, but, rather, a redefinition of one's "community" to include nonhuman inhabitants of the land.

4. In the traditional ethical theories, established hierarchies of duties, rights, virtues, and desired consequences exist so that situations where no single course of action satisfies all of the maxims can still be resolved. Debate continues over where the environment falls in this hierarchy.

5. "Engineers in general, in the pursuit of their profession, affect the quality of life for all people in our society. Therefore, an Engineer... shall participate in none but honest enterprises..."

6. At present, the concept of environmentalism is *widely* held; its future is becoming *deeply* held.

REFERENCES

1. J. Faney, and R. Armstrong, (eds.), *A Peace Reader: Essential Readings on War, Justice, Non-Violence & World Order*, Paulist Press, Mahwah, NJ, 1987.

2. M. W. Martin, and R. Schinzinger, *Ethics in Engineering*, McGraw-Hill, New York, 1989.

3. I. Barbour, *Ethics in an Age of Technology*, Harper, San Francisco, 1993.

4. P. A. Shabecoff, *A Fierce Green Fire*, Hill and Wang, New York, 1993.

5. H. Rolston, III, *Philosophy Gone Wild*, Prometheus Books, Buffalo, NY, 1986.

6. A. Leopold, *A Sand County Almanac*, Oxford University Press, New York, 1949.

7. Presentation by N. Cartusciello, Chief, Environmental Crimes Section, U.S. Department of Justice, May 4, 1994.

8. J. Gleick, *Chaos, Making a New Science*, Viking, New York, 1987.

48

ENVIRONMENTAL JUSTICE

48.1 INTRODUCTION

The need to protect the environment led to the establishment of the U.S. Environmental Protection Agency (EPA) in 1970 in response to growing environmental concerns about unhealthy air, polluted rivers, unsafe drinking water, and waste disposal. Under the mandate of national environmental laws, the EPA is charged with the implementation of these environmental laws and regulations to protect human health and the environment. Although federal regulations to protect human health and the environment are not discriminatory in nature, the combination of several factors such as local zoning laws, permitted emissions, siting requirements, and the establishment of industrial sectors have led to the disproportionate exposure of environmental hazards on particular segments of our society.

Over the past decade, several studies have indicated that minority and low-income communities often bear a disproportionate level of environmental and health effects of pollution. A low-income or racial minority community which is surrounded by multiple sources of air pollution, waste treatment facilities, and landfills and which has lead-based paint in the residences is clearly a community that faces higher-than-average potential environmental risks. Racial minority and low-income populations experience higher-than-average exposures to selected air pollutants, hazardous waste facilities, contaminated fish, and agricultural pesticides in the workplace. Although exposure does not always result in an immediate or acute health effects, high exposure and the possibility of chronic effects are nevertheless a clear cause of health concerns.

This disproportionate distribution of environmental risks have come to be known as environmental justice, environmental equity, and environmental racism. All three terms have been used to describe the belief that poor and minority communities suffer greater exposure to environmental pollution than other communities. The EPA defines *environmental justice* as the fair treatment afforded people of all races, cultures, and incomes regarding the development of environment laws, regulations, and policies. Fair treatment implies that no person or group should shoulder a disproportionate share of negative environmental impact resulting from the execution of environmental programs. The attention to the impact of environmental pollution on particular segments of our society has been steadily growing. Concern that minority populations and/or low-income populations bear a disproportionate amount of adverse health and environmental effects has led President Clinton to issue Executive Order 12898 in 1994 to establish environmental justice as a national priority. This was the first presidential effort to direct all federal agencies with a public health or environmental mission to make environmental justice an integral part of their policies and activities. The Order, entitled "Federal Actions to Address Environmental Justice in Minority Populations and Low-Income Populations," focuses federal attention on environmental and human health conditions of minority populations and low-income populations with the goal of achieving environmental protection for all communities.

48.2 THE ENVIRONMENTAL JUSTICE MOVEMENT

The Environmental Justice movement began in community activism before the program was recognized by the federal government. Evidence of the effects and concentration of environmental pollution in minority communities has fueled a grassroots environmental movement since the early 1980s. The movement calls for grassroots, multiracial, and multicultural activism to redress the distributional inequalities that have resulted from past policies and to prevent the same inequalities from occurring in future policies. The movement advocates that minorities use historically nonexercised political and legal power to push the EPA to address minority concerns and to oppose policies that impoverish the poor. Communities sometimes lack the political clout, economic means, or awareness of rights and opportunities in environmental decision making.

The environmental justice movement is committed to political empowerment as a way to challenge inequities and injustices. Empowerment is the inclusive involvement and education of community members by equipping them with skills for self-representation and defense. Organized activism at the grassroots level could circumvent the power structures that under-represented particular communities in the first place. Community activists want to participate in the decision making that affects their communities. Further, they argue for increased pay for community members who engage in environmentally hazardous labor, for better working conditions in factories, and for the requirement of more on-the-job safety precautions. Activists contend that industries must contribute to community development if they are to detract from the community in other ways.

In 1982, in Warren County, North Carolina, the environmental justice movement grew out of a grassroots protest against the siting of a PCB landfill in a predominantly African-American community. When PCBs were illegally dumped along North Carolina state roads, the state was confronted with a serious toxic waste disposal problem. Warren County, a rural, poor, and 66% Black community, was chosen as the disposal site for the contaminated soil. This should not be dismissed as a product of rural residency; there are a proportionately higher number of White Americans living in rural areas than African- or Hispanic-Americans (Department of Commerce, 1990). Feeling specifically and unjustly targeted, the Warren County citizens organized a protest that attracted national attention and began a movement linking environmental issues with social justice. The protest attracted the attention of civil rights activist and the United Church of Christ's Commission for Racial Justice Director, the Reverend Benjamin Chavis, Jr.

Five years later, the Commission for Racial Justice released the results of a study connecting environmental assaults and racism. The study found that toxic waste dumps were disproportionately located in minority communities. It showed that three out of every five African- and Hispanic-Americans live in areas with controlled toxic waste sites. According to the report, race plays a more significant role than poverty in siting the environmentally dangerous facilities (UCC Commission for Racial Justice, 1987). Plagued by daily struggles for survival, politically unorganized and powerless poor communities have long been a receptacle for toxic waste disposal (1).

In July of 1990, the Administrator of the EPA, William K. Reilly, formed the Environmental Equity Workgroup to explore the emerging consideration in environmental protection: the distribution of environmental problems. The Administrator charged the Workgroup with reviewing the evidence that racial minority and low-income populations bear a higher environmental risk burden than the general population. He also asked the Workgroup to consider what the EPA could do to address any identified disparities.

In its report, "Reducing Risk: Setting Priorities and Strategies for Environmental Protection," the EPA Science Advisory Board urged the EPA to target its environmental protection efforts based on the opportunities for reducing the most serious remaining risks. In response, the EPA began to target its efforts to those environmental problems which pose the greatest risks to human health and the environment. In the context of a risk-based approach to environmental management, the relative risk burden borne by low-income and racial minority communities is a special concern. A low-income or racial minority community which is surrounded by multiple sources of air pollution, waste treatment facilities, and landfills and which has lead-based paint in the residences is clearly a community that faces higher-than-average potential environmental risks. Issues such as these, and how government agencies respond, have come to be known as issues of environmental equity. Environmental equity refers to the distribution of environmental risks across population groups and to our policy responses to these distributions. While there are many types of equity, all of which are important to the EPA, this report focuses on racial minority and low-income populations.

The report to the Administrator reviews existing data on the distribution of environmental exposures and risks across population groups. It also summarizes the workgroup's review of the EPA's programs with respect to racial minority and low-income populations. The following findings are based on the analyses:

- There are clear differences between racial groups in terms of disease and death rates. There are also limited data to explain the environmental contribution to these differences. In fact, there is a general lack of data on environmental health effects by race and income. For diseases that are known to have environmental causes, data are not typically disaggregated by race and socioeconomic group. The notable exception is lead poisoning. A significantly higher percentage of Black children compared to White children have unacceptably high blood lead levels.

- Racial minority and low-income populations experience higher-than-average exposures to selected air pollutants, hazardous waste facilities, contaminated fish, and agricultural pesticides in the workplace. Exposure does not always result in an immediate or acute health effects. Higher exposures, and the possibility of chronic effects, are nevertheless a clear cause for health concerns.

- Environmental and health data are not routinely collected and analyzed by income and race. Nor are data routinely collected on health risks posed by multiple industrial facilities, cumulative and synergistic effects, or multiple and different pathways of exposure. Risk assessment and risk management procedures are not in themselves biased against certain income or racial groups. However, risk assessment and risk management procedures can be improved to better take into account equity considerations.

- Great opportunities exist for the EPA and other government agencies to improve communication about environmental problems with members of low-income and racial minority groups. The language, format, and distribution of written materials, media relations, and efforts in two-way communication all can be improved. In addition, the EPA can broaden the spectrum of groups with which it interacts.

- Since they have broad contacts with affected communities, EPA's program and regional offices are well suited to address equity concerns. The potential exists for effective action by such offices to address disproportionate risks. These offices currently vary considerably in terms of how they address environmental equity issues. Case studies of EPA program and regional offices reveal that opportunities exist for addressing environmental equity issues and that there is a need for environmental equity awareness training. A number of EPA regional offices have initiated projects to address high risks in racial minority and low-income communities.

- Native Americans are a unique racial group that have a special relationship with the federal government and distinct environmental problems. Tribes often lack the physical infrastructure, institutions, trained personnel, and resources necessary to protect their members (2).

Although low-income and minority communities have historically lacked political, legal, and economic power, community activism and mobilization have been effective in combating certain environmental problems. As a result of several major events in the environmental justice movement, environmental justice issues now play an integral part of the EPA's environmental protection policies. The following major events of the environmental justice movement have established environmental justice as a guiding principle in the EPA decision making (3).

- 1971 — CEQ's annual report acknowledges racial discrimination adversely affects urban poor and the quality of their environment.
- 1979 — Robert Bullard's study of an affluent African-American community's attempt to block the siting of a sanitary landfill.
- 1982 — Warren County, North Carolina demonstration against the siting of a PCB landfill in a predominantly African-American community.
- 1983 — GAO report states that 3 of 4 hazardous waste facilities in EPA Region 4 are in African-American communities.

- 1987 — United Church of Christ issues a report entitled "Toxic Wastes and Race in the United States."
- 1990 — Bullard publishes *Dumping in Dixie*, used as the first textbook on environmental justice.
- 1990 — Michigan Coalition Conference releases "Race and the Incidence of Environmental Hazards" report.
- 1990 — The EPA Administrator, William K. Reilly, establishes the Environmental Equity Workgroup. The Administrator charged the Workgroup with reviewing the evidence that racial minority and low-income populations bear a higher environmental risk burden than the general population.
- 1991 — The First National People of Color Leadership Summit in D.C. adopts the "Principles of Environmental Justice."
- 1991 — First of four meetings with the Administrator on environmental justice issues.
- 1992 — The EPA releases "Environmental Equity: Reducing Risks for All Communities" report.
- 1992 — *National Law Journal* report, "Unequal Environmental Protection," alleges EPA discrimination in enforcement.
- 1992 — The EPA establishes the Office of Environmental Justice.
- 1993 — The EPA Administrator, Carol M. Browner, makes environmental justice an Agency priority.
- 1993 — The EPA establishes the National Environmental Justice Advisory Council.
- 1994 — Interagency Symposium on Health Research and needs to Ensure Environmental Justice in Arlington, Virginia.
- 1994 — The President, William J. Clinton, issues Executive Order 12898 designating that 11 agencies are accountable for environmental justice.
- 1994 — Interagency Working Group on Environmental Justice established.
- 1994 — University of Massachusetts issues study challenging siting demographics.
- 1994 — UCC issues "Toxic Waste and Race Revisited," strengthening association between race and waste facilities.
- 1995 — First Interagency Public Meeting on Environmental Justice in Atlanta, Georgia.
- 1995 — The EPA released "Environmental Justice Strategy: Executive Order 12898" in May 1995. The strategy describes environmental justice efforts in six cross-cutting mission areas including health and environmental research; data collection, analysis, and stakeholder access to information; enforcement and compliance assurance; partnerships, outreach, and communication with stakeholders; Native American, indigenous, and tribal programs; and integration of environmental justice into all agency activities.
- 1996 — The EPA published "1996 Waste Programs Environmental Justice Accomplishment Report."

48.3 EXECUTIVE ORDER 12898 ON ENVIRONMENTAL JUSTICE

On February 11, 1994, President Clinton issued Executive Order 12898, "Federal Actions to Address Environmental Justice in Minority Populations," which focused the attention of federal agencies on the environmental and human health conditions of minority and low-income communities. The Executive Order directed Federal agencies to develop environmental justice strategies by April 11, 1995, that identify and address disproportionately high exposure and adverse human health or environmental effects on programs, policies, and activities on minority populations and low-income populations. All agency strategies must consider enforcement of statutes in areas of minority populations and low-income populations, greater public participation, improvement of research and identification of different patterns of subsistence use of natural resources. The Executive Order also requires that agencies conduct activities that substantially affect human health or the environment in a nondiscriminatory manner. In addition, better data collection and research is required by the Executive Order, and it declares that whenever practicable and appropriate, future human health research must look at diverse segments of population and must identify multiple and cumulative exposures. The Executive Order (4) applies equally to Native American programs (5). Highlights from the Executive Order are contained below.

Section 1. IMPLEMENTATION.

1-101. Agency Responsibilities. To the greatest extent practicable and permitted by law, and consistent with the principles set forth in the report on the National Performance Review, each Federal agency shall make achieving environmental justice part of its mission by identifying and addressing, as appropriate, disproportionately high and adverse human health or environmental effects of its programs, policies, and activities on minority populations and low-income populations in the United States and its territories and possessions, the District of Columbia, the Commonwealth of Puerto Rico, and the Commonwealth of the Marian islands.

1-102. Creation of an Interagency Working Group on Environmental Justice.

(a) Within 3 months of the date of this order, the Administrator of the Environmental Protection Agency ("Administrator") or the Administrator's designee shall convene an Interagency Federal Working Group on Environmental Justice ("Working-Group"). The Working Group shall comprise the heads of the following executive agencies and offices, or their designees: (a) Department of Defense; (b) Department of Health and Human Services; (c) Department of Housing and Urban Development; (d) Department of Labor; (e) Department of Agriculture; (f) Department of Transportation; (g) Department of Justice; (h) Department of the Interior; (i) Department of Commerce; (j) Department of Energy; (k) Environmental Protection Agency; (l) Office of Management and Budget; (m) Office of Science and Technology Policy; (n) Office of the Deputy Assistant to the President for Environmental Policy; (o) Office of the Assistant to the President for Domestic Policy; (p) National Economic Council; (q) Council of Economic Advisers; and (r) such other Government officials as the President may designate. The Working Group shall report to the President through the Deputy Assistant to the President for Environmental Policy and the Assistant to the President for Domestic Policy.

(b) The Working Group shall:

(1) provide guidance to Federal agencies on criteria for identifying disproportionately high and adverse human health or environmental effects on minority populations and low-income populations;

(2) coordinate with, provide guidance to, and serve as a clearinghouse for, each Federal agency as it develops an environmental justice strategy as required by section 1-103 of this order, in order to ensure that the administration, interpretation and enforcement of programs, activities and policies are undertaken in a consistent manner;

(3) assist in coordinating research by, and stimulating cooperation among, the Environmental Protection Agency, the Department of Health and Human Services, the Department of Housing and Urban Development, and other agencies conducting research or other activities in accordance with section 3-3 of this order;

(4) assist in coordinating data collection, required by this order;

(5) examine existing data and studies on environmental justice;

(6) hold public meetings at required in section 5-502(d) of this order; and

(7) develop interagency model projects on environmental justice that evidence cooperation among Federal agencies.

1-103. Development of Agency Strategies.

(a) Except as provided in section 6-605 of this order, each Federal agency shall develop an agency-wide environmental justice strategy, as set forth in subsections (b)–(e) of this section that identifies and addresses disproportionately high and adverse human health or environmental effects of its programs, policies, and activities on minority populations and low-income populations. The environmental justice strategy shall list programs, policies, planning and public participation processes, enforcement, and/or rulemakings related to human health or the environment that should be revised to, at a minimum: (1) promote enforcement of all health and environmental statutes in areas with minority populations and low-income populations: (2) ensure greater public participation; (3) improve research and data collection relating to the health of and environment of minority populations and low-income populations; and (4) identify differential patterns of consumption of natural resources among minority populations and low-income populations. In addition, the environmental justice strategy shall include, where appropriate, a timetable for undertaking identified revisions and consideration of economic and social implications of the revisions.

(b) Within 4 months of the date of this order, each Federal agency shall identify an internal administrative process for developing its environmental justice strategy, and shall inform the Working Group of the process.

(c) Within 6 months of the date of this order, each Federal agency shall provide the Working Group with an outline of its proposed environmental justice strategy.

(d) Within 10 months of the date of this order, each Federal agency shall provide the Working Group with its proposed environmental justice strategy.

(e) Within 12 months of the date of this order, each Federal agency shall finalize its environmental justice strategy and provide a copy and written description of its strategy to the Working Group. During the 12 month period from the date of this order, each Federal agency, as part of its environmental justice strategy, shall identify several specific projects that can be promptly undertaken to address particular concerns identified during the development of the proposed environmental justice strategy, and a schedule for implementing those projects.

(f) Within 24 months of the date of this order, each Federal agency shall report to the Working Group on its progress in implementing its agency-wide environmental justice strategy.

(g) Federal agencies shall provide additional periodic reports to the Working Group as requested by the Working Group.

1-104. Reports to The President. Within 14 months of the date of this order, the Working Group shall submit to the President, through the Office of the Deputy Assistant to the President for Environmental Policy and the Office of the Assistant to the President for Domestic Policy, a report that describes the implementation of this order, and includes the final environmental justice strategies described in section 1-103(e) of this order.

Section 2. FEDERAL AGENCY RESPONSIBILITIES FOR FEDERAL PROGRAMS

Each Federal agency shall conduct its programs, policies, and activities that substantially affect human health or the environment, in a manner that ensures that such programs, policies, and activities do not have the effect of excluding persons (including populations) from participation in, denying persons (including populations) the benefits of, or subjecting persons (including populations) to discrimination under, such, programs, policies, and activities, because of their race, Color, or national origin.

Section 3. RESEARCH, DATA COLLECTION, AND ANALYSIS

3-301. Human Health and Environmental Research and Analysis.

(a) Environmental human health research, whenever practicable and appropriate, shall include diverse segments of the population in epidemiological and clinical studies, including segments at high risk from environmental hazards, such as minority populations, low-income populations and workers who may be exposed to, substantial environmental hazards.

(b) Environmental human health analyses, whenever practicable and appropriate, shall identify multiple and cumulative exposures.

(c) Federal agencies shall provide minority populations and low-income populations the opportunity to comment on the development and design of research strategies undertaken pursuant to this order.

3-302. Human Health and Environmental Data Collection and Analysis To the extent permitted by existing law, including the Privacy Act, as amended (5 U.S.C. section 552a):

(a) Each federal agency, whenever practicable and appropriate, shall collect, maintain, and analyze information assessing and comparing environmental and human health risks borne by populations identified by race, national origin, or income. To the extent practical and appropriate, Federal agencies shall use this information to determine whether their programs, policies, and activities have disproportionately high and adverse human health or environmental effects on minority populations and low-income populations;

(b) In connection with the development and implementation of agency strategies in section 1-103 of this order, each Federal agency, whenever practicable and appropriate, shall collect, maintain and analyze information on the race, national origin, income level, and other readily accessible and appropriate information for areas surrounding facilities or sites expected to have substantial environmental, human health, or economic effect on the surrounding populations, when such facilities or sites become the subject of a substantial Federal environmental administrative or judicial action. Such information shall be made available to the public unless prohibited by law; and

(c) Each Federal agency, whenever practicable and appropriate, shall collect, maintain, and analyze information on the race, national origin, income level, and other readily accessible and appropriate information for areas surrounding Federal facilities that are: (1) subject to the reporting requirements under the Emergency Planning and Community Right-to-Know Act, 42 U.S.C. section 11001-11050 as mandated in Executive Order No. 12856; and (2) expected to have a substantial environmental, human health, or economic effect on surrounding populations. Such information shall be made available to the public unless prohibited by law.

(d) In carrying out the responsibilities in this section, each Federal agency, whenever practicable and appropriate, shall share information and eliminate unnecessary duplication of efforts through the use of existing data systems and cooperative agreements among Federal agencies and with State, local, and tribal governments.

Section 4. SUBSISTENCE CONSUMPTION OF FISH AND WILDLIFE

4-401. Consumption Patterns. In order to assist in identifying the need for ensuring protection of populations with differential patterns of subsistence consumption of fish and wildlife, Federal agencies, whenever practicable and appropriate, shall collect, maintain, and analyze information on the consumption patterns of populations who principally rely on fish and/or wildlife for subsistence. Federal agencies shall communicate to the public the risks of those consumption patterns.

4-402. Guidance. Federal agencies, whenever practicable and appropriate, shall work in a coordinated manner to publish guidance reflecting the latest scientific information available concerning methods for evaluating the human health risks associated with the consumption of pollutant-bearing fish or wildlife. Agencies shall consider such guidance in developing their policies and rules.

Section 5. PUBLIC PARTICIPATION AND ACCESS TO INFORMATION

(a) The public may submit recommendations to Federal agencies relating to the incorporation of environmental justice principles into Federal agency programs or policies. Each Federal agency shall convey such recommendations to the Working Group.

(b) Each Federal agency may, whenever practicable and appropriate, translate crucial public documents, notices, and hearings relating to human health or the environment for limited English speaking populations.

(c) Each Federal agency shall work to ensure that public documents, notices, and hearings relating to human health or the environment are concise, understandable, and readily accessible to the public.

(d) The Working Group shall hold public meetings, as appropriate, for the purpose of fact-finding, receiving public comments, and conducting inquiries concerning environmental justice. The Working Group shall prepare for public review a summary of the comments and recommendations discussed at the public meetings.

Section 6. GENERAL PROVISIONS

6-601. Responsibility for Agency Implementation. The head of each Federal agency shall be responsible for ensuring compliance with this order. Each Federal agency shall conduct internal reviews and take such other steps as may be necessary to monitor compliance with this order.

6-602. Executive Order No. 12250. This Executive order is intended to supplement but not supersede Executive Order No. 12250, which requires consistent and effective implementation of various laws prohibiting discriminatory practices in programs receiving Federal financial assistance. Nothing herein shall limit the effect or mandate of Executive Order No. 12250.

6-603. Executive Order No. 12875. This Executive order is not intended to limit the effect or mandate of Executive Order No. 12875.

6-604. Scope. For purposes of this order, Federal agency means any agency on the Working Group, and such other agencies as may be designated by the President, that conducts any Federal program or activity that substantially affects human health or the environment. Independent agencies are requested to comply with the provisions of this order.

6-605. Petitions far Exemptions. The head of a Federal agency may petition the President for an exemption from the requirements of this order on the grounds that all or some of the petitioning agency's programs or activities should not be subject to the requirements of this order.

6-606. Native American Programs. Each Federal agency responsibility set forth under this order shall apply equally to Native American programs. In addition the Department of the Interior, in coordination with the Working Group, and, after consultation with tribal leaders, shall coordinate steps to be taken pursuant to this order that address Federally-recognized Indian Tribes.

6-607. Costs. Unless otherwise provided by law, Federal agencies shall assume the financial costs of complying with this order.

6-608. General. Federal agencies shall implement this order consistent with, and to the extent permitted by, existing law.

6-609. Judicial Review. This order is intended only to improve the internal management of the executive branch and is not intended to, nor does it create any right, benefit, or trust responsibility, substantive or procedural, enforceable at law or equity by a party against the United States, its agencies, its officers, or any person. This order shall not be construed to create any right to judicial review involving the compliance or noncompliance of the United States, its agencies, its officers, or any other person with this order (4).

48.4 IMPLEMENTATION OF ENVIRONMENTAL JUSTICE

The EPA created the Office of Environmental Justice in 1992 and implemented a new organizational infrastructure to integrate environmental justice into EPA's policies, programs, and activities. A Policy Working Group made up of senior managers and policy analysts represents each headquarters office and region. It provides leadership and direction on strategic planning to ensure that environmental justice is incorporated into the agency operations; the most active group is the Environmental Justice Coordinator's Council which serves as the front-line staff specifically responsible to ensure policy input, program development, and implementation of environmental justice throughout the Agency. This new structure has established a clear commitment from EPA's senior management to all personnel that environmental justice is a priority.

In 1993, EPA Administrator Browner made environmental justice an EPA priority. She stated that "Many people of color, low-income and Native American communities have raised concerns that they suffer a disproportionate burden of health consequences due to siting of industrial plants and waste dumps, and from exposure to pesticides or other toxic chemicals at home and on the job and that environmental programs do not adequately address these disproportionate exposures. The EPA is committed to addressing these concerns and is assuming a leadership role in environmental justice to enhance environmental quality for all residents of the United States. Incorporating environmental justice into everyday Agency activities and decisions will be a major undertaking. Fundamental reform will be needed in Agency operations."

To ensure that the EPA would receive significant input from affected stakeholders, the National Environmental Justice Advisory Council (NEJAC) was established as a Federal Advisory Committee and chartered for two years effective September 3, 1993, rechartered September 3, 1995, and rechartered for the second time effective September 3, 1995. During its first two years, the NEJAC Council consisted of 23 members appointed from key environmental justice constituencies, which include community-based groups; business and industry; academic and educational institutions; state and local governments; tribal government; nongovernmental organizations; and environmental groups. The NEJAC Council also had four subcommittees to help develop strategic options for the EPA. Each subcommittee was comprised of approximately 12 individuals knowledgeable in the subject area, from the NEJAC Council as well as from other stakeholder organizations. These were waste and facility siting; enforcement; health and research; and public participation and accountability. In 1995, two new subcommittees were established: the Indigenous Peoples' Subcommittee and the International Subcommittee.

During the 1993–1996 period, the NEJAC produced a number of products and provided advice to help the EPA focus its environmental justice agenda. For example, the initial draft of the EPA's Environmental Justice Strategy required by Executive Order 12898 was reviewed and substantive recommendations made; the Office of Solid Waste and Emergency Response's facility Siting Criteria document was reviewed; a public forum protocol was developed and subsequently used as a model for the first Interagency Public Meeting on Environmental Justice (Atlanta, 1/19/95); in October 1997 the model was used for the first NEJAC/EPA Enforcement Roundtable in San Antonio; the U.S.–Mexico Border XXI program proposal was reviewed; health and research projects to identify high risk communities were developed, reviewed, and commented on the EPA's enforcement and compliance work plan; and public dialogues were conducted in five major cities concerning possible solutions to urban crises resulting from loss of economic opportunities caused by pollution and relocation of businesses.

Southeast Chicago Environmental Initiative

Southeast Chicago is a mosaic of predominately poor or working class, African-American, Hispanic, and White neighborhoods. It is an area of high structural unemployment and multiple environmental problems, including a concentration of disposal sites, countless urban Brownfields, and heavy industries. Located within Southeast Chicago is Altgeld Gardens, a public housing community of thousands of low-income African-Americans which is surrounded by a number of polluting facilities — landfills, incinerators, oil refineries, a paint factory, a steel mill, a sewage treatment plant, a chemical plant, a scrap metal yard, a lagoon, a sludge bed, and a freeway. This community has a high concentration of severe environmental problems and concerns.

The EPA developed the Southeast Chicago Urban Environmental Initiative Action Plan, a framework to improve the environmental conditions of the community. This unique partnership hopes to bring together representatives of the government, industry, community, and environmental groups. Agencies and actions targeted include

- The Agency for Toxic Substances and Disease Registry (ASTDR) is conducing health assessments of the Southeast Chicago community.
- The Department of Housing and Urban Development is developing residential lead-based paint removal projects and other environmental improvements.
- The EPA, Chicago's Department of Environment, and the Illinois Environmental Protection Agency are working together to ensure tougher enforcement and compliance of existing environmental regulations.

Mississippi Delta Project

The Mississippi Delta area has a high concentration of transportation routes, petroleum industries, waste sites, and other facilities. Environmental justice organizations have complained that many of these facilities are sited close to minority communities and that these communities are disproportionately exposed to environmental pollution. An interagency steering committee comprised of ASTDR, the Center of Disease Control (CDC), Occupational Safety and Health Administration (OSHA), the EPA, and the State Health and Environmental departments is working to address these issues.

The goal of this interagency project is to reduce environmental hazards and to prevent them from adversely affecting minority populations and low-income populations residing in the highly industrialized areas along the Mississippi River. This project covers 219 counties in seven States (Arkansas, Illinois, Kentucky, Louisiana, Mississippi, Missouri, and Tennessee), affecting more than 8.3 million people. The project is designed to:

- Identify the key environmental hazards that might affect high risk communities.
- Evaluate the public health impact on high exposure populations.
- Increase health care delivery services in the region, including capacity of State and local health departments to address public health associated with environmental exposures.
- Engage Historically Black Colleges and Universities (HBCUs) and other academic institutions to help increase environmental awareness in these communities.

This project represents the largest geographic-specific public health initiative ever attempted to study the association between hazardous environmental exposure and health effects in minority communities and low-income communities.

New Mexico and Texas Colonias Border Projects

Colonias are Hispanic rural neighborhoods and unincorporated subdivisions in or near cities in Texas, New Mexico, Arizona, and California along the U.S.–Mexico border. Between Texas and New Mexico there are about 1,200 colonias with an estimated population of 300,000 people. Colonias are characterized by substandard housing, inadequate plumbing and sewage disposal systems, and inadequate access to clean water. The common thread is the potential and immediate health threat due to inadequate or lack of safe potable water and sewage disposal.

Under recent grants from New Mexico, nine facility plans and four construction design plans are nearing completion for the thirteen new wastewater collection and treatment systems to serve colonias in New Mexico. This grant program, administered by the New Mexico Environment Department, was made possible through a grant from the EPA.

The Texas Natural Resources Conservation Commission has awarded fifteen grants to provide wastewater collection and treatment systems in Texas Colonias. These projects will affect 64,000 colonias residents. Additional facility plans are being prepared for six colonias which have receive grants for innovative/alternative methods of wastewater collection and treatment. These six projects are designed to identify low-cost methods of wastewater treatment for colonia application.

Pennsylvania Risk and Enforcement Projects

The city of Chester has among the highest concentration of industrial facilities in Pennsylvania. Chester hosts a number of waste processing plants and two oil refineries. All solid waste from Delaware County is incinerated in Chester, and at least 85% of raw sewage and associated sludge is treated there. A large infectious medical waste facility was also recently sited in Chester. Many of the plants are located in close proximity to low-income, minority residential neighborhoods. In fact, a clustering waste treatment facilities have been permitted within 100 feet of over 200 Chester homes.

Chester residents are concerned about the health effects of living and working amid toxic substances and complain of frequent illness. Of cities in the state, Chester has the highest infant mortality rate, the lowest birth rate, and among the highest death rates due to certain malignant tumors.

In response to the Chester community concerns, the EPA has committed to a major initiative involving two studies addressing environmental regulatory and pollutant impact/risk exposure issues. The first was a 30-day study of the EPA's legal authority for existing and proposed facilities in the Chester area. As a result of the 30-day study, the EPA has focused enforcement actions and just recently issued field citations to a number of underground storage tanks located in Chester and the nearby area of Marcus Hook. Other focused enforcement-related activities are proceeding in air toxics reduction and compliance, innovative settlements for toxic emissions violators, and multimedia compliance reviews.

In addition, a 180-day study conducted by a team of toxicologists working with State and local officials is assessing all available environmental media and human exposure pathways. Work products will be displayed through a Regional Geographic Information System overlaying industrial facilities data, NPL sites, and small quantity waste generators, and air emissions data.

Baltimore Urban Environmental Initiative

The Baltimore Urban Environmental Initiative (URI) is an interagency activity being conducted by the EPA in cooperation with the City of Baltimore and the Maryland Department of the Environment. The URI is designed to identify and rank areas of disproportionate risk in Baltimore City for purposes of implementing risk reduction, pollution prevention, public awareness, and other appropriate environmental activities to minimize risks. The Baltimore URI has both short- and long-term tracks. The short-term efforts address issues of immediate concern as well as initial data collection, data analyses, and project planning. The long-term effects will be expanded in areas that warrant continued action.

A project development and problem identification report for the URI will describe the data gathering and risk identification and characterization efforts in support of the overall Initiative. Data has been gathered from a number of existing environmental and demographics-based databases in order to identify and evaluate human health and ecological threats for the purposes of targeting risk reduction/prevention activities. Quantitative and qualitative risk assessment methods have been applied and displayed through the use of a Geographic Information System.

The short-term efforts, based upon preliminary risk analyses, applied knowledge, and experience of an interagency team to jointly target areas of environmental concern that could benefit from immediate action. The six areas of concern identified were:

- Lead.
- Hazardous materials incident.
- Fish consumption/toxics in the harbor.
- Air toxics.
- Ground-level ozone.
- Indoor air and radon.

Individual action teams were formed to address each issue. These teams were responsible for developing action agendas to address the overall goals of risk reduction, pollution prevention, and outreach and education for each risk area (6).

48.5 SUMMARY

1. Racial minority and low-income populations experience higher-than-average exposures to selected air pollutants, hazardous waste facilities, contaminated fish, and agricultural pesticides in the workplace. Although exposure does not always result in an immediate or acute health effect, high exposure and the possibility of chronic effects are nevertheless a clear cause of health concerns. This disproportionate distribution of environmental risks have come to be known as "environmental justice."

2. The environmental justice movement began in community activism before the program was recognized by the federal government. Evidence of the effects and concentration of environmental pollution in minority communities has fueled a grassroots environmental movement since the early 1980s. The movement calls for grassroots, multiracial, and multi cultural activism to redress the distributional

inequalities that have resulted from past policies and to prevent the same inequalities from occurring in future policies.

3 On February 11, 1994, President Clinton issued Executive Order 12898, "Federal Actions to Address Environmental Justice in Minority Populations" which focused the attention of federal agencies on the environmental and human health conditions of minority and low-income communities. The Executive Order directed federal agencies to develop environmental justice strategies by April 11, 1995, that identify and address disproportionately high exposure and adverse human health or environmental effects on programs, policies, and activities on minority populations and low-income populations.

4. The EPA created the Office of Environmental Justice in 1992 and implemented a new organizational infrastructure to integrate environmental justice into EPA's policies, programs and activities.

REFERENCES

1. M. K. Theodore and L. Theodore, *Major Environmental Issues facing the 21st Century*, "Environmental Justice," Prentice-Hall, Upper Saddle River, NJ, 1996.
2. U.S. EPA, *Fact Sheet: Environmental Equity-Reducing Risk for all Communities*, 1992.
3. U.S. EPA, *Environmental Justice 1994 Annual Report*, Office of Environmental Justice, EPA/200-R-95-003, April 1995.
4. Executive Order 12898, President William J. Clinton, The White House, February 11, 1994.
5. U.S. EPA, *OSWER Environmental Justice Action Agenda*, Office of Solid Waste and Emergency Response, EPA/540/r-95/023, 19xx.
6. U.S. EPA, *Environmental Justice Sheet*, Office of Environmental Justice, EPA-300-F-97-003, April 1998.

APPENDICES

APPENDIX A

ENVIRONMENTAL ORGANIZATIONS

Environmental Protection has become a universal phenomena as the citizens of the world move towards protection and better management of their natural resources. The following is a list of environmental organizations whose primary mission is protection of the environment. Leading this charge on a national level is the United States Environmental Protection Agency.

International Organizations
Beauty Without Cruelty International (BWC)
11 Limehill Road
Tunbridge Wells
Kent YNI ILJ, England

Canada–United States Environmental Council
1244 19th Street, NW
Washington, DC, 20037 USA
Tel: 202-659-9510

Caribbean Conservation Corporation (CCC)
P.O. Box 2866
Gainsville, FL 32602 USA
Tel: 904-373-6441

Clean World International (CWI)
c/o Keep Britain Tidy Group
Bostel House
37 West Street
Brighton BNI 2RE, England
Tel: 44-273-23585

International Association on Water Pollution Research and Control (IAWPRC)
   1 Queen Anne's Gate
London SWI H9BT, England

International Board for Soil Resources and Management (INSRAM)
P.O. Box 9-109 Bangkhen
Bangkok 10900, Thailand
Tel: 662-561-1230

The Secretariat for the Protection of The Mediterranean Sea
Place Lesseps 1
E-08023 Barcelona, Spain
Tel: 343-217-1695

The United Nations Development Programme (UNDP)
1 United Nations Plazza
New York, New York 10017 USA
Tel: 212-906-5000

World Environmental Center
605 Third Avenue
New York, New York 10158 USA
Tel: 212-986-7200

U.S. Government Organizations
The United States Environmental Protection Agency (EPA)
401 M Street, SW
Washington, DC 20460
Tel: 202-755-2673

Council on Environmental Quality
722 Jackson Place, NW
Washington, DC 20006
Tel: 202-395-5750

Department of Agriculture
U.S. Forest Service
Soil Conservation Service
Washington, DC. 202
Tel: 202-655-4000

Department of Energy
Washington, DC 205545
Tel: 202-252-5000

Department of Interior
Interior Building
C Street, between 18th and 19th, NW
Washington, DC 20240
202-343-1100

Department of Justice
Land and Natural Resource Division
10th Street and Pennsylvania Avenue, NW
Washington, DC 20530
Tel: 202-633-2701

Department of the Treasury
United States Customs Service
1301 Constitution Avenue, NW
Washington, DC 20229
Tel: 202-566-5104

National Oceanic and Atmospheric Administration (NOAA)
14 Street, NW
Washington, DC 20230
Tel: 202-377-3567

U.S. Nongovernmental Organizations
Endangered Species Committee
Interior Building
Room 4160
Washington, DC 20240
Tel: 202-235-2771

Greenspeace, U.S.A., Inc.
2007 R Street, NW
Washington, DC 20009
Tel: 202-462-1177

Marine Mammal Commission
1625 1 Street, NW
Washington, DC 20006
Tel: 202-653-6237

National Audobon Society
905 Third Avenue
New York, NY 10022
Tel. 212-832-3200

National Wildlife Federation
1412 16th Street, NW
Washington, DC 20036
Tel: 202-797-6800

National Resources Defense Council (NRDC)
Suite 300
1350 New York Avenue, NW
Washington, DC 20005
Tel: 202-783-7800

Sierra Club
530 Bush Street
San Francisco, CA 9418000
Tel: 415-981-8634

APPENDIX B

EPA AND
STATE CONTACTS

RCRA/Superfund Hotline	EPA Small Business Ombudsman	National Response Center
1-800-424-9346	Hotline 1-800-368-5888	1-800-424-8802
(In Washington, D.C.:382-3000)	(In Washington, D.C.:557-1938)	(In Washington, D.C.:426-2675)

Regions

4 — Alabama	5 — Indiana	9 — Nevada	4 — Tennessee
10 — Alaska	7 — Iowa	1 — New Hampshire	6 — Texas
9 — Arizona	7 — Kansas	2 — New Jersey	8 — Utah
6 — Arkansas	4 — Kentucky	6 — New Mexico	1 — Vermont
9 — California	6 — Louisiana	2 — New York	3 — Virginia
8 — Colorado	1 — Maine	4 — North Carolina	10 — Washington
1 — Connecticut	3 — Maryland	8 — North Dakota	3 — West Virginia
3 — Delaware	1 — Massachusetts	5 — Ohio	5 — Wisconsin
3 — D.C.	5 — Michigan	6 — Oklahoma	8 — Wyoming
4 — Florida	5 — Minnesota	10 — Oregon	9 — American Samoa
4 — Georgia	4 — Mississippi	3 — Pennsylvania	9 — Guam
9 — Hawaii	7 — Missouri	1 — Rhode Island	2 — Puerto Rico
10 — Idaho	8 — Montana	4 — South Carolina	2 — Virgin Islands
5 — Illinois	7 — Nebraska	8 — South Dakota	

U.S. EPA REGIONAL OFFICES

EPA Region 1
State Waste Programs Branch
JFK Federal Building
Boston, Massachusetts 02203
(617)223-3468
Connecticut, Massachusetts, Maine,
New Hampshire, Rhode Island, Vermont

EPA Region II
Air and Waste Management Division
26 Federal Plaza
New York, New York 10278
(212)264-5175
New Jersey, New York, Puerto Rico,
Virgin Islands

EPA Region III
Waste Management Branch
841 Chestnut Street
Philadelphia, Pennsylvania 19107
(215)597-9336
Delaware, Maryland, Pennsylvania,
Virginia, West Virginia,
District of Columbia

EPA Region IV
Hazardous Waste Management Division
345 Courtland Street, N.E.
Atlanta, Georgia 30365
(404)347-3016
Albama, Florida, Georgia,
Kentucky, Mississippi, North
Carolina, South Carolina, Tennessee

EPA Region V
RCRA Activities
230 South Dearborn Street
Chicago, Illinois 60604
(312)353-2000
Illinois, Indiana, Michigan,
Minnesota, Ohio, Wisconsin

EPA Region VI
Air and Hazardous Materials Division
1201 Elm Street
Dallas, Texas 75270
(214)767-2600
Arkansas, Louisiana, New Mexico,
Oklahoma, Texas

EPA Region VII
RCRA Branch
726 Minnesota Avenue
Kansas City, Kansas 66101
(913)236-2800
Iowa, Kansas, Missouri, Nebraska

EPA Region VIII
Waste Management Division (8HWM-ON)
One Denver Place
999 18th Street, Suite 1300
Denver, Colorado 80202-2413
(303)293-1502
Colorado, Montana, North Dakota,
South Dakota, Utah, Wyoming

EPA Region IX
Toxics and Waste Management Division
215 Fremont Street
San Francisco, California 94105
(415)974-7472
Arizona, California, Hawaii,
Nevada, American Samoa, Guam,
Trust Territories of the Pacific

EPA Region X
Waste Management Branch—MS-530
1200 Sixth Avenue
Seattle, Washington 98101
(206)442-2777
Alaska, Idaho, Oregon, Washington

STATE HAZARDOUS WASTE MANAGEMENT AGENCIES

ALABAMA
Alabama Department of
  Environmental Management
Land Division
1751 Federal Drive
Montgomery, Alabama 36130
(205)271-7730

ALASKA
Department of Environmental
  Conservation
P.O. Box 0
Juneau, Alaska 99811
Program Manager: (907)465-2666
Northern Regional Office
  (Fairbanks): (907)452-1714
South-Central Regional Office
  (Anchorage): (907)274-2533
Southeast Regional Office
  (Juneau): (907)789-3151

AMERICAN SAMOA
Environmental Quality Commission
Government of American Samoa
Pago Pago, American Samoa 96799
Overseas Operator
(Commercial Call (684)663-4116)

ARIZONA
Arizona Department of
  Health Services
Office of Waste and Water Quality
2005 North Central Avenue
  Room 304
Phoenix, Arizona 85004
Hazardous Waste Management:
  (602)255-2211

ARKANSAS
Department of Pollution Control
  and Ecology
Hazardous Waste Division
P.O. Box 9583
8001 National Drive
Little Rock, Arkansas 72219
(501)562-7444

CALIFORNIA
Department of Health Services
Toxic Substances Control Division
714 P Street, Room 1253
Sacramento, California 95814
(916)324-1826

State Water Resources Control
  Board
Division of Water Quality
P.O. Box 100
Sacramento, California 95801
(916)322-2867

COLORADO
Colorado Department of Health
Waste Management Division
4210 E. 11th Avenue
Denver, Colorado 80220
(303)320-8333 Ext. 4364

CONNECTICUT
Department of Environmental
  Protection
Hazardous Waste Management
  Section
State Office Building
165 Capitol Avenue
Hartford, Connecticut 06106
(203)566-8843, 8844

Connecticut Resource Recovery
  Authority
179 Allyn Street, Suite 603
Professional Building
Hartford, Connecticut 06103
(203)549-6390

DELAWARE
Department of Natural Resources
  and Environmental Control
Waste Management Section
P.O. Box 1401
Dover, Delaware 19903
(302)736-4781

DISTRICT OF COLUMBIA
Department of Consumer and
  Regulatory Affairs
Pesticides and Hazardous Waste
  Materials Division
Room 114
5010 Overlook Avenue, S.W.
Washington, D.C. 20032
(202)767-8414

FLORIDA
Department of Environmental
  Regulation
Solid and Hazardous Waste
  Section
Twin Towers Office Building
2600 Blair Stone Road
Tallahassee, Florida 32301
RE: SQG's
(904)488-0300

GEORGIA
Georgia Environmental
  Protection Division
Hazardous Waste Management
  Program
Land Protection Branch
Floyd Towers East, Suite 1154
205 Butler Street, S.E.
Atlanta, Georgia 30334
(404)656-2833
Toll Free: (800)334-2373

GUAM
Guam Environmental Protection
  Agency
P.O. Box 2999
Agana, Guam 96910
Overseas Operator
(Commercial Call (671)646-7579)

HAWAII
Department of Health
Environmental Health Division
P.O. Box 3378
Honolulu, Hawaii 96801
(808)548-4383

IDAHO
Department of Health and Welfare
Bureau of Hazardous Materials
450 West State Street
Boise, Idaho 83720
(208)334-5879

ILLINOIS
Environmental Protection Agency
Division of Land Pollution
  Control
2200 Churchill Road, #24
Springfield, Illinois 62706
(217)782-6761

INDIANA
Department of Environmental
  Management
Office of Solid and Hazardous
  Waste
105 South Meridian
Indianapolis, Indiana 46225
(317)232-4535

IOWA
U.S. EPA Region VII
Hazardous Materials Branch
726 Minnesota Avenue
Kansas City, Kansas 66101
(913)236-2888
Iowa RCRA Toll Free:
  (800)223-0425

KANSAS
Department of Health and
  Environment
Bureau of Waste Management
Forbes Field, Building 321
Topeka, Kansas 66620
(913)862-9360 Ext. 292

KENTUCKY
Natural Resources and
  Environmental Protection
  Cabinet
Division of Waste Management
18 Reilly Road
Frankfort, Kentucky 40601
(502)4564-6716

LOUISIANA
Department of Environmental
  Quality
Hazardous Waste Division
P.O. Box 44307
Baton Rouge, Louisiana 70804
(504)342-1227

MAINE
Department of Environmental
  Protection
Bureau of Oil and Hazardous
  Materials Control
State House Station #17
Augusta, Maine 04333
(207)289-2651

MARYLAND
Department of Health and Mental
  Hygiene
Maryland Waste Management
  Administration
Office of Environmental Programs
201 West Preston Street, Room A3
Baltimore, Maryland 21201
(301)225-5709

MASSACHUSETTS
Department of Environmental
  Quality Engineering
Division of Solid and Hazardous
  Waste
One Winter Street, 5th Floor
Boston, Massachusetts 02108
(617)292-5589
(617)292-5851

MICHIGAN
Michigan Department of Natural
  Resources
Hazardous Waste Division
Waste Evaluation Unit
Box 30028
Lansing, Michigan 48909
(517)373-2730

MINNESOTA
Pollution Control Agency
Solid and Hazardous Waste
  Division
1935 West County Road, B-2
Roseville, Minnesota 55113
(612)296-7282

MISSISSIPPI
Department of Natural Resources
Division of Solid and Hazardous
  Waste Management
P.O. Box 10385
Jackson, Mississippi 39209
(601)961-5062

MISSOURI
Department of Natural Resources
Waste Management Program
P.O. Box 176
Jefferson City, Missouri 65102
(314)751-3176
Missouri Hotline:
(800)334-6946

MONTANA
Department of Health and
  Environmental Sciences
Solid and Hazardous Waste Bureau
Cogswell Building, Room B-201
Helena, Montana 59620
(406)444-2821

NEBRASKA
Department of Environmental
  Control
Hazardous Waste Management
  Section
P.O. Box 94877
State House Station
Lincoln, Nebraska 68509
(402)471-2186

NEVADA
Division of Environmental
  Protection
Waste Management Program
Capitol Complex
Carson City, Nevada 89710
(702)885-4670

NEW HAMPSHIRE
Department of Health and Human
  Services
Division of Public Health Services
Office of Waste Management
Health and Welfare Building
Hazen Drive
Concord, New Hampshire
  03301-6527
(603)271-4608

NEW JERSEY
Department of Environmental
  Protection
Division of Waste Management
32 East Hanover Street, CN-028
Trenton, New Jersey 08625
Hazardous Waste Advisement
  Program: (609)292-8341

NEW MEXICO
Environmental Improvement
  Division
Ground Water and Hazardous
  Waste Bureau
Hazardous Waste Section
P.O. Box 968
Santa Fe, New Mexico 87504-0968
(505)827-2922

NEW YORK
Department of Environmental
  Conservation
Bureau of Hazardous Waste
  Operations
50 Wolf Road, Room 209
Albany, New York 12233
(518)457-0530
SQG Hotline: (800)631-0666

NORTH CAROLINA
Department of Human Resources
Solid and Hazardous Waste
  Management Branch
P.O. Box 2091
Raleigh, North Carolina 27602
(919)733-2178

NORTH DAKOTA
Department of Health
Division of Hazardous Waste
  Management and Special Studies
1200 Missouri Avenue
Bismarck, North Dakota
  58502-5520
(701)224-2366

**NORTHERN MARIANA ISLANDS,
COMMONWEALTH OF**
Department of Environmental and
  Health Services
Division of Environmental Quality
P.O. Box 1304
Saipan, Commonwealth of Mariana
  Islands 96950
Overseas call (670)234-6984

OHIO
Ohio EPA
Division of Solid and Hazardous
  Waste Management
361 East Broad Street
Columbus, Ohio 43266-0558
(614)466-7220

OKLAHOMA
Waste Management Service
Oklahoma State Department of
  Health
P.O. Box 53551
Oklahoma City, Oklahoma 73152
(405)271-5338

OREGON
Hazardous and Solid Waste
  Division
P.O. Box 1760
Portland, Oregon 97207
(503)229-6534
Toll Free: (800)452-4011

PENNSYLVANIA
Bureau of Waste Management
Division of Compliance
  Monitoring
P.O. Box 2063
Harrisburg, Pennsylvania 17120
(717)787-6239

PUERTO RICO
Environmental Quality Board
P.O. Box 11488
Santurce, Puerto Rico
  00910-1488
(809)723-8184
  –or–
EPA Region II
Air and Waste Management
  Division
26 Federal Plaza
New York, New York 10278
(212)264-5175

RHODE ISLAND
Department of Environmental
  Management
Division of Air and Hazardous
  Materials
Room 204, Cannon Building
75 Davis Street
Providence, Rhode Island
  02908
(401)277-2797

SOUTH CAROLINA
Department of Health and
  Environmental Control
Bureau of Solid and Hazardous
  Waste Management
2600 Bull Street
Columbia, South Carolina
  29201
(803)734-5200

SOUTH DAKOTA
Department of Water and
  Natural Resources
Office of Air Quality and
  Solid Waste
Foss Building, Room 217
Pierre, South Dakota 57501
(605)773-3153

TENNESSEE
Division of Solid Waste
  Management
Tennessee Department of Public
  Health
701 Broadway
Nashville, Tennessee 37219-5403
(615)741-3424

TEXAS
Texas Water Commission
Hazardous and Solid Waste
  Division
Attn: Program Support Section
1700 North Congress
Austin, Texas 78711
(512)463-7761

UTAH
Department of Health
Bureau of Solid and Hazardous
  Waste Management
P.O. Box 16700
Salt Lake City, Utah 84116-0700
(801)538-6170

VERMONT
Agency of Environmental
  Conservation
103 South Main Street
Waterbury, Vermont 05676
(802)244-8702

VIRGIN ISLANDS
Department of Conservation
  and Cultural Affairs
P.O. Box 4399
Charlotte Amalie, St. Thomas
Virgin Islands 00801
(809)774-3320

  –or–

EPA Region II
Air and Waste Management
  Division
26 Federal Plaza
New York, New York 10278
(212)264-5175

VIRGINIA
Department of Health
Division of Solid and Hazardous
  Waste Management
Monroe Building, 11th Floor
101 North 14th Street
Richmond, Virginia 23219
(804)225-2667
Hazardous Waste Hotline:
(800)552-2075

WASHINGTON
Department of Ecology
Solid and Hazardous Waste
  Program
Mail Stop PV-11
Olympia, Washington 98504-8711
(206)459-6322
In-State: 1-800-633-7585

WEST VIRGINIA
Division of Water Resources
Solid and Hazardous Waste/
  Ground Water Branch
1201 Greenbrier Street
Charleston, West Virginia 25311

WISCONSIN
Department of Natural Resources
Bureau of Solid Waste Management
P.O. Box 7921
Madison, Wisconsin 53707
(608)266-1327

WYOMING
Department of Environmental Quality
Solid Waste Management Program
122 West 25th Street
Cheyenne, Wyoming 82002
(307)777-7752
  –or–
EPA Region VIII
Waste Management Division
  (8HWM-ON)
One Denver Place
999 18th Street
Suite 1300
Denver, Colorado 80202-2413
(303)293-1502

APPENDIX C

EPA HAZARDOUS WASTE NUMBERS FOR WASTE STREAMS COMMONLY GENERATED BY SMALL-QUANTITY GENERATORS

The Environmental Protection Agency recognizes that generators of small quantities of hazardous waste, many of which are small businesses, may not be familiar with the manner in which hazardous waste materials are identified. This appendix has been assembled to aid 100- to 1000-kg/month small-quantity generators in determining the EPA Hazardous Waste Numbers for their wastes. These numbers are needed to complete the "Notification of Hazardous Waste Activity," Form 8700-12.

This appendix contains lists of EPA Hazardous Waste Numbers for each waste stream identified in Table 6.1 Chapter 6 of this book. Note that accurately hazardous wastes are identified with an asterisk (*).

To Use This Appendix

1. Locate your business type in Table 6.1 in Chapter 6 of this book. This will help you to identify the waste streams common to your activities.

2. Find each of the waste streams that you identified in Table 6.1 in the more detailed descriptions in this appendix. Review the more detailed descriptions of typical wastes to determine which wastes streams actually result from your activities.

3. If you determine that you actually generate a particular waste stream, report the four-digit EPA Hazardous Waste Number in Item X of Form 8700-12, "Notification of Hazardous Waste Activity."

The specific instructions for completing Item X (Description of Hazardous Wastes) of the notification form are included in the notification package. You should note, however, that specific EPA Hazardous Waste Numbers beginning with:

- "F" should be entered in Item X, Section A.
- "K" should be entered in Item X, Section B.
- "P" or "U" should be entered in Item X, Section C.
- "D" should be entered in Item X, Section E.

The industries and waste streams described here do not provide a comprehensive list, but rather serve as a guide to potential small-quantity generators in determining which of their wastes, if any, are hazardous. Except for the pesticide and wood preserving categories, this appendix does not include EPA Hazardous Waste Numbers for commercial chemical products that are hazardous when discarded unused. These chemicals and their EPA Hazardous Waste Numbers are listed in Title 40 of the Code of Federal Regulations (40 CFR) in Section 261.33.

If the specific EPA Hazardous Waste Number that should be applied to your waste stream is unclear, please refer to 40 CFR Part 261, reprinted in the Notification Form 8700-12 package. In those cases where more than one EPA Hazardous Waste Number is applicable, all should be used. If you have any questions, or if you are unable to determine the proper EPA Hazardous Waste Numbers for your wastes, contact your state hazardous waste management agency, or the RCRA/Superfund Hotline (see Appendix B).

Solvents:

Solvents, spent solvents, solvent mixtures, or solvent still bottoms are often hazardous. This includes solvents used in degreasing (identified as F001) and paint brush cleaning and distillation residues from reclamation. The following are some commonly used hazardous solvents (also see ignitable wastes for other hazardous solvents, and 40 CFR 261.31 for most listed hazardous waste solvents):

Benzene	F005
Carbon Disulfide	F005
Carbon Tetrachloride	F001
Chlorobenzene	F002
Cresols	F004
Cresylic Acid	F004
O-Dichlorobenzene	F002
Ethanol	D001
2-Ethoxyethanol	F005
Ethylene Dichloride	D001
Isobutanol	F005
Isopropanol	D001
Kerosene	D001
Methyl Ethyl Ketone	F005
Methylene Chloride	F001
	F002
Naphtha	D001
Nitrobenzene	F004
2-Nitropropane	F005
Petroleum Solvents	D001
(Flashpoint less than 140°F)	
Pyridine	F005
1,1,1-Trichloroethane	F001
	F002
1,1,2-Trichloroethane	F002
Tetrachloroethylene	
(Perchloroethylene)	F001
	F002
Toluene	F005
Trichloroethylene	F001
	F002
Trichlorofluoromethane	F002
Trichlorotrifluoroethane	
(Valclene)	F002
White Spirits	D001

Acids/Bases:

Acids, bases, or mixtures having a pH less than or equal to 2 or greater than or equal to 12.5, are considered corrosive (for a complete description of corrosive wastes, see 40 CFR 261.22, Characteristic of corrosivity). All corrosive materials and solutions have the EPA Hazardous Waste Number D002. The following are some of the more commonly used corrosives:

Acetic Acid	Nitric Acid
Ammonium Hydroxide	Oleum
Chromic Acid	Perchloric Acid
Hydrobromic Acid	Phosphoric Acid
Hydrochloric Acid	Potassium Hydroxide
Hydrofluoric Acid	Sodium Hydroxide
	Sulfuric Acid

Dry Cleaning Filtration Residues:

Cooked powder residue (perchloroethylene plants only), still residues, and spent cartridge filters containing perchloroethylene or valclene are hazardous and have the EPA Hazardous Waste Number F002.

Still residues containing petroleum solvents with a flashpoint less than 140°F are considered hazardous and have the EPA Hazardous Waste Number D001.

Heavy Metals/Inorganics:

Heavy metals and other inorganic waste materials exhibit the characteristic of EP Toxicity and are considered hazardous if the extract from a representative sample of the waste has any of the specific constituent concentrations as shown in 40 CFR 261.24, Table 1. This may include dusts, solutions, wastewater treatment sludges, paint wastes, waste inks, and other such materials which contain heavy metals/inorganics (note that wastewater treatment sludges from electroplating operations are identified as F006). The following are EP Toxic:

Arsenic	D004
Barium	D005
Cadmium	D006
Chromium	D007
Lead	D008
Mercury	D009
Selenium	D010
Silver	D011

Ignitable Wastes:

Ignitable wastes include any liquids that have a flashpoint less than 140°F, any non-liquids that are capable of causing a fire through friction, absorption of moisture, or spontaneous chemical change, or any ignitable compressed gas as described in 49 CFR 173.300 (for a complete description of ignitable wastes, see 40 CFR 261.21, Characteristic of ignitability). Examples are spent solvents (see also solvents), solvent still bottoms, ignitable paint wastes (paint removers, brush cleaners and stripping agents), epoxy resins and adhesives (epoxies, rubber cements and marine glues), and waste inks containing flammable solvents. Unless otherwise specified, all ignitable wastes have the EPA Hazardous Waste Number of D001.

Some commonly used ignitable compounds are:

Acetone	F003
Benzene	F005
n-Butyl Alcohol	F003
Chlorobenzene	F002[1]
Cyclohexanone	F003
Ethyl Acetate	F003
Ethylbenzene	F003
Ethyl Ether	F003
Ethylene Dichloride	D001
Methanol	F003
Methyl Isobutyl Ketone	F003
Petroleum Distillates	D001
Xylene	F003

Ink Sludges Containing Chromium and Lead:

This includes solvent washes and sludges, caustic washes and sludges, or water washes and sludges from cleaning tubs and equipment used in the formulation of ink from pigments, driers, soaps, and stabilizers containing chromium and lead. All ink sludges have the EPA Hazardous Waste Number K086.

---

[1] Chlorobenzene is listed by EPA as a hazardous waste due to its toxicity and has been assigned EPA Hazardous Waste Number F002. It has a flashpoint, however, of less than 140°F and is therefore included here as an ignitable waste.

Lead–Acid Batteries:

Used lead-acid batteries should be reported on the notification form *only* if they are not recycled. Used lead-acid batteries that *are* recycled do not need to be counted in determining the quantity of waste that you generate per month, nor do they require a hazardous waste manifest when shipped off your premises. (Note: Special requirements do apply if you recycle your batteries on your own premises—see 40 CFR Part 266.)

Lead Dross	D008
Spent Acids	D002
Lead-Acid Batteries	D008

Pesticides:

The pesticides listed below are hazardous. Wastes marked with an asterisk (*) have been designated acutely hazardous. For a more complete listing, see 40 CFR 261.32 and 261.33 for specific listed pesticides, and other wastes, wastewaters, sludges, and by-products from pesticide formulators. (Note that while many of these pesticides are no longer in common use, they are included here for those cases where they may be found in storage.)

* Aldicarb	P070
* Aldrin	P004
Amitrole	U011
* Arsenic Pentoxide	P011
* Arsenic Trioxide	P012
Cacodylic Acid	U136
Carbamic Acid, Methylnitroso-, Ethyl Ester	U178
Chlordane	U036
* Copper Cyanides	P029
1,2-Dibromo-3-chloropropane	U066
1,2-Dichloropropane	U083
1,3-Dichloropropene	U084
2,4-Dichlorophenoxy Acetic Acid	U240
DDT	U061
* Dieldrin	P037
Dimethylcarbamoyl Chloride	U097

Pesticides (Continued)

*Dinitrocresol	P047
*Dinoseb	P020
Disodium Monomethanearsenate	D004
*Disulfoton	P039
*Endosulfan	P050
*Endrin	P051
Ethylmercuric Chloride	D009
*Famphur	P097
*Heptachlor	P059
Hexachlorobenzene	U127
Kepone	U142
Lindane	U129
2-Methoxy Mercuric Chloride	D009
Methoxychlor	D014
*Methyl Parathion	P071
Monosodium Methanearsenate	D004
*Nicotine	P075
*Parathion	P089
Pentachloronitrobenzene	U185
Pentachlorophenol	U242
Phenylmercuric Acetate	D009
*Phorate	P094
*Strychnine	P108
2,4,5-Trichlorophenoxy Acetic Acid	U232
2-(2,4,5-Trichlorophenoxy)- Propionic Acid	U233
*Thallium Sulfate	P115
Thiram	U244
*Toxaphene	P123
Warfarin	U248

Reactives:

Reactive wastes include reactive materials or mixtures which are unstable, react violently with or form explosive mixtures with water, generate toxic gases or vapors when mixed with water (or when exposed to pH conditions between 2 and 12.5 in the case of cyanide or sulfide bearing wastes), or are capable of detonation or explosive reaction when heated or subjected to shock (for a complete description of reactive wastes, see 40 CFR 261.23, Characteristic of reactivity). Unless otherwise specified, all reactive wastes have the EPA Hazardous Waste Number D003. The following materials are commonly considered to be reactive:

Acetyl Chloride	Organic Peroxides
Chromic Acid	Perchlorates
Cyanides	Permanganates
Hypochlorites	Sulfides

Spent Plating and Cyanide Wastes:

Spent plating wastes contain cleaning solutions and plating solutions with caustics, solvents, heavy metals, and cyanides. Cyanide wastes may also be generated from heat treatment operations, pigment production, and manufacturing of anticaking agents. Plating wastes are generally Hazardous Waste Numbers F006-F009, with F007-F009 containing cyanide. Cyanide heat treating wastes are generally Hazardous Waste Numbers F010-F012. See 40 CFR 261.32 for a more complete description of plating wastes.

Wood Preserving Agents:

The wastewater treatment sludges from wastewater treatment operations are considered hazardous (EPA Hazardous Waste Number K001—bottom sediment sludges from the treatment of wastewater processes that use creosote and pentachlorophenol). In addition, unless otherwise indicated, specific wood preserving compounds are:

Chromated Copper Arsenate	D004
Creosote	U051
Pentachlorophenol	F027

INDEX